Print Reading for Industry

11TH EDITION

by **Walter C. Brown / Ryan K. Brown**

Former Associate Professor
Department of Technology
Illinois State University
Normal, Illinois

WRITE-IN TEXT

Publisher
The Goodheart-Willcox Company, Inc.
Tinley Park, IL
www.g-w.com

Manufactured in Canada.

Library of Congress Control Number: 2020939965

ISBN 978-1-64564-672-3

2 3 4 5 6 7 8 9 – 22 – 26 25 24 23 22

Cover image: Marten_House/Shutterstock.com

Preface

Print Reading for Industry is a robust text that focuses on interpreting and visualizing drawings and prints used in industrial settings. It is designed to assist beginning through intermediate students and those learning on-the-job training to build the skills necessary to read and understand "the language of industry." It is ideally suited for teaching semester-long courses. This text may also be used in apprenticeship programs.

Print Reading for Industry uses prints from various industries as examples so that students can become comfortable with real-world common practices. The many prints found in the text are supplemented with the accompanying Large Prints Packet containing 22 C-size prints for even more hands-on learning. These prints are available as PDFs for classroom display and discussion, and can be found in both student and instructor resources.

Additionally, *Print Reading for Industry* provides coverage of several foundational skills needed for print reading success, including basic mathematics, geometry principles, reading engineering drawings, measurement tools, and the design process. Coverage of specialized parts and prints, including applications for fasteners, gears, cams, plastic parts, and precision sheet metal parts, is also included. The role of prints in the digital age is discussed, and coverage reflects the most up-to-date geometric dimensioning and tolerancing, as well as modern trends in the manufacturing industry. New and improved spatial visualization tools and exercises help build students' spatial reasoning skills, making them better print readers.

Print Reading for Industry is organized into sections based on a progression of concepts from simple to complex. Learning objectives and technical terms are listed at the beginning of each unit, providing an overview of the content. Review questions, review activities, industry print exercises, and bonus print reading exercises based on the prints found in the Large Prints Packet provide ample means of assessing student progress. The write-in text workbook format with perforations allows students to tear practice prints out of the textbook, complete the review assignments, and turn them in directly to the instructor.

This textbook will provide the reader with a solid foundation necessary to read prints in an industrial setting and find success in any career where prints are found.

Walter C. Brown
Ryan K. Brown

About the Authors

During his career, **Dr. Walter C. Brown** was a leading authority in the fields of drafting and print reading. He served as a consultant to industry on design and drafting standards and procedures. He authored several books in the fields of drafting, print reading, and mathematics, and was a professor in the Division of Technology at Arizona State University, Tempe, Arizona.

Dr. Ryan K. Brown is a retired faculty member of the Department of Technology at Illinois State University, Normal, Illinois, where, for 26 years, he taught various drafting and design courses in both mechanical and architectural applications. His teaching experiences also include four years as a faculty member at his alma mater, Eastern Kentucky University, and five years teaching drafting at the junior and senior high level. His work experiences cover a variety of short-term consulting positions with a wide range of drafting and graphics-based tasks, such as steel detailing, plant layouts, and architectural drawing and modeling. Throughout his career, he has conducted workshops for companies in the field of print reading and geometric dimensioning and tolerancing. As a service to secondary education in Illinois, he continues to serve as a test author for various drafting and design competitions.

With respect to this text and accompanying materials, feel free to contact the author, Dr. Ryan K. Brown, at the following email address: **rkbrown@ilstu.edu**.

New to This Edition

The 11th edition of *Print Reading for Industry* represents a significant revision, with a new four-color design; new and revised figures, drawings, and industry prints; and significant updates reflecting the latest ASME and AWS standards. Some of the more noteworthy changes to this edition include the following:

- An upgrade to four-color design, with the addition of new images and updated art, provides a more modern, timeless feel and allows new print readers to more readily learn and discern the alphabet of lines necessary to read and interpret industry prints.
- Content related to computer-generated prints and the automatic generation of multiview drawings, section views, auxiliary views, and screw thread representations provides new print readers with knowledge of the most up-to-date practices and representations they are likely to encounter in the field.
- Updated information to match the latest ASME and AWS standards ensures students learn the latest, best-practice methods for drawing and reading industry prints. Content reflects the latest standard (ASME Y14.5-2018) for geometric dimensioning and tolerancing (GD&T), but students also learn enough about discontinued practices to be ready should they encounter older prints in real-world situations.
- Unit 5 now offers greater coverage of multiview drawings and object visualization, including revised and expanded exercises to allow new print readers more practice with the essential but often challenging skills of spatial reasoning and visualization.
- New 3D model PDF files for select parts featured in the text can be used by instructors to assist students with visualization practice, and they can be used for 3D printing, where available.
- Up-to-date pedagogical features provide students a learning edge. New unit summaries and review questions, including critical thinking questions, along with updated print reading questions using industry prints, allow students greater reinforcement of both the unit and cumulative textbook knowledge.

Reviewers

The author and publisher wish to thank the following industry and teaching professionals for their valuable input into the development of *Print Reading for Industry*:

Brian Bennett
Hill College
Cleburne, TX

Rick Calverley
Lincoln College of Technology
Grand Prairie, TX

Peter Fil
Hudson Valley Community College
Troy, NY

Stephanie Gustin
Amarillo College
Amarillo, TX

Bill Hillman
University of Montana
Missoula, MT

Paul Homrich
Lansing Community College
Lansing, MI

Leon Kassler
York County Community College
Wells, ME

Kamyar Khashayar
East Los Angeles College
Monterey, CA

Jack Krikorian
William Rainey Harper College
Palatine, IL

Thomas Looker
Edison State Community College
Piqua, OH

Tony Micallef
Oakland Community College
Auburn Hills, MI

Duane Parrette
Springfield Technical Community College
Springfield, MA

Mark Sewell
Northampton County Community College
Bethlehem, PA

James Shahan
Iowa State University
Ames, IA

John Shepherd
Mt. San Antonio College
Upland, CA

Paul Wanner
Clackamas Community College
Oregon City, OR

Jason Zimmerman
Lenape Technical School
Ford City, PA

Acknowledgments

Dr. Ryan Brown would like to acknowledge and thank Dr. Louis Reifschneider of the Department of Technology at Illinois State University for his review and contributions to the content related to plastic parts. Dr. Reifschneider has experience in the plastics industry as an engineer and designer, and he graciously brought his expertise to this topic.

Images and Prints Used in the Text

Throughout the text are images and illustrations based on actual industry parts and prints supplied by valued contributors. The publisher and authors thank the following companies for their assistance in supplying images and prints for this text and previous editions:

Aerojet-General
AIR Corporation
AISIN Manufacturing Illinois
Barko Hydraulics
Barton Manufacturing
BC Design and Associates
Bell Aerosystems
Boston Gear Group
Brown and Sharpe
Clark Equipment Company
Cleveland Gear
Creatas
Daimler
Deere and Company
DesignJet Division, Hewlett-Packard
Detroit Diesel Allison
Dukane
Engel, Inc.
GenCorp Aerojet
Gleason Works
Grayhill, Inc.
Hydro-Gear
Intermatic, Inc.
Iron-A-Way
J. I. Case
Johnson and Towers, Inc.
Kennametal
Lockmasters, Inc.
MAAC Machinery
Marathon LeTourneau Company
MASCO Corporation
Master Spring and Wire Form Co.
Monarch Sidney
Motorola, Inc.
National Lock
North American Aviation
OMC Lincoln
Perkin Elmer
Polygon Company
Pratt and Whitney Canada
Reed Spectrum
RegO Cryo-Flow Products
Rockwell Manufacturing Company
Skil Corporation
Sono-Mag Corporation
Sperry Phoenix Company
Sterling Precision Corporation
Sunnen Products Company
Talley Industries
The American Welding Society
Unidynamics
Uniloy
United Technologies Otis
Vickers Industrial Division
Wis·Con Total Power Corporation

Features of the Textbook

The instructional design of *Print Reading for Industry* includes student-focused learning tools and print reading exercises to help you succeed. You will find the following features in the textbook to guide you through your learning experience:

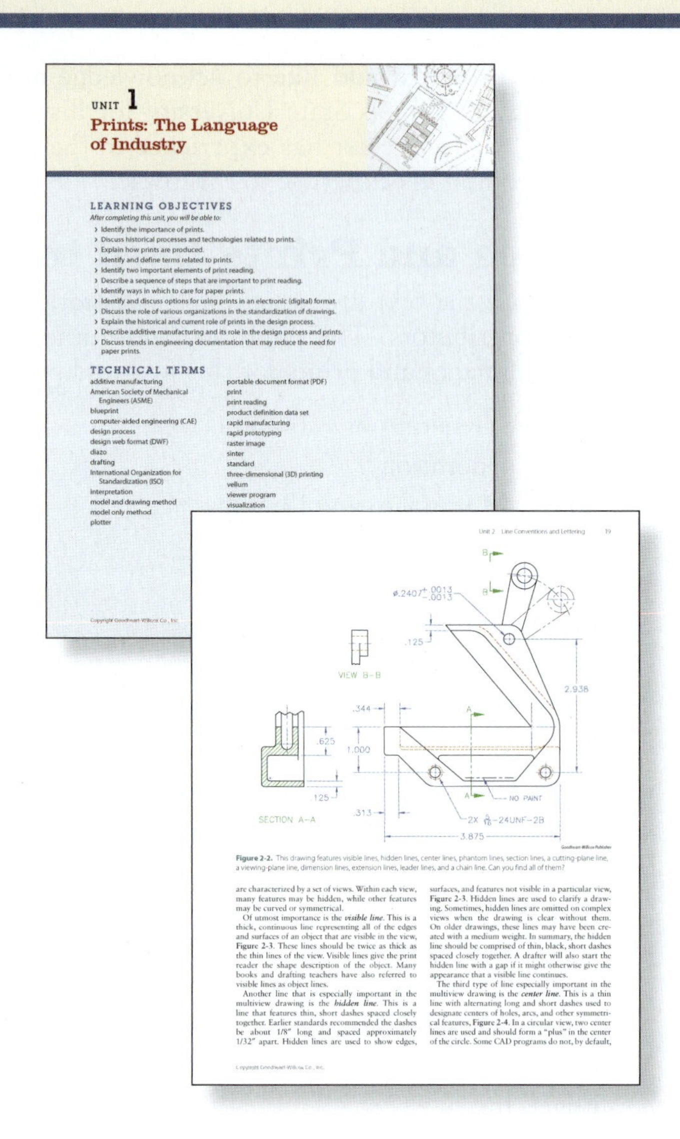
UNIT 1
Prints: The Language of Industry

LEARNING OBJECTIVES

After completing this unit, you will be able to:

- Identify the importance of prints.
- Discuss historical processes and technologies related to prints.
- Explain how prints are produced.
- Identify and define terms related to prints.
- Identify two important elements of print reading.
- Describe a sequence of steps that are important to print reading.
- Identify ways in which to care for paper prints.
- Identify and discuss options for using prints in an electronic (digital) format.
- Discuss the role of various organizations in the standardization of drawings.
- Explain the historical and current role of prints in the design process.
- Describe additive manufacturing and its role in the design process and prints.
- Discuss trends in engineering documentation that may reduce the need for paper prints.

TECHNICAL TERMS

additive manufacturing
American Society of Mechanical Engineers (ASME)
blueprint
computer-aided engineering (CAE)
design process
design web format (DWF)
diazo
drafting
International Organization for Standardization (ISO)
interpretation
model and drawing method
model only method
plotter
portable document format (PDF)
print
print reading
product definition data set
rapid manufacturing
rapid prototyping
raster image
sinter
standard
three-dimensional (3D) printing
vellum
viewer program
visualization

Unit 2 Line Conventions and Lettering

Figure 2-2. This drawing features visible lines, hidden lines, center lines, phantom lines, section lines, a cutting-plane line, a viewing-plane line, dimension lines, extension lines, leader lines, and a chain line. Can you find all of them?

are characterized by a set of views. Within each view, many features may be hidden, while other features may be curved or symmetrical.

Of utmost importance is the ***visible line***. This is a thick, continuous line representing all of the edges and surfaces of an object that are visible in the view, **Figure 2-3**. These lines should be twice as thick as the thin lines of the view. Visible lines give the print reader the shape description of the object. Many books and drafting teachers have also referred to visible lines as object lines.

Another line that is especially important in the multiview drawing is the ***hidden line***. This is a line that features thin, short dashes spaced closely together. Earlier standards recommended the dashes be about 1/8" long and spaced approximately 1/32" apart. Hidden lines are used to show edges, surfaces, and features not visible in a particular view, **Figure 2-3**. Hidden lines are used to clarify a drawing. Sometimes, hidden lines are omitted on complex views when the drawing is clear without them. On older drawings, these lines may have been created with a medium weight. In summary, the hidden line should be comprised of thin, black, short dashes spaced closely together. A drafter will also start the hidden line with a gap if it might otherwise give the appearance that a visible line continues.

The third type of line especially important in the multiview drawing is the ***center line***. This is a thin line with alternating long and short dashes used to designate centers of holes, arcs, and other symmetrical features, **Figure 2-4**. In a circular view, two center lines are used and should form a "plus" in the center of the circle. Some CAD programs do not, by default,

Unit Opening Materials

Each unit opener contains a list of learning objectives and technical terms. **Objectives** clearly identify the knowledge and skills to guide your learning as you progress through the unit. **Technical Terms** are key vocabulary terms you will encounter as you read the content, and their definitions are found in the Glossary.

Illustrations

The text features numerous **illustrations** designed to clearly and simply communicate important concepts. Illustrations and photographic images have been updated for this edition, including the addition of full color.

End-of-Unit Content

End-of-unit material provides an opportunity for review and application of concepts. A concise **Summary** provides an additional review tool that reinforces key learning objectives, helping you focus on important ideas presented in the text.

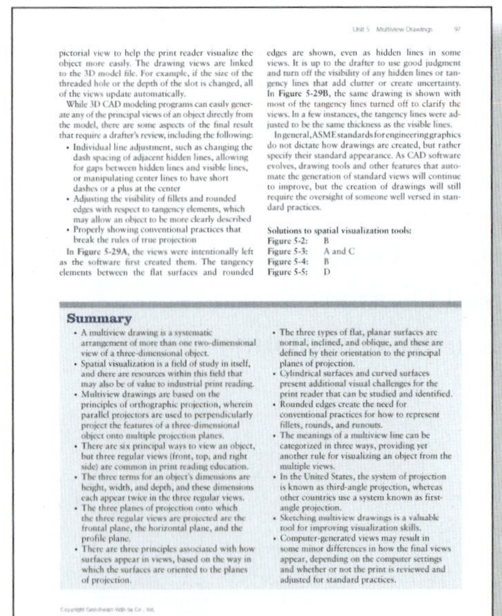
Unit 5 Multiview Drawings 97

pictorial view to help the print reader visualize the object more easily. The drawing views are linked to the 3D model file. For example, if the size of the threaded hole or the depth of the slot is changed, all of the views update automatically.

While 3D CAD modeling programs can easily generate any of the principal views of an object directly from the model, there are some aspects of the final result that require a drafter's review, including the following:

- Individual line adjustment, such as changing the dash spacing of adjacent hidden lines, allowing for gaps between hidden lines and visible lines, or manipulating center lines to have short dashes or a plus at the center
- Adjusting the visibility of fillets and rounded edges with respect to tangency elements, which may allow an object to be more clearly described
- Properly showing conventional practices that break the rules of true projection

In **Figure 5-29A**, the views were intentionally left as the software first created them. The tangency elements between the flat surfaces and rounded edges are shown, even as hidden lines in some views. It is up to the drafter to use good judgment and turn off the visibility of any hidden lines or tangency lines that add clutter or create uncertainty. In **Figure 5-29B**, the same drawing is shown with most of the tangency lines turned off to clarify the views. In a few instances, the tangency lines were adjusted to be the same thickness as the visible lines.

In general, ASME standards for engineering graphics do not dictate how drawings are created, but rather specify their standard appearance. As CAD software evolves, drawing tools and other features that automate the generation of standard views will continue to improve, but the creation of drawings will still require the oversight of someone well versed in standard practices.

Solutions to spatial visualization tools:
Figure 5-2: B
Figure 5-3: A and C
Figure 5-4: B
Figure 5-5: D

Summary

- A multiview drawing is a systematic arrangement of more than one two-dimensional view of a three-dimensional object.
- Spatial visualization is a field of study in itself, and there are resources within this field that may also be of value to industrial print reading.
- Multiview drawings are based on the principles of orthographic projection, wherein parallel projectors are used to perpendicularly project the features of a three-dimensional object onto multiple projection planes.
- There are six principal ways to view an object, but three regular views (front, top, and right side) are common in print reading education.
- The three terms for an object's dimensions are height, width, and depth, and these dimensions each appear twice in the three regular views.
- The three planes of projection onto which the three regular views are projected are the frontal plane, the horizontal plane, and the profile plane.
- There are three principles associated with how surfaces appear in views, based on the way in which the surfaces are oriented to the planes of projection.
- The three types of flat, planar surfaces are normal, inclined, and oblique, and these are defined by their orientation to the principal planes of projection.
- Cylindrical surfaces and curved surfaces present additional visual challenges for the print reader that can be studied and identified.
- Rounded edges create the need for conventional practices for how to represent fillets, rounds, and runouts.
- The meanings of a multiview line can be categorized in three ways, providing yet another rule for visualizing an object from the multiple views.
- In the United States, the system of projection is known as third-angle projection, whereas other countries use a system known as first-angle projection.
- Sketching multiview drawings is a valuable tool for improving visualization skills.
- Computer-generated views may result in some minor differences in how the final views appear, depending on the computer settings and whether or not the print is reviewed and adjusted for standard practices.

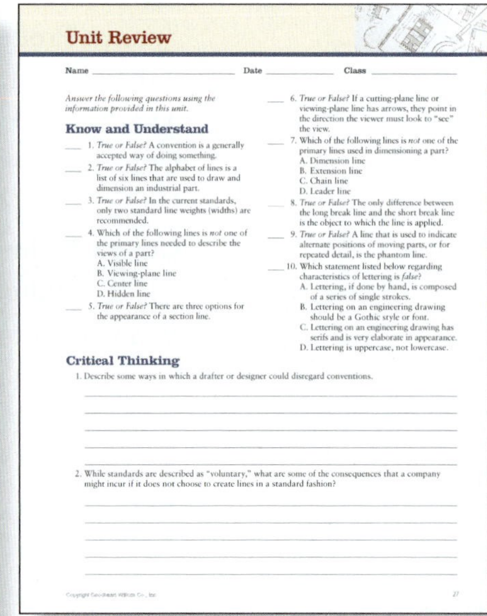
Unit Review

Name ______ Date ______ Class ______

Answer the following questions using the information provided in this unit.

Know and Understand

___ 1. *True or False?* A convention is a generally accepted way of doing something.
___ 2. *True or False?* The alphabet of lines is a list of six lines that are used to draw and dimension an industrial part.
___ 3. *True or False?* In the current standards, only two standard line weights (widths) are recommended.
___ 4. Which of the following lines is *not* one of the primary lines needed to describe the views of a part?
A. Visible line
B. Viewing-plane line
C. Center line
D. Hidden line
___ 5. *True or False?* There are three options for the appearance of a section line.
___ 6. *True or False?* If a cutting-plane line or viewing-plane line has arrows, they point in the direction the viewer must look to "see" the view.
___ 7. Which of the following lines is *not* one of the primary lines used in dimensioning a part?
A. Dimension line
B. Extension line
C. Chain line
D. Leader line
___ 8. *True or False?* The only difference between the long break line and the short break line is the object to which the line is applied.
___ 9. *True or False?* A line that is used to indicate alternate positions of moving parts, or for repeated detail, is the phantom line.
___ 10. Which statement listed below regarding characteristics of lettering is *false*?
A. Lettering, if done by hand, is composed of a series of single strokes.
B. Lettering on an engineering drawing should be a Gothic style or font.
C. Lettering on an engineering drawing has serifs and is very elaborate in appearance.
D. Lettering is uppercase, not lowercase.

Critical Thinking

1. Describe some ways in which a drafter or designer could disregard conventions.

2. While standards are described as "voluntary," what are some of the consequences that a company might incur if it does not choose to create lines in a standard fashion?

27

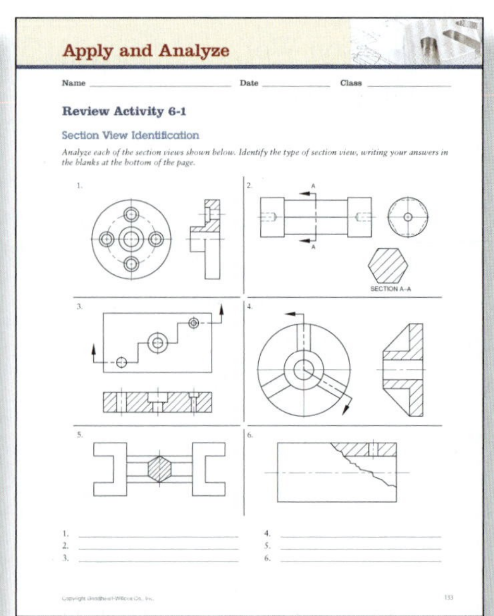
Apply and Analyze

Name ______ Date ______ Class ______

Review Activity 6-1

Section View Identification

Analyze each of the section views shown below. Identify the type of section view, writing your answers in the blanks at the bottom of the page.

1. ______ 4. ______
2. ______ 5. ______
3. ______ 6. ______

133

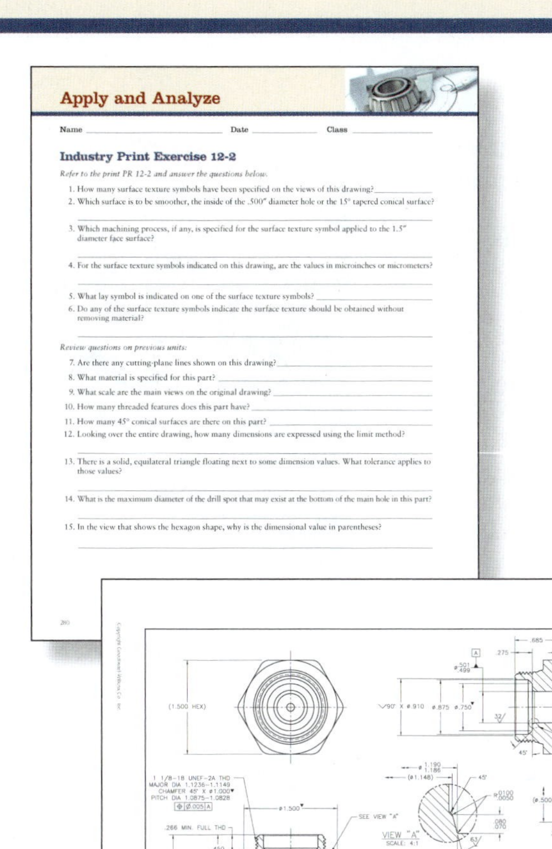
Apply and Analyze

Name ______________ Date ________ Class ________

Industry Print Exercise 12-2

Refer to the print PR 12-2 and answer the questions below.

1. How many surface texture symbols have been specified on the views of this drawing? ______
2. Which surface is to be smoother, the inside of the .500″ diameter hole or the 15° tapered conical surface? ______
3. Which machining process, if any, is specified for the surface texture symbol applied to the 1.5″ diameter face surface? ______
4. For the surface texture symbols indicated on this drawing, are the values in microinches or micrometers? ______
5. What lay symbol is indicated on one of the surface texture symbols? ______
6. Do any of the surface texture symbols indicate the surface texture should be obtained without removing material? ______

Review questions on previous units:

7. Are there any cutting-plane lines shown on this drawing? ______
8. What material is specified for this part? ______
9. What scale are the main views on the original drawing? ______
10. How many threaded features does this part have? ______
11. How many 45° conical surfaces are there on this part? ______
12. Looking over the entire drawing, how many dimensions are expressed using the limit method? ______
13. There is a solid, equilateral triangle floating next to some dimension values. What tolerance applies to those values? ______
14. What is the maximum diameter of the drill spot that may exist at the bottom of the main hole in this part? ______
15. In the view that shows the hexagon shape, why is the dimensional value in parentheses? ______

280

PR 12-2. Back Cap

A **Unit Review** features questions and exercises that reinforce important unit concepts and the cumulative knowledge of previous units. **Know and Understand** questions enable you to demonstrate knowledge, identification, and comprehension of chapter material. **Critical Thinking** questions develop higher-order thinking and problem-solving, personal, and workplace skills. **Apply and Analyze Review Activities** and **Industry Print Exercises** extend learning and help you analyze and apply knowledge by looking at industry prints.

Bonus Print Reading Exercises

End-of-unit **Bonus Print Reading Exercises** correspond to the bonus prints found in the Large Prints Packet that accompanies this textbook. These additional exercises enable you to powerfully apply real-world knowledge and skills.

Apply and Analyze

Name ______________ Date ________ Class ________

Bonus Print Reading Exercises

The following questions are based on the various bonus prints located in the folder that accompanies this textbook. Refer to the print indicated, evaluate the print, and answer the question.

Print AP-001

1. This print does not show a cutting-plane line. Is this acceptable practice or an error? ______

Print AP-002

2. When looking at SECTION A-A, how does the print reader know that this is several parts or pieces and not just one part? ______
3. Most of the sections on this print are full sections. What other method of sectioning is used? ______

Print AP-003

4. What type of section is shown in SECTION B? ______
5. SECTION B uses a center line between the left half and the right half. Is this recommended practice? ______

Print AP-004

6. Based on the shape of the cutting-plane line, what type of section is SECTION B-B? ______
7. Except for SECTION B-B, what other type of section view is used in the main views? ______

Print AP-006

8. Are the section lines for this print representative of the material, or are they simply being shown as general purpose section lines? ______

Print AP-007

9. How many cutting-plane lines are shown on this print? ______

142

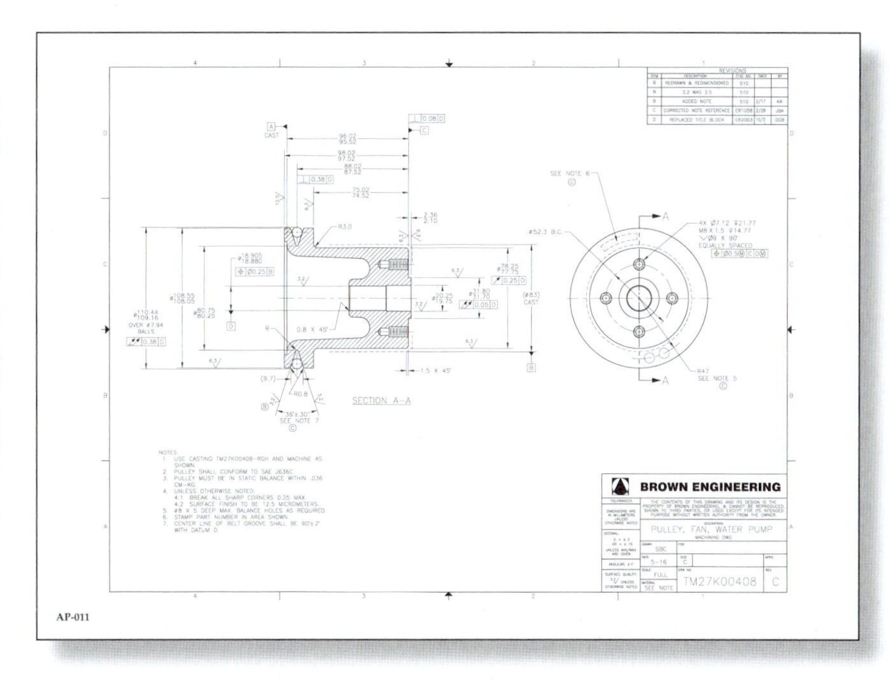

TOOLS FOR STUDENT AND INSTRUCTOR SUCCESS

Student Tools

Student Text

Print Reading for Industry is a robust text that focuses on interpreting and visualizing drawings and prints used in industrial settings, designed to assist beginning through intermediate students and those learning on-the-job training to build the skills necessary to read and understand "the language of industry." Prints from various industries are used to demonstrate real-world common practices. The many prints found in the text are supplemented with the accompanying Large Prints Packet containing 22 C-size prints for even more hands-on learning.

Coverage of specialized parts and prints, including applications for fasteners, gears, cams, plastic parts, and precision sheet metal parts, is also included. The role of prints in the digital age is discussed, and coverage reflects the most up-to-date geometric dimensioning and tolerancing, as well as modern trends in the manufacturing industry. Visualization tools and exercises help build spatial reasoning skills, which are important for print readers. Learning objectives and technical terms listed at the beginning of each unit provide an overview of the content. End-of-unit review questions, review activities, industry print exercises, and bonus print reading exercises based on the prints found in the Large Prints Packet provide ample means of assessing progress. The write-in text workbook format with perforations allows review assignments to be completed and turned in directly to the instructor.

Instructor Tools

LMS Integration

Integrate Goodheart-Willcox content within your Learning Management System for a seamless user experience for both you and your students. LMS-ready content in Common Cartridge® format facilitates single sign-on integration and gives you control of student enrollment and data. With a Common Cartridge integration, you can access the LMS features and tools you are accustomed to using and G-W course resources in one convenient location—your LMS.

G-W Common Cartridge provides a complete learning package for you and your students. The included digital resources help your students remain engaged and learn effectively:

- **eBook content.** G-W Common Cartridge includes the textbook content in an online format. The eBook is interactive, with highlighting, magnification, and note-taking features.
- **Drill and Practice.** Learning new vocabulary is critical to student success. These vocabulary activities, which are provided for all key terms in each chapter, provide an active, engaging, and effective way for students to learn the required terminology.

When you incorporate G-W content into your courses via Common Cartridge, you have the flexibility to customize and structure the content to meet the educational needs of your students. You may also choose to add your own content to the course.

For instructors, the Common Cartridge includes the Online Instructor Resources. QTI® question banks are available within the Online Instructor Resources for import into your LMS. These prebuilt assessments help you measure student knowledge and track results in your LMS gradebook. Questions and tests can be customized to meet your assessment needs.

Online Instructor Resources (OIR)

Online Instructor Resources provide all the support needed to make preparation and classroom instruction easier than ever. Available in one accessible location, the OIR includes Instructor Resources, Instructor's Presentations for PowerPoint®, and Assessment Software with Question Banks. The OIR is available as a subscription and can be accessed at school, at home, or on the go.

Instructor Resources One resource provides instructors with time-saving preparation tools such as answer keys, editable lesson plans, and other teaching aids.

Instructor's Presentations for PowerPoint®
These fully customizable, richly illustrated slides help you teach and visually reinforce the key concepts from each chapter.

Assessment Software with Question Banks Administer and manage assessments to meet your classroom needs. The question banks that accompany this textbook include hundreds of matching, completion, multiple choice, and short answer questions to assess student knowledge of the content in each chapter. Using the assessment software simplifies the process of creating, managing, administering, and grading tests. You can have the software generate a test for you with randomly selected questions. You may also choose specific questions from the question banks and, if you wish, add your own questions to create customized tests to meet your classroom needs.

G-W Integrated Learning Solution

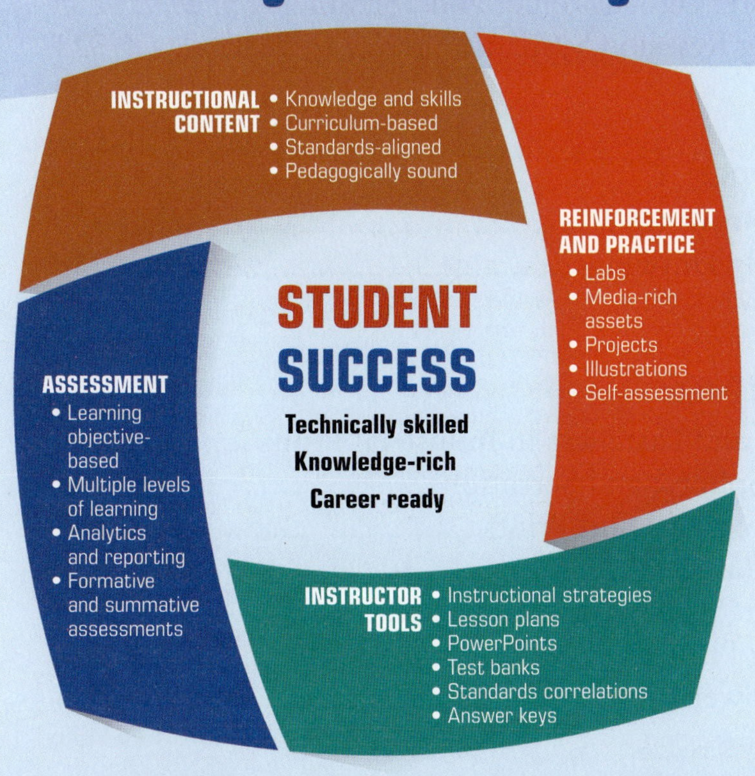

The G-W Integrated Learning Solution offers easy-to-use resources that help students and instructors achieve success.

- **EXPERT AUTHORS**
- **TRUSTED REVIEWERS**
- **100 YEARS OF EXPERIENCE**

EMPLOYABILITY SKILLS · TECHNICAL SKILLS · ACADEMIC KNOWLEDGE · INDUSTRY RECOGNIZED STANDARDS

Brief Contents

Section 1

Introduction to Drafting and Print Reading

Section 2

Fundamentals of Shape Description

Section 3

Fundamentals of Size Description and Annotations

Section 4

Industrial Drawing Types

Section 5

Specialized Parts and Prints

Contents

Section 3
Fundamentals of Size Description and Annotations

Section 4
Industrial Drawing Types

Section 5
Specialized Parts and Prints

SECTION 1

Introduction to Drafting and Print Reading

Bayurov Alexander/Shutterstock.com

UNIT 1

Prints: The Language of Industry

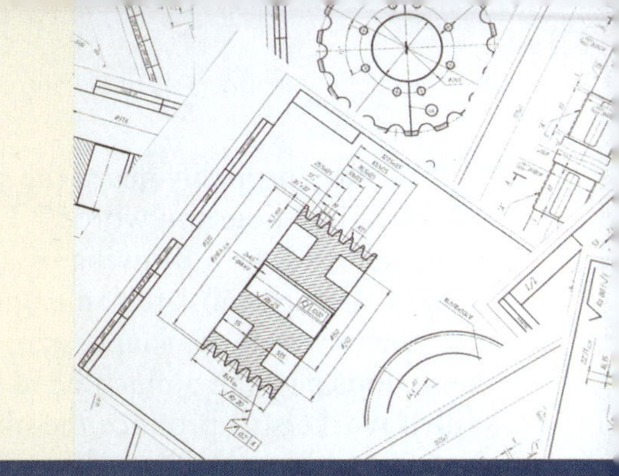

LEARNING OBJECTIVES

After completing this unit, you will be able to:

- Identify the importance of prints.
- Discuss historical processes and technologies related to prints.
- Explain how prints are produced.
- Identify and define terms related to prints.
- Identify two important elements of print reading.
- Describe a sequence of steps that are important to print reading.
- Identify ways in which to care for paper prints.
- Identify and discuss options for using prints in an electronic (digital) format.
- Discuss the role of various organizations in the standardization of drawings.
- Explain the historical and current role of prints in the design process.
- Describe additive manufacturing and its role in the design process and prints.
- Discuss trends in engineering documentation that may reduce the need for paper prints.

TECHNICAL TERMS

additive manufacturing
American Society of Mechanical Engineers (ASME)
blueprint
computer-aided engineering (CAE)
design process
design web format (DWF)
diazo
drafting
International Organization for Standardization (ISO)
interpretation
model and drawing method
model only method
plotter
portable document format (PDF)
print
print reading
product definition data set
rapid manufacturing
rapid prototyping
raster image
sinter
standard
three-dimensional (3D) printing
vellum
viewer program
visualization
wide-format printer

You have probably heard the saying, "a picture is worth a thousand words." This is certainly true when referring to a drawing of a product. It would be next to impossible for an engineer or designer to describe in words the shape, size, and relationship of the various parts of a machine in sufficient detail for skilled workers to produce the object. Drawings are the universal language used by engineers, designers, technicians, and skilled workers to quickly and accurately communicate the necessary information to fabricate, assemble, or service industrial products, **Figure 1-1**.

The Importance of Prints

Within the context of this book, the word ***print*** will simply be defined as a copy of a drawing, but can generally be used the same as the word *drawing*. In many situations, drawings may not be printed, but nevertheless are still a primary means of communication within the manufacturing industry. Many industrial products, such as automobiles, aircraft, and computers, consist of thousands of component parts. These parts may be manufactured in a variety of settings around the globe. The "moment of truth" in the manufacture of these products comes during final assembly or when a spare part is installed in the field. These parts must always fit. Therefore, to meet manufacturing requirements, all industries need workers who can read and understand prints.

A drawing describes what an object should look like when it is completed. Prints provide workers with the details of size, shape, tolerances (allowable variation), materials used, finish, and other special treatments. In many cases, the print is also an important part of the contractual agreement within the industrial setting. The supplier of parts must meet the specifications dictated by the print. Often, purchasing agents have to ensure the print is contractually sufficient to ensure vendors supply quality parts without cutting corners. Quality-control inspectors have to verify that all parts, both those made by suppliers and those made in-house, match the print and continually review the print's role in controlling the precision and quality of the parts needed.

History of Prints

The study of print reading is closely related to the study of ***drafting***, the general term for creating

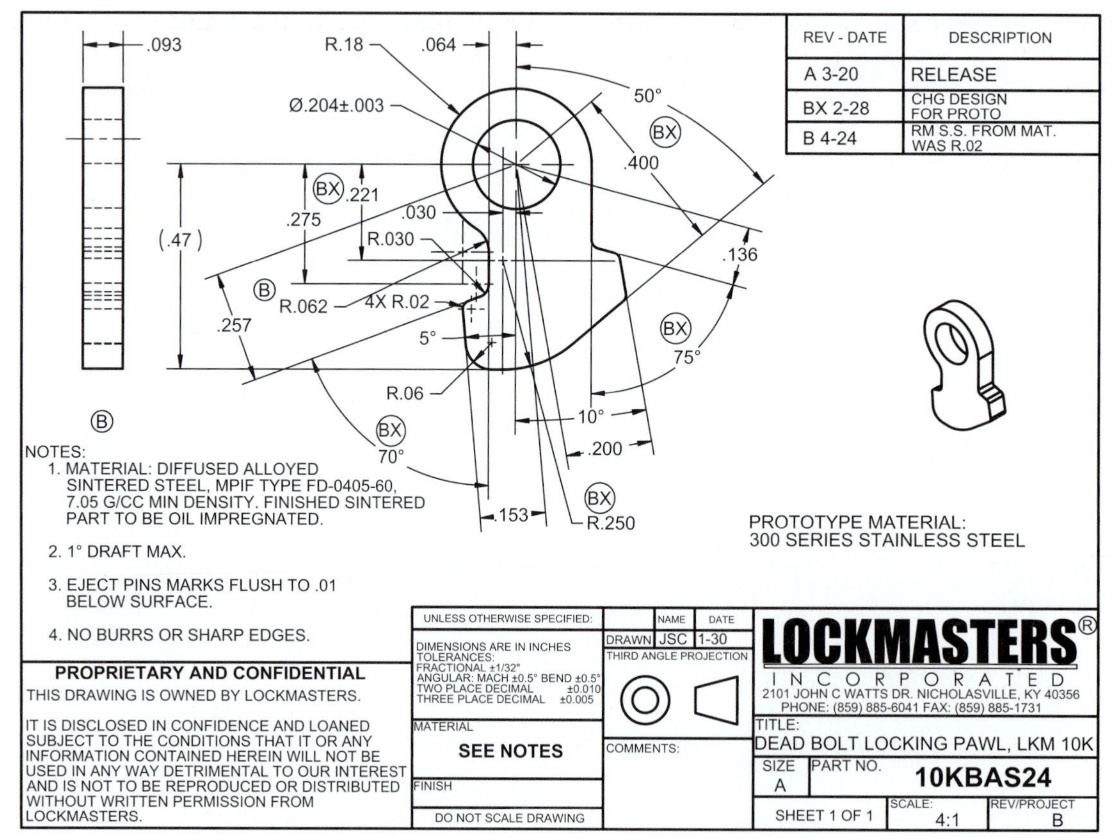

Print supplied by Lockmasters, Inc.

Figure 1-1. Drawings are the universal language used by engineers, designers, technicians, and skilled workers to quickly and accurately communicate the necessary information to fabricate, assemble, or service industrial products.

drawings of objects in technical fields. Of course, those who have studied drafting not only know how to create prints, but also read prints. Courses, curricular programs, and textbooks for drafting have used a multitude of terms throughout the years. Engineering graphics is a common term used within engineering programs, while technical drawing is a common term in drafting programs that train technicians and technologists. In addition, terms such as mechanical drafting or instrumental drafting were used in the past to describe the process of creating industrial drawings. While the focus of this textbook is on reading the drawings, information about creating drawings will also be examined to provide insight into the standards the drafter should be following.

One of the earliest printmaking processes began in the middle of the nineteenth century. It involved exposing treated, photosensitive paper under an original drawing on translucent paper. This early method and the copy paper it used required submersion in a liquid to process the copy. The resulting print had a blue background and white lines. Thus, this type of print was called a ***blueprint***, **Figure 1-2.** Even though the process that created "blue" prints has not been used for decades, the term blueprint has become a part of our common vocabulary to mean "any plan of action or detailed procedure to accomplish a task."

During the 1940s, a dry process evolved that used a paper coated with a type of organic compound referred to as a ***diazo*** compound. Exposure to light evaporated the background area while leaving the coating where lines blocked the light. Ammonia vapor then converted the remaining coating to a permanent blue. This type of print was sometimes called a whiteprint or a blue-line print, but often the term blueprint continued to be used. While the diazo process may still be found in use in a few settings today, most companies have replaced that technology with engineering photocopiers or plotters used to make paper prints.

Mega Pixel/Shutterstock.com

Figure 1-2. Years ago, copies of drawings (prints) were made that featured a blue background and white lines. This is the origin of the term ***blueprint***, which is still in our vocabulary today.

Original drawings were produced on ***vellum*** (transparent, resin-impregnated paper) or plastic film because the copying process required an original that would be durable but also allow light to pass through. Copies of the drawings, or prints, were made and distributed to those who needed them.

Throughout most of the twentieth century, companies produced paper copies of drawings of parts and assemblies, either drawn with traditional drafting equipment or printed from computer-aided drafting (CAD) systems. These original drawings—created by the drafter or printed by the CAD operator—were then stored in a file-drawer system supervised by the engineering or quality control department. Original drawings were seldom used in the plant or field. Instead, prints were made for distribution.

While some companies may still use these traditional systems, for many companies original drawings no longer exist in paper format. Prints can be created directly from the computer as needed. It is also possible for those who need prints to view the drawings in electronic (digital) format using viewing software. There are many software utility programs that allow 2D drawings and 3D models to be viewed without CAD software. At the end of this unit, future trends will be discussed. In some industries today, digital design models already serve as the primary description of parts.

How Prints Are Made

Today, most drawings are produced using a CAD system, **Figure 1-3.** If needed, a hard copy of the drawing can be printed to paper with various technologies. Some printers use standard-size paper fed into the printer as sheets from a tray, while other printers may use a large roll of paper. (Standard paper sizes are covered in Unit 2.) In former days, many CAD output devices capable of printing on larger sheets of paper were referred to as ***plotters***, as the output was vector-based, meaning lines were physically drawn, or plotted, from point to point with pens. Today's CAD drawings are transferred to the paper as a raster image. The term *plotter* continued to be used during the transition from vector-based to raster-based images. A ***raster image*** is composed of tiny pixels or dots. Under high magnification, you can see the dots, but the quality is more than adequate for industrial prints.

With output devices now available in a wide range of sizes and capabilities for paper width, from 24″ wide to several feet wide, the terminology for output devices moved away from *plotters* to ***wide-format printers***, or large-format printers. See **Figure 1-4**. Wide-format printers are designed to accommodate wide media, such as paper, vinyl, or textiles, and also vary with respect to how the ink is applied to the output media, the type of ink, and the purpose of the final output.

In some companies, approved originals may still be produced and filed. Prints can also be made by authorized personnel using a photocopier. In other companies, approved originals may only exist in a protected directory of the computer network, and prints can be sent to output devices by authorized personnel.

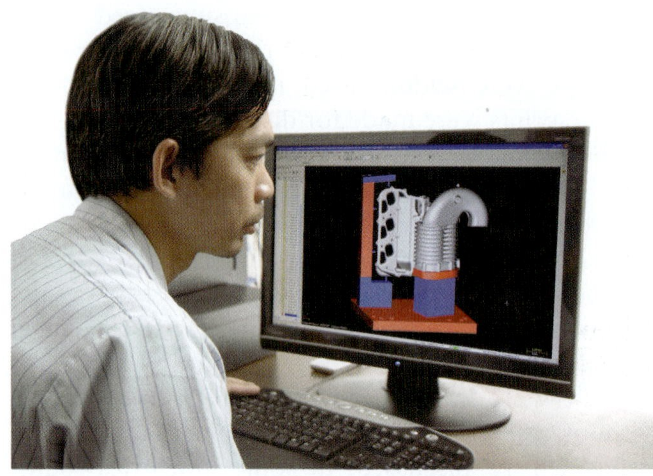

Chuck Rausin/Shutterstock.com

Figure 1-3. Most original drawings are now produced using a computer-aided drafting (CAD) system.

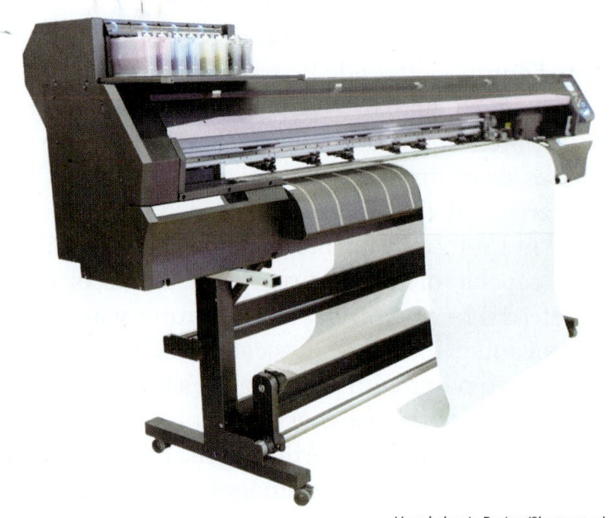

Vereshchagin Dmitry/Shutterstock.com

Figure 1-4. Hard copies are generated from a CAD system using an output device, such as this wide-format printer.

Print Reading

Print reading is the process of analyzing a print to obtain information. This task involves two principal elements—visualization and interpretation. ***Visualization*** is the ability to envision or "see" the shape of the object from the various views shown on a print. See **Figure 1-5**. Every view created by the drafter or designer is based on a projection of the object onto a two-dimensional plane, such as a sheet of paper or a computer screen. Learning the principles of projection will help the reader gain the ability to visualize objects from the views shown on a print. The ***interpretation*** of lines, symbols, dimensions, notes, and other information on a print is also an important factor in print reading. These factors are presented in this text. Actual industrial prints are provided in this text so you can learn print reading using "real-life" drawings.

Print Reading Steps

In addition to visualizing the shape and interpreting the symbols and notes, approaching the print with a sequence of steps is important. The following list is a recommended approach to print reading steps:

1. Read the title of the part.
2. Check the drawing number.
3. Read the title block and notes.
4. Read all specification callouts.
5. Read the revisions and changes.
6. Analyze the part or assembly.

Notice in the sequence above that before analyzing the shape and size of the part, the print reader first observes a few important aspects of the part, such as the name or title and the part number, which may be a flexible number that applies to more than one part on the same print. The notes, callouts, and revisions may also reveal important aspects of the print, such as the angle of projection, the units, the precision level of tolerance, and the year for the standards applied when the print was made.

Care of Prints

Prints are valuable records of information. On many occasions, prints are still output to paper, although newer technologies allow the print reader to use a media device such as a tablet or laptop computer. When working with paper prints, you should observe the following rules:

- Never write on a print in such a way that others may think the print has changed, unless you have been authorized to make changes.

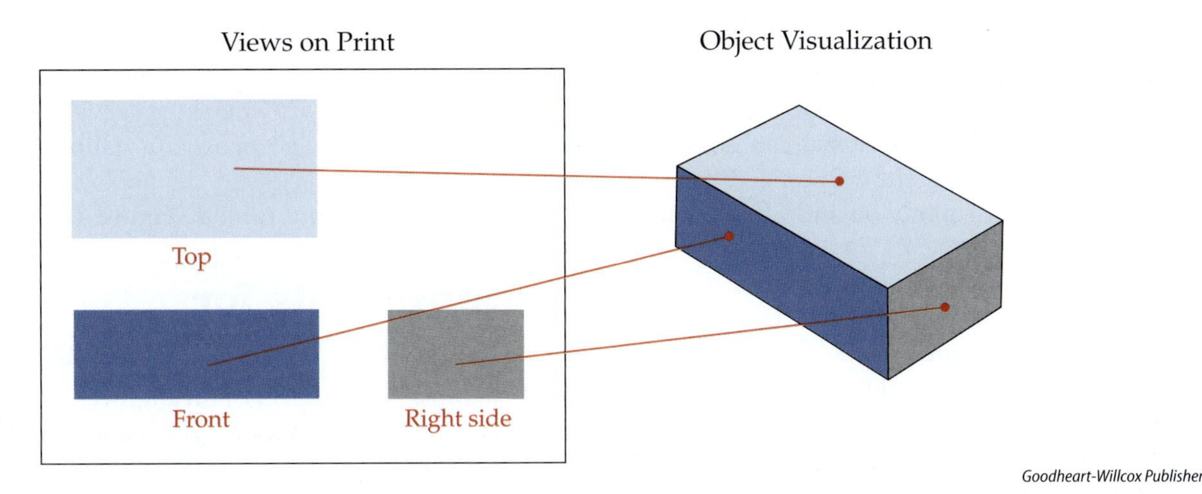

Figure 1-5. Visualization is the ability to "see" the shape of an object from the various views shown on a print.

- Keep prints clean, and especially free of oil and dirt. Soiled prints are difficult to read and contribute to errors.
- If working with prints that are stored in filing cabinets, very carefully fold and unfold the prints to avoid tearing.
- Do not lay sharp tools, machine parts, or similar objects on top of prints.

If the drawing border and title block conform to industry standards, the prints can be folded with the title block showing. A second print number is also often visible when the print is stored in a file folder, **Figure 1-6**. In many companies, prints are easily accessed from a print control area or engineering department. New prints can easily be output for use, and then discarded. Even so, tight control should be maintained over all original computer files that are used to produce the prints, and a naming convention should be in place to ensure that any prints created are the most recent revision. With CAD drawings, protected or "read-only" drawing files may be available through a company computer network. This allows the design and manufacturing teams to make their own prints as needed or to view the design on a computer screen while also ensuring only authorized personnel make changes to the files.

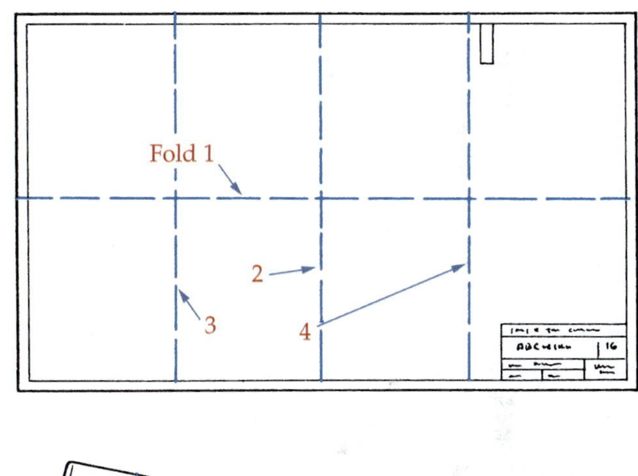

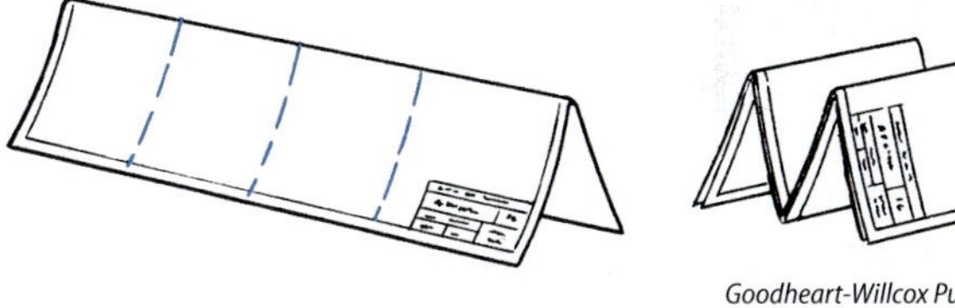

Figure 1-6. If paper prints are filed in a file folder, larger prints should be folded in such a way as to keep the title block showing.

Electronic Formats for Drawings

In our current technological era, digital file formats are available that provide the ability to electronically share prints without jeopardizing or supplying the actual CAD data. The most common electronic format is the ***portable document format (PDF)***, developed by Adobe Inc. Adobe produces Adobe Acrobat®, which is software used to create and alter PDF files, as well as Adobe Reader®, which is free software that allows anyone to read a PDF file. Many third-party software programs exist that enable the creation and editing of PDF files, and most CAD software programs provide tools for exporting drawing files to PDF format. The PDF format was originally designed to allow users to easily share printable documents regardless of computer platform (Windows®, Macintosh®, etc.). The PDF format includes many useful features, such as the ability to combine multiple sheets into one file, mark up and review documents, preserve CAD layer structures, and add password and duplication protection.

In recent years, 3D PDF files have increased in popularity, enabling 3D models to be embedded within a PDF file. With Adobe Reader, PDF files with embedded 3D information can enable the user to dynamically rotate the view, select preset views, turn parts of an assembly on and off, view notes, and more, **Figure 1-7**.

In 2008, Adobe opened the door for the PDF standard to be published and controlled by the ***International Organization for Standardization (ISO)*** as ISO 32000-1. The ISO is responsible for establishing and publishing manufacturing standards worldwide. In short, the PDF format is now the predominant method of creating and sharing electronic documents. Another electronic format is the ***design web format (DWF)*** developed by Autodesk, parent company of several CAD programs, including AutoCAD® and Inventor®. The format was originally designed to allow AutoCAD drawings to be published in a web-based format, but it has evolved into a primary means, within the Autodesk product line, of creating documents that are similar to PDFs in purpose. The DWF file is compressed for faster viewing and printing. The DWF format is not intended to replace the drawing (DWG) file format, but it allows team members such as designers, project managers, and engineers to share design data without requiring every user to have a licensed copy of the CAD software. Autodesk products can also export objects to a 3D DWF format for displaying in 3D manner, with dynamic rotation, view selection, and more. Autodesk's Design Review® software is a logical choice for viewing DWF files, but other third-party programs exist to create, read, and convert DWF files for many purposes.

In summary, there are many ***viewer programs*** that enable users to open, view, and print files in many different formats. Most viewer programs are available as free downloads. Viewer programs typically provide tools for markup, review, and comment tracking. Viewer programs allow drawings and "prints" to be viewed on personal devices such as smartphones and tablets, **Figure 1-8**.

Gorodenkoff/Shutterstock.com

Figure 1-7. PDF files can be viewed on computer, tablet, and smartphone screens, with zoom, pan, and turn-and-tilt (3D orbit) capabilities.

Standards for Engineering Drawings

Almost every aspect of an engineering drawing can be conformed to an established standard. By definition, a ***standard*** is a voluntary guideline. Companies can elect to have their own supplemental standards. While voluntary, standards are often incorporated into regulations and business contracts between manufacturers and their subcontractors and suppliers.

In the United States, the standardization of engineering drawings, including most aspects of prints covered within this text, has long been established through the ***American Society of Mechanical Engineers (ASME)***. This is an independent, not-for-profit organization that, for decades, has defined standards for engineering drawings.

Prior to 1994, the standards were identified by the American National Standards Institute (ANSI) document numbers, even though they were published by ASME. For example, the document titled *Multi and Sectional View Drawings* was identified as ANSI Y14.3-1975. Currently, that document is identified as ASME Y14.3-2012 and is titled *Orthographic and Pictorial Views*.

FERNANDO BLANCO CALZADA/Shutterstock.com

Figure 1-8. Tablets and smartphones can view 2D drawings and 3D models in electronic formats using a variety of apps.

ANSI is the umbrella organization that serves as an overseeing body and is still relevant, but publications are now purchased through ASME, either in paper or protected PDF format. Throughout this text, you will become familiar with some of the ASME standard publications that impact various aspects of print reading. **Figure 1-9** lists a few of the standards most relevant to this text.

For companies with an international presence and product line, the ISO standards may be the more appropriate set of practices. For example, ISO 128-1:2003 is titled *Technical drawings—General principles of presentation—Part 1: Introduction and index*, and ISO 5456-2:1996 is titled *Technical drawings—Projection methods—Part 2: Orthographic representations*. These standards are a major investment for engineering departments, but necessary for success in the twenty-first century.

This text focuses on industrial prints that follow ASME standard practices, both current and recent. It is important to reiterate that not all companies follow all ASME guidelines, sometimes intentionally and sometimes unintentionally. As feasible, this text will identify situations on the "real" exercise prints in the text with respect to their deviation from current ASME standard practice.

ASME Standards	
Y14.1	Decimal Inch Drawing Sheet Size and Format
Y14.1M	Metric Drawing Sheet Size and Format
Y14.2	Line Conventions and Lettering
Y14.3	Orthographic and Pictorial Views
Y14.5	Dimensioning and Tolerancing
Y14.6	Screw Thread Representation
Y14.7	Gear and Spline Drawing Standards (in two parts)
Y14.8	Castings, Forgings, and Molded Parts
Y14.13	Mechanical Spring Representation
Y14.24	Types and Applications of Engineering Drawings
Y14.31	Undimensioned Drawings
Y14.34	Associated Lists
Y14.35	Revision of Engineering Drawings and Associated Documents
Y14.36	Surface Texture Symbols
Y14.37	Composite Part Drawings
Y14.38	Abbreviations and Acronyms for Use on Drawings and Related Documents
Y14.41	Digital Product Definition Data Practices
Y14.43	Dimensioning and Tolerancing Principles for Gages and Fixtures
Y14.46	Product Definition for Additive Manufacturing
Y14.47	Model Organization Practices
Y14.100	Engineering Drawing Practices

Goodheart-Willcox Publisher

Figure 1-9. Listed here are ASME standards related to *Print Reading for Industry*.

Engineering Drawings and the Design Process

While twenty-first-century design processes have been complemented by a new set of computer-based tools, the actual steps of the ***design process*** are not that different from a century ago. In general, to bring an idea to market or to improve a current product, the design process, as illustrated in **Figure 1-10**, happens in the following five steps, with a cycling or looping between the steps:

1. Statement of the problem
2. Generation of possible solutions
3. Refinement of promising solutions
4. Testing and analysis
5. Implementation

Step 1 is to define the problem. This usually involves identifying a problem statement. This statement could be very general, such as identifying a need for "a better scooter," or more specific, such as the desire to "profitably market a low-cost, plastic, one-piece scooter stand for small-wheel scooters."

After the problem statement is established, Step 2 is to generate possible solutions. Step 2 may incorporate brainstorming sessions, research of current products, and the creation of sketches to help communicate possible solutions. These sketches and drawings also help document the dates and creative thoughts for

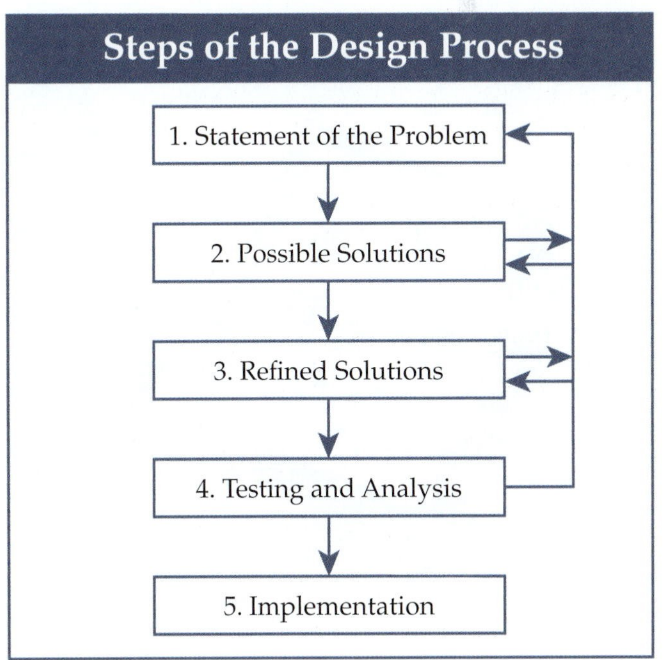

Goodheart-Willcox Publisher

Figure 1-10. The design process involves a number of steps and includes a looping strategy until a successful solution is reached.

future patents. Even at this stage, an understanding of many topics covered by this text can assist in the design process. Understanding spatial dimensions, geometric constructions, and how to use a combination of front, top, and pictorial views can all aid the design team in discovering solutions.

Step 3 involves refining promising solutions. At this stage, sketches may be re-created in a CAD program as precise geometric drawings, either as 2D views or 3D models. An analysis of the strengths and weaknesses of all the possible ideas, as well as visual and aesthetic opinions, may lead to necessary compromises. The results of Step 3 may suggest a return to Step 2 or even Step 1.

Step 4 involves testing and analysis. The design team will build prototypes or create a virtual model in a 3D CAD program. ***Rapid prototyping*** technologies have made it possible to generate prototypes of parts directly from a 3D CAD model without the need for preliminary drawings, **Figure 1-11**. Current 3D modeling software enables the design team to test more ideas, try more variations of shapes and proportions, check for fits and interferences, and conduct strength tests of the component parts. Drawings and prints may or may not be needed at this stage, but they can help communicate design intent to the model-makers and machinists. As with each previous step, depending on the test results, a decision may be made to loop back to a previous step.

The final step, Step 5, can be broadly described as the implementation step. This involves communicating the results to those who will get the product to market. In many cases, this is the step that includes the production of drawings or prints—contract documents that specify much of the information required to manufacture the part. In addition to production drawings, other materials need to be developed, including assembly and operating instructions, specifications, maintenance diagrams, and marketing materials.

Additive Manufacturing and Prints

As discussed in the previous section, rapid prototyping has impacted the role of drawings in the design process, especially the middle stages. In almost all cases, rapid prototyping technologies build parts by adding material in a series of layers, giving rise to the term ***additive manufacturing***. While many processes create a part by subtracting materials, such as drilling a hole or milling a surface, additive manufacturing technologies are based on adding material to the model layer by layer. The resemblance of some additive manufacturing technologies to inkjet printing led to the coining of the term ***3D printing***, **Figure 1-12**.

Goodheart-Willcox Publisher; prototypes courtesy of Envision America

Figure 1-11. The original product on the left was redesigned, and the plastic prototype in the center was generated directly from a CAD file. Although prints were not given to a model-maker, drawings and sketches were useful in the design process. The prototype facilitated assembly, function, and comfort factor testing. The new molded plastic part is on the right.

Monkey Business Images/Shutterstock.com

Figure 1-12. Desktop 3D printers can create plastic prototypes directly from a CAD file. The same technology can also create functioning parts for certain applications.

While rapid prototyping technologies originally created parts from plastic or thin layers of paper that could only function as prototypes, new and different technologies have evolved. Some additive manufacturing technologies can melt, or ***sinter***, metal powder into solid parts. As the processes improved, parts made of plastic and metal proved to be quite functional, and not merely prototypes. Small quantities of parts, as well as special-ordered parts, could be built directly from a CAD model. ***Rapid manufacturing*** is the application of additive manufacturing technologies to the full-scale production of parts and products.

Future developments in additive manufacturing may bring further changes in how prints are used. There are some applications wherein official "prints" or "drawings" will not be needed, at least not in the same role of describing a part to a model-maker or machinist. ASME Y14.46-2017, *Product Definition for Additive Manufacturing*, is a "draft" standard that has been developed to support engineers who are incorporating additive manufacturing technologies. It covers aspects of "model-based product definition" and recommends a standard practice for those exploring and using these new processes. As technologies continue to develop, the role of prints and drawings will continue to be impacted in a variety of ways.

Trends in Engineering Drawing

Newer forms of model documentation and data management are emerging to meet the requirements of modern manufacturing processes. In addition to providing the data included in sophisticated 3D CAD models, ***computer-aided engineering (CAE)*** programs are generating data about manufacturing processes and the properties of parts. Examples of CAE processes include finite element analysis (FEA) for strength analysis, tool path generation software for creating CNC machine tool procedures, and mold flow simulation software, which simulates how well liquid materials fill a mold. This data also assists with managing and sharing a more complete description of a part and its history. Product data management (PDM) software and systems help manage all of this electronic data, and facilitate processing review, revision, and inventory management tasks electronically.

In response to industry demands to incorporate electronic data into the manufacturing process, the ASME Y14.41 *Digital Product Definition Data Practices* standard was developed. This standard establishes practices for communicating design intent and other pertinent information with digital data. A ***product definition data set*** is the total collection of information required to completely define a component part or an assembly of parts, **Figure 1-13**. These data sets can completely define a product for the purpose of design function and analysis, as well as functional test results, manufacturing procedures or tool paths, and inspection data.

The ASME Y14.41 standard addresses two methods of preparing a product definition data set. In the ***model only method***, annotations (such as dimensions, tolerances, and notes) are directly attached to an individual model file. The model in this context is a combination of the design model, annotations, and attributes that describe a product. When using this method, a drawing graphic sheet is not required. The ***model and drawing method*** allows for a hybrid approach to describing a part or assembly. The drawing graphic sheet requires and incorporates the ASME Y14.1 standard specifications for a border line and title block, and allows for 2D geometric views in accordance with other ASME standards. Annotations such as notes and dimensions can still be attached to the model or the graphic sheet and must still conform to other ASME standards, in line with current practices. As conceived, the ASME Y14.41 standard enables model geometry, associated lists, materials, finishes, processes, notes, and analytical data to join a drawing graphic sheet as a means of communication within the manufacturing industry, serving the same role as a print.

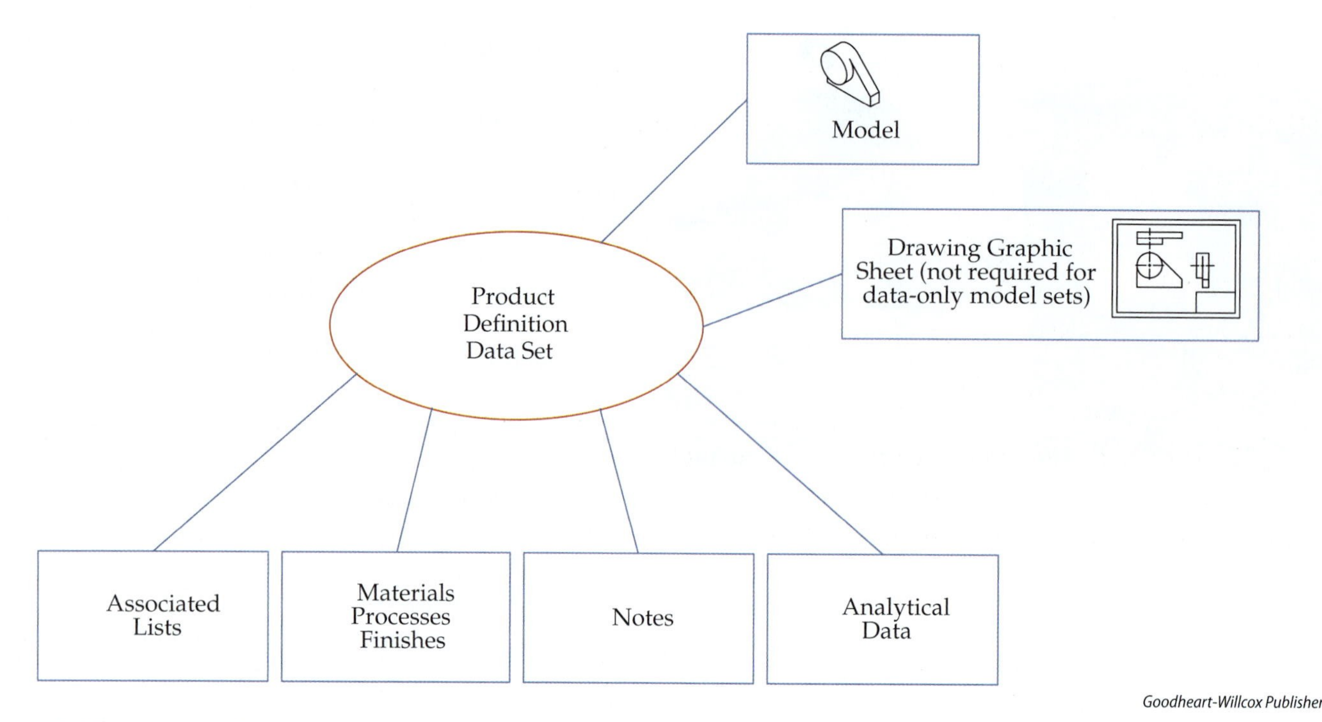

Figure 1-13. Illustrated here are potential contents of a product definition data set.

This is not to suggest that drawings or print reading will no longer be needed. The concepts discussed in the units to follow will continue to be important to the overall design process, and the role of prints as a key communication tool of industry will continue to evolve. Blueprints and whiteprints may no longer be used today, but the principles associated with those prints remain relevant. As John W. Gardner, former US Secretary of Health, Education, and Welfare, once said, "History never looks like history when you are living through it."

Summary

- Prints provide workers with the details of size, shape, tolerances, materials used, finish, and other special treatments.
- The print is an important part of the contractual agreement within the industrial setting.
- Terms such as *blueprint*, *whiteprint*, *diazo print*, and *blue-line drawings* all have their roots in the history of duplicating original drawings as "prints."
- Print reading is the process of analyzing a print to obtain information, and involves two principal elements—visualization and interpretation.
- In addition to visualizing the shape and interpreting the symbols and notes, approaching the print with a sequence of steps is important.
- Care should be taken with paper prints to avoid errors, confusion, and waste.
- Electronic formats such as PDF and DWF have given the print reader many new options for sharing and reading 2D drawings and 3D models.
- Extensive standards for engineering drawings are set forth by the ASME and ISO organizations.
- The design process incorporates the information of prints in various ways and at various stages, depending on the technologies involved.
- Additive manufacturing is impacting the role of prints in the design stages.
- Digital product definition data practices are also impacting the manner in which prints are created for the industrial enterprise.

Unit Review

Name ______________________ Date ______________ Class ______________

Answer the following questions using the information provided in this unit.

Know and Understand

____ 1. *True or False?* A print can simply be defined as a copy of a drawing.

____ 2. *True or False?* Prints are often an important part of a contractual agreement.

____ 3. The study of print reading is closely related to the study of _____.
A. measurements
B. drafting
C. machining
D. art

____ 4. *True or False?* Prints are sometimes called blueprints because the earliest printmaking technology created prints that had blue lines on a white background.

____ 5. For creating large prints, CAD systems can generate drawing files for output devices often referred to as _____.
A. 3D printers
B. photocopiers
C. wide-format printers
D. scanners

____ 6. Identify the two principal aspects of print reading.
A. Visualization and interpretation
B. Models and annotations
C. Views and measurements
D. Part names and part numbers

____ 7. *True or False?* Before analyzing the views of a part on the print, a good practice is to first analyze and read the part name and number in the title block.

____ 8. *True or False?* Prints should never be folded for any reason.

____ 9. What does PDF stand for?
A. Printable document format
B. Plotted drawing file
C. Portable document format
D. Print detail freehand

____ 10. Almost every aspect of an engineering drawing can be conformed to an established _____.
A. method
B. style
C. size
D. standard

____ 11. *True or False?* The "M" in ASME stands for *manufacturing*.

____ 12. Identify the first step in the design process.
A. Creation of sketches and drawings of ideas
B. Statement of the problem
C. Generation of possible solutions
D. Implementation

____ 13. *True or False?* The term *rapid prototyping* was applied to a new technology invented to make parts directly from a 3D CAD model, but the term *additive manufacturing* is more appropriately used today, especially since many of the parts are functional and not just prototypes.

____ 14. *True or False?* CAE and PDM software and systems help manage information about parts, which complements the traditional prints.

____ 15. Which of the following is the total collection of information required to completely define a component part or an assembly of parts?
A. Product definition data set
B. Computer-aided engineering
C. 3D printing
D. Model and drawing method

Critical Thinking

1. Describe a scenario in which a company would lose a great amount of money because someone did not create a print correctly or did not read a print correctly.

2. What are some of the advantages and disadvantages of viewing images or models of parts and assemblies on computer screens, tablets, and other media devices versus having paper prints?

3. While standards are described as "voluntary," what are some of the consequences that a company might incur if it does not choose to create prints in a standard fashion?

Apply and Analyze

Name ______________________ Date ______________ Class ______________

Review Activity

Further apply what you have learned from the unit to answer the following review questions.

1. Briefly explain why a purchasing agent needs to know how to read a print.

____

____

____

2. What common term in our vocabulary means "any plan of action or detailed procedure to accomplish a task"?

____

3. What is one reason it may not be necessary to create a print on paper?

____

____

____

4. How is *visualization* different from *interpretation*?

____

____

____

5. List two reasons that PDF files serve an important role in the print process.

____

____

____

6. List two viewer programs mentioned in the text that enable users to open, view, and print files in different formats.

____

____

7. Explain the roles of ANSI, ASME, and ISO in the field of print reading.

____

____

____

____

8. List an example of how the design process may include a "loop" or "cycle" at some point in the process.

9. Explain the "additive" part of additive manufacturing, and contrast it with other processes.

10. Briefly describe two methods, or approaches, to preparing a product definition data set.

UNIT **2**

Line Conventions and Lettering

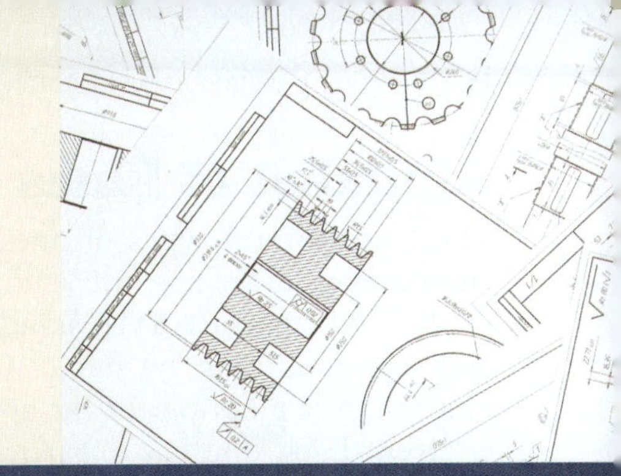

LEARNING OBJECTIVES

After completing this unit, you will be able to:

- Define the alphabet of lines.
- Identify the most common types of lines: visible, hidden, and center.
- Identify the types of lines used in dimensioning.
- Identify the types of lines used in sectional views.
- Understand the ways in which dashes, arrows, and line thickness are used with various lines.
- Describe the types of lines by appearance and purpose.
- Identify the style of lettering recommended for standard industrial drawings.

TECHNICAL TERMS

alphabet of lines
break line
center line
chain line
convention
cutting-plane line
dimension
dimension line
extension line
hidden line
leader line
lettering
phantom line
section line
stitch line
symmetry line
viewing-plane line
visible line

By definition, a ***convention*** is a generally accepted way of doing something. Before computers, prints were created by drafters with pencils and pens, and lettering was primarily a freehand craft. Lines were made with lead pencils, and the ability to create a line of an exact width was a skill that took time to develop. The ASME Y14.2 *Line Conventions and Lettering* standard sets forth the recommended appearance of lines, although in our current age of computers the guidelines no longer specify dash length and line width as precisely as in years past. Because drawings can easily be printed at a variety of scales and sizes, there is more flexibility today.

Alphabet of Lines

There are several types of lines commonly used in engineering drawings, depending on the field of study. The current ASME standard illustrates these and identifies each by name. On some occasions, more than one option for the appearance of the line is given. Each line has a particular meaning to an engineer, designer, or drafter. Individuals who are reading the print must recognize and understand the meanings of these lines in order to correctly interpret an industrial print used in the manufacturing environment.

The list of types of lines, defined in many references as the ***alphabet of lines***, is used throughout industry. See **Figure 2-1**. Each line has a definite form or dash pattern and a standard line weight (width). In older standards, three line weights were recommended—thin, medium, and thick. Current standards recommend two line weights—thick and thin. The standard thick line is recommended to be 0.6 mm in width. The standard thin line is recommended to be 0.3 mm in width. Of course, the actual width of these lines may vary depending on the size of the drawing and whether or not the drawing is reduced or enlarged. In all cases, however, the thin-to-thick ratio should remain 1:2. In most cases, CAD programs can be configured to print drawings at any size or scale while maintaining any desired line thickness. It is important to note that thinner lines should not be lighter. All linework should be black.

Be aware, however, that many drawings throughout the years have been produced without attention to standard line weight. Many of the early CAD systems could not easily print variations in line weight. With on-screen colors, some creators of drawings may not go to the trouble of establishing line weight. This often produces a drawing that is harder to read and interpret. Within your company, you have the opportunity to be the champion for conformance to industry standards, thus promoting drawings that "speak with the proper language."

Lines in a drawing convey information essential to understanding the print. Therefore, to understand the print, you must know and understand the alphabet of lines. Refer to **Figure 2-1** and **Figure 2-2** as you read the remainder of this unit.

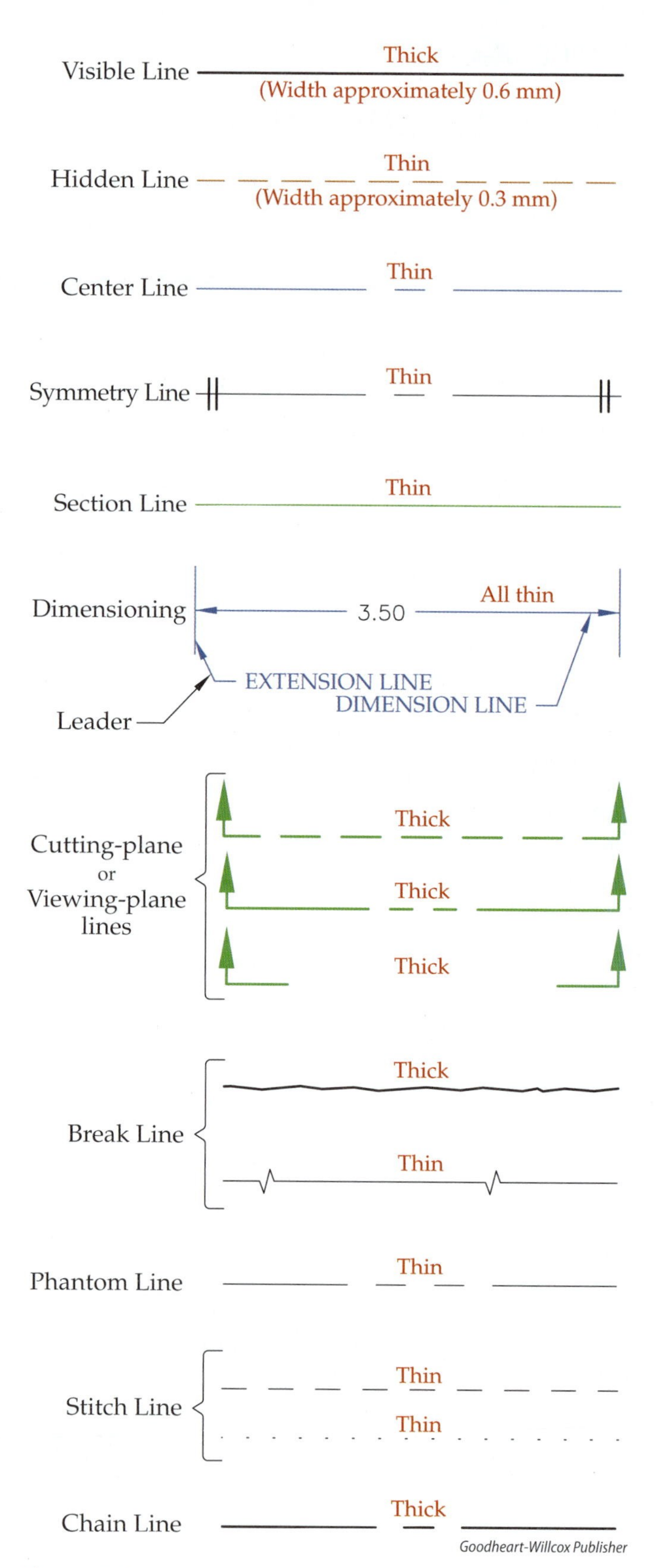

Figure 2-1. The alphabet of lines is a standardized list of lines from which the drafter can choose. Each line has a definite form or pattern and an intended purpose and meaning.

Primary View Lines

Of primary importance to the print reader are the three lines in the alphabet that are used in multiview drawings. Discussed in Unit 5, multiview drawings

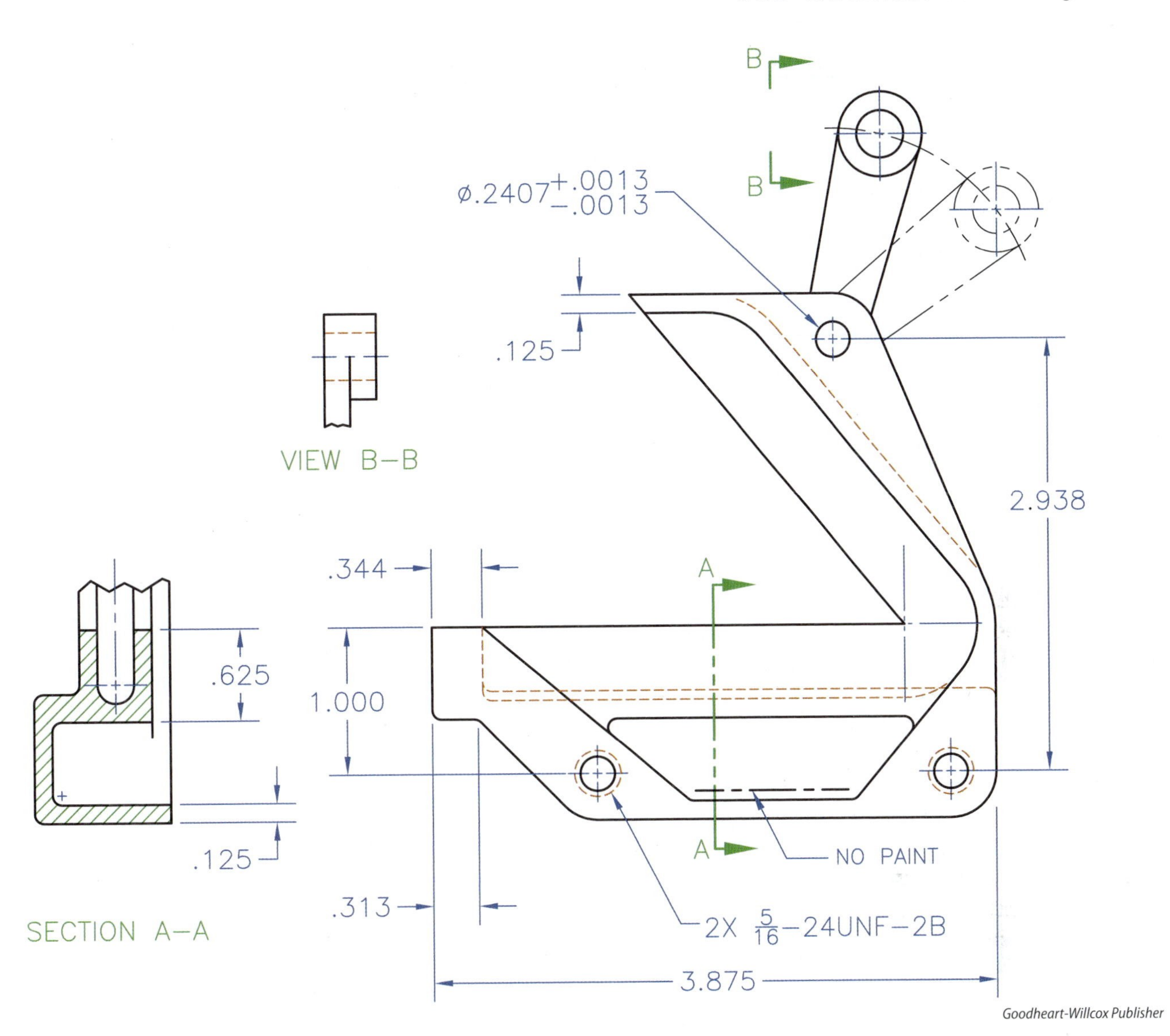

Goodheart-Willcox Publisher

Figure 2-2. This drawing features visible lines, hidden lines, center lines, phantom lines, section lines, a cutting-plane line, a viewing-plane line, dimension lines, extension lines, leader lines, and a chain line. Can you find all of them?

are characterized by a set of views. Within each view, many features may be hidden, while other features may be curved or symmetrical.

Of utmost importance is the ***visible line***. This is a thick, continuous line representing all of the edges and surfaces of an object that are visible in the view, **Figure 2-3**. These lines should be twice as thick as the thin lines of the view. Visible lines give the print reader the shape description of the object. Many books and drafting teachers have also referred to visible lines as object lines.

Another line that is especially important in the multiview drawing is the ***hidden line***. This is a line that features thin, short dashes spaced closely together. Earlier standards recommended the dashes be about 1/8″ long and spaced approximately 1/32″ apart. Hidden lines are used to show edges, surfaces, and features not visible in a particular view, **Figure 2-3**. Hidden lines are used to clarify a drawing. Sometimes, hidden lines are omitted on complex views when the drawing is clear without them. On older drawings, these lines may have been created with a medium weight. In summary, the hidden line should be comprised of thin, black, short dashes spaced closely together. A drafter will also start the hidden line with a gap if it might otherwise give the appearance that a visible line continues.

The third type of line especially important in the multiview drawing is the ***center line***. This is a thin line with alternating long and short dashes used to designate centers of holes, arcs, and other symmetrical features, **Figure 2-4**. In a circular view, two center lines are used and should form a "plus" in the center of the circle. Some CAD programs do not, by default,

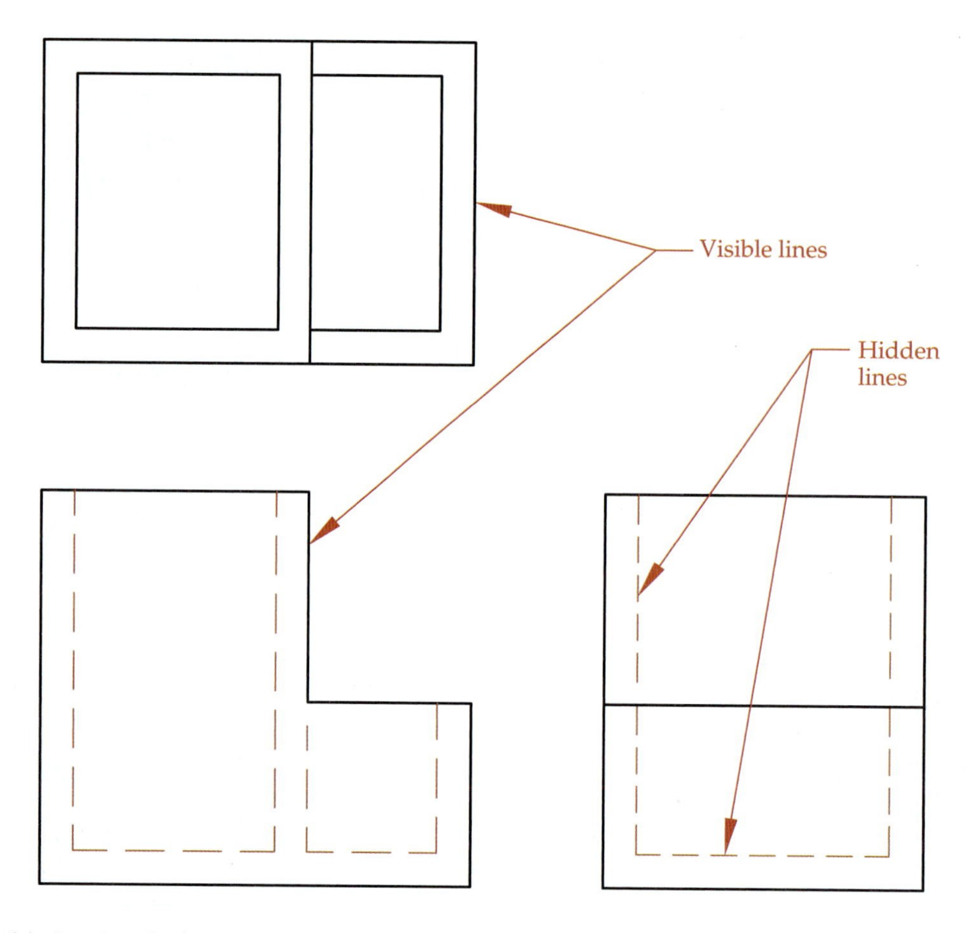

Goodheart-Willcox Publisher

Figure 2-3. A visible line is a thick, continuous line representing all edges and surfaces on an object visible in the view, while hidden lines represent hidden features.

show center lines crossing in the center of a circle quite as well as drawings drawn by hand. In these cases, the drafter should adjust the scale of the center line so the short dashes make a plus. Notice that the center line extends beyond the boundary of the shape description, extending past the visible linework approximately 1/8″ to 1/4″. Center lines are also used to indicate paths of motion, as shown in **Figure 2-2**. In addition, on some drawings, only one side of a part is drawn, and a symbol is placed on each end of the center line to indicate the other side is symmetrical (identical in dimension and shape). When these symbols are added to the center line, the line can be called a ***symmetry line***.

Section View Lines

Some lines are used primarily in section view drawings, which are discussed in Unit 6. Section views are views that show the object as if it has been cut through. Additional conventions are needed to express these views on a drawing.

For surfaces that are assumed to be cut, ***section lines*** are used. These are thin, continuous lines usually drawn at an angle, most commonly 45°. Section lines indicate the surface of an object in a section view that was "cut" by a "cutting plane." Sometimes section lines have dashes to indicate a particular material. General purpose section lining is the same as the cast iron pattern shown in **Figure 2-5**. This type of section lining is commonly used for other materials in section views, unless the drafter or designer wants to indicate the specific material. Some CAD programs refer to the lines within a section-lined area as "hatching" or "cross-hatching."

For section view drawings, a ***cutting-plane line*** is often shown on the view adjacent to the section view to help the print reader know where the "cut" is made. The cutting-plane line should be a thick, dashed line. It usually terminates in a short line at 90° to the cutting plane, with arrowheads in the direction of sight for viewing the section. Letters may be used to indicate the section, but are not required.

There are currently three different choices for the drafter when creating a cutting-plane line. Refer to **Figure 2-1**. The most common cutting-plane line features a long dash and then two short dashes.

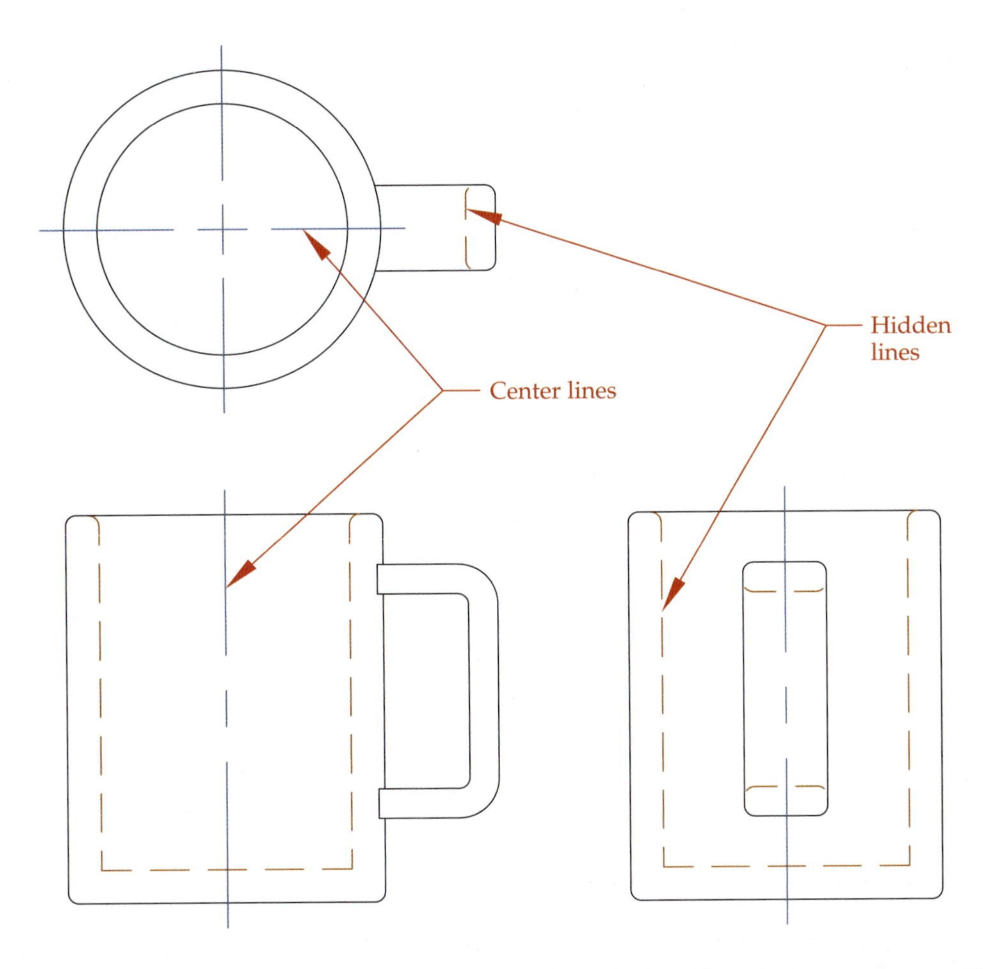

Goodheart-Willcox Publisher

Figure 2-4. Center lines are thin lines with alternating long and short dashes. They designate centers of holes, arcs, and other symmetrical objects.

Another version represented in the standard is a series of medium-size dashes, about twice as long as hidden-line dashes. The current standard also allows for two "elbows" with identifying labels and arrows. In any case, the lines should be the same thickness as visible lines.

Dimensioning Lines

While the visible, hidden, and center lines are used to create the shape description of an object, the size description of an object is indicated in annotations known as ***dimensions***. Dimensions are composed of a variety of lines, all drawn without dashes and all with a thin line weight. Dimensioning is discussed in Unit 9.

The lines that extend the edges of the object out away from the view are called ***extension lines***, **Figure 2-6**. Some books, references, and CAD systems refer to extension lines as "witness" lines. The purpose of the extension lines is to keep the dimensional annotation away from the shape description. Extension lines begin about 1/16″ away from the object's visible corners and extend about 1/8″ beyond the arrows of the dimension line. As shown, extension lines can cross each other without issue.

Between the extension lines are ***dimension lines***, which indicate the extent and direction of the dimensions, **Figure 2-6**. In most engineering drawings, the dimension line is broken in the middle for the dimensional value. Dimension lines are usually terminated by arrowheads against extension lines. While extension lines can cross each other, the drafter tries to avoid dimension lines crossing each other. Dimensioning methods are becoming more diverse in industrial applications, including arrowless coordinate dimensioning and geometric dimensioning and tolerancing. These topics are discussed in greater detail in later units.

In dimensioning, ***leader lines***, commonly known as leaders, are used to point to a feature, such as a hole, or a drawing area to which a local note applies, **Figure 2-6**. Leaders are also used in conjunction with dimension lines if there is insufficient room

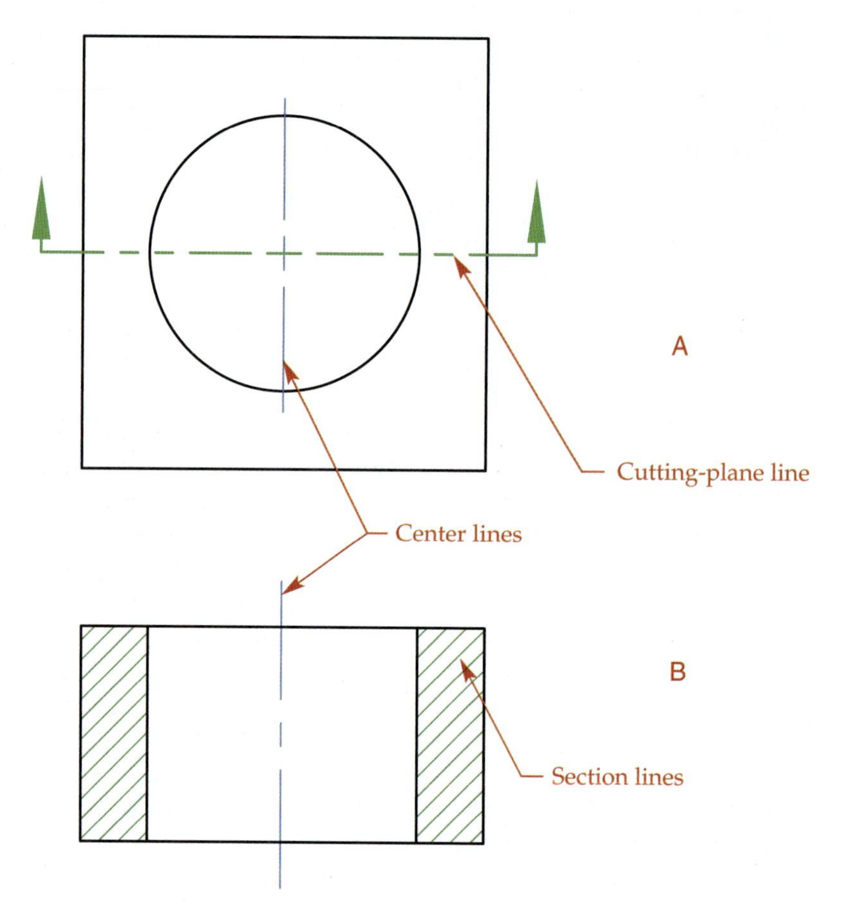

Goodheart-Willcox Publisher

Figure 2-5. Cutting-plane lines can be made in different ways, but the method indicated in A is common. Section lines, shown in B, are thin, continuous lines usually drawn at 45° and about 1/8″ apart.

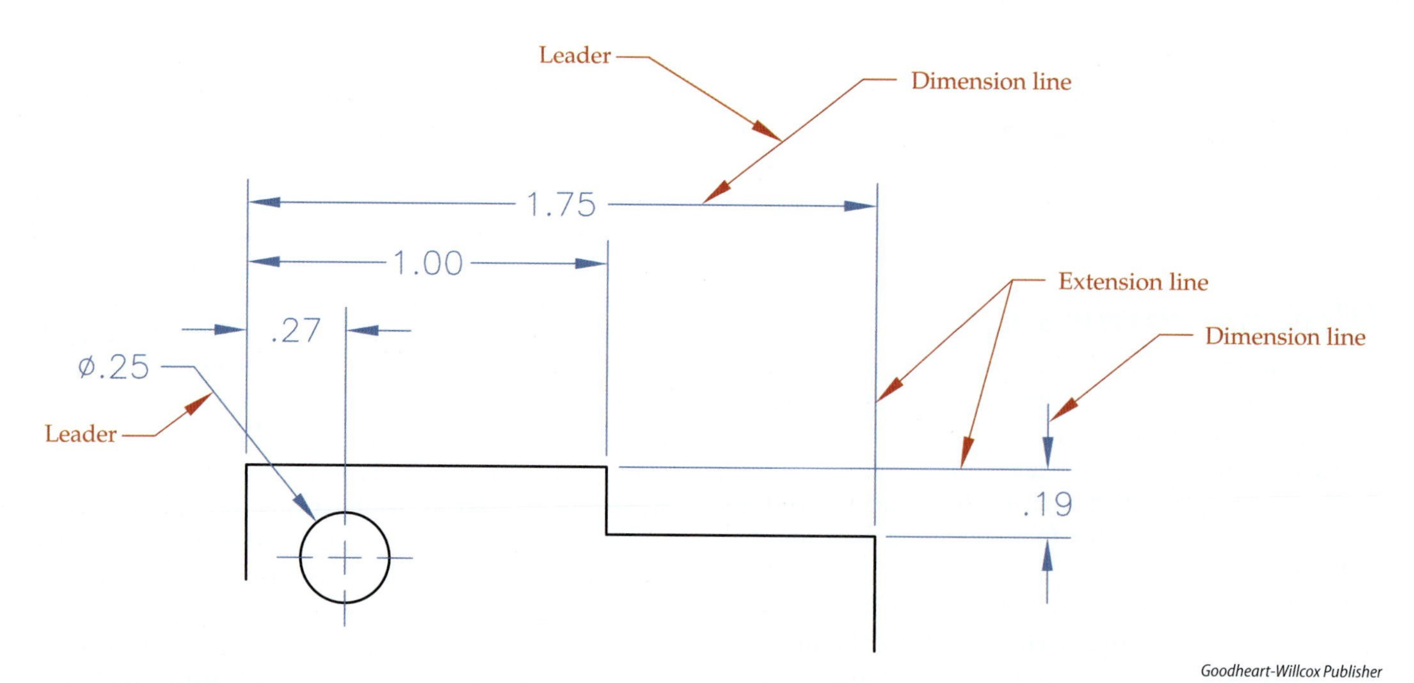

Goodheart-Willcox Publisher

Figure 2-6. Dimension lines are thin lines used to indicate the extent and direction of part dimensions, while the extension lines extend the object. Leaders are characterized by an arrow on one end and a shoulder on the other end, which is vertically centered with the beginning or end of the note.

for dimensional values within the dimension line. Leaders are thin, continuous lines characterized by an arrow on one end and a shoulder on the other. If the leader involves a descriptive note, the shoulder end of the line is vertically centered on the beginning or end of the note. As shown in the figure, the leader line can cross through extension lines, but the drafter tries to avoid crossing lines when possible.

Miscellaneous Lines

A few other lines should be discussed in this unit. As you progress through the textbook, you will encounter applications for these lines. These examples will help to reinforce your ability to discuss the purpose and function of each line.

Sometimes it is not practical to view a particular feature from the normal viewing arrangement. A different viewing direction can be established with a ***viewing-plane line***, **Figure 2-7**. The viewing-plane line is equivalent in appearance to the cutting-plane line, but it simply "floats" outside of the object instead of being placed at a hypothetical cutting position. When used, the view is usually labeled at each elbow with a letter of the alphabet, such as A, and the resulting view is placed off to the side and identified, such as VIEW A-A.

Two types of ***break lines*** are used in drawings to "break out" or "break off" a portion of a view. Long break lines are used to shorten objects that are constant in detail yet too long to fit on the drawing, such as a shovel handle or a long bar, **Figure 2-8**. Long break lines are thin, straight lines with zigzags.

The short break line is used in section view drawings if it is desirable for the cutting plane not to go all the way through an object, **Figure 2-8**. It is also used if a partial view of a part is useful for arranging the views on the page. The short break line, if drawn by hand, should be freehand and thick. This line should not be an exaggerated zigzag like lightning or splintered wood, but should look more like rough granite.

There is an exception when breaking long features such as round tubes or rods. In cases where round stock, such as shafts (solid) or pipe (tubular), needs to be broken, a conventional "S" break is used, **Figure 2-9**. When the part to be broken is not round, a short break line or long break line can be used.

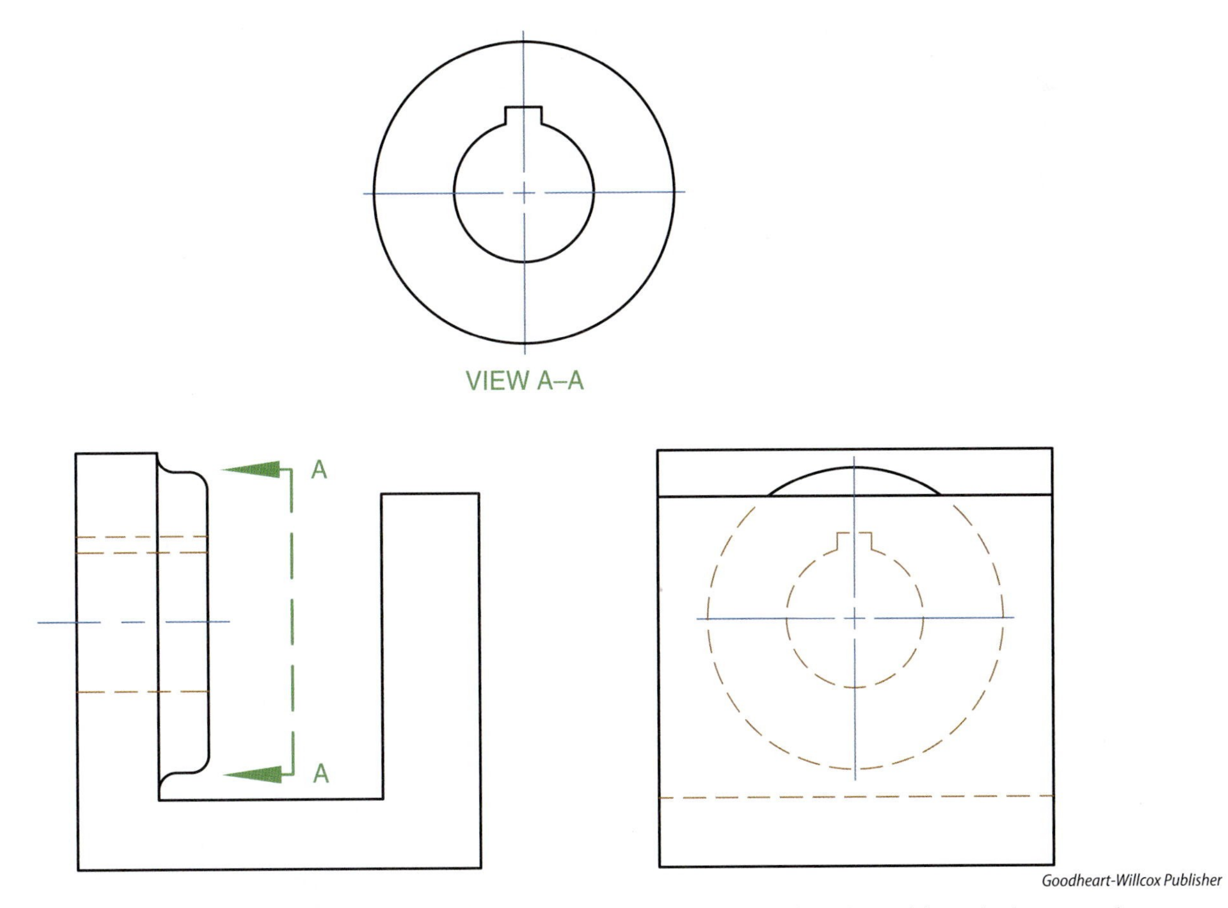

Figure 2-7. A viewing-plane line for a partial view is drawn the same as a cutting-plane line, although there are three standard options. This viewing-plane line indicates the direction for viewing VIEW A-A.

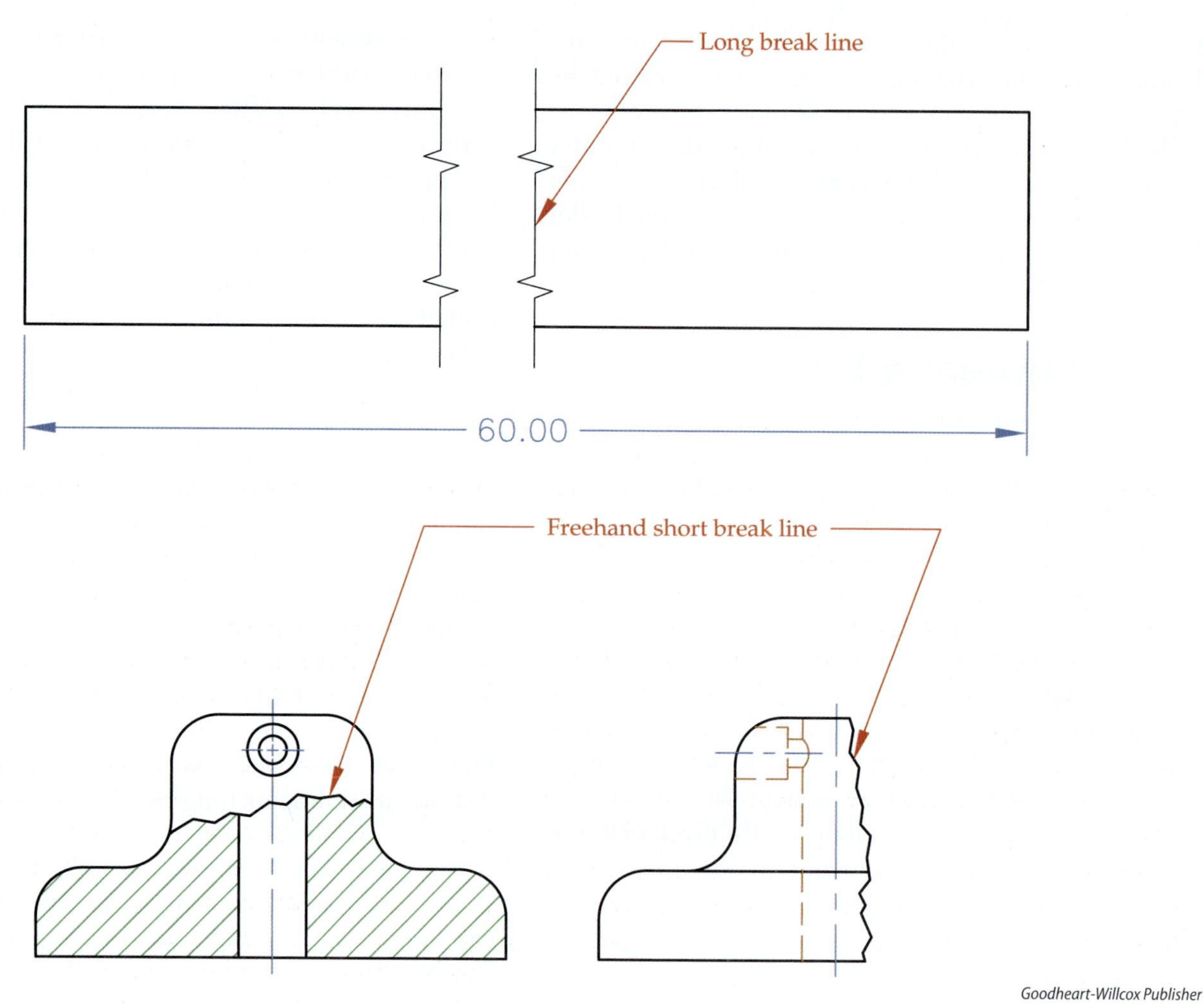

Goodheart-Willcox Publisher

Figure 2-8. Break lines can be used within views to break out sections for clarity or for shortening a view featuring a long and redundant part, such as a handle or tube.

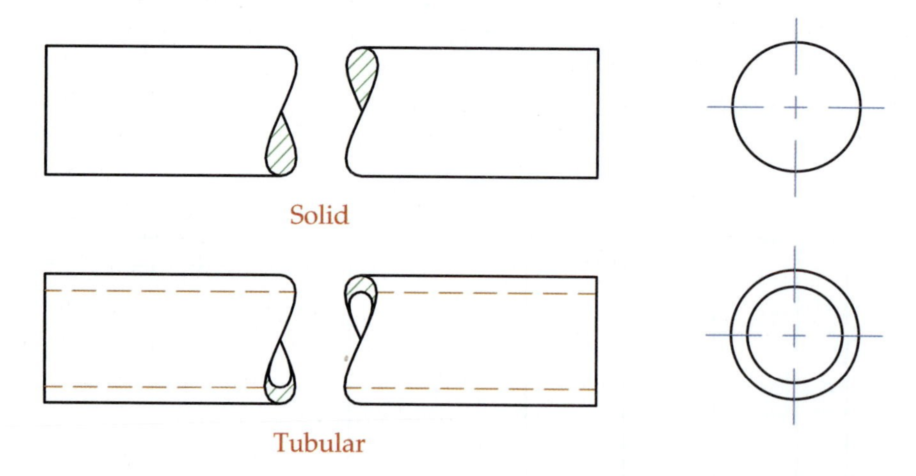

Goodheart-Willcox Publisher

Figure 2-9. Conventional breaks for solid cylindrical and tubular objects use S-shaped break lines.

Phantom lines are thin lines composed of long dashes alternating with pairs of short dashes. The dash pattern is similar to a cutting-plane line, but the line weight should be thin, not thick. Phantom lines are primarily used to indicate alternate positions of moving parts, such as a machine arm, **Figure 2-10A**; adjacent positions of related parts, such as an existing column, **Figure 2-10B**; or for repeated detail, **Figure 2-10C**.

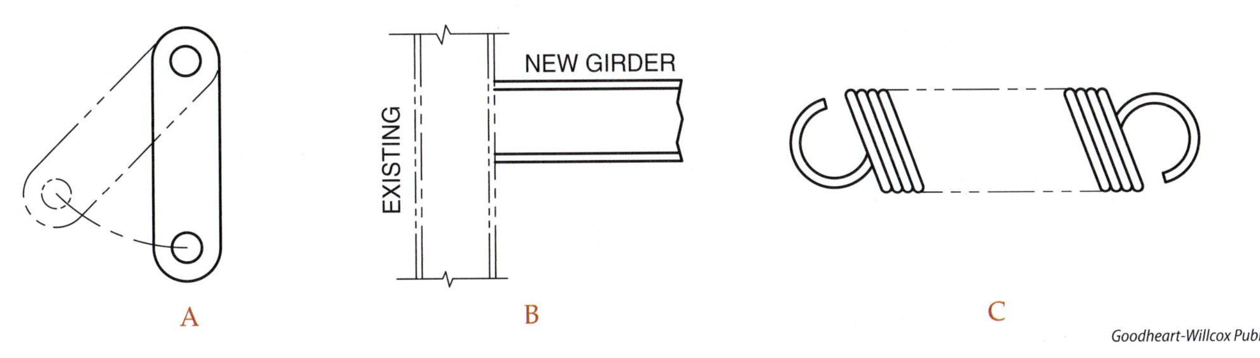

Figure 2-10. Phantom lines are used to show (A) alternate positions of moving parts, (B) adjacent positions of related parts, and (C) repeated detail.

Stitch lines are included in the ASME standard. These lines simply represent the path of a sewing or stitching process. They are comprised of either short dashes with spaces that are the same length, or a series of dots approximately 1/8″ apart.

The ***chain line*** is used to indicate an area on the drawing wherein something special applies, as illustrated by the NO PAINT note in **Figure 2-2**. For example, if the last inch of a rod is to be heat-treated, a 1″ chain line is drawn next to that area of the part, **Figure 2-11**. A local note can point to the chain line. A chain line appears somewhat similar to a center line, with short and long dashes alternating, but it is thick instead of thin.

Standardized Lettering

In engineering drawings, the text and numeric information is referred to as ***lettering***. Within the context of drafting, "lettering" is not only the letters and numerals themselves, but also the process of creating those characters. In the past, when drawing and lettering were both done by hand, it was of the utmost importance that the lettering be legible, uniform, and standard. Even today, for any documents created by hand, such as preliminary sketches, legible lettering helps avoid potentially costly mistakes.

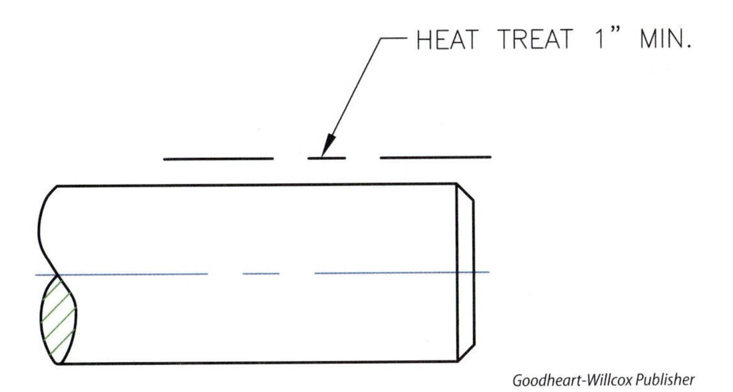

Figure 2-11. A chain line indicates something special applying to an area of a part, such as the heat treatment detail noted here.

Within the context of print reading, lettering is covered in this unit to establish the definition and standard expression recommended by the ASME standard. Lettering on an industrial print is to be uppercase lettering, unless lowercase lettering is specifically required. The recommended minimum height for lettering is 1/8″ (3 mm), with exceptions for drawing block headings, zone letters and numerals, section and view letters, and other information in the title block, such as part name and number. For additional discussion on title block information, see Unit 3.

By definition, engineering drawings created by hand use a font referred to in most drafting resources as "single-stroke Gothic lettering," **Figure 2-12**. "Single-stroke" does not imply each letter is created with a single stroke, but rather refers to the freehand technique of forming each letter from a series of single strokes. Gothic means simple, "sans serif" (without serifs) lettering, as opposed to a Roman font, which contains serifs. Serifs are the small tails that make letters appear fancier. As CAD technologies and computer fonts were developed, not everyone used the same terminology. For example, one popular CAD system has a font called GOTHIC, but it appears like a traditional "old English" font. The same system has a font called ROMAN SIMPLEX that matches the engineering Gothic style rather well, but also a ROMAN DUPLEX font that is a truly Roman-style lettering, complete with serifs. This can be confusing when establishing CAD template drawings and standard CAD management protocols. However, CAD lettering is consistent and accurate, and there are many fonts and styles from which to choose. In summary, it is critical that industrial prints have clear and concise lettering, and computer applications have helped to meet that goal.

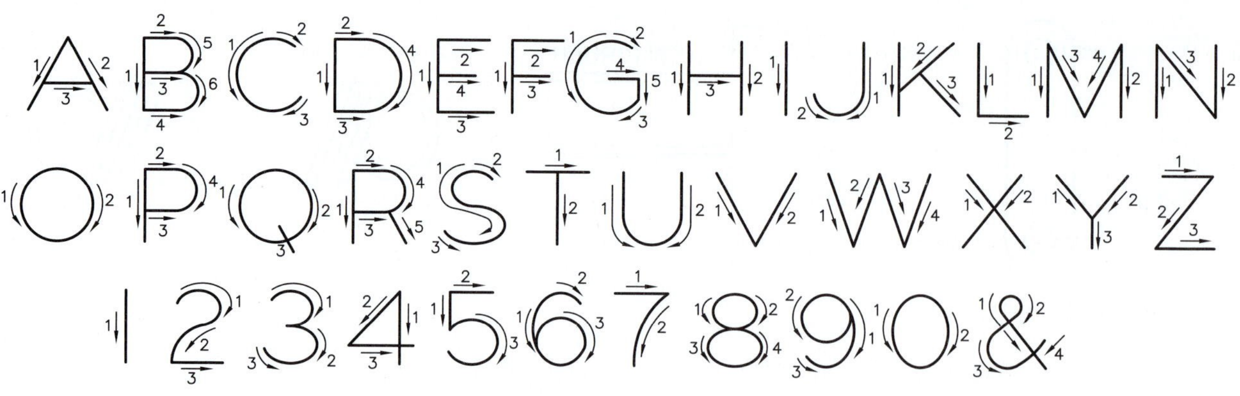

Goodheart-Willcox Publisher

Figure 2-12. Single-stroke Gothic lettering, usually set in uppercase letters, is the established standard for engineering drawings.

Summary

- A convention is a generally accepted way of doing things.
- The list of ASME standard lines is referred to as the alphabet of lines.
- Standard practices recommend two thicknesses for linework in industrial prints: thick (0.6 mm) and thin (0.3 mm).
- The visible line can be described as a thick and continuous line used to show the outline and shape of the part.
- The hidden line can be described as thin, black, short, closely spaced dashes used to describe hidden features of the object in a particular view.
- The center line can be described as a series of thin, black dashes, alternating between medium and long, used to show symmetry, identify center axes, or describe paths of motion.
- Lines used in dimensioning are drawn black, thin, and continuous, including extension lines that extend the shape from the view, and dimension lines that include arrows on each end to show the extent and direction of the dimension.
- Leader lines, drawn thin and black, feature an arrow on one end and a shoulder on the other end, most often in association with a lettered note.
- Lines used in sectional views include section lines, which are thin and black and usually appear in a pattern of angled lines sometimes called hatching, with the possibility of dashes to identify particular materials.
- Cutting-plane lines and viewing-plane lines may be drawn in one of three options, all of which include elbows and arrows to express the direction for viewing.
- Break lines are available in "long break" and "short break" options and are used when drawings or views need to show longer lengths or hidden features in a conventional manner.
- The alphabet of lines includes several miscellaneous types of lines such as the phantom line, stitch line, and chain line, each with a recommended dash pattern and line weight.
- The style of lettering recommended for standard industrial drawings is uppercase single-stroke Gothic lettering, indicating a clear, sans-serif form of lettering that is easy to read.

Unit Review

Name ______________________ **Date** ______________ **Class** ______________

Answer the following questions using the information provided in this unit.

Know and Understand

____ 1. *True or False?* A convention is a generally accepted way of doing something.

____ 2. *True or False?* The alphabet of lines is a list of six lines that are used to draw and dimension an industrial part.

____ 3. *True or False?* In the current standards, only two standard line weights (widths) are recommended.

____ 4. Which of the following lines is *not* one of the primary lines needed to describe the views of a part?
A. Visible line
B. Viewing-plane line
C. Center line
D. Hidden line

____ 5. *True or False?* There are three options for the appearance of a section line.

____ 6. *True or False?* If a cutting-plane line or viewing-plane line has arrows, they point in the direction the viewer must look to "see" the view.

____ 7. Which of the following lines is *not* one of the primary lines used in dimensioning a part?
A. Dimension line
B. Extension line
C. Chain line
D. Leader line

____ 8. *True or False?* The only difference between the long break line and the short break line is the object to which the line is applied.

____ 9. *True or False?* A line that is used to indicate alternate positions of moving parts, or for repeated detail, is the phantom line.

____ 10. Which statement listed below regarding characteristics of lettering is *false*?
A. Lettering, if done by hand, is composed of a series of single strokes.
B. Lettering on an engineering drawing should be a Gothic style or font.
C. Lettering on an engineering drawing has serifs and is very elaborate in appearance.
D. Lettering is uppercase, not lowercase.

Critical Thinking

1. Describe some ways in which a drafter or designer could disregard conventions.

____

____

____

____

____

2. While standards are described as "voluntary," what are some of the consequences that a company might incur if it does not choose to create lines in a standard fashion?

____

____

____

____

____

Apply and Analyze

Name ________________ Date ________ Class ________

Review Activity 2-1

Match the letter of each illustrated line with the correct name. Items are matched only once.

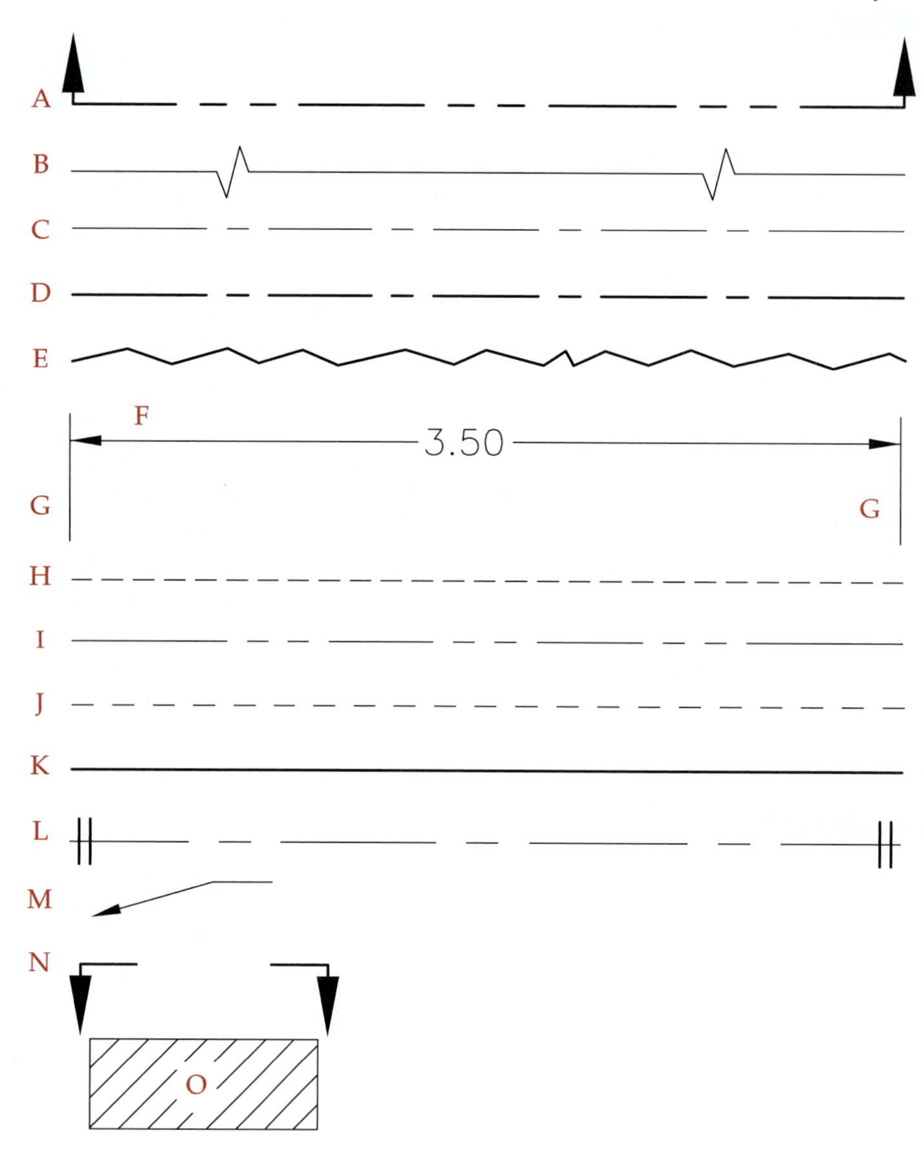

Goodheart-Willcox Publisher

Review Activity 2-1.

____ 1. Short break line
____ 2. Hidden line
____ 3. Center line
____ 4. Phantom line
____ 5. Section line
____ 6. Leader line
____ 7. Dimension line
____ 8. Extension line
____ 9. Cutting-plane line
____ 10. Visible line
____ 11. Viewing-plane line
____ 12. Chain line
____ 13. Long break line
____ 14. Stitch line
____ 15. Symmetry line

Apply and Analyze

Name ______________________ Date ______________ Class ______________

Review Activity 2-2

Study the drawing below and identify the twelve lines by name.

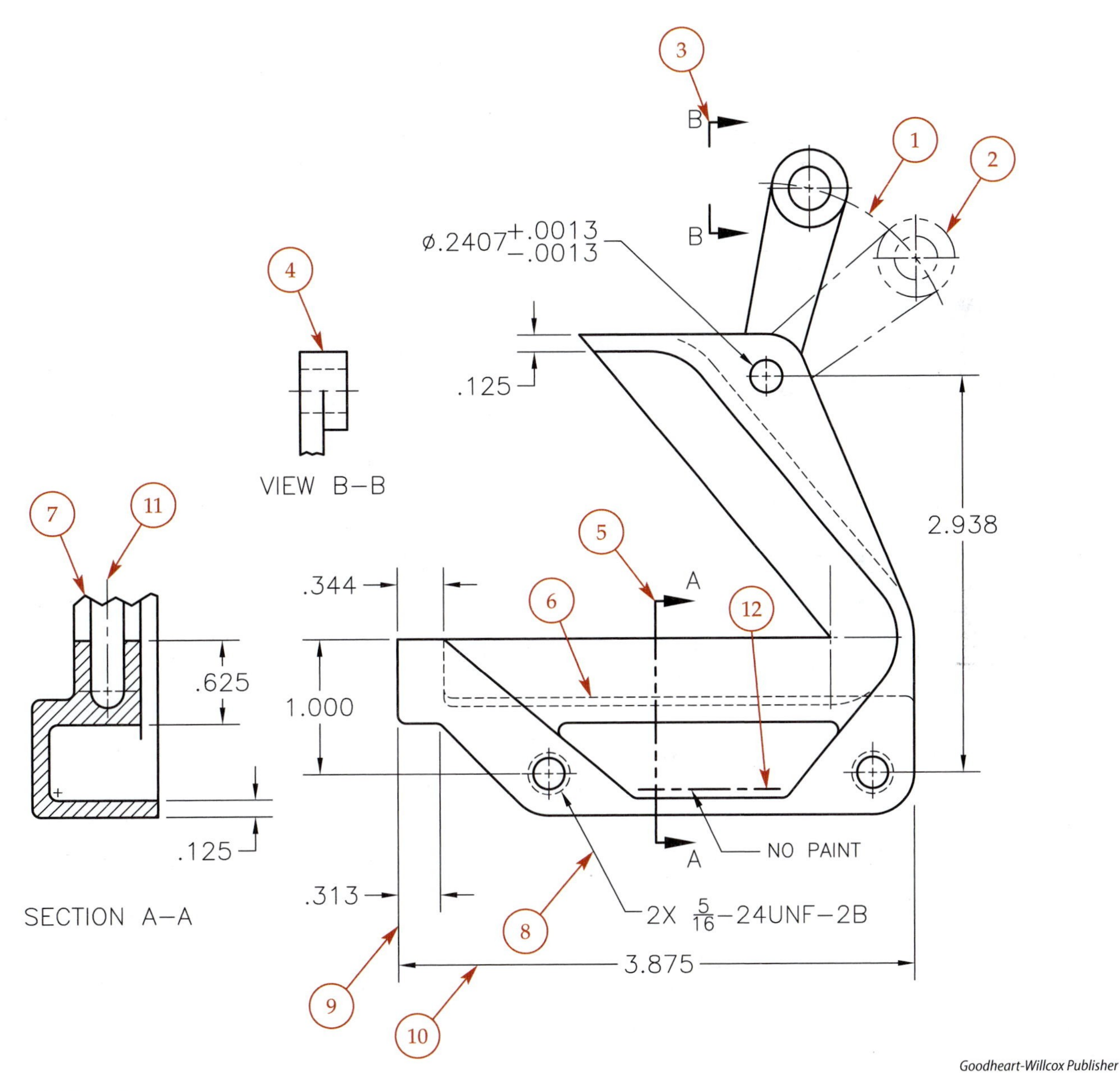

Goodheart-Willcox Publisher

Review Activity 2-2.

1. ______________________
2. ______________________
3. ______________________
4. ______________________
5. ______________________
6. ______________________
7. ______________________
8. ______________________
9. ______________________
10. ______________________
11. ______________________
12. ______________________

Apply and Analyze

Name ______________________ Date ______________ Class ______________

Industry Print Exercise 2-1

Refer to print PR 2-1. Closely examine the print to see which lines from the alphabet of lines are present or absent. For each question, answer P for present or A for absent.

____ 1. Cutting-plane line

____ 2. Dimension line

____ 3. Stitch line

____ 4. Section line

____ 5. Short break line

____ 6. Extension line

____ 7. Leader line

____ 8. Hidden line

____ 9. Chain line

____ 10. Center line

____ 11. Long break line

____ 12. Visible line

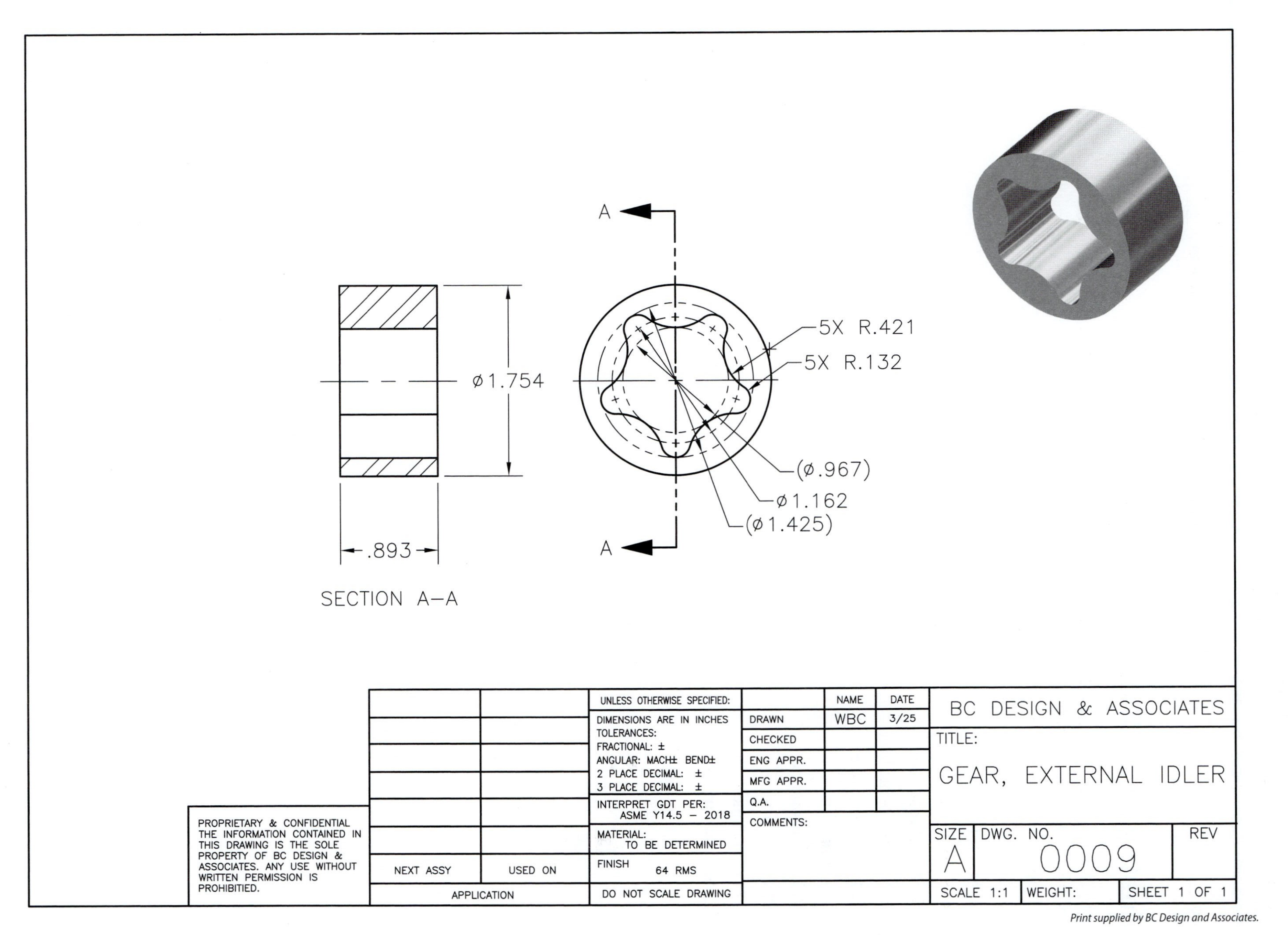

Print supplied by BC Design and Associates.

PR 2-1. External Idler Gear

Apply and Analyze

Name ______________________ Date ____________ Class ______________

Industry Print Exercise 2-2

Refer to the print PR 2-2. Closely examine the print to see which lines from the alphabet of lines are present or absent. For each question, answer P for present or A for absent.

____ 1. Extension line

____ 2. Phantom line

____ 3. Stitch line

____ 4. Visible line

____ 5. Short break line

____ 6. Cutting-plane line

____ 7. Long break line

____ 8. Hidden line

____ 9. Dimension line

____ 10. Center line

____ 11. Leader line

____ 12. Section line

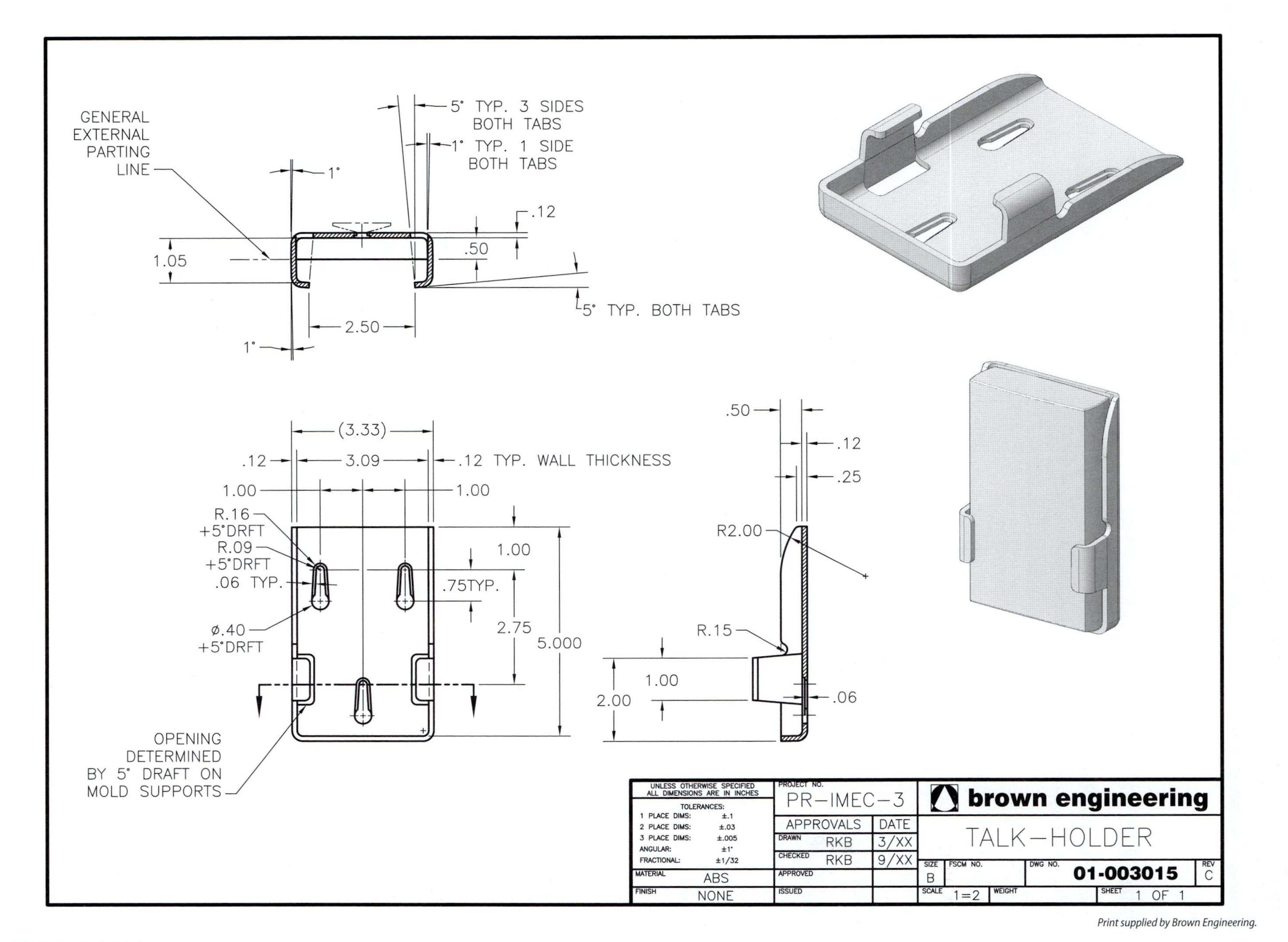

Print supplied by Brown Engineering.

PR 2-2. Talk-Holder

Apply and Analyze

Name ______________________ Date ______________ Class ______________

Bonus Print Reading Exercises

The following questions are based on the various bonus prints located in the folder that accompanies this textbook. Refer to the print indicated, evaluate the print, and answer the question.

Print AP-001

1. Do the visible lines of the object appear to be thicker than the center lines and section lines?

____

2. What type of line is used to indicate the circular area being enlarged for View A?

____

Print AP-003

3. In how many of the four main views are section lines used?

____

4. List the types of lines that appear in Section A-A, including the different types of lines used in dimensioning, if any.

____

____

Print AP-007

5. In the view with one dimension, 8X R.015, what two types of lines are featured besides the leader line?

____

Print AP-008

6. What type of line is used to indicate the cylindrical surfaces that are to be heat treated?

____

7. The involute splines are not shown in full detail. What type of line is used to indicate the involute splines that are specified on each end of this shaft?

____

Print AP-009

8. List four types of lines that are featured in this assembly drawing.

____

____

Print AP-012

9. Of the three options, describe the choice selected for the cutting-plane line.

____

____

10. What type of line is used to show the partial view break off?

____

11. List the types of lines that appear in Section A-A, including the different types of lines used in dimensioning, if any.

____

____

Print AP-016

12. Name the lines that appear in the enlarged detail A, including the different types of lines used in dimensioning, if any.

____

____

Print AP-017

13. What type of line is used in the upper-left round view to indicate the remaining gear teeth without drawing all of them?

____

Print AP-018

14. Not counting the lines used for dimensioning, list the types of lines that appear on this print.

____

____

Print AP-020

15. Do the cutting-plane lines of the object appear to be thicker than the center lines and section lines?

____

Notes

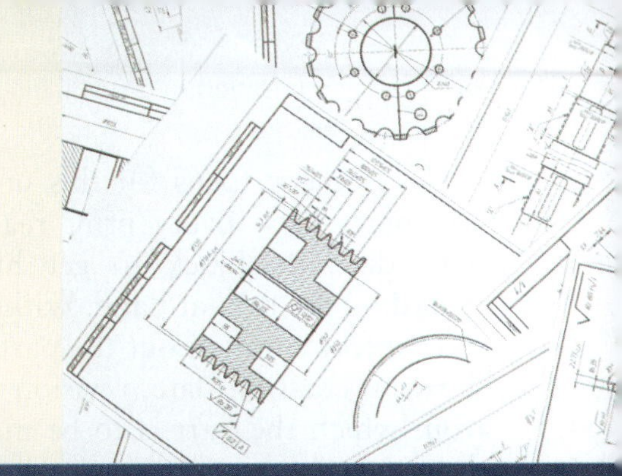

UNIT 3
Title Blocks and Parts Lists

LEARNING OBJECTIVES

After completing this unit, you will be able to:

- Discuss the importance of title blocks and parts lists within print reading.
- Describe paper sizes and systems for engineering, architectural, and ISO standards.
- Describe drawing sheet sizes and formats.
- Identify marginal information and zoning methods for drawing sheets.
- Identify the basic elements of a standard title block.
- Identify additional intermediate title block elements that may be found on an industrial print.
- Describe the function of the revision history block.
- Explain the techniques for identifying parts of an assembly drawing as represented in a basic parts list.

TECHNICAL TERMS

angle of projection block
application block
balloon
CAGE code
drawing number
drawing title
parts list
revision
revision history block
revision status of sheets block
scale
title block
tolerance
tolerance block
zone

Industrial drawings are placed on a variety of sheet sizes, depending on the complexity or size of the part. The drafter uses both of these factors to determine the physical size of the views that will be needed to clearly show detail and describe the part. Several factors are used together to determine the final sheet size and main scale for the majority of the views. More complex parts may even require more than one sheet.

ASME Y14.1, titled *Decimal Inch Drawing Sheet Size and Format*, and ASME Y14.1M, titled *Metric Drawing Sheet Size and Format*, cover the basics of sheet layout and format, title blocks, and revision blocks. ASME Y14.34, titled *Associated Lists*, sets forth recommendations for parts lists and data lists.

The primary focus for this unit is the title block and parts lists. Every print reader needs to begin with the title block to get his or her bearings regarding the task at hand. Within the title block resides information about the part name and number, drawing creation date, revision dates, and material from which the part is to be made. With assembly drawings, the parts list is usually critical to identifying the components that need to be assembled.

Sheet Size and Format

Drawings are prepared on standard-size sheets, but there are two systems for inch-based paper sizes: engineering and architectural. **Figure 3-1** illustrates four of the paper sizes, identified as sizes A through D. Engineering sheet sizes are in multiples of 8 1/2″ x 11″, which is commonly referred to as A-size. If D-size paper, 22″ x 34″, is cut in half, the result is two C-size sheets, each 17″ x 22″. Likewise, C-size paper cut in half results in two B-size sheets, each 11″ x 17″. Roll-size formats G, H, J, and K are also available for large parts and assemblies.

Architectural drawings are usually larger than engineering drawings and are based on rolls of paper manufactured in widths of 24″, 36″, or 48″. The drawing sheet sizes are therefore larger as well. The A-size sheet for architectural applications is 9″ x 12″, B-size sheets are 12″ x 18″, C-size sheets are 18″ x 24″, and D-size sheets are 24″ x 36″. Most professional output devices can accommodate both systems of paper. See **Figure 3-2** for a size comparison.

In the United States, the decimal-inch system is still predominant, especially with respect to paper sheet sizes. Even if drawings are created with metric units, most office supplies are still specified to decimal-inch paper sizes. The ISO 216 standard sets metric paper

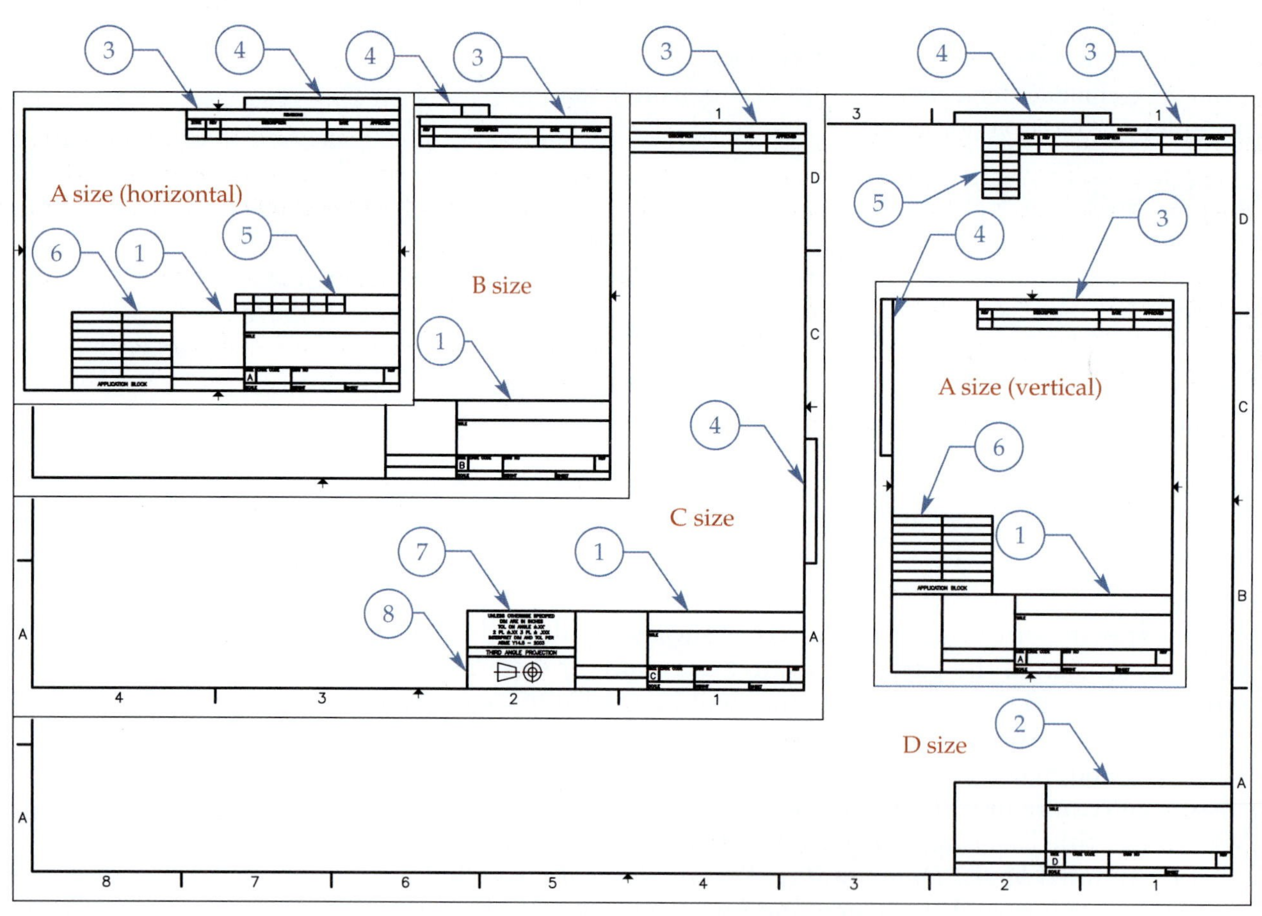

1. Title block for A, B, and C size
2. Title block for D, E, and F size
3. Revision history block
4. Margin drawing number block (optional)
5. Revision status of sheets block
6. Application block (optional)
7. Tolerance block
8. Angle of projection block

Goodheart-Willcox Publisher

Figure 3-1. Flat-sheet formats are standardized by ASME. Sheet formats A through D in landscape mode and A in portrait mode are shown here.

size standards for an "A series" set of paper sizes. Metric paper sizes have the additional benefit that each size, if halved or doubled, maintains the same aspect ratio (length versus height).

Metric sheet sizes commonly used for industrial prints are identified as A0, A1, A2, A3, and A4. See **Figure 3-2** for a size comparison chart. **Figure 3-3** illustrates metric sheet sizes for the ISO 216 "A" series. This unit will not attempt to cover both decimal-inch and metric systems, as the content is virtually the same, simply with different units.

Borders and Zoning

As illustrated in **Figure 3-1**, the margins for a drawing vary depending on sheet size and available space. The recommended minimum space varies from .25″ to 1.00″. Roll-feed sheets for wide-format printers may require more space along one edge. A border line is usually included around the edge of the paper. Border lines are technically not in the alphabet of lines. They are usually drawn thick, perhaps even thicker than visible lines.

ASME standards recommend zoning for sheet sizes larger than B-size, although zoning is permissible even for smaller sheets. Outside of the border, ***zones*** are used to aid in locating details of parts or revision notes, **Figure 3-4**. The zoning system is similar to that used on highway maps. Numbers are used at intervals from right to left. Letters are used at intervals from bottom to top. Zones are noted with the letter first and the number second, for example, zone D3.

Letter Size	Engineering Standard	Architectural Standard	Similar to ISO Standard
A	8.5 x 11	9 x 12	A4
B	11 x 17	12 x 18	A3
C	17 x 22	18 x 24	A2
D	22 x 34	24 x 36	A1
E	34 x 44	36 x 48	A0
E1		30 x 42	
F	28 x 40		

Goodheart-Willcox Publisher

Figure 3-2. The engineering standard for paper is different from the architectural standard. Similar metric paper sizes are also available.

ISO 216 "A" Series	Size in Millimeters	Size in Inches
A4	210 x 297	8.3 x 11.7
A3	297 x 420	11.7 x 16.5
A2	420 x 594	16.5 x 23.4
A1	594 x 841	23.4 x 33.1
A0	541 x 1189	33.1 x 46.8

Goodheart-Willcox Publisher

Figure 3-3. The ISO 216 standard specifies the dimensions of the A series of metric sheet sizes.

Basic Title Block Elements

The ***title block*** provides information that aids in identification and filing of the print. The title block also provides supplementary information about the part or assembly. The title block is usually located in the lower right-hand corner of the print, so that it can be seen when the print is correctly folded. This allows for easy reference and filing.

While many companies have their own variations of the standard title block, the information found in a title block is similar between companies. Title block information included on prints by most industries is explained in this unit. Understanding the information typically found in a title block will help you properly interpret all title blocks.

The name and address of the company is usually placed in the upper part of the title block, **Figure 3-5A**. Often, the company logo also appears in this space. With CAD systems, it is easy to insert a standard logo into a drawing.

The ***drawing number*** is used to identify and control the print. It is also used to designate the part or assembly shown on the print, **Figure 3-5B**. The number is usually coded to indicate department, model,

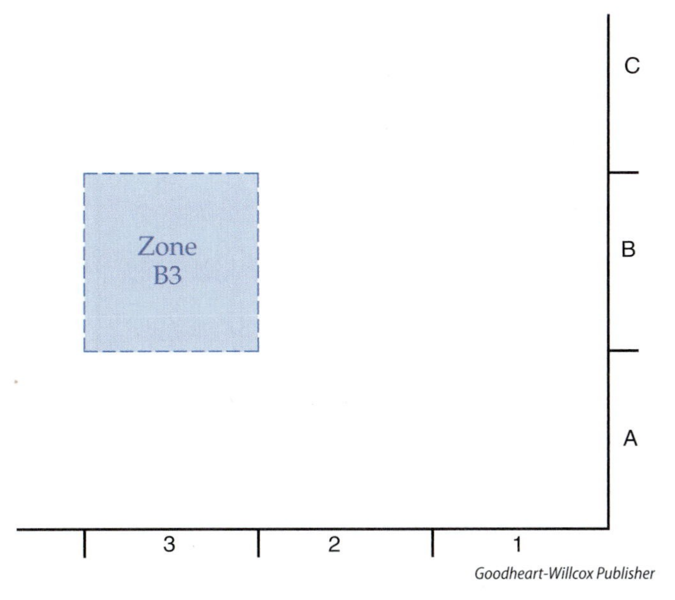

Goodheart-Willcox Publisher

Figure 3-4. Zoning is used to specify an area of a large print, similar to what is used on highway maps.

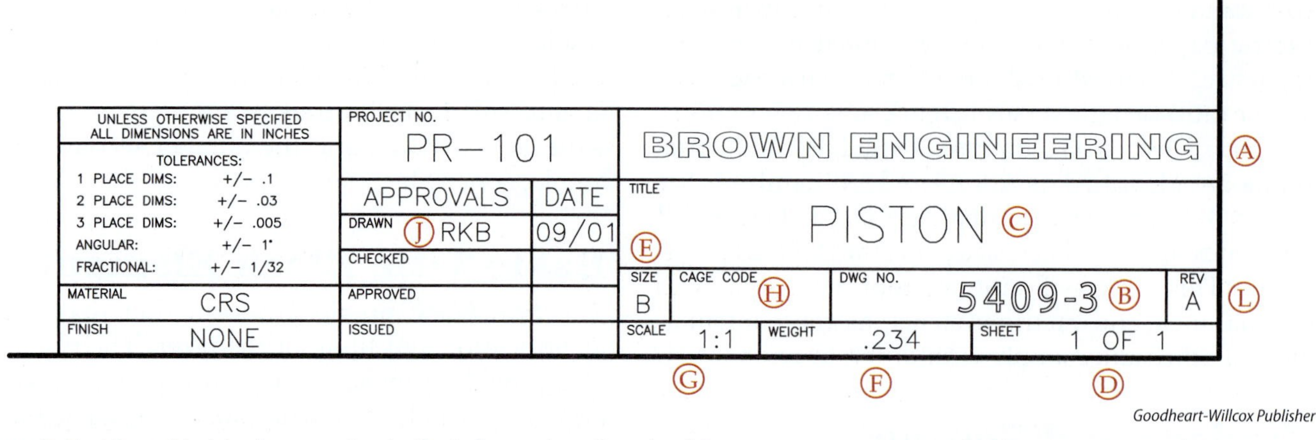

Goodheart-Willcox Publisher

Figure 3-5. Most title blocks contain similar information. Standard formats are recommended but not required.

group, serial number, and dash numbers. The drawing number may also be shown in the margin of the drawing, usually in a rectangular box. This is designed so the drawing number shows along the top edge when the drawing is folded to A-size, **Figure 3-6.**

The ***drawing title*** indicates the name of the part. The title should be descriptive and brief, and should clearly state an identification of the part or assembly. See **Figure 3-5C.** The title starts with the name of the part or assembly, which is followed by descriptive modifiers. When the title is read out loud, the descriptive modifier is read first. For example, the title in **Figure 3-7** is read "pressure regulating valve assembly."

The sheet area of the title block is used for sheet numbering. Sheet numbering is used on multisheet prints to indicate the consecutive order and total number of prints. Sheet numbering is often written as SHEET *x* OF *n*, where *x* is the number of the sheet, and *n* is the total number of sheets for that set. See **Figure 3-5D.**

The size area of the title block is used to indicate which sheet size is used for the print. A letter designation, such as B, C, or D, indicates the standard size of the paper. Refer to **Figure 3-2** or **Figure 3-3** for standard paper sizes.

The weight area of the title block provides the weight of the part, **Figure 3-5F.** The weight can be either the actual or calculated weight, as indicated. Calculated weight is used during design stages to control the weight of the finished part or assembly. Actual weight is obtained after the part or assembly is actually manufactured.

The scale area indicates the scale of the drawing. The ***scale*** is the ratio between the size of the part as drawn and the actual size of the part. It is usually expressed as *paper* = *real* or *paper*:*real*, where the first number (*paper*) indicates the size measured on the paper and the second value represents the size measured on the actual part. See **Figure 3-5G.** Typical scale notations are: 1/2″ = 1″ (half size), 1:2 (half size), FULL (actual size), 1:1 (actual size), and 2:1 (twice size). Sometimes 2X or 3X is used to indicate 2:1 or 3:1, respectively. When several scales are used in the drawing, the scale area of the title block reads AS NOTED, and each scale is indicated below the

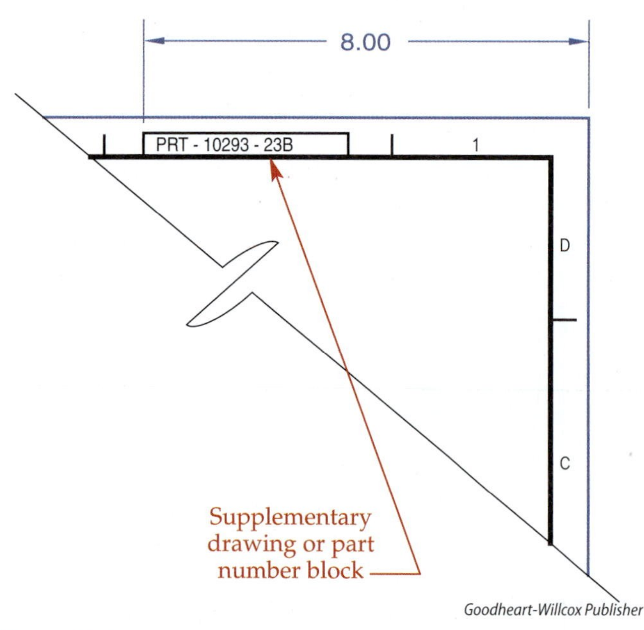

Goodheart-Willcox Publisher

Figure 3-6. Standards allow for the drawing number to be shown in the margin to aid in finding a drawing in the filing drawer.

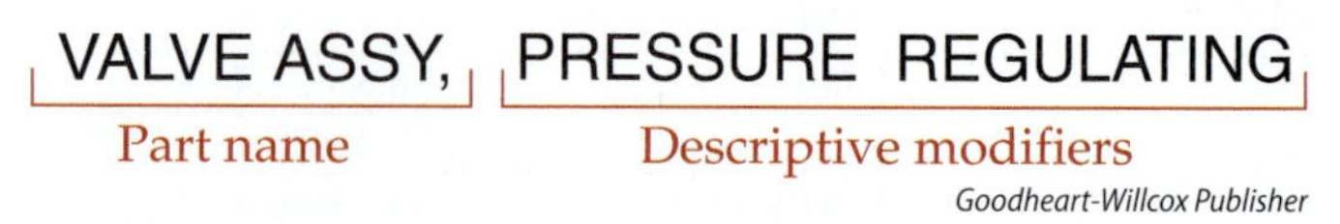

Goodheart-Willcox Publisher

Figure 3-7. A part name often contains descriptive modifiers. When read out loud, the modifiers are read before the part name.

particular view to which it pertains. Some drawings, especially diagrams, have no scale. Pictorial drawings representing a part or assembly in 3D are often not shown to scale, as the object is "turned and tilted" and the object lines are foreshortened. The designation NONE is an appropriate entry in the title block for a drawing with no scale.

It is important to note that even though drawings are usually printed to scale and the scale is indicated in the title block, measurements should *never* be made directly on a print. This is because the print may be reduced in size or stretched. Work from the dimension values given on the print. If you believe these to be in error, report it to the appropriate person in the company.

Some title blocks have an area labeled ***CAGE code***, which is a unique five-digit alphanumeric identifier given to vendors for the government, **Figure 3-5H**. CAGE is an acronym for Commercial and Government Entity. On older drawings, this block may be identified as FSCM (Federal Supply Code for Manufacturers) or NSCM (National Supply Code for Manufacturer). In summary, this area is used to identify part suppliers on drawings. For parts not requiring a CAGE code, this area is left blank or may be omitted from the title block layout.

On the left side of a title block is an area for signatures and date-of-release notations. These areas are not specified in current standards, but may appear as shown in **Figure 3-5J**. Entries in this area are made by those responsible for making or approving certain facets of the drawing or manufacture of the part. There may be many signatures on a given drawing. The following are some areas of responsibility:

- **Drawn.** This area is for the drafter who made the drawing.
- **Checked.** This area is for the engineer or checker who checked the drawing for completeness, accuracy, and clarity.
- **Approved.** This area is to record any other required approvals.
- **Issued.** This area indicates the person who finally issues the drawing as available for general use, making it an "official" drawing.

Intermediate Title Block Elements

There are many other areas of information that will be more clearly understood after some additional units are covered. Within your progression of study, it is important to understand the importance of studying the title block early and often in the process of interpreting the full scope of the product or part. Additional title block information is discussed in future units, but this section is a brief overview of additional information found in some title blocks.

In manufacturing, part dimensions are often within a range. The total amount of the range is the ***tolerance*** for that value. Most title blocks include a ***tolerance block***, which indicates the general tolerance limits for one-, two-, and three-place decimal values and a certain number of degrees for angular dimensions. These limits are to be applied unless the tolerance is otherwise indicated on the drawing. Refer to Unit 13 for a discussion on tolerancing.

The ***angle of projection block*** is usually located below the tolerance block and identifies whether the drawing is a first-angle or third-angle projection. Refer to Unit 5 for a discussion on angles of projection.

The finish area in a title block indicates general finish requirements, such as paint, chemical, or other. Any specific finish requirements called out in a local note can be addressed in the title block with the word NOTED or words AS NOTED in the finish area. Refer to Units 11 and 12 for a discussion on machining specifications and surface texture symbols.

The materials area in a title block indicates the material used to make the part. Some companies have material numbers assigned to all raw materials. The number of the material is then indicated in the title block. See **Figure 3-8**.

The heat treatment area in a title block indicates heat treatment and hardness requirements. See **Figure 3-9**. The entry may be AS REQUIRED or NOTED, which means the part must conform to the specification block notation or to the callout on the drawing detail. If heat treatment is not required, the word NONE or a diagonal line is entered in the block.

An ***application block*** can be added to the title block. The application block lists the assembly or subassembly in which the part or subassembly is used. See **Figure 3-10**. Indicating where the part or

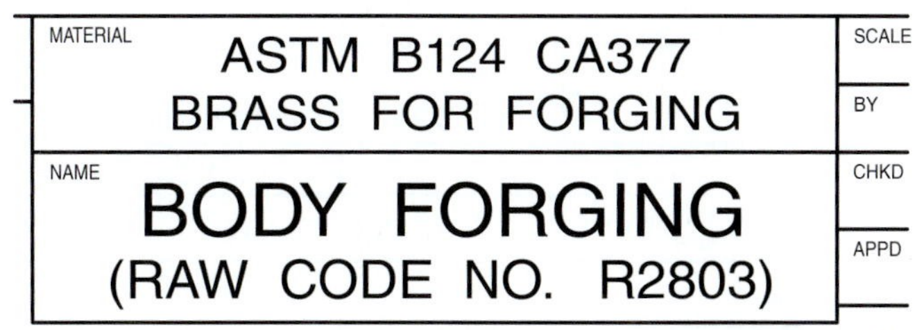

Figure 3-8. The Material area of the title block indicates the material used to make the part.

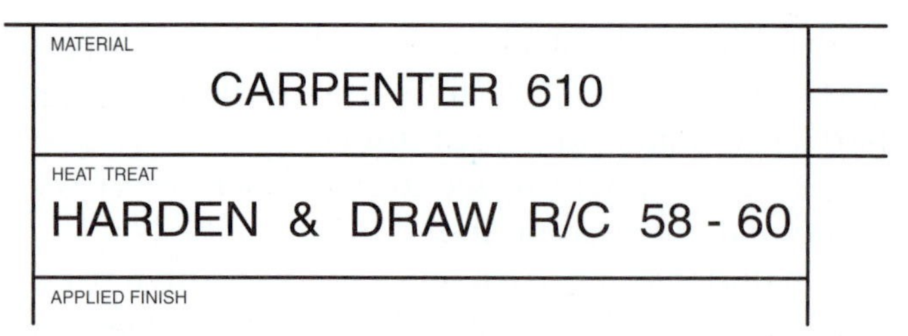

Goodheart-Willcox Publisher

Figure 3-9. The Heat Treatment area in the title block indicates heat treatment and hardness requirements.

assembly is used can help determine the effects of a change in the part or assembly. Refer to Unit 16 for a discussion on assembly drawings.

A few companies include information in the title block indicating DRAWING SUPERSEDED BY or DRAWING SUPERSEDES. For example, if PART 111 supersedes PART 110, then PART 111 is the new part that replaced PART 110 in the system. This enables a drawing to report the previous part number or drawing number as a historical record, which helps the print reader locate information connected to the history of the part. This information can also be applied to a drawing that is being discontinued and archived so individuals can find the new drawing and part that superseded the old part.

Other information may appear in a title block. The customer for whom the product is being produced is sometimes listed, as is the contact number. Standards used for inspection may be used to refer to specific sets of inspection standards used by the company. A diagonal line or X appearing in any part of the title block means the information within that space is not required.

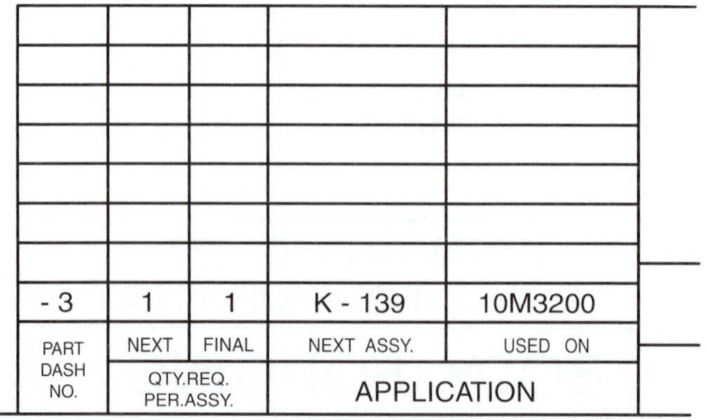

Goodheart-Willcox Publisher

Figure 3-10. The application block, if present, is attached to the title block and lists the assembly or subassembly in which the part is used.

Revision History Block

A ***revision*** is any change made to the original drawing. A system for reporting the history of changes to a part and its drawing is also included on the sheet. Unit 14 discusses revision systems in more detail, but locating the current revision status for a drawing is often a necessary initial step in the print reading process. As a part is revised, the differences are usually recorded on the drawing with alphabetical notations. It is important that the print reader establish the current revision status of the drawing and assess whether the drawing is the current edition. There are times when a revision is pending, which should be indicated somewhere on the print, perhaps as a bold or brightly colored stamp.

While standards recommend the ***revision history block*** be located in the upper-right corner of the sheet layout, **Figure 3-11**, the revision block may reside in other corners or near the title block. This block may include information such as the letter assigned to a change, what change order number was assigned, the revision date, and other useful information. Additional columns can also be added as necessary. For multiple-sheet drawings, there may be a ***revision status of sheets block*** attached near the revision history block or on a separate document to record the revision status of each sheet.

Parts Lists

Lists of information relative to the drawing often take the form of charts or tables. Recommended layouts are covered in ASME Y14.34, titled *Associated Lists*. While some of these are beyond the scope of this text, the recommendations for charting a list of parts for an assembly, subassembly, or welding detail drawing will be discussed. Additional discussion can be found in Unit 16, *Assembly Drawings*, and Unit 22, *Welding Prints*.

REVISION HISTORY			
REV	DATE	DESCRIPTION	APPRVD
A	02/10	REDRAWN	RKB
B	03/10	ADDED GD&T	DCW
C	11/10	REMOVED HOLE AND KEYWAY	MAR

Goodheart-Willcox Publisher

Figure 3-11. The revision block contains a record of the changes that have been made to the original drawing.

The name recommended by ASME for a chart listing the parts within a particular drawing is parts list. The ***parts list*** is a tabular chart or form, usually appearing immediately above the title block on a print, **Figure 3-12**. It provides information with respect to the quantity and description of the parts of a machine, structure, or assembly. Parts lists are used primarily for subassembly, assembly, welding, and installation drawings. Throughout the years, the parts list block has been referred to by many names, including the list of materials, materials list, bill of materials, and schedule of parts.

The following items are usually included in a basic parts list, as illustrated in **Figure 3-12**:

- Find number or mark column
- Quantity required column
- Part or identifying number column
- Nomenclature or description column

The find number or mark column is an optional column used if find numbers or marks have been assigned to a part (in lieu of the actual part number) for the purpose of helping the print reader locate the part on a drawing. The find number is often found within a circle, or balloon, which is discussed later in this unit.

The quantity required column is mandatory and indicates the number of parts required for an assembly. When the letters AR appear in the column in place of a number, it means the number is "as required."

The part or identifying number column is mandatory, indicating part numbers required for the assembly. As discussed in Units 15 and 16, each part normally has its own detail drawing, so this column helps identify additional drawings that contain important and contractual information. All part numbers for company- and subcontractor-manufactured parts are listed. Standard parts, such as bearings, bolts, and screws, are listed by the company-assigned supply number. Sometimes special fasteners, such as rivets, might not be listed in the block. Also, welding prints may not have part names assigned to the individual pieces that are being welded together.

The nomenclature or description column is mandatory and provides "real" names for all subassemblies and parts. These "real" names are usually limited to a basic noun description or, if a main assembly has a description listed, an abbreviated name.

Additional columns for the parts list are discussed in Unit 16, *Assembly Drawings*. Parts lists can become quite involved for assembly drawings that document several different versions of an assembly within one drawing.

Part Identification

A system for identifying the parts appearing in the parts list is necessary. Most assembly or subassembly drawings use leaders to identify the parts. In addition, sometimes the parts are exploded, which means spread out in an aligned fashion, to help the print reader see the relationship between parts. Many companies use balloons neatly arranged on the drawing to identify the parts. ***Balloons*** are circles containing a number or letter and are usually connected to leader lines, **Figure 3-13**. Depending on the complexity of the drawing or part names and numbers, detail callouts can appear adjacent to the part on the drawing. See **Figure 3-14**. A detail callout includes the part name and number, and may eliminate the need for a parts list block.

8	1	8400-356	INSTRUCTIONS
7	1	143-5321-150	LABEL, 100-200 SERIES
6	3	304-5300-101	SPRING, GUIDE
5	3	304-5300-100	PIN, GUIDE
4	1	143-5320-410	BUSHING
3	1	143-5320-407	COLLET, SELF-CENTERING
2	1	143-5321-202	HUB, 200 SERIES
1	1	143-5321-201	BASE, 200 SERIES
MARK	QTY	PART NUMBER	DESCRIPTION
PARTS LIST			

Goodheart-Willcox Publisher

Figure 3-12. This basic parts list provides information about the quantity and names of the parts within an assembly.

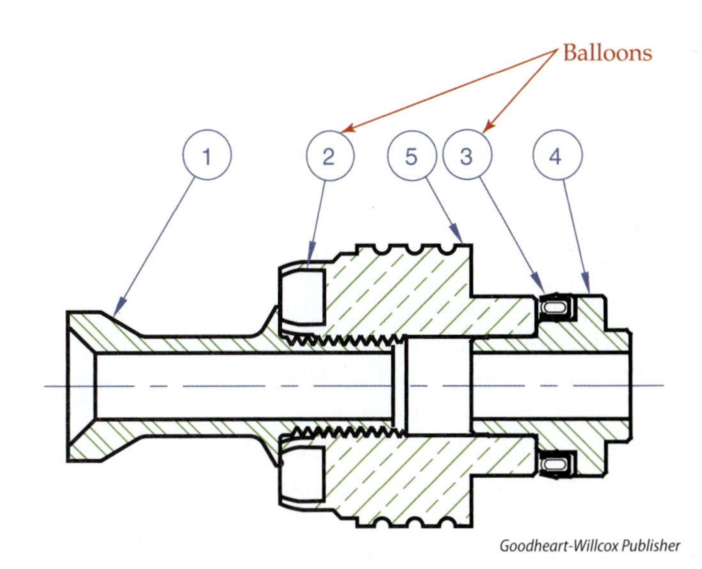

Goodheart-Willcox Publisher

Figure 3-13. Balloons can be used to identify parts.

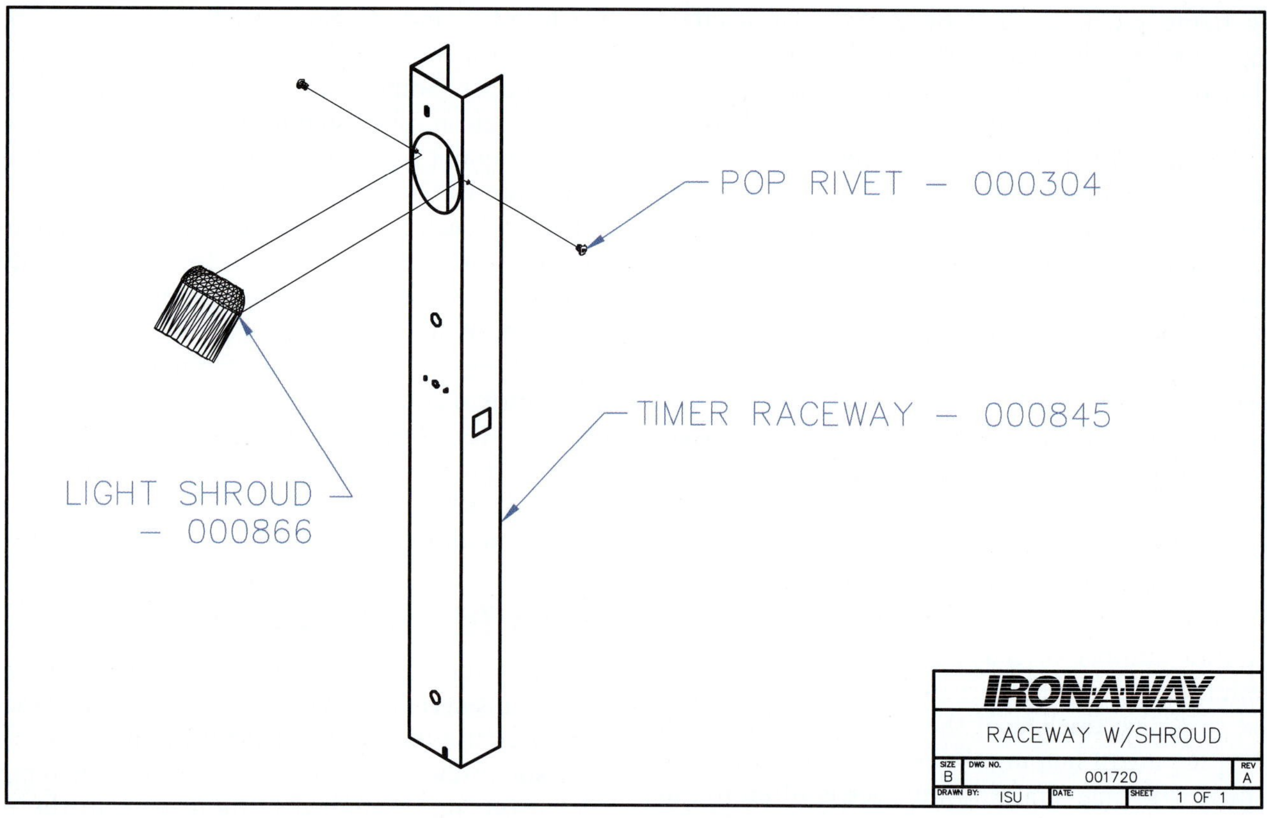

Iron-A-Way

Figure 3-14. A detail callout includes the part name and number without balloons.

Summary

- The title block is an important aspect of print reading and should be the beginning point for reading a print.
- ASME standards are available to guide companies in their use of title blocks and parts lists.
- The engineering and architectural fields both have a letter-based method of describing paper sizes, each with a small A-size basic sheet size.
- Metric paper sizes are standardized by ISO standards and have the advantage of being proportionally the same as the paper doubles in size.
- Thick border lines are used to frame the drawing, and zone letters and numbers are often incorporated on larger sheets to assist with finding information.
- Basic title block elements include drawing name and number, number of sheets, size of paper, part weight, and drawing scale.
- If applicable, title blocks contain a CAGE code for government projects.
- While not specified in standards, drawings usually include spaces for the initials or names of the drafter, the checker, and the approver, and perhaps the issue date.
- Intermediate title block information is often included within a tolerance block, an angle of projection block, and an application block.
- A revision history block, recommended for the upper right-hand corner of the drawing, records and lists the revision information about a drawing.
- Parts lists are common for assembly drawings and take the form of a chart that is connected to the title block and includes information such as identifying mark, quantity, description, and part number for each part listed.
- Balloons connected to leaders are often included in a drawing to help the print reader identify the parts that are listed in the parts list.

Notes

Unit Review

Name ______________________ Date ______________ Class ______________

Answer the following questions using the information provided in this unit.

Know and Understand

____ 1. *True or False?* When reading a print, the title block should be read first.

____ 2. *True or False?* There are no standards for title blocks, so each company creates its own title block with its own company look and feel.

____ 3. C-size paper is either _____ (engineering) or _____ (architectural).
A. 8.5″ x 11″; 9″ x 12″
B. 11″ x 17″; 12″ x 18″
C. 17″ x 22″; 18″ x 24″
D. 22″ x 34″; 24″ x 36″

____ 4. *True or False?* A D-size sheet of paper can be cut into two sheets of C-size paper.

____ 5. *True or False?* The closest metric paper size for A-size paper in the inch system is A1.

____ 6. The numbers and letters found in the margins of larger drawings are related to _____.
A. dates
B. scales
C. material
D. zones

____ 7. *True or False?* For security reasons, the company name and address are usually not placed in the title block.

Match the following hypothetical title block information to the basic title block elements. Answers are used only once.

____ 8. Drawing Number
____ 9. Drawing Title
____ 10. Sheet Size
____ 11. Scale
____ 12. Weight
____ 13. Number of Sheets

A. B
B. 2:1
C. P619405-256
D. PISTON
E. 1 OF 1
F. .45 LB

____ 14. *True or False?* The scale of a print is usually given as "paper = real," so 1 = 2 means one inch on the paper equals two inches on the part.

____ 15. Which of the following is *not* one of the signatures usually found on a print, as discussed in this unit?
A. Drawn by
B. Checked by
C. Supplied by
D. Approved by

____ 16. Which of the following is *not* one of the intermediate title block elements?
A. Application block
B. Tolerance block
C. Angle of projection block
D. General notes block

____ 17. *True or False?* While the revision history block is recommended for the upper right corner, it can be found in other corners of the drawing or attached to the title block.

____ 18. *True or False?* Two types of drawings that may require a parts list are assembly drawings and welding drawings.

____ 19. Which of the following is *not* usually found in a basic parts list?
A. Scale of the part
B. Name of the part
C. Find number of the part
D. Quantity required

____ 20. *True or False?* To assist the print reader in identifying parts on a drawing, find numbers appearing in circles called "bubbles" float near each part in the drawing.

Critical Thinking

1. Why is it important to pay attention to the revision history block when approaching a print as a print reader?

2. Explain why a print reader should never make measurements directly on a print.

3. Why might it be important for a print reader to know who drew, checked, approved, and issued a print?

Apply and Analyze

Name ______________________ Date ______________ Class ______________

Review Activity 3-1

Refer to the print Activity 3-1 and identify the numbered items by name.

1. ______________________
2. ______________________
3. ______________________
4. ______________________
5. ______________________
6. ______________________
7. ______________________
8. ______________________
9. ______________________
10. ______________________

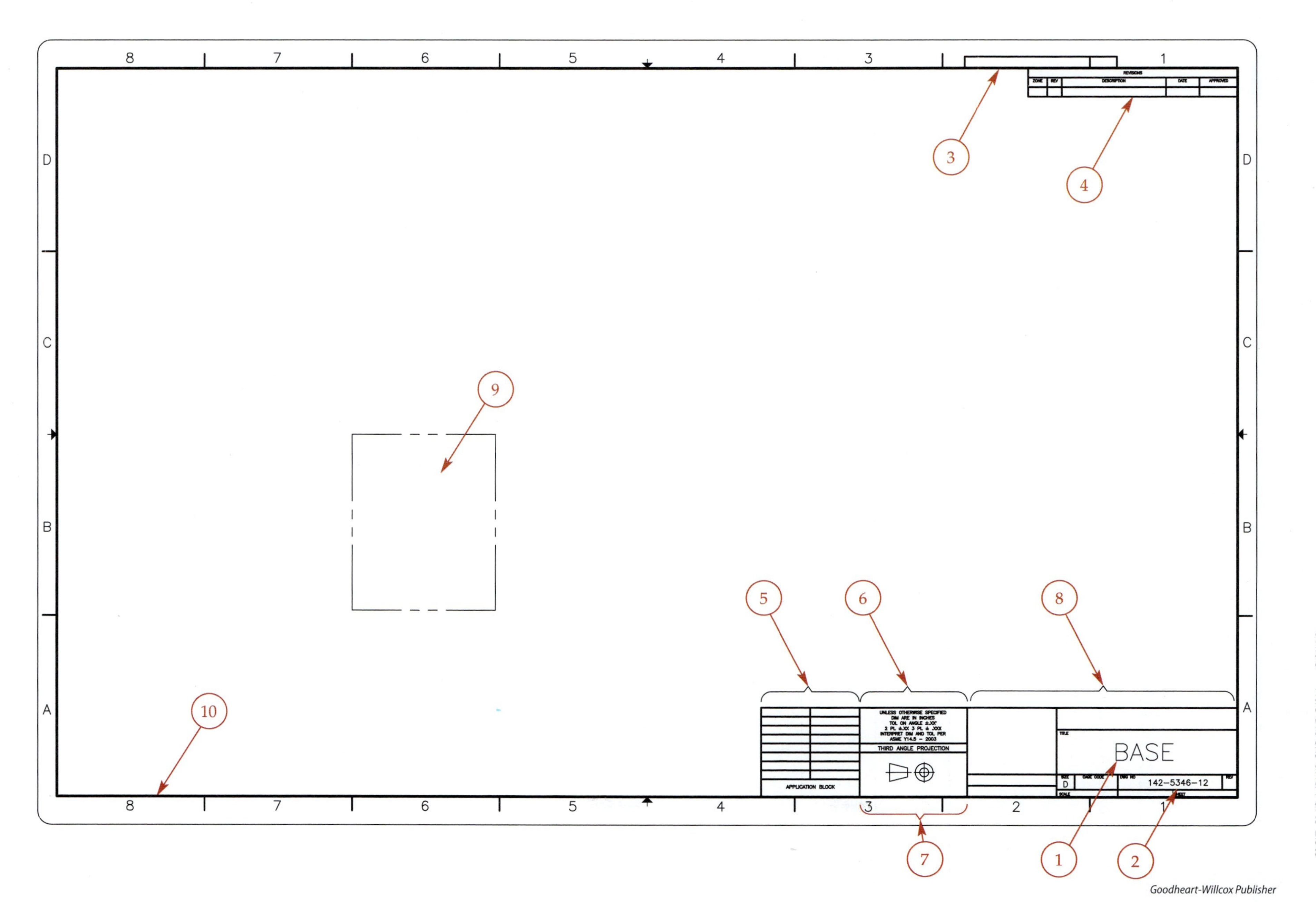

Goodheart-Willcox Publisher

Review Activity 3-1. Base

Apply and Analyze

Name ______________________ Date ______________ Class ______________

Review Activity 3-2

Refer to the print Activity 3-2, evaluate the print, and answer the questions.

1. What is the name of the assembly?

____

2. How many parts are identified by balloons in this drawing?

____

3. How many parts have a quantity of two or more?

____

4. Of the parts shown, what is the name of the smallest part?

____

5. What is the individual part number of the DRAW ROD?

____

6. In the material block, it says SEE LOM. What does LOM stand for?

____

7. For what size sheet is this drawing intended?

____

8. Is this drawing part of a multisheet set?

____

9. What is the name of part number 105-003?

____

10. What are the initials of the person who checked this drawing?

____

ID	QTY	PART NO.	DESCRIPTION
NS	1	105–009	BALL BEARING
NS	1	105–008	SPRING
7	1	105–007	SET SCREW, C105
6	1	105–006	KNOB, 105 SERIES
5	1	105–005	CENTER POST
4	1	105–004	DRAW ROD
3	1	105–003	COLLET
2	1	105–002	PIVOT POST
1	1	105–001	INDEXING BASE

LIST OF MATERIAL

UNLESS OTHERWISE SPECIFIED
ALL DIMENSIONS ARE IN INCHES

TOLERANCES:

1 PLACE DIMS:	+/– .1
2 PLACE DIMS:	+/– .03
3 PLACE DIMS:	+/– .005
ANGULAR:	+/– 1°
FRACTIONAL:	+/– 1/32

MATERIAL SEE LOM
FINISH NONE

PROJECT NO. 08B–105NP

APPROVALS		DATE
DRAWN	DCW	7/20
CHECKED	RKB	7/22
APPROVED		
ISSUED		

BROWN ENGINEERING

INDEXING FIXTURE

SIZE A | CAGE CODE: | DWG NO. 105-60584 | REV A
SCALE NONE | WEIGHT | SHEET 1 OF 1

Goodheart-Willcox Publisher

Review Activity 3-2. Indexing Fixture

Apply and Analyze

Name ______________________ Date ______________ Class ______________

Industry Print Exercise 3-1

Refer to the print PR 3-1, evaluate the print, and answer the questions.

1. What is the name of the part in this drawing?

____

2. What is the drawing number?

____

3. How many parts have a quantity of two or more?

____

4. What is the part number of the NUT?

____

5. What is the part number of the KNOB?

____

6. At what scale was this drawing created?

____

7. For what size sheet is this drawing intended?

____

8. In what state is this company located?

____

9. How many balloons are used to locate parts for the parts list?

____

10. What is the last name of the person who checked this drawing?

____

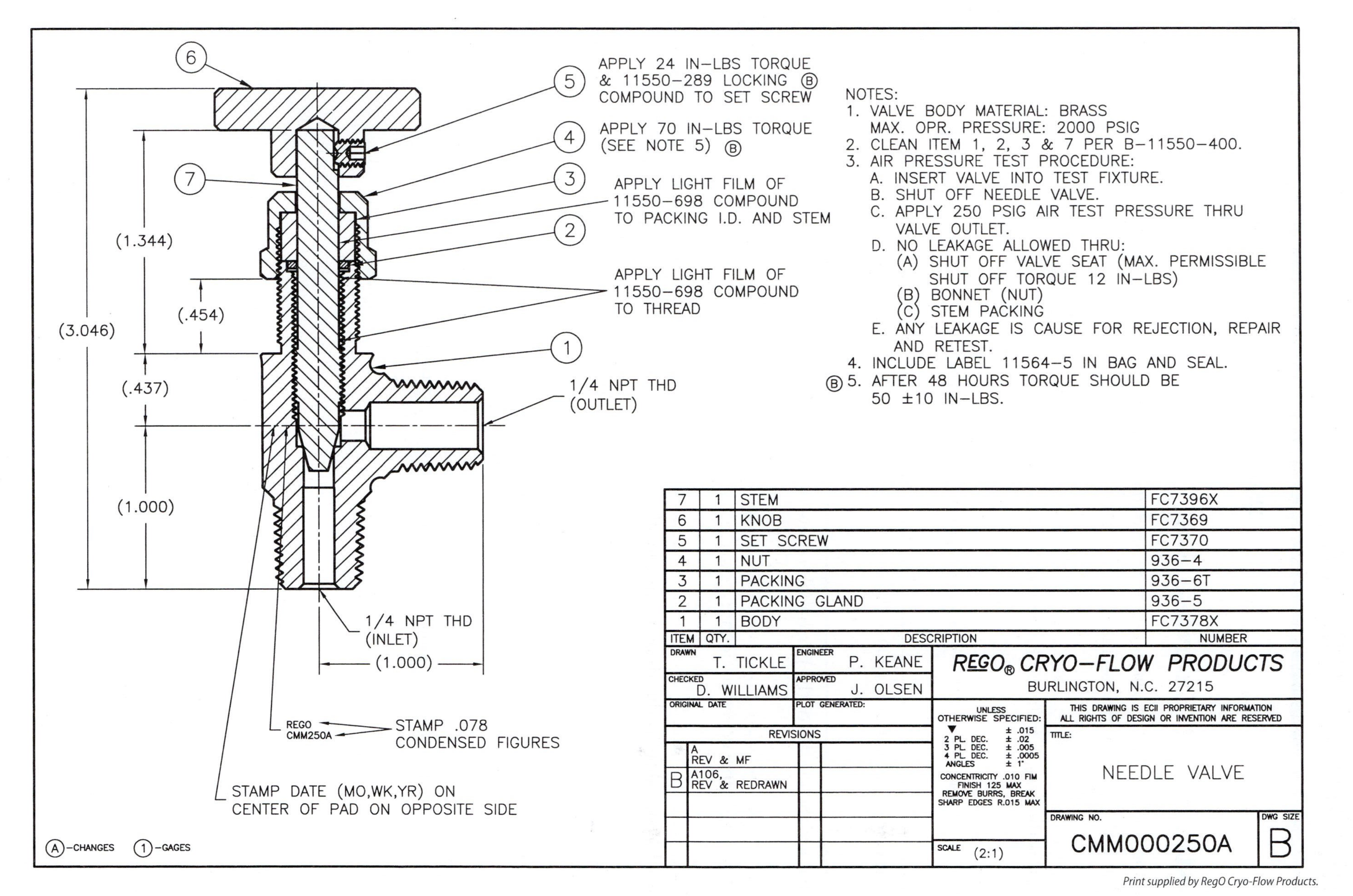

ITEM	QTY.	DESCRIPTION	NUMBER
7	1	STEM	FC7396X
6	1	KNOB	FC7369
5	1	SET SCREW	FC7370
4	1	NUT	936-4
3	1	PACKING	936-6T
2	1	PACKING GLAND	936-5
1	1	BODY	FC7378X

Print supplied by RegO Cryo-Flow Products.

PR 3-1. Needle Valve

Apply and Analyze

Name ______________________ Date ______________ Class ______________

Industry Print Exercise 3-2

Refer to the print PR 3-2, evaluate the print, and answer the questions.

1. What is the name of the assembly?

   ____

2. What is the largest find number in the parts list?

   ____

3. What is the name of the part with the greatest quantity needed?

   ____

4. What is the name of item #14?

   ____

5. What are the initials of the person who drew the drawing?

   ____

6. What is the name of the company?

   ____

7. For what size sheet is this drawing intended?

   ____

8. Is this drawing part of a multisheet set?

   ____

9. What is the name of part number 000412?

   ____

10. What is the individual part number of the WHITE PLASTIC PLUG?

    ____

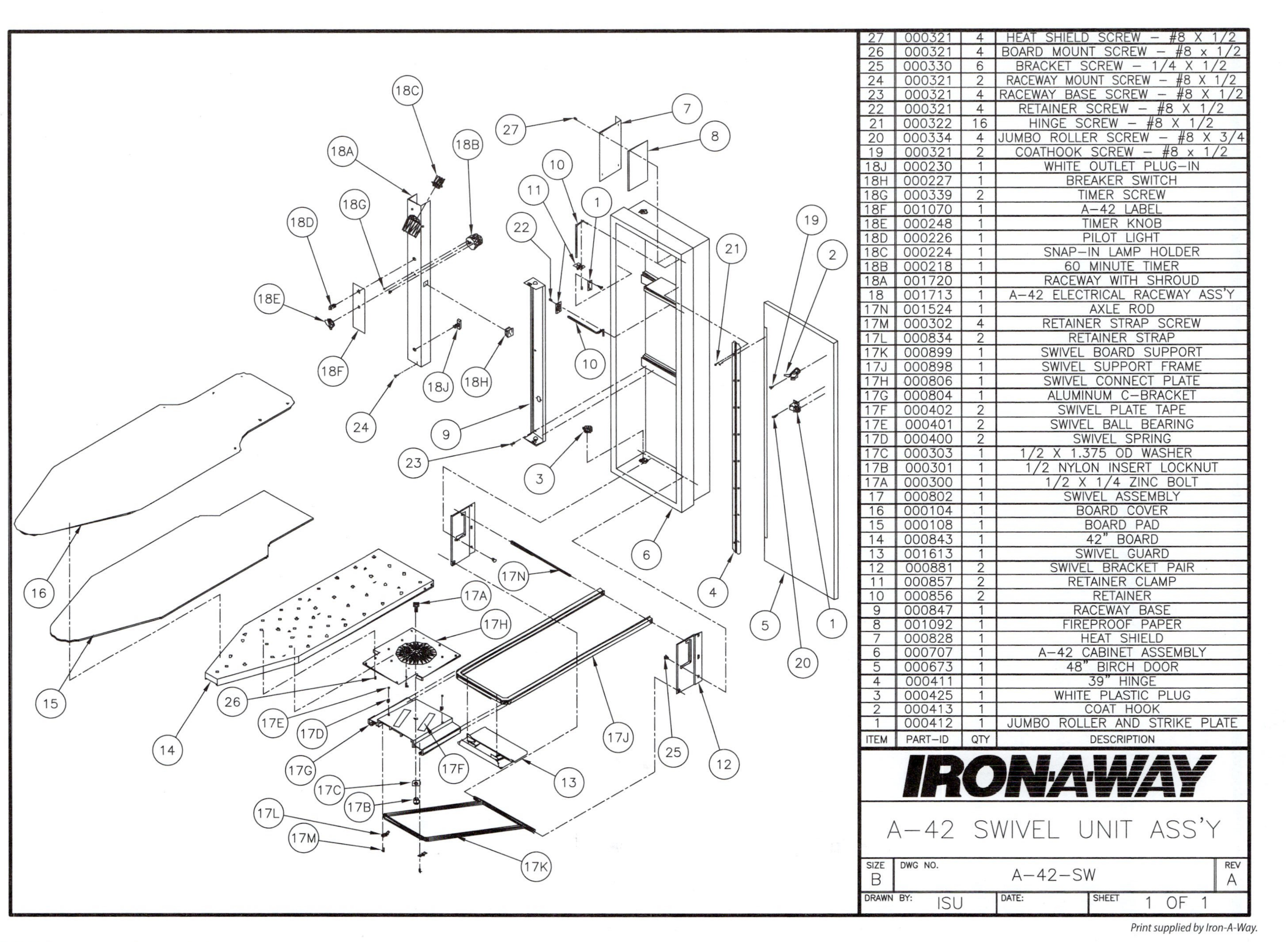

ITEM	PART–ID	QTY	DESCRIPTION
27	000321	4	HEAT SHIELD SCREW – #8 X 1/2
26	000321	4	BOARD MOUNT SCREW – #8 x 1/2
25	000330	6	BRACKET SCREW – 1/4 X 1/2
24	000321	2	RACEWAY MOUNT SCREW – #8 X 1/2
23	000321	4	RACEWAY BASE SCREW – #8 X 1/2
22	000321	4	RETAINER SCREW – #8 X 1/2
21	000322	16	HINGE SCREW – #8 X 1/2
20	000334	4	JUMBO ROLLER SCREW – #8 X 3/4
19	000321	2	COATHOOK SCREW – #8 x 1/2
18J	000230	1	WHITE OUTLET PLUG–IN
18H	000227	1	BREAKER SWITCH
18G	000339	2	TIMER SCREW
18F	001070	1	A–42 LABEL
18E	000248	1	TIMER KNOB
18D	000226	1	PILOT LIGHT
18C	000224	1	SNAP–IN LAMP HOLDER
18B	000218	1	60 MINUTE TIMER
18A	001720	1	RACEWAY WITH SHROUD
18	001713	1	A–42 ELECTRICAL RACEWAY ASS'Y
17N	001524	1	AXLE ROD
17M	000302	4	RETAINER STRAP SCREW
17L	000834	2	RETAINER STRAP
17K	000899	1	SWIVEL BOARD SUPPORT
17J	000898	1	SWIVEL SUPPORT FRAME
17H	000806	1	SWIVEL CONNECT PLATE
17G	000804	1	ALUMINUM C–BRACKET
17F	000402	2	SWIVEL PLATE TAPE
17E	000401	2	SWIVEL BALL BEARING
17D	000400	2	SWIVEL SPRING
17C	000303	1	1/2 X 1.375 OD WASHER
17B	000301	1	1/2 NYLON INSERT LOCKNUT
17A	000300	1	1/2 X 1/4 ZINC BOLT
17	000802	1	SWIVEL ASSEMBLY
16	000104	1	BOARD COVER
15	000108	1	BOARD PAD
14	000843	1	42" BOARD
13	001613	1	SWIVEL GUARD
12	000881	2	SWIVEL BRACKET PAIR
11	000857	2	RETAINER CLAMP
10	000856	2	RETAINER
9	000847	1	RACEWAY BASE
8	001092	1	FIREPROOF PAPER
7	000828	1	HEAT SHIELD
6	000707	1	A–42 CABINET ASSEMBLY
5	000673	1	48" BIRCH DOOR
4	000411	1	39" HINGE
3	000425	1	WHITE PLASTIC PLUG
2	000413	1	COAT HOOK
1	000412	1	JUMBO ROLLER AND STRIKE PLATE

Print supplied by Iron-A-Way.

PR 3-2. A-42 Swivel Unit Assembly

Apply and Analyze

Name ______________________ Date ______________ Class ______________

Bonus Print Reading Exercises

The following questions are based on various bonus prints located in the folder that accompanies this textbook. Refer to the print indicated, evaluate the print, and answer the question.

Print AP-001

1. What size paper was used for the original version of this print?

____

Print AP-002

2. What size paper was used for the original version of this print?

____

3. According to the List of Materials, what is the total quantity of subassemblies required for this assembly?

____

Print AP-003

4. For this print, of what significance is the number 81073?

____

Print AP-004

5. Which of the three parts has a CAGE code, and what is it?

____

6. Within what zone is the revision history block?

____

Print AP-006

7. What material is specified for this part?

____

8. What size paper was used for the original version of this print?

____

Print AP-008

9. What title is given to the revision history block on this print?

____

Print AP-010

10. What is the complete name of this subassembly?

____

11. What is the total quantity of pieces required for this welding assembly?

____

Print AP-011

12. According to the title block, at what scale are the main views on the original print?

____

Print AP-013

13. What is the part number for the label?

____

14. What quantity is required of part number B-9470-12?

____

Print AP-015

15. Analyze the parts list on this print and list the item numbers and part names that are subassemblies within the overall assembly.

____

____

SECTION 2

Fundamentals of Shape Description

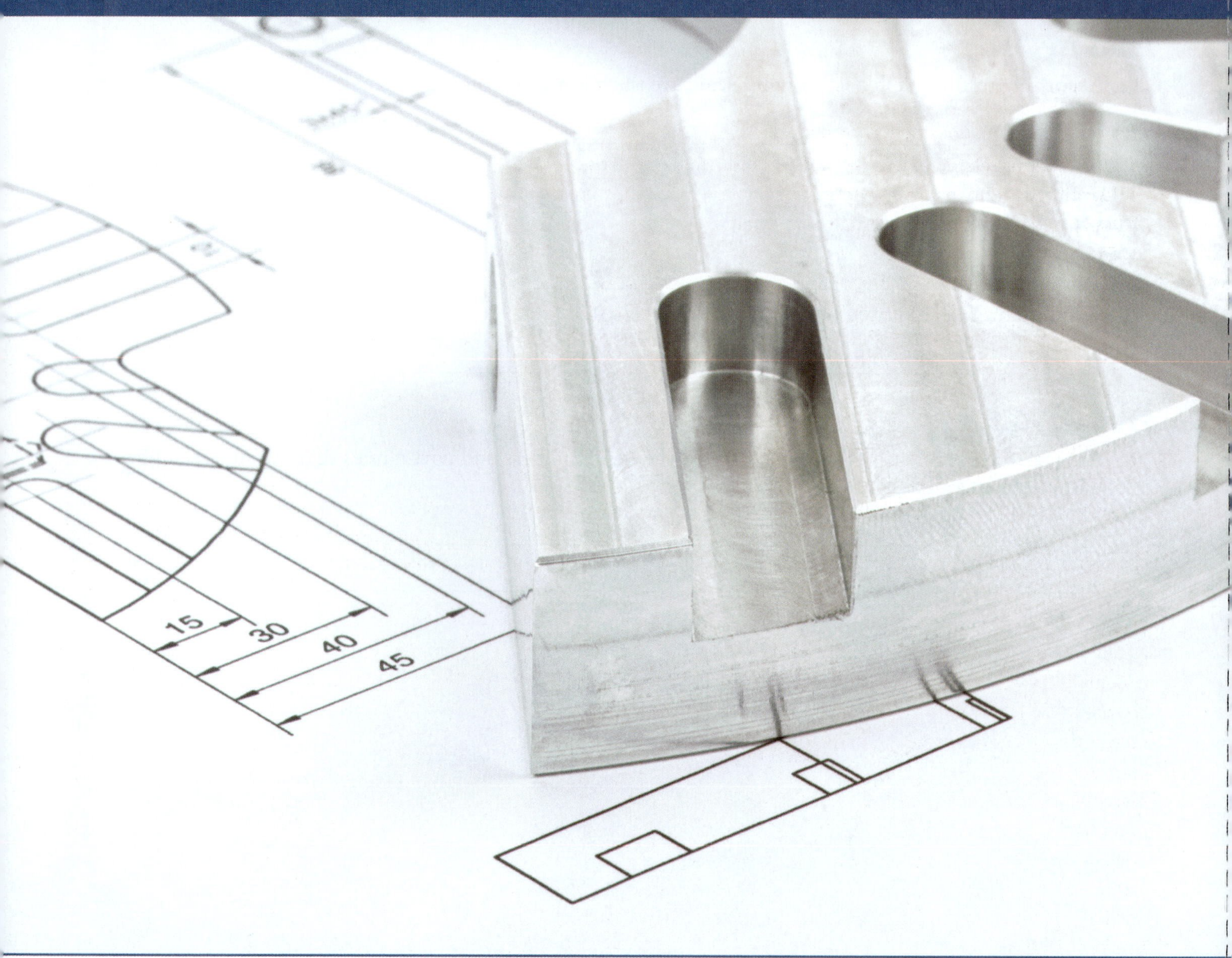

Josef Bosak/Shutterstock.com

UNIT 4

Geometric Terms and Construction

LEARNING OBJECTIVES

After completing this unit, you will be able to:

- Define terms related to the geometry of industrial drawings.
- Identify two-dimensional geometric shapes.
- Identify three-dimensional geometric objects.
- Explain the relationship of a regular polygon to a theoretical circle about which it is circumscribed or inscribed.
- Describe parallelism and perpendicularity as it applies to two- and three-dimensional geometric elements.
- Describe orientation relationships such as concentricity, eccentricity, and coaxiality.
- Describe the nature of tangent relationships in geometric shapes.
- Identify specialized geometric shapes that are used in product design.
- Describe how computer models use polygon meshes to describe an object.
- Describe how freeform surfaces make use of splines to describe an object.

TECHNICAL TERMS

arc
chord
circle
circumscribe
coaxial
concentric
conic section
diameter
eccentric
edge
face
faceted
freeform surface
helix
inscribe
involute
parallel
perpendicular
point of tangency
polygon
polygon mesh
primitive
radius
sector
spline
tangent
vertex

Lines, arcs, circles, ellipses, splines, and polygons are some of the basic two-dimensional building blocks of everything in life. These two-dimensional elements are joined to create planes, curved surfaces, and sculptured surfaces to form three-dimensional models of design ideas. Traditional "pencil" drafters used mechanical equipment such as triangles, T-squares, compasses, and irregular curves to create various geometric shapes. Modern CAD programs automate much of the geometric-design process, but an understanding of geometry is still necessary when interpreting designs and drawings. While someone reading a print does not need to have the same level of understanding as a drafter, knowledge of the terms and geometric constructions is important, especially as the print reader encounters geometric dimensioning and tolerancing (GD&T), which is discussed in Unit 13.

Two-Dimensional Shape Terminology

A line segment is one-dimensional. However, much of the geometry on a print is composed of several straight lines to form two-dimensional shapes, **Figure 4-1**. Line segments can form right angles (90°), acute angles (less than 90°), obtuse angles (greater than 90°), complementary angles (sum of angles equals 90°), or supplementary angles (sum of angles equals 180°). ***Polygons*** are closed shapes composed of straight lines. Triangles are three-sided polygons and can be right (has a right angle), equilateral (all sides equal), isosceles (two sides equal), or scalene (no sides equal). Quadrilaterals are four-sided polygons and can be in the form of a parallelogram, square, rectangle, rhombus, trapezoid, or trapezium. A *regular* polygon has all sides of equal length and all angles of equal measure. Common regular polygons include the equilateral triangle, square, pentagon, hexagon, heptagon, and octagon.

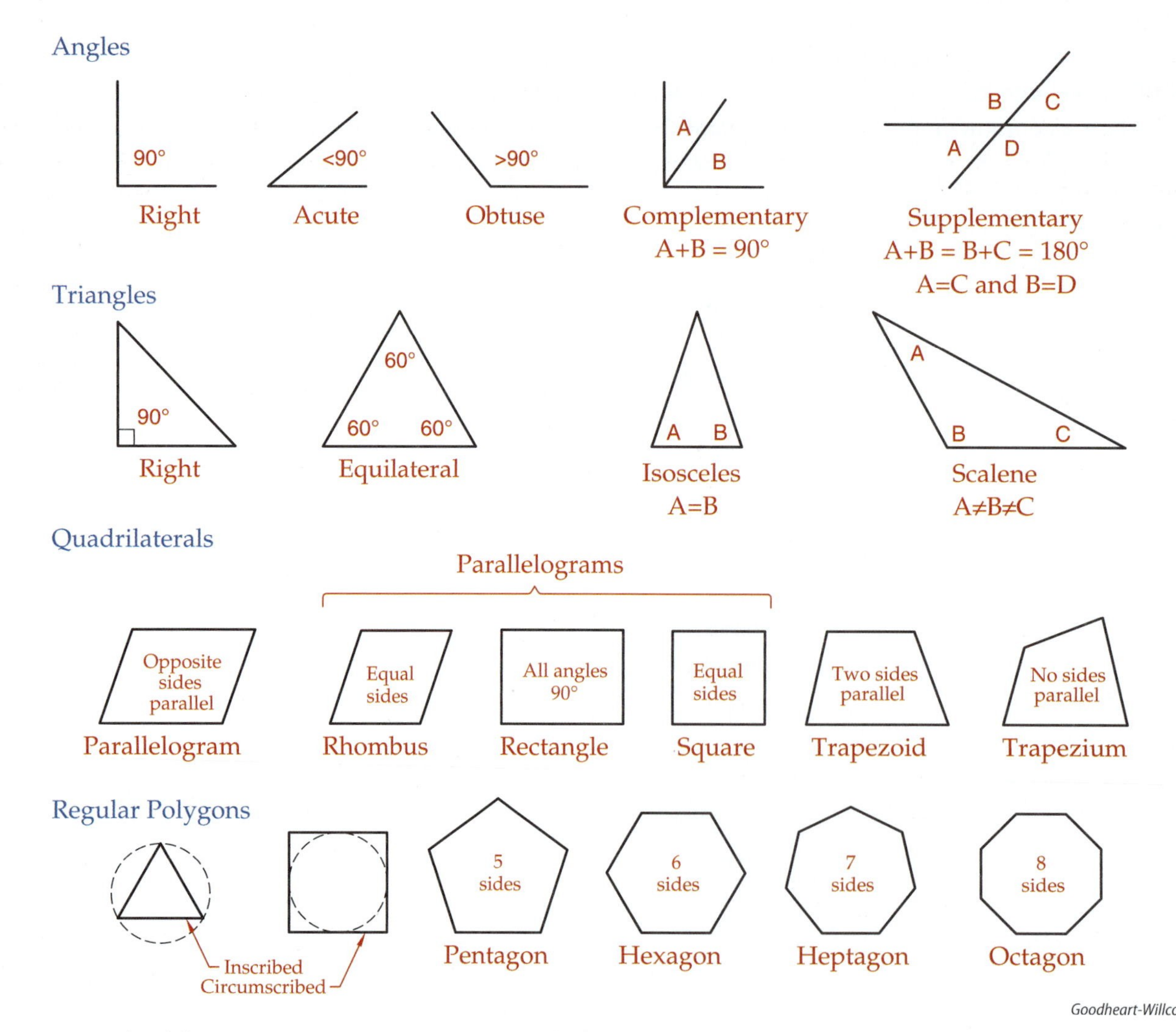

Figure 4-1. Much of the geometry in a print is composed of several straight lines that form two-dimensional shapes.

There are many specific terms associated with lines, circles, and arcs, **Figure 4-2**. A ***circle*** is a closed, 360° curve where all points on the curve are an equal distance from a given point called the center point. The ***radius*** is the distance from the center point to a point on the curve. The ***diameter*** is the distance across the circle through the center point. An ***arc*** is a portion of a circle, or an "open" circle. A line segment within a circle connecting two points other than through the center point is referred to as a ***chord***. A chord is always less than the diameter. A ***sector*** is an area bounded by two radii (plural of radius) and the included arc. If the included angle is 90°, the sector can be identified as a "quadrant," one quarter of a circle. When a line, circle, or arc is bisected, it is divided into two halves.

Regular polygons can be created around the outside or inside of a circle. If the polygon is ***circumscribed*** around the outside of a circle, the measurement "across the flats" is equal to the diameter of the circle. If the polygon is ***inscribed*** within a circle, the measurement "across the corners" is equal to the diameter of the circle. See **Figure 4-3**. A common application of this in industrial prints is a hexagonal bolt head. The bolt head is a hexagon circumscribed about the outside of a circle that is 1.5 times the body diameter of the bolt.

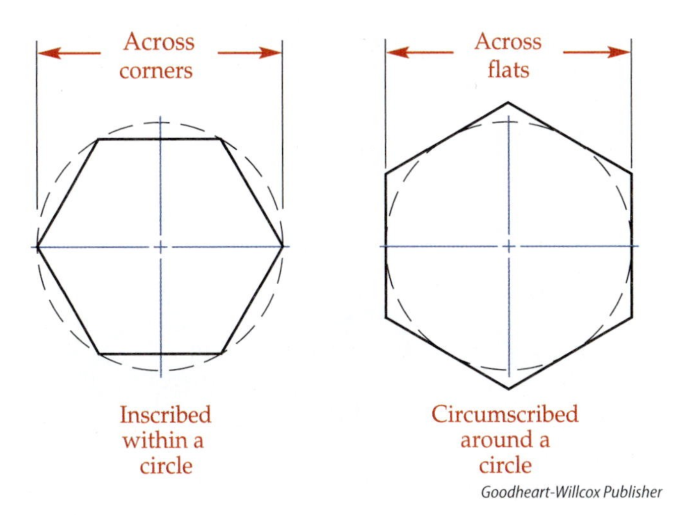

Figure 4-3. Polygons can be created around the outside or inside of a circle.

Three-Dimensional Form Terminology

While the scope of this text is to read prints, which have traditionally been entirely two-dimensional, more and more prints are created from three-dimensional CAD models. Therefore, the print reader may be required to study design solutions that involve an understanding of three-dimensional forms. In most CAD systems, the basic forms used to create complex models are referred to as ***primitives***. Some of the commonly used three-dimensional primitives are prisms, cylinders, pyramids, cones, spheres, and tori (plural for torus), **Figure 4-4**. As shown in the illustration, 3D curved surfaces may be shown with some of the line elements visible. The elements of a cone connect the apex of the cone to points on the base. The elements of a cylinder are the lines that run parallel to the axis, from a point on the base to a point on the top. Technically, there are an infinite number of elements about the curved

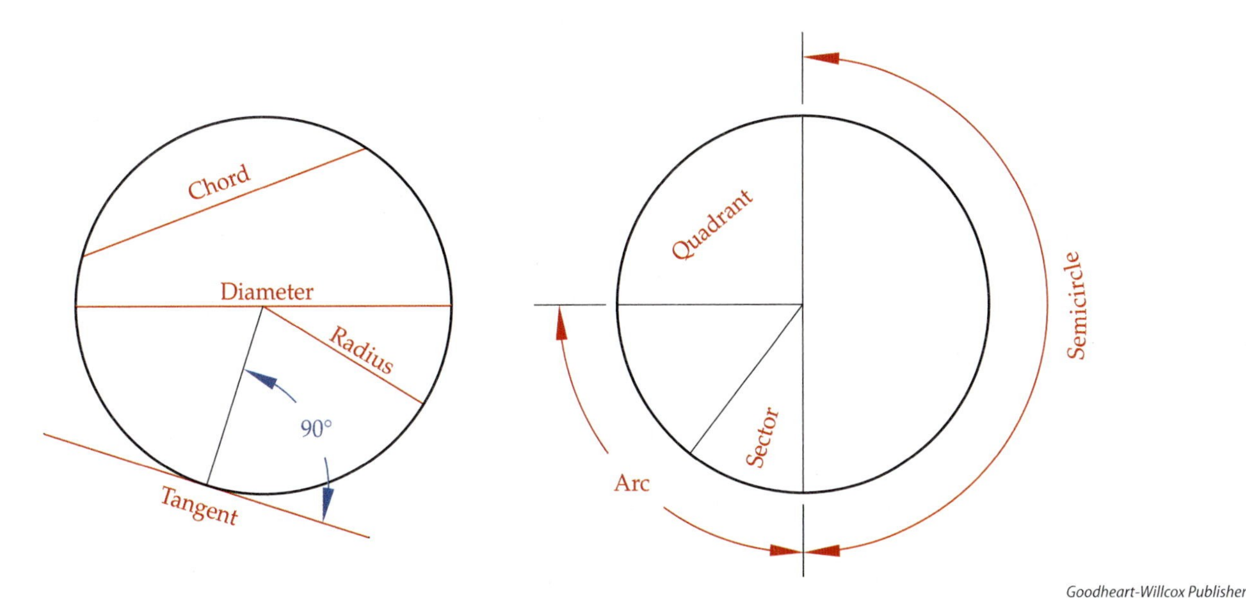

Figure 4-2. There are many terms related to circles and arcs.

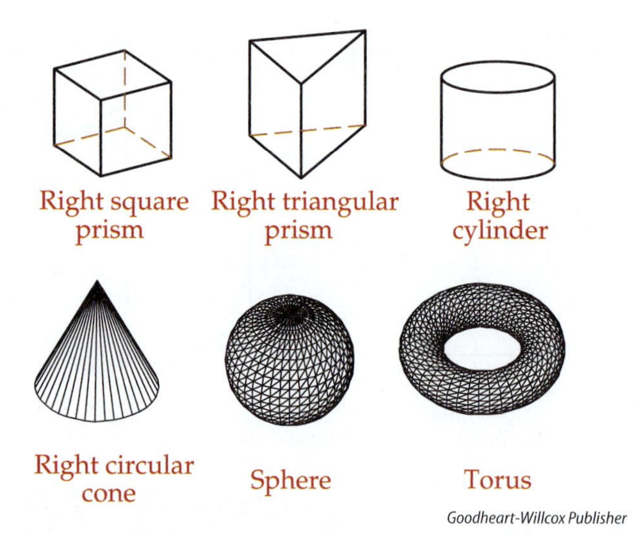

Figure 4-4. Many designs require an understanding of three-dimensional forms.

surface of a cylinder or cone. The term "right" is also applied to three-dimensional forms that have an "axis" squarely oriented to the base. The term "oblique," which means slanted, is applied to those forms with an axis inclined to the base.

Orientation Terminology

Many geometric terms deal with the orientation of one element of geometry to another. Geometric elements are often parallel, perpendicular, or tangent to each other. These terms can be applied to both two-dimensional and three-dimensional geometry.

Parallel means the elements never intersect, even if extended. See **Figure 4-5**. Parallel lines run side by side. Parallel planes likewise maintain a uniform distance, even if extended. Two-dimensional geometry can also be parallel with three-dimensional geometry. Another way of expressing parallelism is to say that all points of one element are "equidistant" (equal distance) from all corresponding points of another element.

Perpendicular means two elements form a right (90°) angle with each other. See **Figure 4-6.** Many geometric principles are based on the right angle. For example, the direction from which you view an object to see the front view is perpendicular to the direction from which you view an object to see the top view. Surfaces can also be perpendicular to each other, and two-dimensional lines or axes can be perpendicular to three-dimensional surfaces. In Unit 13, *Geometric Dimensioning and Tolerancing,* you will learn about specifying the preciseness of parallelism and perpendicularity.

While railroad tracks appear to maintain parallelism as they curve around the bend, there are different terms used for curved elements. Circles and arcs are defined as ***concentric*** if they share a common center point. The three-dimensional extension of concentric is ***coaxial.*** See **Figure 4-7.** For example, two cylinders that share the same center axis are coaxial. Two circles that do not share a center point are said to be ***eccentric.*** Two cylinders with axes that do not align can also be considered eccentric.

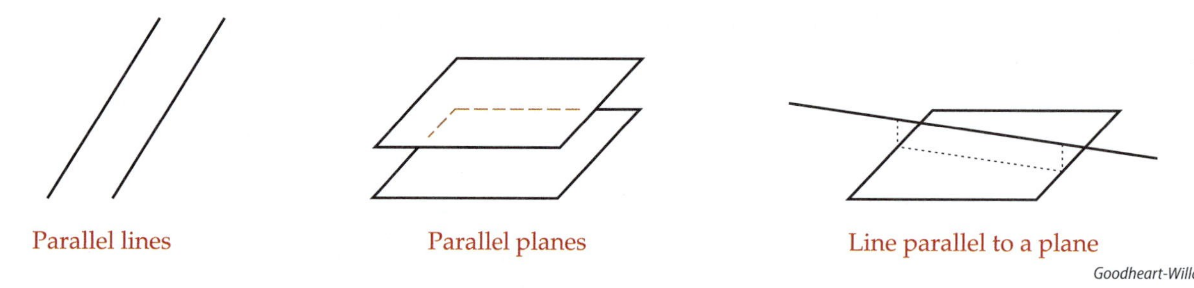

Figure 4-5. Parallel elements never intersect, even if extended. This relationship can exist between 2D and 3D elements.

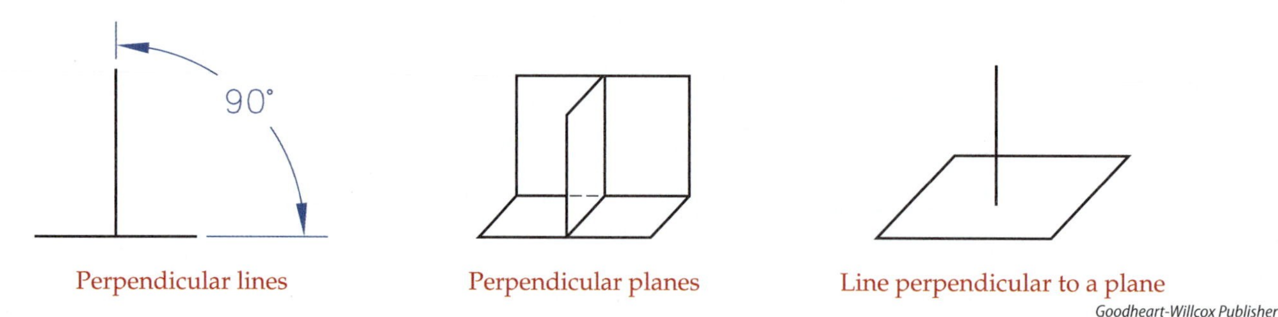

Figure 4-6. Perpendicular elements form a 90° angle to each other.

In the manufacturing department, an analysis of the degree of eccentricity is sometimes a critical aspect of the quality of the part or assembly. Sometimes eccentricity is not desirable, but there are functional designs in industry that require a certain amount of eccentricity.

Tangent is a relationship between two features best described as "sharing one point." In two-dimensional terms, a straight line is tangent to a circle if it touches the circle at only one point and does not intersect the circle again, even if extended, **Figure 4-8**. The shared point is called the ***point of tangency***. For a line and circle, the point of tangency is always found at the end of a line drawn perpendicular to the line from the center point of the arc or circle. Likewise, circles and arcs can be tangent to each other. The point of tangency for two circles or arcs coincides with a segment connecting the center points of the circles or arcs. In three-dimensional models, flat surfaces can be tangent to curved surfaces or curved surfaces can be tangent to curved surfaces, and in these cases we say that the surfaces "share one element." In Unit 5, *Multiview Drawings*, you will learn how to represent views of objects wherein there are *tangency elements*—that element of transition between the curved surface and the flat surface. By default, many CAD systems show the tangency element, but traditional drafters may elect not to show that element of transition unless it clarifies the view.

Specialized Geometric Shapes

Some geometric shapes are based on the way a plane intersects a cone. The resulting shape is called a ***conic section***. See **Figure 4-9**. Depending on the angle by which the plane intersects the cone, a circle, ellipse, parabola, or hyperbola is formed. There are many design applications for conic sections, such as the parabolic shape of headlight lenses and bridge arches. A parabola can also be defined by an algebraic formula.

In addition to passing a plane through a cone, an ellipse results when a circle is viewed at an inclined line of sight. For example, a flat, circular disc appears as a circle if the view is perpendicular to the disc. When the disc is inclined away from the viewer, the circle is foreshortened in one direction and appears as an elliptical shape, **Figure 4-10**. In the CAD system, an ellipse is often defined by the major diameter across the widest part and the minor diameter measured across the narrow part.

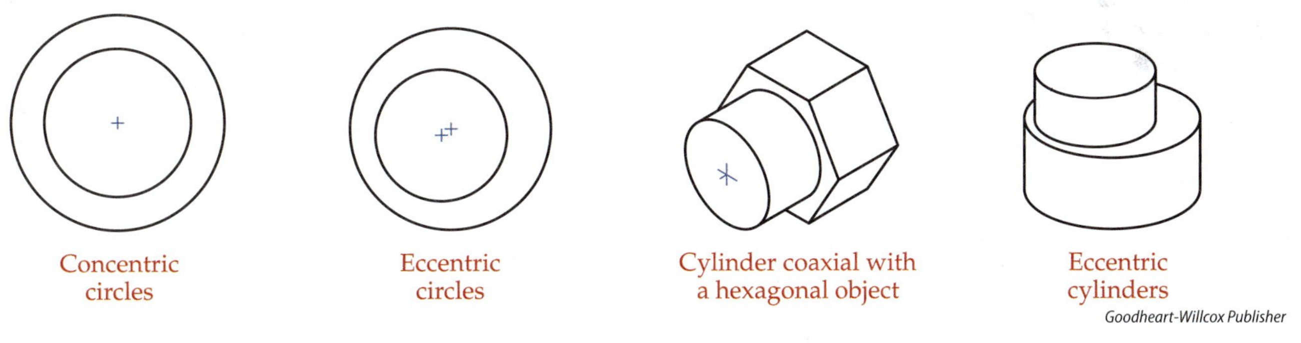

Figure 4-7. Features of a design can exhibit concentric relationships. The extent to which they are not concentric can be called the degree of eccentricity.

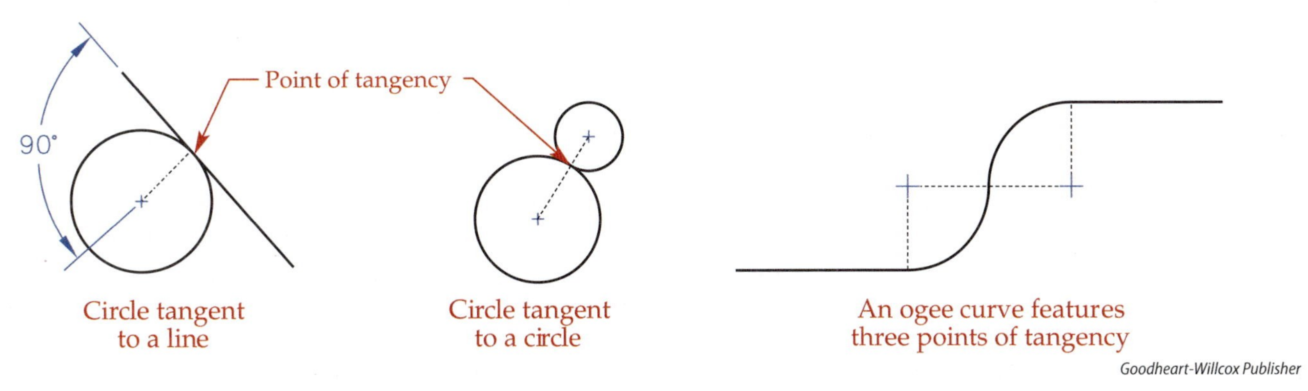

Figure 4-8. Tangent elements share a single point.

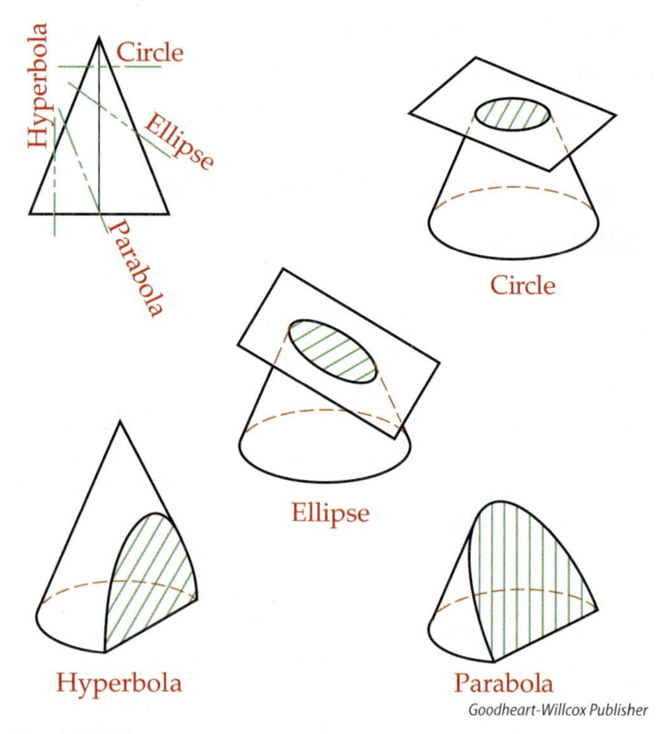

Figure 4-9. Some geometric shapes are based on the way a plane intersects a cone.

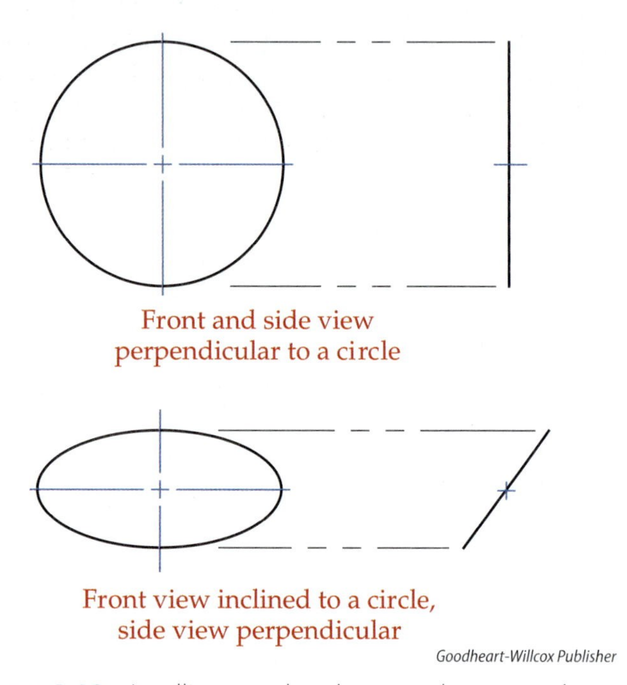

Figure 4-10. An ellipse results when a circle is viewed at an inclined line of sight.

A ***helix*** is generated by rotating a point around an axis while moving it along the axis. This may remind you of a barber pole or candy cane stripe. Screw threads are helical ridges formed on the surface of a cylinder. Springs are also formed in a helical manner. See **Figure 4-11**. An ***involute*** is the path a point follows if it is attached to the end of a taut string and the string is unwound from a cylinder or other geometric form. A gear tooth has a surface based on an involute of a cylinder, which allows it to avoid interference between gear teeth. **Figure 4-12** shows the geometric construction technique a drafter would use to draft out the involute shape with a series of chord steps, but in today's design procedure, standard gear teeth shapes reside within CAD symbol libraries.

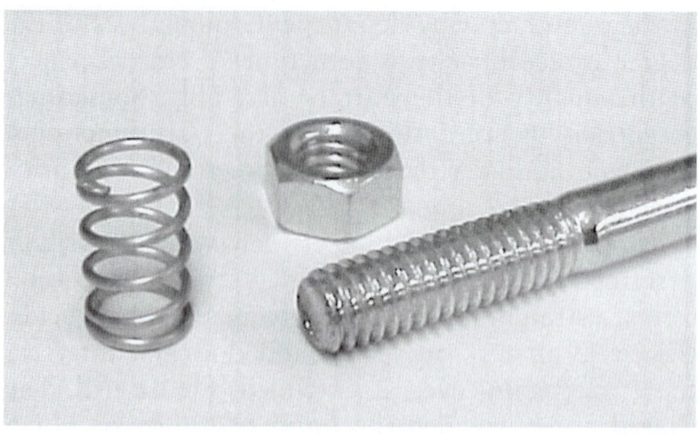

Figure 4-11. Screw threads are helical ridges formed on the surface of a cylinder.

Computer Modeling Terms and Issues

Three-dimensional forms can also be created by combining, extending, or revolving two-dimensional shapes. Within the computer modeling system, two-dimensional shapes are often created first and sometimes referred to as "pre-model" shapes. Computer modeling functions such as **EXTRUDE**, **LOFT**, or **REVOLVE** can be used to create models from these shapes. Computer modeling methods are impacting how parts are designed, and some terms may be found within a product data set that rely on the computer model to communicate information. Some computer models use a ***polygon mesh*** to define the shape of the object. At the basic level, you can see in **Figure 4-4** that the bottom row of primitives is comprised of a mesh of faces. Within a polygon mesh are ***vertices*** (plural of vertex), ***edges***, and ***faces***. Think of vertices as the individual points with X-Y-Z coordinate values, edges as lines that connect the vertices to make a wire-frame model, and faces as surfaces that fill in the area between all the edges, much like covering a wire mesh with foil to create the surface model. From the concept of faces, the term ***faceted*** is derived. Objects that are computer generated

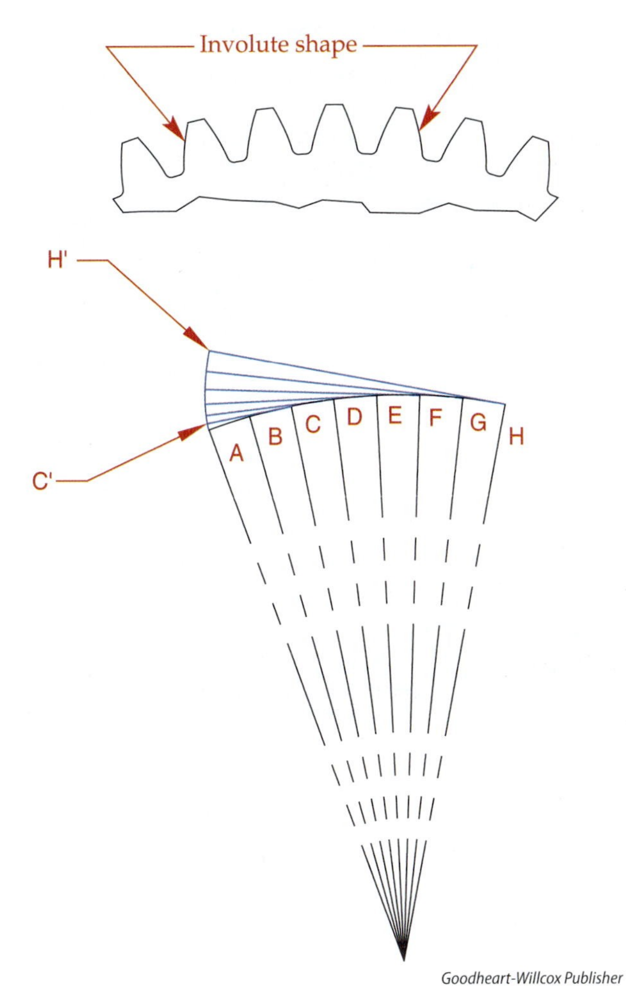

Figure 4-12. A gear tooth has a surface based on an involute.

as polygon meshes are said to be faceted, and the smoothness of the model can be adjusted with a "facet resolution" setting. In summary, the smoothness of a model is determined by the mesh density.

While some 3D geometry is driven by basic geometric shapes and primitives, there are many products in our world today that are more irregular and freeform in nature, such as car fenders, running shoes, and windmill blades. Such objects are more complicated to define with traditional drawing views and dimensions, and computer-aided design systems are assisting the designer with these tasks. The basic building block of this type of computer modeling is the spline. The term ***spline*** can be defined in various ways. In the simplest terms, a spline is a curved "line" that smoothly fits through a series of points. In the days of manual drafting, drafters would specify control points and use an irregular-curve device to form the spline. Today, CAD software can quickly calculate splines using complex mathematical functions.

Splines are sometimes referred to as "B-splines" or "Bézier curves." In early CAD systems, splines were defined by control points through which the spline was "fit." The curve went through those points, just like using the irregular curve of former drafting practice. With Bézier splines, the computer system calculates a spline that does *not* touch the control points, but rather passes near the points, as if drawn toward those points by various settings, **Figure 4-13.** In three-dimensional forms, B-spline surfaces can be referred to as ***freeform surfaces*** or "NURBS" (non-uniform rational B-spline) surfaces. **Figure 4-14** shows an example of a computer model made up of complex surfaces.

While the explanation of these formula-driven shapes is beyond the scope of this text, it is important to be aware that they are impacting the design of various freeform parts and products. With today's technology, prints with traditional dimensions are seldom necessary to define these freeform shapes, as the processes of designing, prototyping, manufacturing, and even inspection, are all being controlled by computer systems.

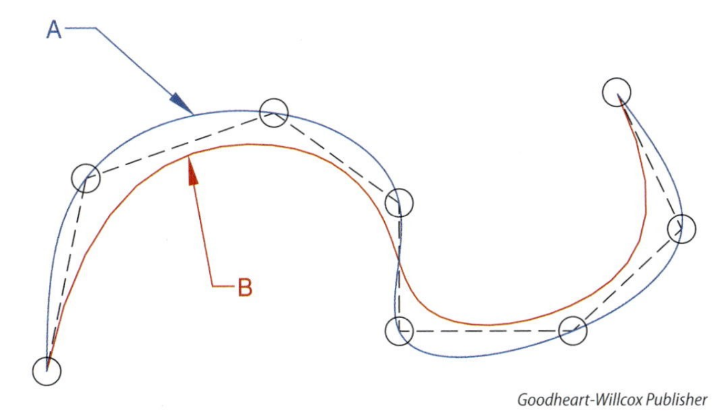

Figure 4-13. Splines drawn with a CAD system. A—Spline calculated through the control points. B—Spline calculated based on the control points.

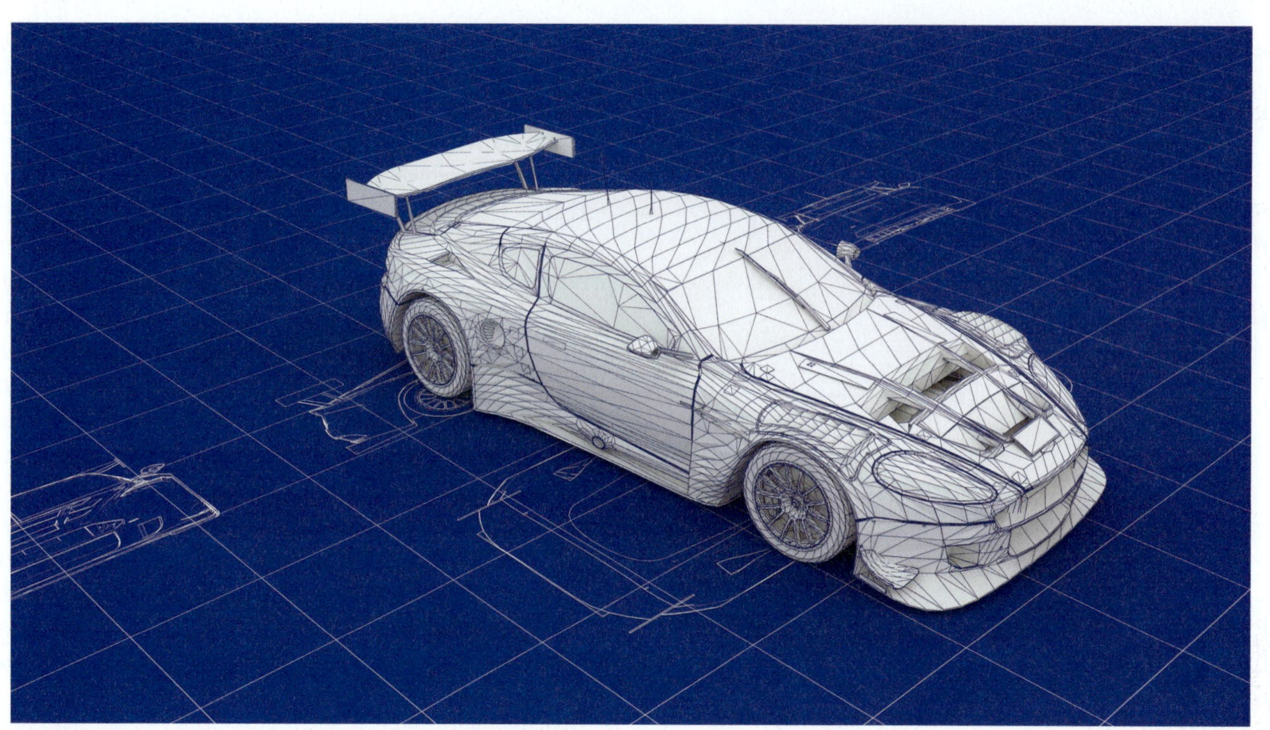

Nuno Andre/Shutterstock.com

Figure 4-14. Pictured here is a computer-generated model rendered as a mesh of small flat surfaces using powerful 3D CAD modeling software.

Summary

- A fundamental understanding of geometry and geometric constructions is complementary to the study of print reading.
- Two-dimensional shape terminology includes angle terms, triangle types, circle terms, and regular polygon terms.
- Regular polygons are either inscribed within a circle or circumscribed about the perimeter of a circle.
- Three-dimensional shape terminology includes primitives such as cylinders, cones, and spheres.
- Orientation terminology includes parallelism, perpendicularity, concentricity, eccentricity, and coaxiality.
- Tangent relationships exist in 2D and 3D geometric forms wherein a shared point or element of tangency is important to the design.
- Specialized geometric shapes found in industrial designs include conic sections (such as the ellipse and parabola), the helix, and the involute.
- Computer modeling terms impacting design include components of polygon meshes (vertices, edges, and faces), and freeform surfaces created within the computer system using complex spline calculations.

Unit Review

Name ______________________ Date ______________ Class ______________

Answer the following questions using the information provided in this unit.

Know and Understand

____ 1. *True or False?* With respect to right angles, right triangles, and right prisms, the word *right* is indicating a 90° relationship is present.

____ 2. Which of the following is *not* a type of triangle?
A. Equilateral
B. Scalene
C. Acute
D. Isosceles

____ 3. Which of the following is *not* a two-dimensional shape?
A. Polygon
B. Circle
C. Pyramid
D. Rhombus

____ 4. *True or False?* The word *chord* describes a line segment that is less than the diameter of the circle.

____ 5. *True or False?* If a hexagon is inscribed within a circle, the "across the flats" measurement of the hexagon is the same as the diameter of the circle that is the basis for the hexagon.

____ 6. *True or False?* The word *primitive* is defined as the two-dimensional shape from which a three-dimensional model is formed by extrusion or revolution.

____ 7. *True or False?* The orientation of the top surface of a cube to the front surface is parallel.

____ 8. *True or False?* The orientation of the top surface of a table to the floor is parallel.

____ 9. Which of the following terms best describes the 3D relationship of a hole and a shaft, wherein the hole is perfectly aligned and centered within the shaft?
A. Parallel
B. Eccentric
C. Coaxial
D. Concentric

____ 10. *True or False?* The point of tangency between two circles tangent to each other is found on a line that connects the two center points.

____ 11. *True or False?* The section derived by slicing a cone with a plane will be either elliptical or circular.

____ 12. The screw thread is defined as a ridge that has the shape of a(n) _____.
A. involute
B. vertex
C. helix
D. parabola

____ 13. Which of the following is *not* a term that is applied to the components of a polygon mesh?
A. Vertex
B. Face
C. Line
D. Edge

____ 14. *True or False?* A freeform surface is comprised of irregular curved elements that are based in part on a geometric shape known as a *spline*.

____ 15. *True or False?* A freeform computer model of something such as a car fender may not have a print with traditional distance dimensions, but will rely on computer systems to define the part, both for manufacturing and inspection.

Critical Thinking

1. Identify three common household objects, and then identify common geometric forms that are exhibited by these objects. You must include at least four different forms.

2. A basic understanding of geometry and geometric shapes is beneficial for reading and interpreting prints. For example, explain how understanding the relationship of a polygon that is circumscribed around or inscribed within a circle might allow you to fill in missing dimensional information on a print.

3. Describe ways in which modern CAD programs have made the geometric-design process and creation of prints easier.

Notes

Apply and Analyze

Name ______________________ Date ______________ Class ______________

Review Activity 4-1

Study the drawing on the following page and answer the questions.

For Questions 1–7, considering the four possibilities of parallel, perpendicular, concentric, and tangent, what relationship exists between the following features?

1. Line A and Line L ______________________
2. Line E and Line G ______________________
3. Line J and Arc K ______________________
4. Line B and Line D ______________________
5. Line L and Arc M ______________________
6. Circle P and Arc M ______________________
7. Line G and Line J ______________________

For Questions 8–12, fill in the blank with the geometric term (arc, chord, etc.) for each of the following features.

8. F ______________________
9. P ______________________
10. U ______________________
11. H ______________________
12. V ______________________
13. If feature V is circumscribed around a circle, what would be the radius of the circle? ______________

For Questions 14–16, how many degrees, measured clockwise along the center-line circle, is it from center point to center point of the following features?

14. Center point R to center point S ______________________
15. Center point S to center point U ______________________
16. Center point U to center point T ______________________
17. What type of angle is formed between D and E? ______________________
18. What type of angle is formed between B and A? ______________________
19. Is the relationship between the angle formed between B and A and the angle formed between D and E complementary or supplementary? ______________________
20. How many degrees are in arc C? ______________________
21. How many degrees are in arc K? ______________________
22. How many degrees are in arc F? ______________________
23. How many degrees are in arc H? ______________________

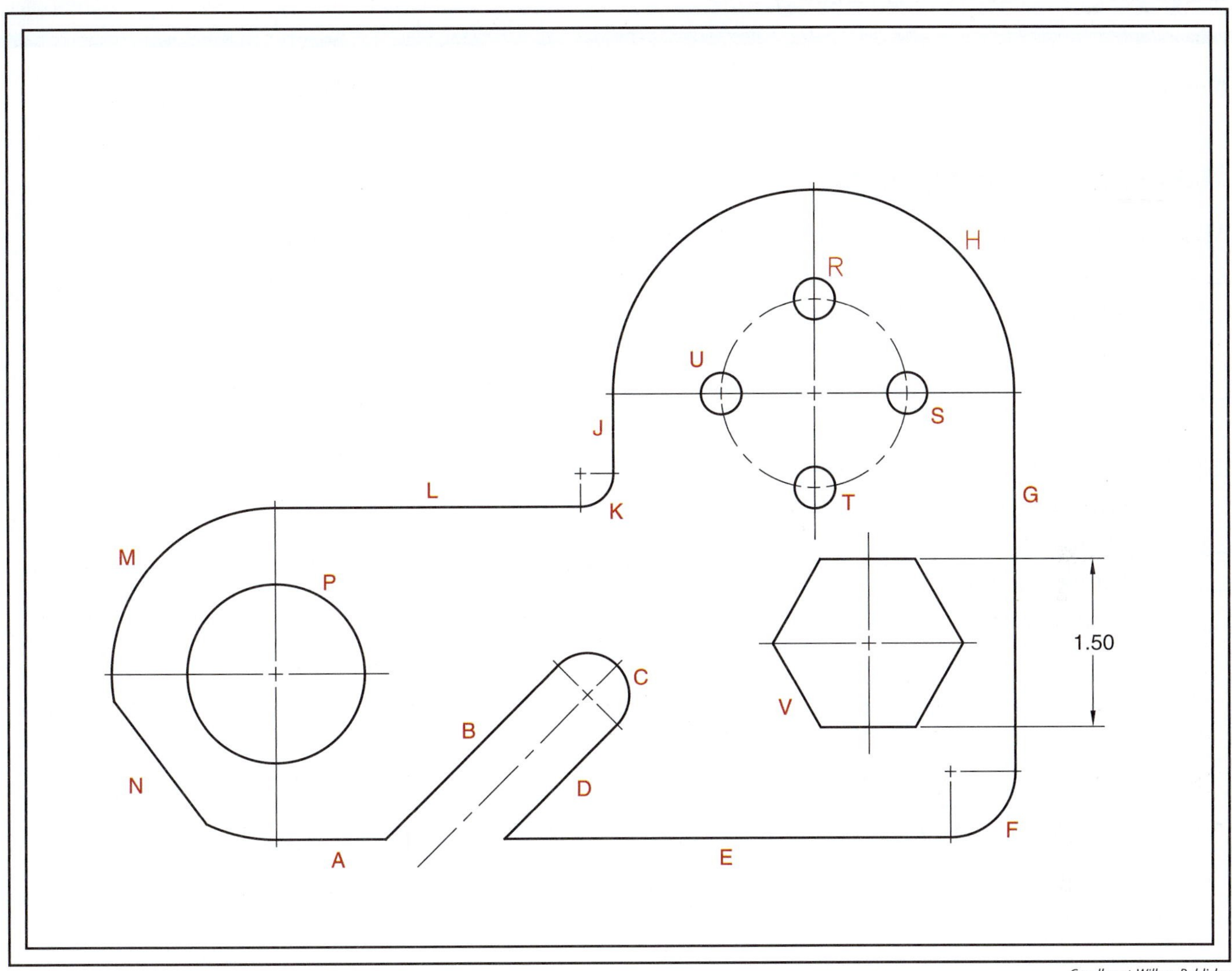

Goodheart-Willcox Publisher

Review Activity 4-1.

Apply and Analyze

Name ______________________ **Date** ______________ **Class** ______________

Review Activity 4-2

Match the letter of each illustrated shape with the correct name. Items are matched only once.

_____ 1. Pentagon
_____ 2. Trapezoid
_____ 3. Octagon
_____ 4. Square
_____ 5. Hexagon
_____ 6. Rhombus
_____ 7. Rectangle
_____ 8. Heptagon
_____ 9. Parallelogram
_____ 10. Trapezium

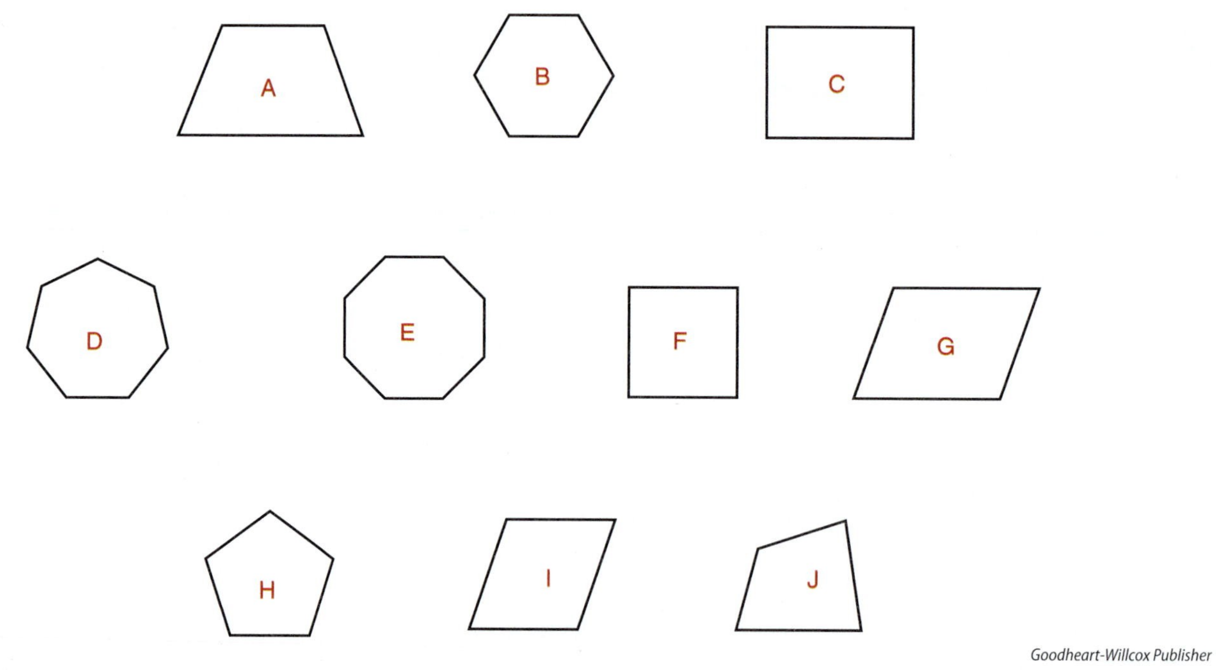

Review Activity 4-2.

Apply and Analyze

Name ______________________________ Date ______________ Class ________________

Review Activity 4-3

Match the letter of each illustrated shape with the correct name. Items are matched only once.

____ 1. Circle
____ 2. Scalene triangle
____ 3. Equilateral triangle
____ 4. Acute angle
____ 5. Semicircle
____ 6. Isosceles triangle
____ 7. Inscribed
____ 8. Right triangle
____ 9. Circumscribed
____ 10. Obtuse angle

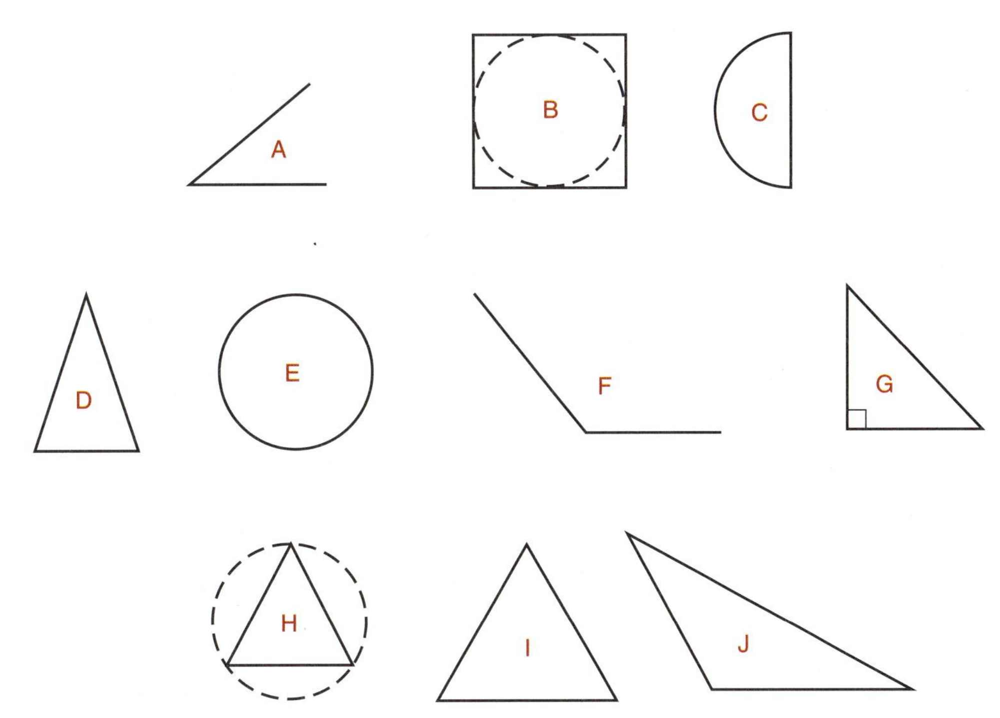

Review Activity 4-3.

Apply and Analyze

Name ______________________ Date ______________ Class ______________

Industry Print Exercise 4-1

Refer to the print PR 4-1 and answer the questions below. The print contains the three regular views, which are (when viewed with the title block in the bottom-right) top (upper-left corner), front (lower-left corner), and right side (lower-right corner).

1. How many circles are featured in the front view?

 ____

2. How many arcs are featured in the front view?

 ____

3. How many points of tangency are there within the geometry of the front view?

 ____

4. How many sets of concentric circles, or circles and arcs, are featured in the front view?

 ____

5. How many circles are featured within the geometry of the top view?

 ____

6. What term describes how the *hidden* lines are oriented to one another in the top view?

 ____

7. How many lines, if any, are tangent with the circle in the top view?

 ____

8. How many vertical visible lines are in the right-side view?

 ____

9. What geometric term describes the general shape of the right-side view?

 ____

10. What geometric orientation does the dimension line have with its extension lines in the right-side view?

 ____

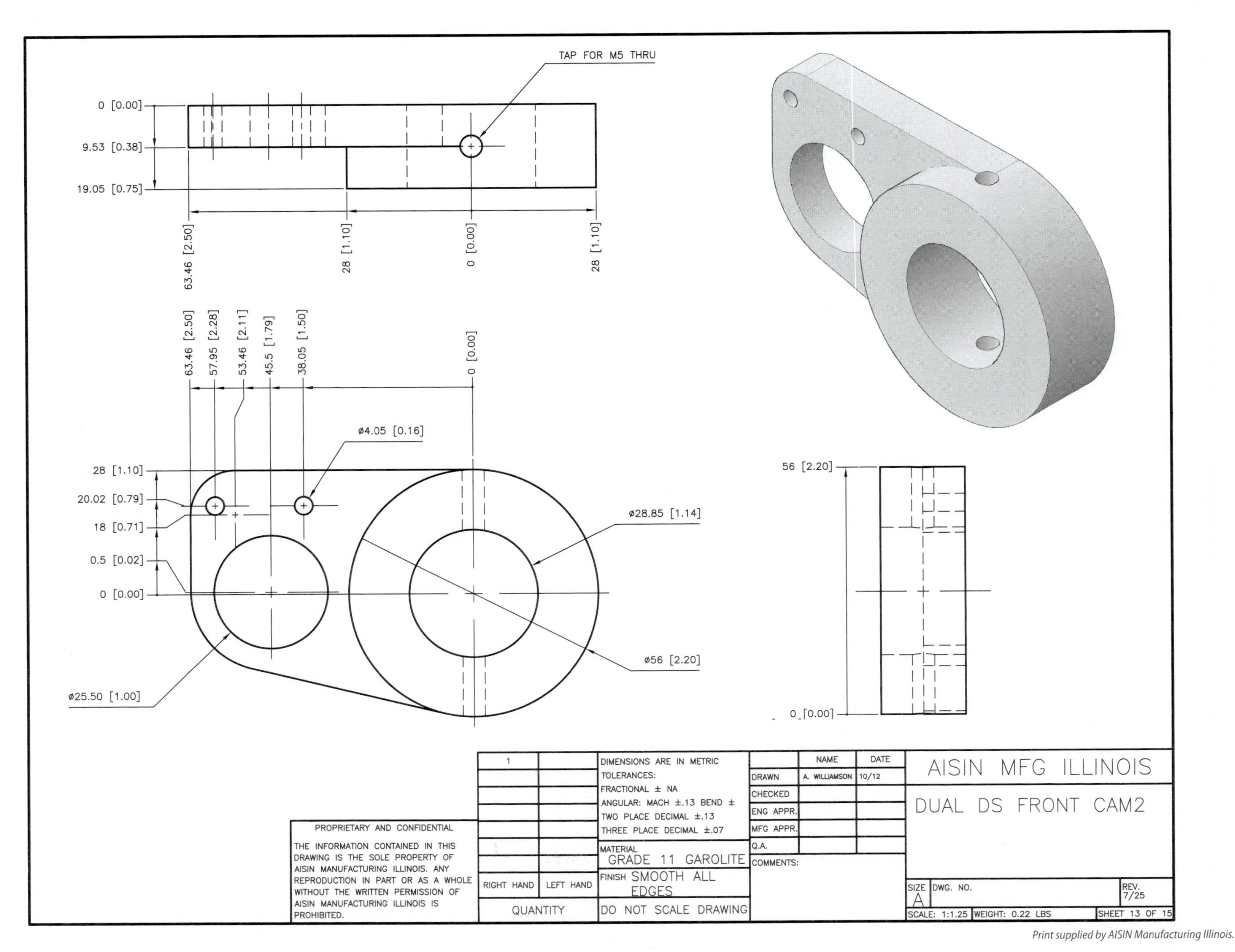

Print supplied by AISIN Manufacturing Illinois.

PR 4-1. Dual DS Front Cam 2

Apply and Analyze

Name ______________________ Date ______________ Class ______________

Industry Print Exercise 4-2

Refer to the print PR 4-2 and answer the questions below.

Note: In this print, the geometric shape shown represents the perimeter surfaces of the Dead Bolt Locking Pawl, and the hole surface going through the part. The symbol Ø means diameter, and a capital "R" means radius.

Analyze the 2D geometry of the front view to answer the following questions:

1. What relationship exists between the circle with a diameter (Ø) of .204″ and the arc with a radius (R) of .18″?

   ____

2. How many straight line segments within the geometric shape of the view are parallel with each other?

   ____

3. *True or False?* The arc with a radius of .250″ is tangent with an arc that has a radius of .06″.

   ____

4. *True or False?* The arc with a radius of .062″ is tangent with an arc that has a radius of .030″.

   ____

5. One line is identified as 50° with respect to the vertical center line. What is the complementary angle that line forms with a horizontal line?

   ____

6. One line is identified as 75° with respect to a vertical line and .136″ away from a center axis. What angle do the extension lines of the .136″ dimension form with a horizontal line?

   ____

7. With respect to Question 6, what supplementary angle could have been given instead of the 75° that was specified?

   ____

8. How many acute angles are specified on the view?

   ____

Analyze the pictorial view to answer the following questions:

9. What 3D orientation term applies to the two curved surfaces that were identified in the 2D geometry of Question 1 above?

   ____

10. What term applies to the straight line elements in the pictorial view that connect the front surface to the back surface? (Hint: The lines show where the curved surfaces turn into flat surfaces, or vice versa.)

    ____

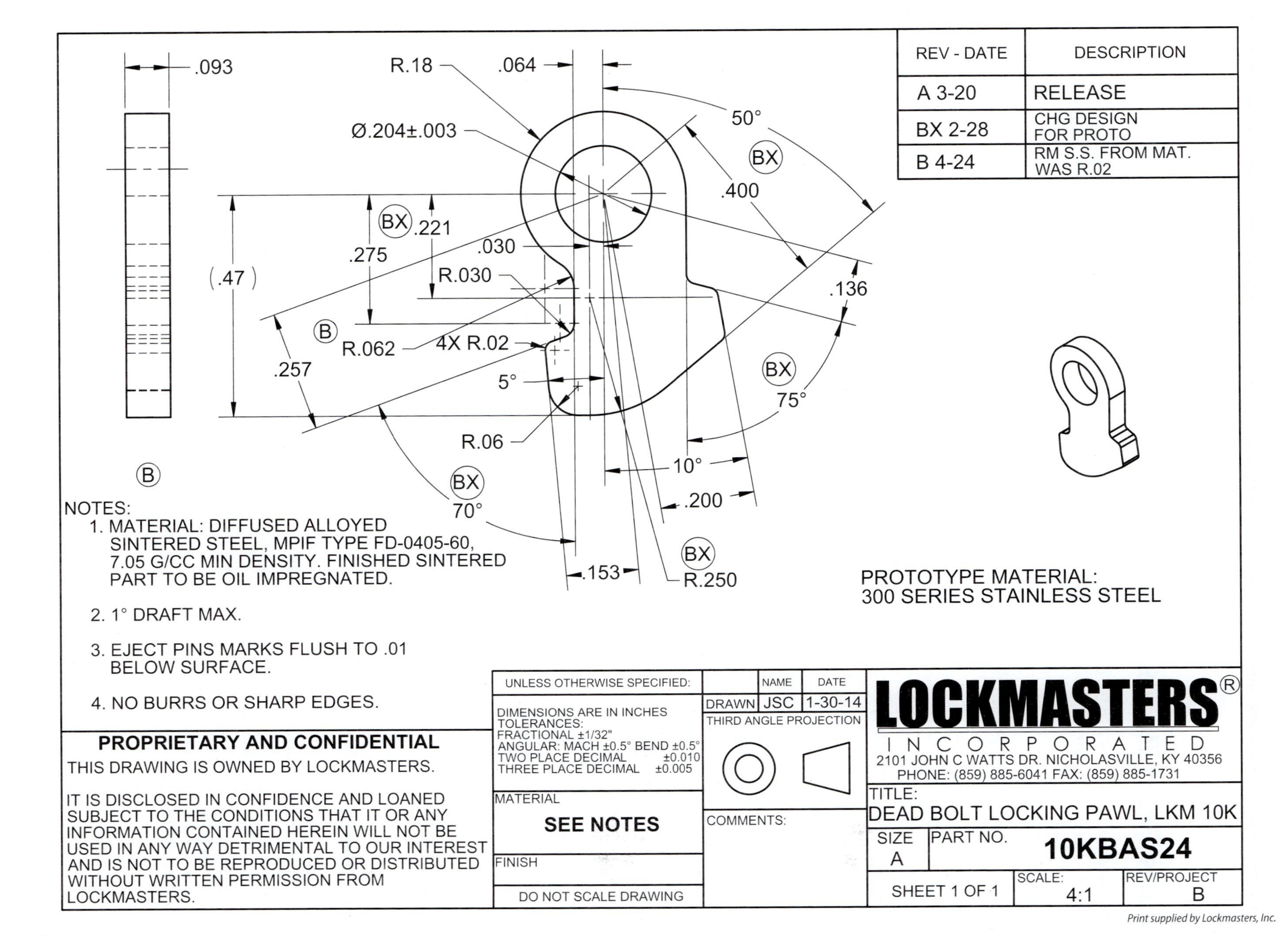

Print supplied by Lockmasters, Inc.

PR 4-2. Dead Bolt Locking Pawl

Apply and Analyze

Name ______________________ Date ____________ Class ____________

Bonus Print Reading Exercises

The following questions are based on the various bonus prints located in the folder that accompanies this textbook. Refer to the print indicated, evaluate the print, and answer the question.

Print AP-002

1. What term describes the relationship of the axis of the 1/2″ NPT pipe plug with the main axis of the bottom assembly?

____

2. With respect to the part of subassembly 1 that attaches to subassembly 2, which word best describes the shape of the material at the connection: round tube, hexagon tube, or square tube?

____

3. The top assembly featured in this drawing appears to have two tall stems, each about 14.5″ tall. What geometric term describes the relationship between the two axes of the stems?

____

Print AP-003

4. This part features some concave arcs that are equally spaced. There are some larger arcs (.072 radius) spaced _____ apart and some smaller arcs (.030 radius) spaced _____ apart.

____

Print AP-007

5. This part features four slots. How many degrees are there between the center planes of adjacent slots?

____

6. The view at the top-right corner of the sheet features two visible circles and a set of center lines. What term applies to the relationship the circles have with each other?

____

Print AP-010

7. In a general way, what geometric shape describes the main objects of this assembly, which is the primary reason only one view is required?

____

Print AP-011

8. This part features a V-groove for a belt. What geometric term applies to the included angle of the V-groove?

____

9. Note 7 specifies what type of relationship between the belt groove and the main axis (identified as datum D)?

____

10. Note 5 specifies some shallow balance holes can be drilled 180° away from what other feature?

____

11. How many degrees are there between each of the M8 × 1.5 threaded holes?

____

Print AP-012

12. What relationship exists between the surface of the pedal pad and the pedal arm, as shown in the auxiliary view in the upper-right corner of the drawing?

____

13. The pedal arm on this part curves at a radius of 11.875. How can the point of tangency between that curve and the 3.000 radius curve be determined?

____

Print AP-016

14. What geometric term is used to describe the manner in which the worm gear ridge is wrapped around the shaft, as shown on the left end of this part?

____

15. What geometric term appears in the Fine Pitch Spur Gear Data table to describe the base circle diameter of 1.17462?

____

Print AP-017

16. What geometric term describes the manner in which the teeth of this gear are considered RH-10°?

____

Print AP-021

17. What geometric term applies to the shape of the opening that is centered approximately 4.310 from the left, assuming the corners were sharp rather than rounded?

____

18. What geometric shape term applies to the curve at each end of the 11 long slots on the right side?

____

Print AP-022

19. What geometric term applies to the relationship the adjacent sides of this part have to each other after the bends are made?

____

20. Are either of the openings in this part square (having equal sides)?

____

Notes

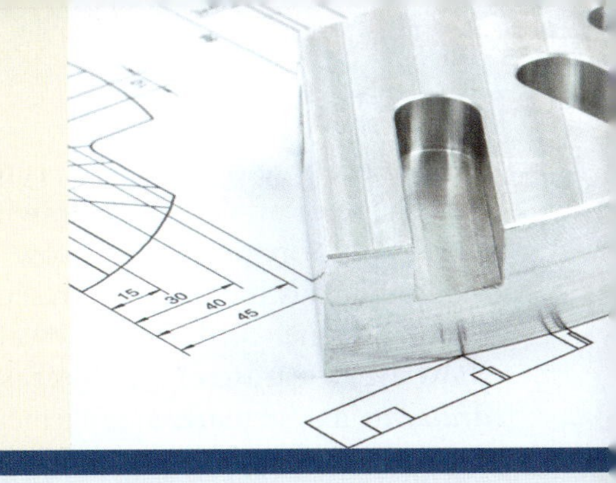

UNIT 5 Multiview Drawings

LEARNING OBJECTIVES

After completing this unit, you will be able to:

- Define spatial visualization and discuss its role in multiview drawings.
- Explain the relationship between an orthographic projection and a multiview drawing.
- Identify and define the three dimensions of an object.
- Identify the six principal views and the three regular views.
- Explain the role of the three principal projection planes in establishing the three regular views.
- Explain three visualization principles for multiview drawings.
- Identify three types of flat surfaces as they relate to the principal projection planes.
- Explain characteristics of cylindrical surfaces in multiview drawings.
- Explain characteristics of fillets, rounds, and runouts in multiview drawings.
- Identify differences between third-angle and first-angle projection.
- Sketch basic multiview drawings of objects.
- Discuss computer-generated views and identify how they may vary from traditional methods.

TECHNICAL TERMS

depth
fillet
first-angle projection
frontal plane
height
horizontal plane
inclined surface
multiview drawing
normal surface
oblique surface
orthographic projection
partial view
profile plane
projectors
removed view
round
runout
spatial visualization
third-angle projection
width

The purpose of a drawing is to show the size and shape of an object. A drawing can also provide certain information about how an object is to be made. Various methods of presentation are available to the designer or drafter. However, the best way to show every feature of an object is to use a ***multiview drawing***, a systematic arrangement of more than one view of an object's features. Multiview drawings are created using the principles of orthographic projection. Many drafting and print reading texts use the terms orthographic projection and multiview drawing interchangeably. The standard practices for multiview drawings are covered in ASME Y14.3, *Orthographic and Pictorial Views*. This unit forms the foundation for many other units. This unit can be described as explaining the science behind the drawings that are used in industry. A print reader may be able to read the views of a drawing without understanding all of the scientific principles, just as a reader of words may not understand all the grammar rules behind sentence structure. Simply stated, it is a goal of this unit to explain how views are projected and to provide various rules that can be applied to the abstract ability to visualize geometric shapes and figures, thereby enhancing the print reading experience.

The Role of Spatial Skills

A skilled technician reading a print must be able to visualize the part as a whole. The print reader must be able to look at the views in a drawing and interpret those into a mental picture, **Figure 5-1**. A related field of study that has been the subject of much research is the field of spatial visualization. ***Spatial visualization*** can be defined as the mental visualization of 2D and 3D shapes and objects, including such tasks as imagining objects in the mind as they are rotated, moved, or reflected in a mirror. While it is beyond the scope of this text to discuss spatial visualization in depth, there is a strong correlation between print reading and spatial visualization. Training in spatial visualization can increase the print reader's ability to read and interpret multiview drawings.

Spatial Visualization Tools

The print reader can take steps other than just looking at prints to develop an ability to visualize. This section presents a few spatial visualization instruments used in spatial research that will give you an idea of the types of mental exercises that are related to this field of study. The exercises in this unit, and in Units 6 and 7, are also designed to help beginning print readers develop the spatial skills necessary to successfully read multiview prints and drawings. Spatial visualization training has also been shown to increase the success rate of students in engineering and technology programs, so you are encouraged to investigate additional resources.

One of the most popular testing instruments is the Purdue Spatial Visualization Test: Visualization of Rotations, which involves observing two pictorial views of a three-dimensional object. The second pictorial view shows the object rotated about a single axis or more than one axis. A different object is then to be mentally rotated in the same fashion. The correct pictorial answer is chosen from five possible answers, **Figure 5-2**. Many examples of this instrument can be found online, and you can conduct your own analysis of your spatial visualization ability.

Another spatial visualization tool involves observing a string of connected cubes drawn pictorially, with a collection of similar strings of cubes turned and tilted at another angle. The altered views are arranged in a

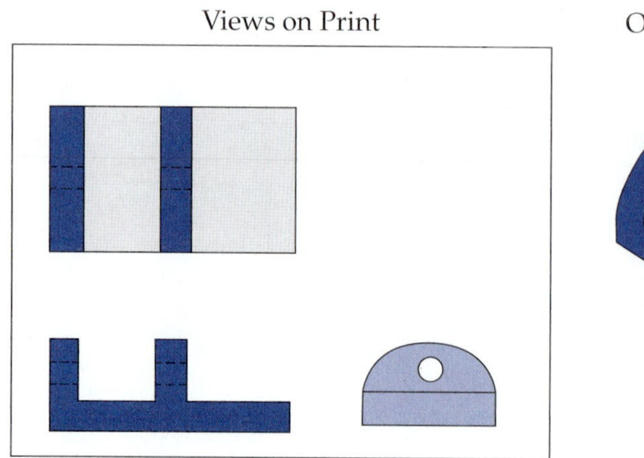

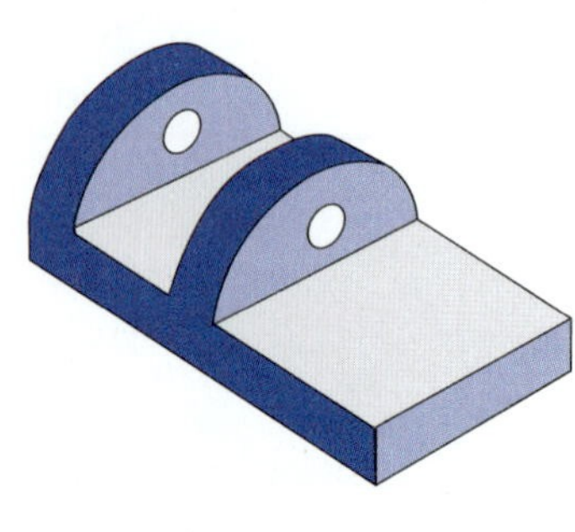

Goodheart-Willcox Publisher

Figure 5-1. The views of a multiview drawing are systematically arranged so anyone can visualize the object.

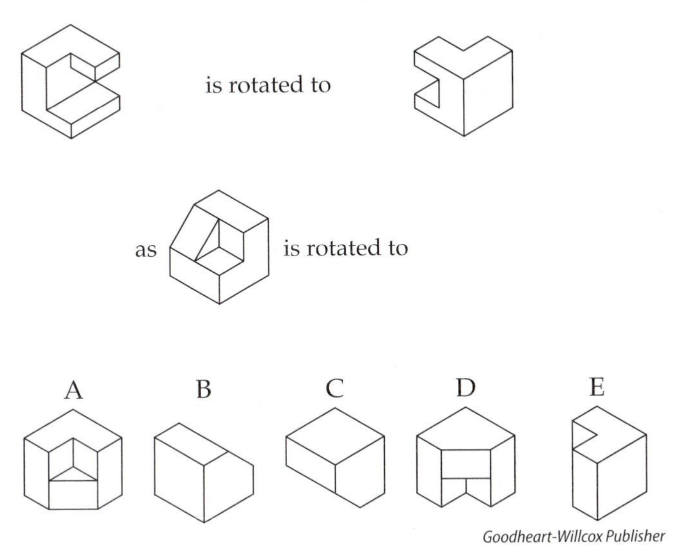

Figure 5-2. The Purdue Spatial Visualization Test: Visualization of Rotations can be used to measure or evaluate an individual's spatial skills. (See page 97 for solution.)

multiple choice format. In the original version of this instrument, two of the choices are the same configuration of cubes, while the other two configurations are different from the original. See **Figure 5-3**.

A surface development test involves the unfolding of a prism with angled surfaces or a cube with graphic patterns on each side, **Figure 5-4**. This test requires selecting a folded object corresponding to the flat pattern. These skills are similar to some of the skills used in Unit 21, *Precision Sheet Metal Parts*. Another test involves illustrations of a sheet of paper being folded multiple times and then punched with one or two holes. The correct unfolded solution is then selected from a set of options, **Figure 5-5**.

These examples are intended as an introduction to different types of spatial visualization instruments that have been used in spatial visualization research. You may wish to pursue additional information about this field of study. The activities at the end of this unit are more traditional to the field of drafting and print reading, and include a variety of exercises for visualizing objects and drawing views. These activities are also designed to help improve spatial visualization skills. Similar activities are presented in Units 6 and 7. Textbooks with titles such as "Technical Drawing," "Engineering Graphics," or "Mechanical Drafting" would also be good resources for exercises and activities related to visualizing three-dimensional objects from two-dimensional views.

In summary, the concepts and skills described in this unit are foundational to print reading. The student who masters these skills will have built a firm foundation for understanding the remaining units in this text. The next few sections of this unit focus on the theory and practice of orthographic projection.

Orthographic Projections of Views

This section could be subtitled "The Science of Technical Drawings." As you will see, any view of an

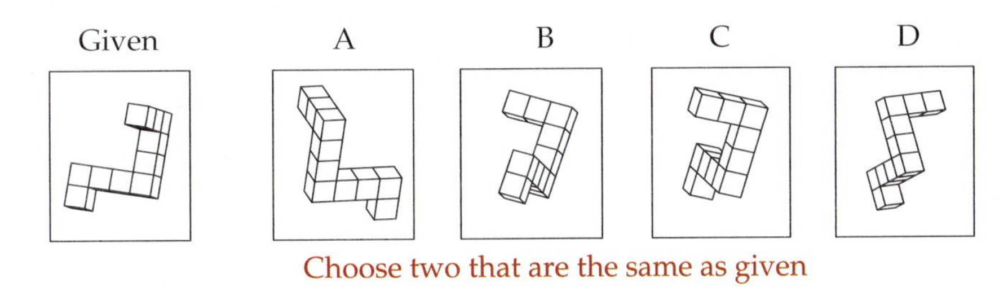

Figure 5-3. This mental rotation spatial visualization instrument challenges the individual to find two additional correct alternative rotations of the given object. (See page 97 for solution.)

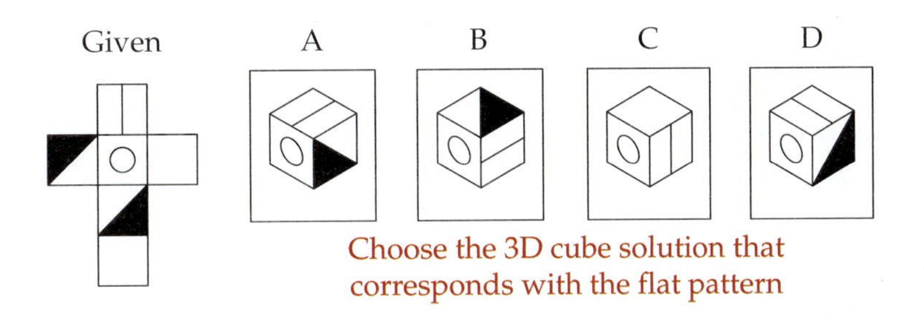

Figure 5-4. Surface development spatial visualization exercises can increase print reading ability, especially within the realm of sheet metal pattern drawings. (See page 97 for solution.)

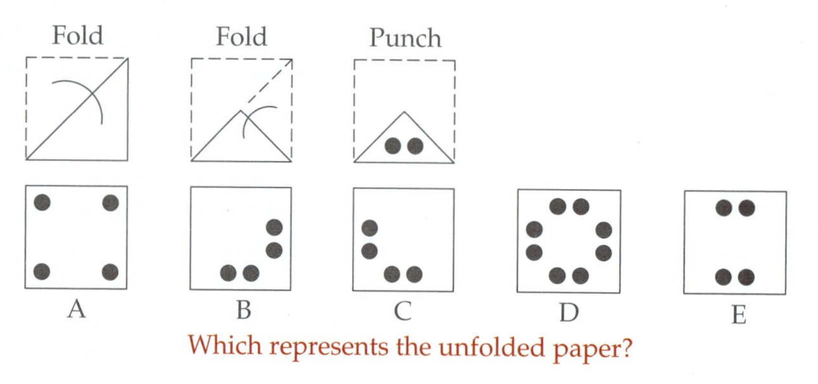

Figure 5-5. This spatial visualization instrument involves mentally visualizing folding and unfolding paper with punched holes. These exercises help develop spatial visualization skills. (See page 97 for solution.)

object can be scientifically explained as the projection of an object's features onto a two-dimensional plane. For a drawing, the plane is often a piece of paper or a computer screen. Simply defined, ***orthographic projection*** is a system wherein parallel lines, called ***projectors***, are used to project the points of a 3D object onto a 2D projection plane. By definition, the "ortho" prefix means "straight" or "right," as exemplified by projectors that are always perpendicular to the projection plane, resulting in an exact and very precise view. If more than one projection plane is used, the result is multiple views, thus the term "multiview projection" or "multiview drawing." In general terms, a multiview drawing is a drawing *based* on the principles of orthographic projection. In the drawing, the different views of a multiview drawing must be systematically arranged according to standard practice. This allows anyone reading the drawing to connect the views together and form a mental picture of the part, **Figure 5-1**.

One way to develop an understanding of the multiview system is to observe how a cardboard box unfolds. Each side of the box is oriented similar to orthographic projection views. The sides are at right angles to each other and have a definite relationship. If the front of the box remains in position, the four adjoining sides unfold similar to how the views of a multiview drawing are arranged. See **Figure 5-6**.

Now think of the cardboard box as made out of glass. Place an object inside the glass box and imagine that the points of the object are projected onto the glass planes as views. See **Figure 5-7**. The surfaces of the glass box represent projection planes. The imaginary projection lines bring the separate views to each projection plane. If the glass box is unfolded like the cardboard box, six views are shown in an orthographic arrangement.

The glass box example is one of several ways that drafters and print readers have been trained to understand the principles of orthographic projection. Through this unit, other principles will be examined. Understanding these principles will help an individual develop the visualization and interpretation skills necessary to read and interpret prints.

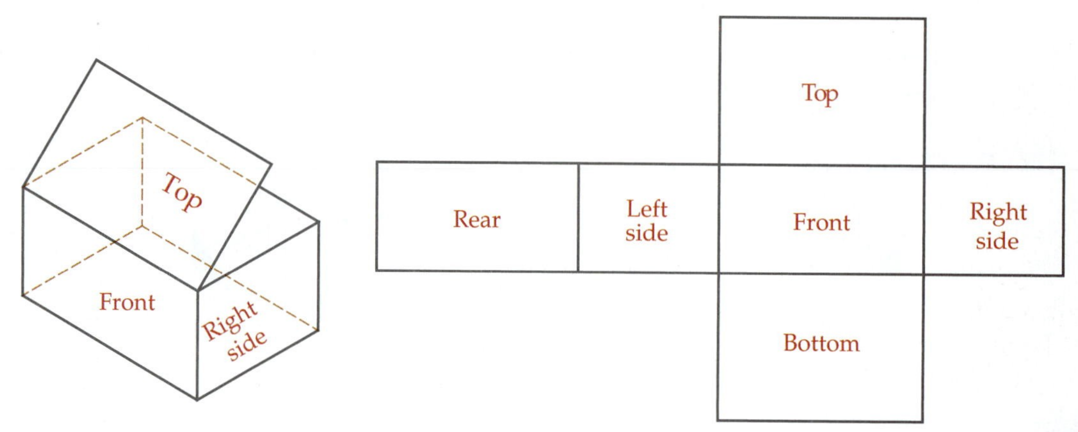

Figure 5-6. Each side of an unfolded cardboard box is oriented similar to orthographic projection views.

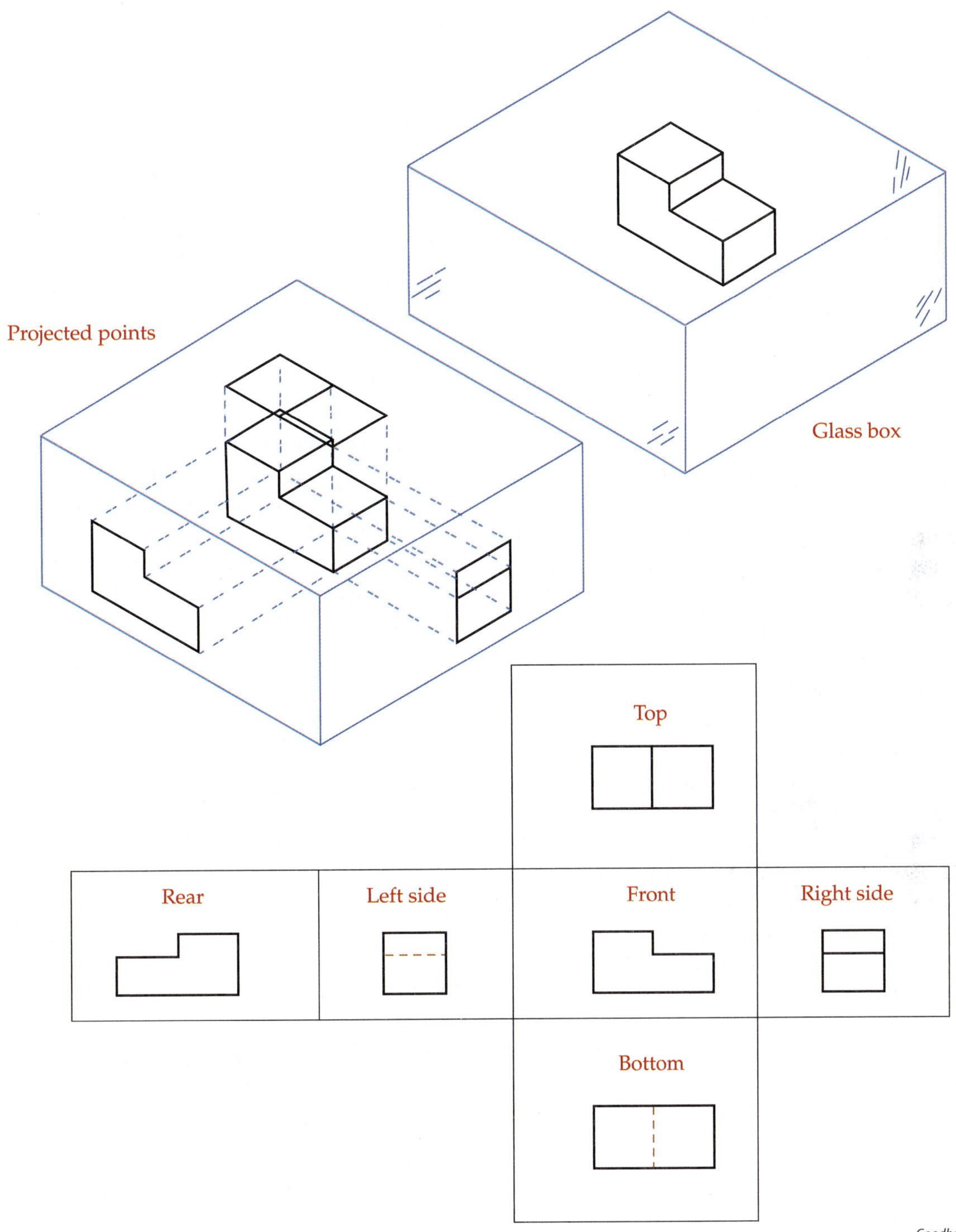

Goodheart-Willcox Publisher

Figure 5-7. If an object is placed inside of a glass box and projected onto each side, when the glass box is unfolded, six views are shown in an orthographic arrangement.

Selection of Views

Since there are three dimensions in space, an object can be viewed from the right or left, top or bottom, and front or back, resulting in six principal views. In the past, the term *normal* was sometimes applied to describe these views. In this text, the term *principal* is used to mean "main" or "primary," as defined by the Y14.3 standard. However, only those views necessary to clearly describe the object need to appear in the drawing. Seldom is an object so complex that all six principal views are required. Usually, the necessary details can be shown in two or three views. Three views will almost always fully describe an object, but more views can be used if there is a lot of detail on the opposite sides of an object. Directions other than the six principal directions are defined as auxiliary views, which are discussed in Unit 7.

In educational settings, the front, top, and right-side views are typically used to describe an object, and may be referred to as the three *regular* views. This type of drawing can also be referred to as a three-view drawing. To be clear, in industrial applications there is nothing that mandates using the three regular views, and views are not given the names that are being applied in this unit for educational purposes. Also, many objects made of flat sheet metal may only require one view, while cylindrical objects may only require two views. The following rules will help in the selection of views:

- Only views clearly describing the shape of the object should be drawn.
- Views containing the fewest hidden lines should be selected. Compare the two side views in **Figure 5-7.**
- If practical, the object should be drawn in its functioning (operating) position.
- If practical, the view best describing the shape of the object should be selected as the front view.

There are situations in which the arrangement of the views necessary to describe the object may be difficult. One feasible solution is to create a view removed out of projection with the usual arrangement. In those cases, a viewing-plane line with labels can be used to identify the viewing direction, and the view can be labeled VIEW A-A (or similar), as discussed in Unit 2. These are called ***removed views***. Another space-saving technique used for regular views or removed views is to draw only a portion of the object. The incomplete view may incorporate short break lines, as if a portion of the object is broken off. These are called ***partial views***. Sometimes, a partial view may be one-half of a symmetrical object wherein the center line is the dividing line.

Dimensions of an Object

One of the keys to reading a multiview drawing is familiarity with the terminology used for the dimensions of an object. Each projected view is two-dimensional, even though the object is three-dimensional. It is critical to discuss the dimensions of an object using standard terms. Throughout the field of drafting education, the three terms predominantly used for the three dimensions of an object are height, width, and depth.

Height is defined as the top-to-bottom measurement for an object from the vantage point of the *front view*, which may or may not be the actual "front" of the object. ***Width*** is defined as the left-to-right measurement of an object, also from the vantage point of the front view. ***Depth*** is defined as the front-to-back distance of the object as based on the orientation of the object for the views. These terms will help you as you go through educational steps to learn about multiview and pictorial drawings.

Even though the object described by a print is three dimensional, each projected view is only two dimensional. The height and width of the object are shown in the front view. The width and depth of the object are shown in the top view. The height and depth of the object are shown in the right-side view. See **Figure 5-8**. Standardizing the terms used for the

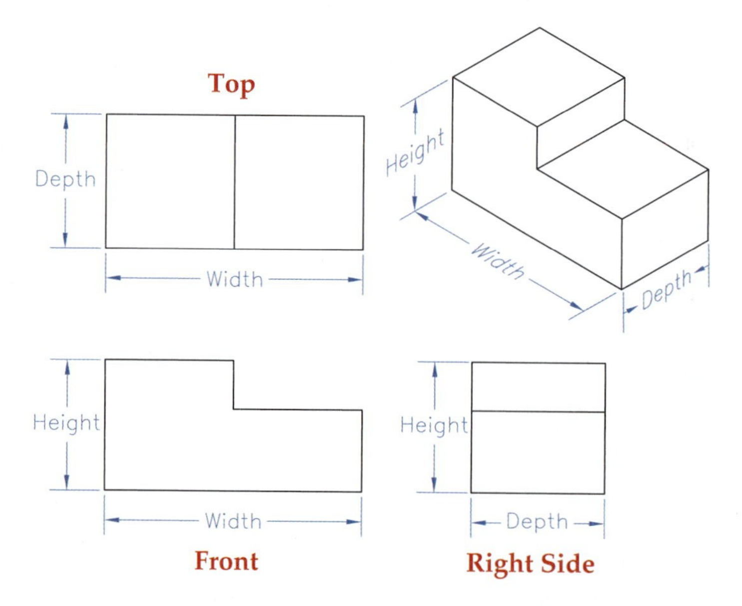

Goodheart-Willcox Publisher

Figure 5-8. The three standard terms for the dimensions of an object are height, width, and depth.

dimensions is critical to studying print reading and completing the exercises in this unit.

Visualization of Objects

Another useful educational tool in studying projection theory is identifying particular "rules" of behavior for surfaces and edges as they project onto projection planes, similar to the sides of the glass box. This section examines how flat, or planar, surfaces are oriented in space and how the orientation of these planes changes how surfaces and edges appear in multiview drawings. For now, three principal projection planes will be established, and only the three regular views will be considered. The front view will project onto a plane called the ***frontal plane***, the top view will project onto a plane called the ***horizontal plane***, and the right-side view will project onto a plane called the ***profile plane***. See **Figure 5-9**.

With these three projection planes in mind, the following principles may be helpful:

- **Principle One.** A flat surface is oriented perpendicular, parallel, or inclined to a projection plane, **Figure 5-10**.
- **Principle Two.** As a result of Principle One, all flat surfaces appear in a multiview drawing as one of the following:
  - A line, if the orientation is perpendicular
  - True size and shape, if the orientation is parallel
  - Foreshortened, if the orientation is inclined (see **Figure 5-11**)

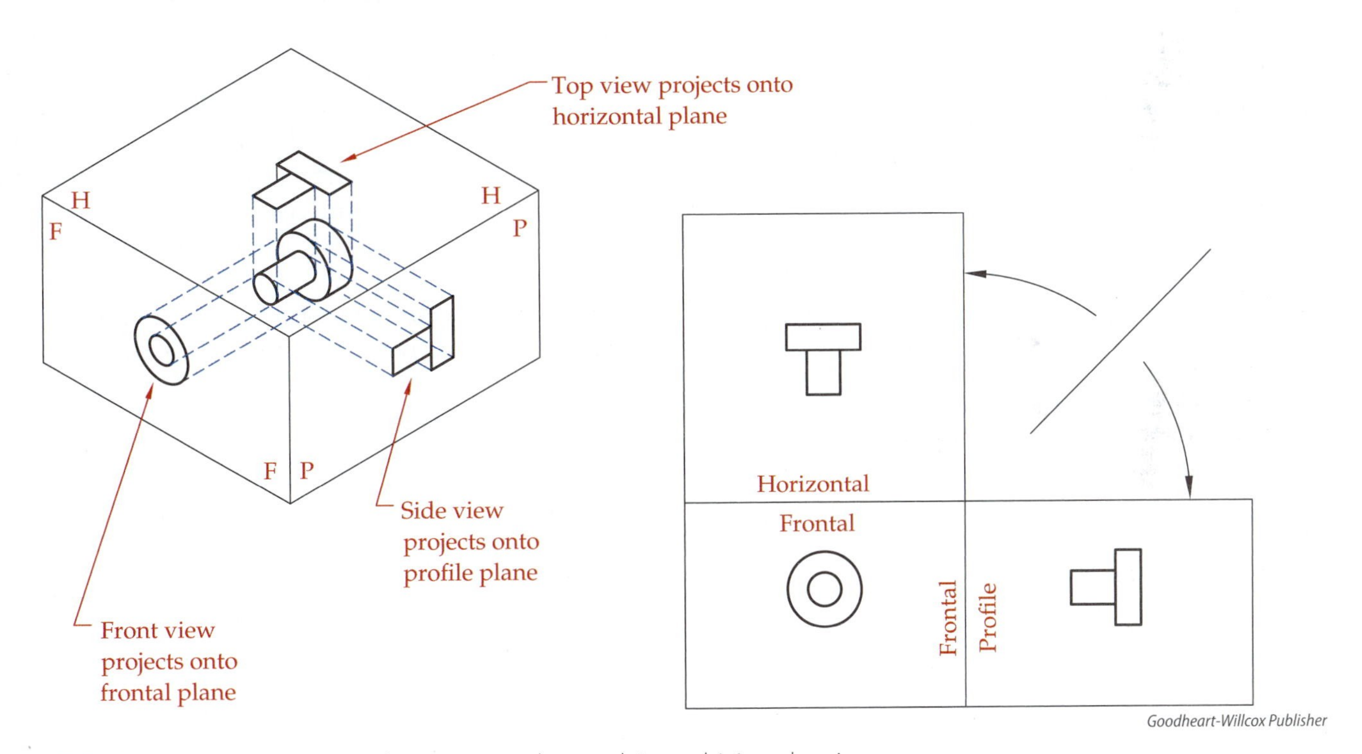

Figure 5-9. Three basic projection planes are used to explain multiview drawings.

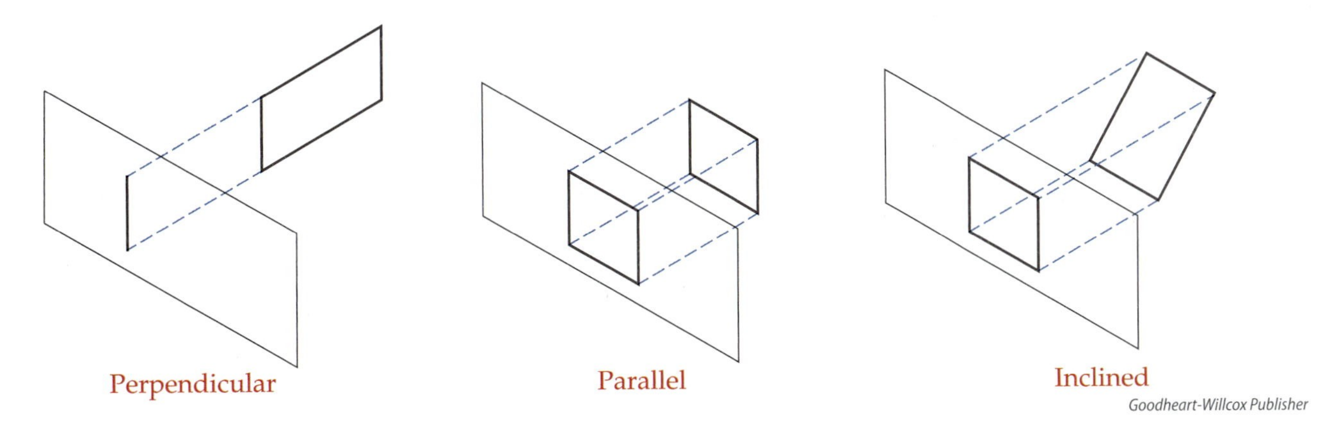

Figure 5-10. A flat surface can be oriented perpendicular, parallel, or inclined to a plane of projection.

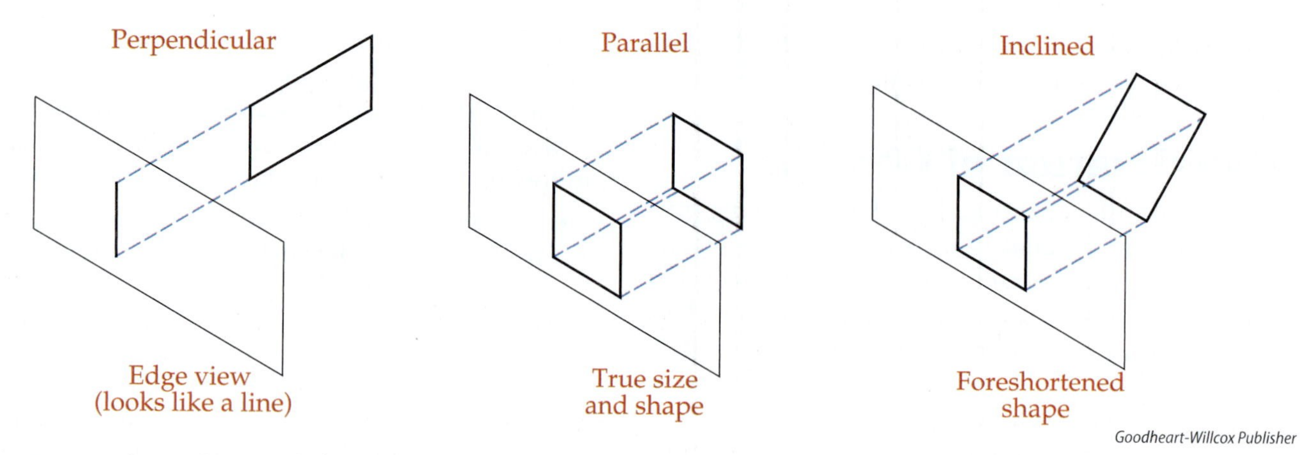

Figure 5-11. A flat surface can appear in a projection as an edge, in true size and shape, or foreshortened.

- **Principle Three.** All surfaces appear in *every* view of a multiview drawing, even if only as a line and even if represented by a hidden line.

Three Types of Surfaces

If the principles previously discussed are further applied, there are three basic types of flat surfaces that can be defined in an orthographic projection. A ***normal surface*** is defined as a surface parallel to one of the three projection planes and, therefore, perpendicular to the other two. For example, each surface of a cube is normal and the top flat surface of a cylinder is normal. If normal surfaces are examined with respect to the three principles stated above, the following will apply:

- The normal surface appears true size and shape in only one view.
- The normal surface appears as a line in two of the three regular views.

Very often, when looking at a line in a multiview drawing, the line represents an edge view of a surface. When looking at the front view of a cube, the top and right-side surfaces appear as lines. Study **Figure 5-12A**. With respect to the top edge of the object in the front view, if you only "see" the front edge along the front surface, you are still thinking in two dimensions. As your visualization skills improve, you will begin to visualize these lines as surfaces that extend back.

A second type of flat surface is the inclined surface. An ***inclined surface*** is defined as a surface that is perpendicular to one projection plane but inclined to the other two projection planes. If inclined surfaces are examined with respect to the three principles stated above, the following will apply:

- An inclined surface appears as a line in only one of the three regular views.
- An inclined surface appears as a foreshortened shape in two of the three regular views.

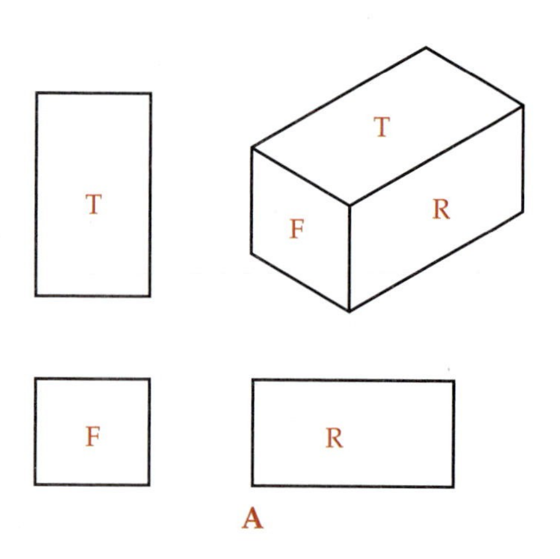

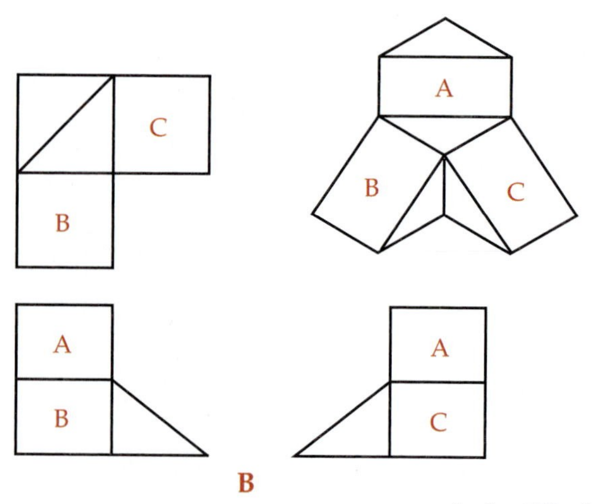

Goodheart-Willcox Publisher

Figure 5-12. These drawings show flat surfaces represented in orthographic projection. A—Normal surfaces appear as either a line or as true size and shape. B—Inclined surfaces appear as a line or as a foreshortened shape.

Study **Figure 5-12B**. Surface C is perpendicular to the frontal plane, so it appears as a line in the front view. However, it is inclined to the horizontal and profile planes, so it appears as a foreshortened shape in the two views projected onto those planes. Surfaces A and B exhibit the same characteristics. In summary, the shape of an inclined surface appears twice in three regular views, but never true size and shape—it is always foreshortened. A normal surface shape appears only once in three regular views, but it appears true size and shape.

The third basic type of flat surface is the oblique surface. An ***oblique surface*** is not only inclined but also rotated. Therefore, it is inclined to all three projection planes and does not appear true shape and size in any regular view. In fact, it may appear a little distorted due to the projection angle it forms with the projection plane. It also does not appear as a line in any regular view. See **Figure 5-13**. Therefore, using only the three regular views, oblique surfaces are harder to visualize than normal and inclined surfaces.

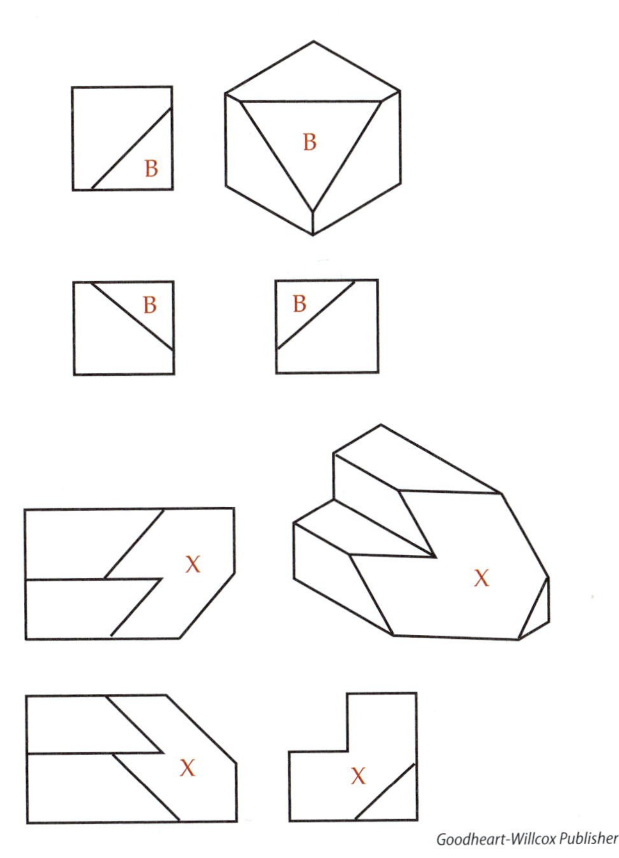

Figure 5-13. An oblique surface is not only inclined but also rotated. Surfaces B and X are oblique surfaces.

Cylindrical and Curved Surfaces

Cylindrical and curved surfaces present another set of visual challenges to the print reader. A flat surface is often tangent to a curved surface, thus making a smooth transition between the curve and the flat, as shown in objects B and D in **Figure 5-14**. In these two cases, no visible lines are shown at the element of tangency, although views generated by the CAD system may create these by default. The projections formed when flat surfaces form intersections and cutouts with cylindrical surfaces can also be tricky. **Figure 5-15** illustrates how cylindrical and curved surfaces are usually projected in multiview drawings.

Conventional Representation

In ASME standard Y14.3, *Orthographic and Pictorial Views*, there is a section titled "Conventional Representation," covering a variety of methods that have been established to enhance views for clarity's sake, as well as simplify some of the views used in industrial prints.

One situation that has always been difficult for drafters is when parts have edges that have been slightly rounded. Molded and cast objects have rounded edges called fillets and rounds, **Figure 5-16**.

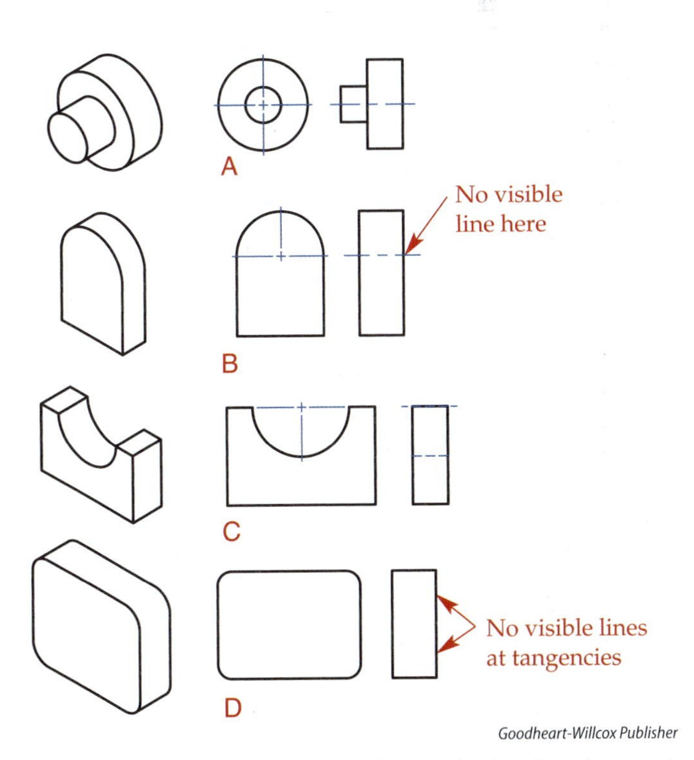

Figure 5-14. These drawings show cylindrical and curved surfaces. A flat surface is often tangent to a curved surface, as shown in B and D.

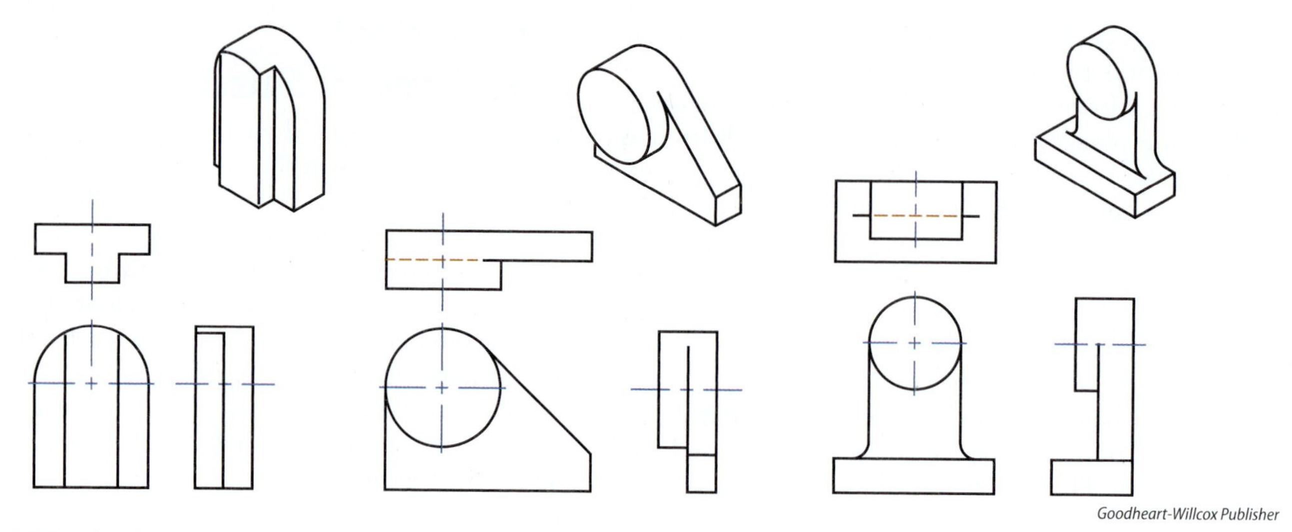

Figure 5-15. This figure helps illustrate how cylindrical surfaces are projected in multiview drawings.

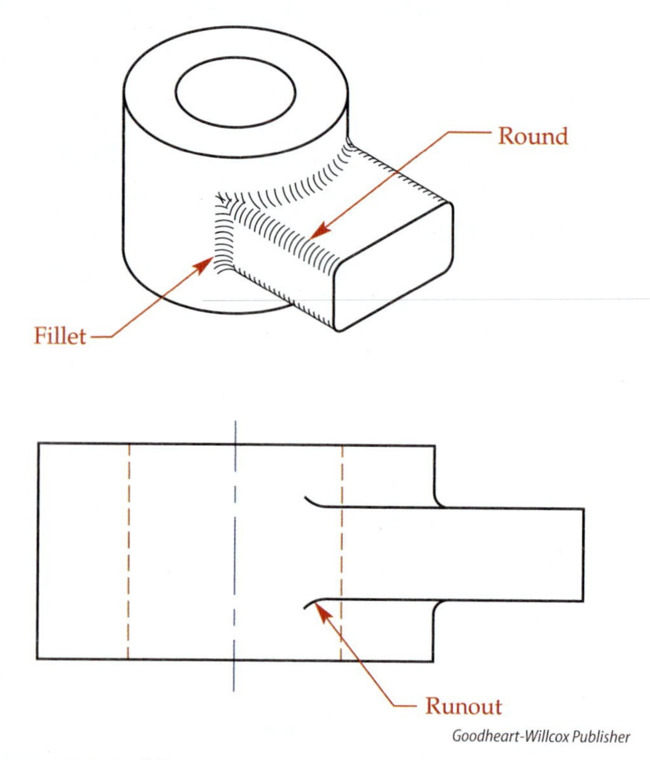

Figure 5-16. Fillets are interior rounded edges. Rounds are exterior rounded edges. When rounded edges intersect curved surfaces, a runout is used to show how the edge tails out.

Fillets are interior rounded edges. ***Rounds*** are exterior rounded edges. Most CAD programs have a command called **FILLET** to perform the task of rounding corners, both interior and exterior.

These rounded corners create situations where the drafter must decide how to represent the change between the two surfaces that do not have a sharp edge separating them. In cases where the rounded edge is very minimal, conventional practice allows for a line to be shown as a phantom line in the location where the edge would be if it were sharp. When a rounded corner intersects a curved surface, the edge fades, or tails out. This is called a ***runout***, and can be represented by an approximated arc, **Figure 5-16.** However, this is different from *runout* as defined in Unit 13, *Geometric Dimensioning and Tolerancing*.

There will be situations in which visible lines coincide with hidden lines or center lines, in which case the visible lines take precedence over the other lines. Hidden lines also take precedence over center lines. **Figure 5-17** shows an object with examples. As you can see, this makes the drawing a little harder to read when a hidden line is omitted because a visible line coincides with it. In similar fashion, even hidden lines that are extremely close to visible lines may be omitted if the lines would bleed together. On parts that are very complicated, with a lot of hidden lines, it is conventional practice to omit some of the hidden lines to avoid distracting the print reader from the important information.

Meanings of a Multiview Line

In summary, there are three meanings a line in a multiview drawing can have. First, a multiview line can represent an edge view of a flat surface or a circular edge view of a cylindrical surface. Second, a line in a multiview drawing may simply be an intersection edge between two flat surfaces or between a flat surface and cylindrical surface where there is not a tangent. Third, a multiview line can represent the maximum contour of a curved surface. See **Figure 5-18.** These meanings apply whether the line is visible or hidden.

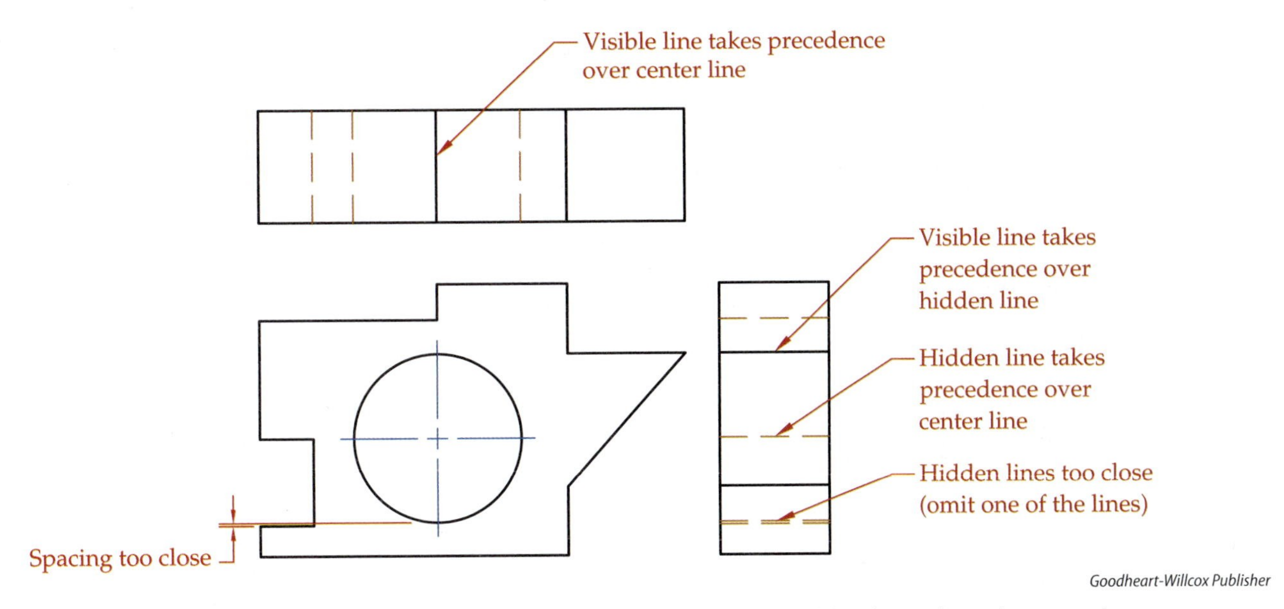

Figure 5-17. Visible lines take precedence over hidden lines and center lines. Hidden lines also take precedence over center lines.

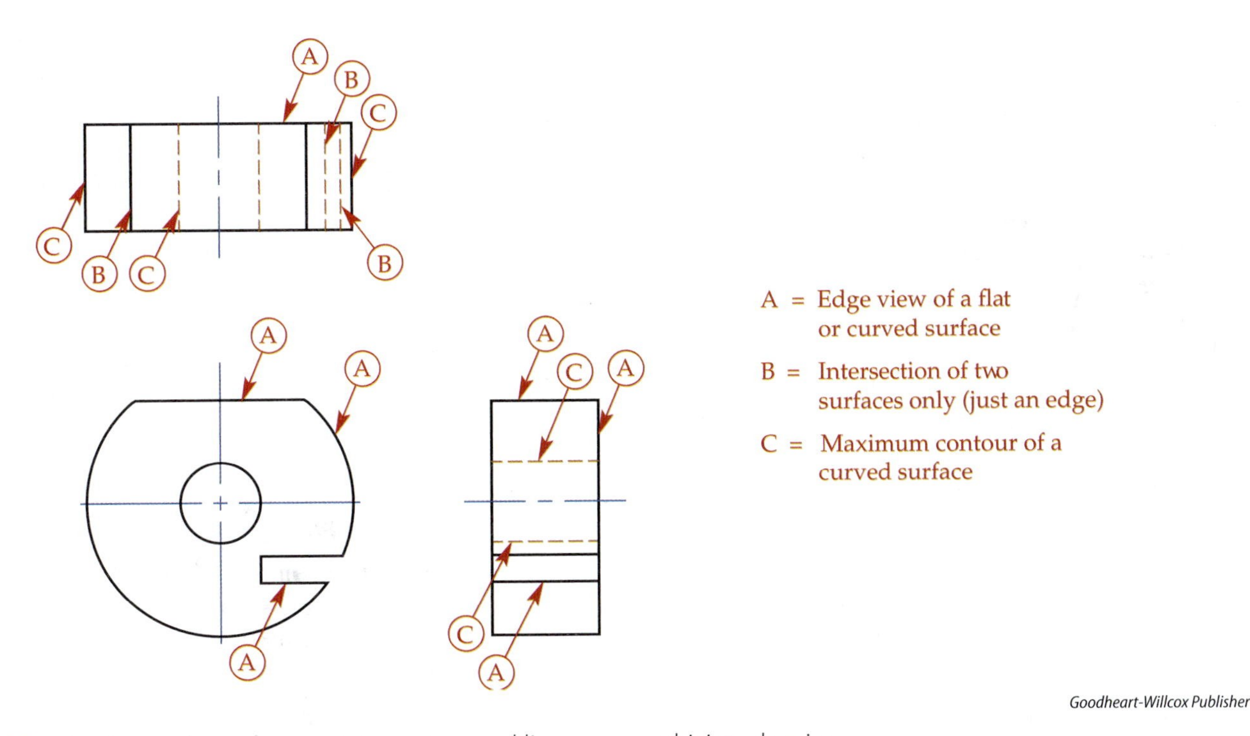

Figure 5-18. This figure shows how to interpret several lines on a multiview drawing.

With respect to the first meaning, be aware there are many different perimeter shapes that a flat surface can have, such as circular, square, or polygonal. Yet, if the surface is perpendicular to the projection plane, it simply projects as a line, or what may be called an edge view. Therefore, many lines in a multiview drawing represent the edge view of planar surfaces that are perpendicular to the projection plane, whether the surfaces are polygonal or curved. As with the other principles of projection discussed previously, understanding these principles will enhance both your visualization and interpretation skills.

First-Angle and Third-Angle Projection

The projection of a view is basically what the viewer sees when looking at the object *through* the glass projection plane. The two projection systems used in industrial drawings are identified as third-angle projection and first-angle projection. These two types of projection result from a theoretical division of space into four quadrants by vertical and horizontal planes, **Figure 5-19**. The viewer of the four quadrants is considered to be in front of the frontal plane and above the horizontal plane. The views are arranged by folding the two planes into one by collapsing the second and fourth quadrants, and the views are then seen from the front. As a result of this, there are no second- or fourth-angle projections. If there were, the views would overlap.

The system of projection explained earlier in the unit is common throughout the United States. The glass box is unfolded in such a way as to place the top view above the front view. This is known as ***third-angle projection***. In third-angle projection, the object resides in the third angle of space, so the projection planes are considered to be located between the viewer and the object. The views are projected toward the viewer onto the planes. See **Figure 5-20**. When quadrants two and four are collapsed, the top view appears above the front view.

In many countries, especially in Europe, a slightly different projection system is used, called first-angle projection. In ***first-angle projection***, the object resides in the first angle of space, so the projection planes are on the opposite sides of the object as the viewer. In other words, the object is between the viewer and the projection planes. The views are projected away from the viewer onto the planes. See **Figure 5-21**. The individual views are exactly the same as those obtained in third-angle projection, but the arrangement on the drawing is different. In essence, the top view is positioned below the front view, and a right-side view is positioned on the left side of the front view. **Figure 5-22** shows a comparison between first-angle and third-angle projection for the six principal views of a simple object.

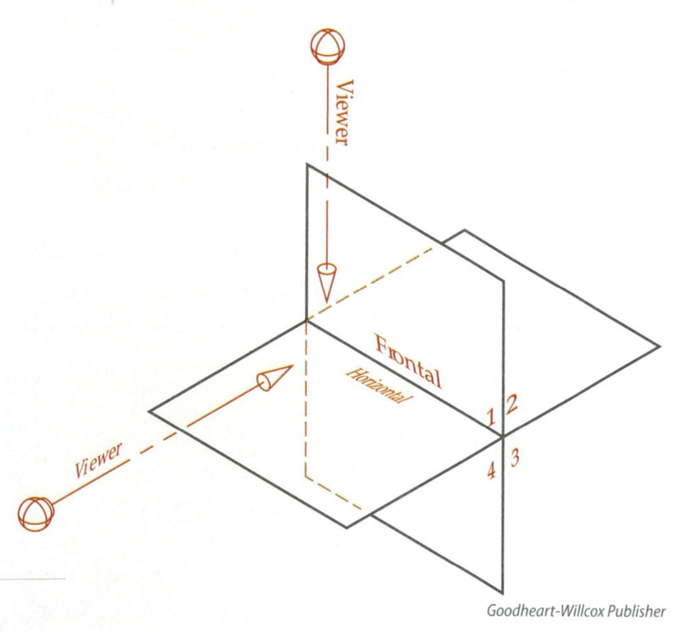

Figure 5-19. The two types of projection are based on a theoretical division of space into four quadrants. Quadrants two and four are not used.

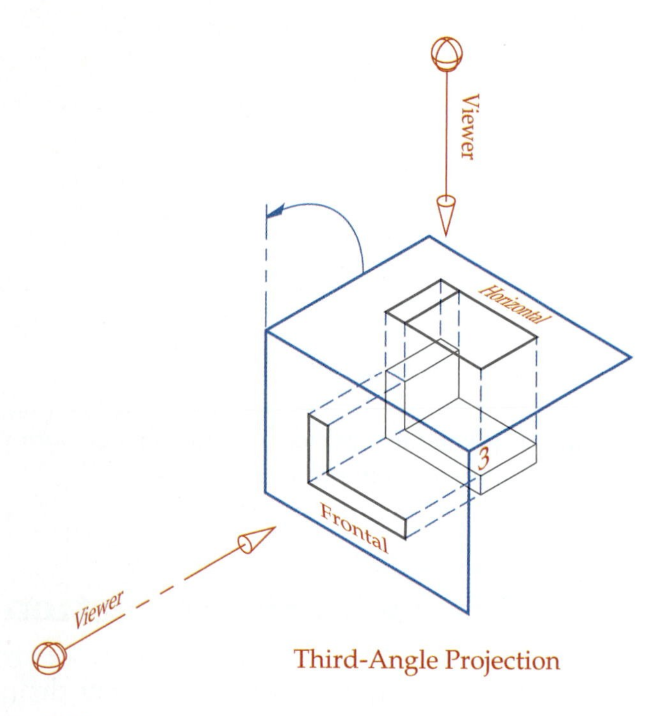

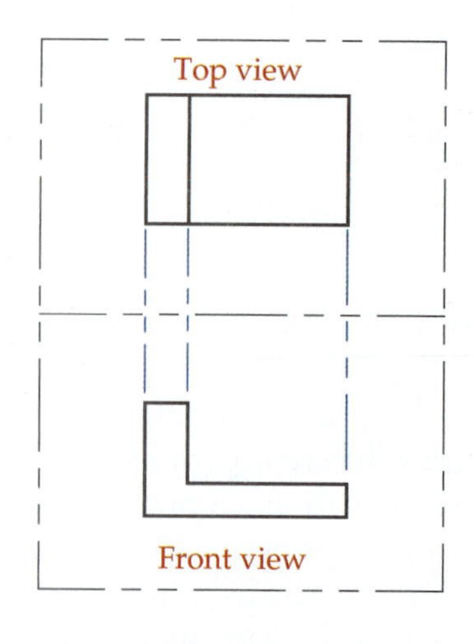

Figure 5-20. In third-angle projection, the projection plane is considered to be between the viewer and the object and the views are projected toward the viewer onto the plane.

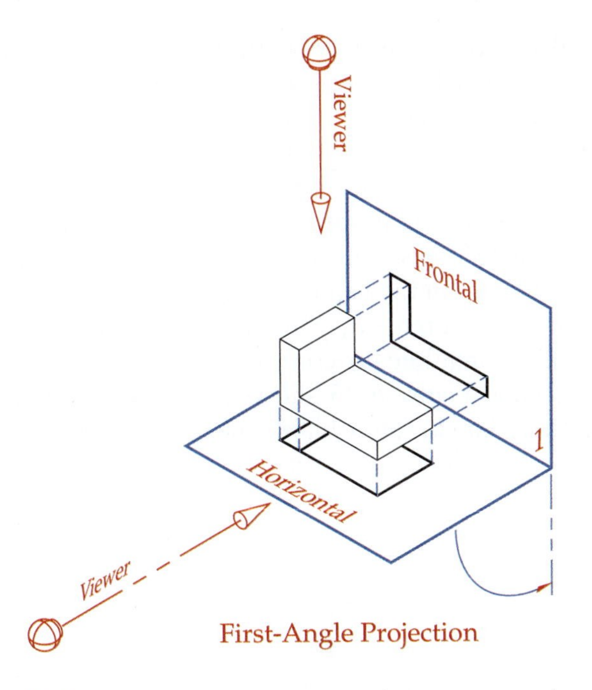

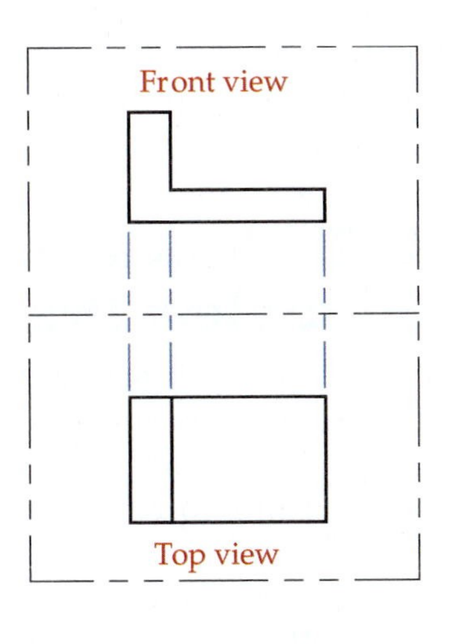

Goodheart-Willcox Publisher

Figure 5-21. In first-angle projection, the projection plane is on the opposite side of the object as the viewer and the views are projected onto the plane on the far side of the object.

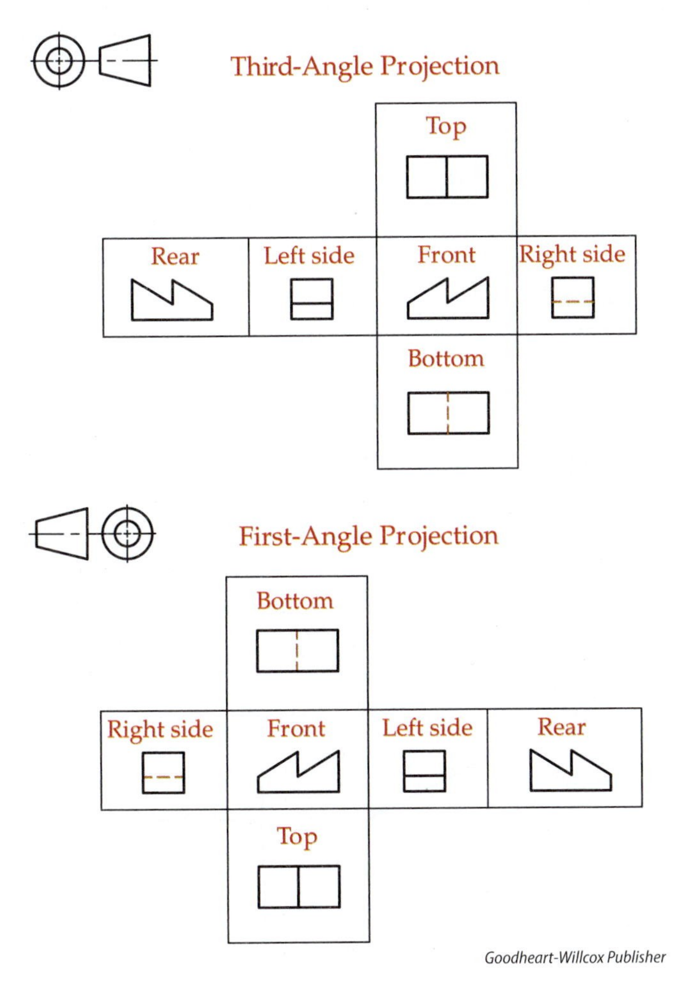

Goodheart-Willcox Publisher

Figure 5-22. This drawing shows the six principal views in third-angle projection compared to first-angle projection.

In summary, the individual views are the same for both angles of projection. The only difference between the two types is the arrangement of views on the drawing. The ASME and ISO standard symbols to indicate third-angle and first-angle projection are shown in **Figure 5-23**. The size of the smaller end of the cone should be about the same size as the main lettering on the print. The circular view can be positioned on

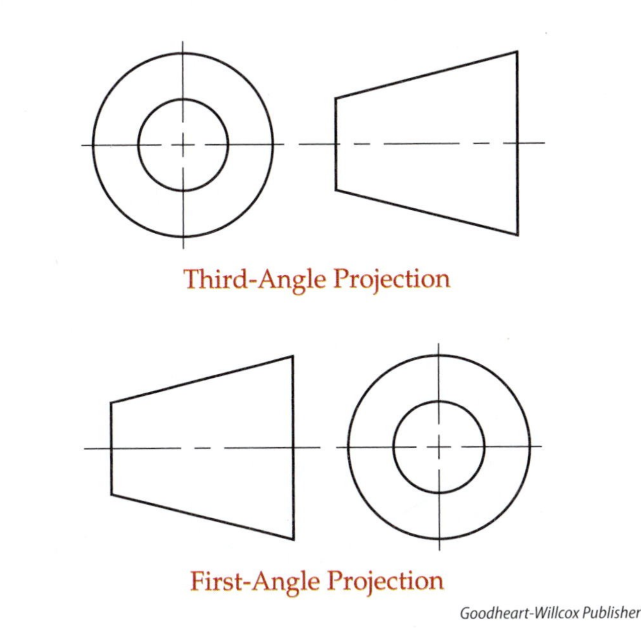

Goodheart-Willcox Publisher

Figure 5-23. The type of projection for a drawing is indicated by one of these two symbols appearing in the title block.

either side of the trapezoidal view, as long as the orientation of the symbol matches the angle of projection illustrated. One of the two versions of the symbol should be included in the title block for drawings that are read within the international community.

Basic Technical Sketching for Multiview Drawings

Freehand technical sketching is a process used by engineers, technicians, and drafters to quickly convey ideas. The ability to freehand sketch can be very helpful in learning to read and interpret multiview drawings. The only materials needed for sketching are a pencil, paper, and eraser. While there are professional drafting pencils available with varying degrees of hardness, a good-quality No. 2 pencil will perform quite well if properly sharpened.

For multiview sketches, construction lines should first be sketched to provide an initial layout of the views. These lines should be created light but still distinct and sharp, not dull and fuzzy. After the initial layout is complete, construct the views, ensuring that height, width, and depth measurements align. Then, darken the final lines with the proper line weight and dash length. Quality lines are better created with a series of short, comfortable strokes, rather than one continuous line. Your eye should be on the point at which the line is to terminate. See **Figure 5-24**. Pull the pencil instead of pushing it. Do *not* use a straightedge to create freehand sketches.

Begin the layout of a multiview drawing with a rectangular box that is sufficient for all three regular views. To help your mind visualize the object, think about all three views simultaneously and not just one view at a time. Within the layout rectangle, you can estimate a height measurement for both front and right-side views, and then estimate a width measurement for both the front view and the top view. For the depth measurement, you can incorporate the use of a 45° miter line to connect the side view with the top view. **Figure 5-25** shows a multiview drawing of a basic block. On a blank sheet of paper, follow the

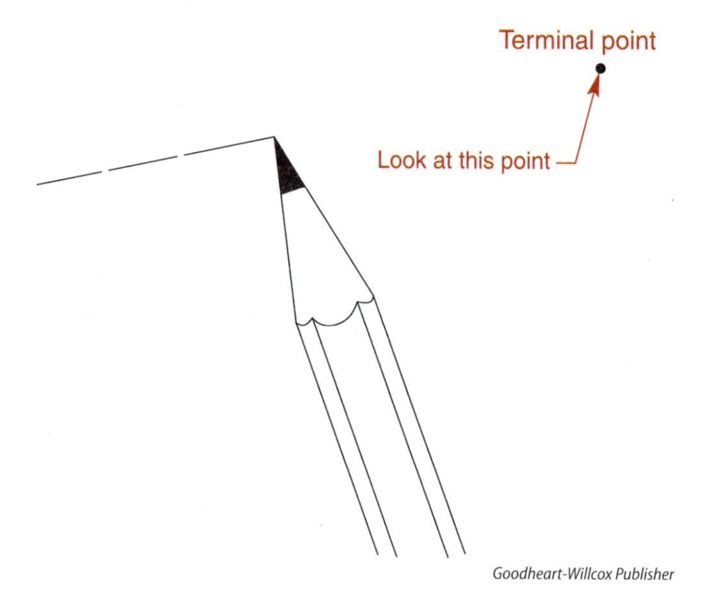

Figure 5-24. When sketching, your eye should be on the point at which the line is to terminate.

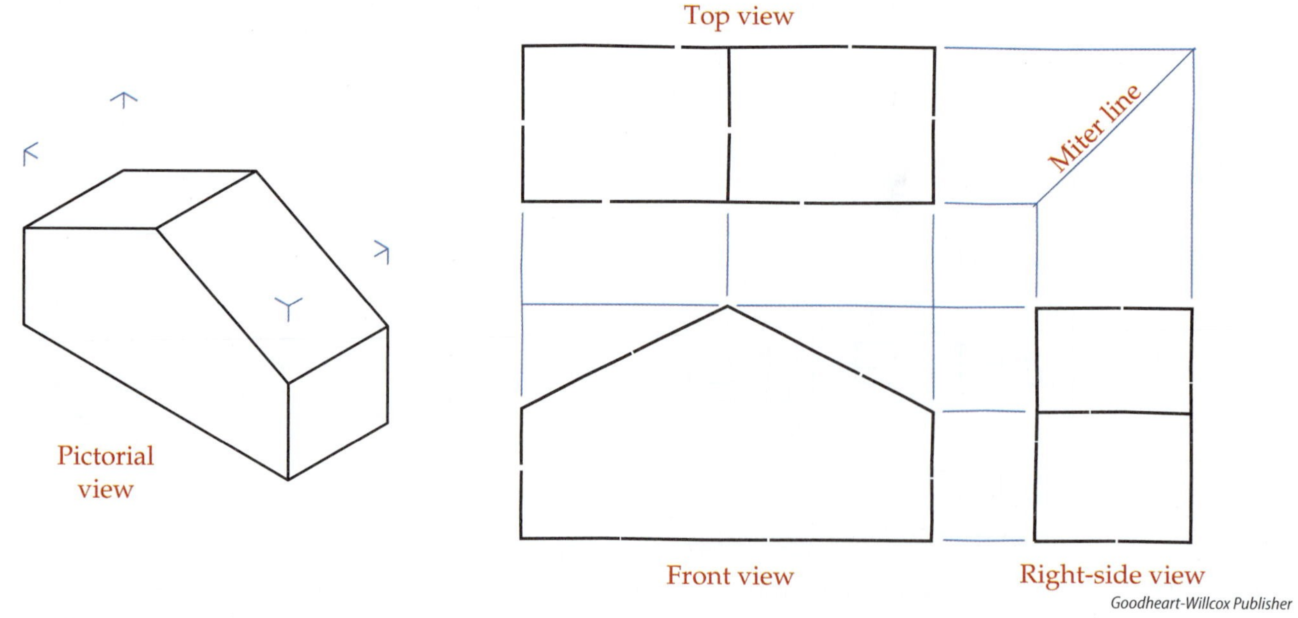

Figure 5-25. The object shown in the pictorial view on the left is sketched as a multiview drawing on the right.

steps shown in **Figure 5-26** to create your own version of this multiview drawing. **Figure 5-27** shows how to construct and project additional features. Some inclined edges and features of the object are not defined with linear distances, so projection techniques ensure the views are properly constructed and aligned from view to view. As mentioned earlier, working on all three views together helps train the mind in visualizing the three-dimensional object using two-dimensional views that are related to each other in very specific ways. Sketching exercises are available at the end of this unit. Appendix C focuses on pictorial sketching, wherein the multiview drawing is given and a picture-like drawing of the object is then constructed.

An additional aid to learning how to visualize an object from the multiview drawing is to label corners (or vertices) of the object with numerals. See **Figure 5-28**. By numbering each corner, you can analyze the multiview and break it down into faces, edges, and vertices. Notice that each numeral is always aligned with its corresponding numeral in the adjacent views, and the miter line is helpful in projecting the depth value. With these numerals, you can identify edges of the object, such as edge

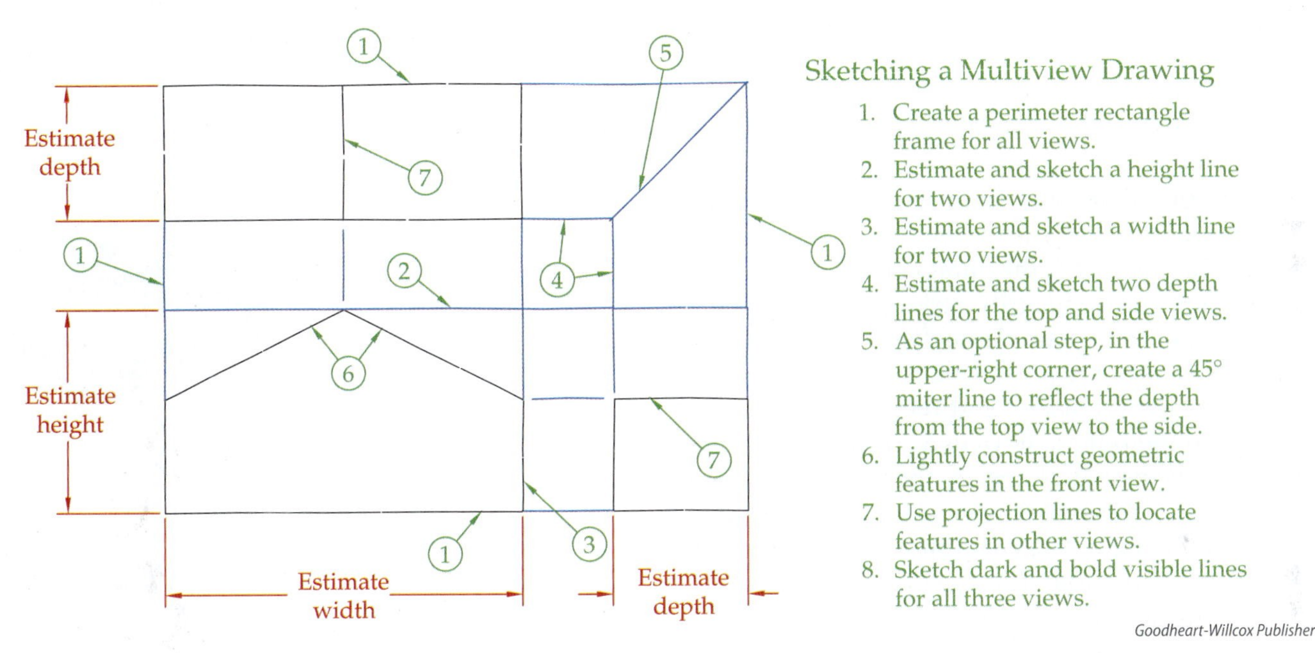

Figure 5-26. The steps for sketching a multiview drawing help guide the drawing to a proper solution. At the basic level, some steps may seem unnecessary, but they establish good practices for more complicated objects.

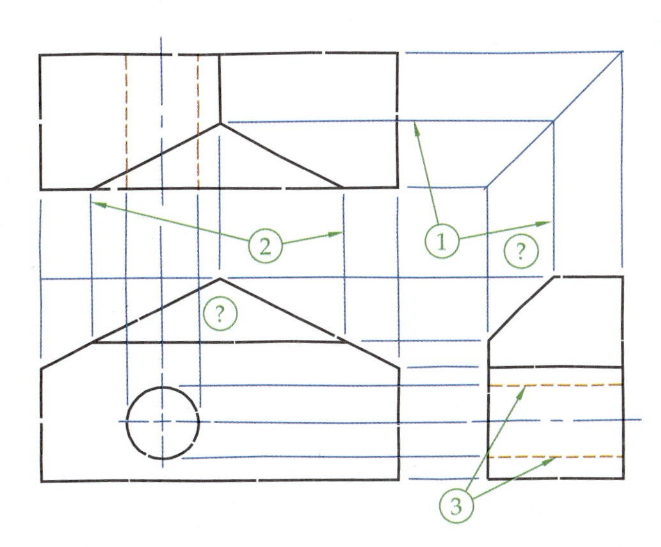

Figure 5-27. Some angular edges and features of the object are not defined by linear dimensions, so projection techniques ensure the views are accurately constructed.

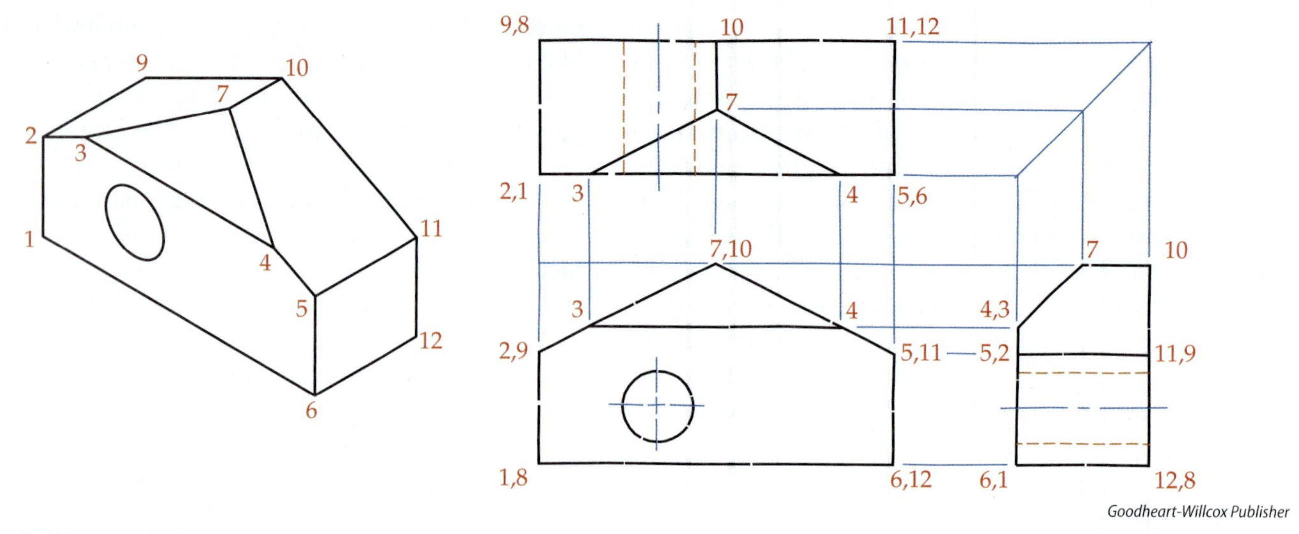

Figure 5-28. Numbering the corners of an object in all three views can be helpful when training the mind to visualize. It also serves as a way to proofread the drawing.

5-11. Notice that edge 5-11 projects as a point in the front view, and since corner 5 is closer to the viewer than point 11, the numbers are shown in that order. You can also identify surfaces by the perimeter numerals. In this example, inclined surface 3-7-4 is perpendicular to the profile plane, so the three points project as a straight line and the surface appears as an edge in the right-side view. Surface 3-7-4 appears as a foreshortened shape in the other two views. As you are learning to visualize, try identifying specific corners, locate them in all views, and mentally visualize various features of the object. This is also a good way to proofread a sketch to make sure that features are aligned properly. Some of the exercises at the end of this unit give you this opportunity.

Computer-Generated Views

In current practice, the views on drawings and prints are often created automatically from 3D models designed with CAD solid modeling software. There are many benefits to creating views in this manner, including the following:

- It is easy to construct views at any scale, using either first-angle or third-angle projection.
- It is easy to update drawings. As the features of the 3D model change, the drawing views are updated automatically.
- The views accurately represent the geometry.
- There is intelligent association between the model and the annotations, such as the dimensions, callouts, and thread notes.

In **Figure 5-29**, a front view was created first, and then two additional views were added, along with a

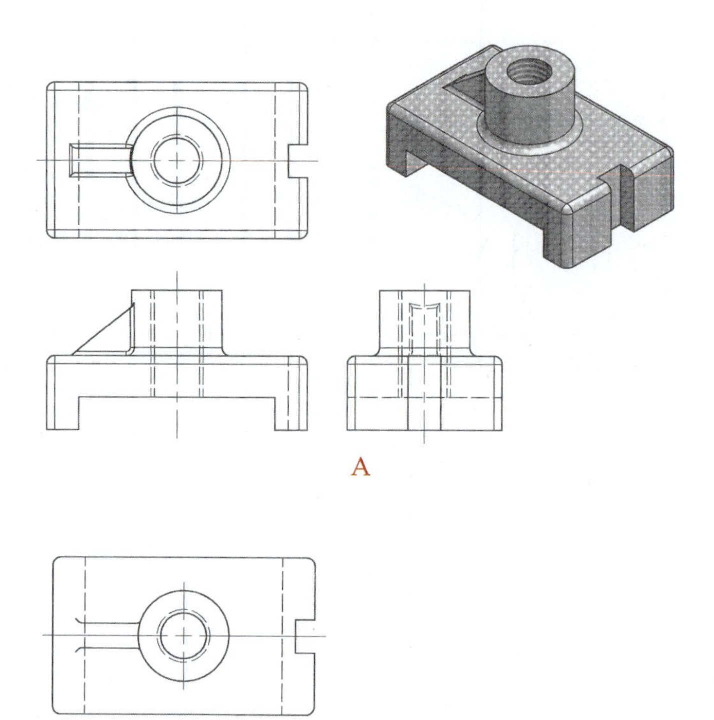

Figure 5-29. CAD software can easily create views of a 3D model. A—The computer-generated views may contain tangency elements for rounded edges. These three views have not been cleaned up and appear cluttered. B—These three computer-generated views have been cleaned up for the sake of clarity.

pictorial view to help the print reader visualize the object more easily. The drawing views are linked to the 3D model file. For example, if the size of the threaded hole or the depth of the slot is changed, all of the views update automatically.

While 3D CAD modeling programs can easily generate any of the principal views of an object directly from the model, there are some aspects of the final result that require a drafter's review, including the following:

- Individual line adjustment, such as changing the dash spacing of adjacent hidden lines, allowing for gaps between hidden lines and visible lines, or manipulating center lines to have short dashes or a plus at the center
- Adjusting the visibility of fillets and rounded edges with respect to tangency elements, which may allow an object to be more clearly described
- Properly showing conventional practices that break the rules of true projection

In **Figure 5-29A**, the views were intentionally left as the software first created them. The tangency elements between the flat surfaces and rounded edges are shown, even as hidden lines in some views. It is up to the drafter to use good judgment and turn off the visibility of any hidden lines or tangency lines that add clutter or create uncertainty. In **Figure 5-29B**, the same drawing is shown with most of the tangency lines turned off to clarify the views. In a few instances, the tangency lines were adjusted to be the same thickness as the visible lines.

In general, ASME standards for engineering graphics do not dictate how drawings are created, but rather specify their standard appearance. As CAD software evolves, drawing tools and other features that automate the generation of standard views will continue to improve, but the creation of drawings will still require the oversight of someone well versed in standard practices.

Solutions to spatial visualization tools:
Figure 5-2: B
Figure 5-3: A and C
Figure 5-4: B
Figure 5-5: D

Summary

- A multiview drawing is a systematic arrangement of more than one two-dimensional view of a three-dimensional object.
- Spatial visualization is a field of study in itself, and there are resources within this field that may also be of value to industrial print reading.
- Multiview drawings are based on the principles of orthographic projection, wherein parallel projectors are used to perpendicularly project the features of a three-dimensional object onto multiple projection planes.
- There are six principal ways to view an object, but three regular views (front, top, and right side) are common in print reading education.
- The three terms for an object's dimensions are height, width, and depth, and these dimensions each appear twice in the three regular views.
- The three planes of projection onto which the three regular views are projected are the frontal plane, the horizontal plane, and the profile plane.
- There are three principles associated with how surfaces appear in views, based on the way in which the surfaces are oriented to the planes of projection.
- The three types of flat, planar surfaces are normal, inclined, and oblique, and these are defined by their orientation to the principal planes of projection.
- Cylindrical surfaces and curved surfaces present additional visual challenges for the print reader that can be studied and identified.
- Rounded edges create the need for conventional practices for how to represent fillets, rounds, and runouts.
- The meanings of a multiview line can be categorized in three ways, providing yet another rule for visualizing an object from the multiple views.
- In the United States, the system of projection is known as third-angle projection, whereas other countries use a system known as first-angle projection.
- Sketching multiview drawings is a valuable tool for improving visualization skills.
- Computer-generated views may result in some minor differences in how the final views appear, depending on the computer settings and whether or not the print is reviewed and adjusted for standard practices.

Unit Review

Name ______________________ Date ____________ Class ____________

Answer the following questions using the information provided in this unit.

Know and Understand

____ 1. Multiview drawings are created using the principles of _____ projection.
A. orthopedic
B. orthodonic
C. orthographic
D. orthodox

____ 2. *True or False?* "Spatial visualization" is synonymous with "print reading."

____ 3. One way to think of the science of projecting views is to place an object inside a _____ box.
A. wooden
B. glass
C. steel
D. foam

____ 4. *True or False?* Although there are six principal ways to view an object, it is common to learn about multiview drawings using three of the six views: front, top, and right-side view.

____ 5. Which of the following dimension terms is *not* used in the unit as a standard term?
A. Length
B. Width
C. Height
D. Depth

____ 6. There are three terms for the projection planes onto which the three regular views are projected. Identify the fictitious or false name below.
A. Frontal
B. Vertical
C. Profile
D. Horizontal

____ 7. *True or False?* All surfaces of the object appear in every view of a multiview drawing, even if only as a line.

Match the following descriptions of how a surface is oriented to a projection plane to the way it appears in the view. Answers are used only once.

____ 8. Parallel
____ 9. Perpendicular
____ 10. Inclined

A. Foreshortened size and shape
B. Edge view (line)
C. True size and shape

Match the following types of surfaces to the descriptions. Answers are used only once.

____ 11. Normal
____ 12. Inclined
____ 13. Oblique

A. Perpendicular to one plane of projection, inclined to the other two
B. Inclined to all three planes of projection
C. Parallel to one plane of projection, perpendicular to the other two

____ 14. *True or False?* If a flat surface is tangent with a curved surface, a visible line is drawn at the element of tangency so the print reader knows the precise line of transition.

____ 15. *True or False?* Interior rounded edges are called rounds, and exterior rounded edges are called fillets.

____ 16. As discussed in the unit, which of the following is *not* one of the three meanings a line in a multiview drawing might have?
A. Edge view of a flat surface
B. Maximum contour element of a curved surface
C. Intersection edge where two surfaces meet
D. Tangency element between a curved surface and a flat surface

____ 17. *True or False?* The United States uses a projection system known as third-angle projection, while Europe uses first-angle projection.

____ 18. *True or False?* When creating a multiview sketch by hand, the front view should be completed first, then the side view, and finally the top view.

____ 19. Computer-generated views may need to be adjusted to conform to standard practices. Which of the following was *not* given as an example in this unit?

A. If a hole is changed in a model, the view will have to be manually changed.

B. Filleted or rounded transition edges may need to be adjusted, or turned off, for clarity.

C. Dash spacing may need to be adjusted on hidden lines that are very close to each other.

D. Center lines may have to be adjusted to create the right dash spacing or to create a plus in the center of the circular feature.

Critical Thinking

1. What are some things people can do to improve their spatial visualization ability?

____

____

____

____

____

2. What do you think are some advantages and disadvantages of creating computer-generated views automatically from a 3D model compared with creating the views of the object manually?

____

____

____

____

____

3. List three principles of orthographic projection that will help you proofread a multiview drawing for correctness.

____

____

____

____

____

Apply and Analyze

Name ______________________ Date ______________ Class ______________

Review Activity 5-1

Normal Surfaces

Study each pictorial (3D) drawing and the identification letters placed on, or pointing to, the normal surfaces. Match the ID letter to the corresponding number for each of the multiview (orthographic) callouts. Answers may be used more than once. Note: The letter I is not used.

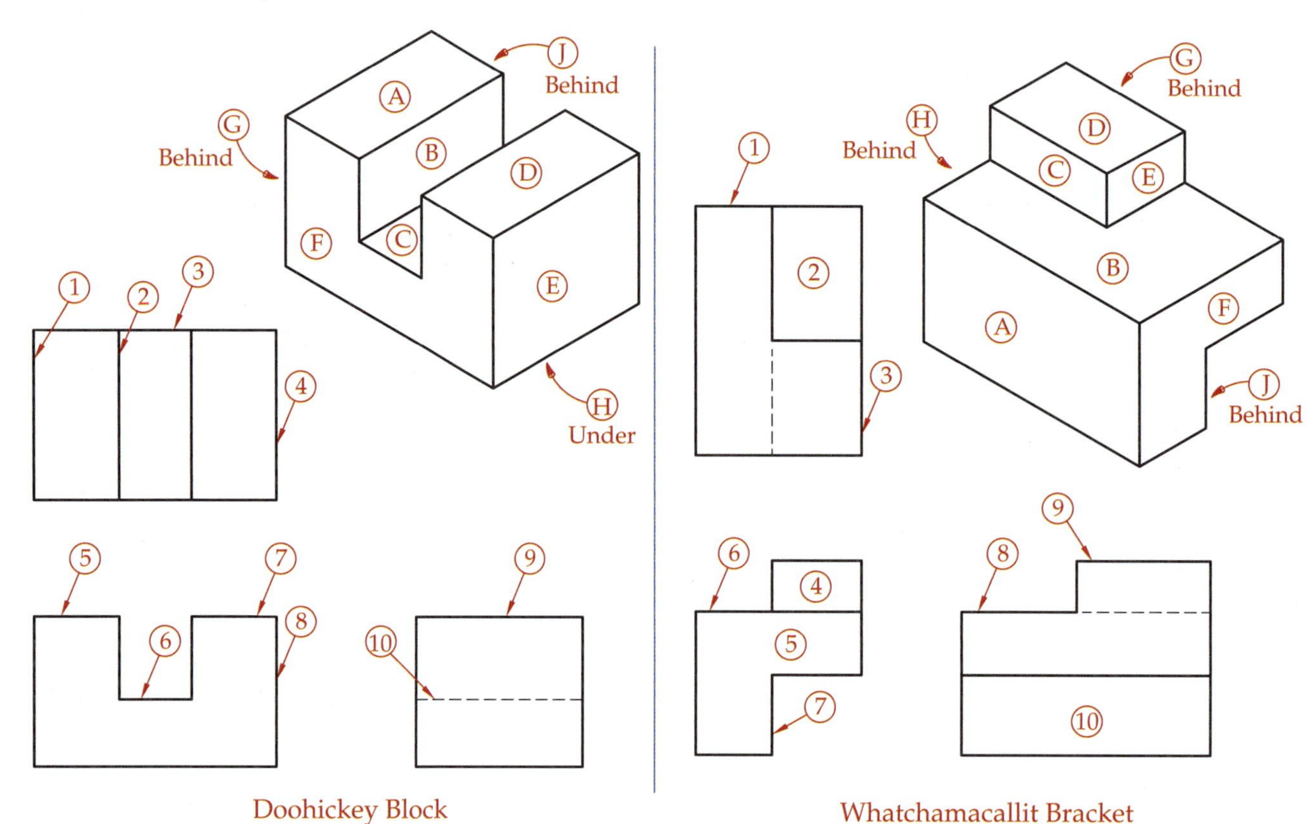

Note: In the pictorial view, arrows pointing directly to a line are referencing a surface that is not visible, but is around the corner of that edge from the viewer's point of view.

Doohickey Block

1. ______________
2. ______________
3. ______________
4. ______________
5. ______________
6. ______________
7. ______________
8. ______________
9. ______________
10. ______________

Whatchamacallit Bracket

1. ______________
2. ______________
3. ______________
4. ______________
5. ______________
6. ______________
7. ______________
8. ______________
9. ______________
10. ______________

Apply and Analyze

Name ______________________ Date ______________ Class ______________

Review Activity 5-2

Normal and Inclined Surfaces

Study each pictorial (3D) drawing and the identification letters placed on, or pointing to, the normal or inclined surfaces. Match the ID letter to the corresponding number for each of the multiview (orthographic) callouts. Answers may be used more than once. Note: The letter I is not used.

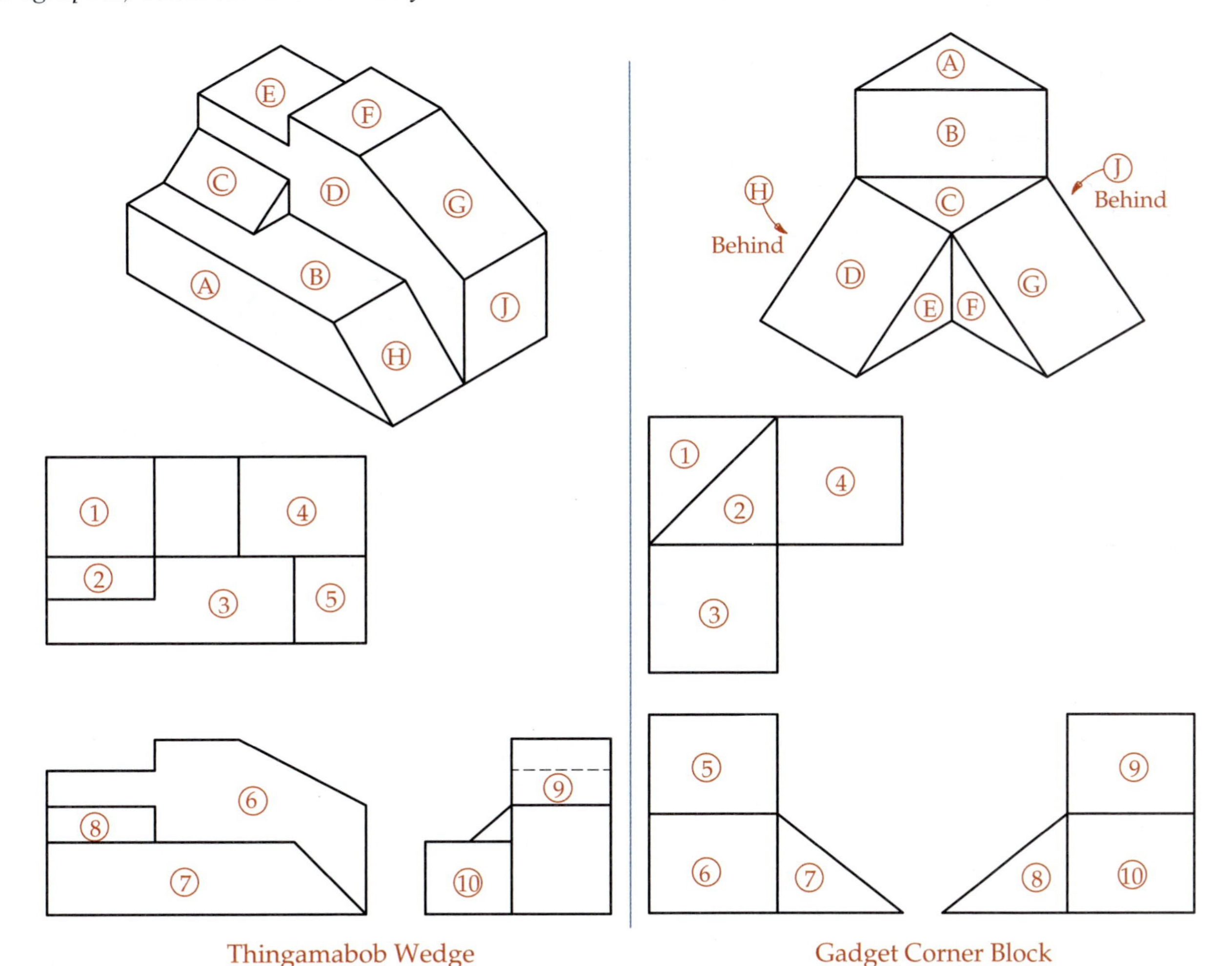

Thingamabob Wedge

Gadget Corner Block

Note: In the pictorial view, arrows pointing directly to a line are referencing a surface that is not visible, but is around the corner of that edge from the viewer's point of view.

Thingamabob Wedge

1. ______________ 6. ______________
2. ______________ 7. ______________
3. ______________ 8. ______________
4. ______________ 9. ______________
5. ______________ 10. ______________

Gadget Corner Block

1. ______________ 6. ______________
2. ______________ 7. ______________
3. ______________ 8. ______________
4. ______________ 9. ______________
5. ______________ 10. ______________

Apply and Analyze

Name ______________________ Date ______________ Class ______________

Review Activity 5-3

True Size and Shape Identification

Analyze each of the multiview drawings. Place a T for true size and shape or an F for foreshortened in the blanks below each multiview. Remember, normal surfaces appear true size and shape in only one view, and inclined surfaces appear foreshortened in shape and size in two views. The first problem is done for you as an example.

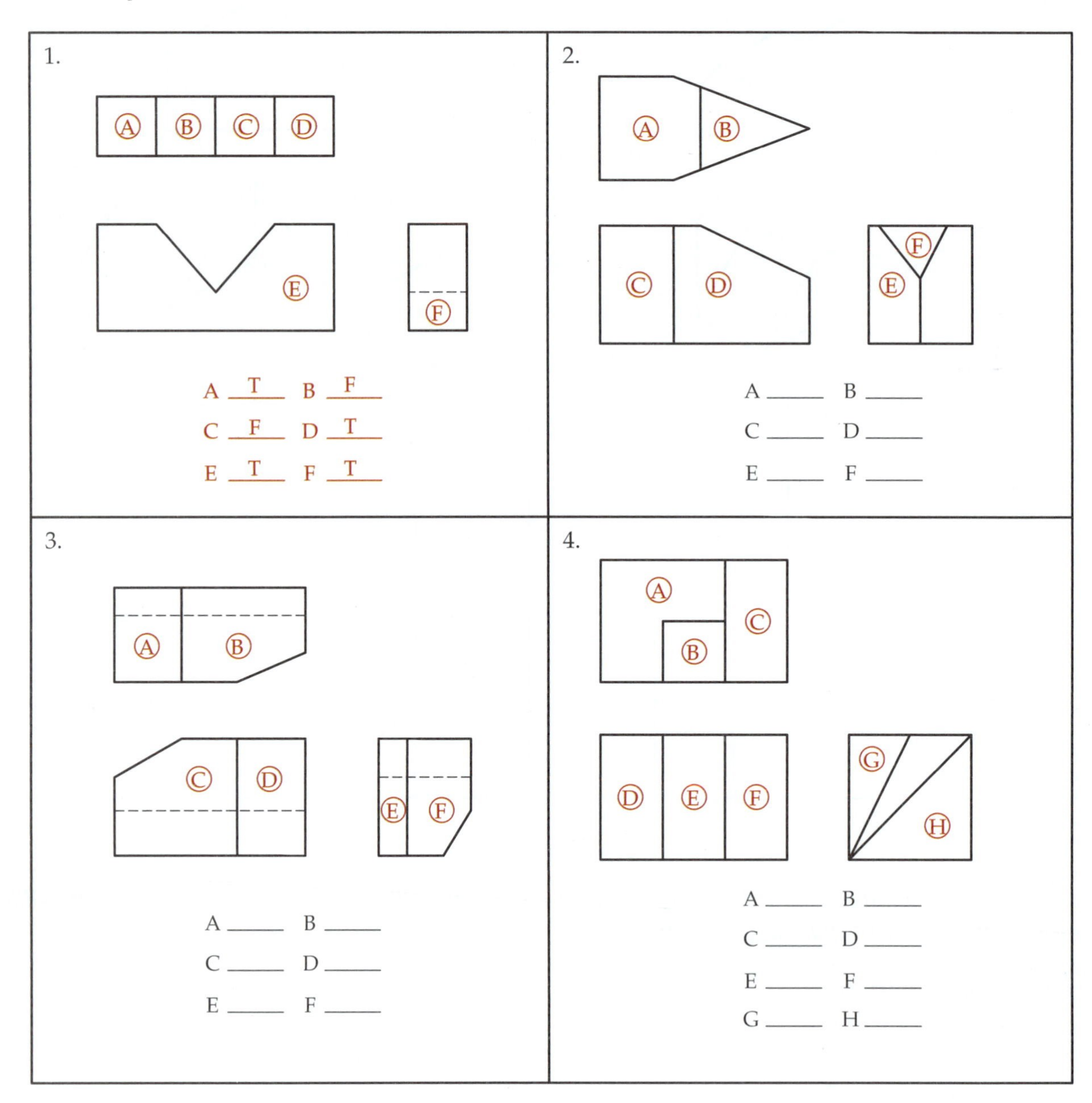

Apply and Analyze

Name ____________________ Date ____________ Class ______________

Review Activity 5-4

Sketching Missing Lines

Sketch the missing line(s) in the following multiview drawings. The final drawing should present a clear view of the object and all views must agree. Include all visible lines, hidden lines, and center lines.

1.

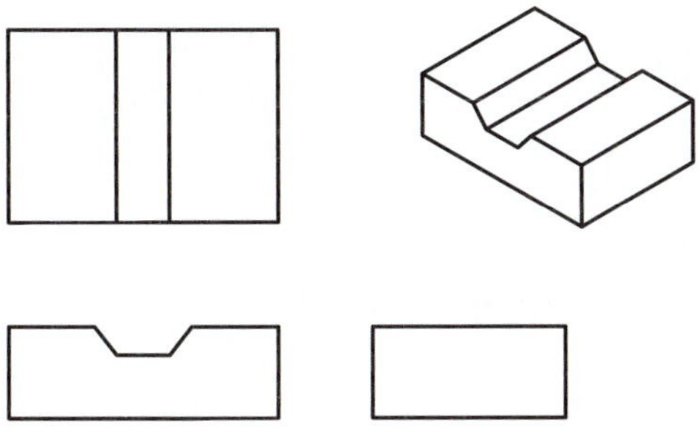

2.

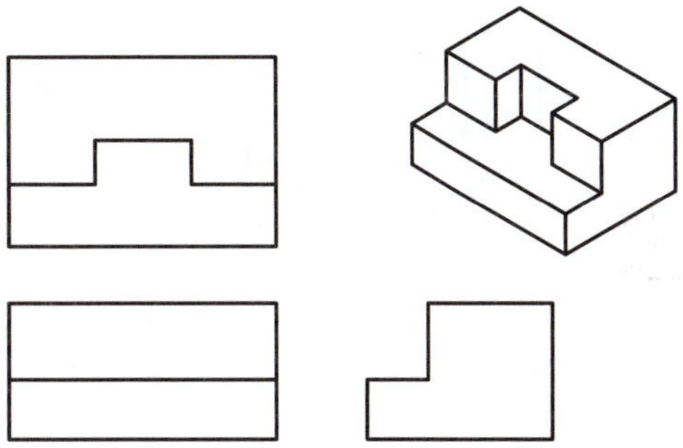

3.

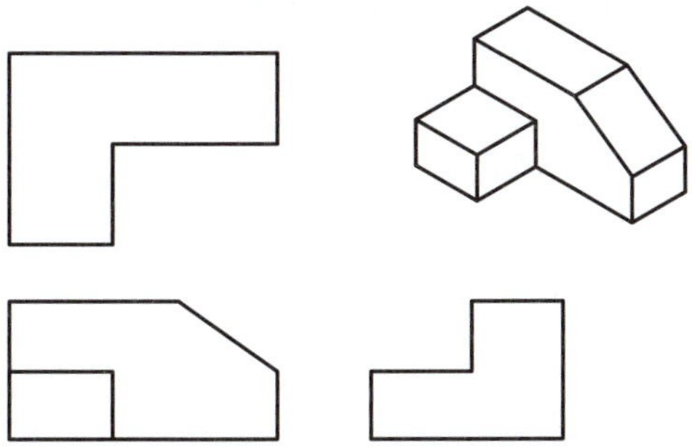

4.

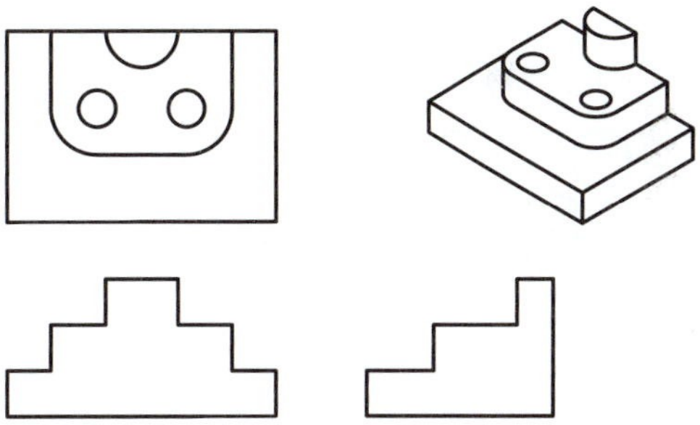

5.

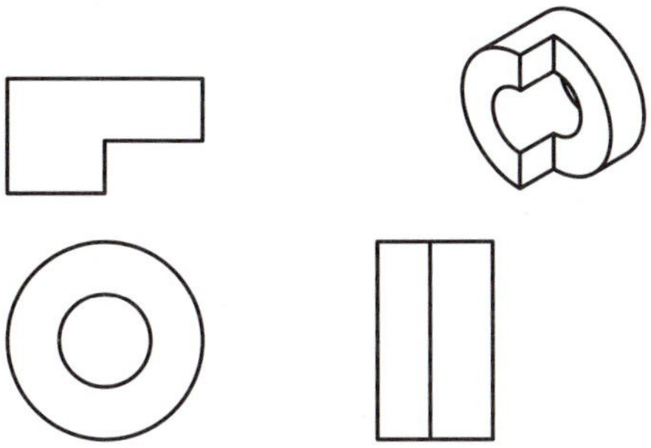

6.

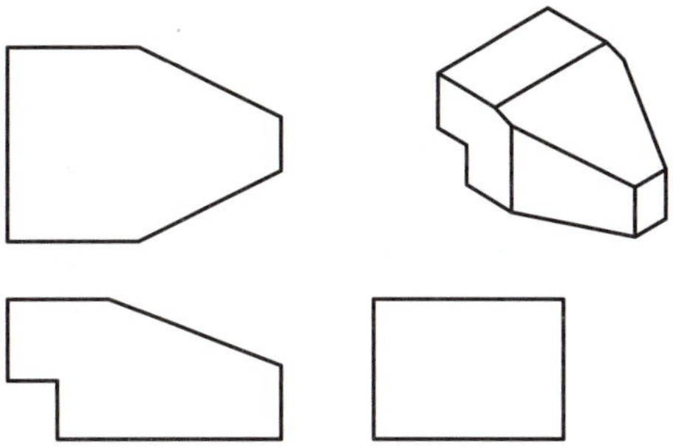

Apply and Analyze

Name ______________________ Date ______________ Class ______________

Review Activity 5-5

Sketching Multiview Drawings

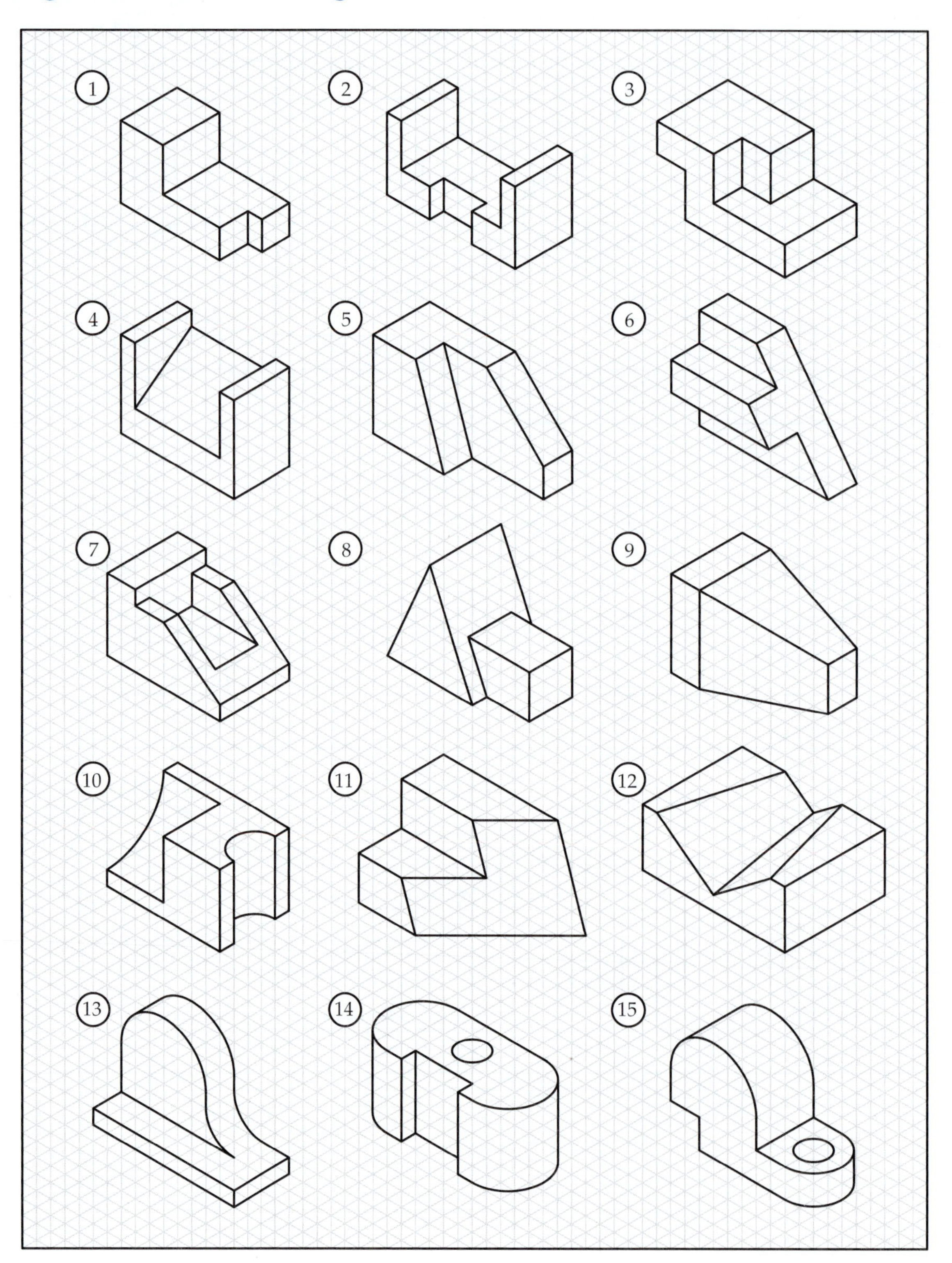

On the previous page, there are 15 pictorial views of various abstract blocks. As assigned, sketch one or more multiview drawings on this sheet and the sheets that follow. Use the steps identified in Unit 5, including a miter line. As an example, the first rectangular layout is provided for you, which establishes the location for each of the three views, as well as the location for the miter line. The initial layout of the views should be done with thin and lighter linework, but the final drawing should present a distinct multiview drawing of the object. Include all visible lines, hidden lines, and center lines as black linework at the appropriate line thickness and dash type. The miter line and other light linework can be left on the page to show proper procedure was used.

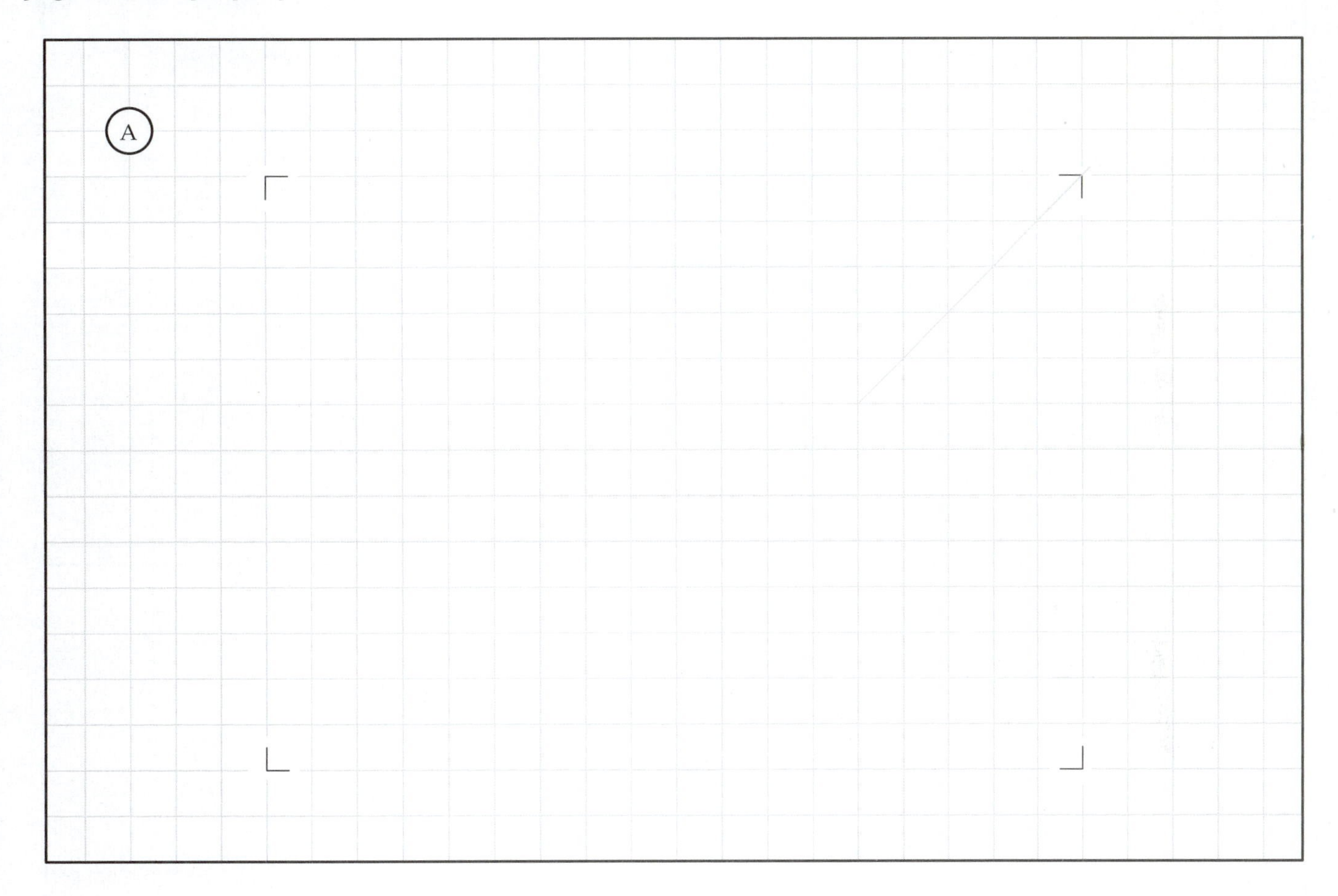

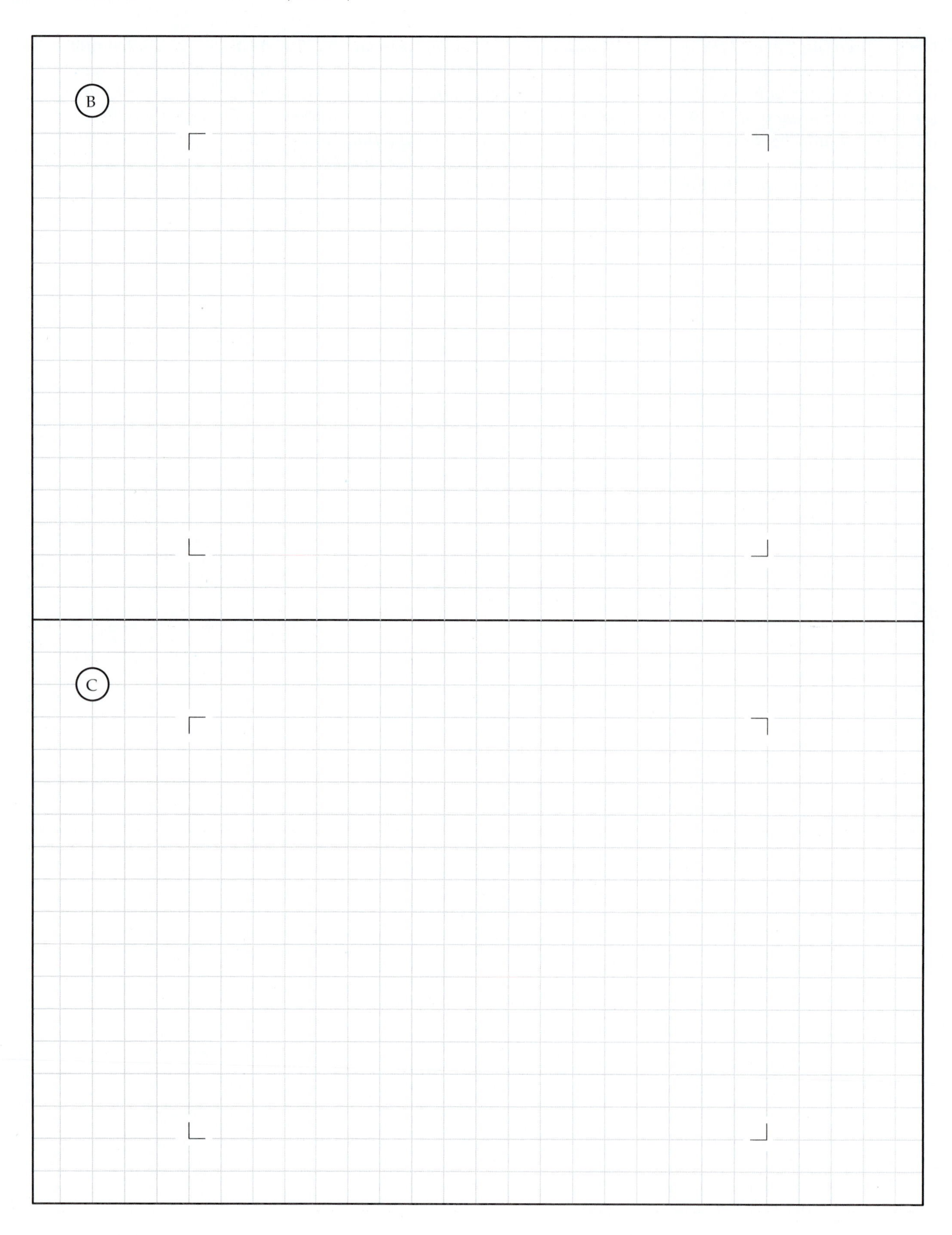
B
C

D

E

Apply and Analyze

Name ______________________ **Date** ______________ **Class** ______________

Review Activity 5-6

Projection Plane Orientation and Surface Type Identification

For each of the four pictorial drawings, imagine a frontal plane in front of the object, a horizontal plane above the object, and a profile plane to the right, as described in this unit. In the charts below, fill in the number of surfaces oriented to each of the three projection planes as indicated for each column (parallel, perpendicular, or inclined). Include all surfaces, even the bottom, back, and left-side surfaces. The first object has been done for you as an example. Also, enter the number of each type of surface in the second chart. Note: The total for each row of the first chart should be the same and should also match the total of the second chart, as shown in the first problem.

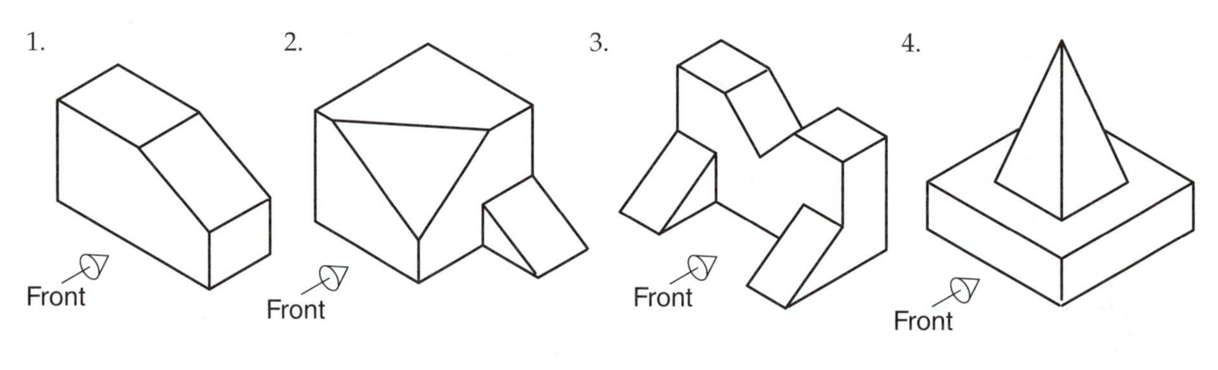

// = Parallel to ⊥ = Perpendicular to ∠ = Inclined to

1.

	//	⊥	∠	
Frontal	2	5	0	7
Horizontal	2	4	1	7
Profile	2	4	1	7
				Total =

Normal	6
Inclined	1
Oblique	0
Total	7

2.

	//	⊥	∠	
Frontal				
Horizontal				
Profile				
				Total =

Normal	
Inclined	
Oblique	
Total	

3.

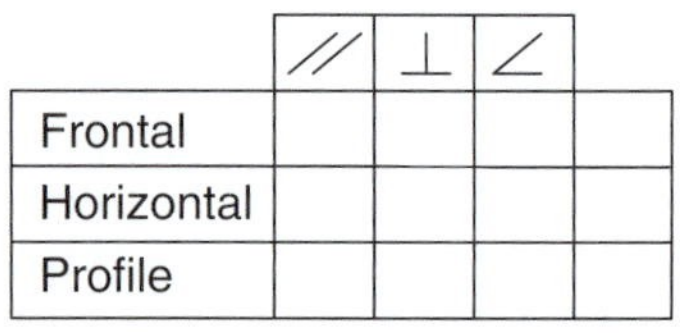

	//	⊥	∠	
Frontal				
Horizontal				
Profile				
				Total =

Normal	
Inclined	
Oblique	
Total	

4.

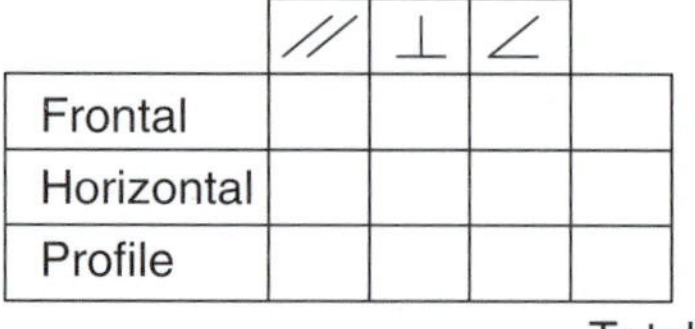

	//	⊥	∠	
Frontal				
Horizontal				
Profile				
				Total =

Normal	
Inclined	
Oblique	
Total	

Apply and Analyze

Name ______________________ Date ______________ Class ______________

Review Activity 5-7

Spatial Visualization and Projection Theory

For each of the questions below, a top view (horizontal plane) is missing. Select the letter of the correct choice for the top view (horizontal plane) and write it in the blank provided. Note: Center lines are also omitted.

1. 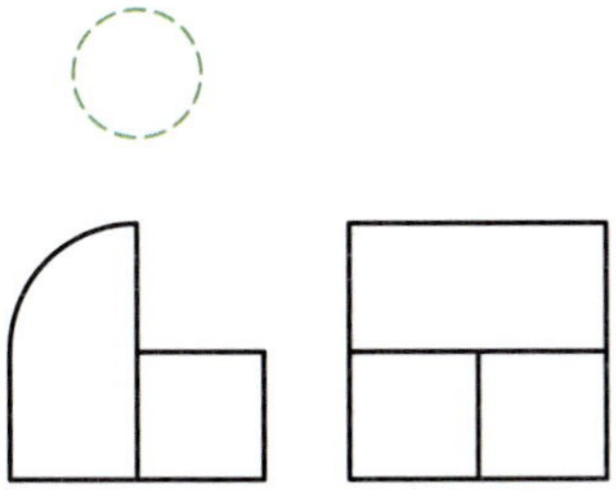 A. B. C. D.

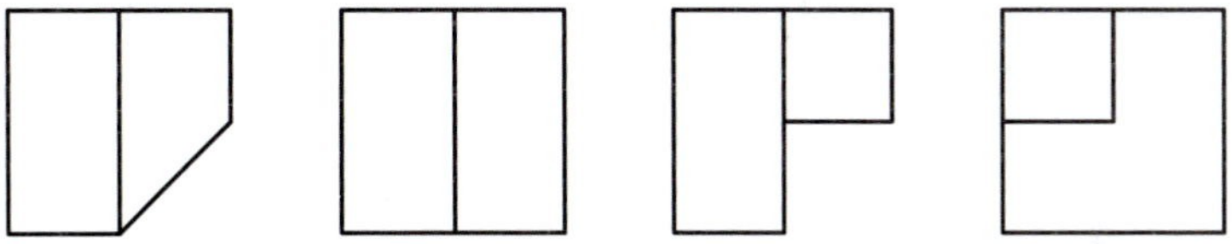

2. 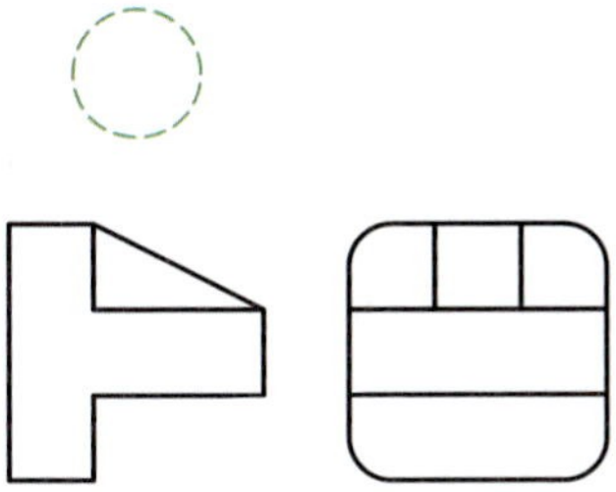 A. B. C. D.

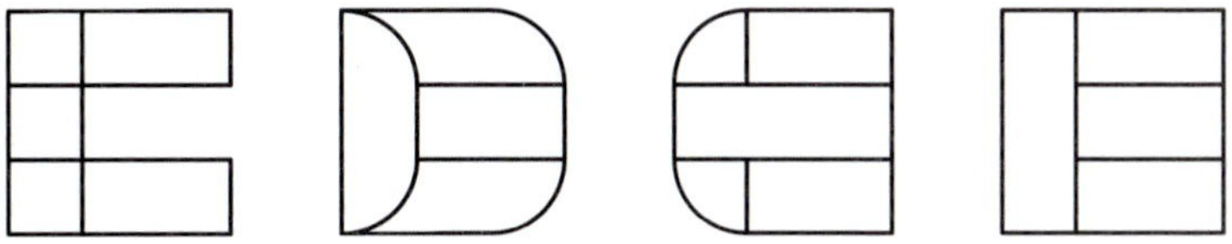

3. 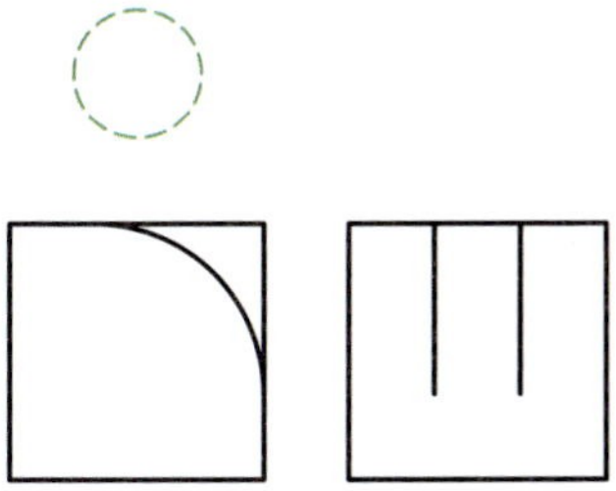 A. B. C. D.

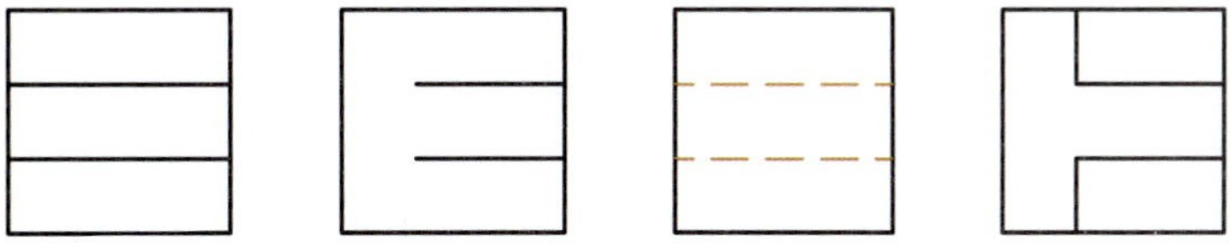

4. 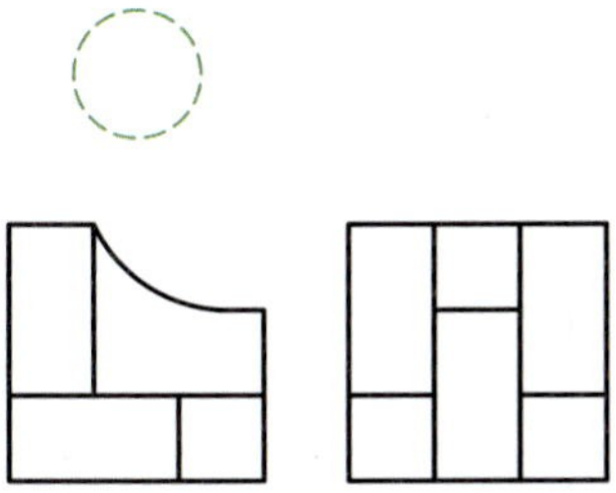 A. B. C. D.

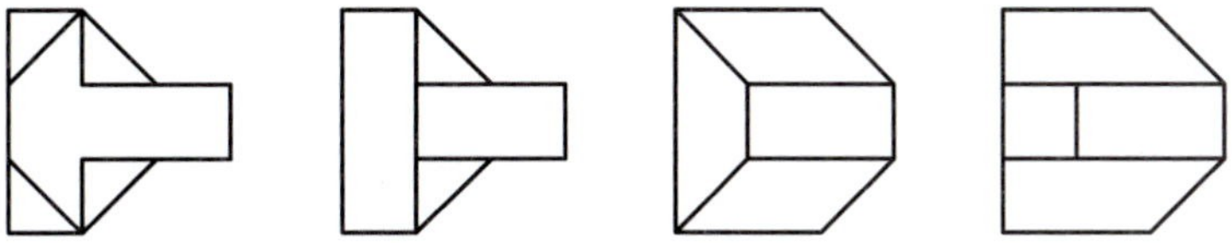

5. 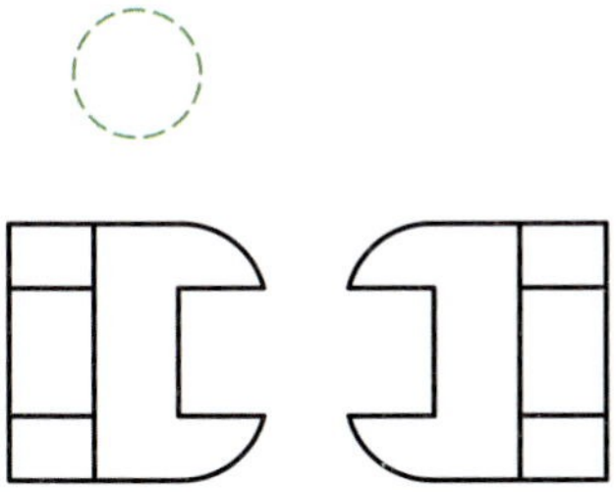 A. B. C. D.

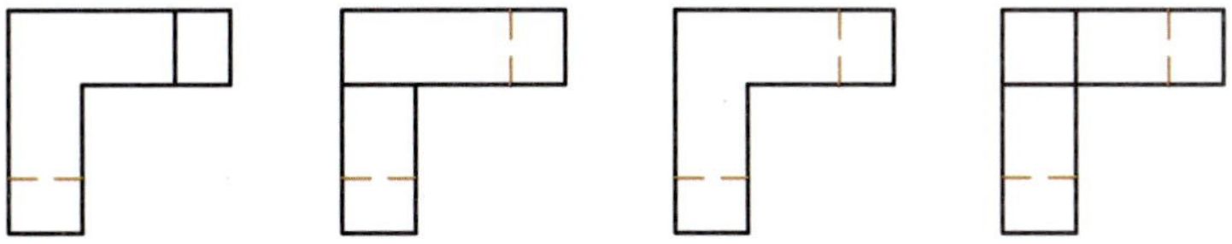

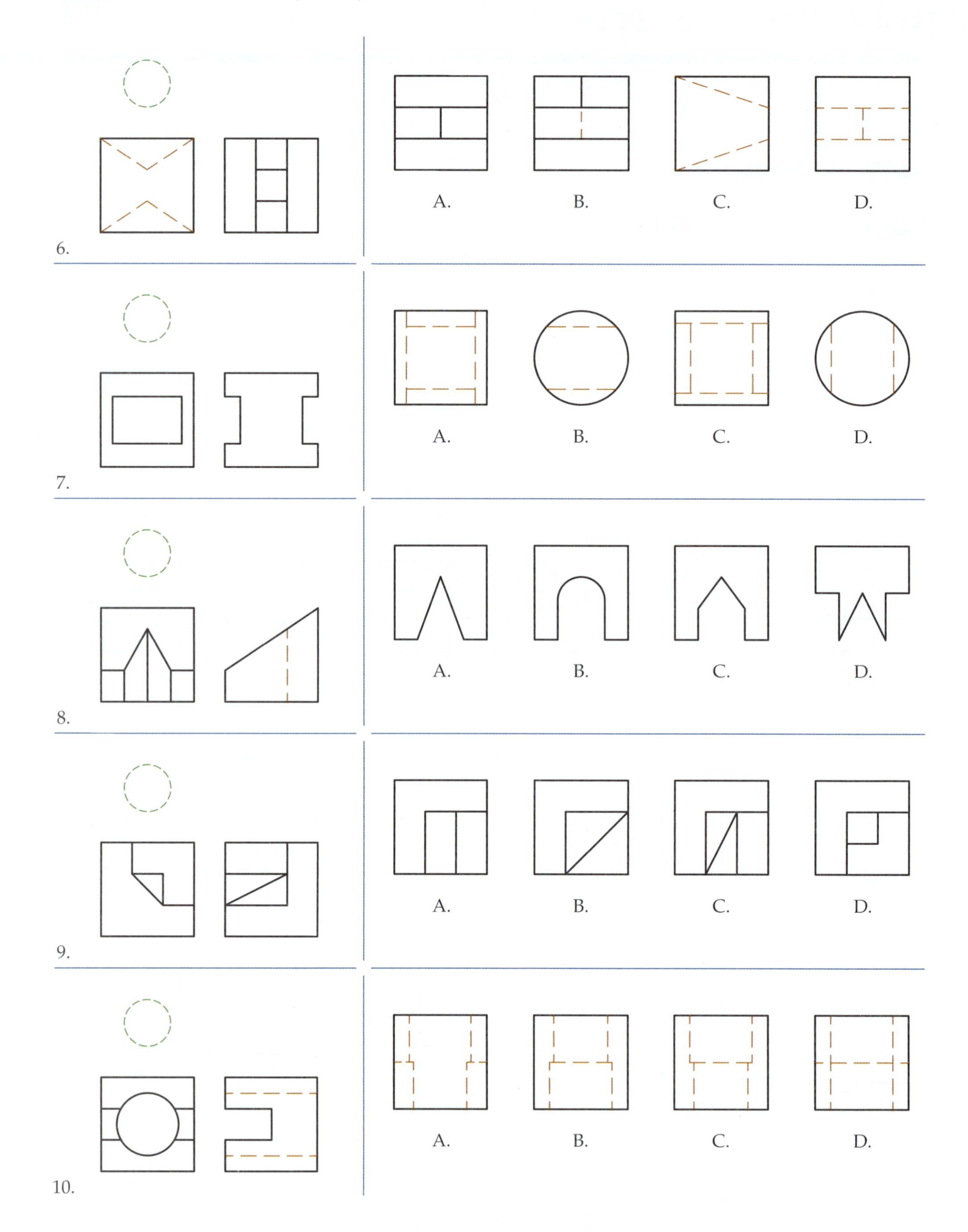
6.
A.
B.
C.
D.
7.
A.
B.
C.
D.
8.
A.
B.
C.
D.
9.
A.
B.
C.
D.
10.
A.
B.
C.
D.

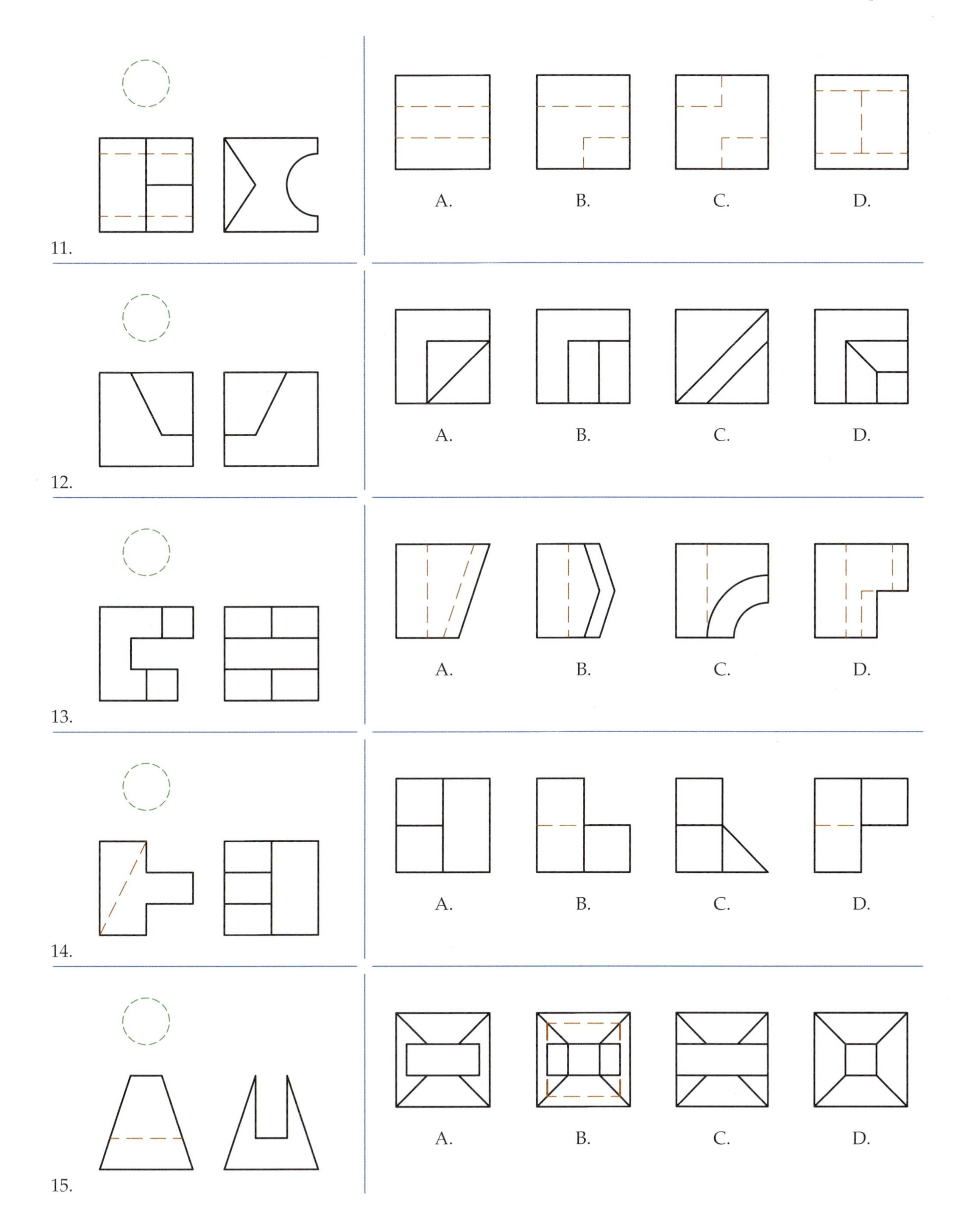
A.
B.
C.
D.
11.
A.
B.
C.
D.
12.
A.
B.
C.
D.
13.
A.
B.
C.
D.
14.
A.
B.
C.
D.
15.

Apply and Analyze

Name ______________________ **Date** ______________ **Class** ______________

Review Activity 5-8

Piston

Study the print and answer the questions.

1. Which regular view name can be applied to each of the views identified as A, B, and C?

   A. ______________________

   B. ______________________

   C. ______________________

2. What are the three overall dimensions of the object?

   Height ______________________

   Width ______________________

   Depth ______________________

3. How many normal surfaces does this object have? ______________________

4. How many inclined surfaces does this object have? ______________________

5. There are no measurements given for the depth (front-to-back) dimension of surface X. In the right-side view, that depth can be determined by analyzing the hidden line of that surface. What is one method of transferring that depth to the top view?

   ____

   ____

6. How wide is surface X? ______________________

7. How wide is surface Z? ______________________

8. Given three choices—edge view, intersection, or maximum contour—answer the following:

   A. What does line 10A represent? ______________________

   B. What does line 10B represent? ______________________

9. The slot (that has surface X as the bottom surface) is not perfectly centered on the part. If the design needed to be adjusted to center that slot, precisely how far would the slot need to be shifted?

   ____

10. With the chamfer on each end at 40°, is the diameter of the end surface greater or less than 1.22”?

    ____

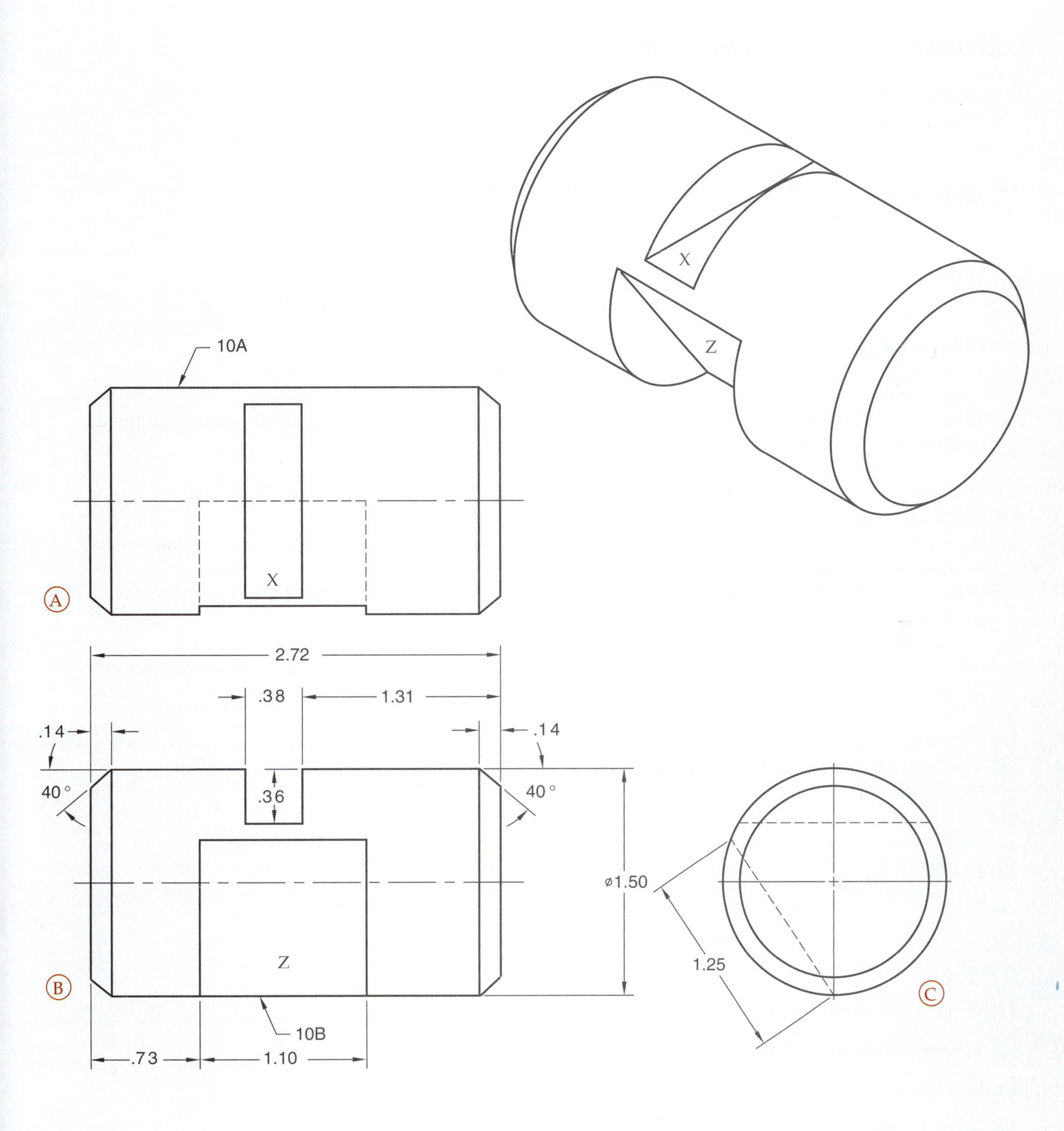
X
Z
10A
X
A
2.72
.38
1.31
.14
.14
40 °
40 °
.36
ø1.50
1.25
Z
B
C
10B
.73
1.10

Apply and Analyze

Name ______________________ Date ______________ Class ______________

Industry Print Exercise 5-1

Refer to the print PR 5-1 and answer the questions below.

1. This drawing has the three principal views. What are they?

____

2. What dimension term is applied to the value of 2.125″?

____

3. Is the circular feature in the right-side view hidden or visible in the front view?

____

4. What is the height of this part?

____

5. What type of surface is the surface that encompasses four holes in the top view, normal, inclined, or oblique?

____

6. What is the total width of this part?

____

7. Which letter of the alphabet, M, H, or L, best describes the general shape of the normal surface facing the frontal plane?

____

8. In the top view, how many lines represent the maximum contour of a curved surface?

____

9. The bottom, flat normal surface of the object appears in the top view as true size and shape. How wide is it?

____

10. In the top view, the right-most vertical line has what meaning: edge view of a surface, intersection of two surfaces only, or maximum contour of a curved surface?

____

Review questions based on previous units:

11. What is the name of this part? ______________________
12. What is the drawing number or part number? ______________________
13. In what state is this company located? ______________________
14. What is the five-digit material code specified for this part? ______________________
15. What is the last name of the person who drew this drawing?

____

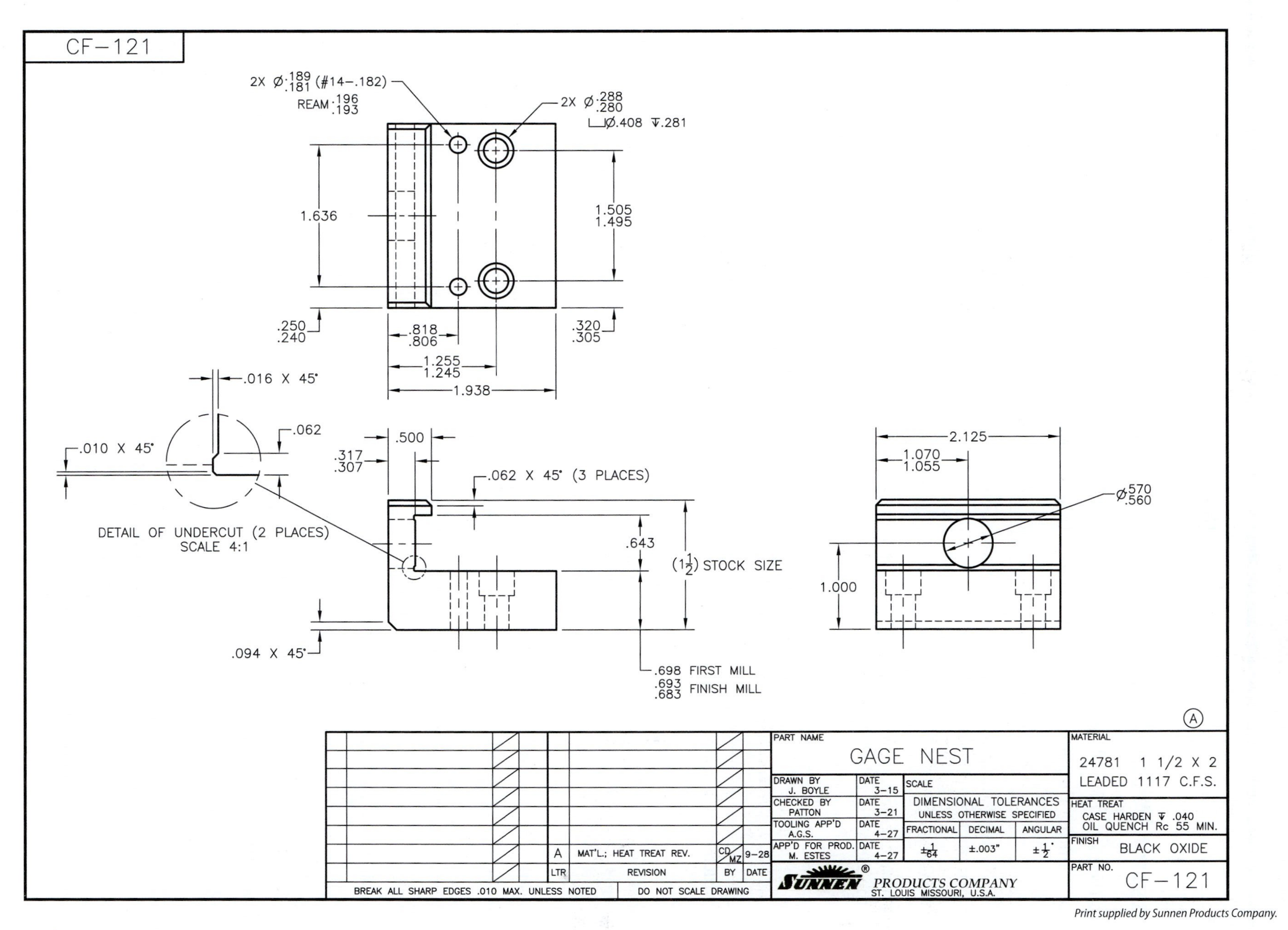

Print supplied by Sunnen Products Company.

PR 5-1. Gage Nest

Apply and Analyze

Name ______________________ Date ______________ Class ______________

Industry Print Exercise 5-2

Refer to the print PR 5-2 and answer the questions below.

1. Not counting the removed view and pictorial views, how many principal views does this object have?

 ____

2. How many of the principal views feature the dimension referred to as height (even if the actual height measurement is not shown)?

 ____

3. For this exercise, if the view that is closest to the center of the sheet is the front view, what is the name of the view directly below it?

 ____

4. The flat surface that shows true size and shape in the top view is parallel to what plane?

 ____

5. In which view is the viewing-plane line shown for the removed view?

 ____

6. What two dimensions, by name, are featured in the left-side view (even though no measurements are given)?

 ____

7. Does this object have any inclined surfaces? ____
8. For this object, which dimension is greater, width or depth?

 ____

9. What is the name of the removed view? ____
10. Are hidden lines shown in the top view? ____

Review questions based on previous units:

11. Which view features cutting-plane lines? ____
12. What is the drawing number or part number? ____
13. How many views feature section lines?

 ____

14. For what size paper was this drawing created (give answer in inches for the engineering system)?

 ____

15. The annotation on the left-side view features a couple of what type of lines?

 ____

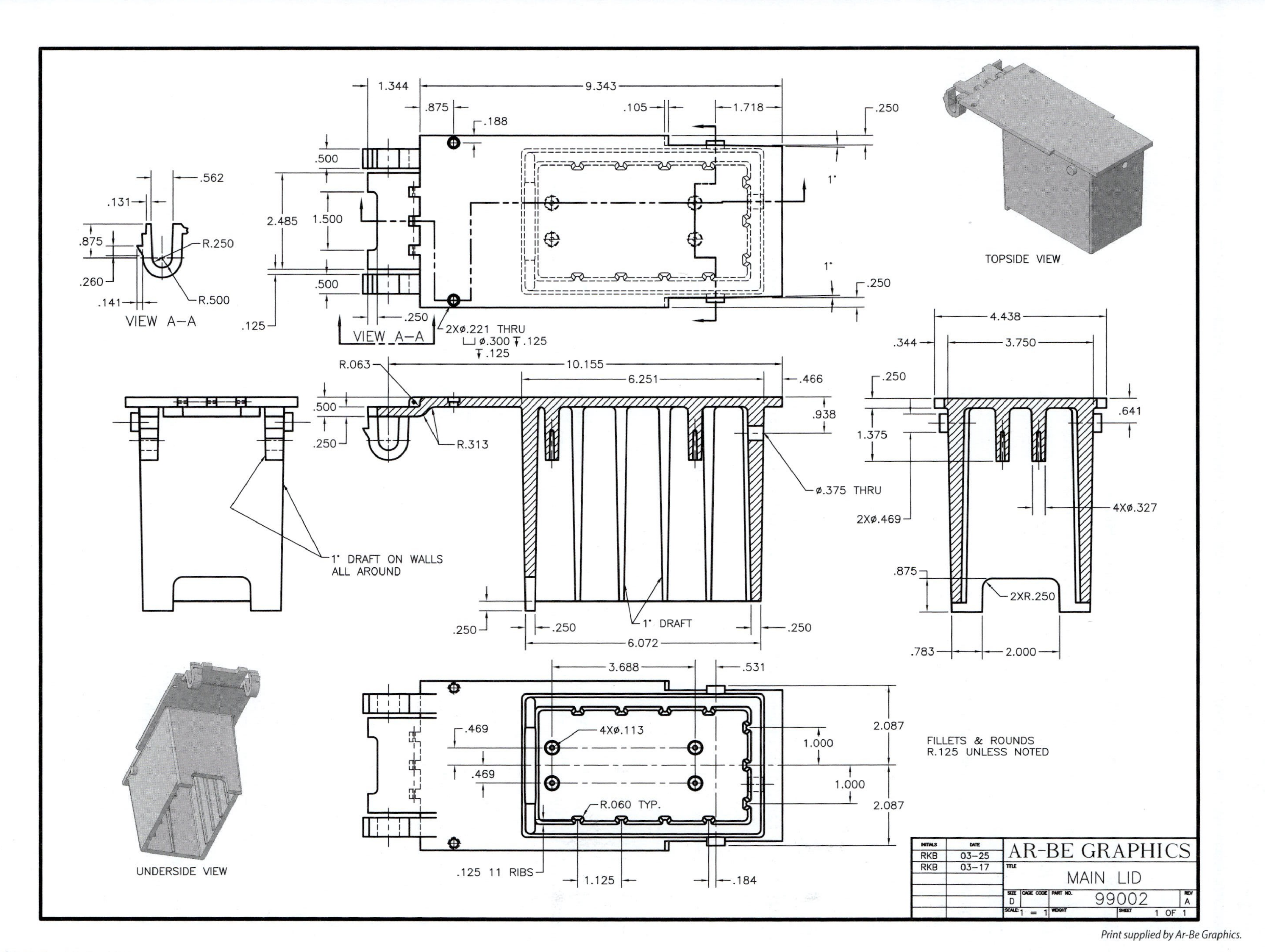

Print supplied by Ar-Be Graphics.

PR 5-2. Main Lid

Apply and Analyze

Name ______________________ Date ______________ Class ______________

Bonus Print Reading Exercises

The following questions are based on the various bonus prints located in the folder that accompanies this textbook. Refer to the print indicated, evaluate the print, and answer the question.

Print AP-001

1. On the left side view of this print, there is a horizontal, straight-line segment that does not connect to other lines. In basic terms, explain why.

____

____

Print AP-004

2. If the view labeled Section B-B is a right-side view, what name would be applied to the other four main views?

____

____

Print AP-005

3. For this print, if the view closest to the pictorial drawings is considered a right-side view, what names would be given to the other three views (list in order from highest to lowest on the page)?

____

____

4. With respect to planar surfaces, versus cylindrical surfaces, does this object have any surfaces that would be considered inclined surfaces?

____

Print AP-006

5. What angle of projection is specified on the print?

____

Print AP-007

6. Carefully examine the drawing. With respect to the enlarged Detail C, how many times larger is it than the other orthographic views?

____

Print AP-009

7. With respect to the enlarged Detail B, how many times larger is it than the other orthographic views?

____

8. This drawing shows a bolt sensor switch for reference only, and it has three labels (NC, NO, and C) identified on a surface. Is that surface oriented as normal, inclined, or oblique?

____

Print AP-012

9. Is the left side view on this print a complete view or a partial view?

____

Print AP-013

10. Are the views of this assembly drawing arranged orthographically in proper third-angle projection?

____

Print AP-018

11. How many of the six principal viewing directions are represented by this print?

____

Print AP-019

12. One view features the depth measurements .300, .250, and .040. How many other main views (not including detail views) on this print feature the depth direction?

____

Print AP-020

13. For the nonpictorial views on this print, how many of the six principal viewing directions are represented?

____

14. In the pictorial views, what do the phantom lines represent?

____

Print AP-021

15. *True or False?* If the lower view is considered a front view featuring width and height, the upper-right view features depth and height.

____

Print AP-022

16. Is the view labeled FORMED VIEW a normal view defined by multiview practice?

____

Notes

UNIT **6**

Section Views

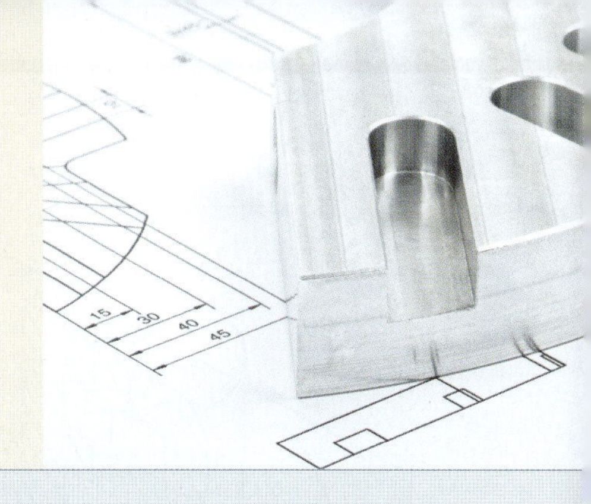

LEARNING OBJECTIVES

After completing this unit, you will be able to:

- Describe the general nature of a section view and explain its purpose.
- Identify the types of lines used in section views.
- Identify basic section line patterns that indicate various materials.
- Explain the characteristics of a drawing that features a full-section view.
- Describe the characteristics of a half-section view.
- Describe the characteristics of an offset-section view.
- Identify the characteristics of a sectional view that includes aligned features.
- Identify the characteristics of a view that features a broken-out section.
- Compare revolved sections and removed sections and identify the characteristics of these sections.
- Explain conventional practices applied to section views.
- Identify the proper representation of partial sections and outline sections.
- Identify the proper representation of separate parts in an assembly section view.
- Discuss computer-generated section views.

TECHNICAL TERMS

aligned section
broken-out section
conventional practice
cutting-plane line
full section
half section
offset section
outline section
partial section
removed section
revolved section
section line
section view

Sometimes the regular views of a drawing cannot clearly show the interior features of an object. A type of view known as a section view is used to help show interior details. A ***section view*** is created by passing an imaginary cutting plane through the object and removing the part nearest to the observer. This allows a direct view of the interior details. Section views are also called sectional views, cross sections, or simply sections. See **Figure 6-1**. The standard practices for sectional views are covered in ASME Y14.3, formerly titled *Multiview and Sectional View Drawings*, but now titled *Orthographic and Pictorial Views*.

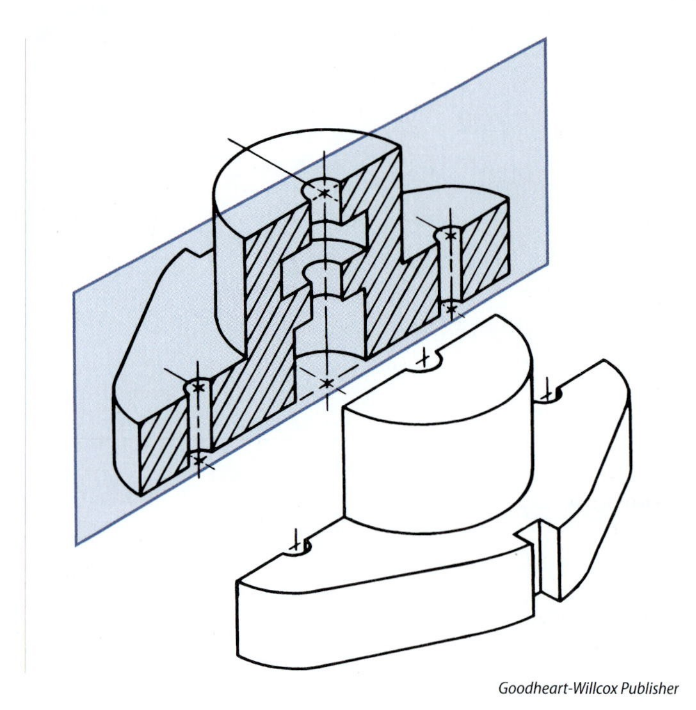

Goodheart-Willcox Publisher

Figure 6-1. A section view allows a direct view of interior detail.

Section View Lines and Principles

The alphabet of lines contains two types of lines used for section views. These are section lines and cutting-plane lines. ***Section lines*** shade the area of material cut by the imaginary cutting plane. The ***cutting-plane line*** represents the edge view of the cutting plane. It is placed in the adjacent view to indicate the location of the cut.

Section lines, also called *crosshatch* or *hatching*, can be created in a variety of patterns. Different patterns can be used to indicate the type of material used for the part, **Figure 6-2**. In practice, however, section lines are often drawn with continuous lines about 1/8″ apart at a 45° angle. These 45° lines, like those shown in **Figure 6-3**, indicate cast iron or malleable iron, but they can be used for general-purpose section lines. The section lines in a section view are not to be the only method designating the material for the object. The material specification should also be listed in the title block, in the materials block, or in a note on the drawing.

While section lines are usually drawn at 45°, they can be drawn at any angle, but preferably not parallel or perpendicular to one of the major visible lines. The angle should be the same throughout the section view of a single part. A section view may serve as one of the regular views on the drawing, or it may appear as an additional view.

As presented in Unit 2, cutting-plane lines are thick and may be drawn as a series of medium dashes or as a series of two short dashes and one long dash. An additional option is to simply show the cutting-plane line with thick "elbows." Arrowheads are placed on each end of the cutting-plane line to indicate the viewing direction for the section view. Most often, capital letters are also placed near each arrowhead (A, B, C, etc.) to help identify the section views on the page with the corresponding cutting plane. The cutting-plane line can be omitted on a simple symmetrical object, but the standards recommend the cutting-plane line be drawn if the cutting plane is bent or offset, or if the section view that results from the cut is not symmetrical.

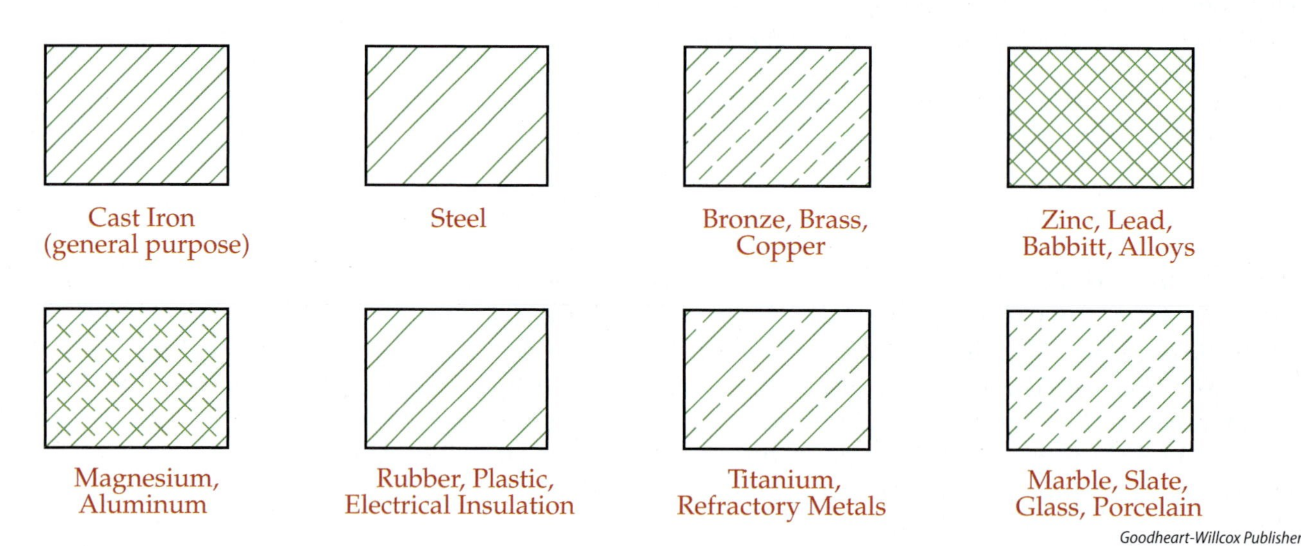

Goodheart-Willcox Publisher

Figure 6-2. Section line patterns are standardized for different types of materials. Most of these are available for industry CAD systems, but general purpose section lining is still common.

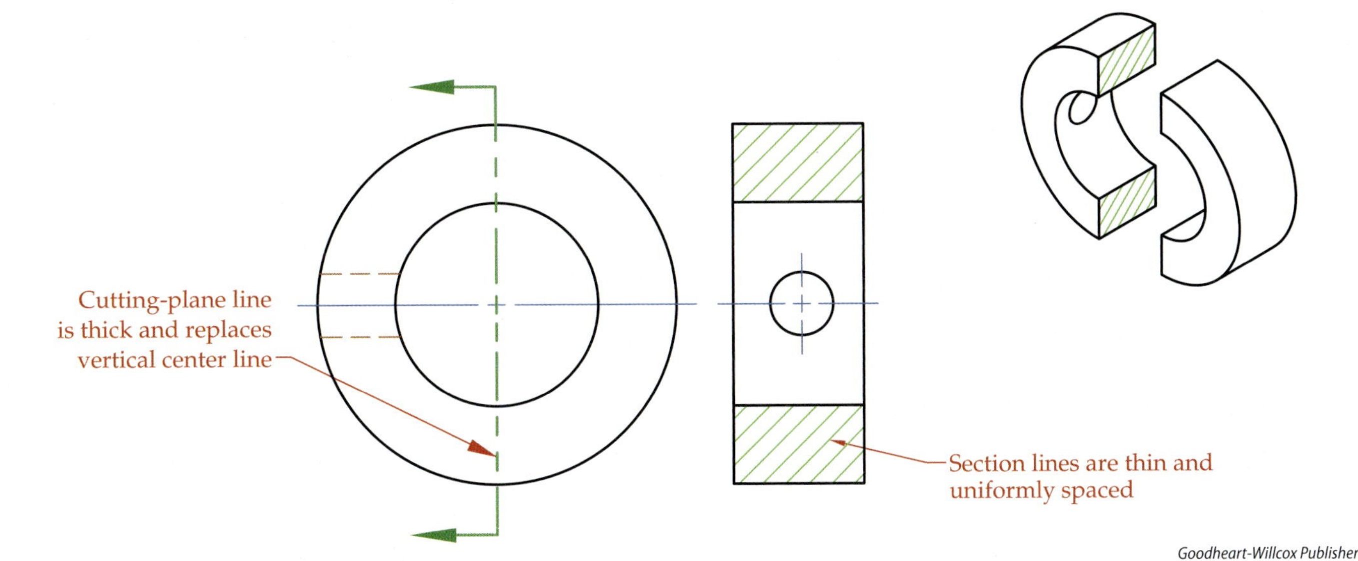

Goodheart-Willcox Publisher

Figure 6-3. Cutting-plane lines and section lines are used in section view drawings.

When possible, the section view should be arranged in orthographic projection with the other views. Scale and sheet size are two factors that are considered when deciding where to place the section view. For multisheet drawings, it is highly desirable to keep the section view on the same sheet as the view that shows the cutting plane, although section names and zoning can assist in cross-referencing sections to other sheets if required.

Section View Types

There are several types of sections. Each is designed for a specific purpose. A ***full section*** is created when the cutting plane passes entirely through the object. For all practical purposes, the object is "cut in half." See **Figure 6-4**. Notice how much more clearly the internal details appear in the section view than in a nonsectioned right-side view, where the internal details would be shown as hidden lines. Lines visible in the section view are shown, but hidden lines usually are omitted. Even if there are hidden features behind the sectioned area, showing hidden lines reduces the clarity of the section view. For complex parts, showing a few hidden lines on a section view is permitted if it eliminates the need to draw another view.

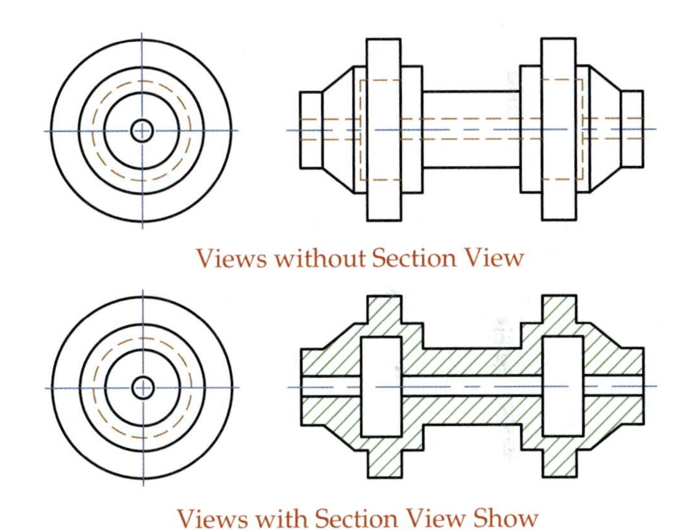

Goodheart-Willcox Publisher

Figure 6-4. A full section is created when the cutting plane passes entirely through the object. The cutting-plane line can be omitted on basic, symmetrical full sections such as this one. A comparison of the views shows the advantage of a section view when internal details need to be clearly visible.

A symmetrical object is the same on both sides of the center line. The ***half section*** is created when a symmetrical object is drawn with one-half of the view as a section and the other half as a regular view. See **Figure 6-5**. In essence, the half section cuts halfway through the object. In your mind, think of this as if one-quarter of the object is removed. The cutting-plane line is shown on the adjacent view, indicating where the part has been cut and the direction from which the section is viewed. Notice there is only one arrowhead on the cutting-plane line for the half-section view. Also note the preferred method of showing the section view, with a center line dividing the sectioned half from the regular half, *not* a visible line. After all, the object is not actually cut. The resulting section is really a "double exposure," with two half views as one.

An ***offset section*** is drawn when the essential internal details do not appear on one flat plane

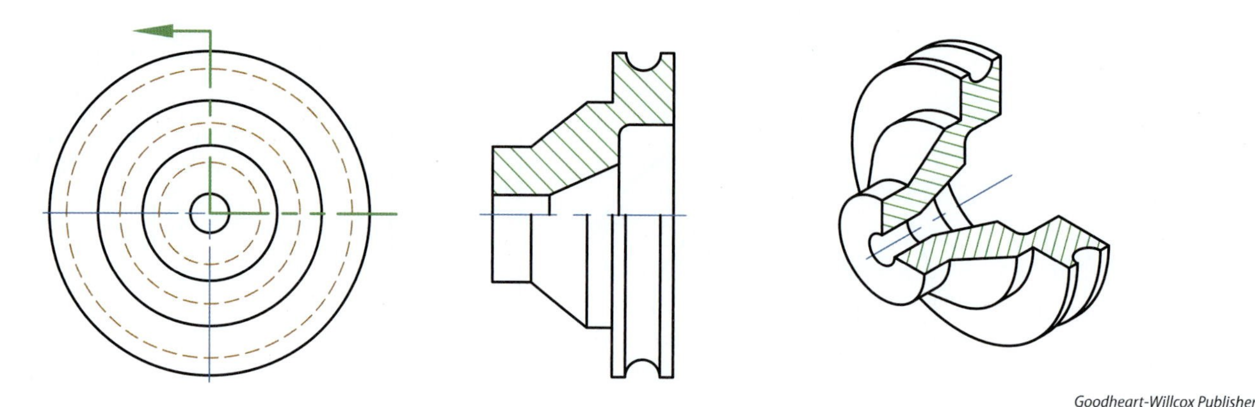

Goodheart-Willcox Publisher

Figure 6-5. The half section is created when a symmetrical object is drawn as a blended view with one-half as a section view and the other half as a regular view.

through the object. The cutting-plane line is drawn offset through the object to include the desired features. Those features are then shown on one plane in the section view. See **Figure 6-6**. The section view itself does *not* show the bends in the cutting plane and appears as if it was cut by a flat plane.

An ***aligned section*** is usually drawn for a cylindrical object with an odd number of features. The cutting-plane line is offset through the features such that they can be "rotated" to a normal vertical or horizontal plane and projected to the section view. The result is equivalent to the feature being "aligned" with a normal full-section cutting plane. See **Figure 6-7**. Notice the upper arrow in this figure is perpendicular to the cutting-plane line, since the cut profile will be rotated into a vertical position and not foreshortened.

A ***broken-out section*** is created when a small portion of a part is exposed to show the interior construction. This is like starting to cut the object with a plane, but then breaking off a piece of the object, leaving the rest of the object shown in a regular way. A cutting-plane line in an adjacent view is unnecessary. The sectioned portion of the view is separated by a short break line, as presented in Unit 2. See **Figure 6-8**. This type of section view can also be used to show exterior and interior details on the same view, similar in fashion to a half section. As in other cases, the hidden lines in the nonsectioned portion of the view may be drawn if needed to explain other interior details, or they may be omitted for clarity.

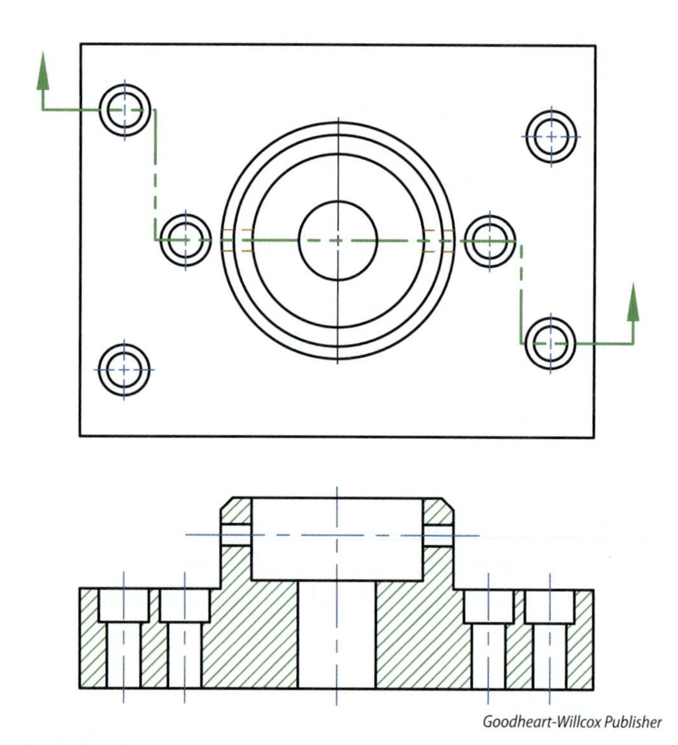

Goodheart-Willcox Publisher

Figure 6-6. An offset section is drawn with a cutting-plane line offset through the object to include the desired features.

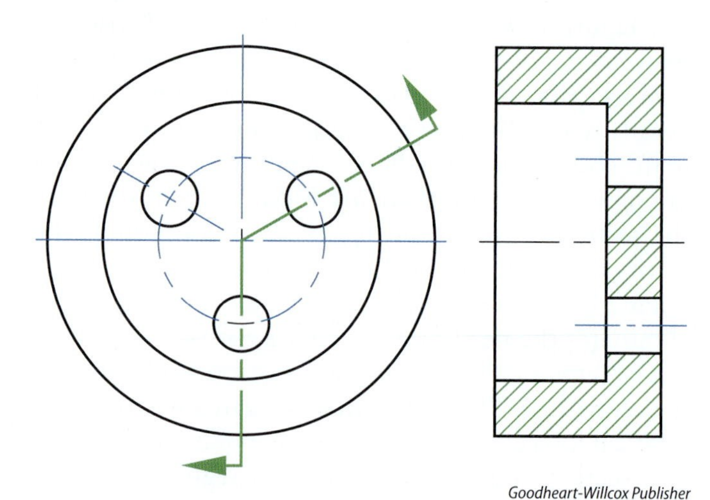

Goodheart-Willcox Publisher

Figure 6-7. The cutting-plane line of an aligned section is offset through the features so they can be "rotated" to a normal vertical or horizontal plane and projected to the section view.

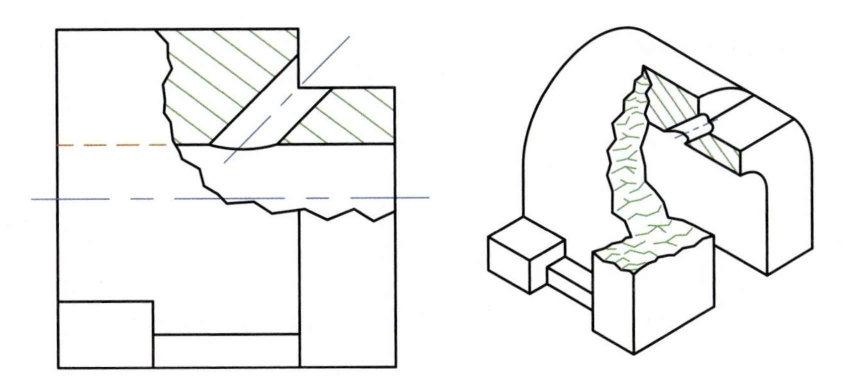

Goodheart-Willcox Publisher

Figure 6-8. A broken-out section is created when a small portion of a part is exposed to show the interior construction.

A ***revolved section*** is used for a feature such as a wheel spoke or long steel bar. The cutting plane slices through the feature parallel to the line of sight, but the "cut" shape is revolved 90° directly on the regular view as if it were an overlay, as shown in **Figure 6-9A**. To further clarify the view, the part may be broken on each side of the section, as shown in **Figure 6-9B**.

A ***removed section*** is similar to a revolved section, but the section view is shown in another place on the drawing. In addition, the removed section should be placed "out of projection" with the other views. See **Figure 6-10**. Removed sections are frequently used as detail sections. Each removed view is labeled with uppercase letters corresponding to the letters at each end of the cutting-plane line. Refer to SECTION A-A and SECTION B-B in **Figure 6-10**. Removed sections may also be shown at a different scale, usually enlarged to clarify detail. In the case of multiple-sheet drawings, if possible, the removed section should be on the same sheet as the corresponding "cut" view.

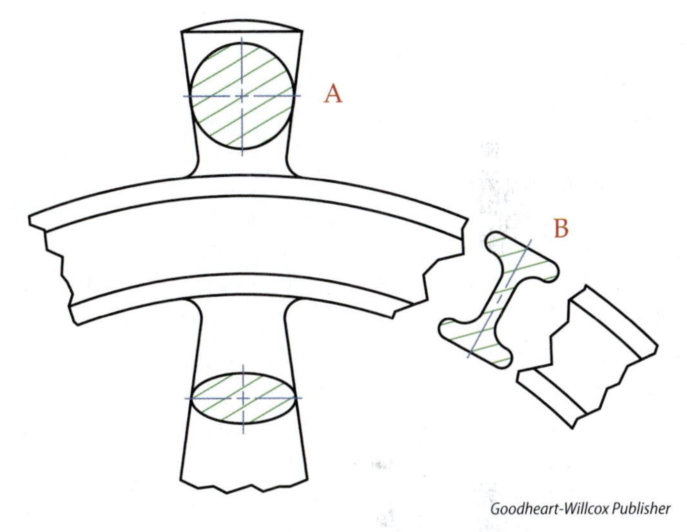

Goodheart-Willcox Publisher

Figure 6-9. A—A revolved section has a cutting plane slicing through the feature parallel with the line of sight, but the "cut" shape is rotated 90° directly on the regular view. B—The regular view can also be broken on each side of the revolved section.

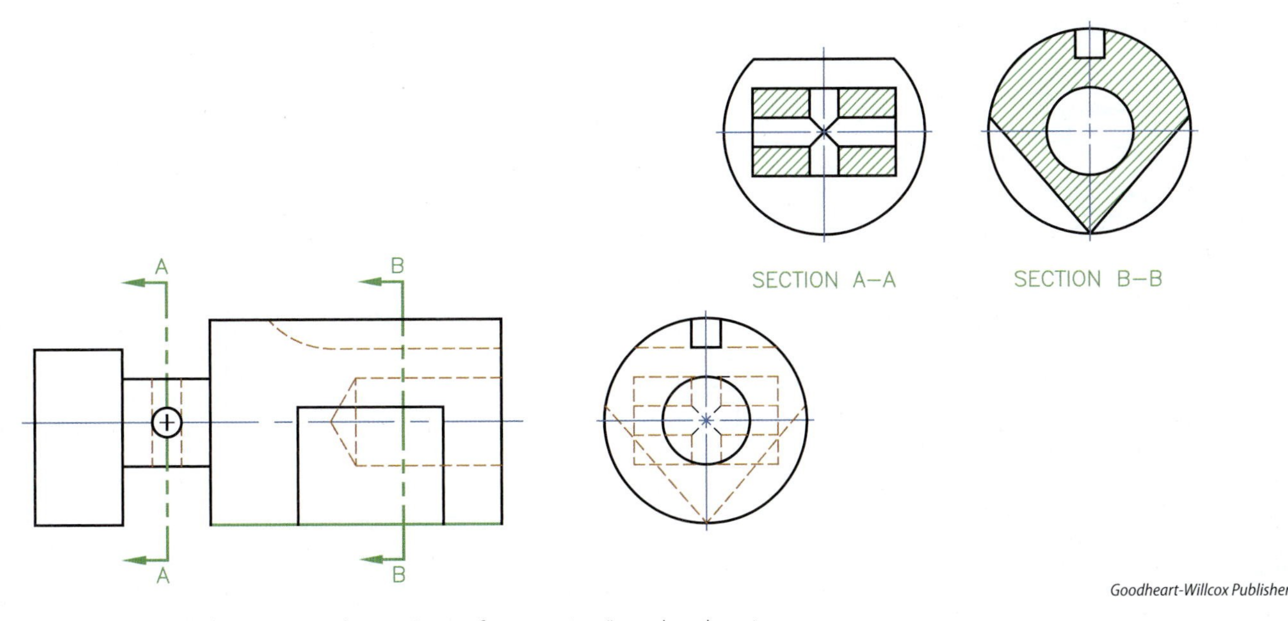

Goodheart-Willcox Publisher

Figure 6-10. A removed section is shown "out of projection" on the drawing.

Conventional Practice in Section Views

Section views present many opportunities for conventional practice. A ***conventional practice*** is one in which drafters break the principles or rules of projection theory for the sake of clarity. Standards establish the rules for a particular exception. One example of a conventional practice can be called the "rib rule." Many sectioned objects have thin walls, webs, or ribs that help support a feature of the object, as in **Figure 6-11**. To avoid a false impression of thickness, it is a conventional practice to leave the web or rib without section lines. Also, the rib is outlined, even though the cutting plane passes through it. This rule applies when the cutting plane is parallel to the thin material, not when the cutting plane passes across it in a perpendicular fashion. In some cases, it may be clearer to show the rib or web sectioned with alternating lines to distinguish it from the rest of the sectioned area.

The aligned section illustrated in **Figure 6-12** also shows the conventional practice for spokes of a wheel. As with the rib rule, spokes and arms in a section view are not section lined, because it would give the impression there is a solid web.

Other Section View Practices

There are a few other practices specific to section views that should be explained in this unit. A section can be a ***partial section***. This practice allows for a removed section to show details of an object without drawing complete views. In **Figure 6-13**, the partial

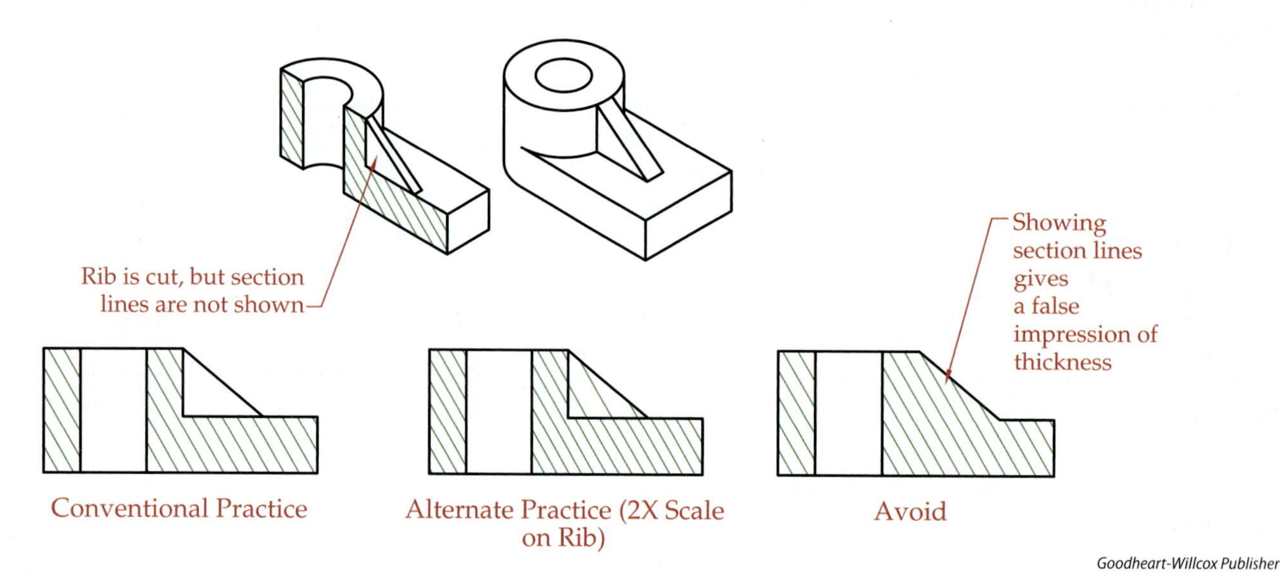

Figure 6-11. It is conventional practice to leave webs or ribs without section lines.

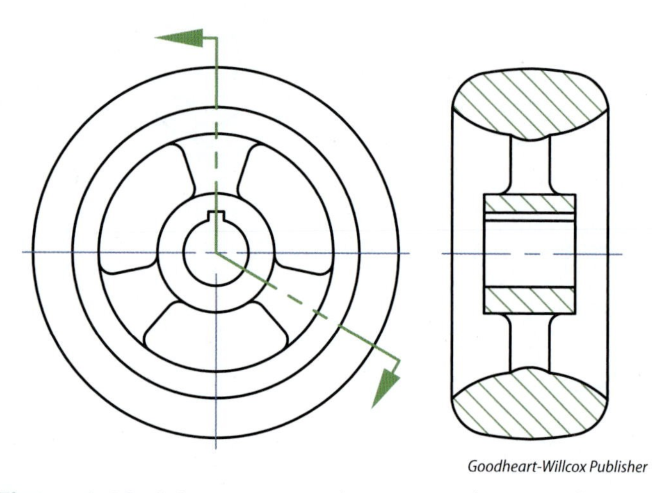

Figure 6-12. It is conventional practice to leave spokes without section lines.

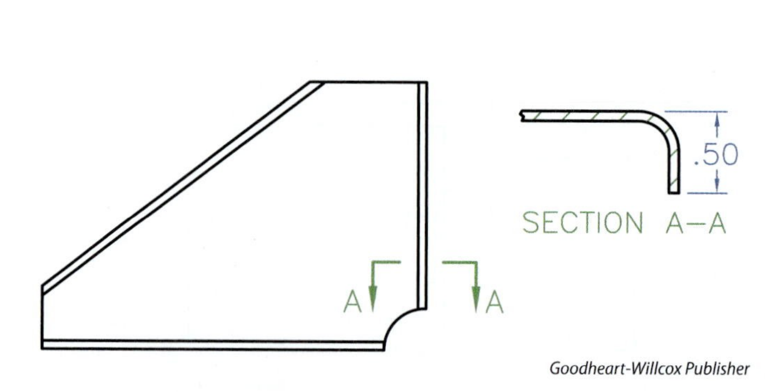

Figure 6-13. A partial section view shows details of an object without drawing complete views.

section is labeled SECTION A-A and is removed to the side. It could have also been drawn at a different scale to enlarge the view, if desired.

If a drawing can be clearly expressed without complete section lines, the section lines can be segments drawn only along the visible object lines. This is called an ***outline section*** and is only used on large views with a lot of surface area to be section lined. See **Figure 6-14**. Large parts are often drawn with outline sections, making the drawing easier to read.

Auxiliary views are projected in directions other than the six principal directions. Auxiliary views are discussed in Unit 7. The principles of section views can logically be applied to auxiliary views when the cutting plane is oriented in an auxiliary-view direction, **Figure 6-15**. Any type of section found on regular views—full, half, broken-out, etc.—may also be found on auxiliary views. When an auxiliary view is sectioned and the portion of the view extending beyond the section is not fully shown, the view should follow the guidelines for any partial view of an object.

Sections in Assembly Drawings

When multiple parts are shown assembled in a drawing, the drawing is called an assembly drawing. Assembly drawings are discussed in Unit 16. With respect to assembly drawings of multiple parts, many of the parts may be shown in full section and distinguished by orienting the section lines at different angles or spacing. For example, the largest part may feature section lines that slope to the right, the second part may feature section lines that slope to the left, and then additional parts may have section lines at other angles or with closer spacing, **Figure 6-16**.

Also note in **Figure 6-16** the conventional practice of leaving fasteners and shafts uncut as the cutting plane passes through the assembly. Fasteners and shafts are usually more recognizable by their exterior features.

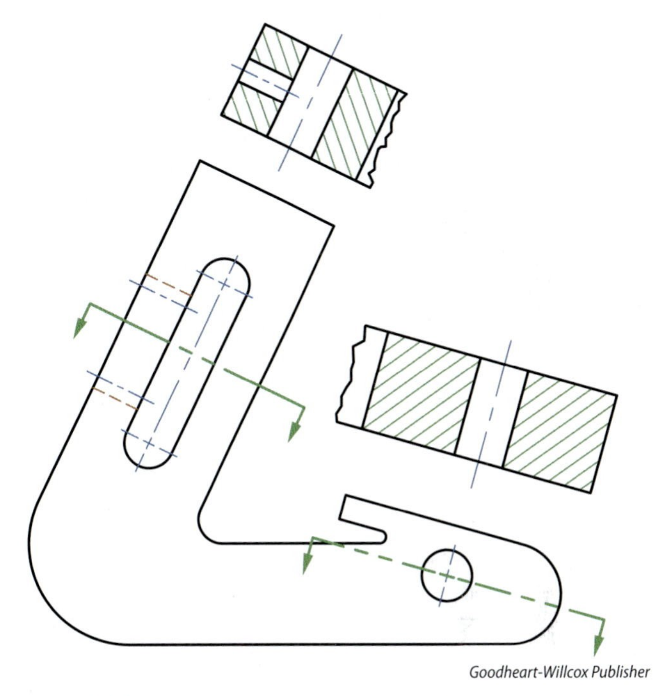

Goodheart-Willcox Publisher

Figure 6-15. An auxiliary view can also be a section view.

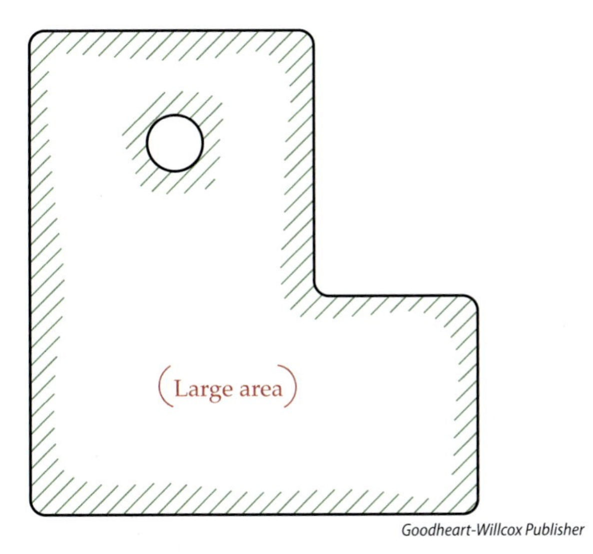

Goodheart-Willcox Publisher

Figure 6-14. The lines of an outline section are drawn only along the visible lines.

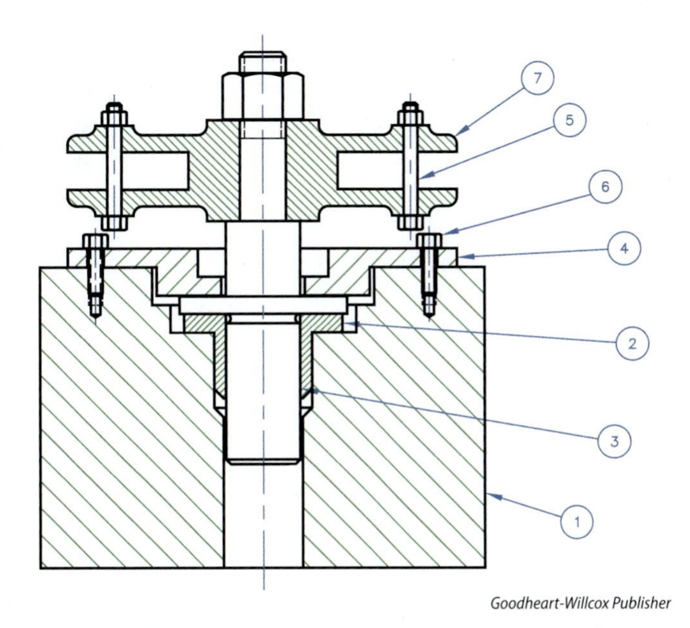

Goodheart-Willcox Publisher

Figure 6-16. Fasteners and shafts are not sectioned in assembly drawings. The section lines should have a unique appearance for each part.

In assembly drawings, thin parts such as rubber inserts, gaskets, and thin plates are often shown differently in section than other parts. As the linework in thin shapes would be so close together and hard to see, thin parts may simply appear as a solid black or shaded area in assembly drawings, **Figure 6-17**.

Computer-Generated Section Views

As with the principal orthographic views, 3D CAD programs can quickly and easily generate section views directly from a model. In **Figure 6-18**, the front, top, and right-side views were established first. In this example, the first removed view, SECTION A-A, was created by designating where the cutting plane would "cut" on the front view. The software automatically placed the section view in a projected alignment and inserted a cutting-plane line in the front view. The section view was then manually moved out of the projected alignment. The section labels, scale, and direction of the arrows on the cutting-plane line are also editable.

As discussed in the previous unit, while computer-generated views can be created quickly, they often require some adjusting for the sake of clarity. For **Figure 6-18**, some of the rounded edge tangency elements were left showing and some were hidden. Computer-generated section views usually work well for basic objects and sections, but broken-out, aligned, and revolved sections may be more difficult to create depending on the capabilities of the software. Also, CAD programs do not always take into consideration the conventional practices associated with ribs and spokes. However, with computer-generated views, additional sections that serve the same purpose can easily be added to the product definition data set.

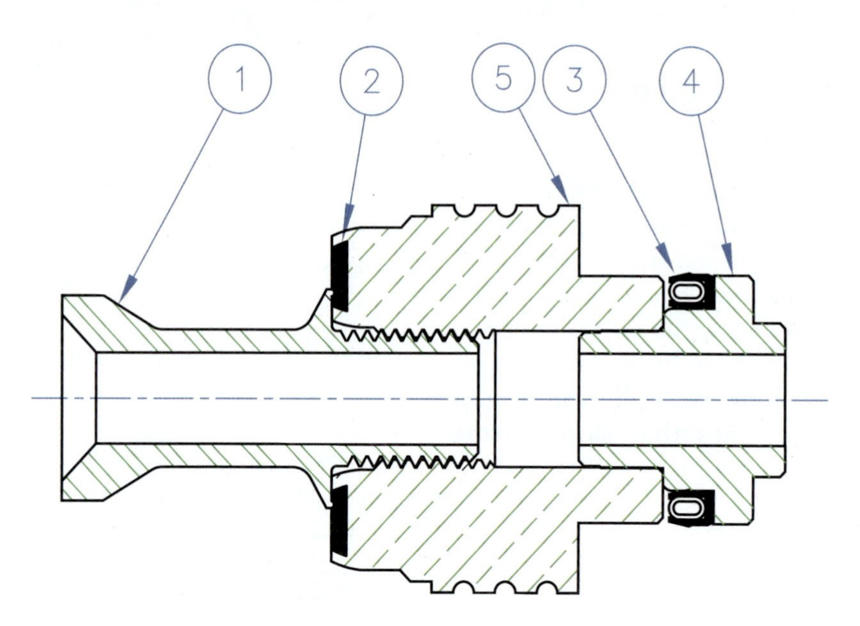

Goodheart-Willcox Publisher

Figure 6-17. Thin parts in an assembly section view are often simply shaded solid.

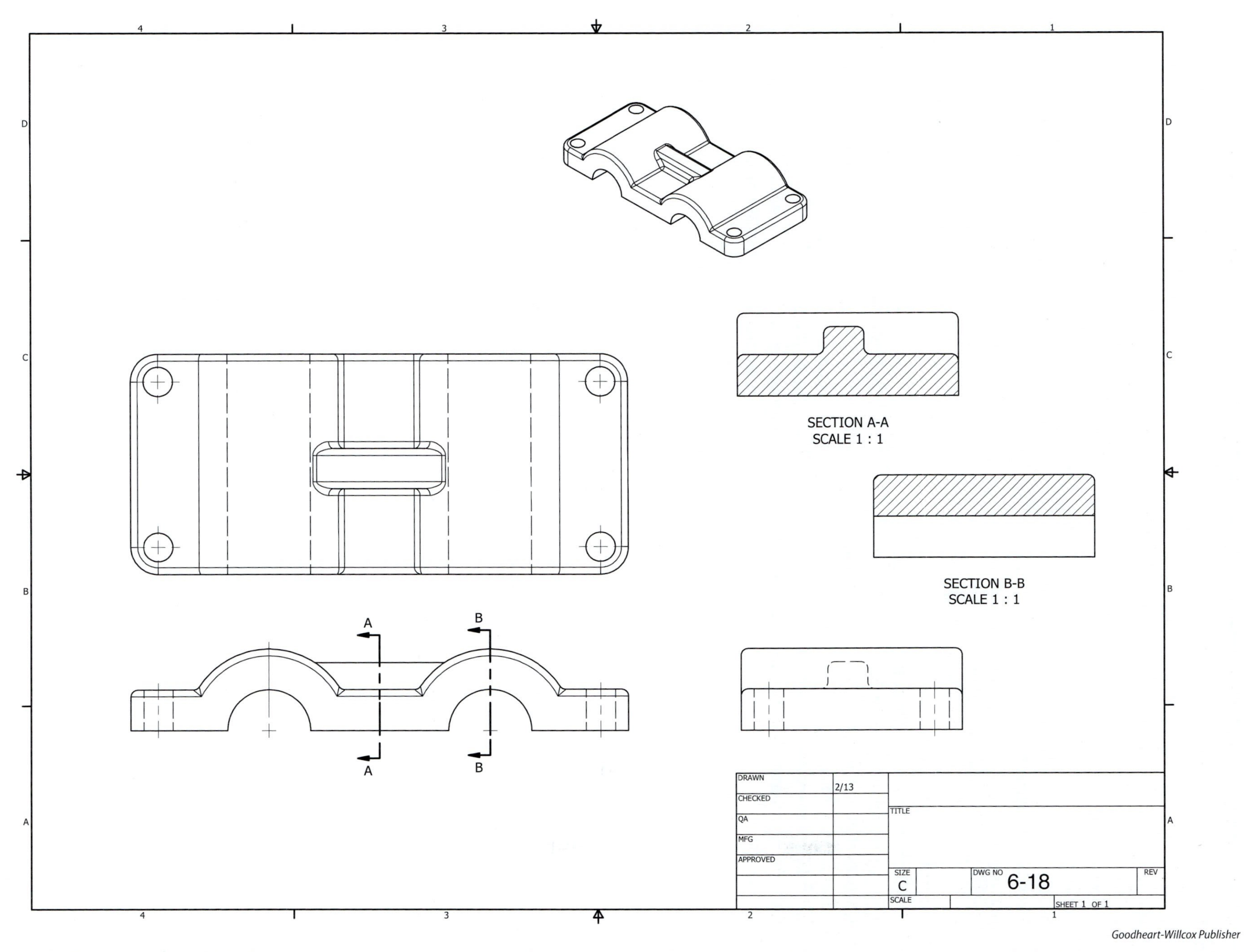

Goodheart-Willcox Publisher

Figure 6-18. Three-dimensional CAD modeling systems can create section views directly from the model in the same manner that principal views are created.

Summary

- Section views are based on an imaginary cut of the object in some fashion and are created to help show interior detail.
- Two types of lines exclusive to section views are the cutting-plane line and the section line.
- Section lines can also incorporate dashed lines and patterns to symbolize a particular material or categories of material, such as steel, alloys, plastic, or marble.
- Full sections are those in which the object is cut fully through, while half sections are those in which the object is cut halfway through.
- The cutting-plane line can be offset through features that are not aligned, resulting in an offset section or an aligned section.
- A broken-out section allows for a partial cut into the object and incorporates the short break line.
- Removed and revolved sections allow for a "cut and revolve" cutting-plane technique, sometimes directly on the view, but sometimes removed out of projection.
- For clarity's sake, various conventional practices are used when cutting through features such as webs, ribs, and spokes.
- Partial sections and outline sections can be implemented to make drawings less cluttered and for clarity's sake.
- Section techniques can also be applied in partial views and auxiliary views.
- When multiple parts are represented in an assembly drawing, section-view patterns are implemented at different angles and spacing to help clarify the assembly.
- Computer-generated sectional views can be easily generated from the 3D model but may show some features in a slightly different fashion than traditional drawing practices.

Unit Review

Name ____________________ Date ____________ Class ____________

Answer the following questions using the information provided in this unit.

Know and Understand

____ 1. *True or False?* The primary purpose of section views is to help feature exterior detail without using hidden lines.

For the next three terms, match the three types of lines to the three descriptions. Answers are used only once.

____ 2. Cutting-plane line
____ 3. Section line
____ 4. Short break line

A. Not exclusive to section views, but needed for broken-out sections.
B. Drawn thick in a view adjacent to the section view.
C. Drawn thin, and usually continuous, but sometimes dashed.

____ 5. Of the following, which is *not* a type of section view featured in this unit?
A. Full
B. Half
C. Quarter
D. Offset

____ 6. *True or False?* The preferred dividing line between the two halves of a half-section view is the visible line.

____ 7. An object with a number of odd features, such as spokes or holes, often has a(n) _____ section so the odd features can be rotated around the axis into a vertical cutting plane, resulting in a view that is easier to read and more informative.
A. aligned
B. revolved
C. removed
D. rotated

For the next three terms, match the three types of section views to the three descriptions. Answers are used only once.

____ 8. Broken-out section
____ 9. Removed section
____ 10. Revolved section

A. Section view will behave like an overlay on the same view where the cut is imagined.
B. Section view will not cut fully through the object.
C. Section view will be moved out of projection with the regular views.

____ 11. For all of the following situations *except one*, there is an applicable conventional practice that states that even though that part of the object was cut by the cutting plane, section lines are not usually shown. Identify the exception.
A. Thin rib cut flatwise
B. Spoke
C. Web cut flatwise
D. Thin rib cut across perpendicular to the center plane

____ 12. *True or False?* An edge section is a section view with a large area to be sectioned, and the section lines are just drawn around the edges.

____ 13. Identify the *false* statement about section views of assembled parts.
A. Each part has section lines at an angle unique to that part.
B. Shafts, bolts, and nuts are often not shown cut by the cutting plane.
C. The angle of section lines on one side of the center line should be 90° to the section lines on the other side of the center line.
D. Very thin parts such as gaskets or rubber inserts may feature solid shading instead of section lines (hatching).

____ 14. *True or False?* An auxiliary view (projected from a regular view in a direction other than the regular views) can also be a section view.

____ 15. *True or False?* Computer-generated section views usually work well for basic objects and sections, but broken-out, aligned, and revolved sections may be more difficult to create depending on the capabilities of the software.

Critical Thinking

1. List some of the key issues with respect to why section views are used.

____

____

____

____

____

____

2. Explain why conventional practices are so important, and some of the ways in which print readers can be knowledgeable about those conventional practices.

____

____

____

____

____

____

Apply and Analyze

Name ______________________ **Date** ______________ **Class** ______________

Review Activity 6-1

Section View Identification

Analyze each of the section views shown below. Identify the type of section view, writing your answers in the blanks at the bottom of the page.

1.

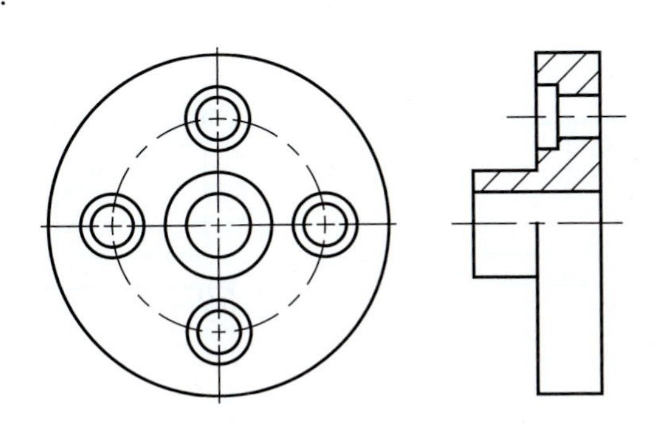

2.

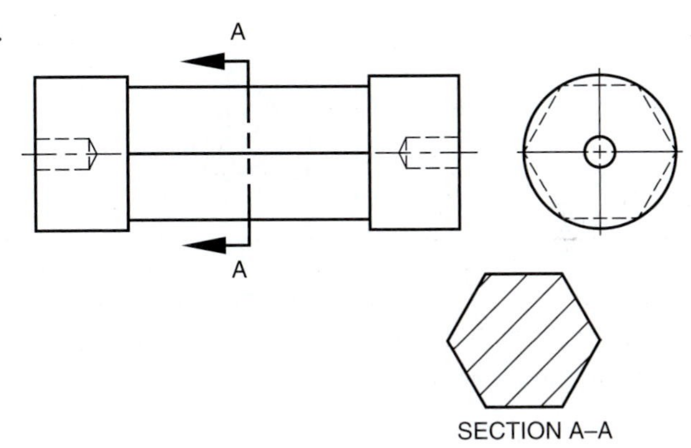

3.

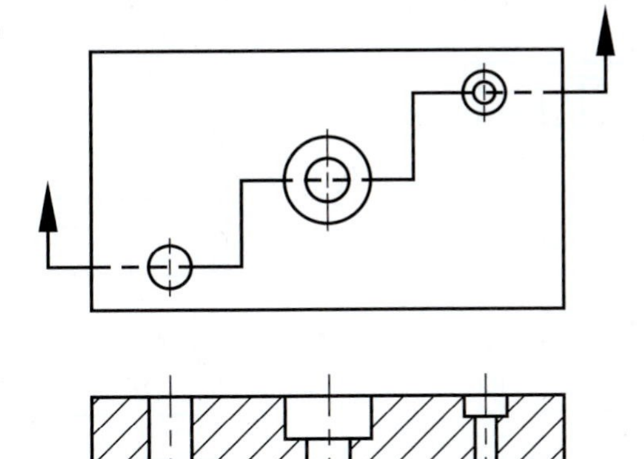

4.

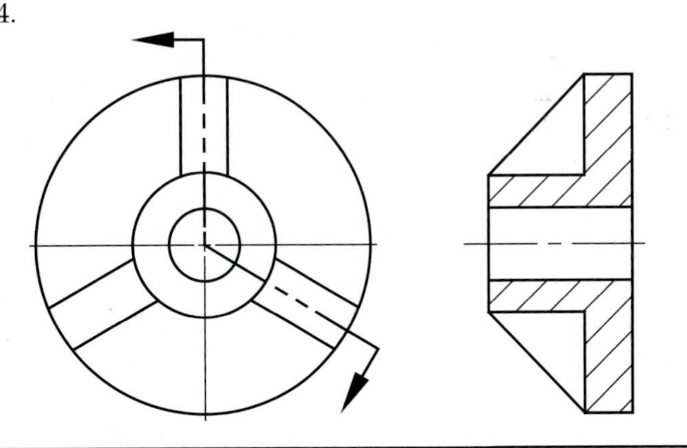

5.

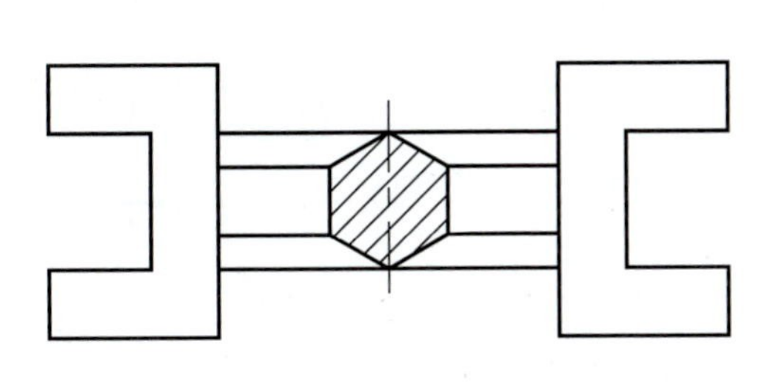

6.

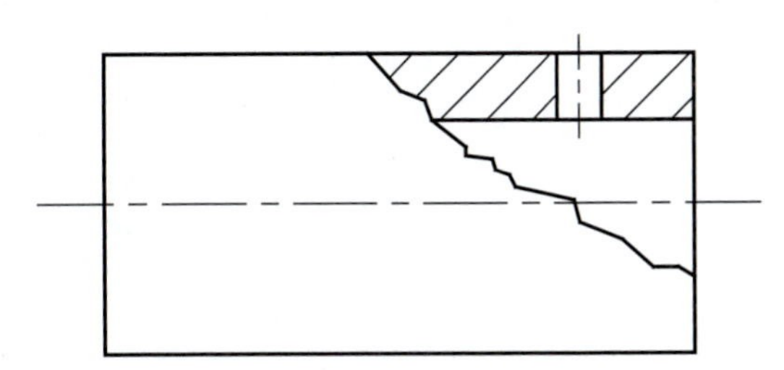

1. ______________________
2. ______________________
3. ______________________
4. ______________________
5. ______________________
6. ______________________

Apply and Analyze

Name ______________________ Date ______________ Class ______________

Review Activity 6-2

Section Views

For each of the numbers below, select the correctly drawn section view (A, B, C, or D), as cut by a cutting plane parallel with a frontal plane at the midpoint of the object. Place the letter of the correct section view in the blank provided.

______________ 1. 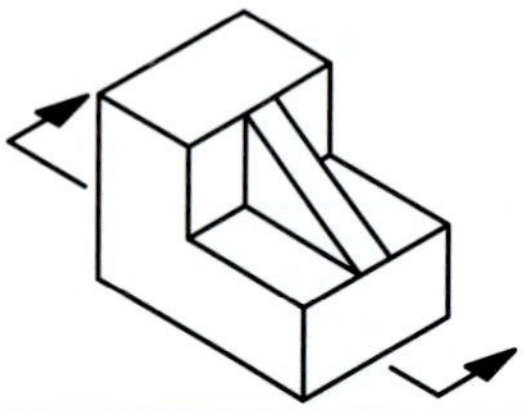 A. 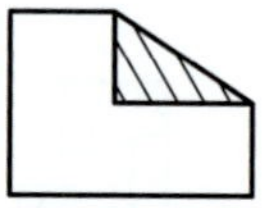 B. 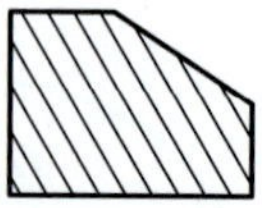 C. 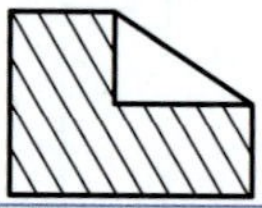 D.

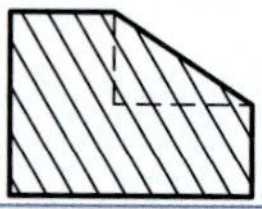

______________ 2. 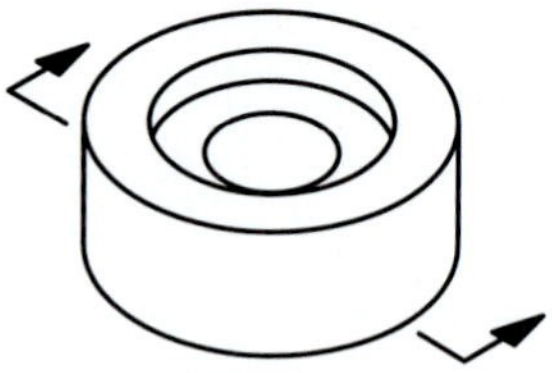 A. 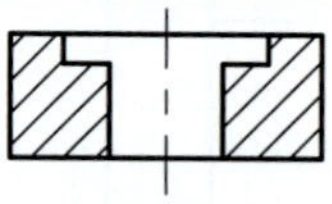 B. 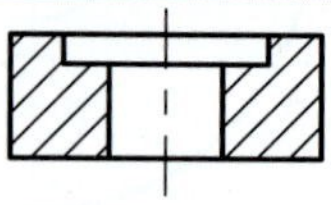 C. 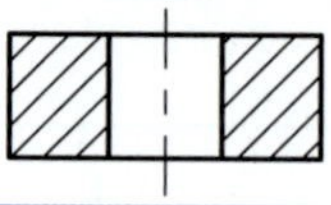 D.

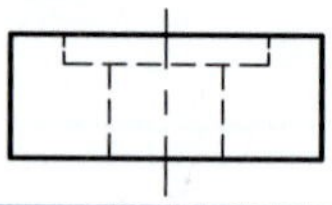

______________ 3. 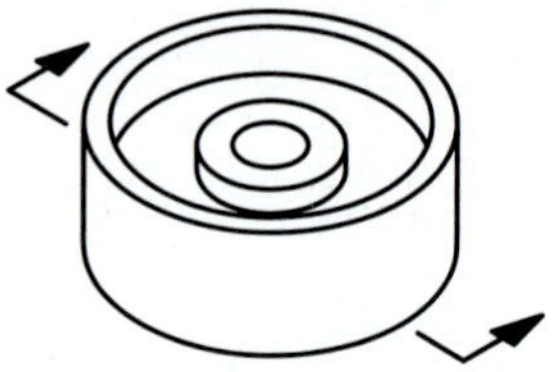 A. 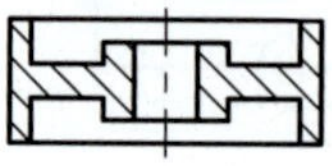 B. 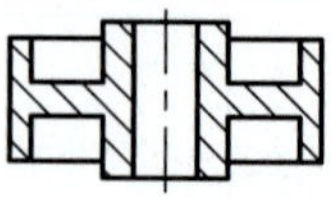 C. 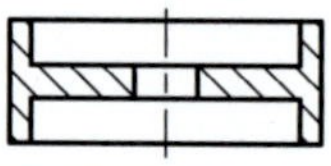 D.

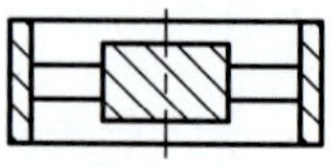

______________ 4. 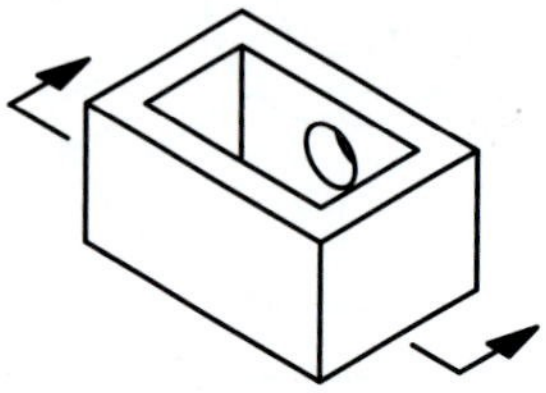 A. 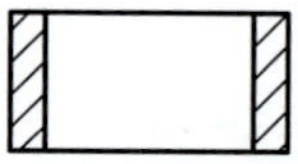 B. 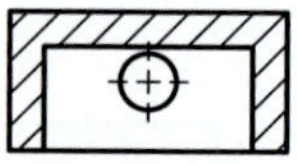 C. 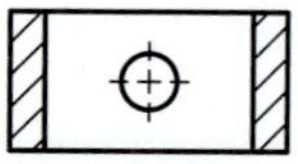 D.

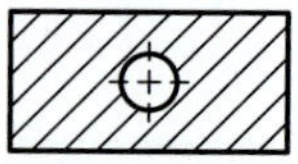

______________ 5. 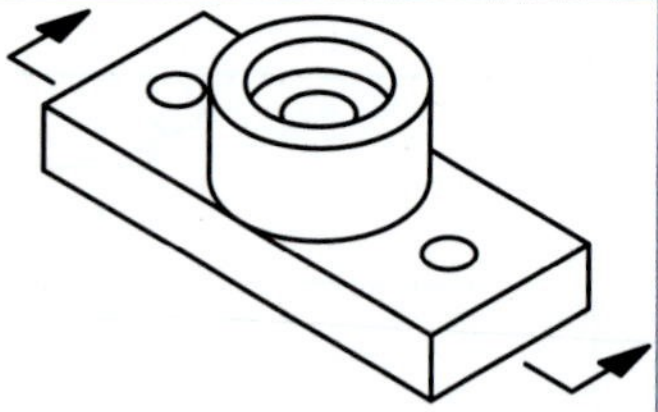 A. 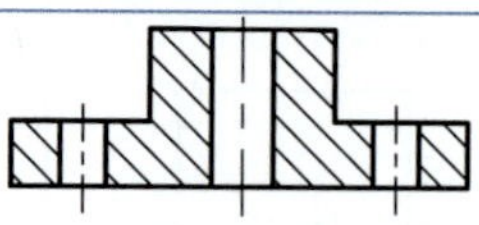 B. 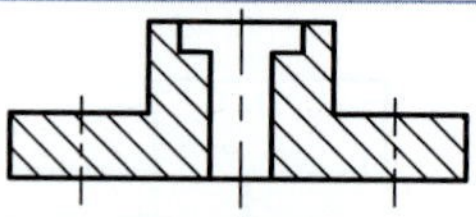 C. 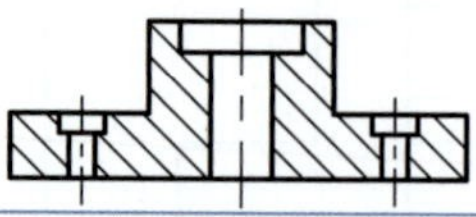 D.

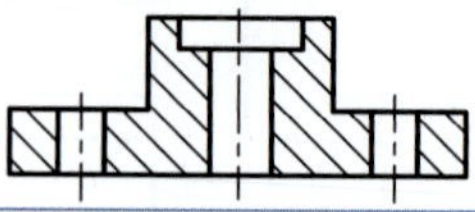

______________ 6. 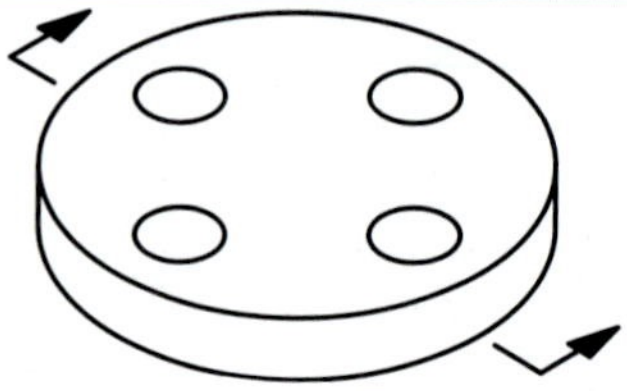 A. 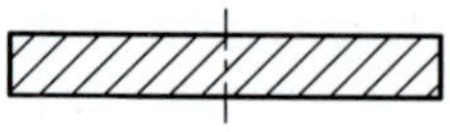 B. 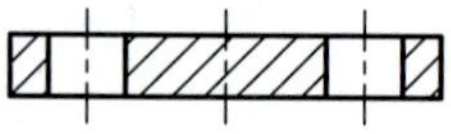 C. 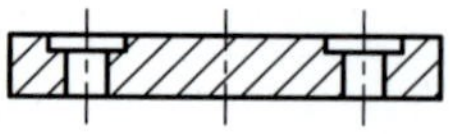 D.

Apply and Analyze

Name ______________________ Date ______________ Class ______________

Review Activity 6-3

Section Views

For each of the numbers below, a front section view is missing. Place the letter of the correct section view in the blank provided. Note: To make these problems a little more challenging, center lines have been omitted!

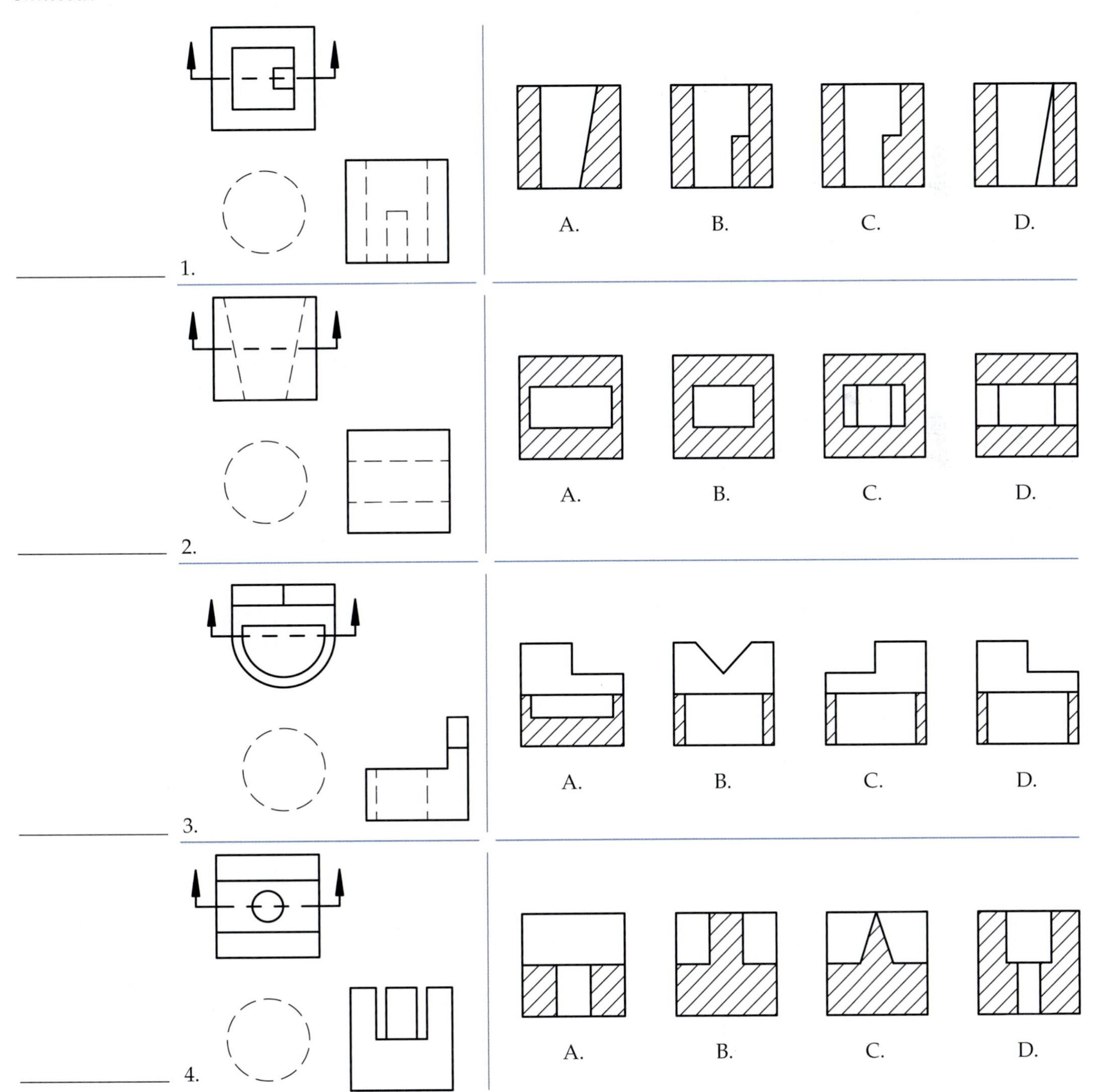

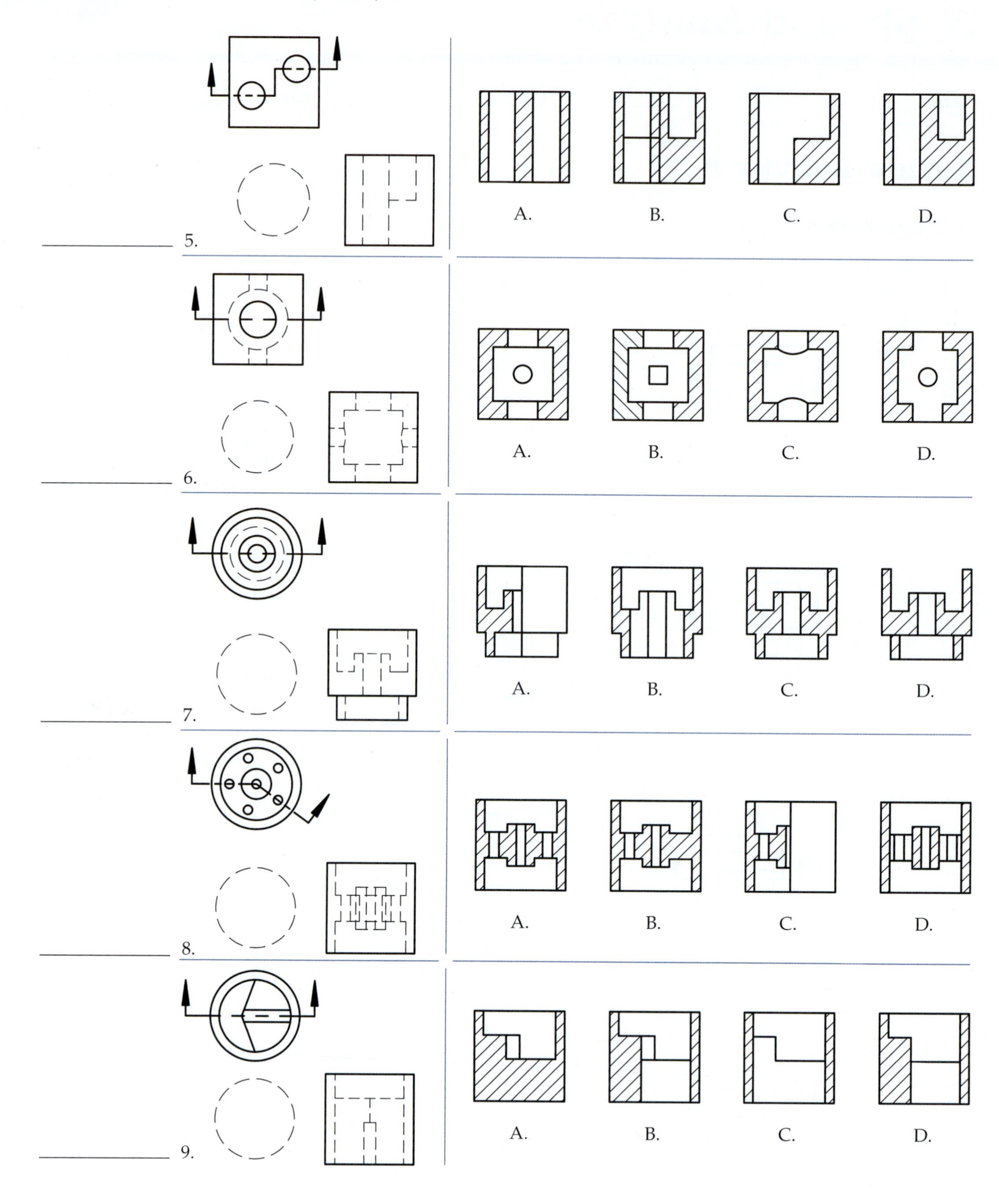
5.
A.
B.
C.
D.
6.
A.
B.
C.
D.
7.
A.
B.
C.
D.
8.
A.
B.
C.
D.
9.
A.
B.
C.
D.

Notes

Apply and Analyze

Name ______________________ Date ______________ Class ______________

Industry Print Exercise 6-1

Refer to the print PR 6-1 and answer the questions below.

1. What type of section view is shown on this print?
2. Is the cutting-plane line shown?
3. Are there any hidden lines shown in the section view?
4. The section line pattern indicates cast iron. Is this the material for the part? If not, what is the material?
5. Does this drawing feature a top view?

Review questions based on previous units:

6. What is the drawing number or part number?
7. What is the name of this part?
8. Was this drawing checked by someone?
9. Does this drawing have center lines?
10. According to one of the general notes, in what state is the supplier located?
11. On what size paper was the original drawing created?
12. On what part number(s) will this part be used?
13. At what scale was the original drawing created?
14. In the section view, the leftmost vertical object line has what meaning?
15. In the section view, what numerical dimension value is given for the overall depth?

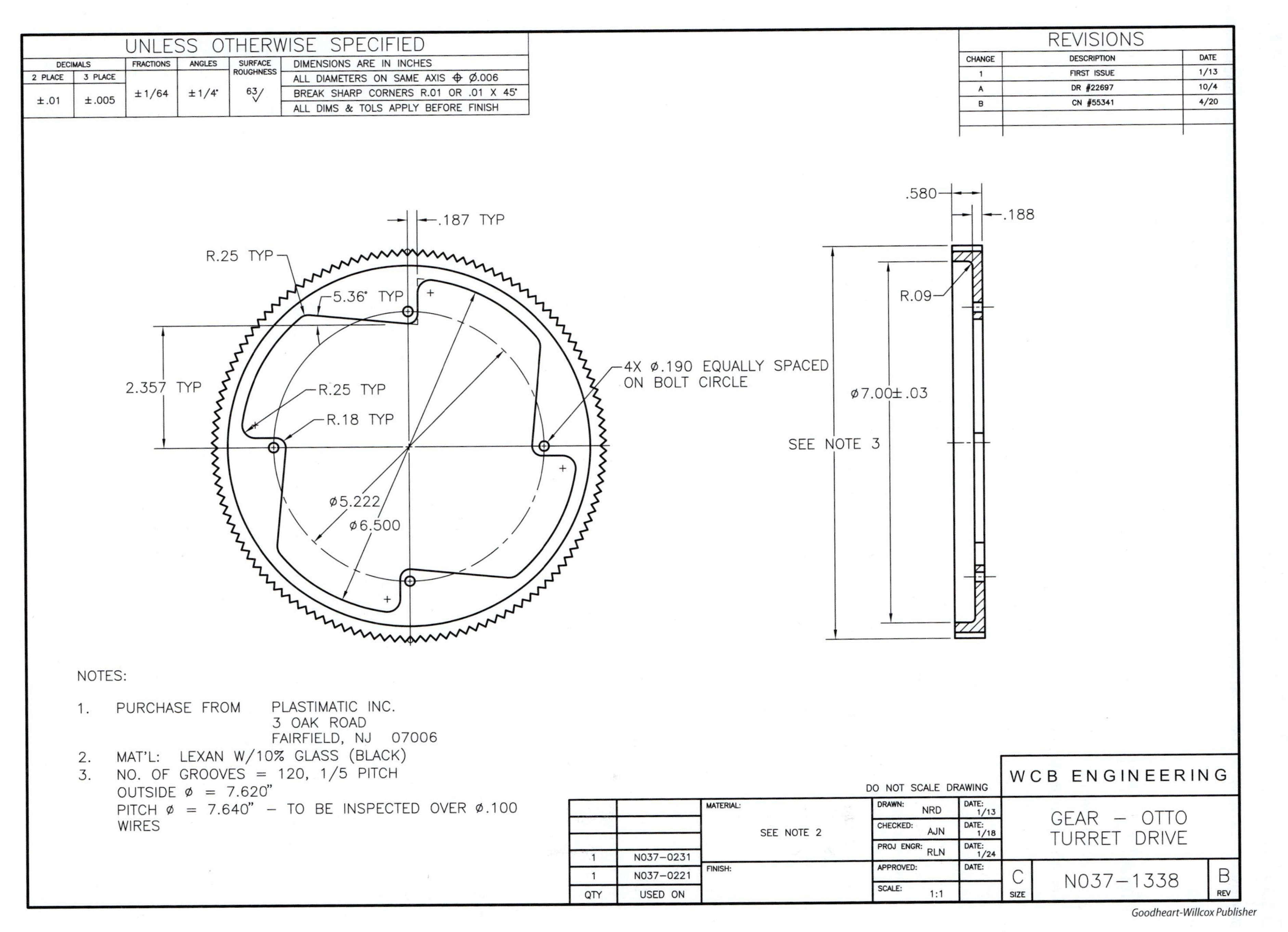

PR 6-1. Otto Turret Drive Gear

Apply and Analyze

Name ______________________ Date ______________ Class ______________

Industry Print Exercise 6-2

Refer to the print PR 6-2 and answer the questions below.

1. For the sake of this exercise, if the main section view is considered the front view, what term would describe the circular view?

____

2. Is the cutting-plane line shown?

____

3. Does the section line pattern shown in the section view match the type of material specified in the title block? (Refer to **Figure 6-2** in the unit.)

____

4. What is the name of the jagged line in the partial section view labeled Detail A?

____

5. What type of section view is represented by this drawing, full or half?

____

Review questions based on previous units:

6. What is the drawing number or part number?

____

7. What is the name of this part?

____

8. Was this drawing checked by someone?

____

9. Does this drawing have center lines?

____

10. In what state is the company that owns this print located?

____

11. What scale is the view labeled Detail A?

____

12. In addition to cylinder forms, what other 3D geometric form is featured in the body shape of this object?

____

13. At what scale was the original drawing created?

____

14. In the section view, the uppermost horizontal object line has what meaning?

____

15. In the round view, is the smallest visible circle an internal feature (hole) or an external feature (post)?

____

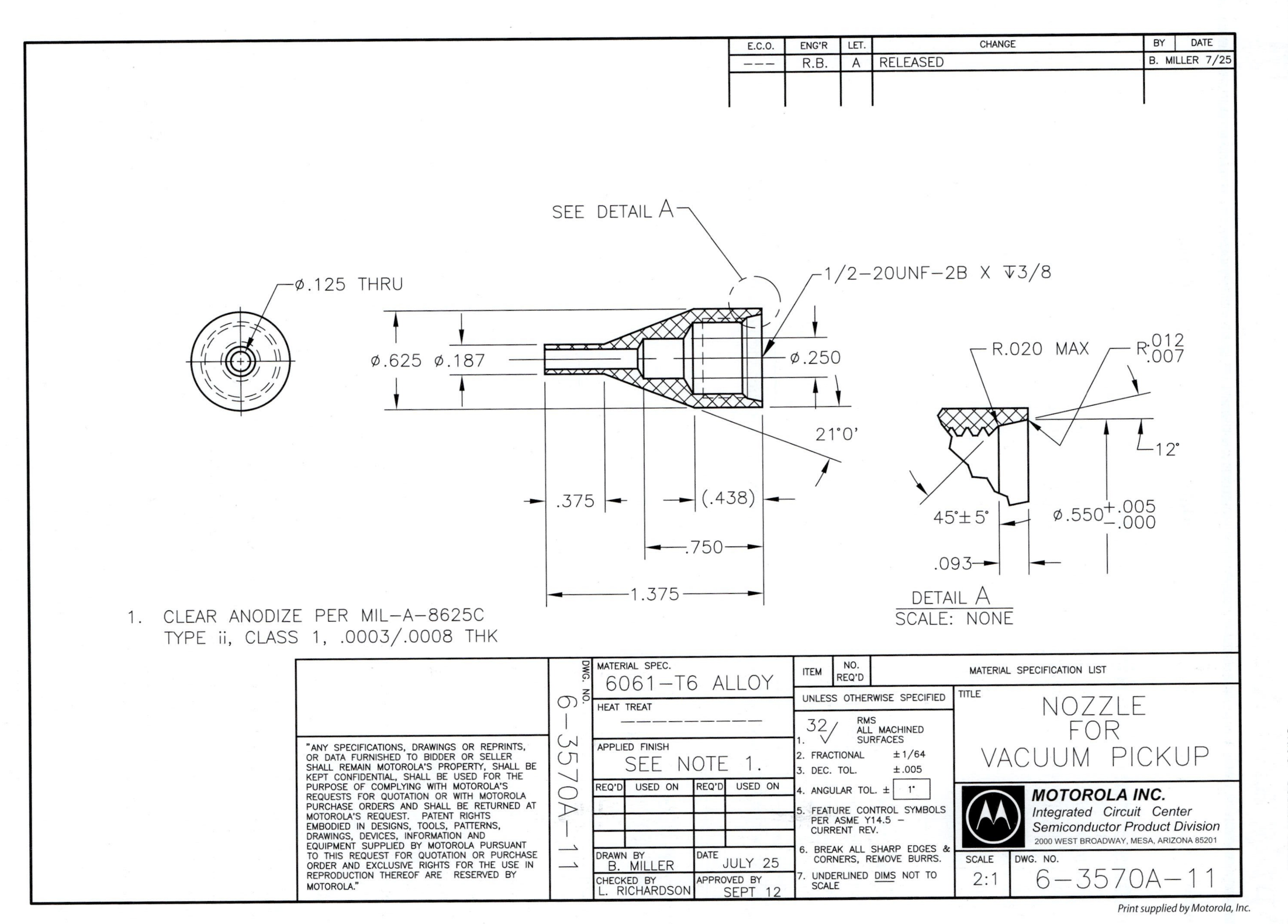

Print supplied by Motorola, Inc.

PR 6-2. Nozzle for Vacuum Pickup

Apply and Analyze

Name ______________________ Date ______________ Class ______________

Bonus Print Reading Exercises

The following questions are based on the various bonus prints located in the folder that accompanies this textbook. Refer to the print indicated, evaluate the print, and answer the question.

Print AP-001

1. This print does not show a cutting-plane line. Is this acceptable practice or an error?

____

Print AP-002

2. When looking at SECTION A-A, how does the print reader know that this is several parts or pieces and not just one part?

____

3. Most of the sections on this print are full sections. What other method of sectioning is used?

____

Print AP-003

4. What type of section is shown in SECTION B?

____

5. SECTION B uses a center line between the left half and the right half. Is this recommended practice?

____

Print AP-004

6. Based on the shape of the cutting-plane line, what type of section is SECTION B-B?

____

7. Except for SECTION B-B, what other type of section view is used in the main views?

____

Print AP-006

8. Are the section lines for this print representative of the material, or are they simply being shown as general purpose section lines?

____

Print AP-007

9. How many cutting-plane lines are shown on this print?

____

Print AP-009

10. Based on the cutting-plane line, what type of section is Section A-A?

____

Print AP-010

11. What type of section describes this one-view drawing?

____

Print AP-011

12. This print features a section view labeled SECTION A-A. In this particular case, is this label necessary?

____

Print AP-012

13. What type of section is featured on the main front view of this object?

____

Print AP-013

14. List the names of the two parts located on the central cutting plane that are not shown as cut, but are instead still shown whole.

____

15. Are there any section lines on this part that are material specific?

____

Print AP-016

16. What type of section is featured to show the hole on the worm gear end of this part?

____

Print AP-017

17. What conventional practice is applied to the full section view with respect to the gear tooth?

____

____

Print AP-018

18. For the section views shown on this print, are there any revolved or aligned sections?

____

Print AP-020

19. Besides a full section, what other type of section is used on this print?

____

20. Is the section line pattern appropriate for this part, and what materials can be represented by the pattern shown?

____

____

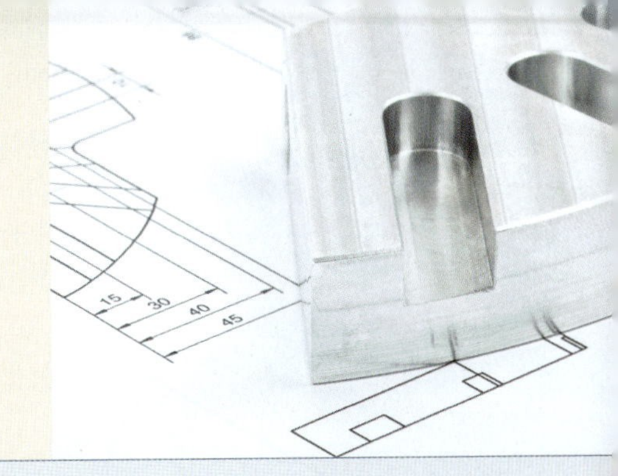

UNIT 7 Auxiliary Views

LEARNING OBJECTIVES

After completing this unit, you will be able to:

- Define auxiliary views as compared to principal views.
- Explain the purposes and benefits of auxiliary views.
- Explain the relationship of auxiliary projection planes to the object and to the principal projection planes.
- Discuss the purpose and use of reference lines in auxiliary view construction.
- Describe techniques that may be helpful to visualizing auxiliary views.
- Discuss terms that help describe or define different types of auxiliary views.
- Identify the differences between primary and secondary auxiliary views.
- Discuss characteristics of computer-generated auxiliary views.

TECHNICAL TERMS

auxiliary view
depth auxiliary view
dihedral angle
folding line
height auxiliary view
primary auxiliary view
secondary auxiliary view
width auxiliary view

Drafters sometimes find it necessary to use auxiliary views to fully describe an object. The word *auxiliary* simply means "additional" or "supplemental," as in "ways of helping beyond the regular." ***Auxiliary views*** are still orthographic projections, with parallel projectors perpendicular to a projection plane, but auxiliary views are projected in some angular direction. One reason to create an auxiliary view is to show the true shape and size of features that would not be shown true size and shape in the principal views. In a previous unit, the six principal views were identified as the front, top, right side, left side, rear, and bottom. Simply put, any projection of an object that is not a principal view is an auxiliary view. Standards for creating auxiliary views are covered by ASME Y14.3, *Orthographic and Pictorial Views*.

Auxiliary View Principles and Purposes

As explained in Unit 5, an inclined or oblique surface will not project true size and shape in any of the principal views. The block in **Figure 7-1** features an inclined surface labeled 1-2-3-4. In the front view of the block, the edge view of the inclined surface is featured because the surface is perpendicular to the frontal plane. The two other regular views show a foreshortened shape of surface 1-2-3-4 because the surface is inclined to the horizontal and profile planes.

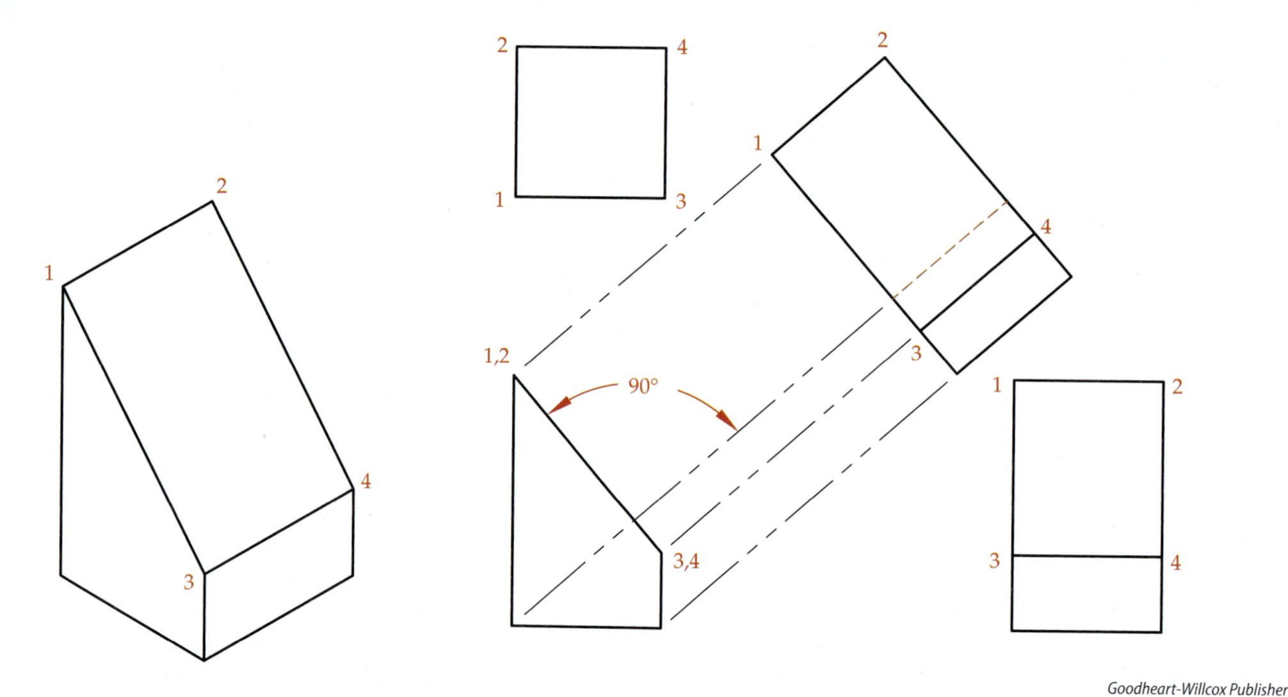

Goodheart-Willcox Publisher

Figure 7-1. In this case, an auxiliary view is needed to show the true shape and size of the inclined surface.

An auxiliary view of the surface can be projected onto an auxiliary projection plane strategically oriented to create a view that will show the true size and shape of the surface. Applying orthographic principles, a line of sight perpendicular to the inclined surface is chosen, and a plane of projection parallel to the inclined surface is used. When the inclined surface is projected with parallel projectors perpendicular to the projection plane, the resulting projection will be true size and shape. This is a common reason to use an auxiliary view. Also, if the inclined plane has additional features, such as rounded corners or holes located perpendicular to the edges of the surface, the auxiliary view showing the true size and shape of that inclined plane will provide a better option for where to locate the dimensions for those features.

As illustrated in **Figure 7-2**, the auxiliary plane is "hinged" to the frontal plane of projection and swung around into alignment with the other planes. While not standard practice, the ***folding lines*** between the views may be present as reference lines. If so, they should be phantom lines and may be marked with projection plane labels. Reference lines such as these

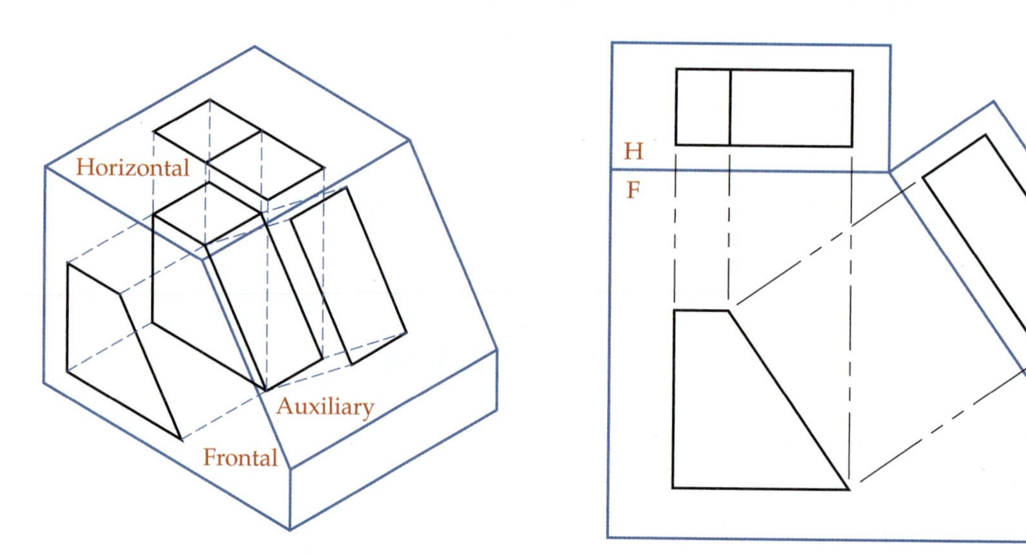

Goodheart-Willcox Publisher

Figure 7-2. If a plane of projection is placed parallel to an inclined surface, the projection onto that plane is true size and shape.

are more common when reading ASME standards or drafting texts, **Figure 7-3**. Accordingly, it is helpful in visualizing the spatial relationship between two views if you think of the phantom reference line as a 90° bend line between two adjacent planes. Also notice in the figure that the depth distances from the reference line to the object are equal, which is useful if the auxiliary views are being constructed manually by the drafter.

Another purpose for an auxiliary view is to create a view that shows the true angle between two surfaces, known more technically as the ***dihedral angle***. The dihedral angle between two planes is shown true only in a view that features the point view of the edge of intersection between the two surfaces. In cases where a V-groove runs in a normal direction, there is likely a principal view that shows the true angle. In **Figure 7-4**, the V-groove is inclined to the normal directions of sight, so an auxiliary view is required to show the true angle between the two surfaces of the groove. In this situation, the auxiliary view provides the best choice for where to place the dimension for the dihedral angle.

Visualizing Auxiliary Views

Visualizing auxiliary views should not be that much different from visualizing regular views, but the angles do seem to play tricks on the mind. As with regular views, learning how to visualize the views comes with practice. One technique is to picture the object being turned while your line of sight remains fixed. Another technique is to picture the object remaining fixed while you move yourself around the object to view it from different vantage points. With auxiliary views, you can imagine that your eye is located in such a way as to squarely look at the inclined surface or to look squarely at the point view of an inclined V-groove.

As described above and in earlier units, views of an object are related to each other and should be lined up with each other in ways that help the print reader visualize the part. Remember, any view adjacent to another view is like turning the object 90°. In the case of auxiliary views, the turn axis is not vertical or horizontal, but rather at an angle. To help with this, sometimes a print reader can simply rotate the print when reading an auxiliary view so the lines of projection between the principal view and auxiliary view are vertical or horizontal, **Figure 7-5**. This technique is helpful for those who are accustomed to reading regular views.

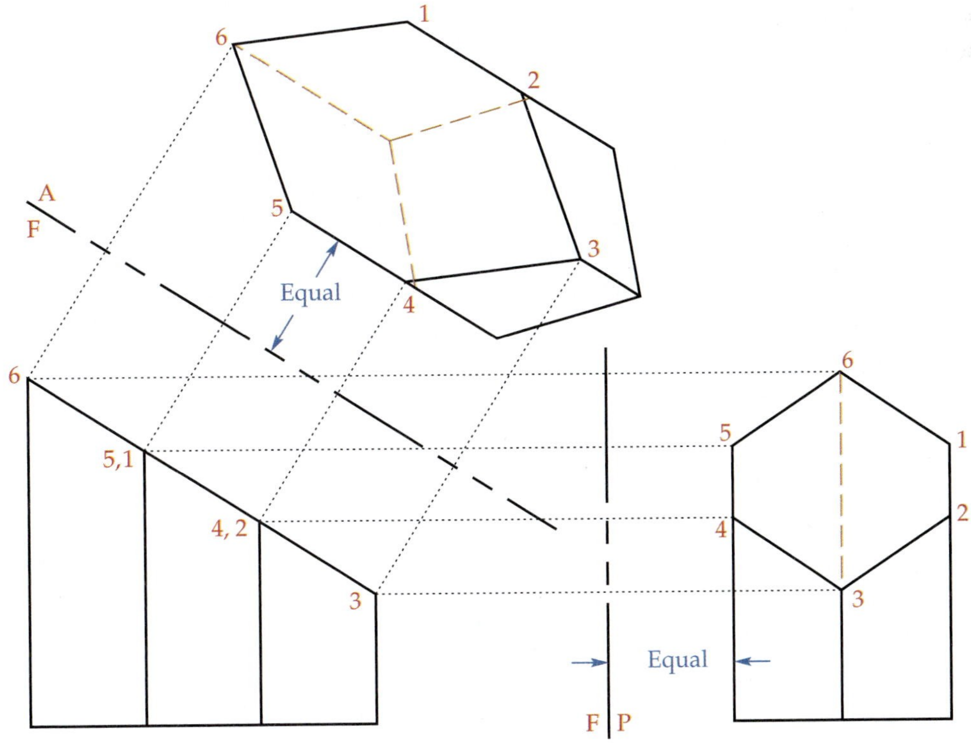

Figure 7-3. Reference lines may be indicated in some textbooks or standards. These lines represent the hinge line of the projection planes: A = auxiliary, F = frontal, and P = profile. These lines are not usually part of an industry print. Distances from the reference lines to the views are equal, as indicated.

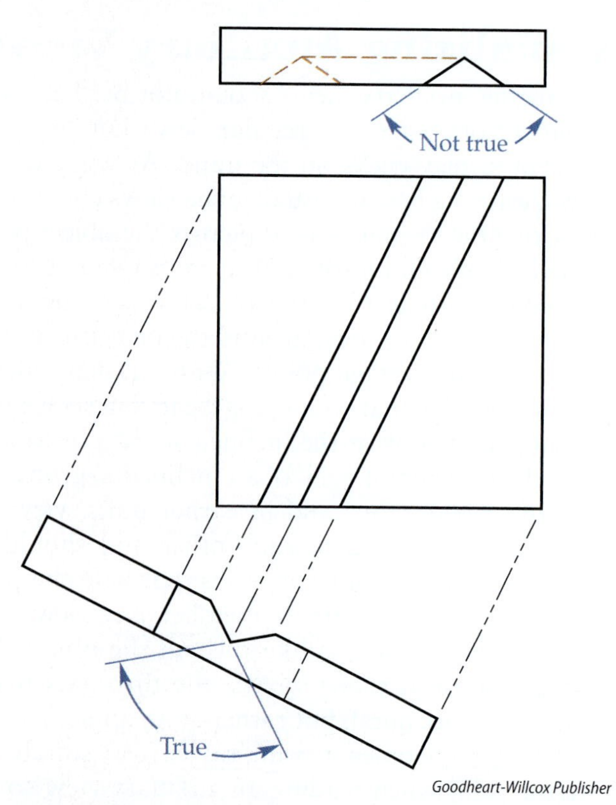

Goodheart-Willcox Publisher

Figure 7-4. An auxiliary view projected to show the point view of the intersection edge is needed to show the true dihedral angle of the groove.

Also, as discussed in Unit 5, numbering corners (vertices) of the object can help you read a drawing. As you are learning how to read prints, experiment with numbering some corners of objects, as illustrated in some of the figures in this unit.

In the case of inclined surfaces, the auxiliary views may be projected from any view in which the inclined surface appears as a line. Depending on the angle, auxiliary views can be difficult to locate on the sheet. Some may choose to connect an auxiliary view to an adjacent view with one or more projection lines, but that is optional. A center line can also be extended to connect two views if the views feature a hole or symmetrical feature. The complexity of the part, the scale of the drawing, and the available drawing space all present challenges in locating auxiliary views. As with regular views, viewing-plane lines can also be used. Auxiliary views can be placed out of projection, but they must remain turned and tilted in the auxiliary-view orientation dictated by the viewing-plane line, **Figure 7-6**. As with this figure, partial views may also be used to simplify the views, as long as the critical features are shown.

Auxiliary View Types

Auxiliary views should be projected and aligned with principal views. They are not labeled as auxiliary views, just as front, top, and right-side views are not labeled. In drafting textbooks, and perhaps in industry standards, there are terms used to help explain the auxiliary views and the system of projection that allows them to be properly constructed.

With that in mind, an auxiliary view projected from a principal view is called a ***primary auxiliary view***. In some cases, only the inclined or special feature is shown, which can be achieved with a partial primary auxiliary view. **Figure 7-7** shows a partial primary auxiliary view projected from the front view to provide a true representation of the inclined, rectangular feature of the object. The location dimensions for the hole pattern and the size of this inclined surface are foreshortened in the regular top and right-side views. The partial primary auxiliary view provides not only a true size and shape description of that portion of the object, but also the best view in which to locate the dimensions.

Some textbooks define auxiliary views by the primary dimension that is featured in the auxiliary

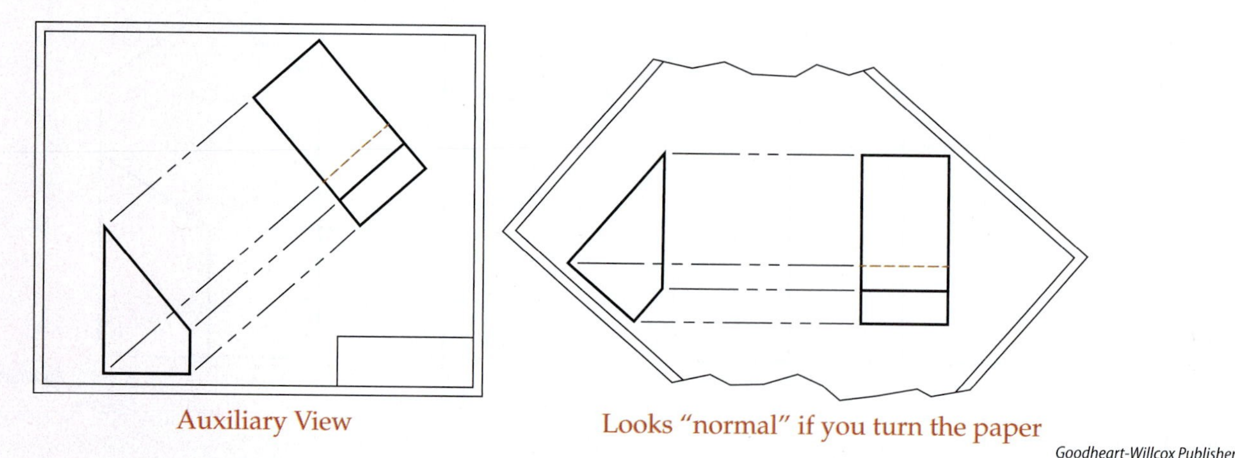

Goodheart-Willcox Publisher

Figure 7-5. Rotating the print when reading an auxiliary view can help the reader visualize the view.

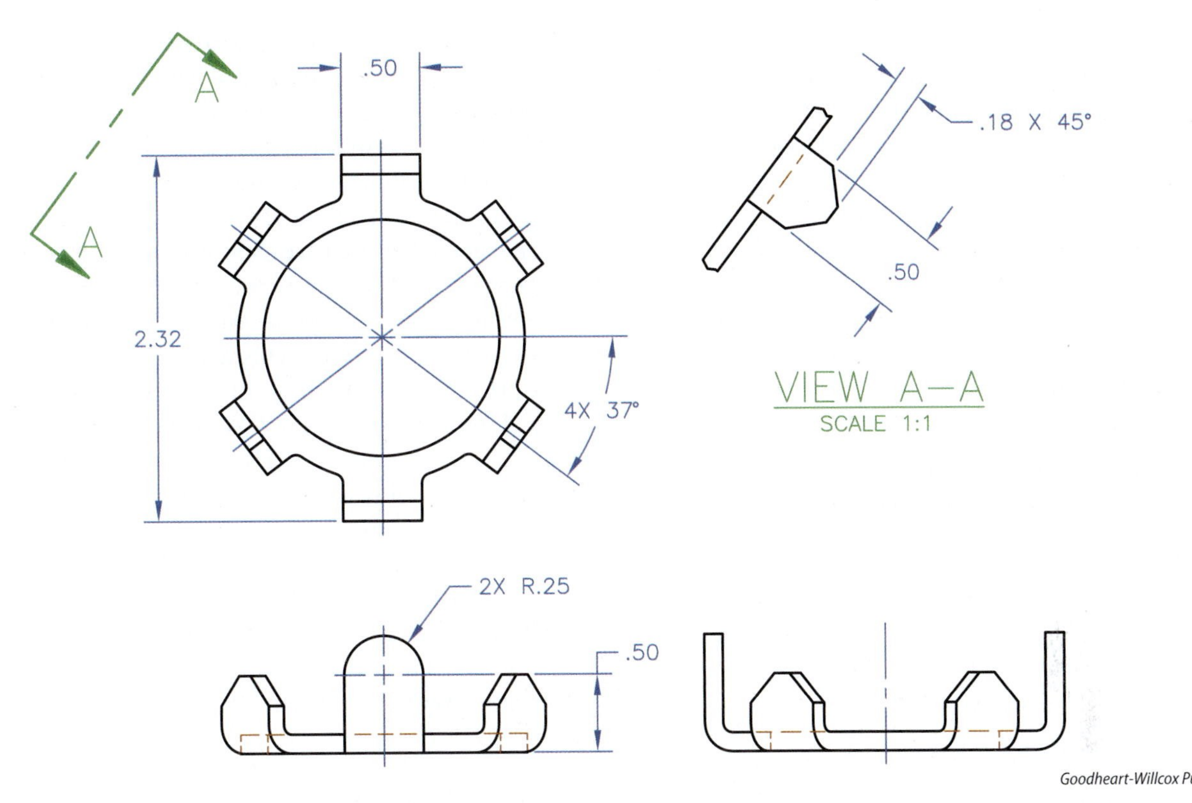

Figure 7-6. This auxiliary view was moved out of projection but still maintains the auxiliary angle represented by the viewing-plane line.

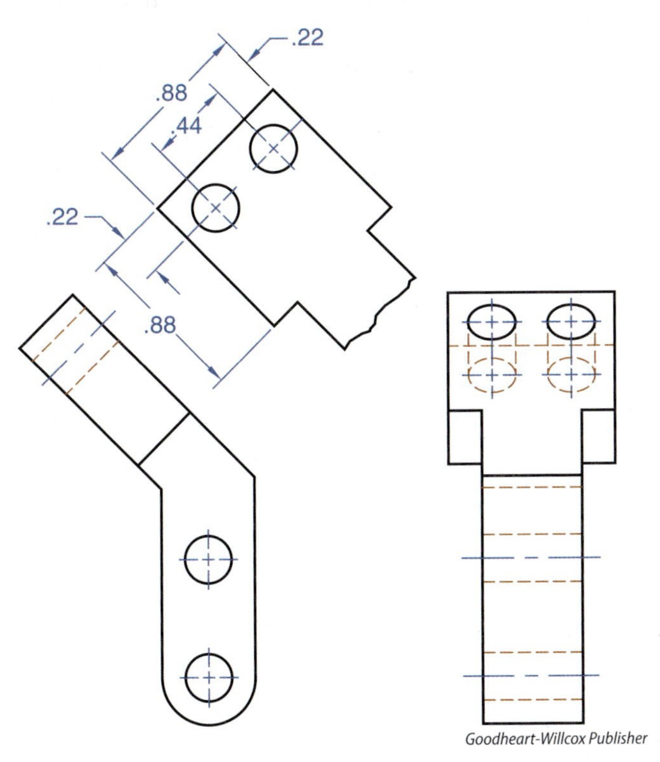

Figure 7-7. The partial auxiliary view provides a true representation of the inclined feature of this object.

view. For example, in multiview projection, the front view of an object does not feature or show the depth dimension of the object, only width and height. By the principles of projection, *any* view projected from the front view will show the depth (front-to-back dimension) because it is a 90° turn of that view. So, just as the right-side view and top view feature the depth dimension of an object, any auxiliary view projected from a front view also features the true depth of the object. Therefore, these particular auxiliary views could be identified as ***depth auxiliary views***. When manually constructing depth auxiliary views, one dimension is constructed by projection from the front view, but the depth measurement is obtained from some other view projected from the front view, such as the top or a side view, or simply by the known measurements of the object.

Under this system of defining auxiliary views, an auxiliary view projected from a top view is called a ***height auxiliary view***. Any auxiliary view projected from a right-side view is called a ***width auxiliary view***. In summary, the auxiliary view shows the dimension (height, width, or depth) that is missing from the principal view (top, front, or right side) from which it is projected. These terms are not likely to be used on a print, but they do help describe the system of auxiliary view projection. An understanding of these terms may help the print reader determine the dimensional values and how they apply to the object.

A primary auxiliary view is often projected from a principal view that shows the edge view of an inclined surface. However, if the surface is oblique, then it is not only inclined, but also rotated. Creating a view that shows the true size and shape of an oblique surface first requires the drafter to create a primary auxiliary view that shows the oblique surface as a line, or edge view. After this, a ***secondary auxiliary view*** can be projected from the primary auxiliary view, showing the true size and shape of the oblique surface. See **Figure 7-8.**

Computer-Generated Auxiliary Views

As with the principal orthographic views, 3D CAD programs can easily generate an auxiliary view from a model. In **Figure 7-9**, the auxiliary view was projected from the front view by selecting the short edge line parallel with the angle of projection. As discussed in previous units, while these views can be quickly generated from a model, they often require some adjusting for a clearer result. For example, if a partial primary auxiliary view was desired in **Figure 7-9**, it may require significant adjusting, depending on the software. Also, to clarify the rounded edges of the model in **Figure 7-9**, the visibility of some tangent elements was turned on, while other tangency elements were turned off. The projection lines connecting the front view with the auxiliary view are optional.

Three-dimensional CAD tools have considerably increased the designer's ability to communicate with auxiliary views. What used to be an intermediate or advanced concept is now as easy to create as any principal view. Print readers should expect to see more auxiliary views to help describe the shape of parts and features. As software tools improve, the ability to communicate complex designs to the print reader will continue to be enhanced.

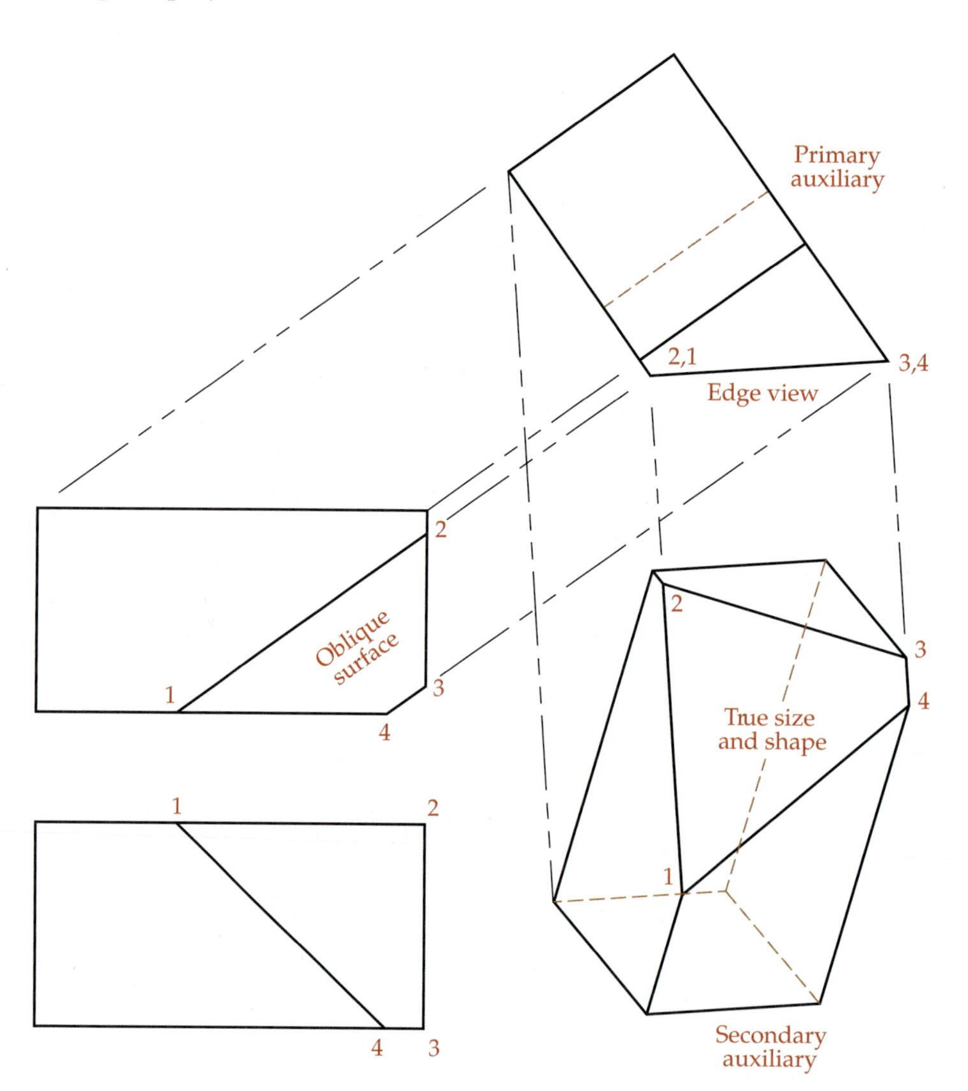

Figure 7-8. Secondary auxiliary views are necessary to show the true size and shape of an oblique surface.

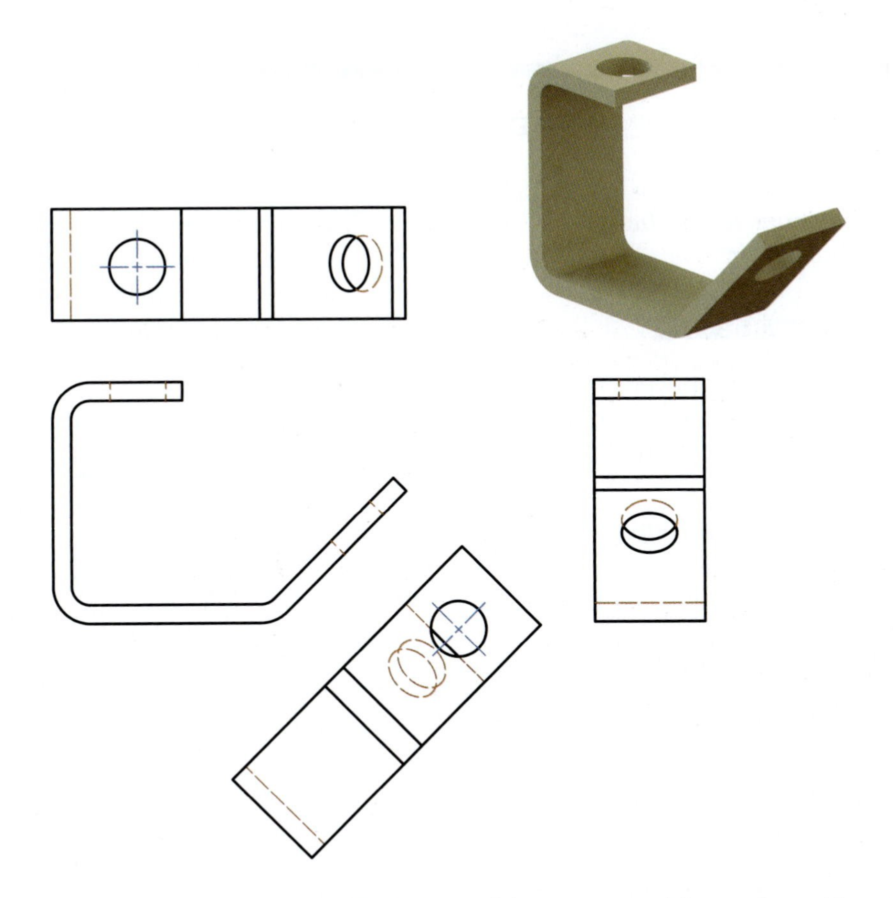

Goodheart-Willcox Publisher

Figure 7-9. Auxiliary views created directly from the CAD model are projected from selected features within the principal views. In this example, the visibility of the tangent element lines and hidden lines are turned off in some cases, while others have been left visible. The automatic center lines for the true-shape holes were created, but the center lines for the foreshortened holes were not. The drafter will need to finish some of the detailing.

Summary

- Auxiliary views are those views projected at angles other than the six principal views.
- Auxiliary views are created to show the true size and shape of inclined surfaces, or to show a dihedral angle where the intersecting surfaces are inclined to the principal viewing direction.
- Auxiliary projection planes are strategically placed parallel with an inclined surface or perpendicular to the shared edge of two surfaces, so that the orthographic projection features the true size and shape of the surface or the true dihedral angle between the two surfaces.
- In standard manuals and textbooks, reference lines that help establish distances and projection relationships are often shown between the principal and auxiliary views.
- The visualization of an auxiliary view can be aided with corner numbering techniques, mental rotation techniques, and paper turning to orient the views in a more regular fashion.
- Primary auxiliary views are projected from, and therefore adjacent to, regular views and can be defined as height, depth, or width auxiliary views.
- To feature the true size and shape of an oblique surface, a secondary auxiliary view is projected from a primary auxiliary view that features the edge view of the surface.
- Computer-generated auxiliary views are easily generated from the model but may need to be adjusted for clarity.

Unit Review

Name ______________________ Date ____________ Class ____________

Answer the following questions using the information provided in this unit.

Know and Understand

____ 1. *True or False?* The primary purpose of auxiliary views is to show the less common views of the object, such as the rear view and the bottom view.

____ 2. Of the following, which is a common reason to use an auxiliary view?
- A. To show a turned and tilted view of a normal surface
- B. To show the true size and shape of an inclined surface
- C. To show the object as if it has been cut in half
- D. When the title block gets in the way of a normal right-side view

____ 3. *True or False?* The purpose for a particular auxiliary view will determine the angle that is chosen for it.

____ 4. Of the following, to which principal projection plane can an auxiliary projection plane be hinged?
- A. Frontal
- B. Horizontal
- C. Profile
- D. Any of the above

____ 5. Choose the *true* statement about reference lines.
- A. Reference lines should be created with the same dash style as hidden lines.
- B. A reference line represents the "hinge," or folding line, of two projection planes.
- C. Reference lines should always be shown and labeled in industry drawings.
- D. A reference line will be perpendicular to the edge view of the inclined surface.

____ 6. *True or False?* A dihedral angle is the angle formed between two surfaces, and is only seen in the "point view" of the intersection between the two surfaces.

____ 7. For the auxiliary view, what orientation does the viewer's line of sight have with the inclined surface being featured?
- A. Perpendicular
- B. Parallel
- C. Tangent
- D. Inclined

____ 8. *True or False?* An auxiliary view is still a 90° "turn and tilt" of the object with respect to the adjacent view.

____ 9. *True or False?* If a viewing-plane line is used to indicate a direction for the auxiliary view, the result will be a partial view, but oriented in the same manner as the principal views, and not the auxiliary orientation.

Match the following four auxiliary view terms to the descriptions. Answers are used only once.

____ 10. Width

____ 11. Height

____ 12. Secondary

____ 13. Depth

- A. A primary auxiliary projected from a regular top view
- B. An auxiliary projected from a primary auxiliary view
- C. A primary auxiliary projected from a regular side view
- D. A primary auxiliary projected from a regular front view

____ 14. *True or False?* If the secondary auxiliary view features the true size and shape of a surface, the primary auxiliary view features the edge view of the surface.

____ 15. *True or False?* Due to the angles, a computer-generated auxiliary view requires the CAD operator to create the view manually, unlike regular views, which are generated automatically.

Critical Thinking

1. Describe some reasons an auxiliary view might be needed on the print for a part.

2. Why might you be likely to see more auxiliary views in prints made today (and in the future) than in older prints?

Apply and Analyze

Name ______________________ Date ______________ Class ______________

Review Activity 7-1

Examine the drawing on the next page. Given a complete front view, a partial top view, and two partial auxiliary views, answer the following questions.

1. Which of the three dimension terms—height, width, or depth—is represented by each of the following dimensions?

   A. ______________________________

   B. ______________________________

   C. ______________________________

   D. ______________________________

   E. ______________________________

2. If a regular right-side view would have been shown to the right of the front view, what geometric shape would describe the appearance of hole H?

   ____

For questions 3–10, place a check (or X) next to T for top, F for front, or A for auxiliary to indicate which view or views meet the stated criteria.

3. Shows the true height of hole F

   T _____ F _____ A _____

4. "Looks" straight through hole H

   T _____ F _____ A _____

5. "Looks" straight through hole G

   T _____ F _____ A _____

6. Shows the axis-to-axis distance from hole F to hole G

   T _____ F _____ A _____

7. Shows a depth location distance for hole J

   T _____ F _____ A _____

8. Shows the true shape of the inclined surfaces

   T _____ F _____ A _____

9. Shows the true angle between the inclined surfaces and the horizontal surfaces

   T _____ F _____ A _____

10. Shows the depth location difference between hole F and hole G, if any

    T _____ F _____ A _____

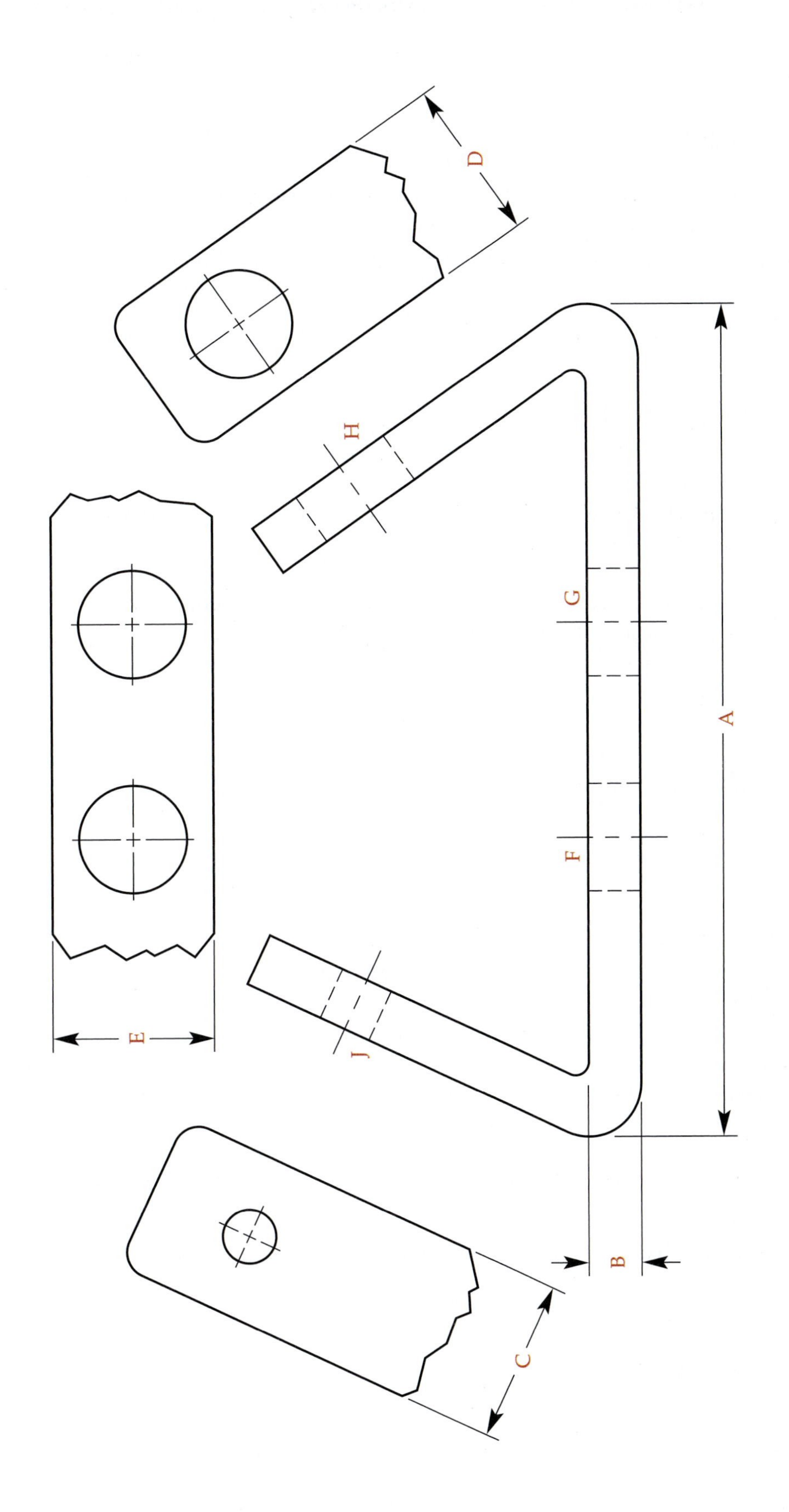
D
H
E
G
F
A
J
B
C

Apply and Analyze

Name ____________________ Date ____________ Class ____________

Review Activity 7-2

Examine the drawing and answer the following questions. This drawing features threaded holes and machining specifications that have not yet been covered, but it provides a good example of auxiliary views.

1. For the sake of this exercise, if the section view of this drawing is considered the front view, what other three views are shown? Include the word *partial* if applicable.

   ____

____ 2. *True or False?* There are some hidden lines that were omitted from the right-side view.

3. In the auxiliary view, what is the name of the jagged and irregular line that indicates this is a partial view?

   ____

4. What two dimensions are exhibited in the partial primary auxiliary view?

   ____

____ 5. Of the following, which is a *true* statement about the partial auxiliary view?
   A. The view was projected from the right-side view.
   B. The view shows the true size and shape of an inclined surface.
   C. The main purpose of the view is to show a true dihedral angle.
   D. The 3.125″ dimension is *not* a true dimension in this view.

6. If the auxiliary view were to be labeled by the dimension (height, width, or depth) it features that is *not* featured in the front view, what dimension would that be?

   ____

7. What geometric shape would describe the inclined hole in the right-side view?

   ____

____ 8. If the drawing page is turned in such a way so the longest center line of the auxiliary view is horizontal, the front view could still be called the front. If so, the auxiliary view could now be called the right-side view. In this case, what would the former right-side view now be called?
   A. Secondary auxiliary view
   B. Primary auxiliary view
   C. Left-side view
   D. Partial width auxiliary view

____ 9. *True or False?* The auxiliary view is oriented to provide the viewer with the *point view* of the axis of the .453″ diameter hole.

10. List four different lines from the alphabet of lines that are featured in the auxiliary view.

   ____

   ____

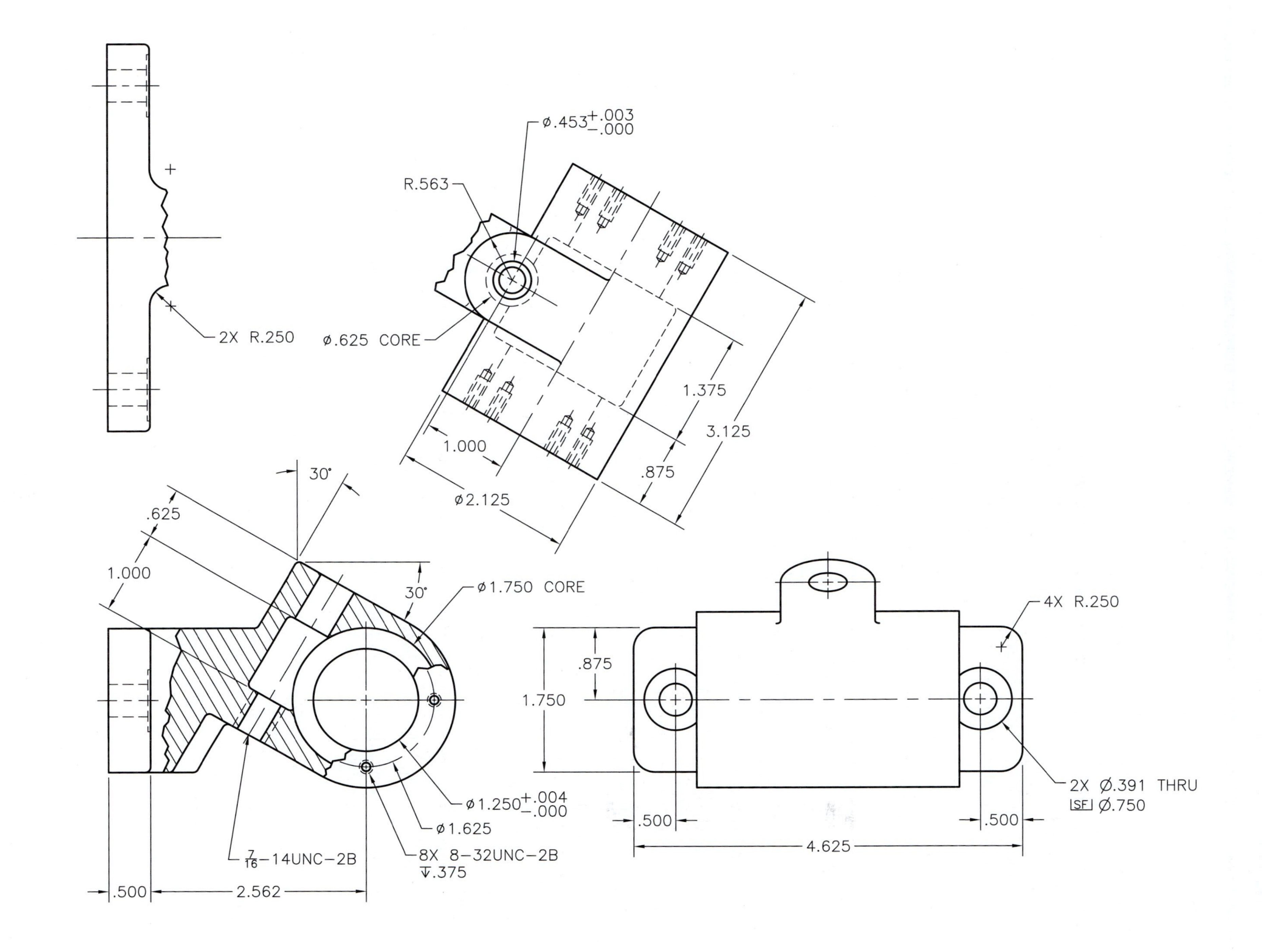

2X R.250
Ø.453 +.003 -.000
R.563
Ø.625 CORE
1.375
3.125
1.000
.875
Ø2.125
30°
.625
1.000
30°
Ø1.750 CORE
4X R.250
.875
1.750
2X Ø.391 THRU
SF Ø.750
Ø1.250 +.004 -.000
Ø1.625
7/16-14UNC-2B
8X 8-32UNC-2B
↧.375
.500
.500
.500
4.625
2.562

Apply and Analyze

Name ______________________ Date ______________ Class ______________

Industry Print Exercise 7-1

Refer to the print PR 7-1 and answer the questions below.

1. For this print, is SECTION A-A an auxiliary view? ______________________
2. Notice the view that features the circular view of the hole through this part. Not counting Detail A, how many auxiliary views are projected from that view?

   ____
3. Are there any center lines shown in the auxiliary views? ______________________
4. If the lower-left auxiliary view is considered a *depth auxiliary*, which value shown in the view is a depth dimension, the .63 measurement or the .57 measurement?

   ____
5. In the view identified as SECTION A-A, how many depth dimensions are given? ______________
6. Detail A is an enlarged partial view. Is it considered to be an auxiliary view also?______________

Review questions based on previous units:

7. What is the drawing/part number? ______________________
8. What is the name of this part? ______________________
9. Is there a cutting-plane line on this drawing?______________________
10. Is the view located in the lower-left area of the drawing a section view? ______________
11. At what scale is the enlarged Detail A? ______________________
12. On what size paper was the original drawing created? ______________________
13. Do the visible lines of this drawing appear to be two times thicker than the other lines?__________
14. As you visualize this part, does it have any holes in it?______________________
15. In this drawing, the material is specified outside of the title block as a note. In inches, how thick is the material used to make this part?

    ____

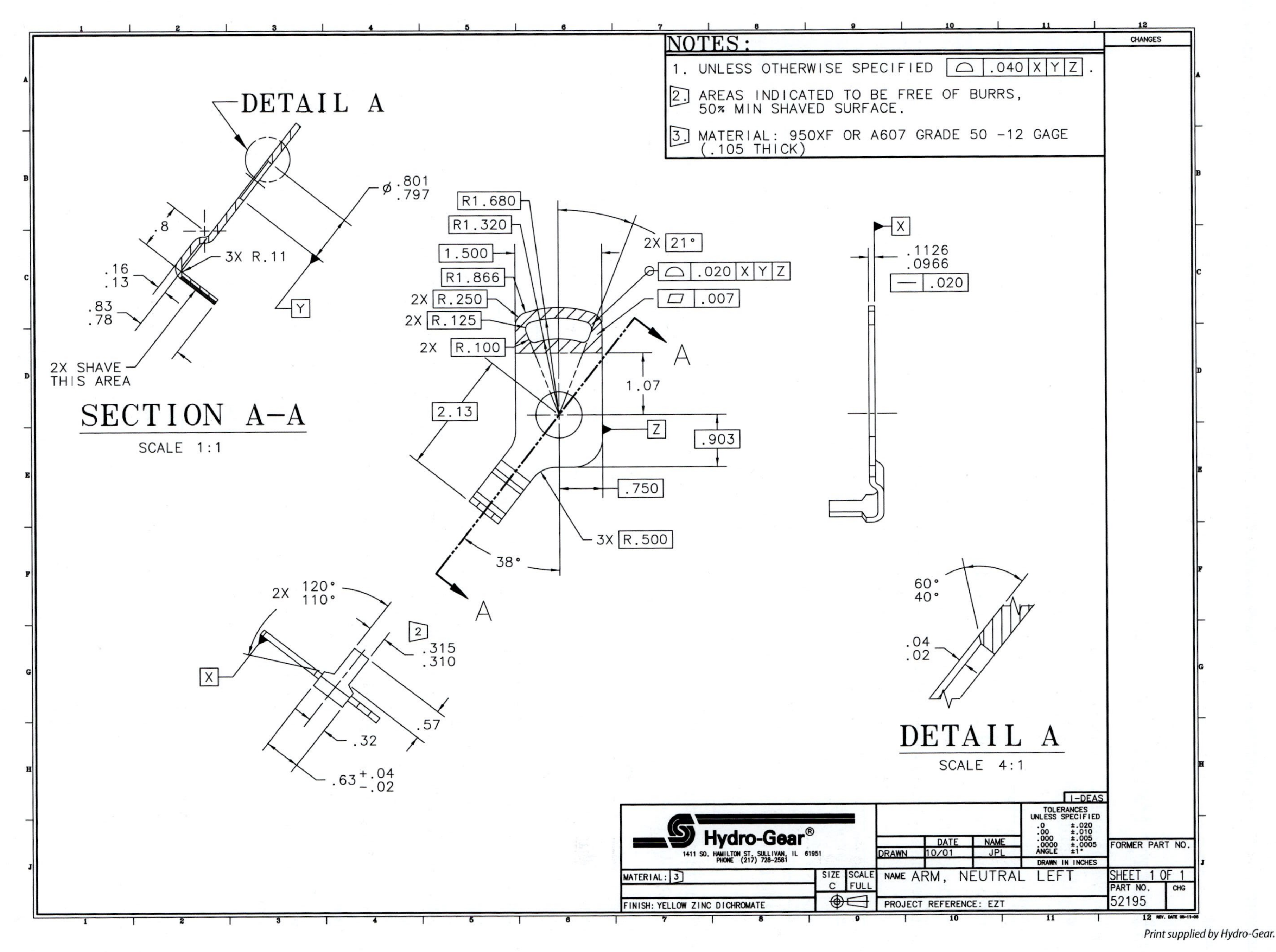

Print supplied by Hydro-Gear.

PR 7-1. Neutral Left Arm

Apply and Analyze

Name ______________________ Date ______________ Class ______________

Industry Print Exercise 7-2

Refer to the print PR 7-2 and answer the questions below.

1. For this print, the auxiliary view is also identified as _____.

   ____

2. How many additional auxiliary views, if any, are there on this print? ______________

3. In the lowermost view, the true thickness of the material is shown and is specified by the toleranced dimension with a range of .046″–.053″. Is this true thickness also featured in the auxiliary view?

   ____

4. Is the purpose of the auxiliary view to show the true size and shape of an inclined surface?

   ____

5. If the main view could be considered a front view, what name could be given to the auxiliary view?

   ____

6. If the upper-left view is considered the top view, what angle does cutting-plane A-A form with the horizontal plane?

   ____

Review questions based on previous units:

7. Does this drawing feature extension lines? ______________
8. Are any of the views at a different scale than others? ______________
9. What material is specified in the title block? ______________
10. The drawing layout has a place for changes (revisions) to be recorded. Are there any changes recorded on this print? ______________
11. What is the name of the company that owns this print? ______________
12. What are the initials of the person who drew this print? ______________
13. What is the name of this part? ______________
14. This part features a hole that has vertical and horizontal sides (not counting the little rectangular notch). Is the larger hole of that feature dimensioned as a square, or as a rectangle?

    ____

15. If the main view is considered the front view, what is the overall width of the part?

    ____

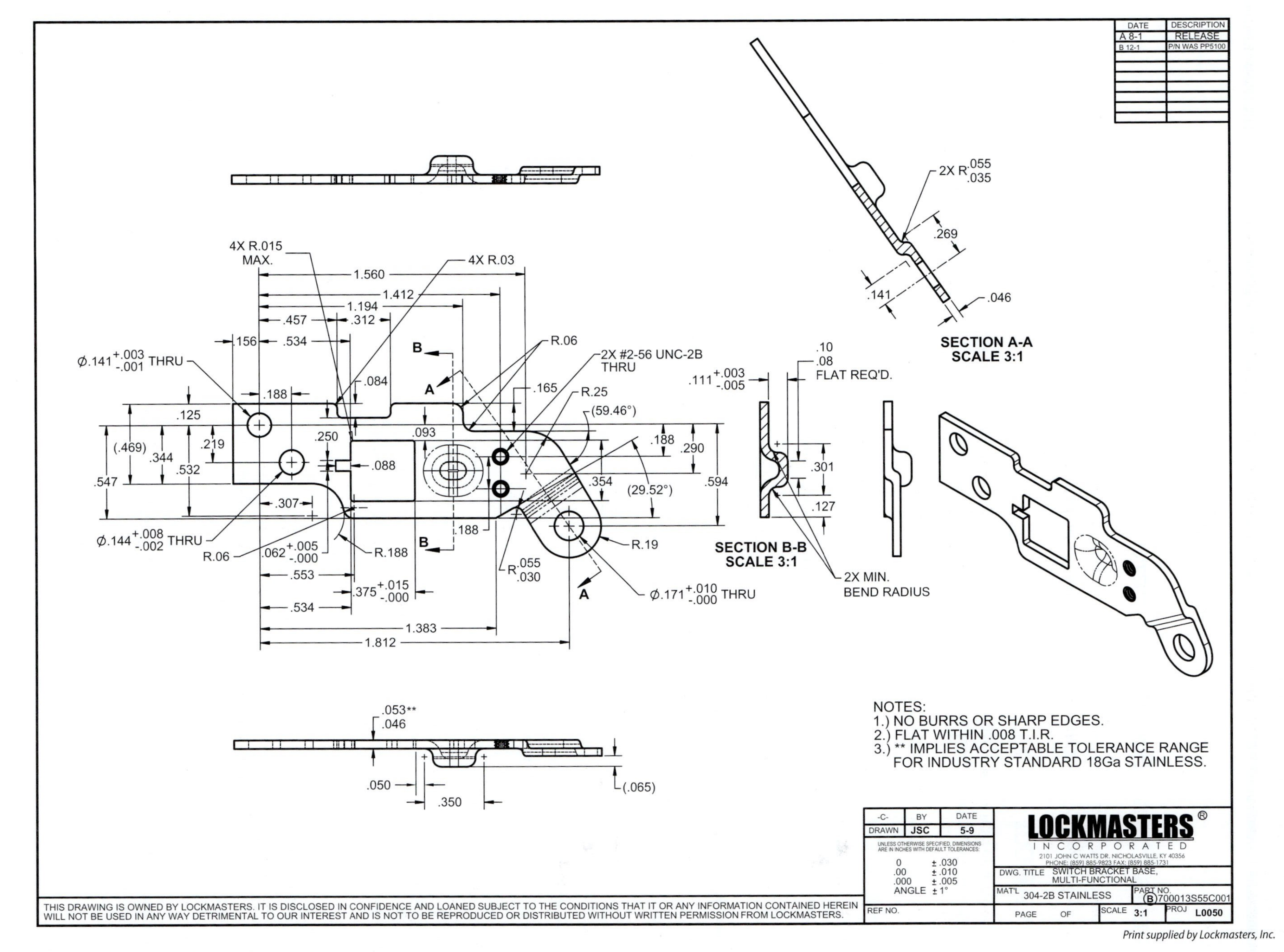

Print supplied by Lockmasters, Inc.

PR 7-2. Multifunctional Switch Bracket Base

Apply and Analyze

Name ______________________ Date ____________ Class ____________

Bonus Print Reading Exercises

The following questions are based on the various bonus prints located in the folder that accompanies this textbook. Refer to the print indicated, evaluate the print, and answer the question.

Print AP-001

1. If an auxiliary view were used to show a view looking into the hole that is inclined 15°, would the auxiliary view be projected above the left side view, above the section view, or below the section view?

____

Print AP-002

2. On this print, there is a stem and knob assembly at an angle to the main view. If an auxiliary view were used to show the round knob as a true circle, could it be projected off this same view?

____

Print AP-006

3. The major diameter (100 mm) of this part is interrupted by a flat surface on top. Is an auxiliary view required to show the true size and shape of that flat surface?

____

Print AP-007

4. How many auxiliary views are found on this print?

____

5. Is the SECTION B-B auxiliary view of this print in projection or removed out of projection?

____

6. At what apparent angle, relative to a horizontal line, is this auxiliary view projected?

____

Print AP-012

7. Based on a note that was added during revision B, what is the name of the 2″ portion of this part featured in the upper-right auxiliary view?

____

8. One of the auxiliary views in this print is jogged out of position so it can fit on the page. What type of line is used to show the jogged alignment?

____

9. In the two auxiliary views, the break lines indicate the views are also _____ views.

____

10. One auxiliary view shows a hole in a true circular position. How many other views include a projection of that hole as an elliptical shape?

____

Notes

UNIT 8

Screw Thread Representation

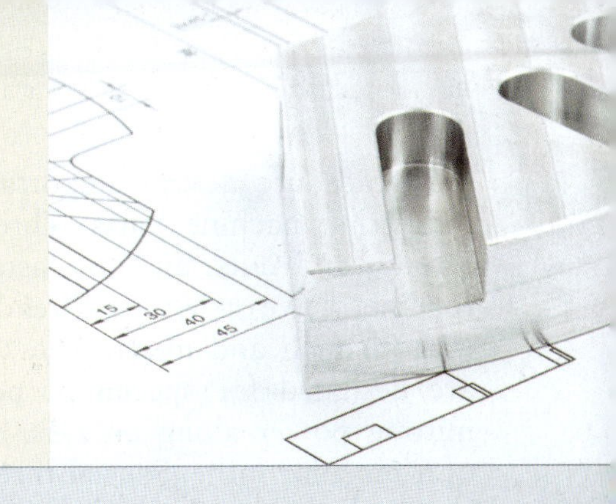

LEARNING OBJECTIVES

After completing this unit, you will be able to:

- Define terms related to screw threads.
- Identify common screw thread forms.
- Describe three methods for representing screw threads in industrial prints.
- Describe the characteristics of the four common Unified thread series.
- Describe the nature of thread classes in industrial applications.
- Explain the nature of multiple threads with respect to appearance and lead.
- Explain the nature of left-hand threads with respect to appearance and use.
- Explain the different parts of a screw thread specification or callout.
- Discuss the differences between metric threads and inch threads.
- Identify standard pipe thread representation and designations.
- Describe unique characteristics of computer-generated threads and their respective representation in industrial prints.

TECHNICAL TERMS

crest
detailed representation
external thread
flank
flank angle
internal thread
lead
left-hand thread
major diameter
minor diameter
multiple-start thread
nominal size
pitch (P)
pitch diameter
right-hand thread
root
schematic representation
screw thread
simplified representation
single-start thread
thread
thread angle
thread class
thread form
Unified Thread Standard (UTS)

A *screw thread* or *thread* is a ridge of a particular shape, often a V, which follows a helical path around a cylindrical surface. A helix can be seen in common items such as a barber pole or candy cane. A thread can also be compared to wrapping a rope around a cylinder, keeping each coil snug against the cylinder and the previous coil. By definition, a helix is formed by a point uniformly progressing along a path with a constant orientation to the axis of a cone or cylinder. Threads can be formed by hand, ground or turned on a lathe, or formed using thread-cutting machines.

Threads are a very important feature of many industrial machine parts. Threads used for bolts, screws, and nuts, and for fastening machine parts together have a V-type thread form standardized in both inch and metric sizes. Other thread forms are designed for special purposes, such as transmitting power along an axis, as in the lead screw of a machine lathe. Special threads are also used to hold light bulbs securely in place, or for bottle caps on aluminum and plastic bottles. Pipe threads are machined according to standards and are designed for a variety of specific functions.

Thread Terms and Definitions

ASME B1.7, *Screw Threads: Nomenclature, Definitions, and Letter Symbols*, is the best resource for screw thread terms and definitions. A few of the terms are given here. Refer to **Figure 8-1** as you read the following definitions.

- An ***external thread*** is a thread on the outside of a shaft. The threads on a bolt are external threads.
- An ***internal thread*** is a thread on the inside of a hole. The threads in a nut are internal threads.
- The ***crest*** is the top edge or surface of a ridge.
- The ***root*** is the bottom or "bottomland" between two ridges.
- The ***flank*** is the surface joining the crest and root. It could be considered the "side" of the ridge or groove.
- The ***flank angle*** is the angle formed between the flank and a line perpendicular to the axis. For example, a sharp-V groove has a flank angle of 30°.
- The ***thread angle*** is the angle formed between two adjacent flanks. For example, a sharp-V groove has a thread angle of 60°.
- The ***major diameter*** is the largest diameter of a thread, measured crest-to-crest across the diameter of the cylinder.
- The ***minor diameter*** is the smallest diameter of a thread, measured root-to-root across the diameter of the cylinder.
- The ***pitch diameter*** is an imaginary diameter where the width of the thread ridge is equal to the width of the groove between ridges. This is important for designing the fit of threaded parts.
- The ***pitch (P)*** is the distance from one point on a screw thread to the same point on the next thread. American inch threads are specified by the number of threads per inch. If there are 16 threads per inch, the pitch is 1/16″. The pitch for metric threads is specified in the thread note.
- The ***lead*** is how far a thread advances when the threaded part is rotated one complete revolution. For most threads, the lead and pitch are equivalent. For multiple threads, the lead is double the pitch for double threads or triple the pitch for triple threads. Multiple threads are covered later in this unit.

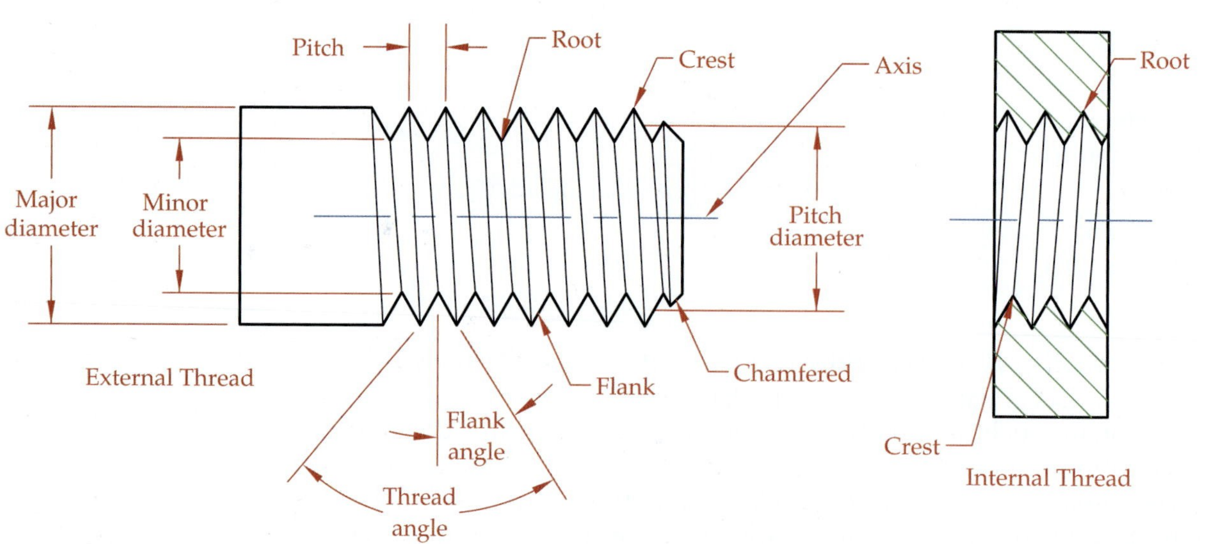

Figure 8-1. A few of the terms associated with screw threads are illustrated here.

Screw Thread Forms

The *thread form* describes the shape of the ridges (or grooves) that form the threads, **Figure 8-2**. Most thread forms used in industry today are based on approved American standards. Historically, thread standardization originated in the nineteenth century. By the mid-twentieth century, a movement to resolve differences and improve uniformity between American National threads in the United States and Whitworth threads in Great Britain led to a unified system of threads that is standard for inch-based applications today. While the thread forms before this were similar in nature, and based on a sharp-V pattern, they differed in thread angle and various other subtle ways. Should specifications for these older thread forms show up on current prints, the print reader should ask for a clarification on how to address those thread forms.

There are over a dozen ASME standards related to the manufacture of threads. ASME B1.1 sets forth the parameters for a ***Unified Thread Standard (UTS)*** thread form, often referred to generally as Unified threads. Unified threads are the recognized standard in the United States for straight threads used on screws, bolts, nuts, and other threaded parts. Unified threads are similar to the older American National thread form, but are easier to manufacture. As mentioned, these Unified threads have the general appearance of a sharp V with a 60° thread angle. However, the Unified thread form features a profile with crests and roots that are flattened or rounded to break off the sharpness, **Figure 8-3**.

Metric threads have the same parameters for thread form, crest and root truncation, and proportions, so the appearance on the print is identical to inch threads. Of course, the pitches and diameters are based on metric dimensions rather than fractional inch measurements. The design principles of metric threads are set forth in ASME B1.13M and ISO 68-1.

There are various uses for threads other than for fasteners. Thread forms in use for purposes other than fasteners include Acme, Stub Acme, buttress, knuckle, and square. Acme and buttress threads are used in applications that exert force, such as vises

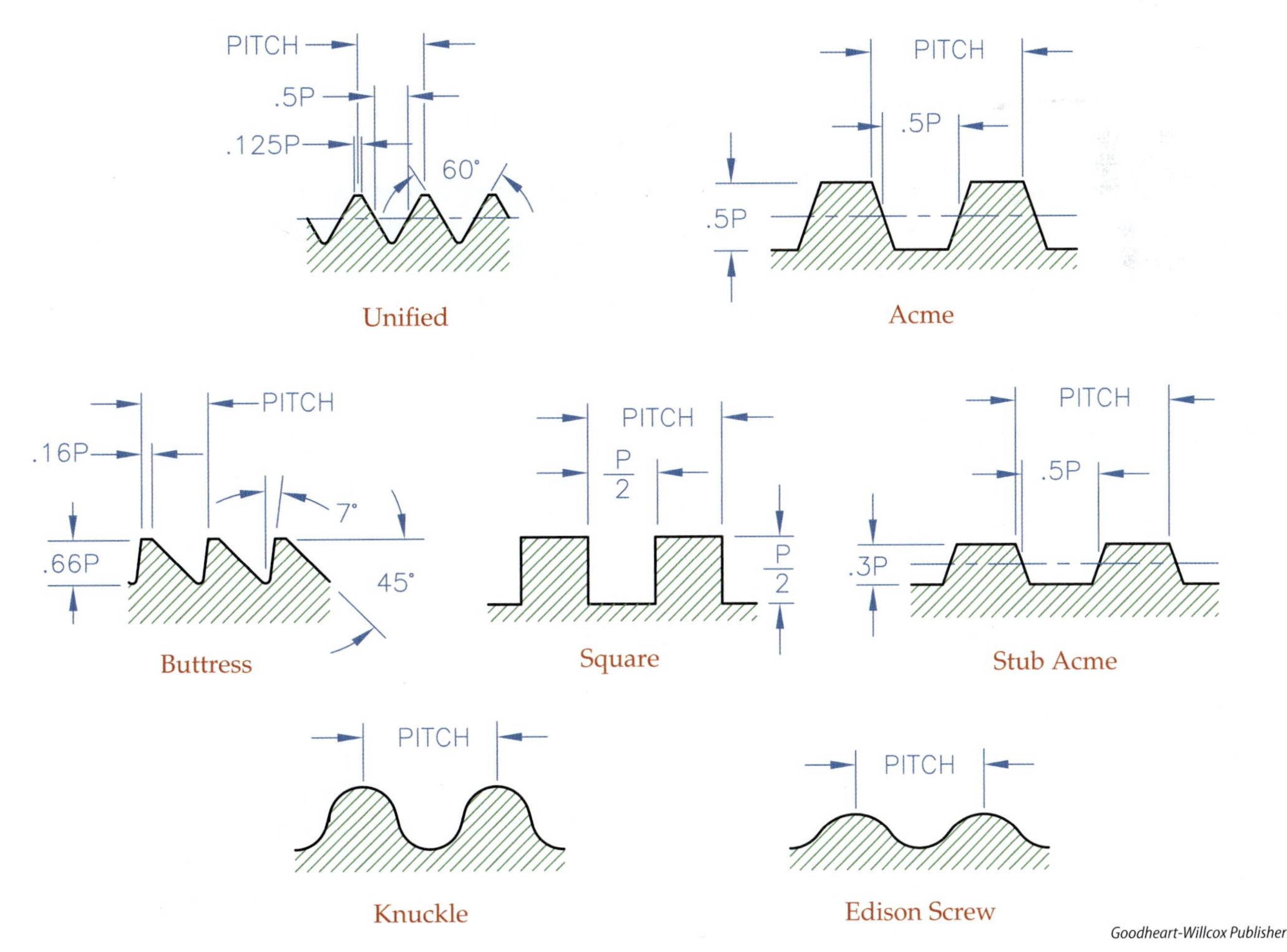

Figure 8-2. The thread form describes the shape of the ridge or groove that forms the threads.

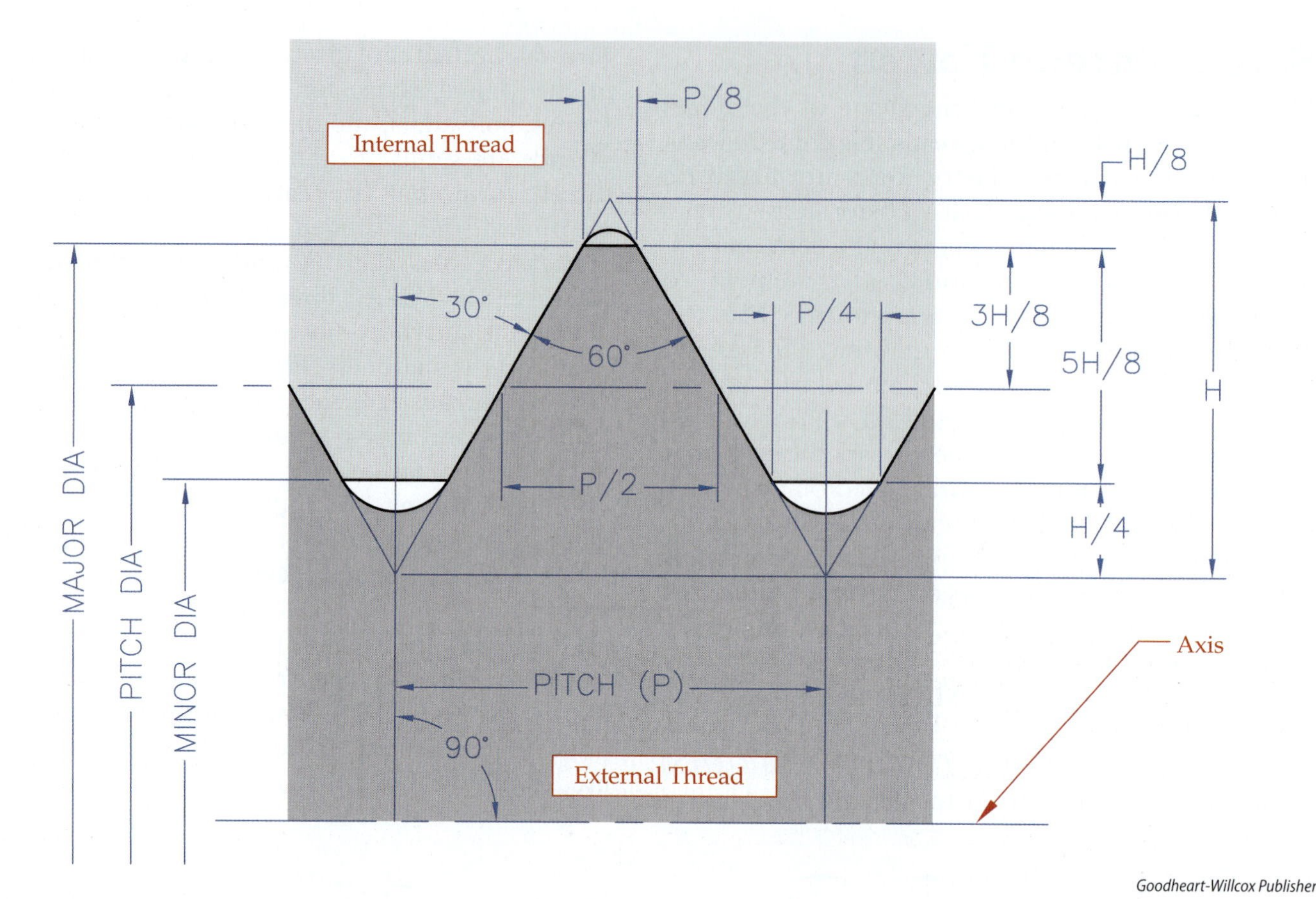

Figure 8-3. The Unified thread form features a profile with crests and roots that are flattened or rounded to break off the sharpness.

and jacks, or in equipment with a lead screw that demands high strength and accuracy. A version of knuckle threads has been standardized by the German Institute for Standardization, and features a 15° flank angle. These standard knuckle threads are similar to the Edison screw used on light bulbs, although the Edison standard is shallower. Light bulb threads are standardized by the International Electrotechnical Commission and fall into 10 categories, or sizes, such as E12 for candelabra bulbs and E26 for standard 120V lamps. In summary, each of the thread forms may or may not have an American standard. The table in **Figure 8-4** shows applicable ASME standards for some common threads.

ASME Standards for Selected Common Threads

Screw Thread Form	ASME Standard
Unified inch threads (UN & UNR)	ASME B1.1
Acme screw threads	ASME B1.5
Stub Acme screw threads	ASME B1.8
Buttress inch screw threads	ASME B1.9
Metric screw threads (M profile)	ASME B1.13M
Pipe threads, general purpose (inch)	ASME B1.20.1
Dryseal pipe threads (inch)	ASME B1.20.3
Unified inch threads (UNJ)	ASME B1.15
Metric (MJ profile)	ASME B1.21M

Goodheart-Willcox Publisher

Figure 8-4. This table shows ASME standards that apply to some common threads.

Thread Representations

A true projection view of a screw thread is very complex due to the helical shape, which projects as an irregular curve. Therefore, threads are usually represented on a print in conventional ways. In an earlier unit, conventional practice was defined as a standard way of doing something, even if this method breaks the principles on which views are based. The three conventional methods in which threads may be represented on drawings are known as the detailed, schematic, and simplified representations. See **Figure 8-5**. While the standards shown in the previous table set forth the guidelines for manufacturing threads, ASME Y14.6 establishes the standards for representing threads in industrial drawings and prints.

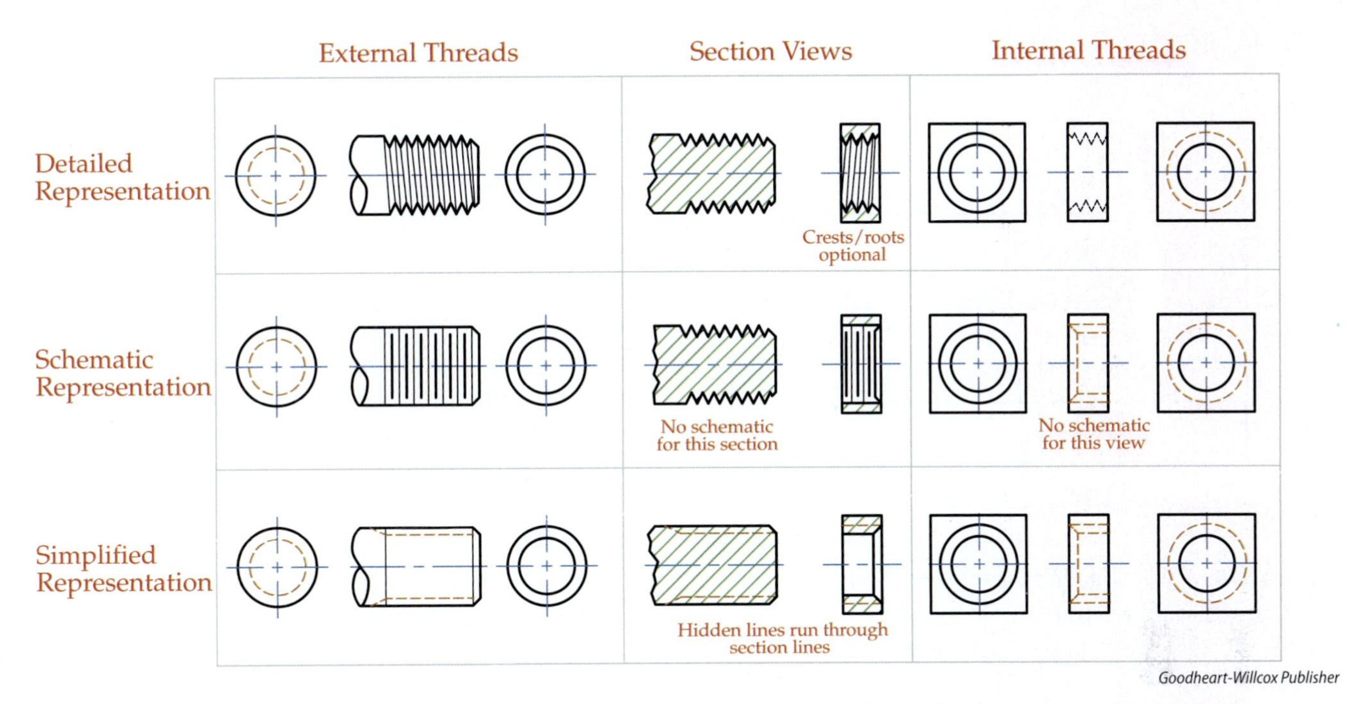

Figure 8-5. The three conventional methods for representing threads are the detailed, schematic, and simplified representations.

The ***detailed representation*** method most closely represents screw threads as they would appear in true projection. This convention is sometimes used to show the geometry of a thread form as a portion of a greatly enlarged detail on a drawing. A rule of thumb for drafters is to draw detailed threads if the major diameter appears larger than 1″ on the drawing. Under this rule, if a 3″ diameter is drawn quarter scale, the actual diameter on the drawing is .75″, so a detailed representation would usually not be applied. If the 3″ diameter thread is placed on the drawing at half scale, a detailed representation might be a better choice.

For smaller diameters and reduced-scale views of threaded features, the ***schematic representation*** and ***simplified representation*** methods are more quickly and easily drawn, are easy to read, and help describe the threaded feature. A simplified representation is perhaps the easier of these two conventions to use and has been more common in recent years. It is also recommended for assembly drawings. One disadvantage of the schematic representation is that the spacing looks good at one particular scale, but the spacing may require adjustment if the view is enlarged. With a simplified representation, this is not an issue. The ASME standard recommends that simplified and schematic representations not be mixed in the same drawing unless there is a pressing reason to do so.

Certain thread elements can be confusing on a drawing. For example, if the end of an external thread is chamfered, the resulting visible circle may coincide with the minor diameter. As a result, a visible circle will be shown instead of a hidden circle. On threaded holes that are also countersunk, two visible circles may result. If the countersink diameter is close but not exactly the same as the major diameter, the hidden circle may be omitted for the major diameter. For special conditions, an enlarged detail may be appropriate. Study **Figure 8-5** again to see the various ways the three thread representations appear.

In real life, threads do not just start or stop suddenly. For the end of a bolt, the root of the thread technically forms a spiral appearance as it intersects with the chamfered end, but this is not shown in the end view. In recent standards, a technique was introduced to show the fadeout or runout of the threads beyond the last fully cut thread in the side view, which allows dimensions to be added. On the drawing, dimensions can be applied to show the amount of external or internal surface that must have a complete, fully cut thread. Likewise, dimensions can be added to show that even partial threads must not infringe on a certain area or feature. See **Figure 8-6**.

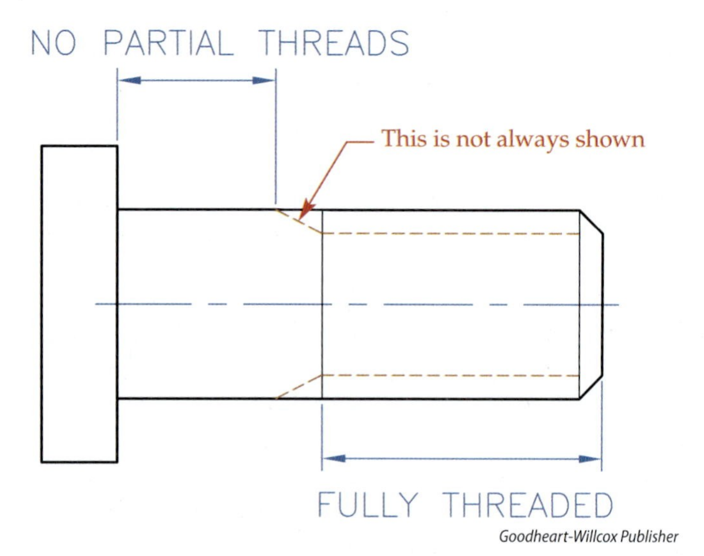

Goodheart-Willcox Publisher

Figure 8-6. Partial threads may run beyond the design length needed for full threads. Dimensions may be added to clarify the boundaries of full and partial thread requirements.

Thread Characteristics

There are several other characteristics besides thread form associated with threads. These include the thread series, the class of fit (precision level), whether or not the thread is a multiple thread, and the direction of the thread. The following sections describe these additional characteristics.

For schematic and simplified thread representations, there is no differentiation for the thread form, pitch, thread direction, or multiple-start status. In other words, the drawing looks the same regardless of the characteristics discussed in the next sections. In addition, the spacing for a schematic representation is usually just a factor of the major diameter and nothing more. Larger diameters can have more generous spacing. Some drafters may attempt to space the schematic lines based on the pitch, but this is not required, nor practical in many cases.

Thread Series

There are four main thread series of inch-based Unified screw threads: coarse, fine, extra fine, and constant pitch. See **Figure 8-7** for standard values such as basic diameter, threads per inch, and recommended tap drill size. Notice that screw numbers are designated for diameters smaller than 1/4″. Each of the following thread series has its own designated number of threads per inch:

- **Unified coarse.** Designated UNC, this series of threads is used for nuts, bolts, screws, and general uses where fine threads are not required.

Nominal Diameter	Basic Diameter	Coarse UNC		Fine UNF		Extra Fine UNEF	
		Thds per in	Tap drill dia	Thds per in	Tap drill dia	Thds per in	Tap drill dia
0	.060			80	.0469		
1	.073	64	No.53	72	No.53		
2	.086	56	No.50	64	No.50		
3	.099	48	No.47	56	No.45		
4	.112	40	No.43	48	No.42		
5	.125	40	No.38	44	No.37		
6	.138	32	No.36	40	No.33		
8	.164	32	No.29	36	No.29		
10	.190	24	No.25	32	No.21		
12	.216	24	No.16	28	No.14	32	No.13
1/4	.250	20	No.7	28	No.3	32	.2189
5/16	.3125	18	F	24	I	32	.2813
3/8	.375	16	.3125	24	Q	32	.3438
7/16	.4375	14	U	20	.3906	28	.4062
1/2	.500	13	.4219	20	.4531	28	.4688
9/16	.5625	12	.4844	18	.5156	24	.5156
5/8	.625	11	.5313	18	.5781	24	.5781
11/16	.6875	...	...	...	...	24	.6406
3/4	.750	10	.6563	16	.6875	20	.7031
13/16	.8125	...	...	...	...	20	.7656
7/8	.875	9	.7656	14	.8125	20	.8281
15/16	.9375	...	...	...	...	20	.8906

Nominal Diameter	Basic Diameter	Coarse UNC		Fine UNF		Extra Fine UNEF	
		Thds per in	Tap drill dia	Thds per in	Tap drill dia	Thds per in	Tap drill dia
1	1.000	8	.875	12	.922	20	.953
1–1/16	1.063	...	...	...	...	18	1.000
1–1/8	1.125	7	.904	12	1.046	18	1.070
1–3/16	1.188	...	...	...	...	18	1.141
1–1/4	1.250	7	1.109	12	1.172	18	1.188
1–5/16	1.313	...	...	...	...	18	1.266
1–3/8	1.375	6	1.219	12	1.297	18	1.313
1–7/16	1.438	...	...	...	...	18	1.375
1–1/2	1.500	6	1.344	12	1.422	18	1.438
1–9/16	1.563	...	...	...	...	18	1.500
1–5/8	1.625	...	...	...	...	18	1.563
1–11/16	1.688	...	...	...	...	18	1.625
1–3/4	1.750	5	1.563	...	...	...	...
2	2.000	4.5	1.781	...	...	...	...
2–1/4	2.250	4.5	2.031	...	...	...	...
2–1/2	2.500	4	2.250	...	...	...	...
2–3/4	2.750	4	2.500	...	...	...	...
3	3.000	4	2.750	...	...	...	...
3–1/4	3.250	4	...	...	...	...	...
3–1/2	3.500	4	...	...	...	...	...
3–3/4	3.750	4	...	...	...	...	...
4	4.000	4	...	...	...	...	...

Goodheart-Willcox Publisher

Figure 8-7. Standard thread specifications for Unified coarse, fine, and extra-fine threads are illustrated in this table. See Appendix D for more tables detailing screw thread series.

- **Unified fine.** Designated UNF, this series of threads is used where the length of the threaded engagement is short and where a small lead is desired.
- **Unified extra fine.** Designated UNEF, this series of threads is used for very short lengths of thread engagement and for thin-wall tubes, nuts, ferrules, and couplings.
- **Unified constant-pitch.** Designated UN with the number of threads per inch preceding the designation, such as 8UN, this series of threads is for special purposes, such as high-pressure applications. Constant-pitch threads are also used for large diameters where the other thread series do not meet the requirements.

Additional Unified threads have joined the standards in recent years. Technically, Unified threads are defined as having a "flat root contour," but thread tool wear results in some rounding. A rounded root actually results in a stronger part. For these reasons, there are a couple of Unified designations that specify the nature of the rounded root. The UNR designation indicates a root radius of $0.10825P$, but this can only be applied to an external thread. The UNJ designation is another rounded-root form originating in the military standards. It uses a slightly larger radius ($0.15011P$) as the minimum mandatory radius. In addition to these, a designation of UNS indicates a Unified form that is special or nonstandard.

Thread Classes

After classification by form and series, threads are further classified by manufacturing tolerance. This is a general way of specifying the level of precision or the quality of the surfaces involved. These ***thread classes*** are 1A, 2A, and 3A for external threads and 1B, 2B, and 3B for internal threads. On some older drawings, classes 2 and 3 may appear without a letter designation, so the print reader may wish to ask for clarification in those instances.

Classes 1A and 1B replace the older American Standard class 1. They are used in applications requiring minimum binding. The tolerance of these classes allows for frequent and quick assembly or disassembly of parts and represents the loosest class of fit.

Classes 2A and 2B are threads with tighter tolerances. They are used for general purposes such as for nuts, bolts, screws, and normal applications by mass-production industries.

Classes 3A and 3B are threads with very stringent and close tolerances. They are used for applications in industries requiring tighter tolerances than the preceding classes of 1A and 1B or 2A and 2B.

Multiple Threads

The vast majority of screw threads are comprised of a single ridge wrapped around the cylindrical surface, and are technically referred to as ***single-start threads***. Threads can also be manufactured with two or three ridges running side by side. These have been commonly referred to as double threads or triple threads, but more recently are referred to as ***multiple-start threads***. Double threads can be called *two-start threads*, and triple threads can be called *three-start threads*. Due to the increased slope angle of the thread, double or triple threads have less holding power.

Conceptually, it may help to think of a rope wrapped around a wooden rod. If one rope is wrapped, it is like a single thread. If two ropes are wrapped, they are like double threads. Three ropes wrapped around the wooden rod are like triple threads. Looking at three ropes wrapped around a rod, it may appear as one rope, but the angle of progression is steeper than it would be if one rope is wrapped around the rod. This steeper angle corresponds to a different lead, defined earlier as the distance a thread advances when the threaded part is rotated one complete revolution.

The print reader should not expect to have a visual indication of multiple threads. If the drafter properly applies a detailed representation, the slope of the crest line is inclined 1/2 pitch in the side view of a single-start thread. On the "back side" of the cylinder, the crest slopes the other 1/2 pitch. Study the hidden lines in **Figure 8-8** that represent the crest lines on the back side of the cylinder. A detailed double thread should indicate a crest directly above a crest. The slope of a crest line is then one pitch. On the back side of the cylinder, the crest line slopes an additional pitch. Remember, the *lead* of a double thread, or two-start thread, is twice the pitch. One revolution of the threaded cylinder advances two pitches because there are two ridges. Likewise, a triple thread, or three-start thread, advances three pitches. For a detailed representation, the slope of a crest line on the side view is tilted 1.5 pitches.

With schematic and simplified representations, there is no difference in the graphic symbolization, regardless of the thread characteristics. For those cases, the print reader should rely solely on the thread specification note, and *not* on the visual representation, as a means of determining if the thread is single, double, or triple.

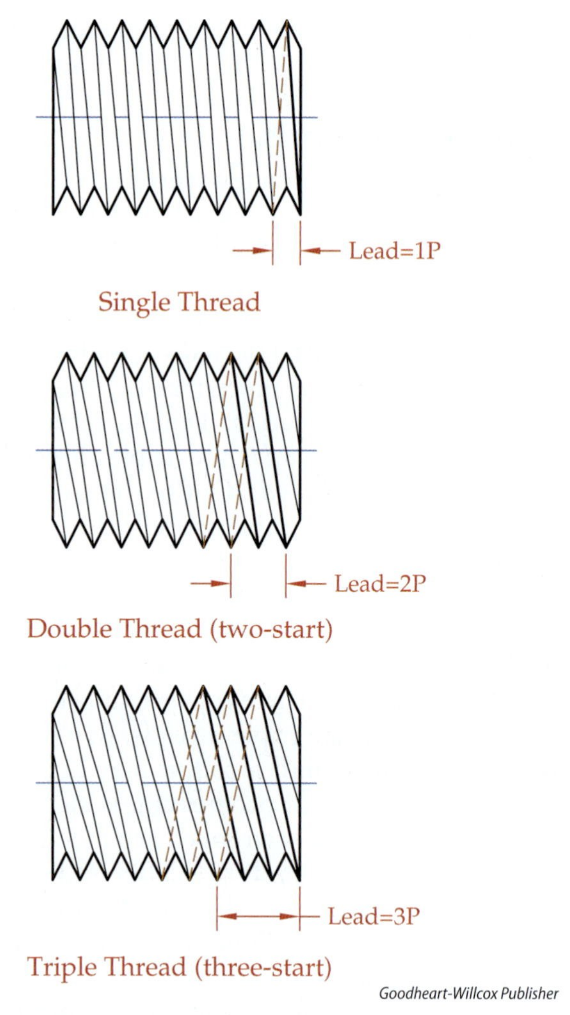

Goodheart-Willcox Publisher

Figure 8-8. If drawn in detailed representation, multiple-start threads should show the crest lines sloping accordingly. The hidden lines in the illustration represent the slope on the "back side," but are not shown on industrial prints. The double thread crest slopes one pitch on the near side and an additional pitch on the back side, which explains the lead being equal to twice the pitch.

Left-Hand Threads

Threads are most commonly created so that clockwise revolutions advance the threaded parts together. Common nuts and bolts tighten with clockwise turns. These threads are called ***right-hand threads***. However, some applications require threads that tighten with counterclockwise movement. These threads are referred to as ***left-hand threads*** or reverse threads.

Left-hand threads are usually found on items where right-hand threads may produce unwanted loosening, perhaps due to motion. Examples include threads on bicycle pedals, threads on a toilet tank flush handle, the arbor nut on a table saw, and similar applications. Sometimes left-hand threads are used to ensure components are connected properly, such as the fittings on oxyacetylene welding equipment. Left-hand threads are indicated on a print using a callout. Detailed representations should also reflect a slope angle opposite that of right-hand threads, **Figure 8-9**. As a tip to help drafters remember the details of right-hand threads, some drafting textbooks taught that the crest and root lines slope in the same direction as the thumb tilts as you are looking at the back of your right hand.

Specification of Screw Threads

A screw thread and all of its characteristics are specified on a drawing using a standard note with a leader and an arrow pointing to the thread. The note contains information describing the specific thread and may be composed of a variety of details. In **Figure 8-10**, the note on the right is interpreted as:

A. The nominal size (major diameter or screw number for smaller diameters) is 1″.
B. The number of threads per inch is five (5) threads per inch.
C. The thread form is Acme.

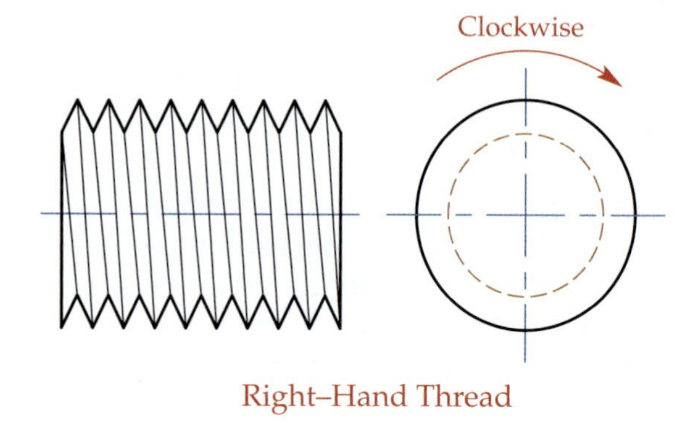

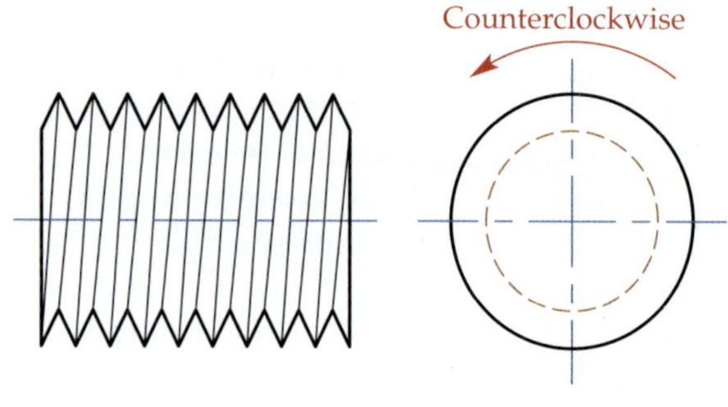

Goodheart-Willcox Publisher

Figure 8-9. Right-hand and left-hand threads are illustrated here.

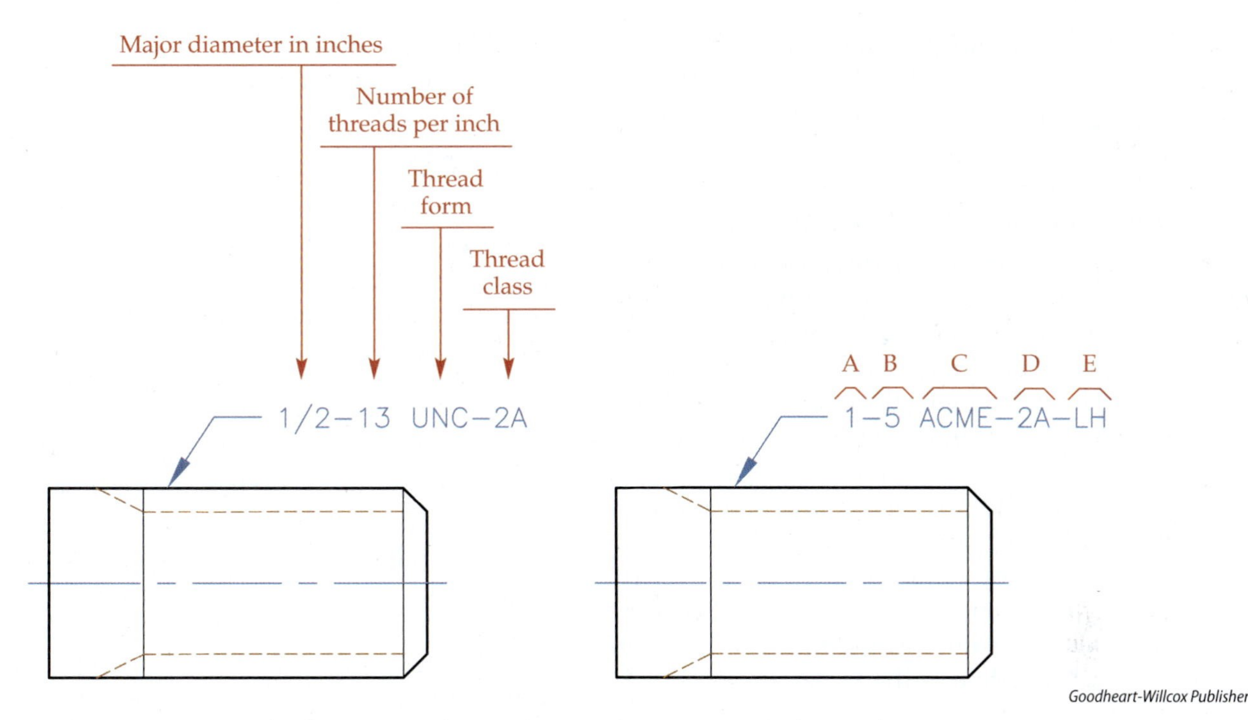

Figure 8-10. A screw thread is specified in a local note that may be composed of several components.

D. The thread class number or symbol and internal/external designation is 2A, which signifies external threads.

E. The threads are left-hand (LH) threads. This designation is only included for left-hand threads. No designation is used for right-hand threads.

Since fractional numbers are widely used, they are acceptable for the nominal size, even though the rest of the drawing is using decimal values. The ***nominal size*** is a general classification term used to designate the size of a commercial product. If decimal values are used for the nominal size, they should be shown to four places (unless the fourth place is a zero). For common numbered screw threads, usually those smaller than 1/4″, the decimal equivalent should be shown in parentheses to three places, as in this example: #10(.190)-32 UNF-2A.

In newer standards, additional elements of the thread note include the designation MOD to indicate there is a special modification or qualifying information. If MOD is placed in the thread note, another line is added to explain the modification. For inspection, there are standard thread gaging systems guided by ASME B1.3 that determine the acceptability of screw threads. System 21, for example, allows the inspector to accept screw threads with a *go* or *no-go* ring-and-plug system. This gaging system number can also be added in parentheses at the very end of the thread note.

- **Example 1:**
  - {$\frac{3}{8}$}-24 UNF-3A MOD (21)
  - MAJOR DIA .3648-.3720 MOD
- **Example 2:**
  - {$1\frac{1}{2}$}-10 UNS-3B MOD (21)
  - MINOR DIA 1.398-1.409 MOD
  - PD 1.4350-1.4413
  - MAJOR DIA 1.500 MIN

In Example 1, the modification is described in line two. In this case, the print specifies the major diameter should be between .3648 and .3720, a specification that requires a slight modification to the crests of the thread. Thread gaging system 21 is also specified. In Example 2, the Unified special threads callout means a special diameter and pitch diameter are specified in the following three rows, but gaging system 21 is still the method of inspection.

Unless otherwise specified, threads are right hand and single start. The letters LH after the class symbol specify the thread as left hand. In former practice, the word DOUBLE may also have been shown to indicate a double thread, or the word TRIPLE may have been shown to indicate a triple thread. Single-start threads and right-hand threads have never required RH or SINGLE to be specified.

Newer standards have implemented some additional techniques for multiple-start threads. Multiple threads can be described by replacing the number of threads per inch with the pitch and lead given in the note. The following are two examples of a thread note for a standard Unified multiple-start thread.

- **Former practice:**
  - .750-16 UNF-2A TRIPLE
- **Current practice:**
  - .750-.0625*P*-.1875*L*(3 STARTS)UNF-2A

Other specifications for a threaded hole may be given in the same local note along with the thread note. See **Figure 8-11.** These additional specifications may include thread length, hole size (for internal threads), or chamfer. Giving the pitch diameter below the thread note is a common option adopted by many companies. Additional manufacturing specifications are discussed in Unit 11, *Machining Specifications and Drawing Notes*.

If the tolerance, or range of size, for the thread pitch diameter is given, it should be placed below the note. However, on some prints, the tolerance may appear on the same line as the thread note.

{$\frac{3}{4}$}-10 UNC-2B

PD .6850 TO .6927

The specification for a constant-pitch series thread with a tolerance for the pitch diameter is listed as follows.

{2$\frac{1}{4}$}-8 UN-3A

PD 2.1688 TO 2.1611

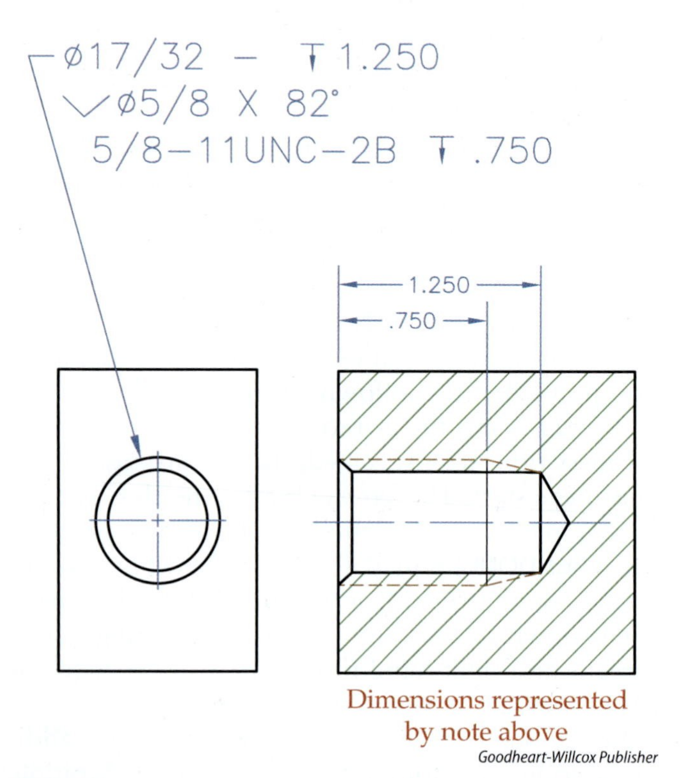

Figure 8-11. Additional specifications for a thread may include thread length, hole size, depth, and chamfer specification.

Additional information about tolerances is covered in Unit 10, *Tolerancing*.

Metric Threads

Metric threads are graphically represented in the same manner as Unified threads, but the format of the specification note is different. In a note for metric threads, **Figure 8-12**, the letter M is followed by "diameter × pitch," and the tolerance (class of fit) follows. The M designates the thread as a metric series. In former practice, if the pitch was not indicated, the print reader assumed the standard coarse thread pitch applied, but metric fine threads always had the pitch designated. Today, the preferred standard is to always specify the thread pitch. Notice the specification for metric series threads does not indicate the number of threads per unit of length, such as threads per millimeter or threads per centimeter. Instead, the pitch is simply stated in the specification. In **Figure 8-12**, the M8 represents a nominal thread diameter of 8 mm with the standardized fine thread pitch of 1.0 mm. For an 8 mm diameter, the standard coarse thread pitch is 1.25 mm. If the thread note were specified simply M8 (older practice), then the standard coarse pitch could be assumed. Current standards recommend the diameter, pitch, tolerance class, and gaging system be specified. However, in actual practice, many prints may not indicate the tolerance and gaging system. **Figure 8-13** features a chart of standard metric sizes and corresponding pitches. As shown, there are not a significant number of fine or extra fine options in the metric standard.

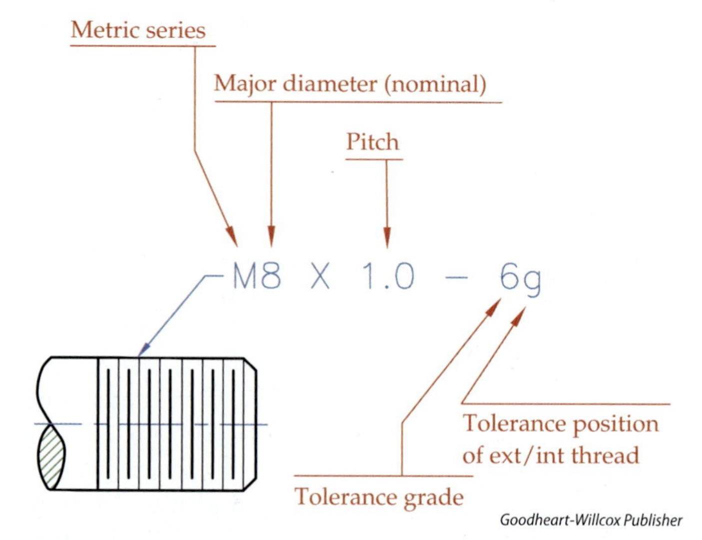

Figure 8-12. Metric threads are specified in a different fashion than inch-based threads.

Metric Thread Sizes			
Nominal Diameter (mm)	Thread Pitch (mm)		
	Coarse	Fine	Extra Fine
1.6	0.35		
2	0.4		
2.5	0.45		
3	0.5		
3.5	0.6		
4	0.7		
5	0.8		
6	1		
8	1.25	1	
10	1.5	1.25	1
12	1.75	1.25	1.25
14	2	1.5	
16	2	1.5	
18	2.5	1.5	
20	2.5	1.5	
22	2.5	1.5	
24	3	2	
27	3	2	
30	3.5	2	
33	3.5	2	
36	4	3	
42	4.5		
48	5		
56	5.5		
64	6		
72	6		
80	6		
90	6		
100	6		

Goodheart-Willcox Publisher

Figure 8-13. This chart shows some standard metric thread sizes and corresponding pitches. See Appendix D for a more extensive table detailing metric screw thread series.

The tolerance and class of fit for metric threads are designated by adding numbers and letters in a certain sequence to the callout. In **Figure 8-12**, the 6g specification indicates a middle level of tolerance with a small allowance for external threads. The numeral choices range from 3 to 9 for external threads and 4 to 8 for internal threads. Alphabetical characters "g" or "h" (external) or "G" or "H" (internal) are used in the United States for the position of the tolerance. An H specification is more precise with zero allowance, and a G specification has a small allowance. In brief, the 2A/2B classification for inch threads is approximately equivalent to 6g/6H for metric threads.

The thread designation in **Figure 8-14A** calls for a 20 mm diameter fine thread with a pitch of 1.5 mm. In addition, the thread note specifies a tolerance class of 6h6g. In this example, the first part, 6h, is a tolerance grade applied to the pitch diameter. The second part, 6g, is a tolerance grade applied to the crest diameter. If these two diameters have the same tolerance class, then only one is given. The note can also cover the expected class of fit between two threaded parts, in which case the note specifies both the internal and external tolerance class. In **Figure 8-14B**, the thread note indicates a tolerance class specification of 4H/6g. This indicates the nut (internal thread of mating part) for the threaded feature should be a precise tolerance class 4H fit, while the pitch diameter and major crest diameter of the external threads both have a medium grade fit (6g). Companies that stock metric gages for inspecting threads usually stock class 6H plug gages for internal threads and 6g ring and set plug gages for external threads, because this comprises a large percentage of the metric thread applications.

Pipe Threads

The three forms of pipe threads used in industry are regular, aeronautical, and Dryseal pipe threads. The regular pipe thread is the standard for the plumbing trade. The aeronautical pipe thread is the standard in the aerospace industry. Regular and aeronautical pipe thread forms must be filled with a lute or sealer to prevent leakage in the joint. The Dryseal pipe thread is the standard for automotive, refrigeration, and hydraulic tube and pipe fittings. Dryseal pipe threads do not allow leakage, even without the use of sealer. This is due to the metal-to-metal contact at the crest and root of the threads.

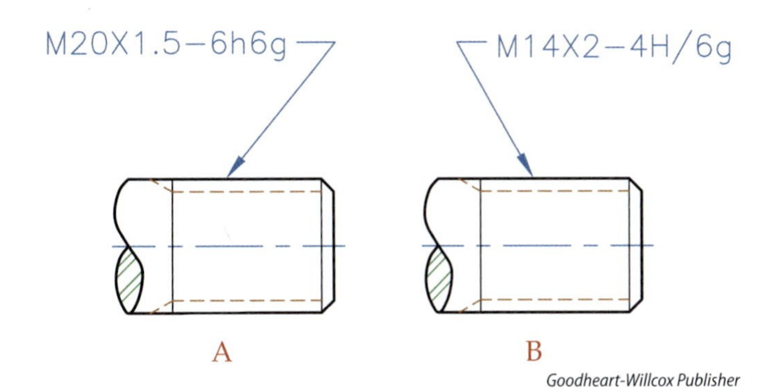

Goodheart-Willcox Publisher

Figure 8-14. The tolerance class for metric threads can be specified in different ways. A—The tolerance class for the pitch diameter is a little more precise than the major diameter. B—The internal thread is specified to be more precise than the external thread.

Representation and Specification

One of the more common pipe threads is the NPT designation and is defined by ASME B1.20.1. **Figure 8-15** shows common NPT sizes (in nominal diameters) from 1/4″ to 2″. As expressed by the chart, a pipe referred to as a 1″ pipe has an outside diameter of 1.315″, and the length of threads cut into the end of the pipe is 7/8″, so at 11.5 threads per inch, about 10 threads are cut. NPT threads have a taper of 3/4″ on the diameter, per foot, although in drafting practice, to help show the taper, the taper may be exaggerated, as in **Figure 8-16**. Some pipe threads are not tapered and are referred to as straight pipe threads. Straight and taper pipe threads are graphically represented in a manner similar to other screw threads. Even on features that have tapered pipe threads, some companies may elect to not show the taper.

The specifications for American Standard Pipe Threads are listed in sequence of nominal pipe size, number of threads per inch, and the symbols for the thread series and form. For example, the thread specification 1/2-14 NPT designates a 1/2″ nominal pipe size, 14 threads per inch, American Standard taper pipe thread.

Pipe Thread Designations

The following pipe series symbols are used for American Standard pipe threads.

- **NPT.** American Standard taper pipe thread for general use
- **NPS.** American Standard straight pipe thread
- **NPTR.** American Standard taper pipe thread for railing joints
- **NPSC.** American Standard pipe thread for general couplings
- **NPSM.** American Standard pipe thread for various free-fitting mechanical joints
- **NPSL.** American Standard pipe thread for loose-fitting mechanical joints with locknuts
- **NPSH.** American Standard pipe thread for hose couplings

Pipe series symbols used for Dryseal pipe threads are designated as follows.

- **NPTF.** Dryseal American Standard pipe thread in many types of service, especially fuel connections
- **PTF-SAE SHORT.** Dryseal SAE short taper pipe thread
- **NPSF.** Dryseal American Standard fuel internal straight pipe thread
- **NPSI.** Dryseal American Standard intermediate internal straight pipe thread

A typical specification for a Dryseal pipe thread is shown in **Figure 8-16**, and includes the following:

A. Nominal size (1/4″)

B. Number of threads per inch (18)

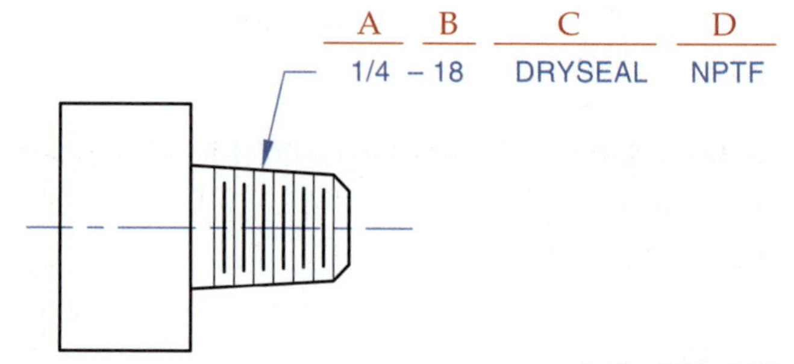

Goodheart-Willcox Publisher

Figure 8-16. Pipe threads are drawn in a fashion similar to that of normal threads, but the major diameter is tapered at a slight angle to the center axis.

NPT - American Standard Taper Pipe Thread				
Nominal Pipe Size	Threads per Inch	Threaded Length	Number of Threads Cut	Nominal Outside Dia.
1/4"	18	5/8"	11	.540"
3/8"	18	5/8"	11	.675"
1/2"	14	3/4"	10	.840"
3/4"	14	3/4"	10	1.050"
1"	11-1/2	7/8"	10	1.315"
1-1/4"	11-1/2	1"	11	1.660"
1-1/2"	11-1/2	1"	11	1.900"
2"	11-1/2	1"	11	2.375"

Goodheart-Willcox Publisher

Figure 8-15. This chart shows some common NPT sizes, in nominal diameters, from 1/4″ to 2″.

C. Form (Dryseal)
D. Pipe series symbol (NPTF = Dryseal American Standard pipe thread)

Computer-Generated Thread Representations

The representation of screw threads in a CAD system will vary depending on the software and the user options selected. Three-dimensional solid models with threaded features are often created without actually modeling the true helical screw thread. For some basic parts with small threaded holes, it may not be an issue for the model to contain true geometry, and an actual representation of the screw thread might be shown in the views. It will be up to the engineer, designer, or drafter to determine if modeling the threaded holes might create an unnecessarily complex model. In some cases, the appearance of threads on the 3D pictorial view is produced by a shaded image applied to a cylindrical surface. If the designer needs a true thread model, this can be accomplished by lofting a thread form shape along a helical path about the surface of the cylinder.

Computer-generated orthographic views of threads often use a simplified representation. **Figure 8-17** shows a pictorial view of a 3D model with a shaded image applied in place of true threads, while the orthographic views appear in a simplified representation. In many programs, thread specifications for ASME and ISO standards are stored within the CAD data set. Thread notes can be generated automatically as the views are annotated. Different software packages may or may not feature all the options the drafter needs.

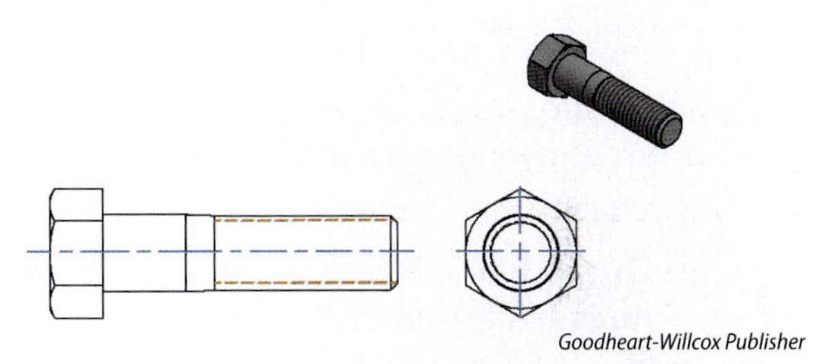

Goodheart-Willcox Publisher

Figure 8-17. Most 3D CAD modeling programs do not represent threads as true helical thread models. Orthographic views are automatically generated based on user settings. In this case, a simplified representation is shown.

Summary

- A screw thread is a ridge of a particular shape that follows a helical path around the surface of a cylinder.
- A fundamental understanding of screw thread terminology (crest, root, flank, lead, major diameter, etc.) is important to the print reader.
- The shape of the screw thread ridge (or groove) is known as the thread form and is the basis for different types of threads.
- The Unified Thread Standard (UTS), or Unified, form is the most common thread form standardized for general nuts, bolts, and industrial applications.
- Threads are often represented on a drawing in one of three ways: detailed, schematic, or simplified.
- There are conventional practices that deal with the manner in which the helical ridge starts, or how a thread fades out as a partial thread.
- For Unified threads, there are coarse, fine, extra-fine, and constant-pitch series that establish the standard pitch for available body diameters.
- Threads are manufactured at different classes of fit, some being more precise than others.
- While most threads are single-start threads with a single ridge, multiple-start threads are also used in industrial applications.
- While most threads are right-hand threads that advance when turned clockwise, there are applications in industry wherein left-hand threads are utilized.
- Standard notes with leader lines are used to specify information about a screw thread, such as the diameter, form, pitch, and class of fit.
- While metric threads are primarily designed in the same manner as inch threads, the specification callouts and class-of-fit specifications are different.
- Pipe threads are represented in similar fashion to regular threads, but standardized forms are varied in nature, and thread note designations are based on piping applications.

Unit Review

Name ______________________ Date ______________ Class ______________

Answer the following questions using the information provided in this unit.

Know and Understand

____ 1. *True or False?* The geometric shape that is the basis for screw threads is the *spiral*.

____ 2. *True or False?* Threads have one purpose, which is holding parts together.

Match the following screw thread terms to their definitions. Answers are used only once.

____ 3. The bottom between ridges

____ 4. Distance of advance when a threaded feature turns 360°

____ 5. The side of a ridge

____ 6. The top edge or surface of a ridge

A. Lead
B. Root
C. Crest
D. Flank

Match the following screw thread terms to their definitions. Answers are used only once.

____ 7. Measured root-to-root

____ 8. Measured crest-to-crest

____ 9. Measured flank-to-flank

____ 10. Measured thread-to-thread (any same point)

A. Major diameter
B. Pitch
C. Minor diameter
D. Thread angle

____ 11. Which of the following thread forms are the most similar in appearance?
A. Metric and Unified
B. Square and Stub Acme
C. Knuckle and buttress
D. Unified and Acme

____ 12. On industrial prints, which of the following is *not* a way that screw threads usually appear?
A. Detailed
B. Schematic
C. Simplified
D. True projection

____ 13. *True or False?* Detailed representation is the most realistic of the three methods of representation, but is usually only used if the major diameter of the screw thread is 1″ or larger on the drawing.

____ 14. If a UNC thread for a 1/4″ body diameter is 20 threads per inch, then UNF threads for that body diameter would be _____ threads per inch.
A. 12
B. 16
C. 20
D. 28

____ 15. *True or False?* Inch-based screw threads are classified by their precision, or manufacturing tolerance, with 2 being most common, 1 being less precise, and 3 being more precise.

____ 16. *True or False?* A two-start thread is when there are two ridges that form the screw thread, resulting in a lead equal to half the pitch.

____ 17. Which of the following terms describes threads that advance with counterclockwise motion instead of the normal clockwise motion?
A. Lead
B. Series
C. Left hand
D. Backward

Match the following parts of the thread specification 5/8-18 UNEF – 2A *to the description. Answers are used only once.*

____ 18. Thread form

____ 19. Thread series

____ 20. Number of threads per inch

____ 21. External thread

____ 22. Major diameter

____ 23. Class of fit

A. 5/8
B. 18
C. UN
D. EF
E. 2
F. A

____ 24. For a thread note that states M24 X 3, which of the following is *false*?
A. This is a metric thread.
B. The pitch is 3 mm.
C. The minor diameter is 24 mm.
D. The thread is right hand.

____ 25. For a thread note that states 3/4 – 14 NPT, which of the following is *false*?
A. The pitch is 14.
B. The major diameter is 3/4″.
C. The threads are pipe threads.
D. The diameter is tapered.

____ 26. *True or False?* Three-dimensional CAD models of screw threads are not possible due to the complex helical path of the crests and roots.

Critical Thinking

1. Describe three advantages for having a system of standardized screw threads.

____

____

____

____

____

____

2. Discuss three advantages or disadvantages of the standardized ways of showing screw threads on the print.

____

____

____

____

____

____

Apply and Analyze

Name ______________________ **Date** ______________ **Class** ______________

Review Activity 8-1

Examine the chart that follows. Fill in the blanks for the sample threads shown in views A–D. Either read the thread specification note or use the paper scale provided at the bottom to measure or transfer distances with a separate scrap sheet of paper. The guidelines are to assist you with neat capital lettering. Also, answer question 10 related to the metric thread specification.

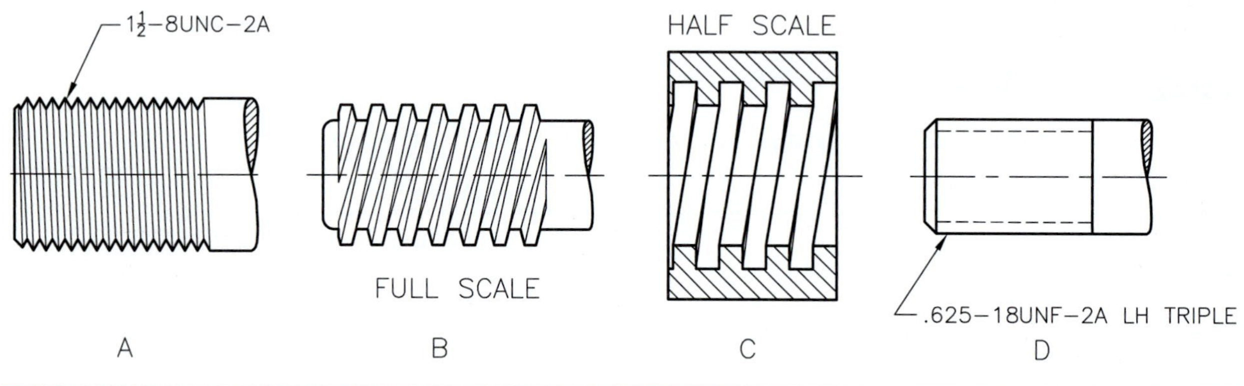

	A	B	C	D
1. Number of threads/inch				
2. Pitch of thread				
3. Thread form				
4. RH/LH?				
5. Single or which?				
6. Lead of thread				
7. Major diameter				
8. Internal/external?				
9. How represented?				

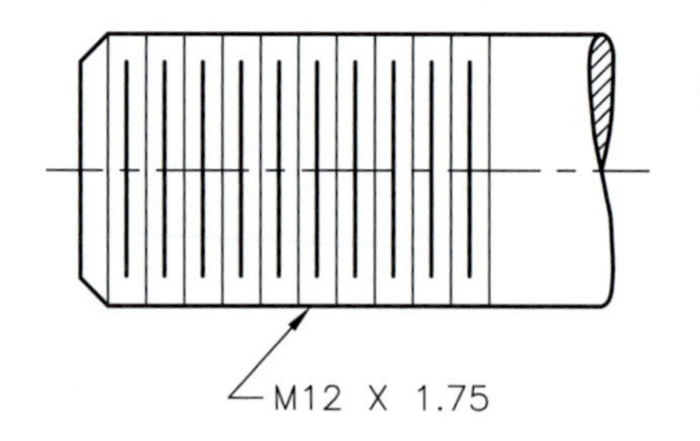

10. For the metric thread:
    A. What is the major diameter? ______________
    B. What is the pitch? ______________
    C. How is it represented? ______________

0 1 2

Copy or transfer distances to scale to answer questions about B and C above

Apply and Analyze

Name ______________________ Date ______________ Class ______________

Review Activity 8-2

For each of the thread notes below, write out the description using complete words. The first one has been done as an example.

1. 3/8-16 UNC-2A

   Example: 3/8″ major diameter, 16 threads per inch, Unified coarse thread form, with a Class 2 medium fit, external thread

2. 1-20 UNEF-3B

   ____

   ____

3. .625-8 ACME-LH

   ____

   ____

4. M10 X 1.5-6H/5g6g

   ____

   ____

5. 3/4-14 NPT

   ____

   ____

For each of the thread descriptions below, create a standard thread note specification.

6. 1 1/2″ major diameter, 12 threads per inch, Unified fine thread form, with a Class 2 fit, internal thread, left hand

   ____

7. #5 or 1/8″ major diameter, 40 threads per inch, Unified coarse thread form, with a Class 2 fit, external thread

   ____

8. 1/2″ major diameter, 10 threads per inch, Stub Acme thread form, internal thread, right hand

   ____

9. Metric thread, 20 mm major diameter, 2.5 mm pitch, and “6g” grade for external pitch diameter and external major diameter

   ____

10. 1/4″ pipe size, 18 threads per inch, Dryseal American standard pipe thread

    ____

Apply and Analyze

Name ______________________ Date ____________ Class ______________

Industry Print Exercise 8-1

Refer to the print PR 8-1 and answer the questions below.

1. On the left end of this object, the thread note indicates 10UNC. What is the pitch for this thread?

____

2. Is either of the threads on this part left hand?

____

3. What type of thread representation is used on this print?

____

4. On the right end of the object, what is the major diameter of the thread?

____

5. What thread form is used for the threads on this sheet, and is it coarse, fine, or extra fine?

____

6. If a nut were screwed onto the right end of this part, one 360° turn of the nut would advance the nut what distance?

____

7. Does there appear to be a chamfer on one or both ends of this part?

____

Review questions based on previous units:

8. Is there a cutting-plane line on this print?

____

9. Are there hidden lines featured on this print?

____

10. What is the part number for this drawing?

____

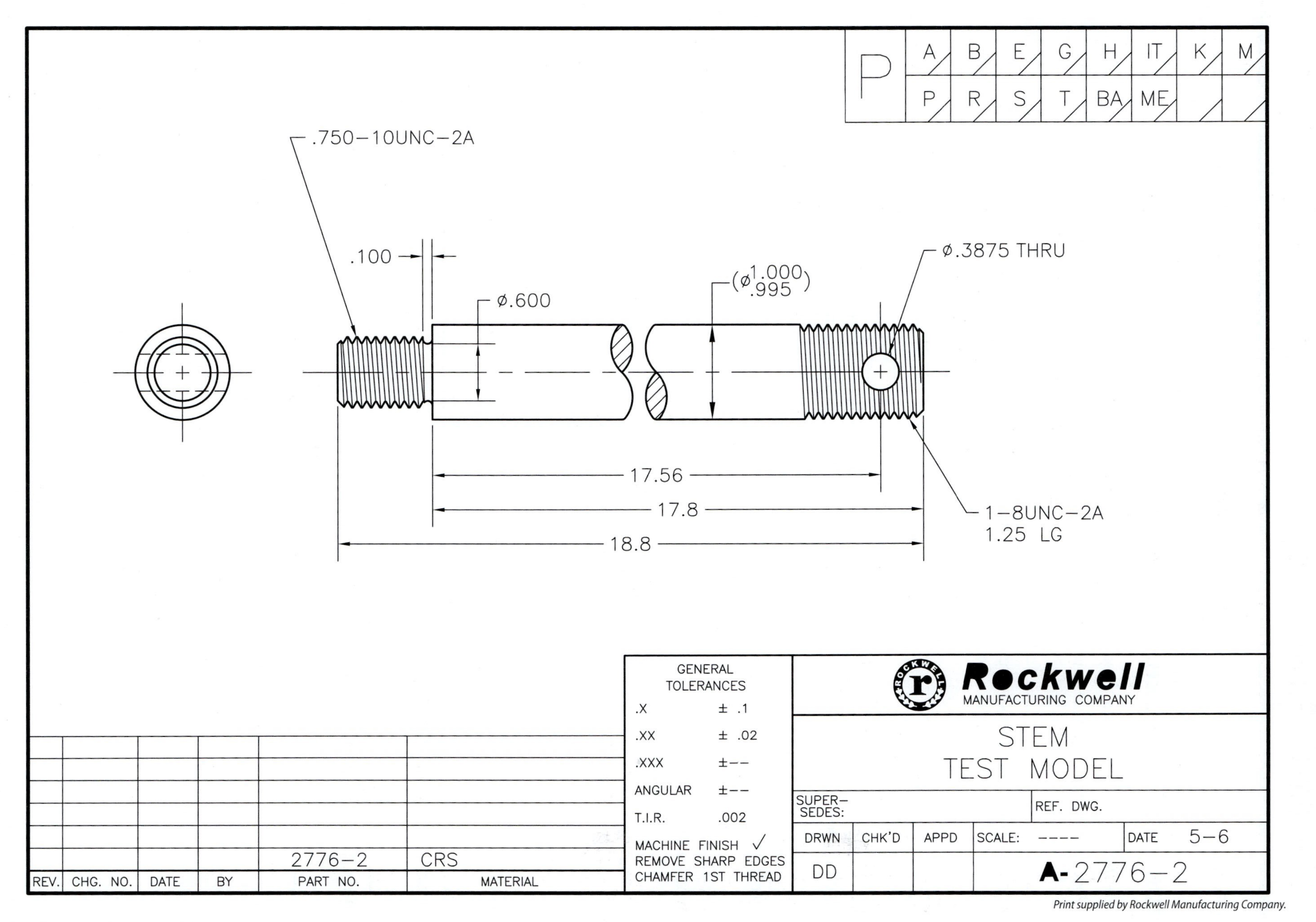

Print supplied by Rockwell Manufacturing Company.

PR 8-1. Stem Test Model

Apply and Analyze

Name ______________________ Date ______________ Class ______________

Industry Print Exercise 8-2

Refer to the print PR 8-2 and answer the questions below.

1. How many threaded holes are there for the part on this print?

____

2. For this print, the views were generated from holes that were modeled with true geometry. Which method of representation presented in this unit is the closest to how these threads are represented?

____

3. What is the pitch of the threads for this part?

____

4. Use Figure 8-7 in this unit to determine the major diameter of the threads for this part.

____

5. Use Figure 8-7 to determine if these threads are coarse, fine, or extra fine.

____

6. Within the thread note, what does the "2B" indicate?

____

Review questions based on previous units:

7. What scale are the views on this print?

____

8. What is the material for this part?

____

9. What is the name of this part?

____

10. Does this drawing feature an auxiliary view?

____

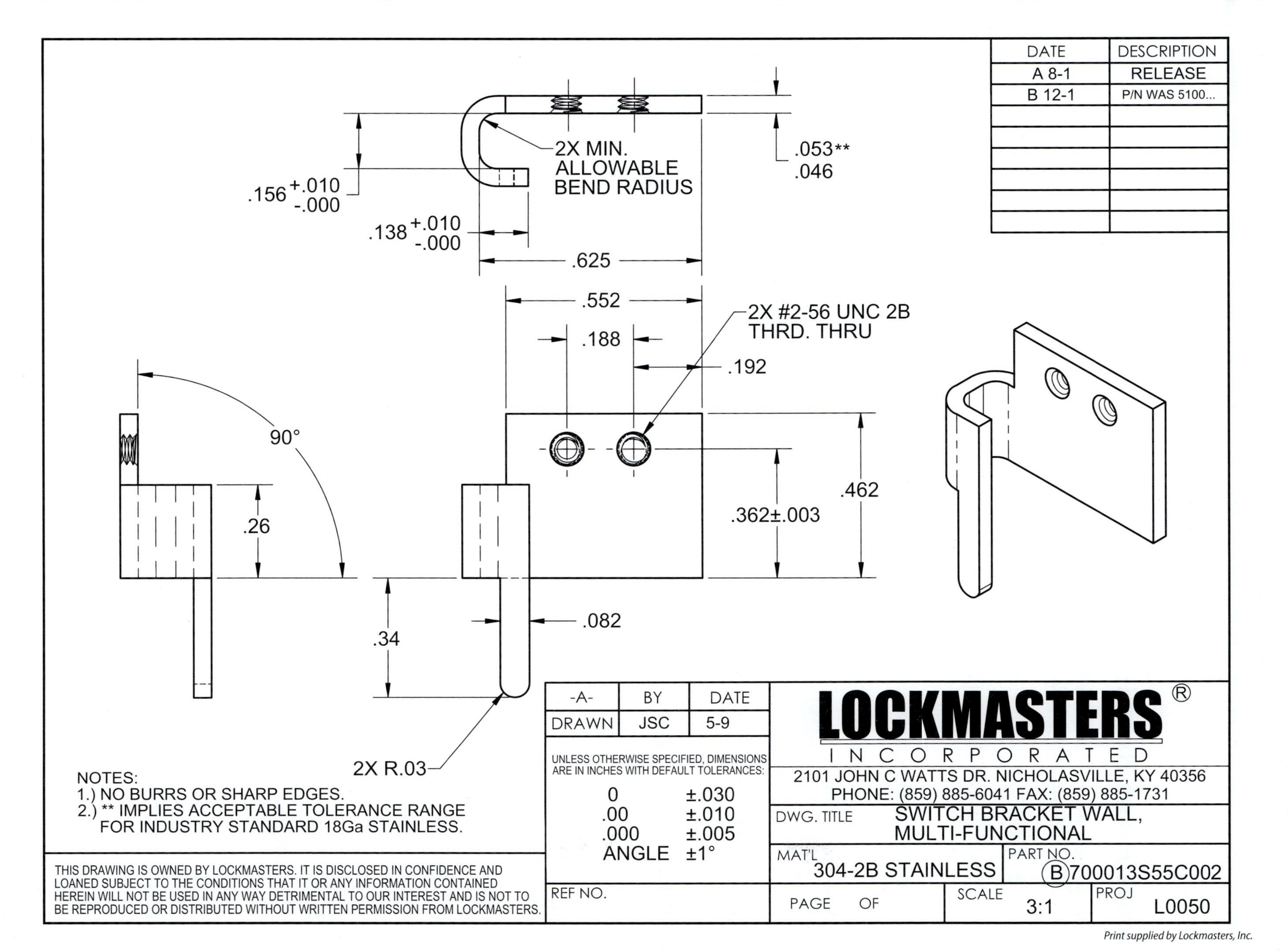

Print supplied by Lockmasters, Inc.

PR 8-2. Multifunctional Switch Bracket Wall

Apply and Analyze

Name ______________________ Date ______________ Class ______________

Bonus Print Reading Exercises

The following questions are based on various bonus prints located in the folder that accompanies this textbook. Refer to the print indicated, evaluate the print, and answer the question.

Print AP-001

1. Is the National Gas Outlet (NGO) thread shown on this print right-handed or left-handed?

____

2. What is the major diameter of the National Gas Tapered (NGT) thread shown on this print?

____

3. What is the smallest thread pitch specified on this print?

____

Print AP-003

4. What does UNEF stand for in the thread note on this print?

____

5. In the thread note on this print, what does 2A specify?

____

6. The major diameter for the one threaded feature is 3/8″, but the thread is shown detailed because the scale of the drawing is 4:1 and, therefore, the diameter on the print will be at a size of _____.

____

Print AP-004

7. This assembly drawing includes a threaded stud. By what method are the threads represented?

____

Print AP-005

8. How many threaded holes are there in this subassembly?

____

Print AP-006

9. What is the major diameter of the threaded feature on this print?

____

10. What is the pitch of the threaded feature on this print?

____

Print AP-008

11. How deep are the threads specified on this print?

____

Print AP-010

12. What method of thread representation is used on this print?

____

Print AP-011

13. What is the thread specification for the threaded holes on this part?

____

14. What method of thread representation is used on this print?

____

Print AP-012

15. What thread note is specified for a through hole on this print?

____

SECTION 3

Fundamentals of Size Description and Annotations

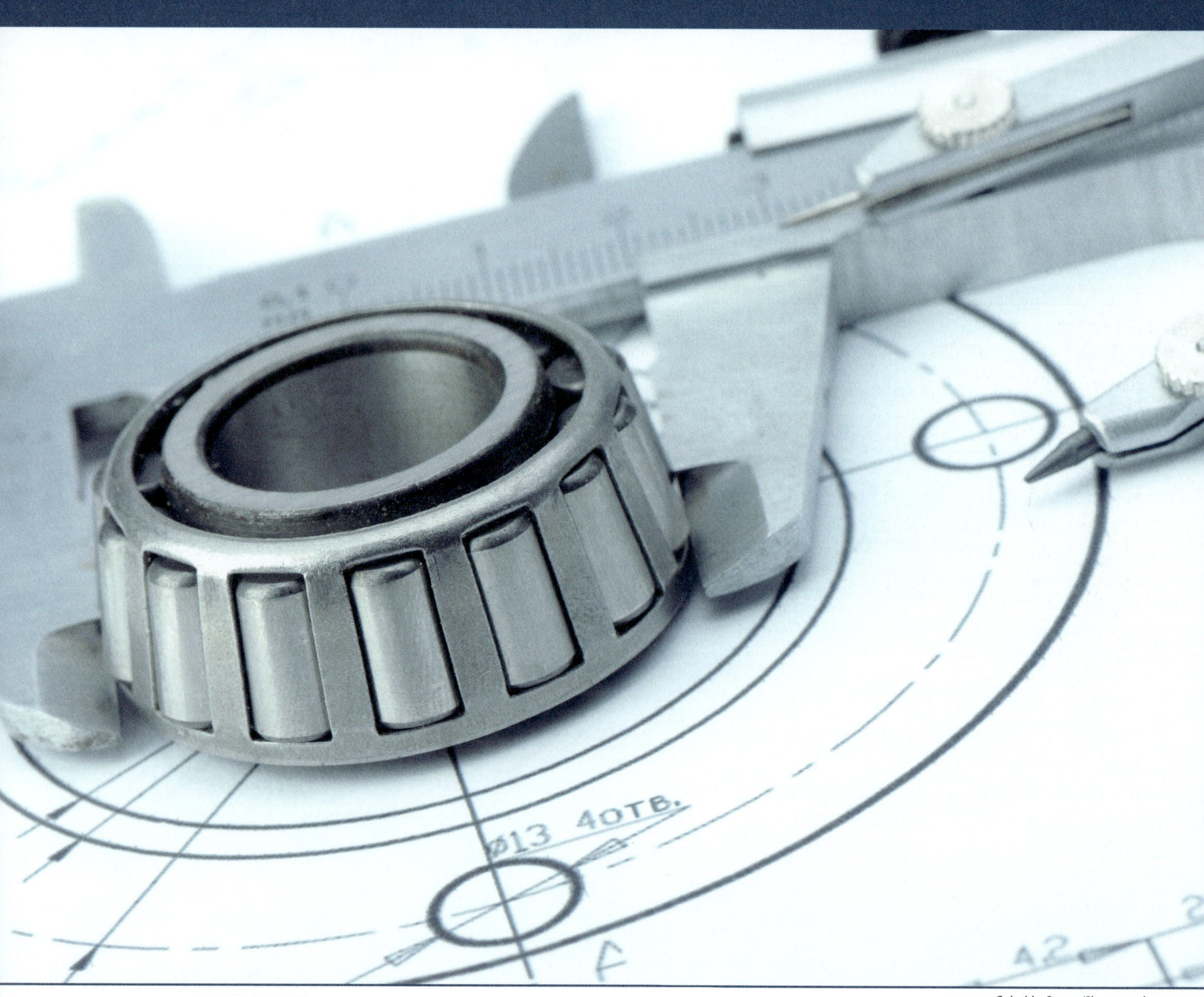

Galushko Sergey/Shutterstock.com

UNIT 9
Dimensioning

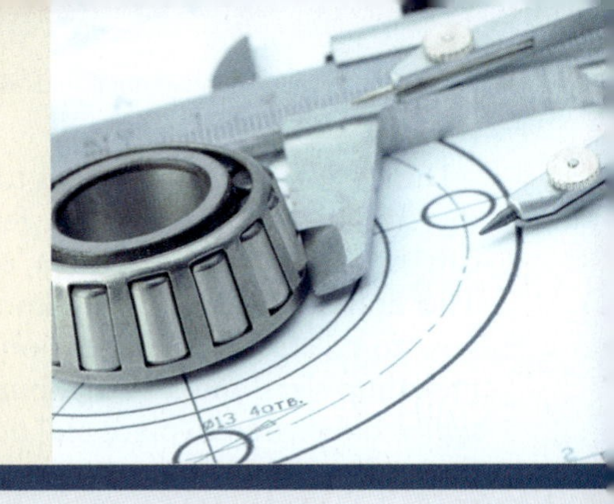

LEARNING OBJECTIVES

After completing this unit, you will be able to:

- Define dimensioning and its role in industrial prints.
- Describe fundamental rules affecting the application of dimensions to a part or assembly.
- Identify the manner in which units of measure are applied and formatted on an industrial print.
- Identify terms and measurements associated with dimensioning mechanics, including line types, symbols, arrows, and spacing.
- Identify symbols that have been standardized for use in dimensions and dimensioning notations.
- Explain the choice and placement rules drafters use when creating dimensions.
- Identify and discuss various systems, notes, and methods for dimensioning.
- Describe additional types of dimensions used in special ways.
- Discuss alternative methods for describing size and applying dimensions within a digital product definition.
- Identify examples of industrial prints that may be undimensioned.

TECHNICAL TERMS

aligned dimensioning
baseline dimensioning
bolt circle
chain dimensioning
contour dimensioning
coordinate dimensioning without dimension lines
datum
dimensioning
dimensioning mechanics
dimension origin symbol
drawing graphic sheet
dual dimensioning
International System of Units (SI)
location dimension
reference dimension
size dimension
tabular dimension
true geometry view
undimensioned drawing
unidirectional dimensioning
US customary units

Print reading requires the reader to understand the shape description of the object. This is called visualization. Previous units primarily focused on describing the shape of an object or part with different view systems, sectional techniques, and special symbols for screw threads. The second major component of print reading is the *size* description. The reader must understand the size of a part as specified on the print. The size description can simply be called ***dimensioning***.

Dimensions are annotations to the views describing the size or location of each feature. Dimensioning can be a very difficult area of drafting to learn. The drafter must not only know the *mechanics* of how to create all the dimension and extension lines, including sizes and spacing, but also how to choose *which* dimension to place and *where* to best place the dimension. Sometimes this can be a real challenge. Unfortunately, many drafters and engineers have not had sufficient training in how to properly dimension a drawing. Therefore, some prints may not follow standard recommendations for dimensioning as well as they should, making them harder to read.

This unit explains some of the rules a drafter should use to choose and place dimensions. This will assist the print reader in knowing which dimensions to expect and where to look for a dimension. As a print reader, you can be a valuable team member by identifying ways in which to improve a drawing with more concise or appropriate dimensioning.

The primary standard that covers dimensioning practices is ASME Y14.5, titled *Dimensioning and Tolerancing*. Units 10, 11, and 13 of this text also rely heavily on this standard. The ASME Y14.41 *Digital Product Definition Data Practices* standard addresses the use of computer models and data sets to organize, control, and manage the data associated with parts. With this newer method of dimensioning, there will be the opportunity to place dimensions in the 3D model environment. Machines can now manufacture, check, and analyze parts without many of the dimensions and notes previously defined by two-dimensional drawings and prints. The ASME Y14.31 *Undimensioned Drawings* standard helps set forth guidelines for undimensioned drawings, such as undimensioned printed circuit drawings and flat pattern drawings. The Y14.41 and Y14.31 standards are assisting industry with new ways to perform some of the tasks defined in this unit. These standards are discussed in more detail later in this unit.

Fundamental Rules

Before discussing dimensioning at a detailed level, an examination of some fundamental rules expressed in ASME Y14.5 is appropriate. In industrial prints, a fundamental principle is that a complete size description must be given either in a drawing or in a product definition data set. Think of the product definition data set as a computer model that contains information that can be passed on to a machine for manufacturing or inspection. With respect to prints, one fundamental rule is that no measuring with a scale or other instrument will ever be necessary, nor will assumptions need to be made. In addition, no extra dimensions will be given that would provide more than one way to interpret a distance or size.

There are a couple of logical assumptions of which print readers should be aware. In standard practice, 90° is implied for center lines, features located on center planes and axes, and corners of objects and features that appear to be perpendicular on the print or model. A zero basic dimension is implied for parts on a drawing that appear to be aligned with center planes and axes. The print reader is still advised to ask for clarification if the print is unclear with respect to features being aligned with each other or with center axes and center planes.

Dimensioning, if done properly, should involve the selection and arrangement of dimensions that suit the function of a part or one of its features. The mating relationship of parts to be assembled is also key to a properly dimensioned drawing. In recent decades, standards have evolved alongside new technologies, and therefore manufacturing methods should not be specified on the print as in former practice. A good drawing also features dimensions that are arranged for optimum readability, with dimensions shown in profile views where outlines and shapes are visible. In this unit, you will see these fundamental principles applied.

Units of Measure

There are two primary units of measure for American industrial prints. While many companies still design and function with inches, companies whose products are widely used in an international market have adopted the metric system. Due to many reasons, including the extreme cost of changing over equipment and standard-size materials and components, many companies have continued to use ***US customary units***. Length or distance units include the inch, foot, yard, and mile. This system is very similar to what is known as *Imperial units*, but there are some differences.

The metric system is known as the ***International System of Units***, abbreviated as ***SI***, from the French *Système Internationale d'Unités*. SI is the modern form of the metric system and is the most widely used system of measurement outside the United States, not just for linear distances but across the spectrum of weights, volumes, and other measurements.

Dimensioning Mechanics

Dimensioning mechanics can be defined as the instructions or guidelines for what size and spacing to use for all of the components of the dimension. For example, arrowhead appearance, lettering size, and spacing for extension lines are all part of dimensioning mechanics. Even CAD systems that automatically create dimensions need to be set up and managed so that the automatic dimensioning conforms to standards. Simply put, mechanics is the part of dimensioning concerned with what dimensions look like rather than where dimensions are placed, which dimensions are chosen, or whether or not a particular dimension is even needed.

Lines Used in Dimensioning

The standard lines used in dimensioning are presented in Unit 2, including the extension line, dimension line, and leader line. **Figure 9-1** not only illustrates these lines, but also illustrates spacing and size guidelines for gaps and extensions. Technically, the *chain line* is also a line used in dimensioning, and it will be covered in a future unit on specifications. Dimension lines are always parallel to the distance dimensioned, with an arrowhead on each end. They are usually broken in the middle for the dimensional value.

Extension lines extend from the part or feature being dimensioned. The line extends approximately 1/8″ (3 mm) beyond the arrowhead. About 1/16″ (1.5 mm) should be allowed between the extension line and the object to clearly separate the shape description from the size description. Center lines can be used as extension lines, in which case they cross visible lines without a gap. The first dimension should be at least 3/8″ (10 mm) away from the view. Subsequent dimensions should be another 1/4″ (6 mm) or more beyond the previous one. These distances are recommended in the ASME standards, but they are suggested guidelines and not a basis for rejection or acceptance of the drawing.

Arrowheads should be about 1/8″ (3 mm) long, drawn sleek and slender, sharply formed, and precisely terminated at the extension line. Lettering is usually 1/8″ (3 mm) tall with 1/16″ (1.5 mm) gaps on each side to allow clearance around the number value.

Leader lines are used for local notes that point to a particular feature on the drawing, **Figure 9-2**. Leader lines are commonly used to point to a hole or screw thread. In the case of a hole, the leader should line up with the center of the hole and be at an angle. The leader line typically has a shoulder of about 1/4″ at the end leading into the beginning or end of a note. Normally, leader lines terminate with an arrowhead, but in cases where the leader refers to an outlined surface area, perhaps indicated by chain lines, the preferred termination is a dot. Also notice in **Figure 9-2** that for dimensions placed within the 3D model environment, a leader line may be attached to a surface with a dot termination, but if the leader points to the edge of a feature, such as a hole, the arrowhead is still the preferred termination.

Standard Dimensioning Symbols

Dimensional values often require symbols, **Figure 9-3**. In previous practices, it may have been common to express the machining process for a hole, for example, as 1/2″ DRILL or .750″ REAM. It also may have been common to simply state the word RADIUS or DIAMETER. In current standards, however, it is recommended the dimensioning callouts use symbols

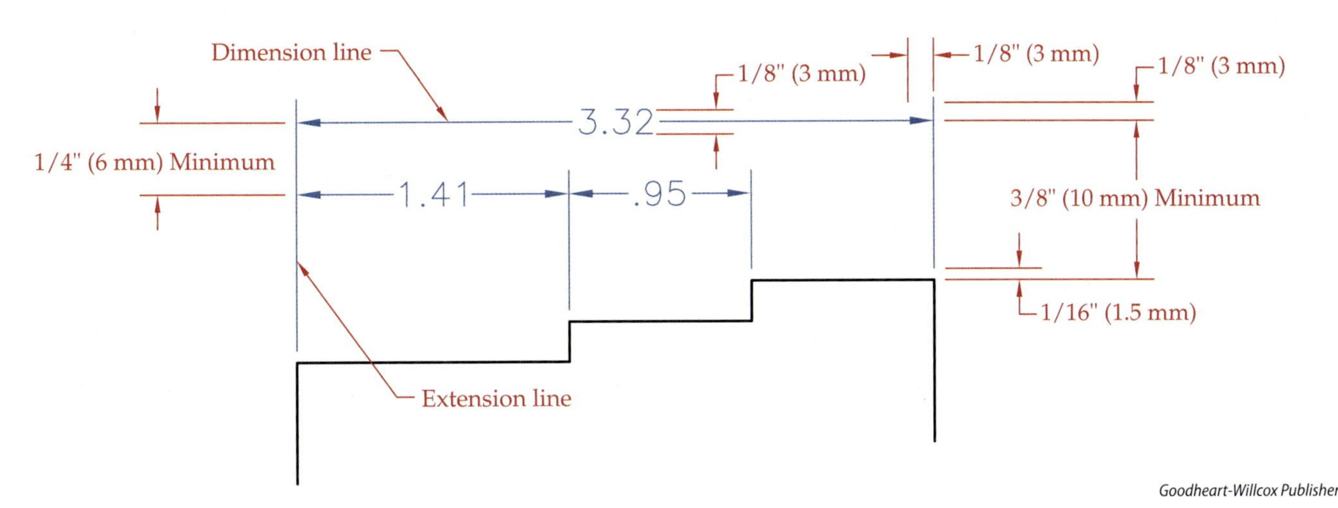

Goodheart-Willcox Publisher

Figure 9-1. This illustration describes some of the dimensioning mechanics the drafter uses to place dimension lines and extension lines in a clear and organized manner.

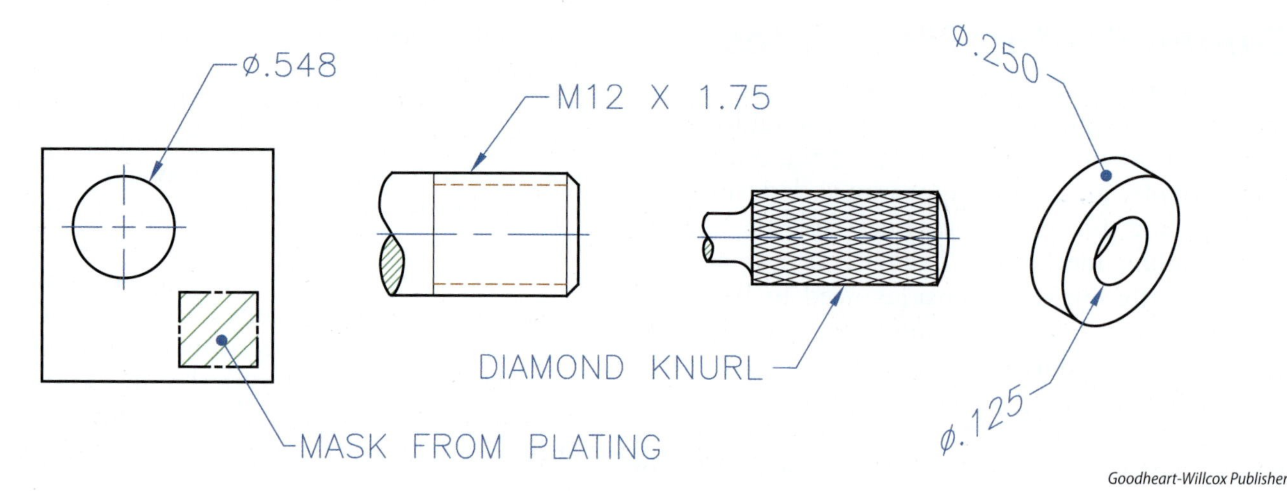

Goodheart-Willcox Publisher

Figure 9-2. Leader lines are used for local notes that point to a particular feature on the drawing or model.

for diameter (Ø) and radius (R), placing the symbol in front of the dimension value. Stating the machining process should be avoided unless there is a pressing reason to do so. **Figure 9-3** shows symbols for spherical radii and diameters, the "X" for "by" or "times," parentheses for reference dimensions, and an arc symbol for use with measuring along an arc. One of the newer dimensioning symbols is the *dimension origin symbol.* For a dimension between two surfaces or features, the dimension origin symbol can replace an arrowhead, and in such cases the dimension origin symbol clarifies which surface or feature defines the plane from which to measure. These symbols will be covered later in the unit, and additional symbols not shown in the figure will be covered in an upcoming unit.

Ø	Diameter
R	Radius
SR	Spherical radius
SØ	Spherical diameter
X	Places or By
( )	Reference (around numeral)
⌒	Arc length (above numeral)
	Dimension origin

Goodheart-Willcox Publisher

Figure 9-3. ASME recommends standard symbols that avoid the use of words when possible.

Dimensional Values

There are two methods of placing dimensional values on a drawing, print, or sketch. These are the unidirectional and aligned dimensioning methods. In ***unidirectional dimensioning***, dimensions are placed so they read from the bottom of the drawing. With ***aligned dimensioning***, dimensions are placed so they read either from the bottom or right-hand side of the drawing, with the dimension value aligned with the dimension line, **Figure 9-4**. Unidirectional dimensions are the most common type in industrial prints. The aligned method is more common on architectural drawings, woodworking drawings, and other drawings that use fractional expressions.

On drawings where all dimensions are either US customary or SI units, it is not necessary to identify the units with "mm" for millimeters or "in" for inches. In these cases, a note such as UNLESS OTHERWISE SPECIFIED, ALL DIMENSIONS ARE IN INCHES should be located on the drawing, perhaps in the title block. For US customary drawings, the standard practice is to use the same number of decimal places as the tolerance precision level, even if zeros need to be added. For example, many of the dimensional values for mass-produced parts are to the third decimal place (thousandths), so 1.000 would be a proper expression for one inch. For values less than one inch, a zero is

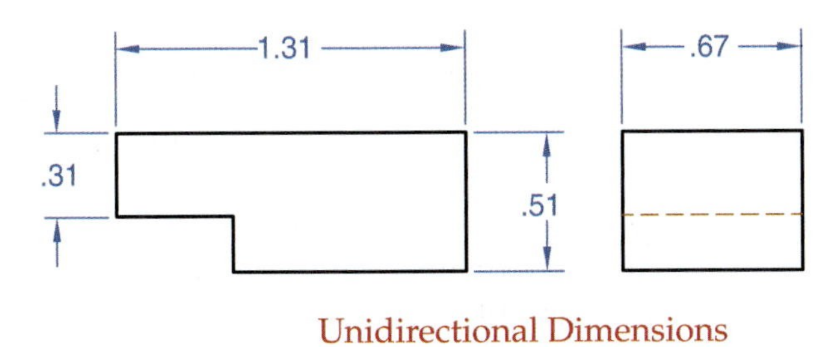

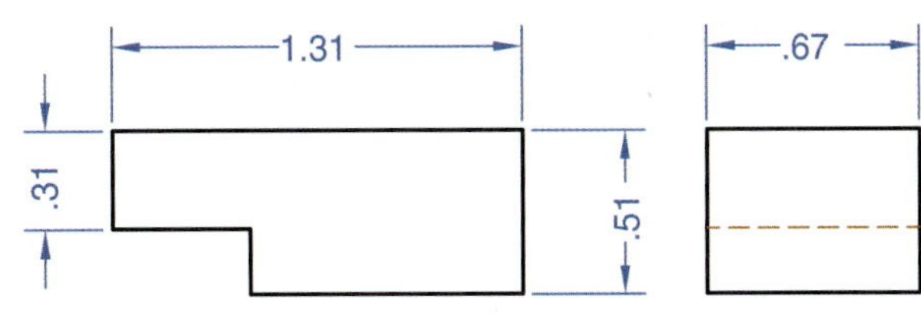

Goodheart-Willcox Publisher

Figure 9-4. There are two methods of placing dimensions on a drawing: unidirectional and aligned.

not placed in front of the decimal, so .750 would be proper, whereas 0.75 would be improper. For millimeter expression, however, the standard practice is different. With SI units, a zero does precede a value less than 1 mm. Whole numbers generally are specified without a decimal, and decimal fractions of a millimeter do not feature any trailing zeros.

Print readers should also be aware that not all countries of the world use a decimal point as the decimal separator between the whole number and the fractional value. Many countries use a decimal comma. For example, the decimal point is used in US practice and in the ASME standards, so 45.25 would be the proper expression for 45 1/4 millimeters, whereas countries that use a decimal comma would express this value as 45,25. Similarly, there are differences in the approach to the separator for thousands. In 2003, at the General Conference on Weights and Measures, it was confirmed that the decimal marker "shall be either the point on the line, or the comma on the line," while it was reaffirmed that "numbers may be divided in groups of three in order to facilitate reading," but neither dots nor commas should be used to separate thousands. For millimeter expression, ASME standards for industrial prints state that neither commas nor spaces should be used to separate larger millimeter values into groups.

Dimensional Linework and Arrangement

When creating the dimensions for a drawing, the drafter or designer creating the print has one global rule that supersedes all suggestions: *Dimension the drawing so it is clear and not confusing to the reader.* With this in mind, the following recommendations are also part of the dimensioning mechanics:

- It is permissible for extension lines to cross each other, and they should not be broken in those cases where they do cross. In cases where an extension line crosses through or near an arrowhead, it may be broken across the arrowhead.
- Dimensions are arranged with shorter dimensions closer to the views so that, if at all possible, extension lines do *not* cross dimension lines.
- It is also desirable that dimension lines do *not* cross each other.
- Subdimensions should line up with each other.
- Crossing dimension lines with leader lines should be avoided.
- Small dimensions create the need to break the dimension line into two pieces, **Figure 9-5.**
- As shown in **Figure 9-5**, the dimension origin symbol can replace an arrowhead if it helps clarify the surface or feature from which a dimension is measured.
- Oblique extension lines can be used in tight quarters, but in these cases, only the extension lines are slanted, and the dimension line direction is still the same, **Figure 9-6.**

Angular dimensions are used on prints to indicate the size of angles in degrees (°) and fractional parts of a degree. Fractional parts of a degree can be

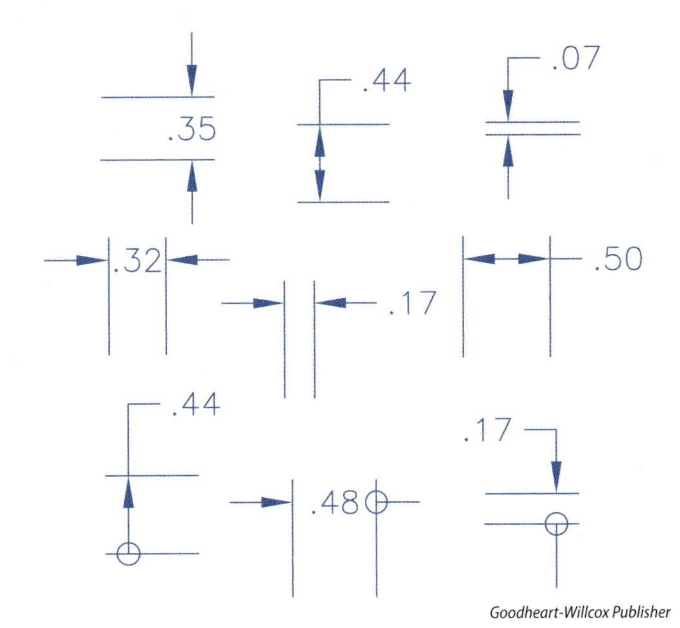

Goodheart-Willcox Publisher

Figure 9-5. The dimension line must be broken and separated in various ways for small dimensional values.

expressed as decimal degrees or as minutes (′) and seconds (″). A complete circle contains 360°. One degree contains 60′ (minutes). One minute contains 60″ (seconds). For an angular dimension, the dimension line is a curve with its center at the vertex of the dimensioned angle, **Figure 9-7.**

While diameter dimensions are used for cylinders and holes, radius dimensions are used for surfaces and features that are not fully round. A radius dimension features one arrowhead, with the dimension line drawn radial to the center point or axis of the curved feature. The radius symbol (R) should precede the dimensional value, **Figure 9-8.** In general, the radius of many rounded features is self-locating, so a location dimension for the arc center is not required. In cases where the radii have a fixed location dimension, a center "plus" should be shown from which those location dimensions can extend. For smaller radii, if space does not permit the dimensional value to be placed between the center point and the arc, various techniques that appear similar to a leader line are used to arrange the arrowhead and dimension value. As illustrated in **Figure 9-8,** on those occasions wherein the radius is very large and the center point may fall off the graphic sheet, the dimension line may be foreshortened with a jog. In those cases, the portion of the dimension line that has the arrowhead is still radial relative to the arc's center. If there is a coordinate location dimension to the center, it will also need to

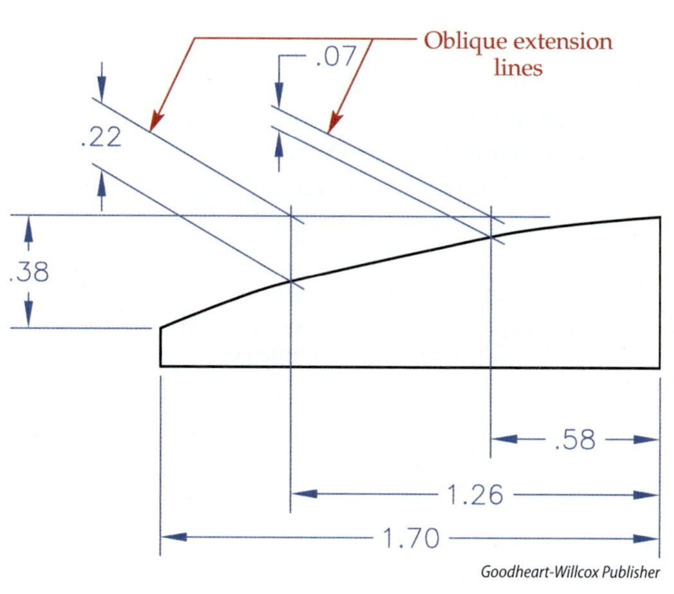

Figure 9-6. In tight situations, extension lines can be drawn at an oblique (slanted) angle, but the dimension lines must maintain their direction.

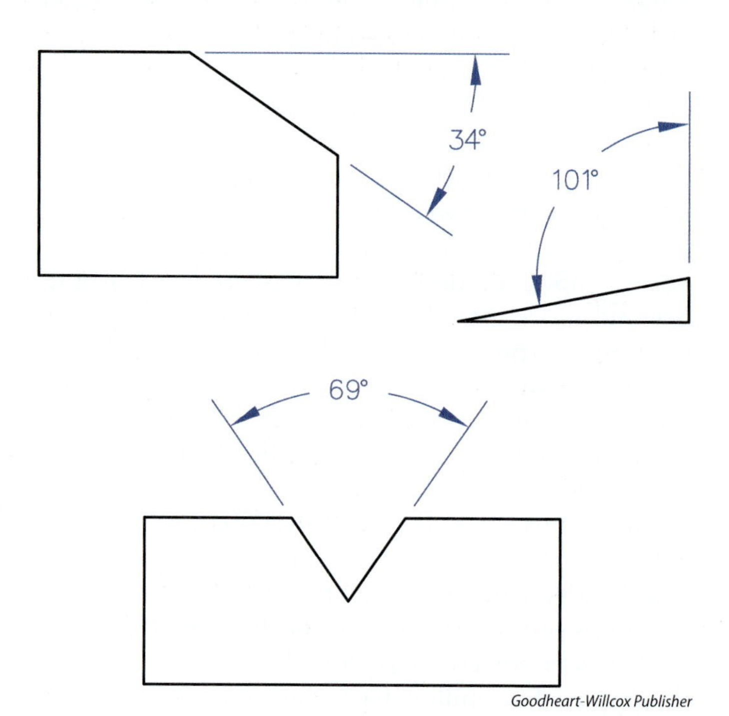

Figure 9-7. For an angular dimension, the dimension line is a curve with its center at the vertex of the dimensioned angle.

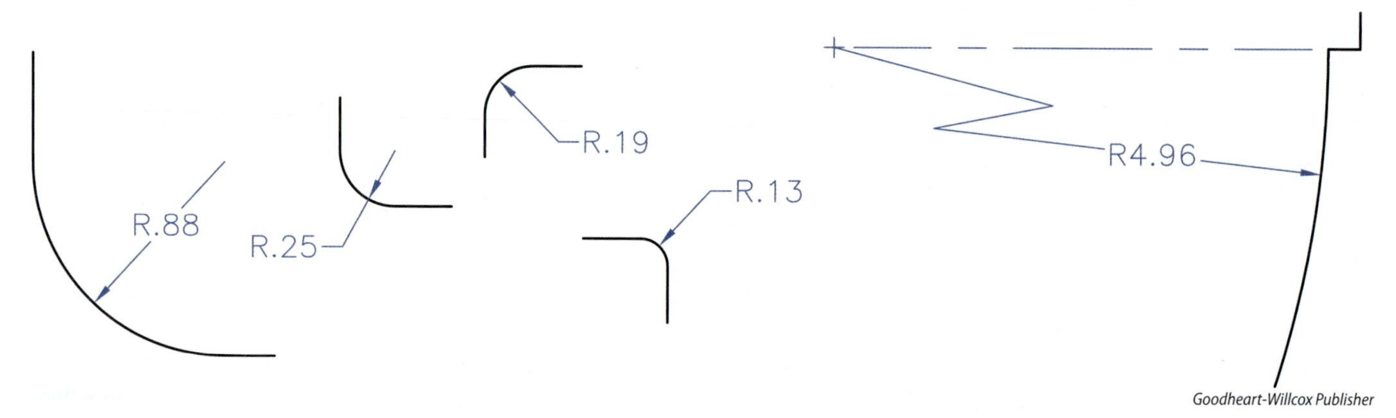

Figure 9-8. For a radius dimension, the dimension line only has one arrowhead and may be constructed in various ways depending on the radius size.

be foreshortened. When a radius dimension appears in the 3D model environment, it should not be foreshortened. Spherical features can be dimensioned by radius or diameter, with an "S" inserted prior to the radius or diameter symbol, **Figure 9-9**.

Dimensioning along an arc can be accomplished with a dimension line that is parallel with the arc, together with a small arc symbol above the dimension value, **Figure 9-10**. Notice that if a linear dimension is given directly across the ends of the arc, a chordal distance is implied.

With respect to dimensioning symbols, the "X" symbol can be applied in two different ways, **Figure 9-11**. The X is commonly used to express the quantity of features being dimensioned, so in this case X can be read as "times." In **Figure 9-11**, both holes have the same size, so 2X precedes the diameter value, with a space after the X. Another common use of the X symbol is to separate two values with the word "by," as in 2 X 4 or .125 X 45°, with a space on each side of the X.

Dimensioning Rules of Choice and Placement

In addition to following the mechanics of dimensioning, the drafter or designer must choose which dimensions to use and then select the best view or location for those dimensions. In industrial mass production, parts are dimensioned in such a way that there is only one way to determine the location or size of a feature, so multiple options will exist for a feature.

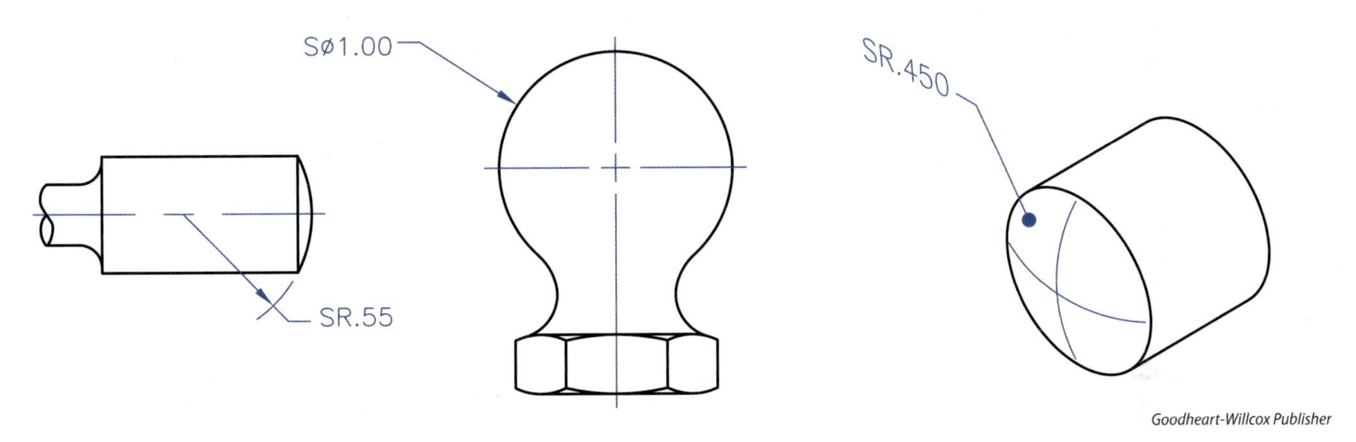

Figure 9-9. For spherical features, the "S" symbol is added before the radius or diameter symbol.

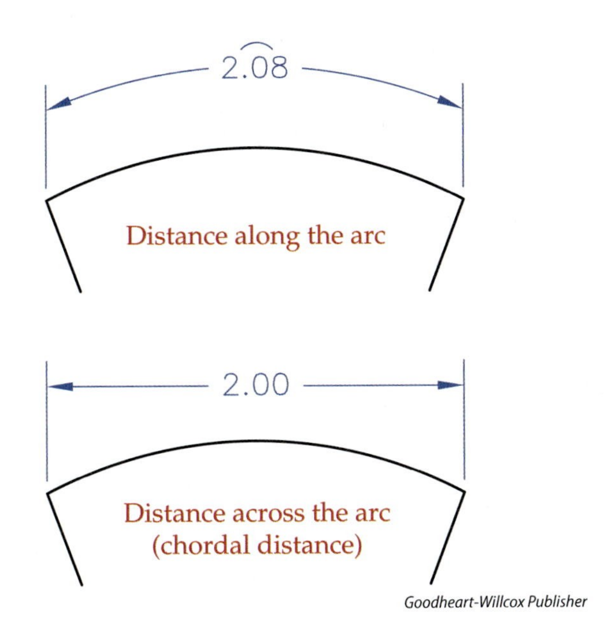

Figure 9-10. For dimensioning the length along an arc, versus the chordal distance across the arc, the arc symbol is placed above the dimension value.

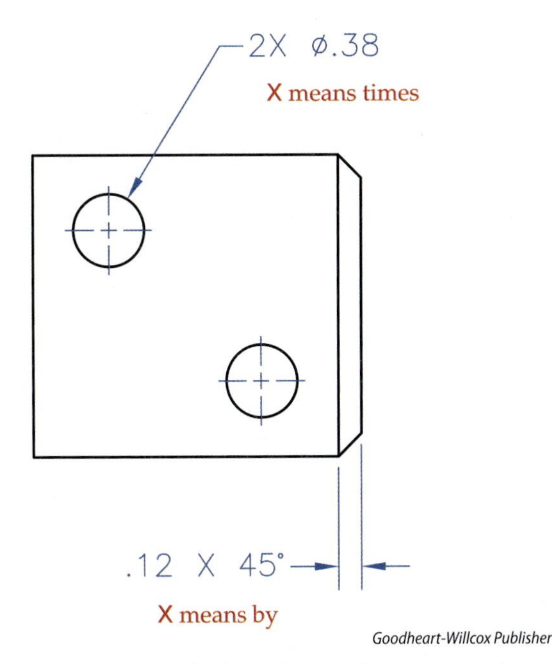

Figure 9-11. The "X" symbol can be used to express the word "times" or the word "by."

The elements of *choice* and *placement* are the real difficult aspects of dimensioning. Knowing the guidelines for choice and placement may help the print reader know where to look for dimensions on a print. However, the function of the part, the purpose of a feature, or the machining and inspection processes may dictate dimension locations that do not follow these guidelines. In summary, the guidelines for choice and placement are just that—guidelines.

In keeping with the global rule "to clearly express the size," dimensions should be located off the views but still central to the drawing. Therefore, you should find most dimensions between the views. Notes and leader lines, however, are likely located around the perimeter of the view arrangement.

A feature should be dimensioned in the view that is most descriptive of the feature's shape. This is known as ***contour dimensioning.*** A feature such as a slot or notch has its shape, or contour, shown in one particular view. Closely related to this principle is a drafter's rule to avoid dimensioning to a hidden line or the center line of a hidden hole. **Figure 9-12** shows two versions of a drawing with contour dimensioning, one properly constructed and one poorly constructed.

For objects comprised of external cylindrical features, the diameter of the cylinder should be dimensioned in the "rectangular" view. See **Figure 9-13.** Even if a cylindrical feature is milled flat across one or more sides, the diameter of the cylinder is usually given. Older standards allowed the diameter symbol to be omitted if there was a "round" view. However, current standards request the symbol to be used in all cases. In summary, in orthographic views, leader

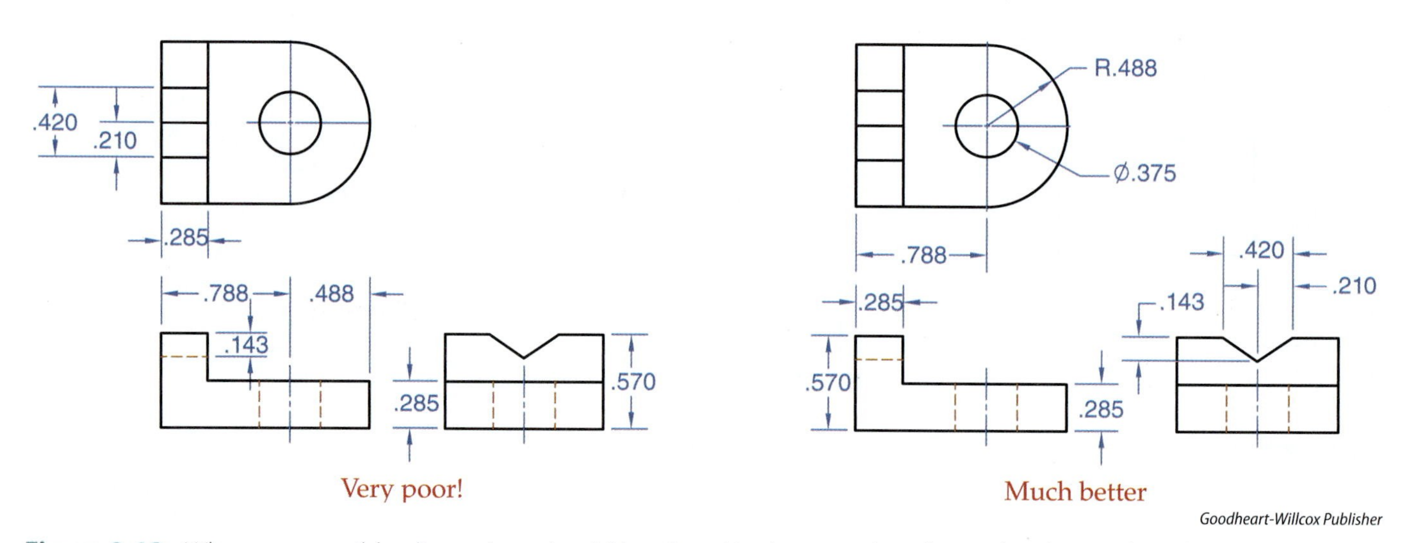

Goodheart-Willcox Publisher

Figure 9-12. Whenever possible, dimensions should be placed in the view that shows the shape of the feature being dimensioned. Also, try to avoid dimensioning to a hidden line. This is not always possible.

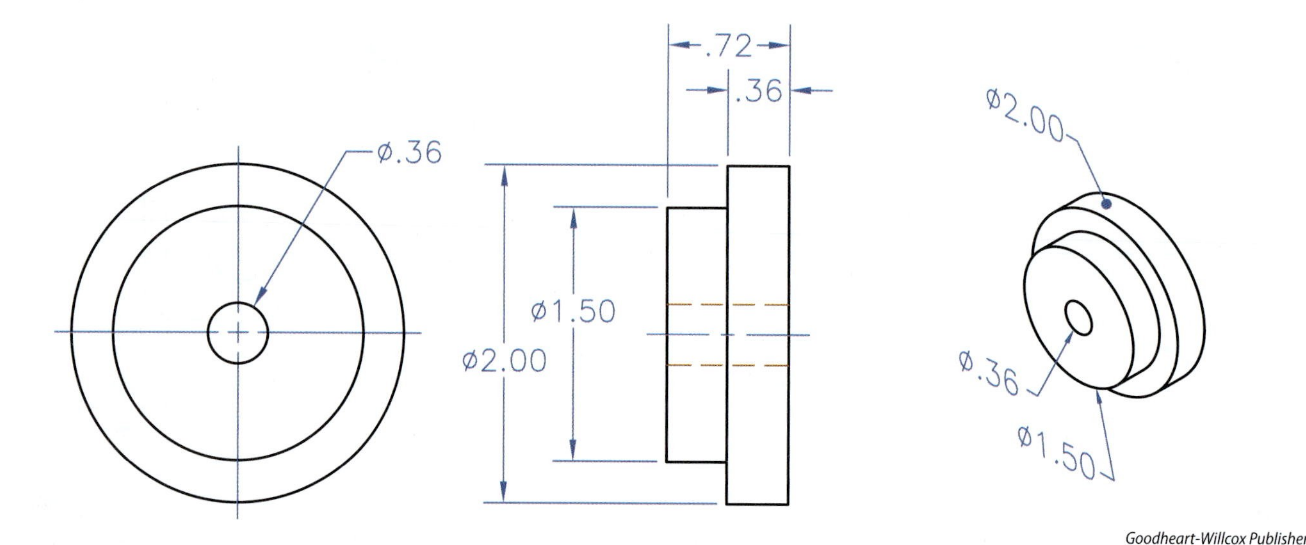

Goodheart-Willcox Publisher

Figure 9-13. External cylindrical features should be dimensioned in the rectangular view.

lines should only be used to point to a circle if it is a hole. One exception to this is for dimensions placed within the 3D model environment. For these applications, leader lines may be used to specify cylinder sizes if that is as clear as a linear dimension option. Holes located around a ***bolt circle***, or "circle of centers," can be located in several ways. Some common methods are shown in **Figure 9-14**.

An angled surface can be dimensioned by two offset measurements or by an angular value and an offset measurement, **Figure 9-15**. Sometimes an angular surface, by the nature of the angle, creates an edge that does not need a size dimension. If a print seems to be missing a dimension, check to see if this condition exists. Again, hopefully the drawing is dimensioned with the function and purpose of the angled surface in consideration. Manufacturing or inspection methods often dictate a particular choice of dimensions. In some cases, of course, the angled feature may not be critical. Unit 13, *Geometric Dimensioning and Tolerancing*, discusses various ways that geometric features, such as angled features, can be controlled with more precision and intent.

Chain dimensioning, wherein dimensions are linked together end to end, can result in the accumulation of tolerances, if applied. To prevent this accumulation, features can be dimensioned from a common datum surface or feature, **Figure 9-16**. This is called ***baseline dimensioning***. A ***datum*** is the origin from which the location of features is established, or the reference plane or axis to which a geometric control is referenced. Tolerancing and datum references are covered in detail in Unit 10.

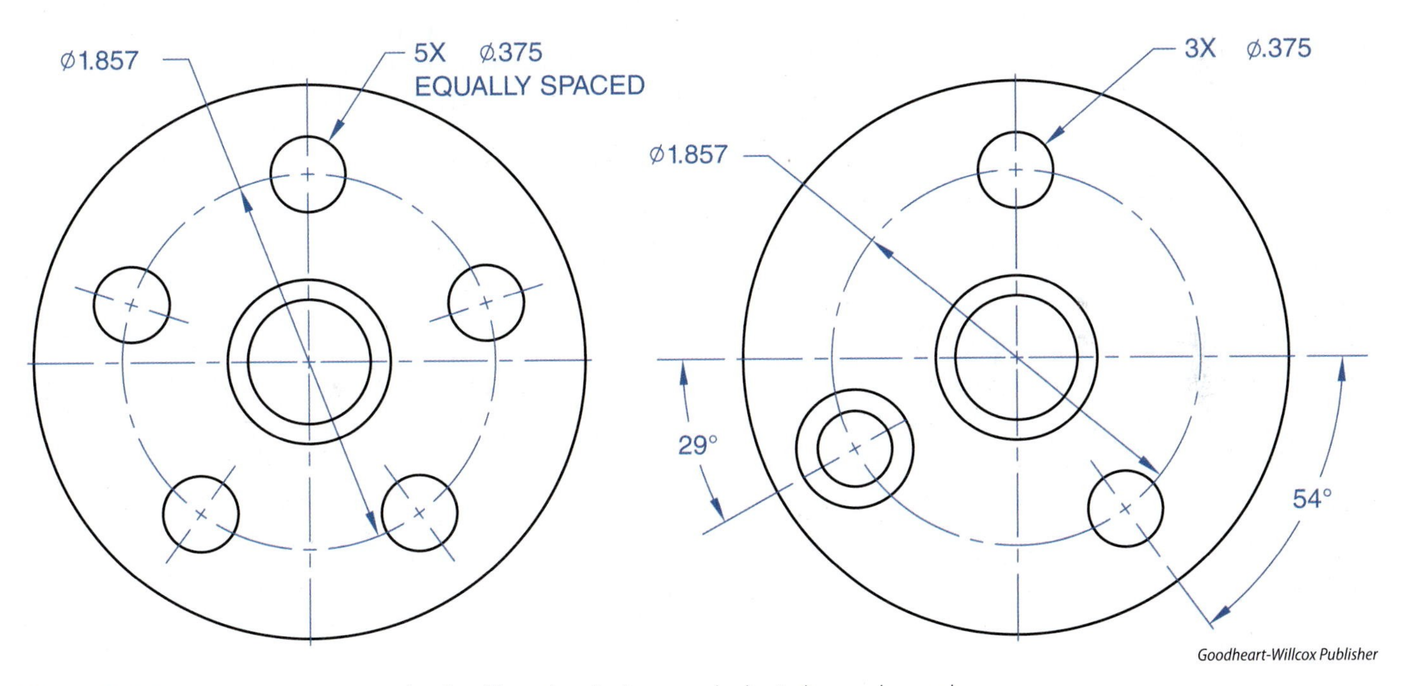

Figure 9-14. Some common methods of locating holes on a bolt circle are shown here.

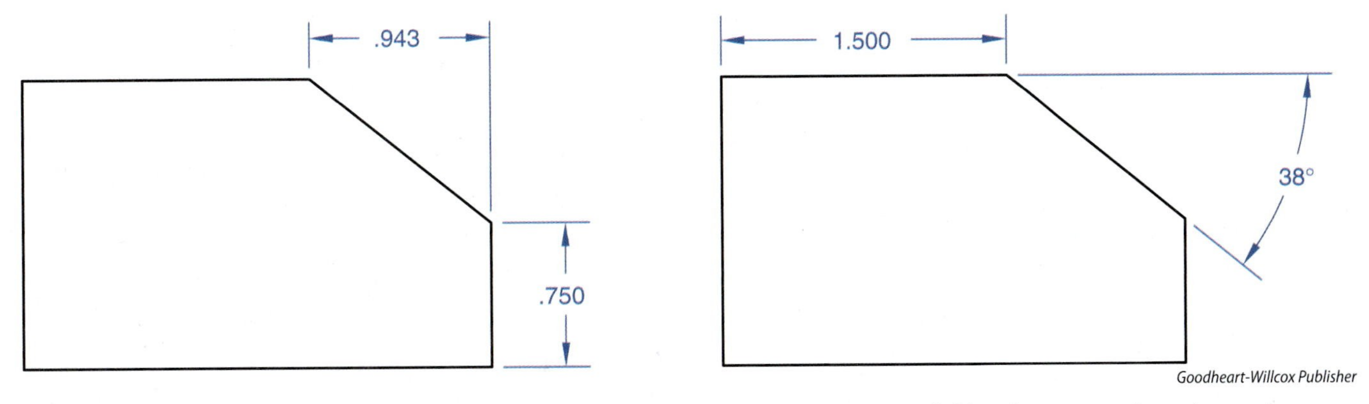

Figure 9-15. An angled surface can be dimensioned by two offset measurements (left) or by an angular value and an offset measurement (right).

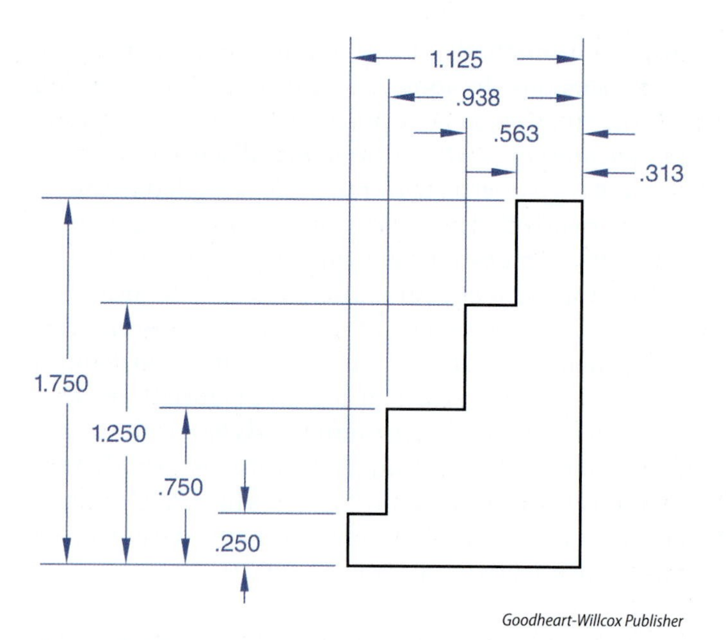

Figure 9-16. Features are dimensioned from a common datum surface or feature to prevent the accumulation of tolerances.

Features with rounded ends provide many choices for dimensioning. See **Figure 9-17** for examples of external features. In these cases, an overall dimension may not be present. The "correct" choice is often based on the accuracy required or the machining process used to create the feature. See **Figure 9-18** for examples of rounded-end slots and standard dimensioning techniques for interior features. The rounded ends of internal slots are often dimensioned as the distance across the slot instead of the radius at the end of the slot. If the slot will be machined with a cutting bit, then the diameter of the bit is the more valuable dimension. In the case of a full radius at the end of the slot, a radius is indicated but no dimensional value is given. Sometimes the word FULL may be added in front of the R symbol. If the radius is not FULL, then a dimension is specified.

In summary, the guidelines in this section have been recommended throughout the years to help drafters create clear, concise size descriptions on drawings. There are too many special cases to cover within the

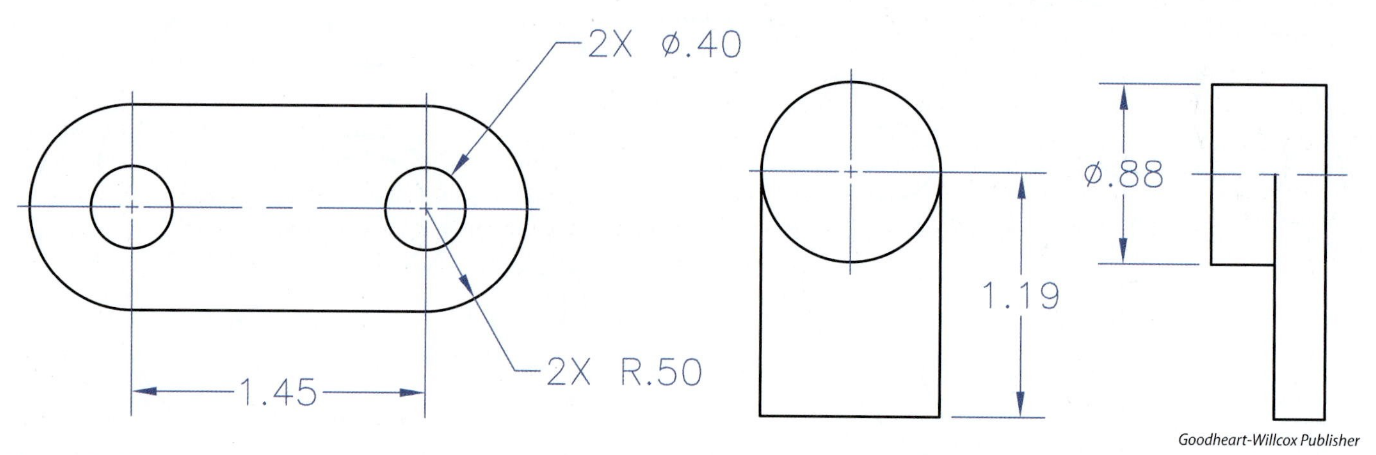

Figure 9-17. Exterior features with rounded ends may be dimensioned center to center with a radius given rather than stating the overall dimension.

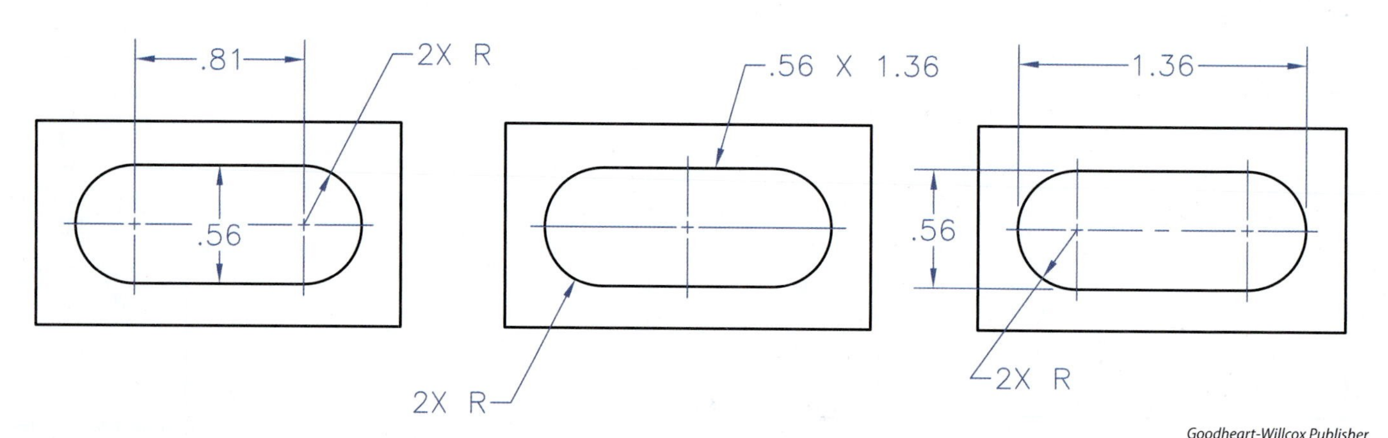

Figure 9-18. Interior features with rounded ends may be dimensioned with a distance across the slot, with a radius indicated but not stated.

scope of this text. Therefore, if a feature does not have a dimension or it appears a dimension is missing, the print reader should *not* make an assumption or measure the drawing for critical information. As discussed in the "Fundamental Rules" section, dimensions should *never* be calculated or assumed. Omissions should be reported to the department responsible for the prints.

Additional Dimensioning Techniques

On a dimensioned drawing, there are usually both size and location dimensions. ***Size dimensions*** indicate the size of the part and the size of its various geometric features, such as holes, fillets, and slots. ***Location dimensions*** indicate the location of features such as holes, slots, and grooves.

Reference dimensions are occasionally given on drawings to assist in knowing certain general distance information, like the basic total size of a feature or sum of dimensions. ***Reference dimensions*** are not toleranced and are *not* to be used for layout, machining, or inspection operations. These dimensions are marked by parentheses. In older standards, the dimension value was followed by REF, **Figure 9-19**. According to ASME Y14.5, the use of reference dimensions on a drawing should be minimized.

Tabular dimensions are placed on the drawing as reference letters within the dimension lines. A table on the drawing lists the corresponding dimensions, **Figure 9-20**. Tabular dimensions are useful when a company manufactures a series of sizes of an assembly or part. With this system, more than one part can be featured on a single drawing.

Another use for tabular dimensions is dimensioning a part with a large number of repetitive features, such as holes, **Figure 9-21**. The chart in this application is sometimes referred to as a hole chart. Running extension and dimension lines to each hole would make the drawing difficult to read. To make the drawing clearer, each hole or feature can be assigned a letter, number, or letter with a subscript number. The dimensions of the feature and its location along the X and Y axes can then be given in a table on the drawing.

Coordinate dimensioning without dimension lines is frequently used on drawings containing datum lines or planes, **Figure 9-22**. This practice eliminates numerous dimension and extension lines and improves the clarity of the drawing. Coordinate dimensioning without dimension lines is also called *ordinate dimensioning*, *arrowless dimensioning*, or *datum dimensioning*. This system of dimensioning is especially useful on drawings of parts that are to be machined on computer numerically controlled (CNC) equipment. Coordinate dimensioning without dimension lines requires zero-value reference planes be established, from which all values are given in the

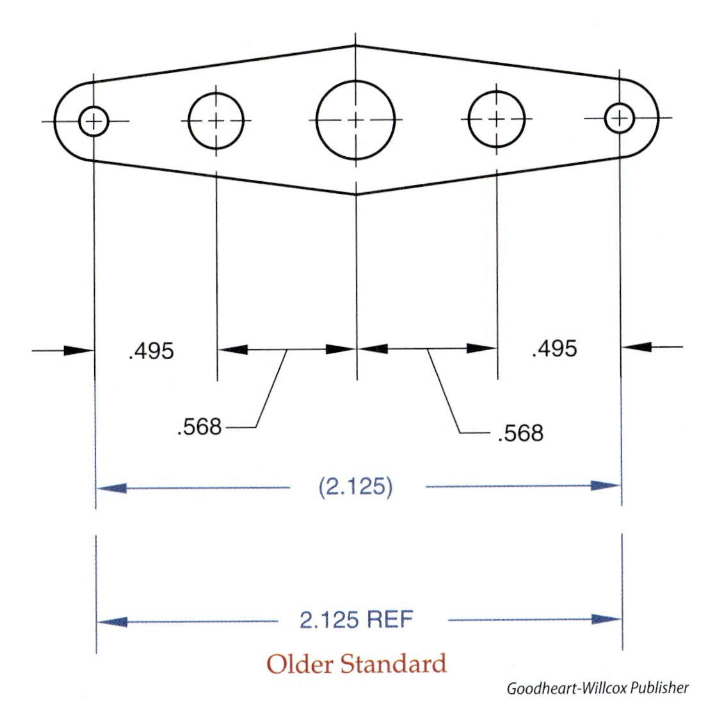

Goodheart-Willcox Publisher

Figure 9-19. Reference dimensions are marked by parentheses or followed by REF.

PART NO.	A	B	C	D
41–8706	.750	.500	1.312	.875
41–8707	1.125	.750	1.812	1.000
41–8708	1.500	1.000	2.062	1.125

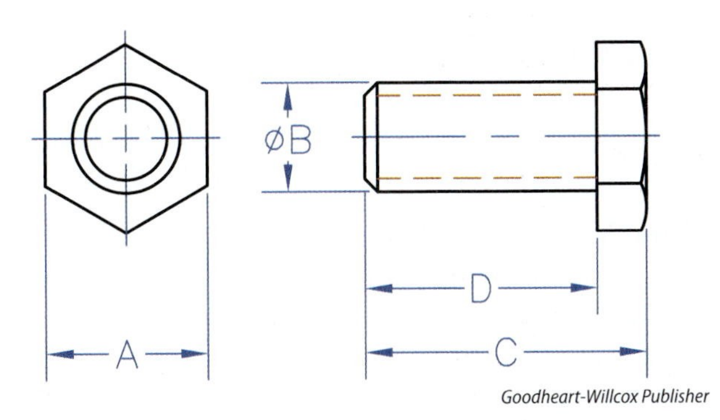

Goodheart-Willcox Publisher

Figure 9-20. In tabular dimensioning, a table on the drawing may list the corresponding dimensions for more than one part number.

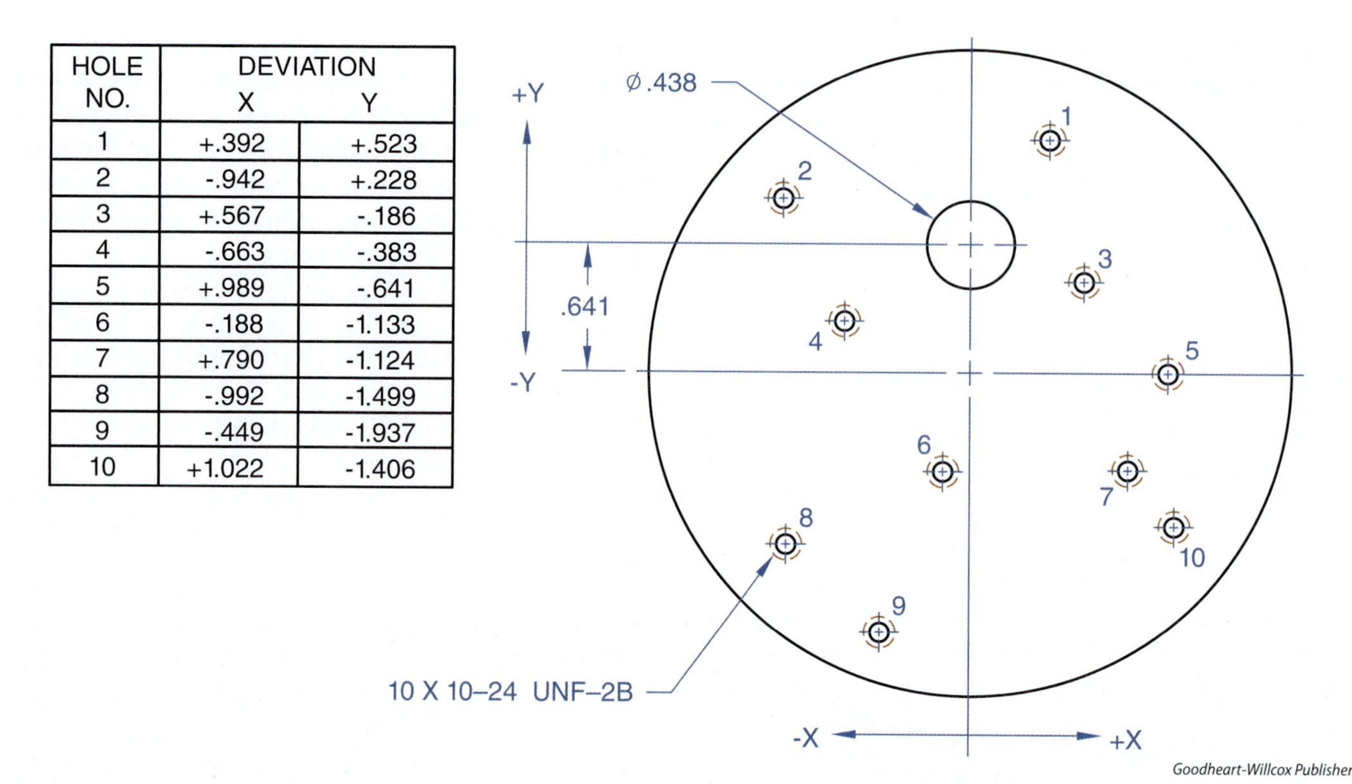

HOLE NO.	DEVIATION X	Y
1	+.392	+.523
2	-.942	+.228
3	+.567	-.186
4	-.663	-.383
5	+.989	-.641
6	-.188	-1.133
7	+.790	-1.124
8	-.992	-1.499
9	-.449	-1.937
10	+1.022	-1.406

Goodheart-Willcox Publisher

Figure 9-21. A hole chart is a type of tabular dimensioning for a part with a large number of repetitive features, such as holes.

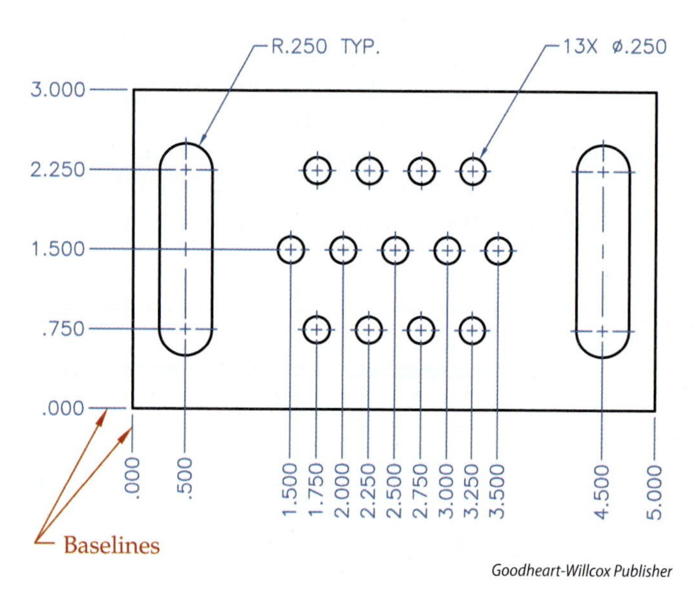

Goodheart-Willcox Publisher

Figure 9-22. Coordinate dimensioning without dimension lines eliminates numerous dimension and extension lines and improves the clarity of the drawing.

X and Y directions. These lines are sometimes referred to as *baselines*, as this dimensioning falls into the category of baseline dimensioning discussed earlier.

Sometimes on a print it may be desirable to have dimensions that are not the same as the primary unit of measure. In these cases, ASME standards dictate that, for an inch-based drawing, the abbreviation "mm" should follow those metric values that are given. Likewise, if an inch value is shown on a millimeter-based drawing, the abbreviation "IN" should be applied. ***Dual dimensioning*** is a practice that has been used on industrial prints to show both inch and millimeter values for the dimensional values throughout the drawing. In these cases, the bracket method is a common method that places the alternate value in brackets following the primary value. In **Figure 9-23**, the primary value is in inches and the alternate value is in millimeters. The method used should be consistent throughout the drawing, and a note should appear on the print to explain the method used. It is important to note that dual dimensioning is not provided for in the ASME Y14.5 standard. Technically, if both dimensional values are considered of equal importance and tolerances are to be applied, this breaks a fundamental rule that each size and location dimension has only one interpretation. Therefore, dual dimensioning has been a subject of debate throughout the years. In summary, the print reader should be aware of the intention and purpose of any dual dimensioning that is expressed, and should know whether or not one unit of measure has priority over the other value with respect to inspection.

Notes and Special Features

Notes are almost always necessary to complement views and dimensions. Local notes use leader lines to

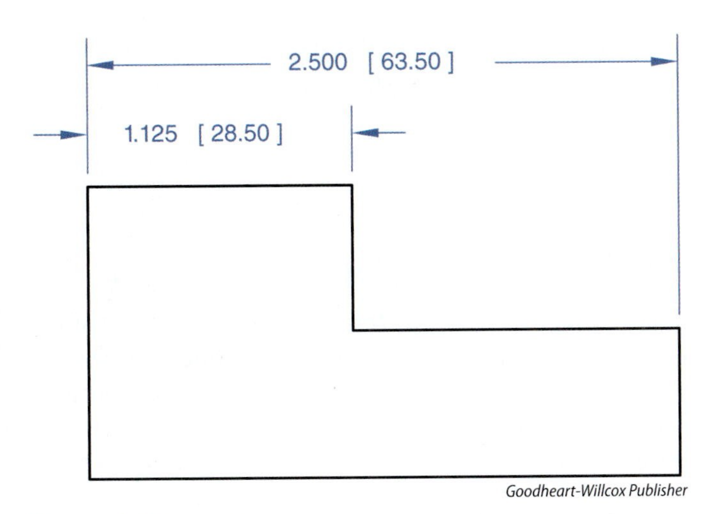

Figure 9-23. Dual dimensioning is sometimes used on a drawing to show both inch and millimeter values.

point to special features, holes, and threads. Special features often include machining specifications, including counterbores, countersinks, tapers, knurls, keyways, and chamfers. General notes usually apply to the entire part. Examples of general notes include FILLETS AND ROUNDS R.125 UNLESS SPECIFIED, FINISH ALL OVER, and BREAK ALL SHARP CORNERS. Many of these special features, machining specifications, and other notes will be examined in detail in Unit 11, *Machining Specifications and Drawing Notes*.

Digital Product Definition

The ASME Y14.41 *Digital Product Definition Data Practices* standard was developed to address alternative practices for describing size and applying dimensions. Within these practices, an electronic model of the part or assembly can serve as the contract information, and the use of projected orthographic views is not required. Still included with the model file data are items you would expect to find on a standard drawing sheet, such as the title, approval dates, part number, originator's name, and other pertinent data.

Annotations, such as dimensions and notes, can also be included. However, rather than attaching the annotations to 2D views, they are attached to the 3D model on 2D planes, **Figure 9-24**. In reference books and standards, the annotations appear similar

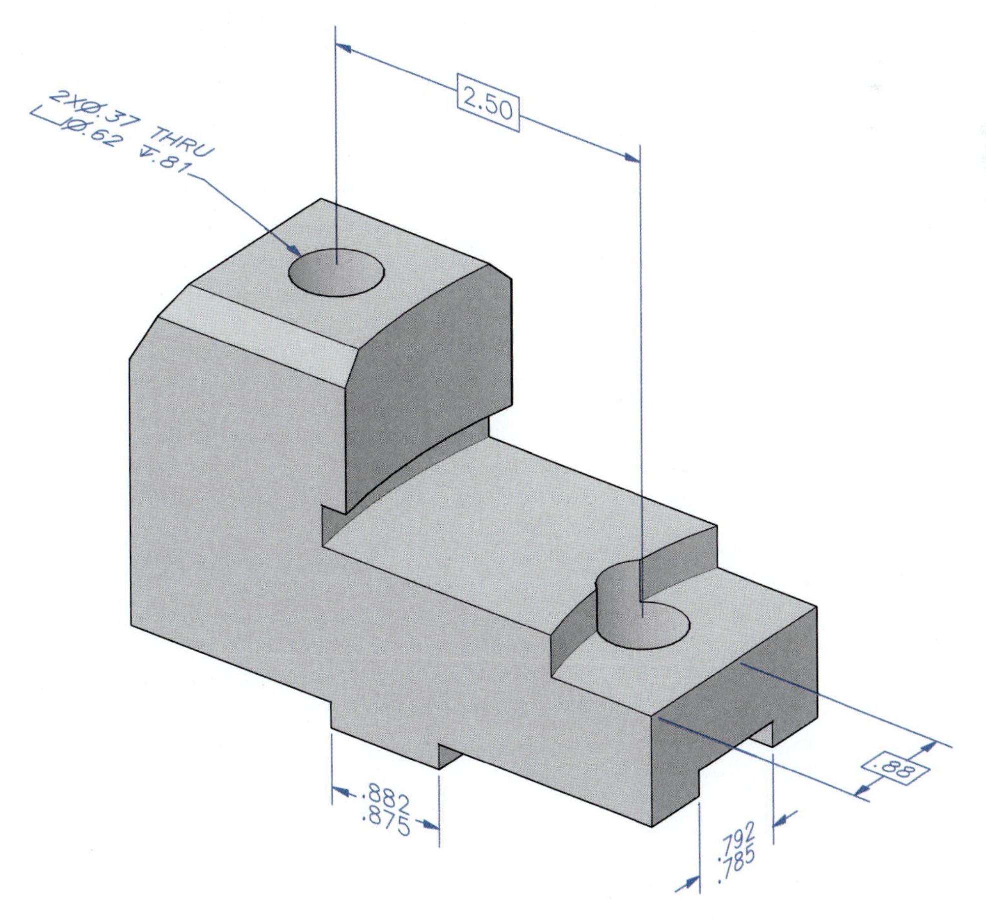

Figure 9-24. Some companies are now incorporating digital product definition, wherein the 3D model data provides most of the size information, but critical dimensions and annotations can still be added.

to how dimensioned pictorial drawings appear on paper. However, when displayed electronically, the dimensions rotate and turn with the model as the model is rotated. Because the 3D model provides much of the size description, annotations may only be necessary for critical measurements and tolerances, features that are not modeled (such as thread notes), or geometric size tolerances and datum identification.

The Y14.41 standard also allows for a ***drawing graphic sheet*** to be included. The drawing graphic sheet may include orthographic and axonometric (rotated) views that are generated directly from the model, **Figure 9-25**. The views are not required to provide a complete product definition, although that is allowed. Requirements for creating a data set that uses both a model and a drawing graphic sheet include the following:

- Product definition data created or shown in the model must be in agreement with the drawing graphic sheet, and vice versa.
- The drawing graphic sheet must include a border and title block. Hard copy outputs must be obtainable.
- When complete product definition is not contained on the drawing graphic sheet, it must be noted.
- When complete product definition is not contained on the model, it must be noted.
- Dimensions, tolerances, datums, and notes may appear in orthographic views or axonometric views.
- The use of color is acceptable on drawing graphic sheets.

Undimensioned Drawings

The ASME Y14.31 *Undimensioned Drawings* standard was developed to set forth recommendations for drawings that describe the size of an object without annotations, much as if the drawing itself were a size template or overlay. By definition, an ***undimensioned drawing*** defines the item with a true geometry view. The ***true geometry view*** serves as a true profile of the part. Within this system, drawings are usually printed or plotted full scale on a strong and durable media, such as polyester film. While some dimensions may be included, they are usually of a nominal reference nature or for features held to tolerances different from the undimensioned feature tolerances.

Undimensioned drawings are suitable for applications such as parts with numerous irregular contours, printed circuit boards, wire harnesses, flat patterns, art layout for instruction plates, and formed sheet metal sections. Undimensioned drawings must incorporate dimensional accuracy methods so that the accuracy of the drawing can be verified as true size. Acceptable methods could include grid lines, dimensional accuracy points, or registration marks.

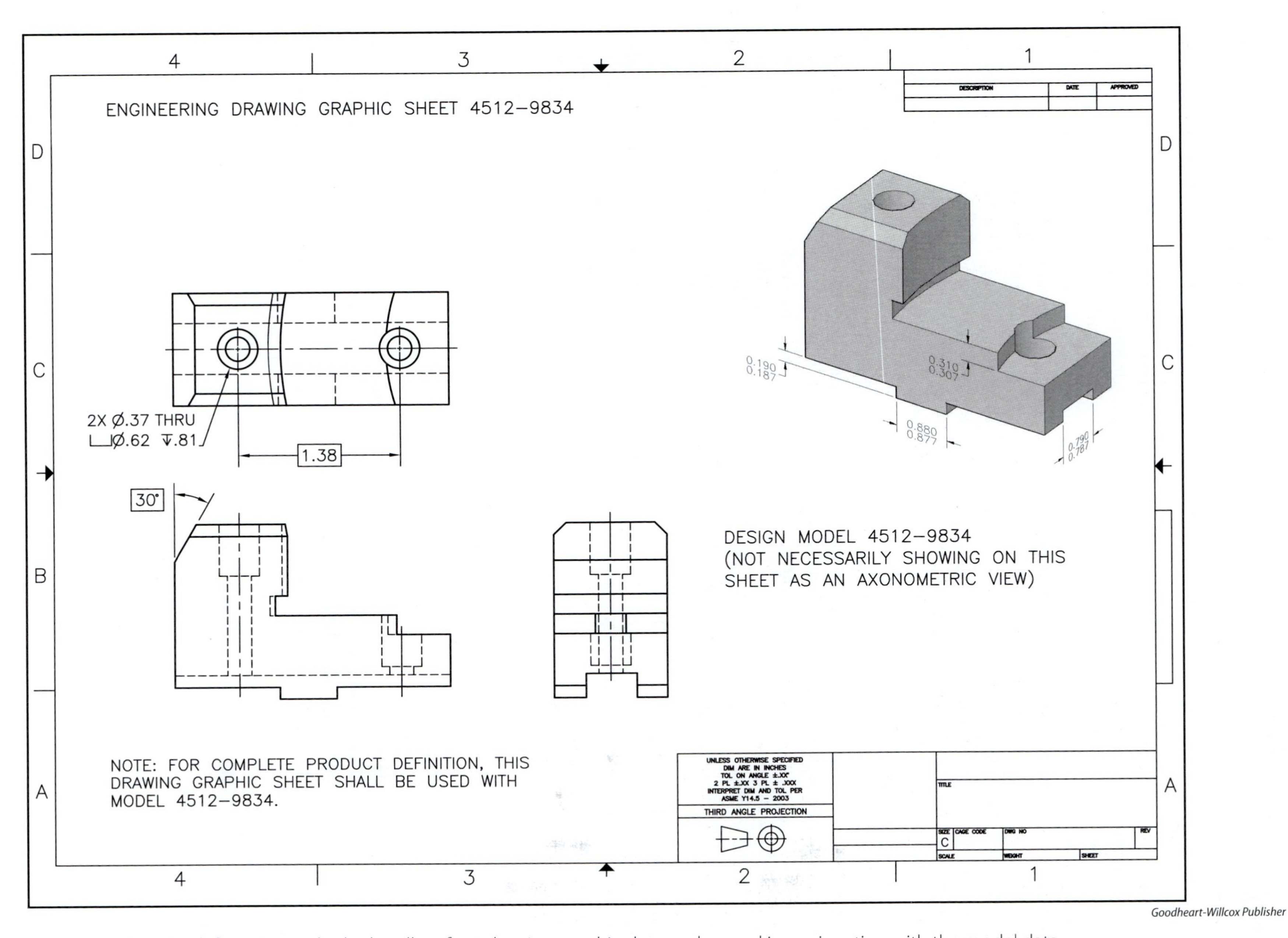

Goodheart-Willcox Publisher

Figure 9-25. Digital product definition standards also allow for a drawing graphic sheet to be used in conjunction with the model data.

Summary

- Dimensioning is defined as the annotations to the *shape* description that comprise the *size* description of the part or assembly.
- The primary standard that covers dimensioning practices is ASME Y14.5, titled *Dimensioning and Tolerancing*, while other standards cover newer accepted practices for dimensions within a product definition data set or for undimensioned drawings.
- There are fundamental rules that guide the application of dimensions, including expectations and assumptions.
- There are two units of measure that are common to industrial prints, US customary units and SI units.
- Dimensioning mechanics is concerned with the rules for constructing dimensions, including arrowhead appearance, lettering size, and spacing for extension lines.
- Lines used in dimensioning include extension lines, dimension lines, leader lines, and chain lines.
- Standard symbols for dimensioning include symbols for diameters, radii, spherical features, reference values, arc length, and dimension origin.
- Dimension values are placed on a drawing in either an aligned or unidirectional method, and there are rules that govern the use of both US customary units and SI units, which are applied in different ways.
- One aspect of dimensioning is choice and placement, which involves the study of rules that guide the print creator on which dimensions are selected and where the dimensions should be located to best describe the size and location of the object and its features.
- Contour dimensioning is the principle of applying dimensions in the view in which the shape is shown.
- Standard practice for dimensioning cylindrical features in orthographic views requires the diameter to be given in the rectangular view for external features and leader lines to be used for internal features.
- Angled features may be dimensioned in various ways, either with angular dimensions or offset dimensions.
- Dimensioning choices must be made with respect to chaining dimensions together versus dimensioning from a baseline.
- Dimensioning rounded-end features requires choices to be made, and the "correct" choice is often based on the accuracy required or the machining process used to create the feature.
- Reference dimensions are sometimes used to provide summary information in addition to the manufacturing dimensions.
- There are several dimensioning techniques that assist with creating a clear and concise size description, including tabular dimensions, hole charts, and coordinate dimensions without dimension lines.
- Although not endorsed by standards, dual dimensioning is a technique of providing both US customary unit values and SI unit values for each dimension.
- Newer alternate methods for describing the size of an object are set forth in a standard titled *Digital Product Definition Data Practices*, which includes methods for applying dimensions within the 3D model environment, with or without drawing graphic sheets.
- Undimensioned drawings are suitable for applications such as parts with numerous irregular contours, printed circuit boards, wire harnesses, flat patterns, art layout for instruction plates, and formed sheet metal sections.

Unit Review

Name ______________________ Date ______________ Class ______________

Answer the following questions using the information provided in this unit.

Know and Understand

____ 1. *True or False?* Dimensioning can be described as the *size description*.

____ 2. *True or False?* Dimensioning is the easiest part of creating an industrial print.

____ 3. Which of the following is *not* a true fundamental rule of dimensioning?
- A. An industrial print should be a *complete* size description.
- B. Measuring of a print is not allowed, nor should it be expected.
- C. It is helpful to give the manufacturing process (such as DRILL or REAM) when possible.
- D. The arrangement of the dimensions should be related to function, or mating relationships.

____ 4. *True or False?* Industrial prints for small parts will most likely be dimensioned in inches (US customary units) or millimeters (SI units).

____ 5. Which of the following is the best definition for dimensioning mechanics?
- A. What dimensions look like
- B. Where dimensions are placed
- C. Which dimensions are chosen
- D. All of the above

____ 6. Which of the following is *not* recommended to be 1/8″ (3 mm)?
- A. The height of the lettering
- B. Amount of gap between extension lines and the object
- C. The length of the arrowhead
- D. Amount of extension line beyond the arrowhead

____ 7. *True or False?* Leader lines have an arrowhead on one end and a shoulder on the other end, although the arrowhead could be replaced with a dot for some situations.

Match the following dimensioning symbol clues to a description of the symbol. Answers are used only once.

____ 8. Looks like a circle with an arrow

____ 9. Looks like parentheses

____ 10. Looks like the letter R

____ 11. Looks like a circle with a slash through it

____ 12. Looks like the letter X

- A. Diameter
- B. Radius
- C. Times
- D. Reference dimension
- E. Dimension origin

____ 13. *True or False?* Dimension values on industrial prints are always created in such a way that the values line up with the dimension lines and read from the bottom or from the right.

____ 14. *True or False?* The rules and methods for the format of millimeter values on the print are the same as the format of inch values.

____ 15. *True or False?* It is permissible for extension lines to cross each other but, if possible, dimension lines should not cross each other.

____ 16. Which of the following will have a curved dimension line?
- A. A reference dimension
- B. A diameter dimension
- C. A rounded-end radius dimension
- D. An angular dimension

____ 17. Identify the *false* statement.
- A. Angular values less than one degree must be given as minutes and seconds.
- B. The radius value, when given, is preceded by the R symbol.
- C. A spherical radius or diameter is preceded by the S symbol.
- D. An X symbol can mean the word "by" or the word "times."

____ 18. Which of the following is a term related to placing dimensions in the view where the shape is shown?
- A. Baseline dimensioning
- B. Chain dimensioning
- C. Contour dimensioning
- D. Reference dimensioning

____ 19. *True or False?* Rounded-end shapes, both interior and exterior, should be dimensioned center-to-center with a radius value given.

Match the following dimensioning terms to the descriptions. Answers are used only once.

____ 20. Dimensions where the values may be A, B, C, etc.

____ 21. Dimension values are not to be used for inspection.

____ 22. Dimension values are in brackets.

____ 23. A form of baseline dimensioning

A. Reference dimensions
B. Tabular dimensions
C. Arrowless dimensioning
D. Dual dimensioning

____ 24. *True or False?* General notes apply to the entire part, whereas local notes are usually placed at the end of a leader line.

____ 25. What is the name of the ASME standard that sets forth guidelines for including size information with a 3D model, perhaps in conjunction with 2D views?
A. Undimensioned Drawings
B. Geometric Dimensioning and Tolerancing
C. Digital Product Definition Data Practices
D. Orthographic and Pictorial Views

____ 26. *True or False?* There are objects such as printed circuit boards, wire harnesses, or parts with irregular contours that may need to be checked with full-size templates or overlays, rather than dimensions from a drawing.

Critical Thinking

1. Choose two of the fundamental rules for dimensioning that were presented in Section 9.1, and explain why each of those rules is important.

____

____

____

____

2. Discuss when you think dual dimensioning might be appropriately used and why dual dimensioning has been the subject of debate among print readers and drafters.

____

____

____

____

____

3. Discuss future trends in dimensioning with respect to how a 3D computer model can be the way that a machinist or quality control inspector could make or check a part without a fully dimensioned print.

____

____

____

____

____

Apply and Analyze

Name ______________________ Date ______________ Class ______________

Review Activity 9-1

Sketch the dimensions required to completely describe the size of the block shown below without overdimensioning. If feasible, discuss the choice and placement with others to evaluate the available options. Your instructor will indicate whether or not you should use real number values and measurements or simply use "X" for the number values to focus on the mechanics and the choice and placement options.

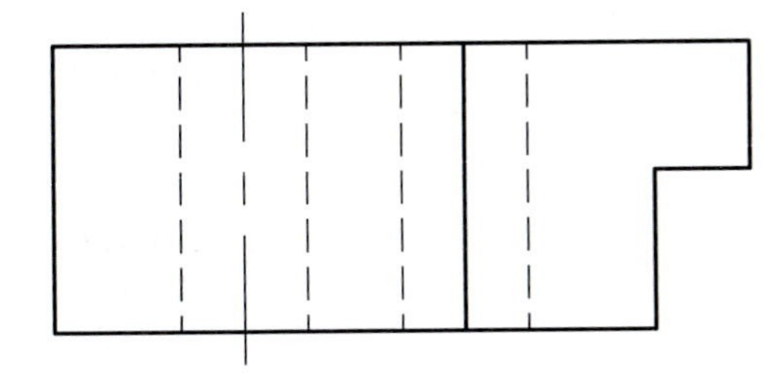

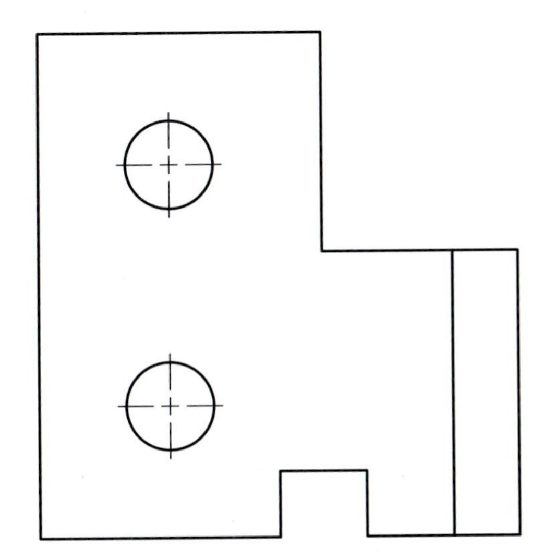

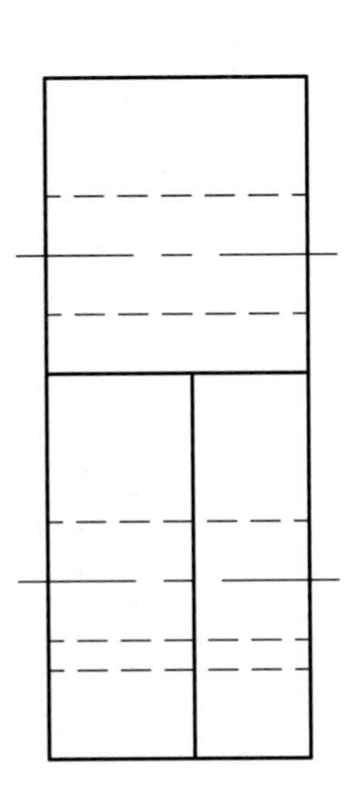

Apply and Analyze

Name ______________________ Date ______________ Class ______________

Review Activity 9-2

Sketch the dimensions required to completely describe the size of the block shown below without overdimensioning. If feasible, discuss the choice and placement with others to evaluate the available options. Your instructor will indicate whether or not you should use real number values and measurements or simply use "X" for the number values to focus on the mechanics and the choice and placement options.

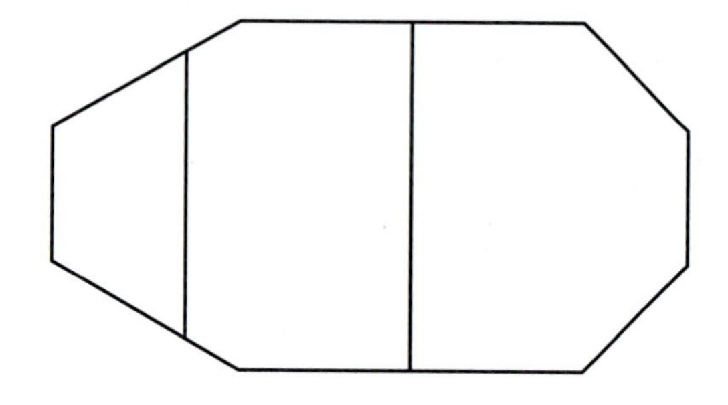

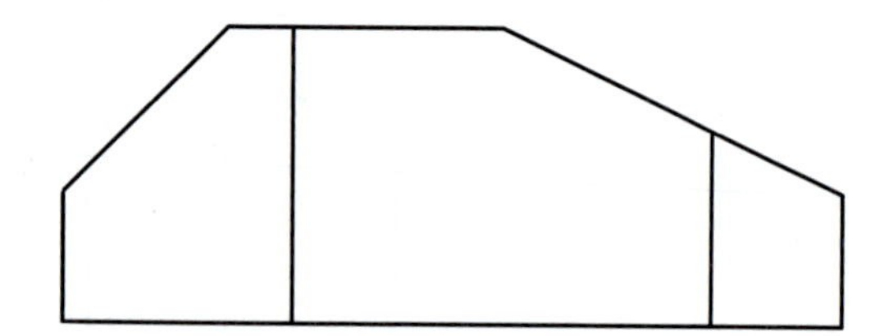

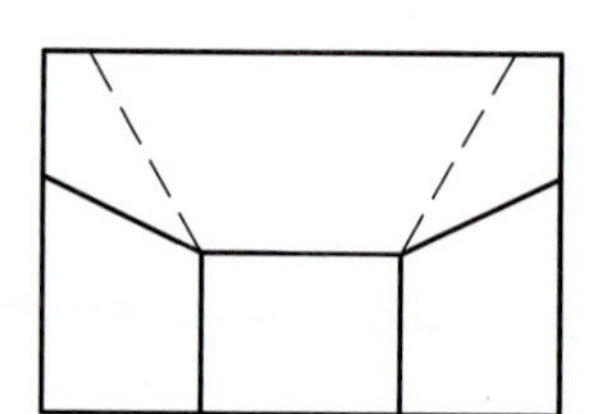

Apply and Analyze

Name ______________________ **Date** ______________ **Class** ______________

Review Activity 9-3

Sketch the dimensions required to completely describe the size of the block shown below without overdimensioning. If feasible, discuss the choice and placement with others to evaluate the available options. Your instructor will indicate whether or not you should use real number values and measurements or simply use "X" for the number values to focus on the mechanics and the choice and placement options.

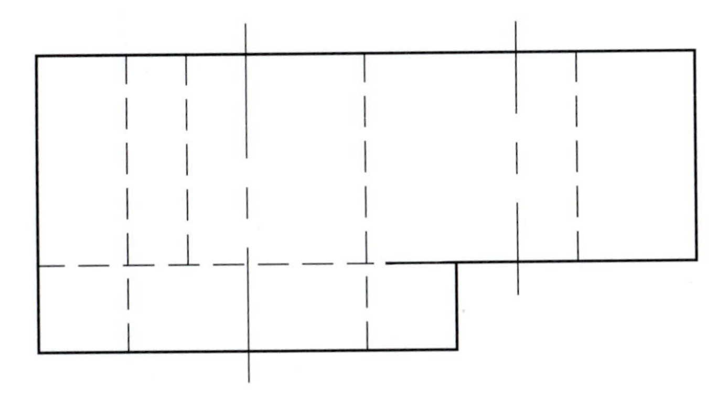

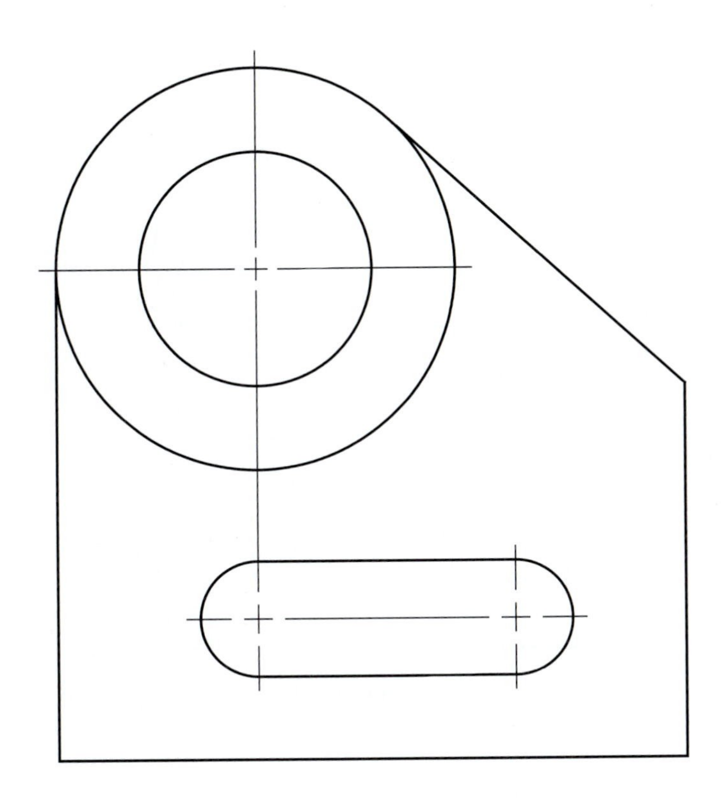

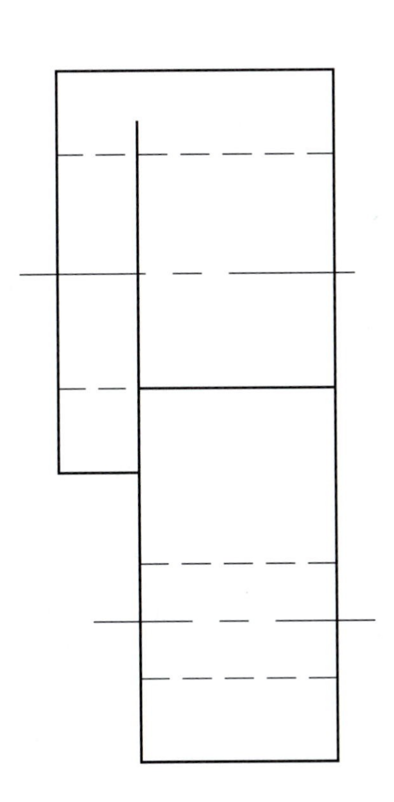

Apply and Analyze

Name ______________________ Date ____________ Class ______________

Review Activity 9-4

For each drawing, determine the minimum number of size and location dimensions required to completely dimension the given views.

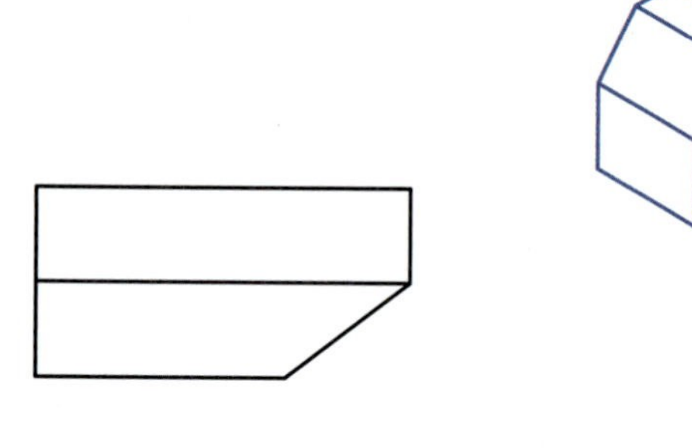

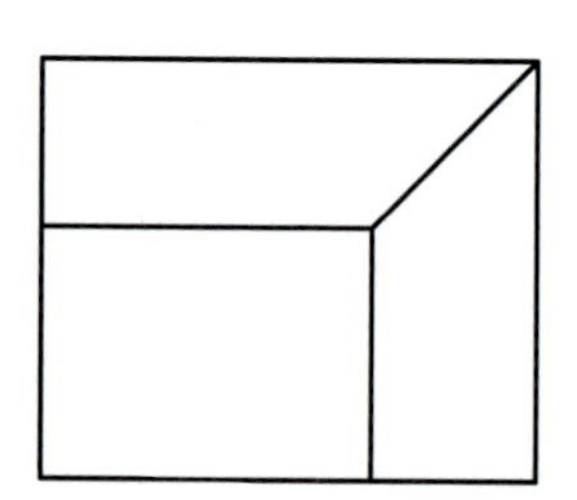

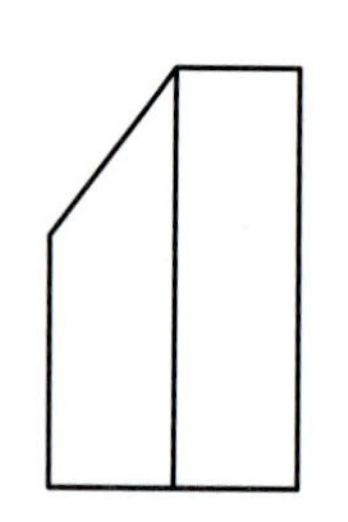

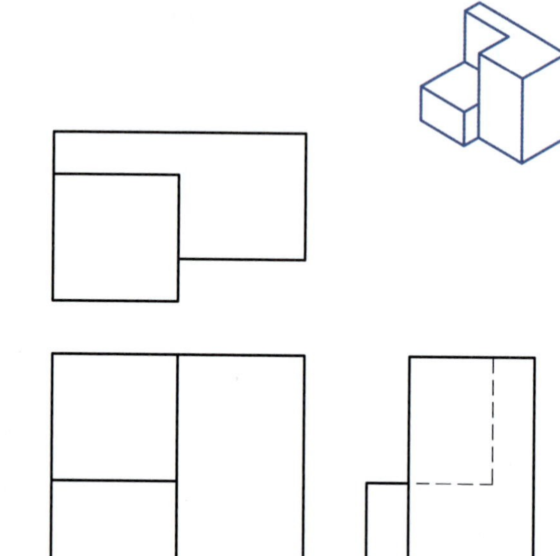

____ 1. Minimum number of dimensions required
A. 6
B. 7
C. 8
D. 9

____ 2. Minimum number of dimensions required
A. 6
B. 7
C. 8
D. 9

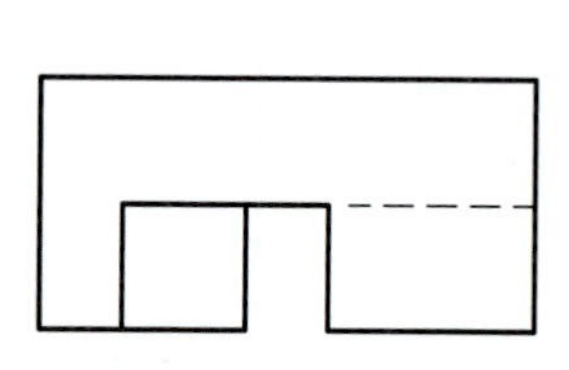

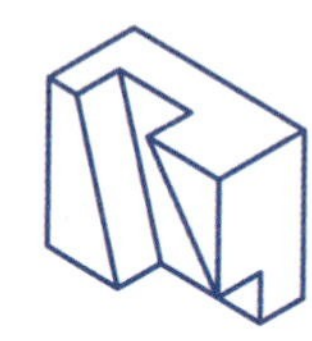

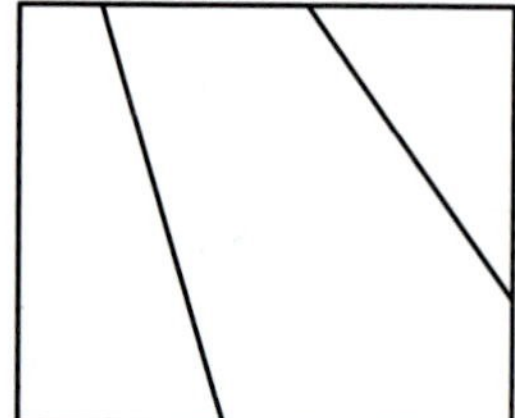

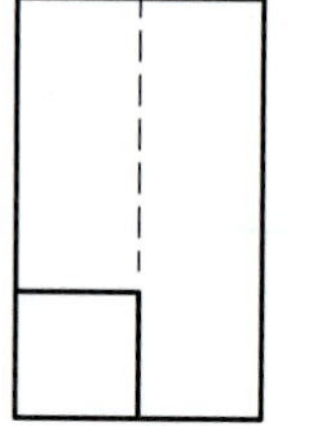

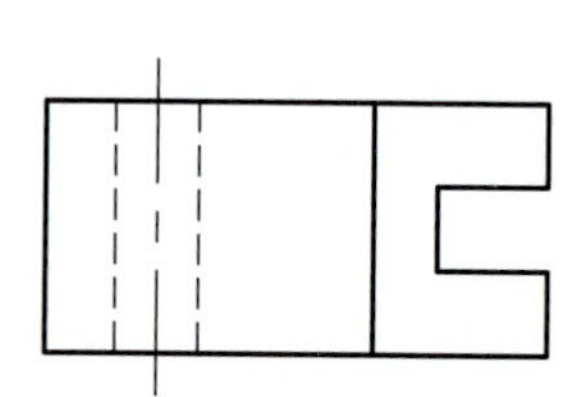

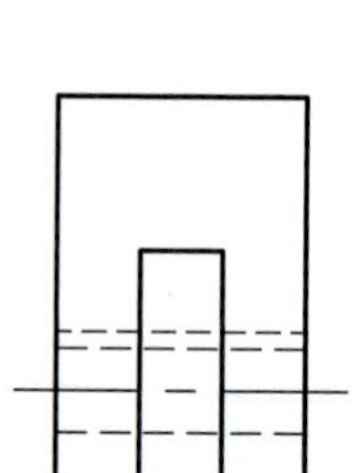

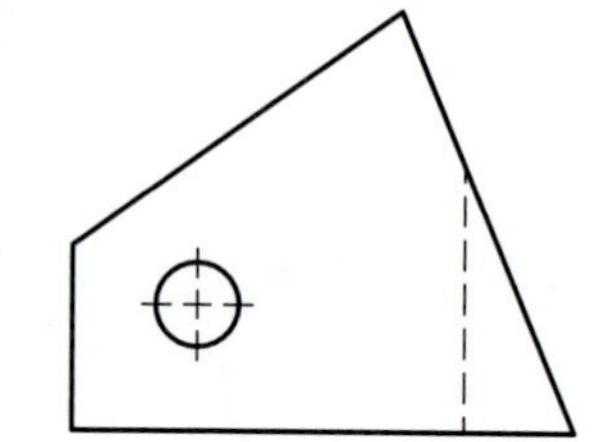

____ 3. Minimum number of dimensions required
A. 6
B. 7
C. 8
D. 9

____ 4. Minimum number of dimensions required
A. 9
B. 10
C. 11
D. 12

Apply and Analyze

Name ______________________ Date ____________ Class ____________

Review Activity 9-5

For each drawing, determine the minimum number of size and location dimensions required to completely dimension the given views.

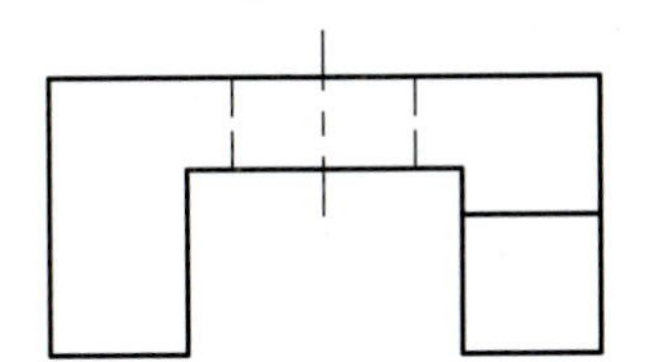

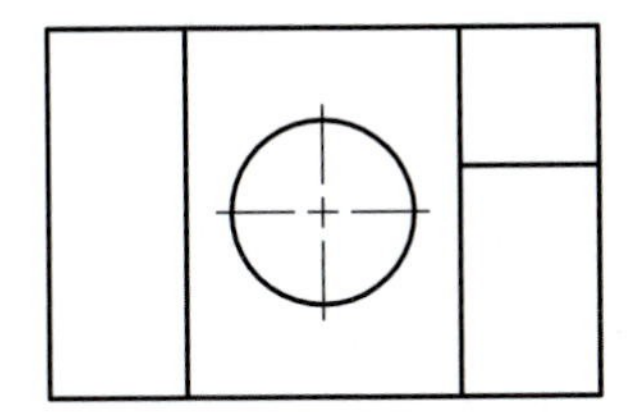

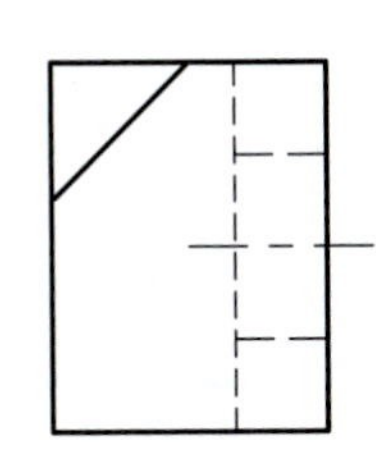

____ 1. Minimum number of dimensions required
A. 8
B. 9
C. 10
D. 11

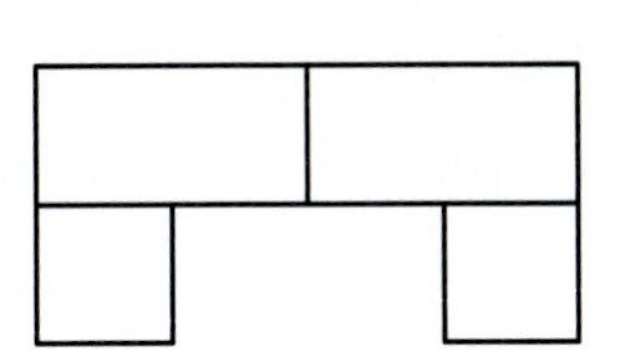

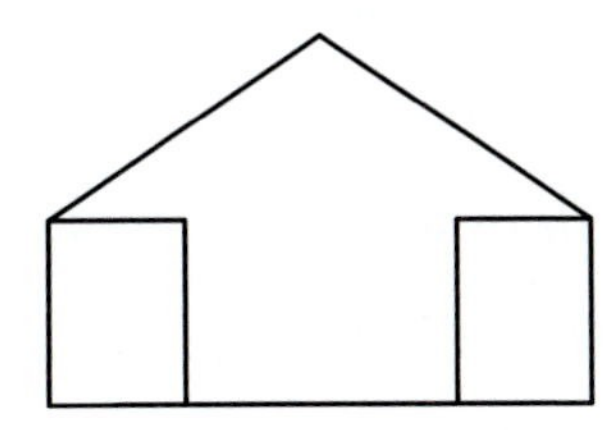

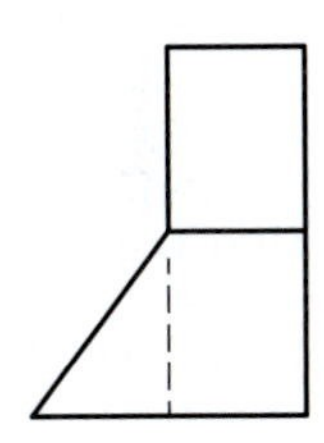

____ 2. Minimum number of dimensions required
A. 6
B. 7
C. 8
D. 9

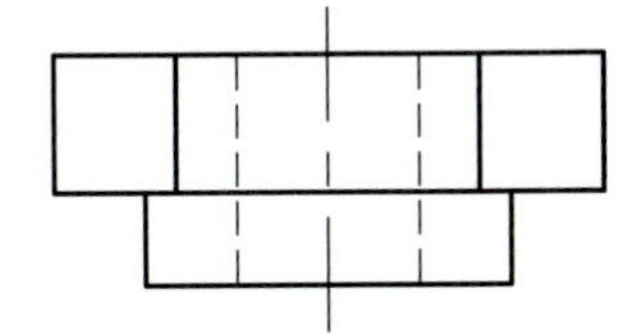

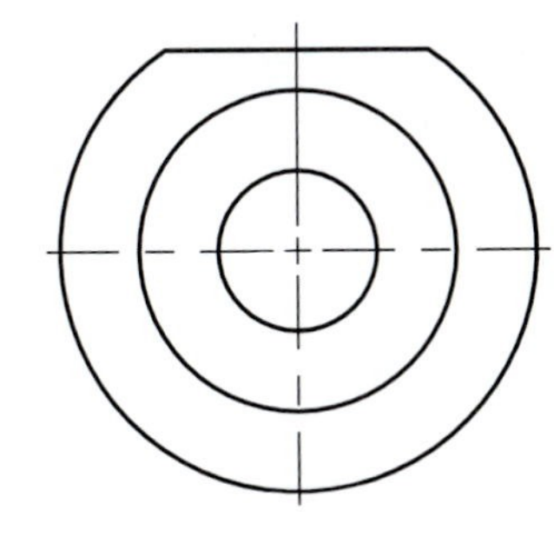

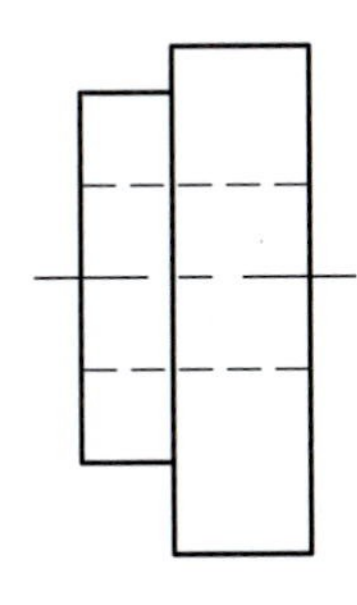

____ 3. Minimum number of dimensions required
A. 6
B. 7
C. 8
D. 9

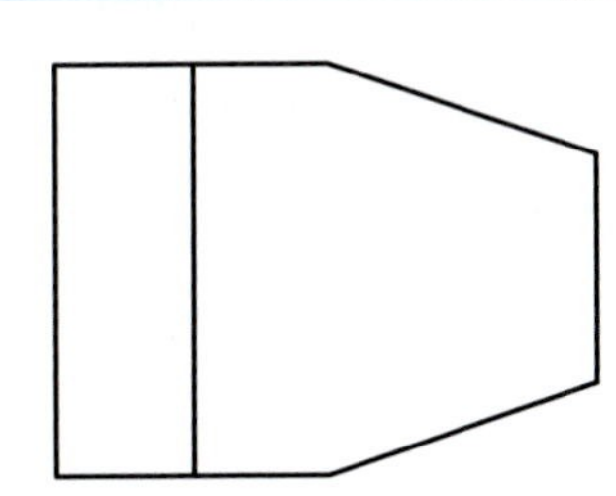

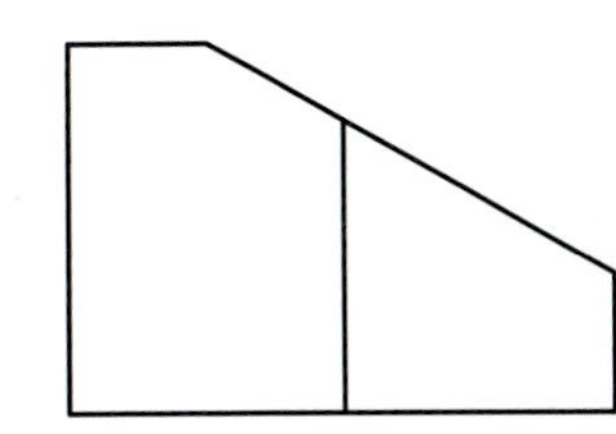

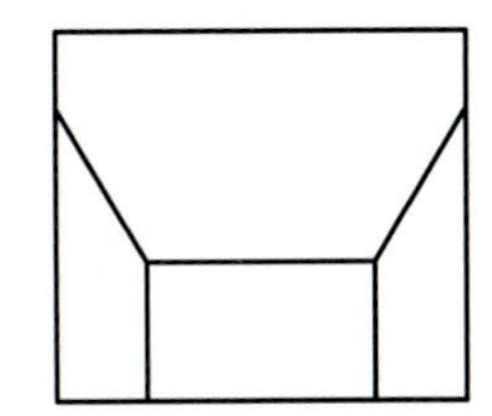

____ 4. Minimum number of dimensions required
A. 7
B. 8
C. 9
D. 10

Apply and Analyze

Name ______________________ Date ____________ Class ____________

Industry Print Exercise 9-1

Refer to the print PR 9-1 and answer the questions below.

1. How many angular dimensions are shown on this print? ______________________
2. How many reference dimensions are shown on this print? ______________________
3. Did the creator of this print follow the cylinder rule for the overall diameter? ______________________
4. Does this print use the aligned system or the unidirectional system? ______________________
5. How many local notes with leader lines are there on this print? ______________________
6. Counting reference dimensions, how many diameters are specified? ______________________

Review questions based on previous units:

7. What paper size is the original version of this print? ______________________
8. What is the material for this part? ______________________
9. What is the name of this part? ______________________
10. What is the part number of this print? ______________________
11. To what scale is the part drawn on the original drawing? ______________________
12. Does this print use SI or US customary units? ______________________
13. How many "through holes" does this object have? ______________________
14. Counting all surfaces of the object, how many are noncylindrical planar surfaces? ______________________
15. Does this drawing feature any section views? ______________________

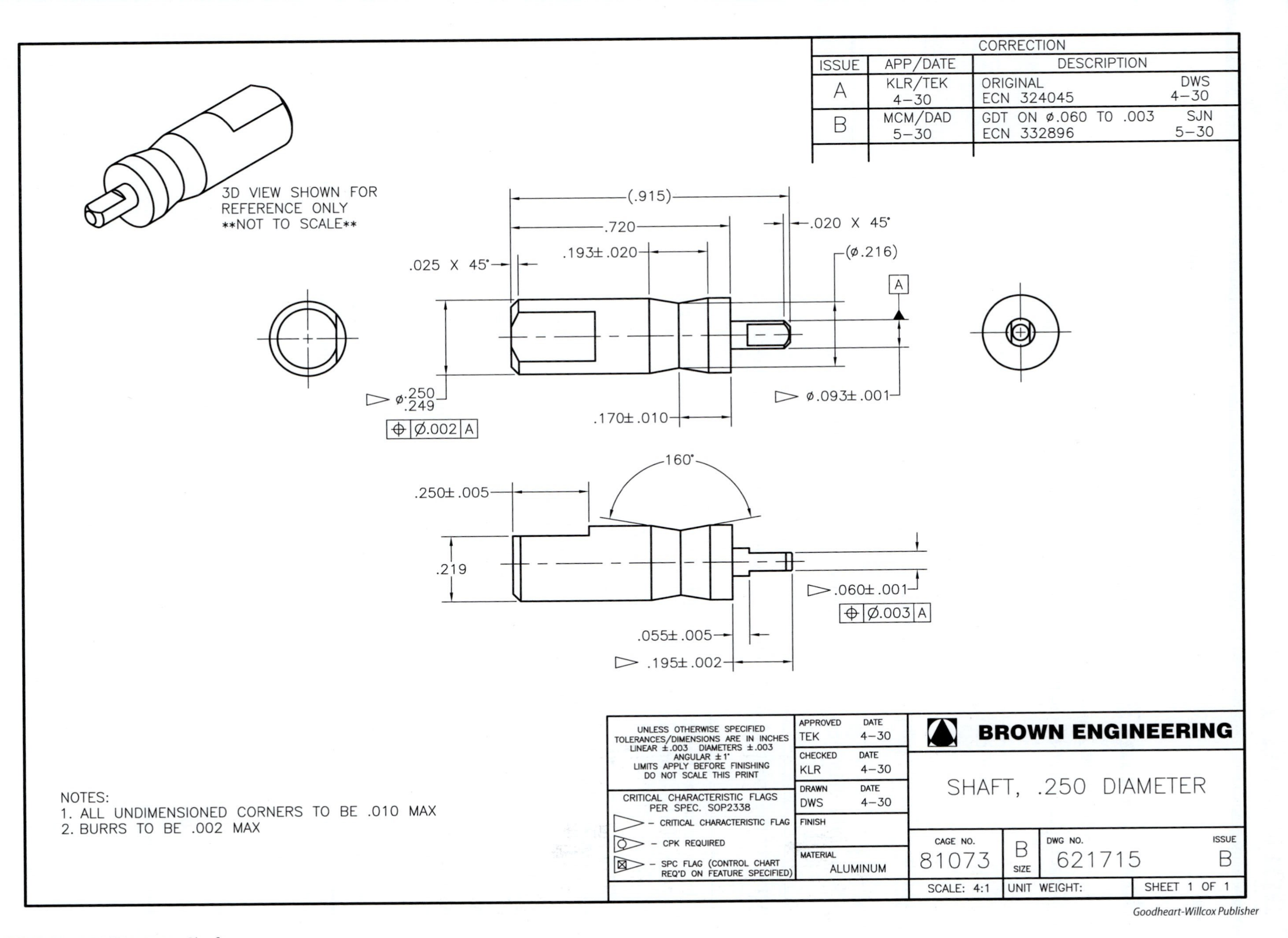

Goodheart-Willcox Publisher

PR 9-1. .250 Diameter Shaft

Apply and Analyze

Name ______________________ **Date** ____________ **Class** ____________

Industry Print Exercise 9-2

Refer to the print PR 9-2 and answer the questions below.

1. What is the center-to-center width measurement between the two "posts" of this object?

____

2. On the enlarged Detail A, two dimensional values begin with 4X. What does that mean?

____

3. On the enlarged Detail A, what dimensional value is repeated as a reference measurement?

____

4. Of all of the diameters specified on this print, which is the largest?

____

5. Of all of the arcs that are specified by radius, which is the smallest?

____

6. The main stem of each post is tapered with a smaller diameter at the tip. What angle does the taper form with the center axis?

____

Review questions based on previous units:

7. Are there any section lines shown on this drawing? ______________________
8. What is the name of this part? ______________________
9. What is the number of this part? ______________________
10. What geometric 3D shape is created due to the 4° dimension shown in Section A-A?

____

11. Does this object have any features that could be described as cylindrical? ______________________
12. *True or False?* The view closest to the general notes is viewed from a direction 180° from the direction for viewing Section A-A.

____

13. What scale is used for Detail A? ______________________
14. Does this drawing feature any auxiliary views? ______________________
15. Does this part feature any threads? ______________________

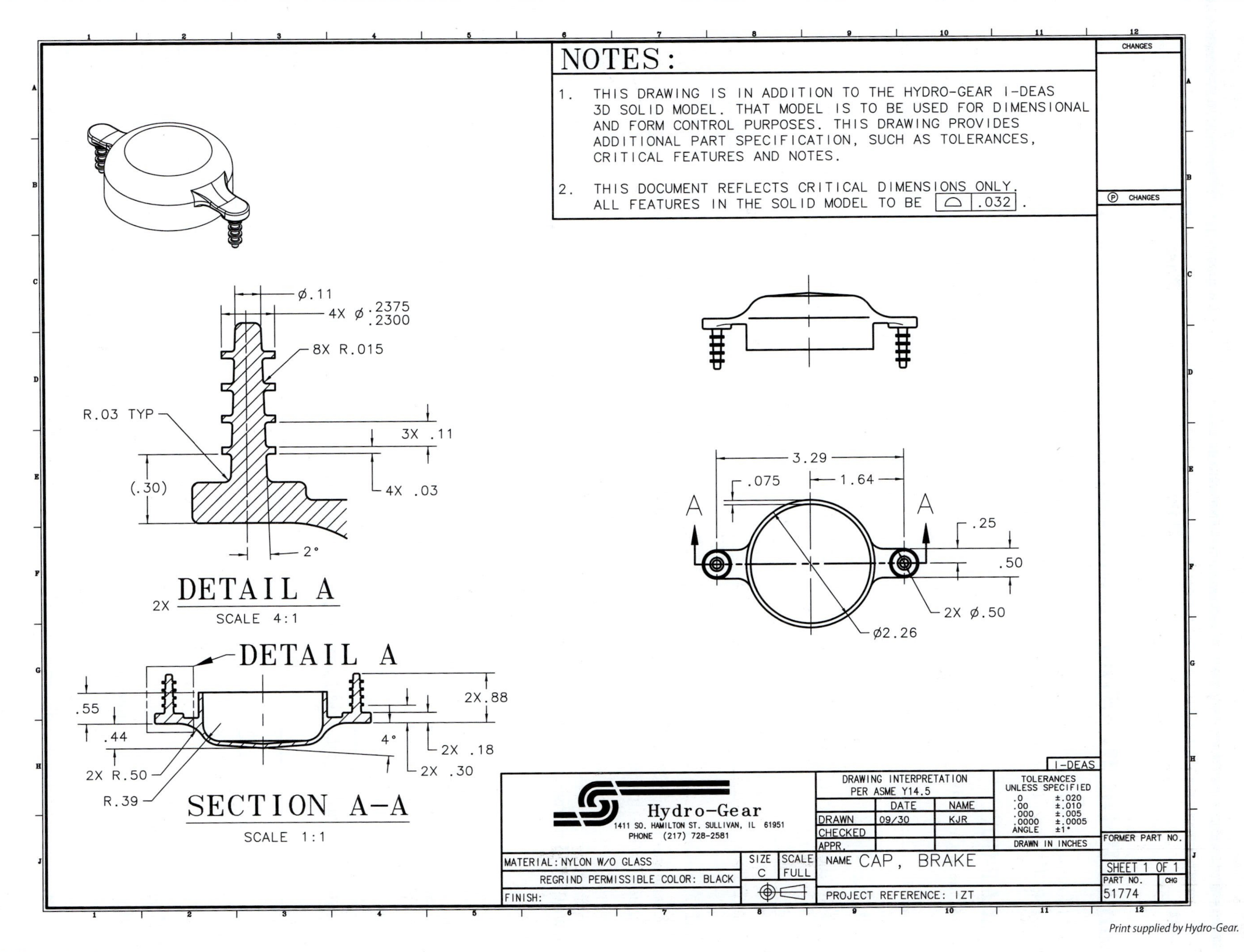

Print supplied by Hydro-Gear.

PR 9-2. Brake Cap

Apply and Analyze

Name ______________________ **Date** ______________ **Class** ______________

Bonus Print Reading Exercises

The following questions are based on various bonus prints located in the folder that accompanies this textbook. Refer to the print indicated, evaluate the print, and answer the question.

Print AP-001

1. What is the diameter of the hole that is inclined 15°?

____

Print AP-002

2. This drawing only has nine dimensions. Why are those all in parentheses?

____

____

Print AP-003

3. Many companies use symbols to point out critical dimensions. How many critical dimensions are specified on this print?

____

4. For the dimensions, does this print use the aligned method or unidirectional method?

____

Print AP-004

5. Based on a guideline for dimensioning discussed in the unit, why is the total width not given in the front view (central on drawing)?

____

6. List the overall height, width, and depth of this part. Some calculation may be necessary.

____

____

____

Print AP-007

7. How many reference dimensions are specified on this print?

____

8. Not counting reference dimensions, how many angular dimensions are specified on this print?

____

Print AP-008

9. Not counting the holes in each end, and not counting the spline data, how many different diameter sizes are specified that are smaller than 17.00 mm?

____

10. Does this print employ mostly chain dimensioning or baseline dimensioning?

____

Print AP-012

11. How many radius dimensions in this print are specifying a radius value larger than 1″?

____

12. What is the radius of the semicircular rounded slot?

____

Print AP-016

13. What diameter is specified for the circle of centers (bolt circle)?

____

14. Does this print feature unidirectional or aligned dimensioning?

____

Print AP-018

15. There is a leader line on this print calling out a cavity number. Why does this leader line *not* feature a triangular arrowhead?

____

____

16. How many dimensions of this print are specifying an angular measurement?

____

Print AP-021

17. One dimension indicates a distance of .300 TYP. What does TYP stand for?

____

18. For some holes, this print also implements a type of location dimensioning known as _____.

____

Print AP-022

19. What letters are used to specify the datums on this print?

____

20. What method is used to specify the diameters of the holes?

____

Notes

UNIT 10

Tolerancing

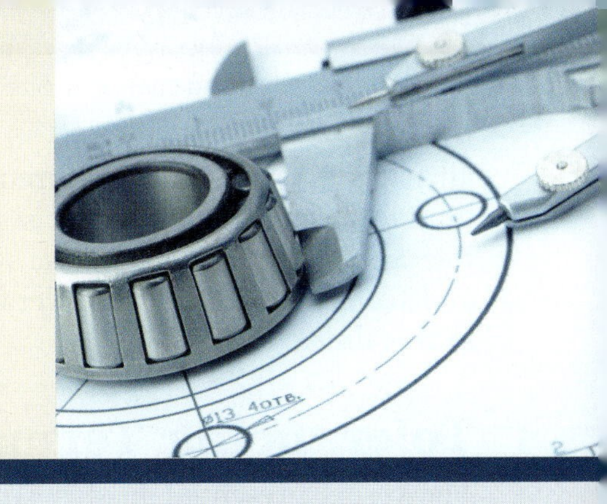

LEARNING OBJECTIVES

After completing this unit, you will be able to:

- Discuss the need for tolerances within the manufacturing enterprise.
- Define terms related to tolerancing.
- Explain how tolerances are expressed on a drawing.
- Identify tolerance values for dimensions on the drawing, regardless of the tolerancing expression.
- Describe how designers and engineers determine tolerances for mating parts using a basic hole system.
- Identify standard symbols that guide or clarify the use of tolerances to define size ranges.
- Calculate tolerances or limits for mating parts based on maximum material conditions and allowance.
- Describe basic characteristics of American standard classes of fits for inch-based parts.
- Describe basic characteristics of standard metric tolerance classes for millimeter-based parts.

TECHNICAL TERMS

allowance
basic dimension
basic hole system
basic size
bilateral tolerance
conical taper
continuous feature
datum feature symbol
dimension origin symbol
interchangeable manufacturing
least material condition (LMC)
limit dimensioning
limits
maximum material condition (MMC)
nominal size
plus and minus dimensioning
single limit
statistical process control (SPC)
statistical tolerance
tolerance
unilateral tolerance

Most companies do not manufacture all of the parts and subassemblies required in their products. Frequently, the parts are manufactured by specialty industries or subcontractors that work from specifications provided by the company. The parts and subassemblies are then assembled into the final product. The key to the successful operation of the final product is that duplicate parts must be interchangeable and still function satisfactorily. A common term for this is ***interchangeable manufacturing***. To achieve this, all parts must be manufactured to within specified limits of size. The range of size is defined by the tolerance. The various terms, symbols, procedures, and techniques related to tolerancing are presented in this unit.

Tolerancing Terms

The terms presented in this section are frequently used in industry. In order to satisfactorily read and interpret prints, it is necessary to understand their true meaning and application. Other terms relating to conditions and applications of tolerancing can be found in the glossary of this text.

The ***tolerance*** is the total amount by which a dimension can vary. You can think of this as how much you can "tolerate" the dimension being less than 100% precise. Tolerance may be expressed as the desired size followed by a bilateral or unilateral tolerance expression or as a set of limits, **Figure 10-1**. Notice that all four of the examples given create the same range of sizes for the hole. Tolerances can be applied to all feature dimensions, including size dimensions and location dimensions, **Figure 10-2**.

A ***unilateral tolerance*** is a tolerance that permits variation from the dimensional value in one direction only. A ***bilateral tolerance*** is a tolerance that permits variation from the dimensional value in both directions. The permitted variation in a bilateral tolerance can have equal or unequal values. An example of a bilateral tolerance is ±.004″. A common error in reading a bilateral tolerance, such as this example, is to say that the tolerance is .004″. In reality, the tolerance is .004″ + .004″, or .008″ total. The method of

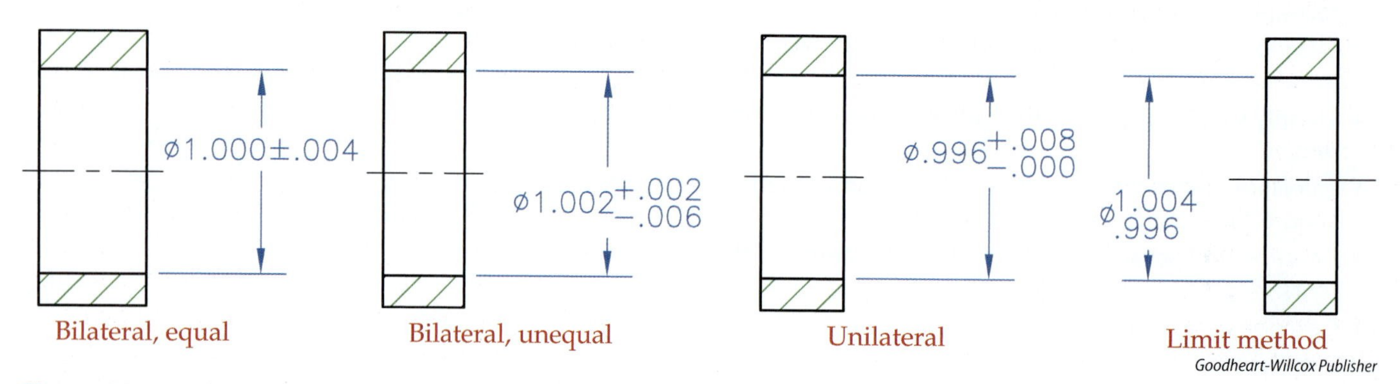

Figure 10-1. Tolerance may be expressed as the desired size followed by a bilateral or unilateral tolerance or as a set of limits. In all four examples shown here, the tolerance is .008″. Therefore, the limits are the same.

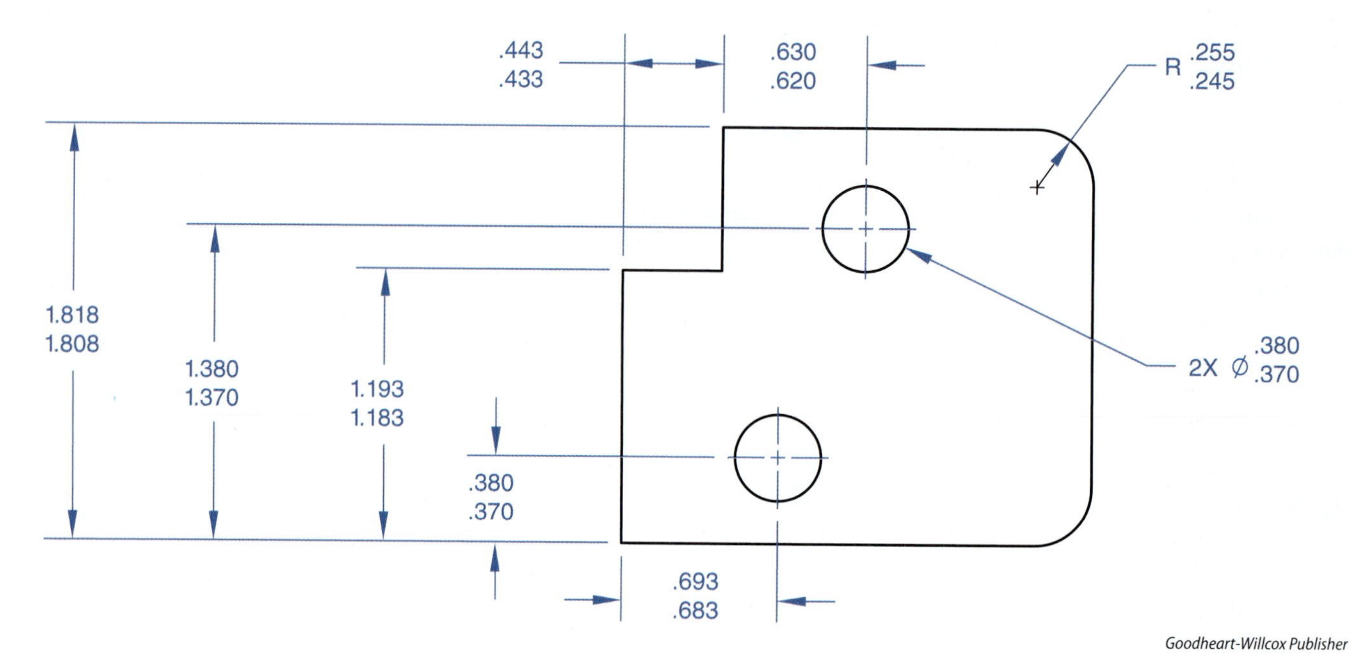

Figure 10-2. Tolerances can be applied to all feature dimensions, including size dimensions and location dimensions.

expressing tolerances with a basic size and a plus and minus value is called ***plus and minus dimensioning***.

Limits are the extreme maximum and minimum dimensions of a part allowed by the application of a tolerance. Two limit dimensions are always involved, which are a maximum size and a minimum size. For example, the design size of a part may be 1.375. If a tolerance of plus or minus two thousandths of an inch (±.002″) is applied, then the two limits are a maximum limit of 1.377 and a minimum limit of 1.373. This method of expressing a tolerance with limits is called ***limit dimensioning***.

Nominal size is most often a general term used to designate size for a commercial product, **Figure 10-3**. It may or may not express the true numerical size of the part or material. For example, a seamless wrought-steel pipe of 3/4″ (.75″) nominal size diameter has an actual inside diameter of .824″ and an actual outside diameter of 1.050″. Another example of the use of nominal size is with dimensional lumber used in construction. A 2 × 4 lumber size actually measures 1 1/2″ × 3 1/2″. In some cases, such as cold-finished, low-carbon steel rounds, the nominal 1″ size comes within two-thousandths (.002″) tolerance of actual size. Therefore, nominal size may or may not be an accurate numerical size of a material.

In general, the term ***basic size*** means the size determined by engineering and design requirements for a part or set of mating parts. For a part that does show a dimensional value with a plus and minus tolerance, the basic size is the target value indicated. Basic size also means the size from which allowances and tolerances are calculated and applied. Be aware that terms are not used the same in all regions of the country or from company to company. In certain texts or standards, the term "nominal size" may also refer to the basic size as defined in this paragraph.

For example, strength and stiffness may require a 1″ diameter shaft and hole for two parts to work well together. Using a system known as the ***basic hole system***, the basic 1″ size is applied to the hole as the smallest value. For a clearance or sliding fit, a slightly smaller size is applied to the shaft's larger value. Tolerances are also applied to both parts to determine the looser fit. See **Figure 10-4**. In actuality, the sizes of both parts are calculated based on the basic size of 1″. Each part is then dimensioned with tolerances expressed using one of the methods discussed above. In the metric standards, the system is referred to as the *hole basis system*. A shaft basis system is also available but is less common.

Another more recent development in tolerancing terminology is the basic dimension. ***Basic dimensions*** are defined as dimensions that are theoretically exact, and they are most often symbolized as a dimension value within a rectangular box. See **Figure 10-5**. Drawings that incorporate geometric dimensioning and tolerancing (GD&T) use basic dimensions to locate features such as hole patterns or describe irregular profiles. In these cases, the tolerance is specified in a feature control frame, so the tolerance is applied in a different fashion. The GD&T application of the term *basic dimension* is similar, but unrelated to, the term *basic size* as discussed in this unit. Geometric dimensioning and tolerancing will be covered in more detail in Unit 13.

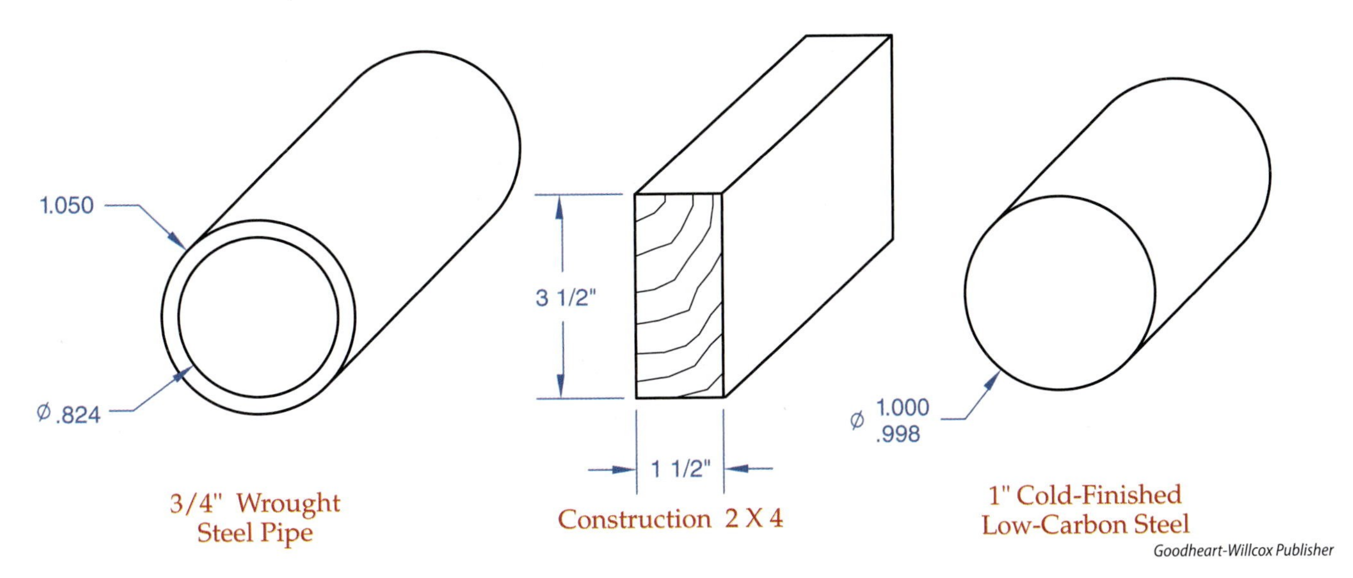

Figure 10-3. Nominal size may or may not express the true, actual size of the part or material.

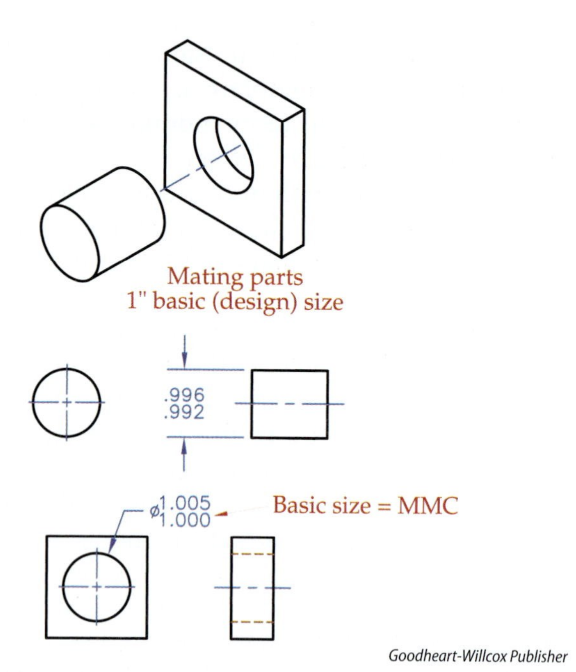

Goodheart-Willcox Publisher

Figure 10-4. With the basic hole system, the nominal or basic size is used as the MMC of the hole, and tolerances are calculated for both the hole and the shaft based on the requirements of the fit.

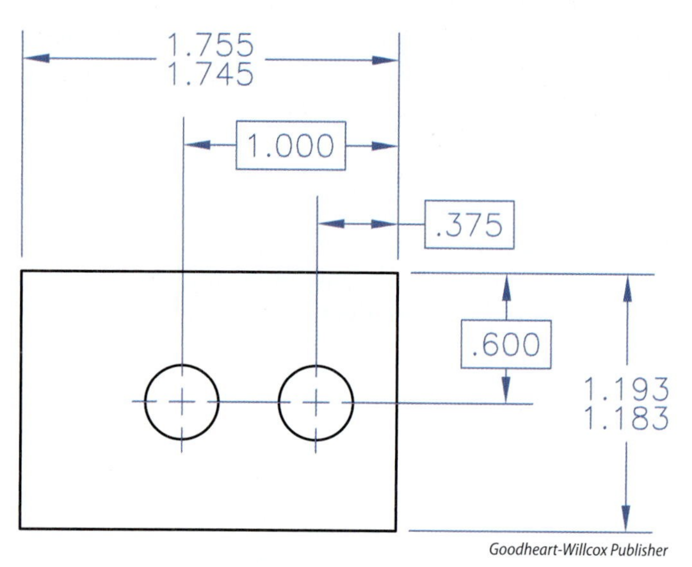

Goodheart-Willcox Publisher

Figure 10-5. Basic dimensions are used in a variety of ways and indicate a theoretical value wherein the tolerance will be determined in another manner, such as a feature control frame.

Material Conditions and Mating Parts

With respect to parts being assembled, certain terms are useful for understanding the relationship of mating parts. Designers and engineers must establish not only a range for the sizes of the two parts, but also for the clearance space between the two parts.

Maximum material condition (MMC) is a standard term that is very useful in determining how precisely parts fit together. The MMC of a feature is the size limit value representing the feature when the most material is present. For an external feature, such as a pin or post, MMC is the *largest* value. However, for an internal feature, such as a slot or hole, MMC is the *smallest* value. At first, this may seem like an unusual way to express sizes. Be careful to not talk about "maximum size," but rather "maximum material." Using the terms "upper limit" and "lower limit" to speak about the larger and smaller values allows the word "maximum" to be reserved for MMC. The MMC of an external feature is the upper limit, or larger value, because there is more material when the part is larger. The MMC of an internal feature is the lower limit, or smaller value, because there is more material when the hole is smaller. Making the hole larger requires removing material. One of the advantages of using the term MMC is the ability to determine the tightest fit between two parts simply by comparing the MMC values.

While not as critical to design calculations, the ***least material condition (LMC)*** of a feature is also an appropriate term. The least material condition of an internal feature is the larger size. The least material condition of an external feature is the smaller size. The loosest fit can be determined by comparing the least material conditions. LMC also has implications in GD&T, which will be covered in Unit 13.

Allowance is the intentional difference found when comparing the MMC values of mating parts. There are different ways of describing allowance. Perhaps the most appropriate is "tightest fit," whether the parts have clearance, or even if they must be forced or pressed together. Another way to describe allowance is the MMC of the external feature subtracted from the MMC of the internal feature. Therefore, if allowance is a positive number, the fit is a clearance fit. If allowance is a negative number, the fit is an interference fit. In **Figure 10-4**, the MMC of the hole is 1.000″ and the MMC of the pin is .996″. Therefore, the allowance (tightest fit) is +.004″. As the hole size departs from MMC (larger hole) or as the pin size departs from MMC (smaller pin), there is more clearance. In this example, if the parts pass inspection based on their size tolerances, the loosest fit is a clearance of +.013″. This combines the largest possible hole diameter, 1.005″, with the smallest possible pin diameter, .992″ (1.005 – .992 = .013).

Symbols

In Unit 9, several ASME symbols were introduced for dimensioning, and these can be reviewed in **Figure 9-3**. In this unit, additional symbols are introduced, as illustrated in **Figure 10-6**. This section will introduce these symbols with a brief explanation of each.

You will recall from Unit 9 that a *datum* is the origin from which the location of features is established, or the reference plane or axis to which a geometric control is referenced. A datum is established by real features or surfaces of a part. Datums are essential to many of the GD&T controls discussed in Unit 13. A datum is assumed to be exact for the purpose of reference. The ***datum feature symbol***, per the ASME Y14.5 standard, is indicated by a letter in a small rectangle or box attached to the views in some manner by a stem that features a triangle, usually filled, as shown in **Figure 10-7**. On older drawings, the datum may be indicated in a different manner. The older datum symbol featured a letter preceded and followed by a dash and enclosed in a small rectangle or box, without a connecting stem, and it was often connected to the part by an extension line or attached to the bottom of a feature control frame.

As introduced in Unit 9, sometimes it is important to specify from which surface a dimension originates. By definition and unless otherwise specified, the surface points of either surface can be used to establish a measurement plane for a dimension. The surface points of the opposite surface are determined to be a certain distance away from the plane. With some features, such as illustrated in **Figure 10-8**, it may be that a shorter surface is the mounting surface and the longer surface must be within a tolerance range as measured from the shorter surface. The ***dimension origin symbol*** provides a concise and clear way of stating which surface acts as the origin on the print without establishing a datum reference or using GD&T controls.

CR Controlled radius

Dimension origin

Conical taper

Slope

ST Statistical tolerance

CF Continuous feature

Goodheart-Willcox Publisher

Figure 10-6. ASME standard symbols are used with dimensions to help clarify the interpretation of tolerances applied in certain situations.

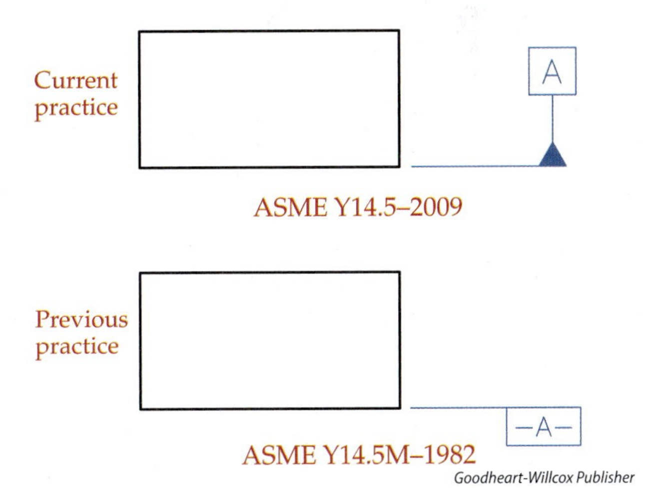

Goodheart-Willcox Publisher

Figure 10-7. A datum is identified with a datum feature symbol, and the feature can then be used as the origin from which the location of other features is established or the reference plane or axis to which a geometric control is referenced.

Tolerancing an angled surface dimensioned with an angular dimension results in a fan-shaped tolerance zone. **Figure 10-9** illustrates the tolerance zone of an angled surface dimensioned in that fashion. For more critical control of a surface wherein the surface needs to be controlled between two parallel planes, GD&T will offer an angularity control. For tolerancing the amount of taper with offset values, the taper symbol can be incorporated, as shown in **Figure 10-10**. ***Conical taper*** is the ratio of the diameters on each end of a cone-shaped surface, and it can also be toleranced more precisely with the assistance of the conical taper symbol, as shown in **Figure 10-11**. Notice the use of the basic dimension to establish the tolerance zone. For noncritical tapers, such as a simple transition between two cylinders, toleranced diameters and a toleranced length can be quite adequate. If the conical taper is a standard machine taper used throughout the tooling industry, the taper can also be dimensioned and toleranced by specifying the taper name and number. ASME B5.10, titled *Machine Tapers*, is a resource for additional information.

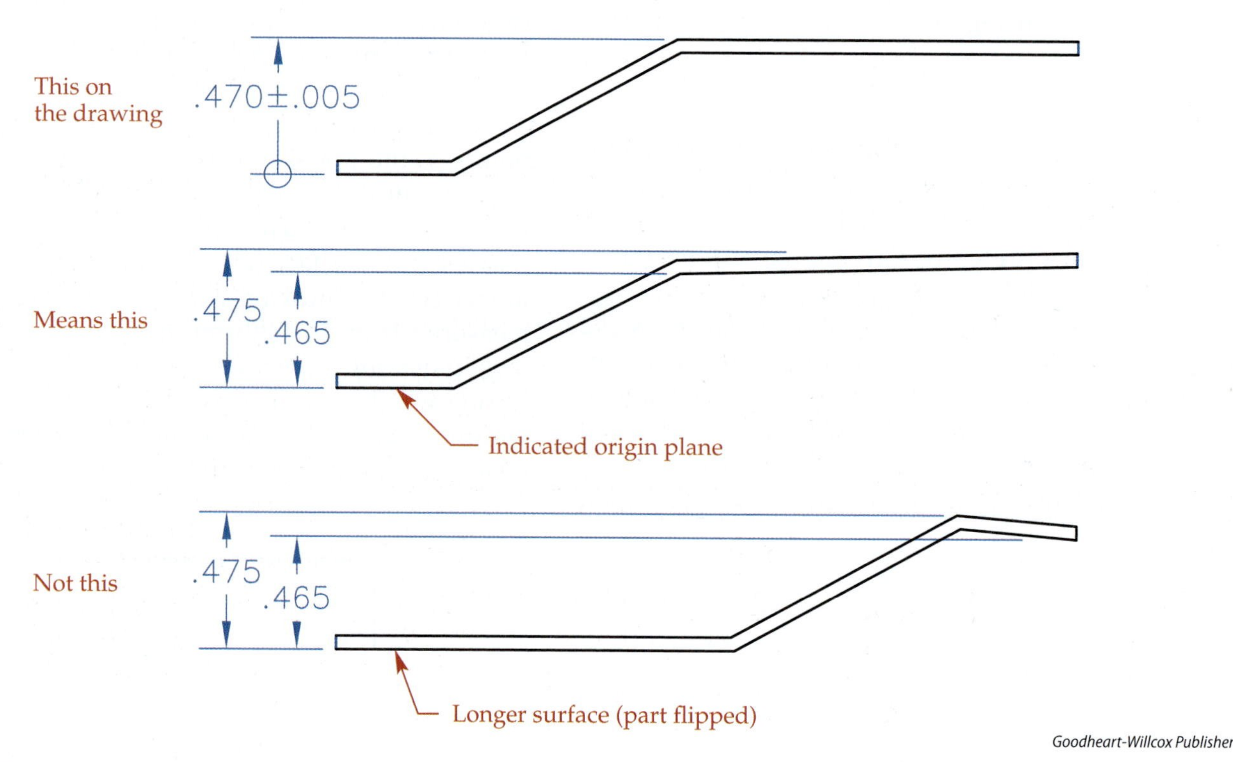

Figure 10-8. A dimension origin symbol is sometimes used to specify which surface should be used for the measuring plane. Especially in cases without a datum reference, this could make a difference when inspecting the size measurement to see if it is within tolerance.

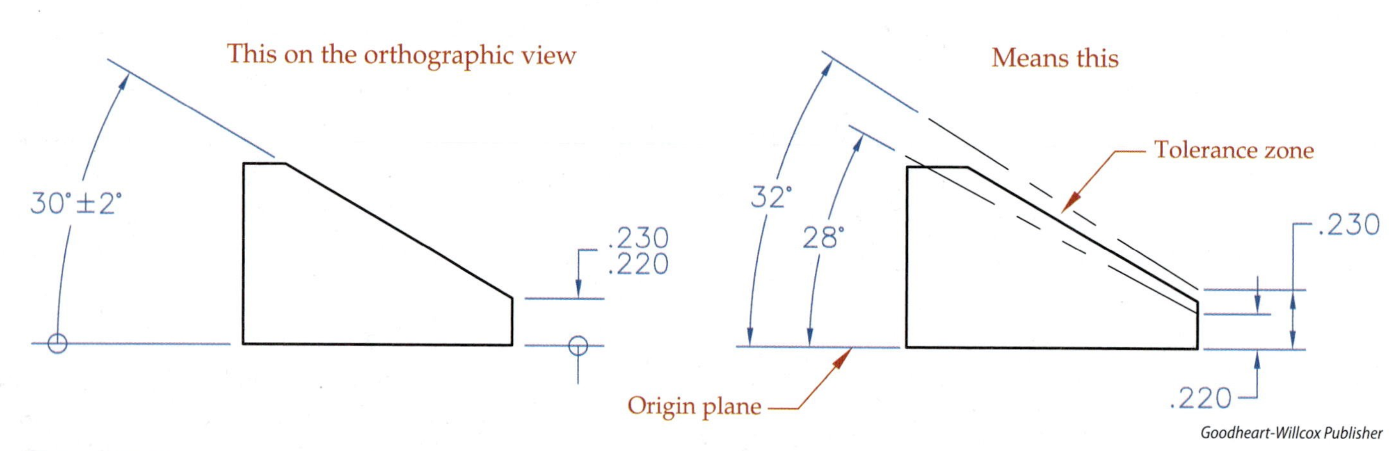

Figure 10-9. Angular dimensions given in degrees result in a fan-shaped tolerance zone.

This on the orthographic view

Means this

Tolerance zone maintains slope

.25:1

.558
.548

1.318
1.308

.558
.548

Goodheart-Willcox Publisher

Figure 10-10. The taper symbol can be used as a way of defining a sloped surface's location.

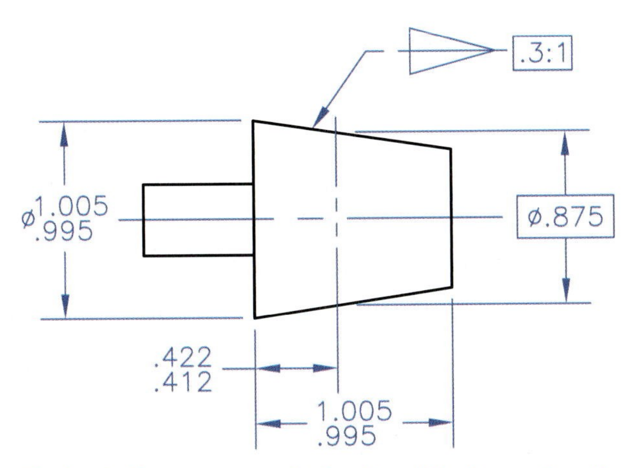

The basic diameter controls the size of the tapered section as well as its position in relation to the surface from which it is located.

Goodheart-Willcox Publisher

Figure 10-11. Cone-shaped features, depending on their precision requirements, can be dimensioned with a conical taper symbol to help define their location and range of size.

When tolerancing a radius, there are two situations that need clarification. If the location of the radius is given with dimensions, the tolerance zone of the arc is two concentric arcs. When the center of the radius is not located by dimensions, it is expected that the curved surface remains tangent to the adjoining surfaces, so the tolerance zone becomes a crescent shape, **Figure 10-12.** In addition to these clarifications, ASME standards allow for a controlled radius symbol to be used in the form of a CR symbol before the radius value. The controlled radius is to be a "fair curve without reversals." In these more tightly controlled cases, standards recommend clarification be given through a general note or a reference to a company standard or other engineering specification.

Another tolerancing technique that sometimes appears on prints is referred to as ***single limit***, and incorporates the MIN or MAX note. Single limits may be used on such things as hole depth, thread length,

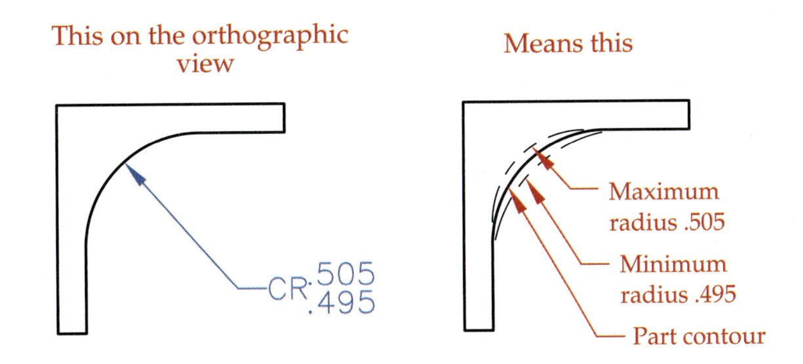

For unlocated (floating) radius dimensions of curved features, the tolerance zone is crescent-shaped. For more precision, a controlled radius specification (shown using the CR symbol) is used to indicate "a fair curve without reversals."

Goodheart-Willcox Publisher

Figure 10-12. Curved surfaces defined by radius values may create tolerance zones that are crescent shaped. A controlled radius option is for critical curve control.

corner radii, and chamfers. In a case where the dimension value is small, such as rounding or breaking off corners, it may be convenient for the designer to indicate the maximum value for the radius or break, such as R.03 MAX. This is a unilateral expression that designates the largest value. The smallest amount is left up to the machinist. Usually, in these cases, it is the intent of the designer that the radius or break *not* be zero, although the specification indicates that zero is the lower limit, with the value given as the upper limit.

Statistical process control (SPC) is a system used in industry to monitor a manufacturing process, and it uses statistical analysis to ensure a higher level of quality. For example, particular dimensions may be monitored to observe dimensional change as a tool or fixture device experiences wear. A symbol used to flag those dimensions as ***statistical tolerances*** is shown in **Figure 10-13.** Notice a standard convention allows for a dimension to be given as a statistical tolerance alongside more restrictive arithmetic limits. In either case, a note such as FEATURES IDENTIFIED WITH THE STATISTICALLY TOLERANCED SYMBOL SHALL BE PRODUCED WITH STATISTICAL PROCESS CONTROLS should also be given on the drawing.

Another symbol that helps clarify how dimensions and tolerances are to be interpreted is the ***continuous feature*** symbol, **Figure 10-14**. This symbol is used to establish that two surfaces that are interrupted, perhaps by a groove, should still be treated as one feature. While extension lines can be shown or omitted, extension lines by themselves do not indicate that two features should be measured and inspected as a single continuous feature.

As you progress through this text, additional uses and applications of the symbols discussed in this unit will be presented. While there are too many examples

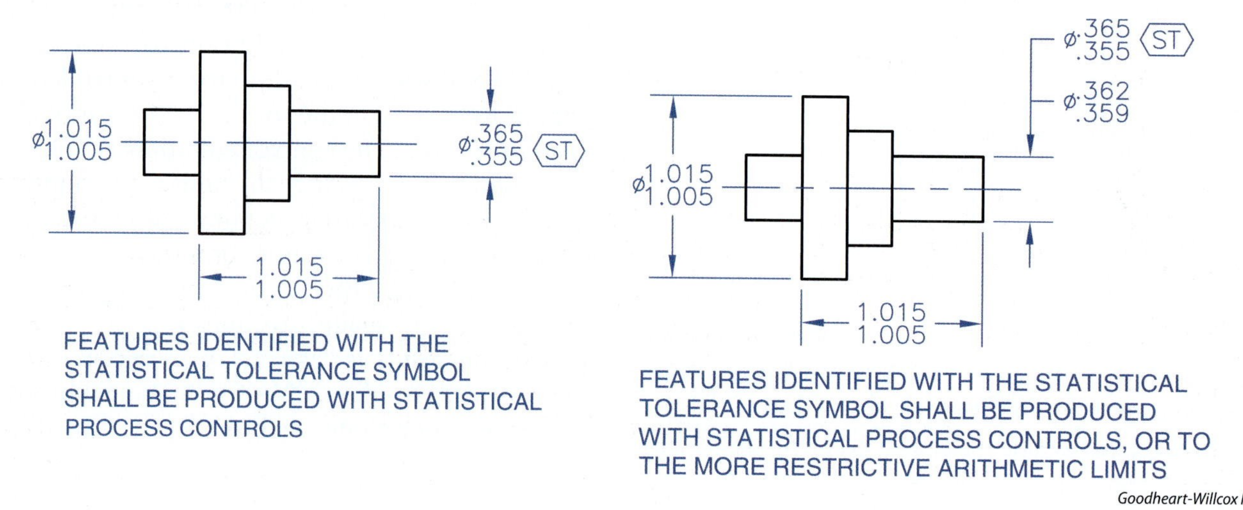

Figure 10-13. The statistical tolerance symbol is used to indicate certain dimensions are involved in the statistical process control plan.

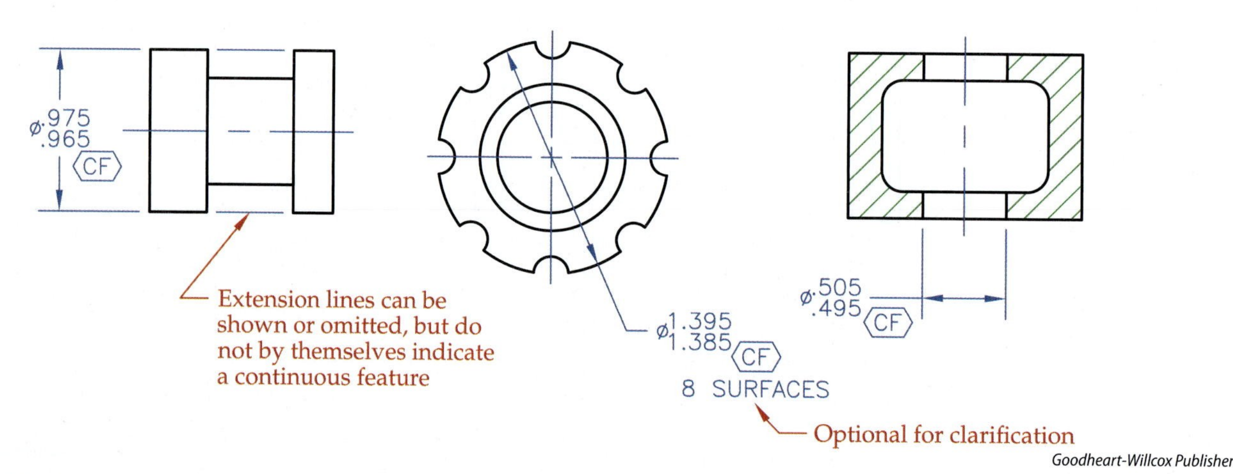

Figure 10-14. The continuous feature symbol is used to indicate that surfaces on each side of an interruption should still be considered as a single feature.

to illustrate all uses in this text, the ASME Y14.5 document is recommended as a thorough resource for interpreting the symbols used on industrial prints.

American Standard Tolerances

Engineers and designers are responsible for assigning tolerances to parts in an assembly so the assembly can function properly. Tolerances should always be as large as possible to reduce manufacturing costs, but they should also ensure that any random selection of two parts does not have a situation that is too loose or too tight. Therefore, engineers and designers must know the appropriate level of accuracy for a given application and process. Fits between plain (non-threaded) cylindrical parts are recommended within the ASME B4.1 standard, which sets forth preferred limits and fits based on application. Within this standard, there are five types of fit. The following are the fits and the letter symbols used to identify them:

- Running or Sliding Clearance Fits (RC)
- Locational Clearance Fits (LC)
- Transitional Clearance or Interference Fits (LT)
- Locational Interference Fits (LN)
- Force or Shrink Fits (FN)

The abbreviations for each of these fits are not intended to be shown on industrial prints. Therefore, this section provides a general understanding of tolerances and how they may be derived, but does not explain what the print reader will see on the drawing.

Specifying one of the standard fits is a quick and easy way for an engineer or designer to communicate a desired amount of tolerance to a drafter or technician. For example, the engineer may simply specify on a sketch that a hole and its mating part should be a 3/4″ RC3 class fit. The technician then uses tables to calculate the limits for each of the mating parts. See **Figure 10-15**. Finally, a drawing is created using

Basic Hole System
Limits are in thousandths of an inch. Apply limits to the basic size.

Nominal Size Range Inches	RC1			RC2			RC3			RC4		
	Limits of Clearance	Standard Limits		Limits of Clearance	Standard Limits		Limits of Clearance	Standard Limits		Limits of Clearance	Standard Limits	
Over – To		Hole	Shaft		Hole	Shaft		Hole	Shaft		Hole	Shaft
0.00 – 0.12	0.1 0.45	+0.2 −0	−0.1 −0.25	0.1 0.55	+0.25 −0	−0.1 −0.3	0.3 0.95	+0.4 −0	−0.3 −0.55	0.3 1.3	+0.6 −0	−0.3 −0.7
0.12 – 0.24	0.15 0.5	+0.2 −0	−0.15 −0.3	0.15 0.65	+0.3 −0	−0.15 −0.35	0.4 1.12	+0.5 −0	−0.4 −0.7	0.4 1.6	+0.7 −0	−0.4 −0.9
0.24 – 0.40	0.2 0.6	+0.25 −0	−0.2 −0.35	0.2 0.85	+0.4 −0	−0.2 −0.45	0.5 1.5	+0.6 −0	−0.5 −0.9	0.5 2.0	+0.9 −0	−0.5 −1.1
0.40 – 0.71	0.25 0.75	+0.3 −0	−0.25 −0.45	0.25 0.95	+0.4 −0	−0.25 −0.55	0.6 1.7	+0.7 −0	−0.6 −1.0	0.6 2.3	+1.0 −0	−0.6 −1.3
0.71 – 1.19	0.3 0.95	+0.4 −0	−0.3 −0.55	0.3 1.2	+0.5 −0	−0.3 −0.7	0.8 2.1	+0.8 −0	−0.8 −1.3	0.8 2.8	+1.2 −0	−0.8 −1.6
1.19 – 1.97	0.4 1.1	+0.4 −0	−0.4 −0.7	0.4 1.4	+0.6 −0	−0.4 −0.8	1.0 2.6	+1.0 −0	−1.0 −1.6	1.0 3.6	+1.6 −0	−1.0 −2.0
1.97 – 3.15	0.4 1.2	+0.5 −0	−0.4 −0.7	0.4 1.6	+0.7 −0	−0.4 −0.9	1.2 3.1	+1.2 −0	−1.2 −1.9	1.2 4.2	+1.8 −0	−1.2 −2.4
3.15 – 4.73	0.5 1.5	+0.6 −0	−0.5 −0.9	0.5 2.0	+0.9 −0	−0.5 −1.1	1.4 3.7	+1.4 −0	−1.4 −2.3	1.4 5.0	+2.2 −0	−1.4 −2.8
4.73 – 7.09	0.6 1.8	+0.7 −0	−0.6 −1.1	0.6 2.3	+1.0 −0	−0.6 −1.3	1.6 4.2	+1.6 −0	−1.6 −2.6	1.6 5.7	+2.5 −0	−1.6 −3.2
7.09 – 9.85	0.6 2.0	+0.8 −0	−0.6 −1.2	0.6 2.6	+1.2 −0	−0.6 −1.4	2.0 5.0	+1.8 −0	−2.0 −3.2	2.0 6.6	+2.8 −0	−2.0 −3.8
9.85 – 12.41	0.8 2.3	+0.9 −0	−0.8 −1.4	0.8 2.9	+1.2 −0	−0.8 −1.7	2.5 5.7	+2.0 −0	−2.5 −3.7	2.5 7.5	+3.0 −0	−2.5 −4.5
12.41 – 15.75	1.0 2.7	+1.0 −0	−1.0 −1.7	1.0 3.4	+1.4 −0	−1.0 −2.0	3.0 6.6	+2.2 −0	−3.0 −4.4	3.0 8.7	+3.5 −0	−3.0 −5.2

Goodheart-Willcox Publisher

Figure 10-15. A table similar to this is used to calculate tolerances for standard inch-based fits. These calculations are not usually done by the print reader, but rather by the drafter or technician creating the drawing.

the proper values and an appropriate tolerancing method, such as limit or plus and minus dimensioning. These sets of mating dimensions are sometimes referred to as precision fits and may have a tolerance value carried to four decimal places. Less critical dimensions are typically to three-place decimal precision on drawings measured in inches.

To use the table in **Figure 10-15**, find the row that matches the size range of the nominal diameter. If the nominal diameter is 1″, the size range is 0.71–1.19. Notice that the values are given as thousandths of an inch, so the decimal would move three places to the left, with zeros added as needed. If the class of fit specified by the designer or engineer is RC2, then use the "Hole" column to determine the tolerance for the hole. Using the basic hole system described earlier, the MMC of the hole should be 1.0000″, while the other limit is 1.0005″ (+0.5 means add .5 thousandths, which equals .0005″). The "Shaft" column indicates that 0.3 and 0.7 should be subtracted from the basic size to derive limits of .9997″ to .9993″ for the shaft. The MMC of the shaft is, therefore, .9997″, and the MMC of the hole is 1.0000″. This means the allowance for this RC2 class fit is .0003″. The loosest fit for a 1.0005″ hole and a .9993″ shaft would be .0012″. As you can see, this is a very precise fit. The maximum clearance is only 1.2 thousandths of an inch, which is a very tight range.

Metric Tolerances and Fits

As discussed in earlier units, the international scope of many companies has moved them toward the metric system in many applications. For metric dimensions, standards from the International Organization for Standardization (ISO) have been incorporated into ASME B4.2, titled *Preferred Metric Limits and Fits*. This standard does not use the same notations as ASME B4.1, nor are the tables quite the same, but the concepts are the same. In **Figure 10-16**, a partial table is shown. Since the precision of millimeters is usually two or three decimal places, the tables do not require decimal manipulation. Also, preferred basic sizes are given, so designers can choose directly from

Preferred Metric Hole Basic Clearance Fits – Partial Table

Basic Size		Loose Running			Free Running			Close Running			Locational Clearance		
		Hole H11	Shaft c11	Fit	Hole H9	Shaft d9	Fit	Hole H8	Shaft f7	Fit	Hole H7	Shaft h6	Fit
1	Max	1.060	0.940	0.180	1.025	0.980	0.070	1.014	0.994	0.030	1.010	1.000	0.016
	Min	1.060	0.880	0.060	1.000	0.955	0.020	1.000	0.984	0.006	1.000	0.994	0.000
1.2	Max	1.260	1.140	0.180	1.225	1.180	0.070	1.214	1.194	0.030	1.210	1.200	0.016
	Min	1.200	1.080	0.060	1.200	1.155	0.020	1.200	1.184	0.006	1.200	1.194	0.000
1.6	Max	1.660	1.540	0.180	1.625	1.580	0.070	1.614	1.594	0.030	1.610	1.600	0.016
	Min	1.600	1.480	0.060	1.600	1.555	0.020	1.600	1.584	0.006	1.600	1.594	0.000
2	Max	2.060	1.940	0.180	2.025	1.980	0.070	2.014	1.994	0.030	2.010	2.000	0.016
	Min	2.000	1.880	0.060	2.000	1.955	0.020	2.000	1.984	0.006	2.000	1.994	0.000
2.5	Max	2.560	2.440	0.180	2.525	2.480	0.070	2.514	2.494	0.030	2.510	2.500	0.016
	Min	2.500	2.380	0.060	2.500	2.455	0.020	2.500	2.484	0.006	2.500	2.494	0.000
3	Max	3.060	2.940	0.180	3.025	2.980	0.070	3.014	2.994	0.030	3.010	3.000	0.016
	Min	3.000	2.880	0.060	3.000	2.955	0.020	3.000	2.984	0.006	3.000	2.994	0.000
4	Max	4.075	3.930	0.220	4.030	3.970	0.090	4.018	3.990	0.040	4.012	4.000	0.020
	Min	4.000	3.855	0.070	4.000	3.940	0.030	4.000	3.978	0.010	4.000	3.992	0.000
5	Max	5.075	4.930	0.220	5.030	4.930	0.090	5.018	4.990	0.040	5.012	5.000	0.020
	Min	5.000	4.855	0.070	5.000	4.970	0.030	5.000	4.978	0.010	5.000	4.992	0.000
6	Max	6.075	5.930	0.220	6.030	5.970	0.090	6.018	5.990	0.040	6.012	6.000	0.020
	Min	6.000	5.885	0.070	6.000	5.940	0.030	6.000	5.978	0.010	6.000	5.992	0.000
8	Max	8.090	7.920	0.260	8.036	7.960	0.112	8.022	7.987	0.050	8.015	8.000	0.024
	Min	8.000	7.830	0.080	8.000	7.924	0.040	8.000	7.972	0.013	8.000	7.991	0.000
10	Max	0.090	9.920	0.260	10.036	9.960	0.112	10.022	9.987	0.050	10.015	10.000	0.024
	Min	0.000	9.830	0.080	10.000	9.924	0.040	10.000	9.972	0.013	10.000	9.991	0.000
12	Max	12.110	11.905	0.315	12.043	11.950	0.136	12.027	11.984	0.061	12.018	12.000	0.029
	Min	12.000	11.795	0.095	12.000	11.907	0.050	12.000	11.966	0.016	12.000	11.989	0.000

Goodheart-Willcox Publisher

Figure 10-16. A table similar to this is used to calculate basic clearance fits for metric applications.

the table values that are preset. For example, a close running fit for a 5 mm diameter could simply be specified on the print as 5.000–5.018 (or 5.000 + 0.018) for the hole and 4.990–4.978 (or 4.990 – 0.012) for the shaft. This means the allowance is 0.01 mm and the loosest fit is 0.04 mm. In metric applications, it is sometimes common to specify the MMC and then a unilateral tolerance for mating features.

Some companies may prefer to use the symbolic representations of metric precision fits. As you can see in **Figure 10-17**, hole or shaft relationships fall into 10 different levels of tightness, and each level has an alphanumeric designation. A free running fit for a hole has an H9 designation, while the mating shaft has a d9 designation. One method of calling out the hole measurement is to specify the size of the hole but then place the basic size and tolerance symbol in parentheses: 8.000–8.036 (8 H9). Some companies may choose to reverse those components and specify the hole as 8 H9 (8.000–8.036). In the case of 8 H9 being the only callout, the print reader needs to use the standard charts or a machinist handbook to determine the limits for that feature.

Tolerancing Methods

Each company determines the best practice for tolerancing. At one time, it was common to use the limit method for every dimension. To reduce clutter on the drawing, other companies may elect to use limit dimensions only for important and critical dimensions. In these cases, basic size values are given for most dimensions. The print reader should examine title block notes or general notes to determine the tolerance amount. Most title block notes use decimal precision to indicate tolerance precision. For example, two-place decimals may have a tolerance of ±.01″, while three-place decimals are ±.005″. Many drawings combine tolerancing methods, depending on the processes for checking or manufacturing the parts or features. As you will see in Unit 13, *Geometric Dimensioning and Tolerancing*, GD&T also impacts the way in which tolerances and dimensions are expressed on a drawing.

ISO Symbol		
Hole Basis	Shaft Basis	Description
H11/c11	C11/h11	Loose–Running Fit Wide commercial tolerances or allowances on external members.
H9/d9	D9/h9	Free Running Fit Not for use where accuracy is essential, but good for large temperature variations, high running speeds, or heavy journal pressures.
H8/f7	F8/h7	Close Running Fit For running on accurate machines and for accurate location at moderate speed and journal pressures.
H7/g6	G7/h6	Sliding Fit Not intended to run freely, but to move and turn freely and locate accurately.
H7/h6	H7/h6	Locational Clearance Fit Provides snug fit for locating stationary parts; but can be freely assembled and disassembled.
H7/k6	K7/h6	Locational Transition Fit For accurate location, a compromise between clearance and interference.
H7/n6	N7/h6	Locational Transition Fit For more accurate location where greater interference is permissible.
H7/p6	P7/h6	Locational Interference Fit For parts requiring rigidity and alignment with prime accuracy of location but without special bore pressure requirements.
H7/s6	S7/h6	Medium Drive Fit For ordinary steel parts or shrink fits on light sections, the tightest fit usable with cast iron.
H7/u6	U7/h6	Force Fit Suitable for parts which can be highly stressed or for shrink fits where the heavy pressing forces required are impractical.

Goodheart-Willcox Publisher

Figure 10-17. Symbols made up of letters and numerals that classify the tolerance of a metric feature can be used in the local note. These are based on ISO standards but are incorporated into ASME standard B4.2.

Summary

- Interchangeable manufacturing requires that all parts be manufactured within specified limits of size, and so all dimension values have a tolerance, or range of size, specified.
- Tolerance is the total amount by which a dimension can vary, and it can be expressed in a variety of ways.
- In general, the basic size of a feature or pair of mating parts is the designer's value before tolerances are applied.
- Basic dimensions are theoretically exact dimensions with numerical values in boxes and are used in situations wherein the tolerance is applied differently, most often in geometric dimensioning and tolerancing.
- Maximum material condition (MMC) is an important term that helps the designer understand the tightest fit, and it is represented by the larger value of an external feature and the smaller value of an internal feature.
- Allowance is the intentional difference found when comparing the MMC values of mating parts, simply defined as the tightest fit between two features, expressed as a positive value for clearance fits.
- Datums are the reference planes or axes from which measurements and geometric controls are measured, and they are identified with alphabet characters within a datum feature symbol.
- Dimension origin symbols help clarify surfaces from which dimensions should originate without establishing datums.
- Symbols are available to clarify how the taper of flat and conical surfaces can be toleranced, how radii can be controlled, and how single limits can be applied.
- Statistical tolerances and continuous features also can be clarified with standard symbols.
- Engineers and designers use standard classes of fits to establish precise values for inch-based mating parts. These include clearance fits, transitional fits, and interference fits.
- Metric-based design also uses established and standardized classes of fits to establish relationships for mating parts.
- The print reader often does not need to know the standard classes of fits, but rather only needs to read the values established for the dimensions on the print.
- Each company will determine the best practices for tolerancing, so the print reader should be aware of company standards and title block default tolerances and be ready for a variety of tolerance expressions, symbols, and practices.

Unit Review

Name ______________________ **Date** ______________ **Class** ______________

Answer the following questions using the information provided in this unit.

Know and Understand

____ 1. What term describes the assurance that when you select one of part A from among hundreds of those parts and likewise a mating part B, the two parts will fit together properly?
A. Selective manufacturing
B. Controlled manufacturing
C. Interchangeable manufacturing
D. Flexible manufacturing

____ 2. *True or False?* Tolerance is the total amount by which a dimension can vary.

For the next four auxiliary view terms, match the terms to the descriptions. Answers are used only once.

____ 3. Plus and minus method—bilateral but not equilateral	A. 4.875 / 4.885
____ 4. Plus and minus method—unilateral	B. 4.880±.005
____ 5. Limit method	C. $4.877^{+.008}_{-.002}$
____ 6. Plus and minus method—equilateral	D. $4.888^{+.000}_{-.010}$

____ 7. *True or False?* A dimension that specifies 1.500″ ±.005″ has a tolerance of .010″.

____ 8. The _____ size of an object is the size by which we identify it and may not be the actual size.
A. maximum
B. basic
C. nominal
D. general

____ 9. The _____ size of a feature or object is the designer's intended value, with tolerances applied in one or both directions.
A. maximum
B. basic
C. minimum
D. allowable

____ 10. *True or False?* Basic dimensions will be represented with a box around the dimension value, and tolerances are not applied to these values because they are theoretically exact.

____ 11. *True or False?* The maximum material condition (MMC) of a feature or a part is another way of saying the "upper limit" or the "largest size."

____ 12. _____ is another term for tightest fit.
A. Allowance
B. Running
C. Interference
D. Clearance

____ 13. *True or False?* The loosest fit of two parts can be determined by subtracting the least material condition (LMC) of the shaft from the LMC of the hole.

____ 14. The term _____ applies to a surface that has been identified as a reference surface with a(n) _____ feature symbol. *(Same word for both blanks)*
A. plane
B. origin
C. reference
D. datum

____ 15. *True or False?* The dimension origin symbol should always be used with a datum feature symbol.

____ 16. *True or False?* The tolerance specified for an angled surface may result in a fan-shaped tolerance zone.

____ 17. *True or False?* The tolerance for a radius value specified for a curved feature (such as a rounded edge) may result in a crescent-shaped tolerance zone.

____ 18. What term, symbol, or expression would you expect to see near the dimension value for a "single limit" tolerance expression?
A. MAX
B. SL
C. ST
D. CF

____ 19. *True or False?* The statistical tolerance symbol and the continuous feature symbol are both applied to the drawing by leader line.

____ 20. Identify the one classification of fit that may be either clearance or interference in nature.
A. RC—Running or Sliding
B. LC—Locational
C. LT—Transitional
D. FN—Force or Shrink

____ 21. *True or False?* Standard classes of fits such as RC, LC, and FN are often not identified on the industrial print but may only have been used by the engineers to communicate with the drafter creating the print.

____ 22. *True or False?* The ASME table for running or sliding clearance fits used in the text was based on a basic hole system, so the MMC value for the hole will be the basic design size, and the LMC value of the hole will be slightly larger.

____ 23. Identify the *false* statement with respect to metric-based fits.
A. Metric fits are identical to inch-based fits but use millimeters instead of inches.
B. Metric fits are the same in concept to inch-based fits, but the terms and symbols are different.
C. ASME B4.2 is the primary resource for metric-based fits.
D. With the metric system, there are alphanumeric symbols for the classes of fit, and these sometimes appear on the prints.

Critical Thinking

1. If parts were made with very precise tolerance requirements, then the assembly of the parts would be of higher quality. What would be some of the consequences?

____

____

____

____

____

2. Some companies elect to show limit dimensions for every dimension on the print. What are some reasons that others may elect *not* to do that?

____

____

____

____

____

3. What are some specific things that would qualify an individual to be the person who determines the amount of tolerance for a mating relationship of two parts?

____

____

____

____

____

Apply and Analyze

Name ______________________ Date ______________ Class ______________

Review Activity 10-1

For each nominal size given below, calculate the maximum material condition and least material condition for the hole and the shaft. Then, calculate the allowance.

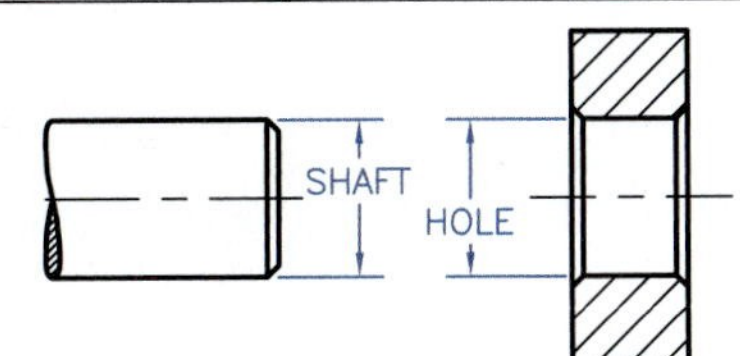

Using the basic hole system, calculate the dimensions as established for standard fits. Use the table in Figure 10-15 for values, and use four (.XXXX) decimal places of accuracy for all values.

Nominal Fit "Basic" Size		Hole Limits		Shaft Limits		Allowance
		MMC	LMC	MMC	LMC	
Ø A	5/8″ RC3					
Ø B	1 – 1/2″ RC2					
Ø C	1/4″ RC4					
Ø D	2 – 3/8″ RC1					
Ø E	1″ RC2					

Apply and Analyze

Name ____________________ Date ____________ Class ____________

Review Activity 10-2

For each of the mating part pairs below, calculate the upper and lower limit for each part. For A and B, express the final dimension using limit dimensioning. For C and D, express the final dimension using plus and minus dimensioning (equal bilateral expression).

A.

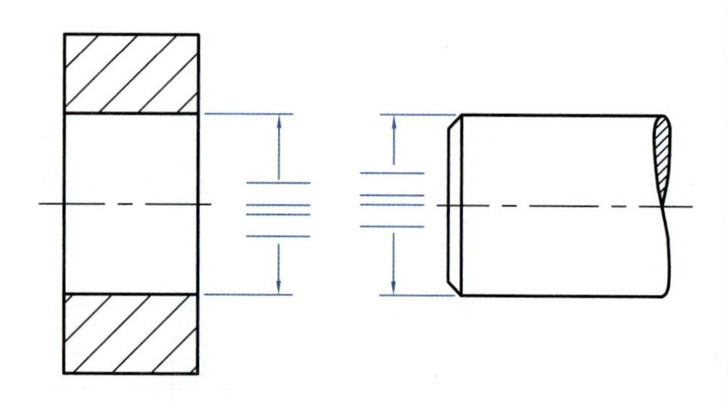

Basic hole size [MMC] is 5/8″
Allowance [A] = .005
Shaft tolerance = .011
Hole tolerance = .006

B.

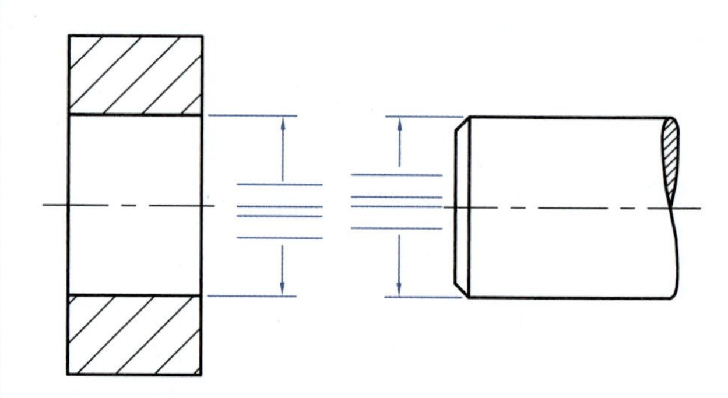

Basic hole size [MMC] is 1 1/2″
Allowance [A] = .010
Shaft tolerance = .010
Hole tolerance = .010

C.

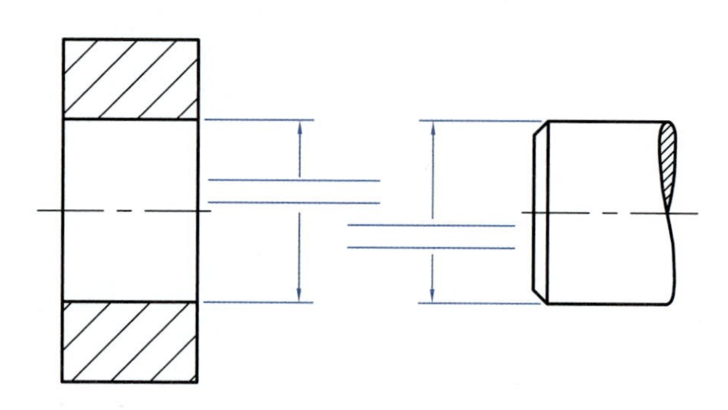

Basic hole size [MMC] is 3/4″
Allowance [A] = .008
Shaft tolerance = .008
Hole tolerance = .008

D.

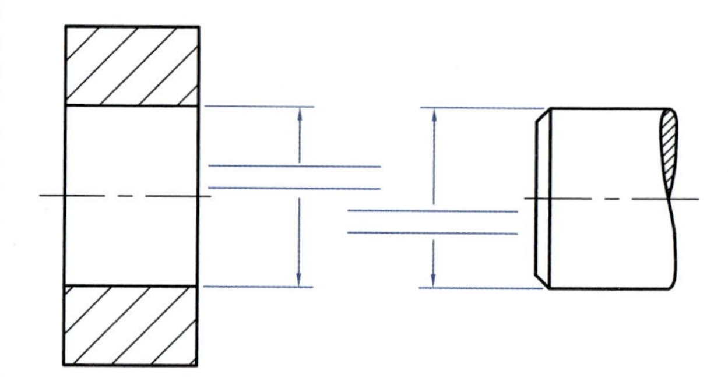

Basic hole size [MMC] is 1″
Allowance [A] = –.006
Shaft tolerance = .002
Hole tolerance = .002

Notes

Apply and Analyze

Name ______________________ Date ______________ Class ______________

Industry Print Exercise 10-1

Refer to the print PR 10-1 and answer the questions below.

1. Note 3 in the lower-right area of the drawing is referring to the crosshatched area on the front view. What is the maximum material condition of the hole in the middle of that area?

____

2. Where on the drawing is the tolerance specified for the R.50 and the 2X R.12 dimensions, and how much is this tolerance?

____

3. How many dimensions on this print are expressed with the limit method? ______________
4. This drawing shows a fastener inserted into a hole, as indicated by note 5. What is the maximum material condition of that pin? *(Note: The front view indicates the size of the hole.)*

____

5. With reference to question 4, the hole in the part is to be .451–.453. What is the allowance for the pin and the hole?

____

6. Two of the dimensions, 3.165 and 2.114, have rectangular boxes around them. As discussed in this unit, what are those dimensions called?

____

7. Note 1 indicates 10 GAGE 1010-1020 sheet metal can be used to make the main body of this part. According to the left side view, how much tolerance is given to the thickness of the metal?

____

8. The box symbols that indicate an A, B, or C are called _____ feature symbols.

____

9. The top view shows a bend radius of the thin metal. Would a bend radius of 1/8″ meet the requirements?

____

10. What is the tolerance of the single limit dimension? ______________

Review questions based on previous units:

11. Are there any section views shown on this print? ______________
12. What is the name of this part? ______________
13. What is the part number of this print? ______________
14. How many chamfered edges are noted by leader lines on this drawing? ______________
15. Which system for placing dimension values is used, aligned or unidirectional? ______________

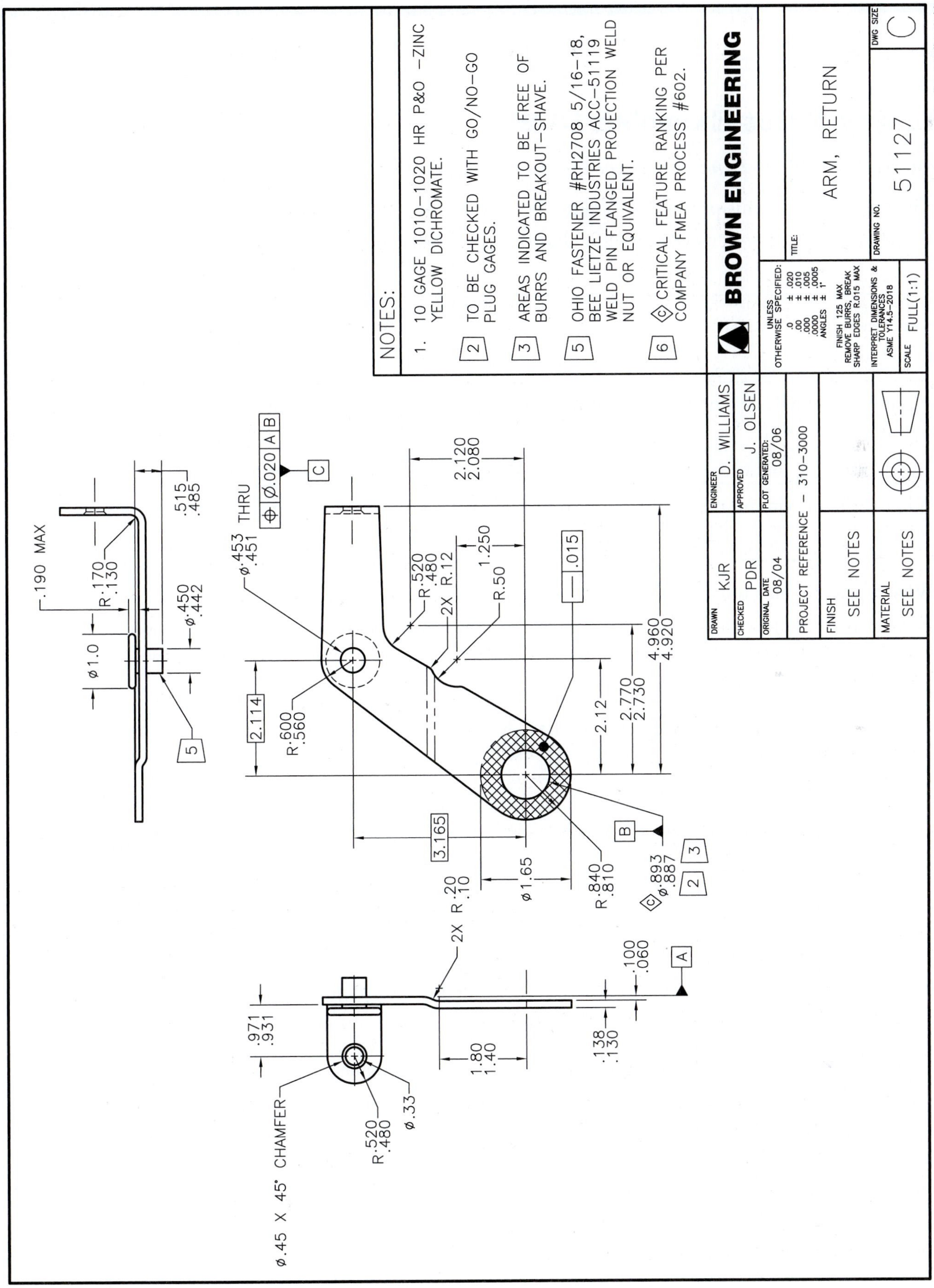

NOTES:

1. 10 GAGE 1010–1020 HR P&O –ZINC YELLOW DICHROMATE.
2. TO BE CHECKED WITH GO/NO-GO PLUG GAGES.
3. AREAS INDICATED TO BE FREE OF BURRS AND BREAKOUT-SHAVE.
5. OHIO FASTENER #RH2708 5/16–18, BEE LIETZE INDUSTRIES ACC–51119 WELD PIN FLANGED PROJECTION WELD NUT OR EQUIVALENT.
6. ◇C CRITICAL FEATURE RANKING PER COMPANY FMEA PROCESS #602.

DRAWN KJR	ENGINEER D. WILLIAMS	BROWN ENGINEERING	
CHECKED PDR	APPROVED J. OLSEN		
ORIGINAL DATE 08/04	PLOT GENERATED: 08/06	UNLESS OTHERWISE SPECIFIED: .0 ± .020, .00 ± .010, .000 ± .005, .0000 ± .0005, ANGLES ± 1°	TITLE: ARM, RETURN
PROJECT REFERENCE – 310–3000		FINISH 125 MAX REMOVE BURRS, BREAK SHARP EDGES R.015 MAX	
FINISH SEE NOTES		INTERPRET DIMENSIONS & TOLERANCES ASME Y14.5–2018	DRAWING NO. 51127 / DWG SIZE C
MATERIAL SEE NOTES		SCALE FULL(1:1)	

Goodheart-Willcox Publisher

PR 10-1. Return Arm

Apply and Analyze

Name ______________________ Date ______________ Class ______________

Industry Print Exercise 10-2

Refer to the print PR 10-2 and answer the questions below.

1. The basic depth of this part is .715″. What is the tolerance on that dimension? ______________
2. In the auxiliary section view, there is a chamfer note. What is the tolerance on the chamfer?

   ____

3. The central, smallest hole has a target size of .1577″. What is the tolerance on that hole?

   ____

4. What is the MMC of a mating part for a clearance fit within this same hole featuring an allowance of .0006″?

   ____

5. What is the maximum material condition of the largest body diameter of this part?

   ____

6. What is the angular tolerance for angles, unless otherwise specified? ______________
7. What is the tolerance of the diameter dimension that has a tolerance expressed in a bilateral expression, but not equilateral?

   ____

8. There are four symbols on this print that have letters enclosed in squares, each with a stem leading to a filled triangle. What do these symbols represent?

   ____

9. How many notes indicate a unilateral tolerance for rounding a corner "up to" a certain amount?

   ____

10. In the view that is furthest to the right, a 1.323 diameter dimension is given. What is the *least material condition* of that feature? ______________
11. Based on tolerances given, calculate the deepest value for the counterbore. ______________

Review questions based on previous units:

12. Is Section B-B shown in alignment with other views, or is it shown as a removed view? ______________
13. Use charts in Appendix D of this text to identify the major diameter of the four threaded holes.

    ____

14. How many cutting planes are indicated? ______________
15. Are there any reference dimensions on this print? ______________

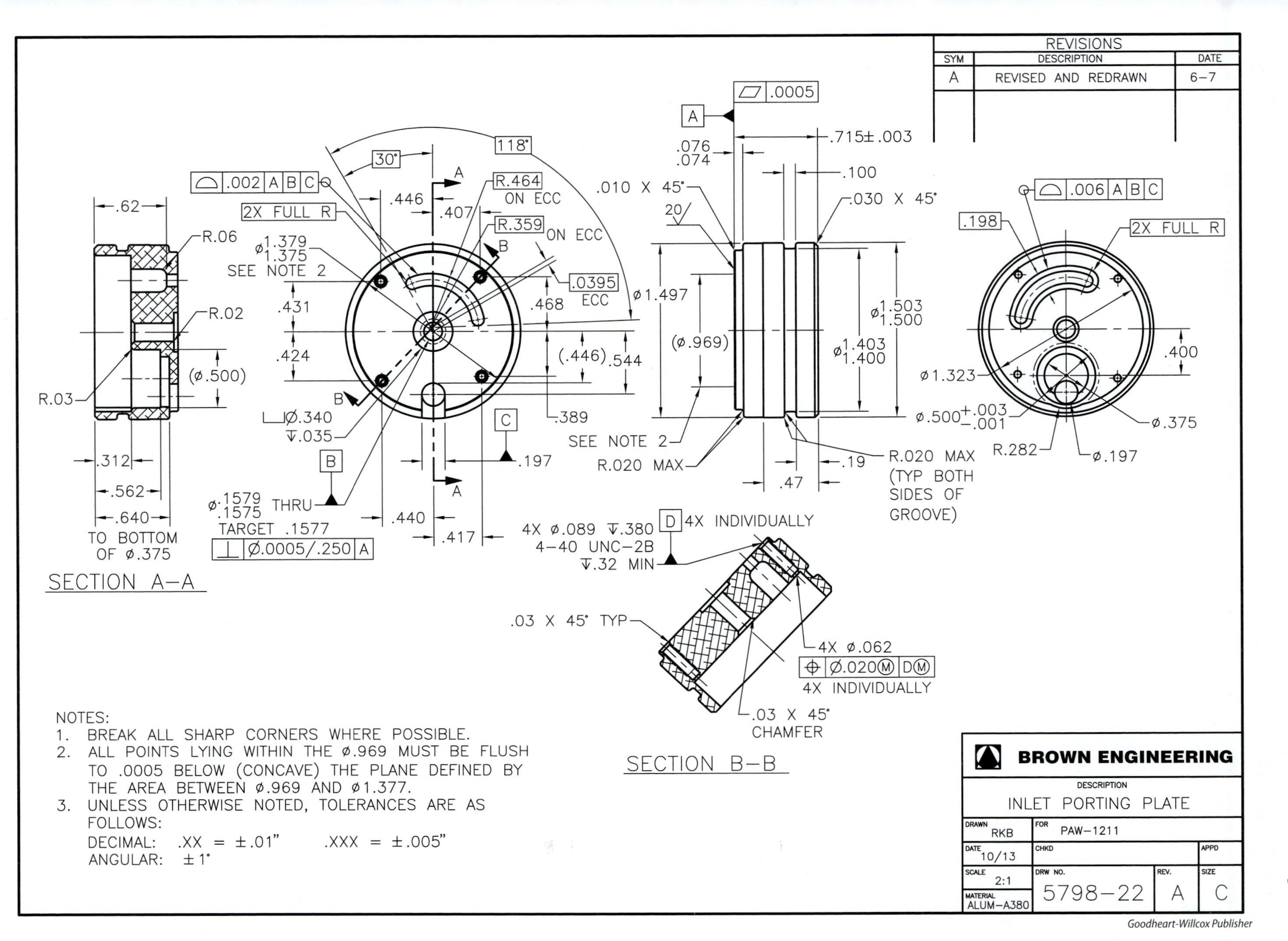

PR 10-2. Porting Plate Inlet

Apply and Analyze

Name ______________________ Date ______________ Class ______________

Bonus Print Reading Exercises

The following questions are based on various bonus prints located in the folder that accompanies this textbook. Refer to the print indicated, evaluate the print, and answer the question.

Print AP-001

1. For this print, there are special triangles floating next to a few dimensions. What tolerance is applied to those dimensions?

____

2. How many dimensions on this print express the tolerance using the limit method? *(Note: Exclude thread notes.)*

____

Print AP-003

3. For the overall (largest) diameter dimension, what name is given to the tolerancing method used?

____

4. In detail A, the angular measurements are given in degrees and minutes. How much tolerance is specified for the angles?

____

Print AP-004

5. For the three holes located .875″ apart, what is the total size tolerance specified for those holes?

____

6. What tolerance is applied to the total height of 3.000″?

____

Print AP-006

7. What is the MMC of the smallest hole on the main central axis?

____

8. Do the dimensions and tolerances apply before the finish is applied or after?

____

9. Is it possible for the grooves featured in the enlarged details to have parallel sides?

____

Print AP-011

10. How much tolerance is allowed when checking the diameter over balls?

____

11. Calculate the total width of the sectional view. *(Specify your answer as a range.)*

____

Print AP-012

12. For the few dimensions that have a tolerance specified locally, what method of tolerance representation is used?

____

Print AP-016

13. Is the tolerance on the 1.656″ length specified as unilateral, equilateral, or bilateral?

____

14. What is the maximum material condition of the .3120″ diameter feature?

____

Print AP-017

15. What tolerance is specified for the total depth of this object, as featured left to right in the section view?

____

16. What is the MMC of the cylindrical surface that will establish datum A?

____

Print AP-018

17. What tolerance is specified for dimensions without a local tolerance specified?

____

Print AP-019

18. What tolerance is specified for the overall width of .860″?

____

Print AP-022

19. *True or False?* The holes marked A have a tighter size tolerance than the holes marked B.

____

20. The holes marked B are located 9.8 mm away from datum Y. How much tolerance is allowed for that linear dimension value?

____

Notes

UNIT 11 Machining Specifications and Drawing Notes

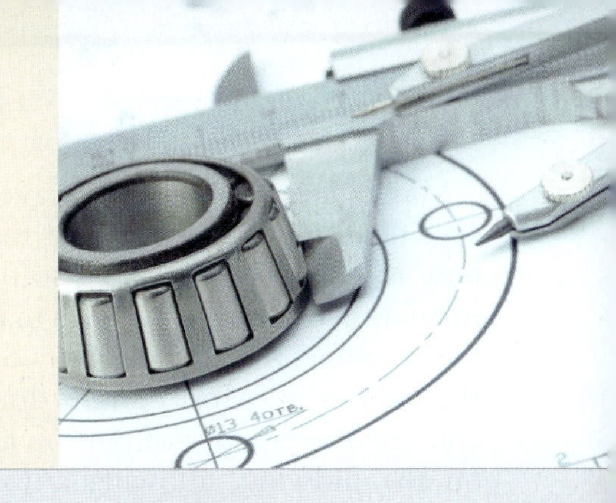

LEARNING OBJECTIVES

After completing this unit, you will be able to:

- Identify and interpret general notes on a print.
- Identify and interpret local notes on a print.
- Identify standard symbols that guide or clarify the specification of features that are typical of common machining processes.
- Read and interpret specifications for holes that feature counterbores and spotfaces.
- Read and interpret specifications for holes that feature countersinks and counterdrills.
- Read and interpret dimensions with respect to chamfered edges.
- Read and interpret callouts and other dimensioning methods for features such as necks and keyways.
- Identify and interpret callouts for knurls.
- Define the manufacturing process known as broaching.

TECHNICAL TERMS

broaching
chamfer
counterbore
counterdrill
countersink
flag
general note
key
keyseat
keyway
knurling
local note
machining center
reaming
spotface
typical (TYP)

Additional annotations must often appear on a drawing to provide information and instructions beyond the title block information, parts list, graphic shape description, and basic dimensioning. These additional annotations are usually classified as *notes*, *specifications*, or *callouts*. Notes can be used to eliminate repetitive information or to give more information about the size of holes, fastener types, or other particular specifications for the removal of machining burrs.

Sometimes notes can contain so much information that placing the notes on the drawing makes the drawing unreadable. This is often the case for architectural and structural drawing specifications. For these notes, the information is published on separate sheets. The sheets are then included with the set of drawings. This is how the term "drawings and specifications" originated.

Many large industries have internal process-specification manuals. These manuals may specify how to annotate a drawing so the machinist has information on how to perform machining processes—the machine, tools, and cutters to be used, as well as the tolerances. Most current drafting standards discourage the practice of putting all of these processes on the drawing, allowing any vendor or supplier to choose their own processes, as long as the part matches size and location dimensions. Older drawings commonly will have used different standards, and they must be read by the print reader. You should become familiar with the standards used by your company and the processes involved in the work you will be required to perform. This unit covers common machining processes and methods for specifying processes on drawings.

Notes

Basically, notes are classified as either *general* or *local*, but both types of notes can contain similar information. The type of note is determined by the application of the note and how it is placed on the drawing or within the 3D model environment. ***General notes*** apply to the entire drawing or model. They are usually placed in a horizontal position above or to the left of the title block. General notes are not referenced in the parts list or tied to specific areas of the drawing. Some examples of general notes are given in **Figure 11-1**.

Sometimes there are exceptions to general notes. In this case, the general note is followed by the phrase EXCEPT AS SHOWN or UNLESS OTHERWISE SPECIFIED. The general note then applies to the entire part or to the entire drawing, except where a difference is noted by a local note.

Local notes, also referred to as *specific notes* or *callouts*, apply only to certain features or areas. They are positioned near the feature or area to which the note applies. A leader is used to connect the note to the feature or area. See **Figure 11-2**. In some cases, a chain line can be used to indicate the note applies only to a specific portion of the part.

1. THIS PART SHALL BE PURCHASED ONLY FROM SOURCES APPROVED BY THE ENGINEERING DEPARTMENT.
2. BREAK SHARP EDGES R.030 MAX UNLESS OTHERWISE SPECIFIED.
3. REMOVE BURRS.
4. FINISH ALL OVER.
5. METALLURGICAL INSPECTION REQUIRED BEFORE MACHINING.

Goodheart-Willcox Publisher

Figure 11-1. General notes apply to the entire drawing.

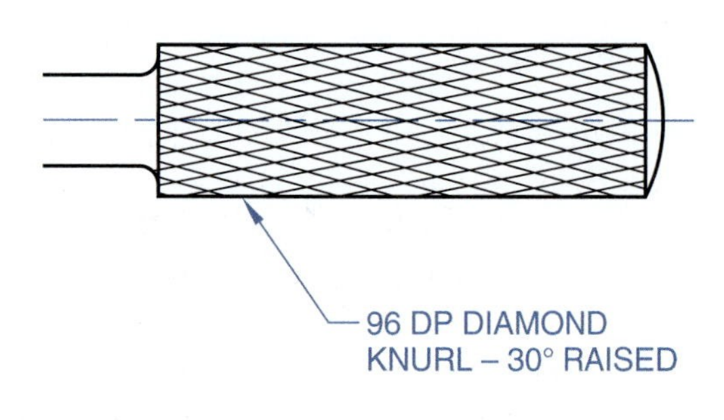

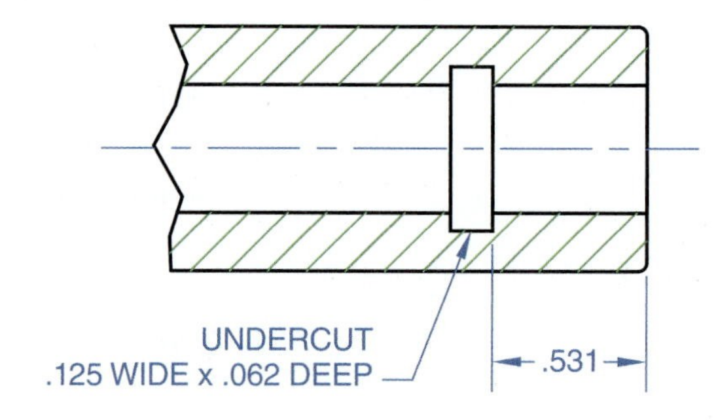

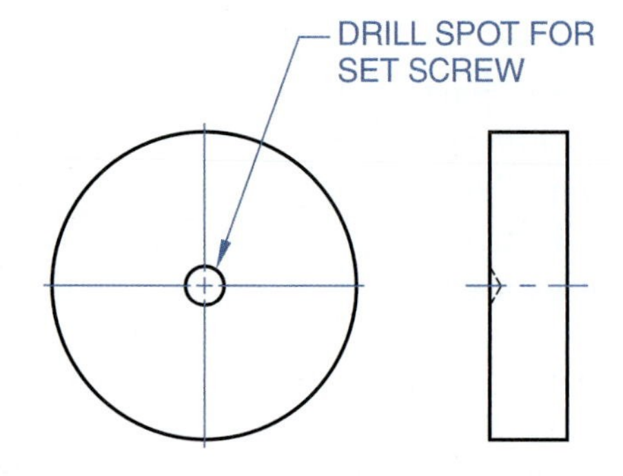

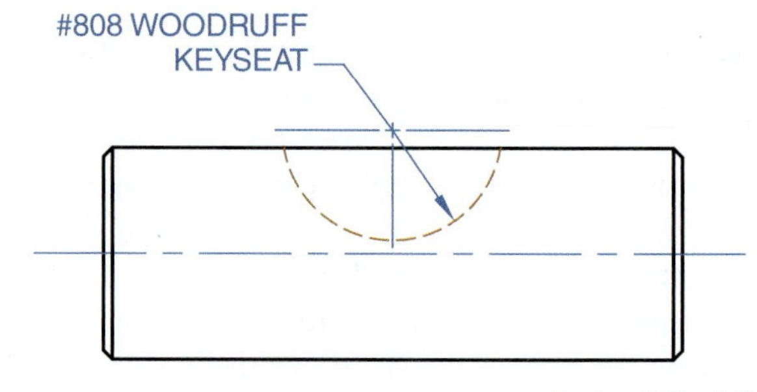

Goodheart-Willcox Publisher

Figure 11-2. Local notes often describe special manufacturing processes or treatments. Sometimes these processes do not need to be described and can be replaced with symbols or linear dimensions that do not refer to the process.

Sometimes local notes use numbers enclosed in a geometric shape, such as an equilateral triangle. This type of note is called a ***flag***. See **Figure 11-3**. The actual note text appears in a central location with other flagged local note text, or perhaps with general notes. This technique can help keep the area near the views free from clutter.

Symbols

In previous units, several ASME symbols were introduced for dimensioning and tolerancing, and those can be reviewed in **Figure 9-3** and **Figure 10-6**. This section will introduce additional symbols with respect to machining specifications, **Figure 11-4**. As with previous units, these symbols are referenced and explained in ASME Y14.5, complete with recommended size for each symbol. The symbols in current standards explain common terms symbolically or with alphabetic characters and should be used on drawings in lieu of words when feasible. In former practice, the local note for two holes may have specified .250 DRILL, TWO PLACES, and the word DEEP, if used, would have been placed after the value instead of placing the depth symbol in front of the value. For additional symbols, see Appendix D.

Holes

One of the most common features to be specified by note is the hole, which may be counterbored or countersunk, or partially threaded. On older drawings, the type of operation was often specified, such as whether or not the hole was to be punched, drilled, or reamed. These specifications were usually identified in a note that gave more information than just the size.

1 .030 X 45° CHAMFER

THIS SURFACE TO BE COPLANAR WITH SURFACE MARKED "Y"

RUBBER STAMP PART NUMBER HERE

MOUNT IN CHUCK USING THIS SURFACE

Goodheart-Willcox Publisher

Figure 11-3. A flagged local note is a number enclosed in a geometric shape, in this case an equilateral triangle, that will then be located at the end of a leader line. The textual content of the note may appear in a convenient location, in similar fashion to general notes.

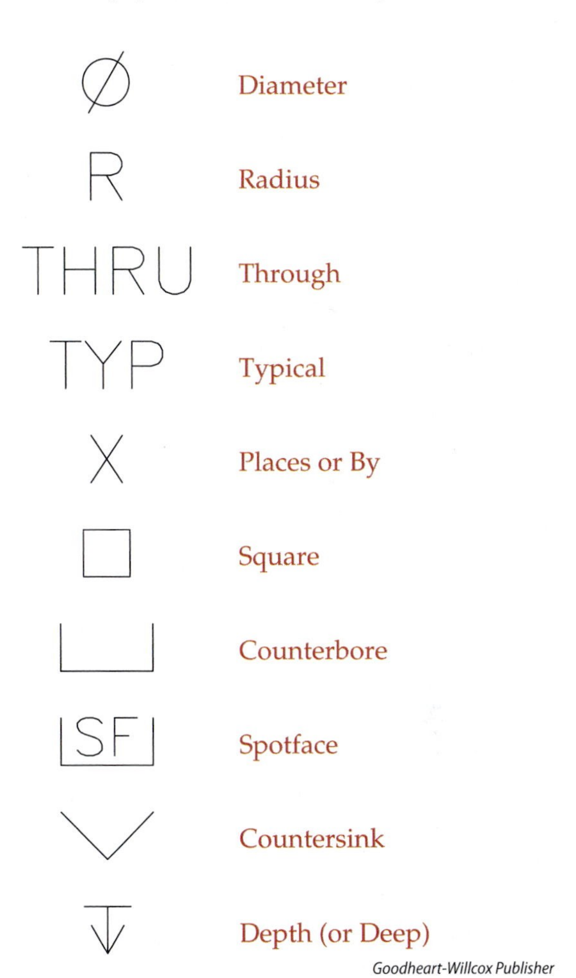

Goodheart-Willcox Publisher

Figure 11-4. These are the ASME recommended standard symbols used in local notes for holes.

Drilled and Reamed Holes

Drilled holes are created with a twist drill, not a mill cutter. Current ASME standards discourage specifying whether or not the hole must be manufactured by drilling. If details are included, drilled holes are usually specified by the diameter of the drill bit. The number of holes and depth may also be specified. **Figure 11-5** shows how symbols are used in local callouts for holes. For multiple holes of the same size, standard practice recommends specifying the holes with the leader line only pointing to one hole. For example, if there is a four-hole pattern, one hole could exhibit the callout 4X ∅.250. When used, the depth symbol is placed in front of the depth value. In situations where it may not be obvious, the depth can be specified as THRU (for through) to clarify that the hole passes entirely through the feature. The depth can also be expressed with a linear dimension, especially when the hole enters a curved surface or other special shape that makes the interpretation of depth less clear. **Figure 11-6** shows how local notes can be applied within the 3D model environment.

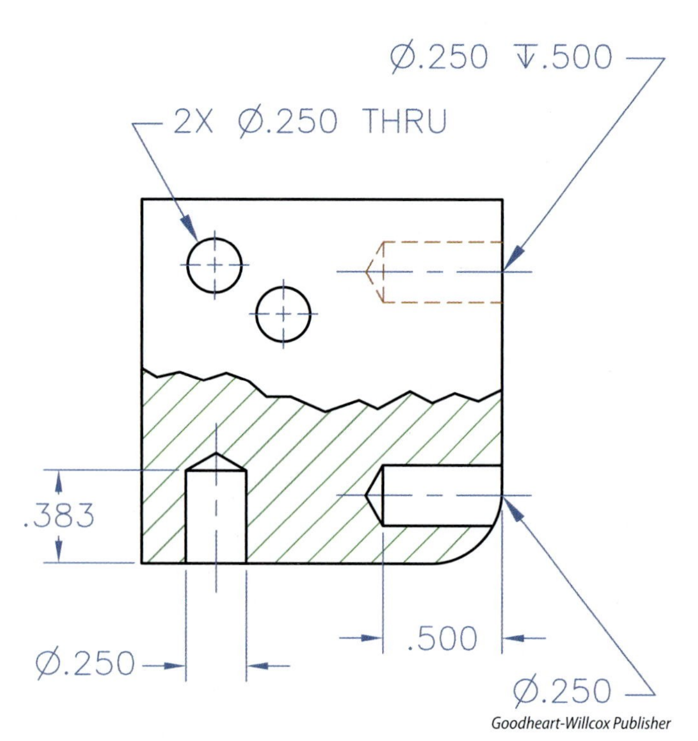

Figure 11-5. These hole callouts use standard symbols for number of places, diameter, and depth, but linear measurements are sometimes required, or are more clear.

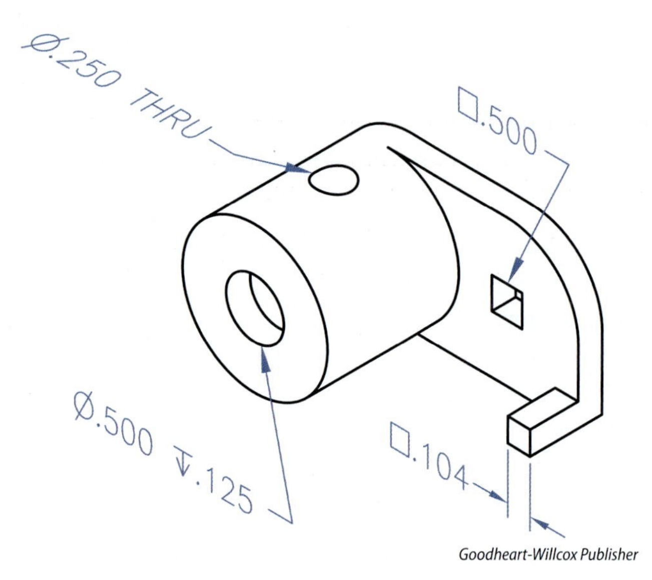

Figure 11-6. Local notes and callouts can be incorporated into the 3D model environment.

Square holes can incorporate the square symbol, although that symbol can also be applied to external features such as square posts.

Some hole notes may also contain the term TYP, which stands for ***typical***. When this term appears in the specification, the note applies to all similar features on the print, unless otherwise noted. Usually the similar features will be obvious and will not require the print reader to make assumptions that could be in error.

Reamed holes are created with a machine tool called a ream or reamer. ***Reaming*** primarily applies to metal manufacturing and creates a very true, smooth, and accurately sized hole. To create a reamed hole, the hole is initially created as a drilled hole slightly smaller (.010″ to .025″) than the finished size. Then, the hole is reamed to the specified finished size. The drilled hole size may be omitted, but if the process to create the hole is critical, a callout for both the drilling and reaming may be shown. Additional symbols for surface finish quality will be discussed in the next unit, and will allow for additional clarification of the machining process.

Counterbores and Spotfaces

Counterbored holes have been cylindrically enlarged on one or both ends to form a recessed flat shoulder. A ***counterbore*** is often used to recess a bolt or machine screw head below the surface of the part. A counterbore specification is shown in **Figure 11-7.** Current standards provide a symbol for counterbore

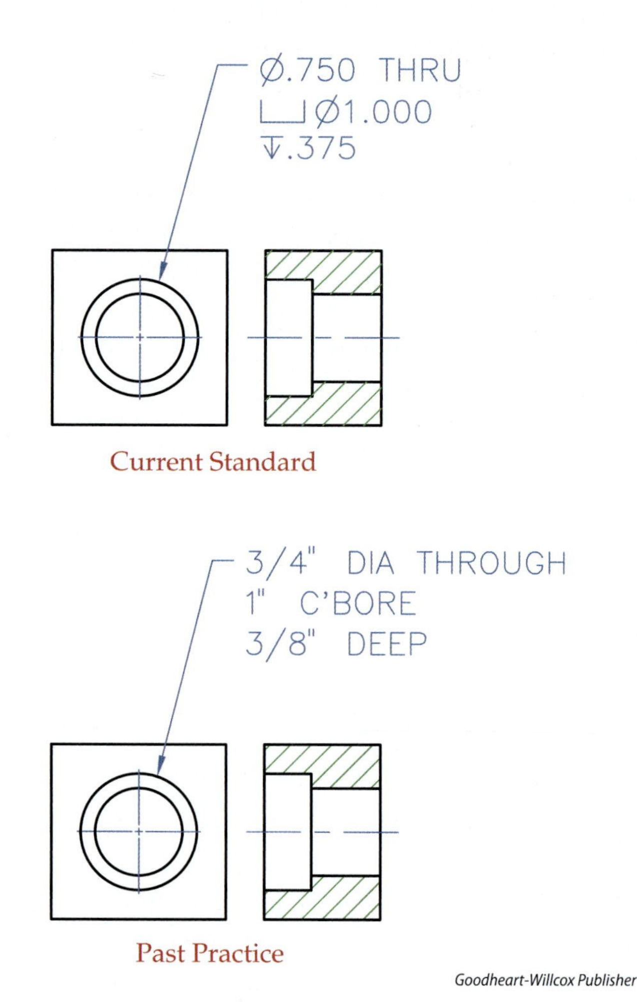

Figure 11-7. The top figure shows the current practice for specifying a counterbored hole.

and depth. In former practice, the abbreviation C'BORE was used in the callout instead of the counterbore symbol. The counterbore diameter, the through hole, and the depth of the counterbore are often given in one note that points to the outermost circle representing a plan view of the hole. Separate notes can be used when it may be beneficial to separate out the information. The note should be in the order of the processes, so the hole diameter and depth are given first, and the counterbore diameter and depth are given second. Additionally, a radius can be specified at the end of the note. On some occasions, the remaining material may have more significance, so a linear dimension can be given for the thickness rather than the depth. Some holes may have more than one counterbore. In those cases, all depth distances are referenced from the same surface. **Figure 11-8** shows additional counterbore notes.

Spotfaces are simply very shallow counterbores. The ***spotface*** provides a flat surface on rough stock for the purpose of a bearing or seating surface (for a bolt head, nut, or washer, for example). See **Figure 11-9**. The depth is often omitted, but it may be given in similar fashion to the counterbore. When the depth is omitted, the spotface should be created with the minimum depth needed to clean up the indicated surface. The machinist understands to simply "break off" the rough surface at a depth of roughly .062″ or less. The ASME standard symbol features an "SF" inside the counterbore bracket symbol and is a newer symbol that may not be found on older prints. Another newer practice is the allowance for a radius to be specified. In the case of a spotface, the diameter of the spotface specifies the flat surface within the fillet radius.

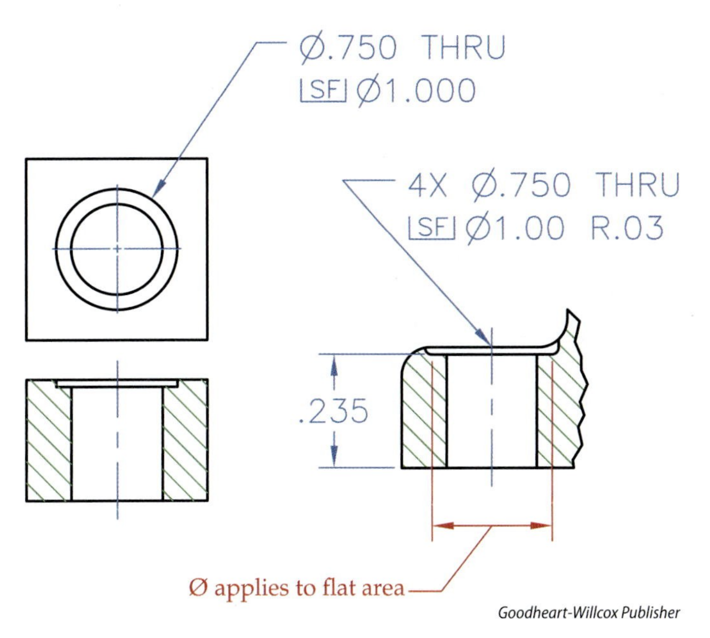

Figure 11-9. The specification for a spotfaced hole can point to the circle view or be incorporated into the section view, as shown on the right. It can include a radius value, and in those cases, the flat area must meet the diameter specified.

Countersinks and Counterdrills

Another process to improve the function of a hole is the countersink. A ***countersink*** is defined as the cone-shaped enlargement on the end of a hole. Countersunk holes are often used to provide a seat for conical screw heads and rivets. Current standards provide a symbol for countersinking holes. See **Figure 11-10**. Older practices specified countersink with C'SINK, and this may still be found on some drawings. A countersunk hole is often drawn with an included angle of 90°, even if the angle is 82°. There are different ways to specify the size of a countersink. Usually the included angle is called out in the local

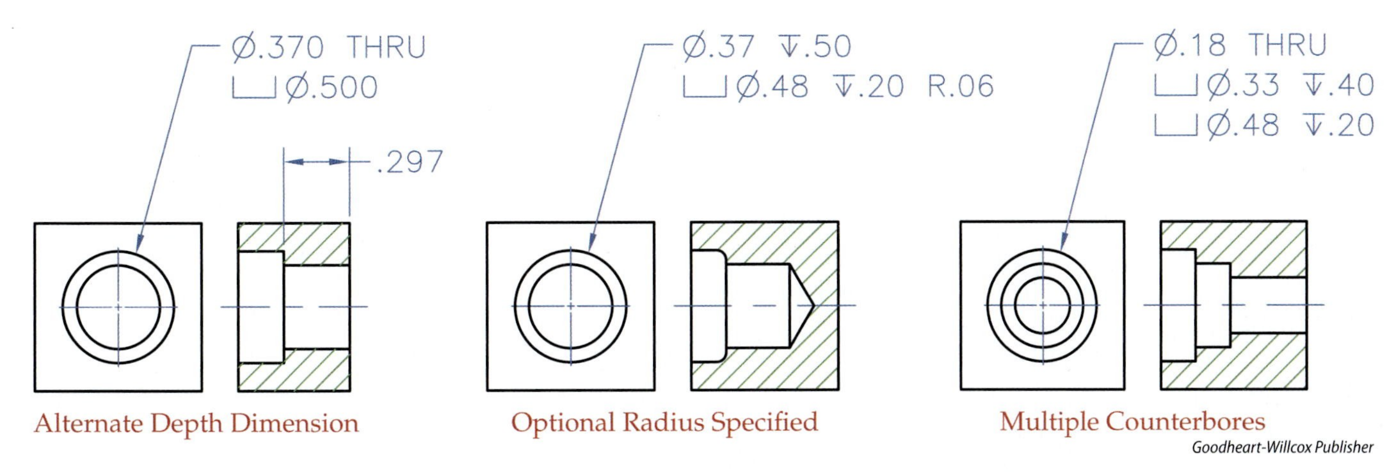

Figure 11-8. Specifications for counterbored holes can incorporate linear dimensions, a fillet radius specification, and multiple counterbores in one callout.

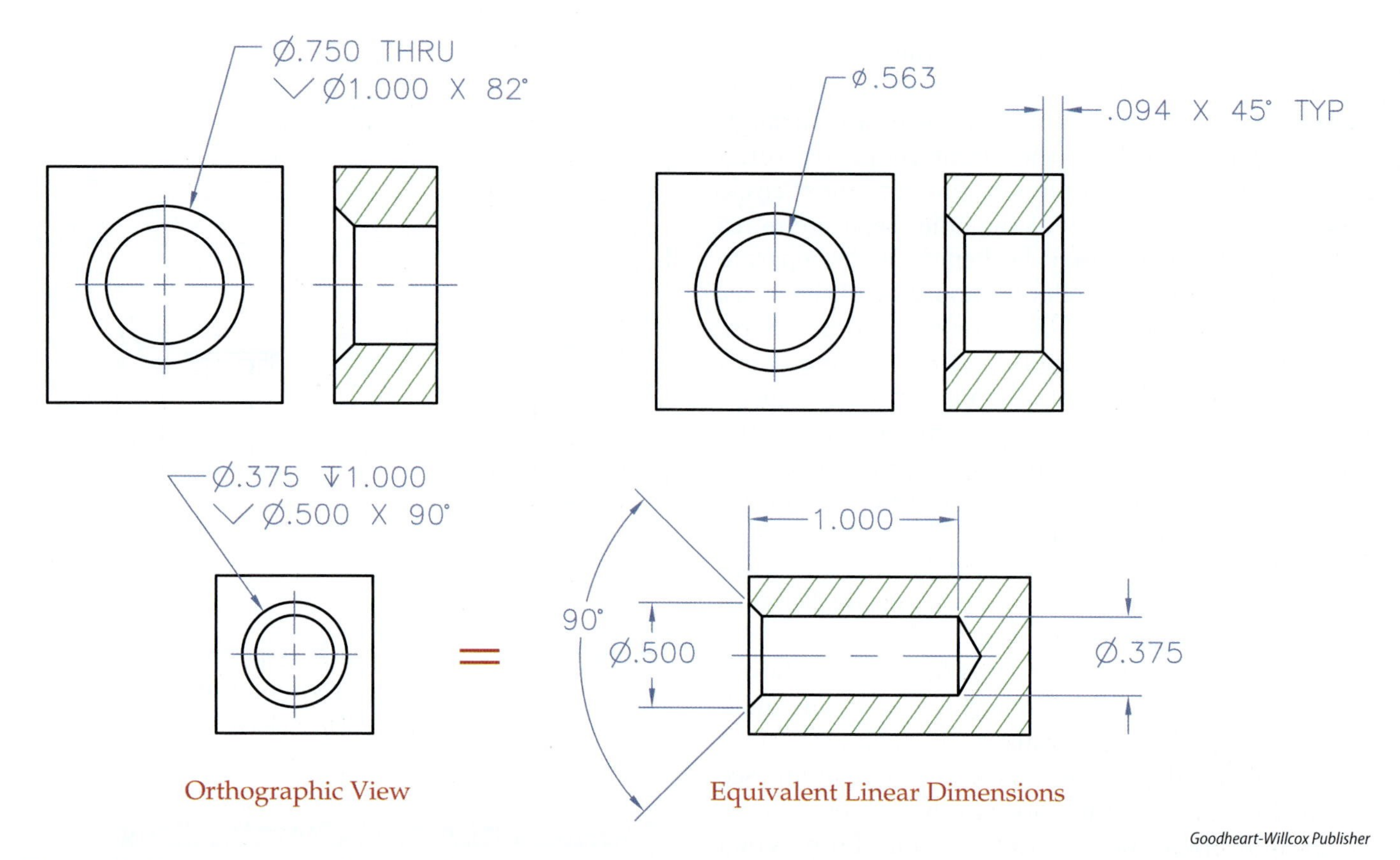

Goodheart-Willcox Publisher

Figure 11-10. Countersunk holes are holes with a cone-shaped enlargement, and they can be dimensioned by note. Chamfer depth can also be specified as shown, or other linear measurements may be given if that is more clear and concise.

note, which is determined by the bit. The machinist also then needs a specification for the larger diameter of the countersink, or the depth of the countersink can be given in some situations. If the countersink is on a curved surface, the diameter specified applies at the minor diameter of the resulting oval-like shape, **Figure 11-11.**

When a larger hole has a conical transition to a smaller hole, this is called a ***counterdrill.*** The transition is usually shown at the same angle as a drill bit tip (120°), and dimensioning the angle is considered optional. The depth of the counterdrill is considered to be the portion of the larger hole that is full diameter. See **Figure 11-12.** As with holes and counterbores, linear dimensions can be given in lieu of symbols if it helps clarify the size description.

Holes are sometimes drilled and countersunk for the purpose of holding the part between lathe centers or in a machining fixture. These are called ***machining centers*** and have no function other than providing a place to hold the part. If the machining center remains on the finished part, it can be indicated by local note or callout. Machining center surfaces can also be identified as datum surfaces to establish a datum axis.

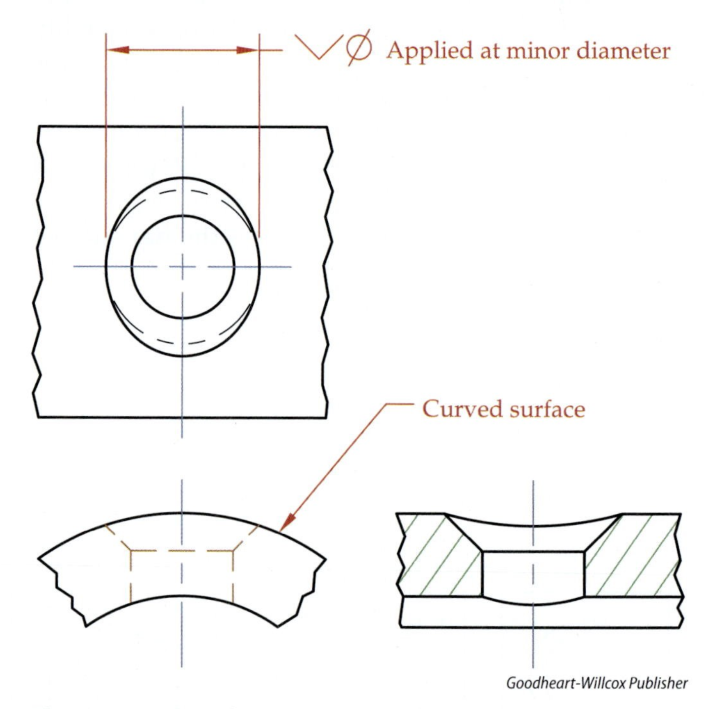

Goodheart-Willcox Publisher

Figure 11-11. When a countersink emerges from a curved surface, the resulting intersection forms an oval-like profile. In these cases, the specified countersink diameter applies at the minor diameter.

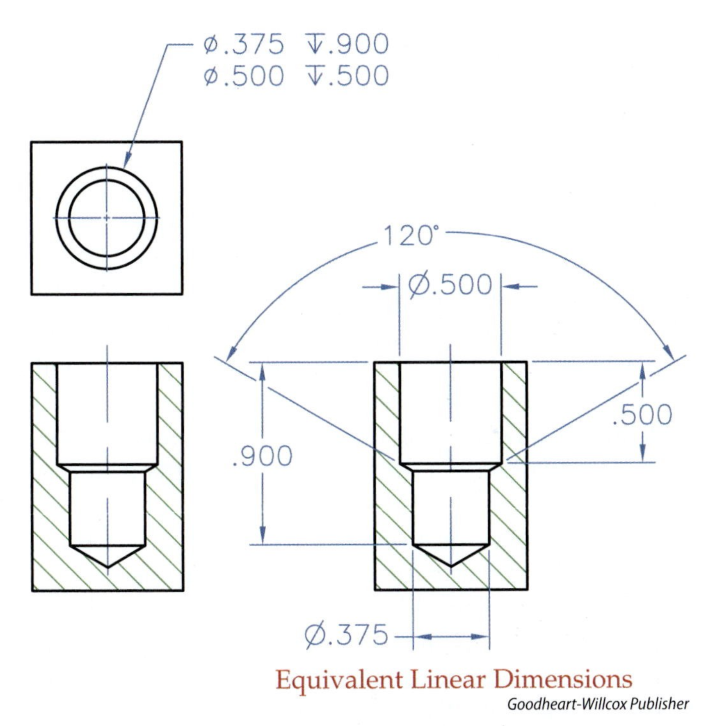

Figure 11-12. A counterdrilled hole has a conical transition that does not require an angular dimension. The depth measurements are for the full-diameter distances, as shown at right. As with counterbores and countersinks, linear dimensions can be used when they more clearly describe the feature.

Chamfered Edges

Beveled edges on parts are common for a variety of reasons and are often called ***chamfers***. In essence, a chamfered edge around a hole is the same as a countersink, but the term *countersink* is more commonly used with smaller holes. A countersink bit can perform the operation on smaller holes. The term *chamfer* is more often applied for larger holes. It also describes an edge that has been beveled. In many cases, the end of a cylinder is also chamfered to allow for easier assembly.

Figure 11-13 illustrates different methods for specifying a chamfer on the print, either by using a local note or by using linear and angular dimensions. If a chamfer is 45°, the leg distance is the same on both sides. (A *leg* is one of the sides that form a triangle.) For chamfer angles other than 45°, a leg distance must be specified (the .225 dimension in **Figure 11-13**). **Figure 11-14** shows recommended standard practice for chamfered edges where surfaces are not 90° to each other.

One reason for chamfering an edge is to remove burrs. One general note for this could be stated as BREAK ALL EDGES .031 MAX, which would allow a slight chamfer to each edge, as long as the legs of the

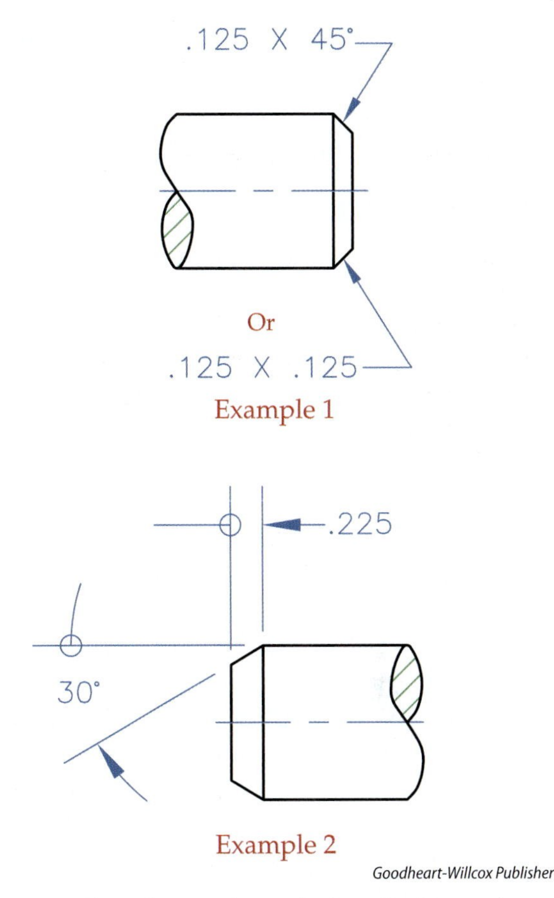

Figure 11-13. Chamfers on the end of a cylinder can be specified by note (Example 1) or by dimensions (Example 2).

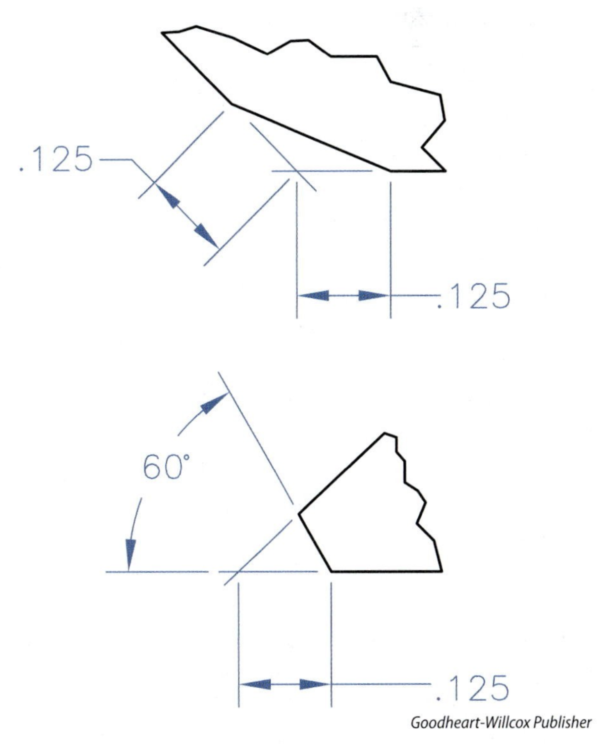

Figure 11-14. Chamfers on nonperpendicular surfaces should *not* be specified by note, but rather with dimensions.

chamfers were both less than the maximum distance specified.

Common Machining Operations

There are various machining operations that can be performed on a part. In the past, these operations were performed using manual machining equipment. Today, in many cases the machining process has been automated with computer-controlled equipment. This section covers a few special machining processes.

Necks and Undercuts

Sometimes a groove or neck is cut into a cylindrical surface for a retaining ring or to provide a better transition between two features. In former practice, the groove was usually dimensioned by a local note, such as .06 WIDE × .03 DEEP. Current standards recommend linear measurements that specify the resulting diameter within the groove, providing the inspector with an easier checking distance, **Figure 11-15**. In similar fashion, sometimes an undercut groove is used at the head of a cylindrical part to eliminate the chance for a fillet being created due to tool tip wear or a rounded cutting tool. While local notes can be used to specify the undercut, linear measurements that are easily checked are preferred. Ideally, the dimensions given should take into consideration the measuring equipment commonly used, such as calipers for checking the dimensions in **Figure 11-16**.

Keyways

A ***key*** is a fastener used to prevent the rotation of gears, pulleys, and rocker arms on rotating shafts. The key is a small piece of metal that fits into a ***keyseat*** in the shaft. A ***keyway*** in the hub allows the hub to slide over the key, preventing rotation. There are a variety of keys used in industry for different situations, as shown in **Figure 11-17**.

While a common practice at one time was to indicate a keyway with a note, newer standards promote the use of linear dimensions to specify exact distances that are logical inspection distances. See **Figure 11-18**. This technique eliminates confusion regarding the size of the keyway. The older method allowed the depth to be interpreted as the height above the center of the curve, but it was also sometimes interpreted to be the keyway sidewall. To eliminate confusion, keyways should not be dimensioned by a callout, but rather should be dimensioned as illustrated.

Knurls

Knurling is a pattern of ridges and grooves machined on a cylindrical surface for the purpose of providing a better grip or increasing the diameter of a part. The pattern is often crisscrossed, leaving a diamond-like series of ridges. A metal part may feature a knurl so that it can be press fit into a plastic part. When a knurl is required for a press fit, a toleranced diameter

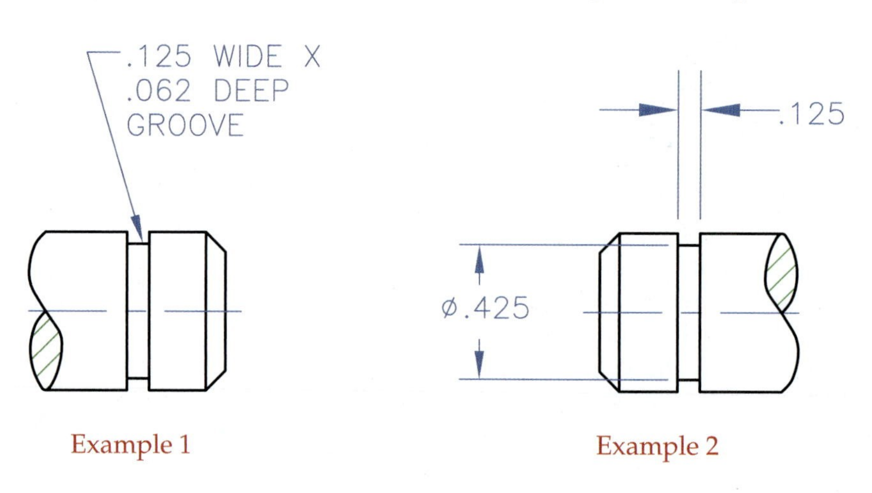

Figure 11-15. A ring groove or neck may be specified by note (Example 1) or by dimensions (Example 2).

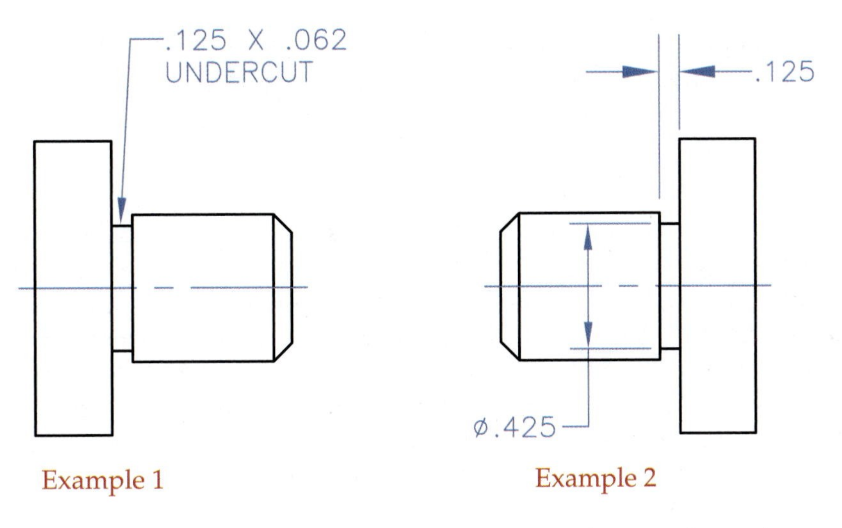

Figure 11-16. An undercut is sometimes used to help prevent corner interference where a larger cylinder intersects a smaller cylinder.

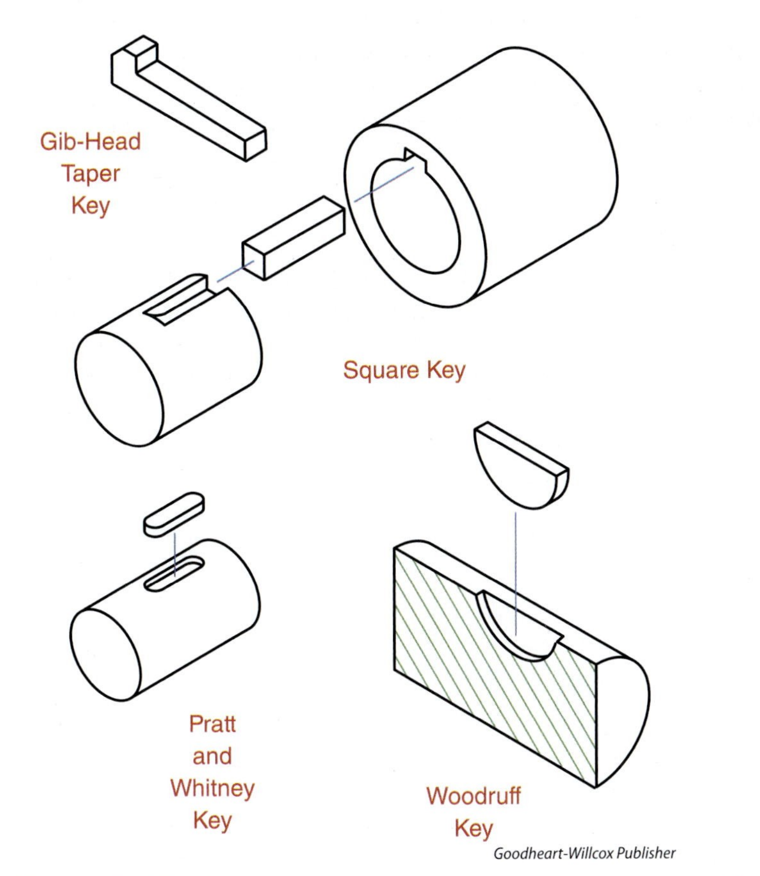

Figure 11-17. There are a variety of keys used in industry for different situations.

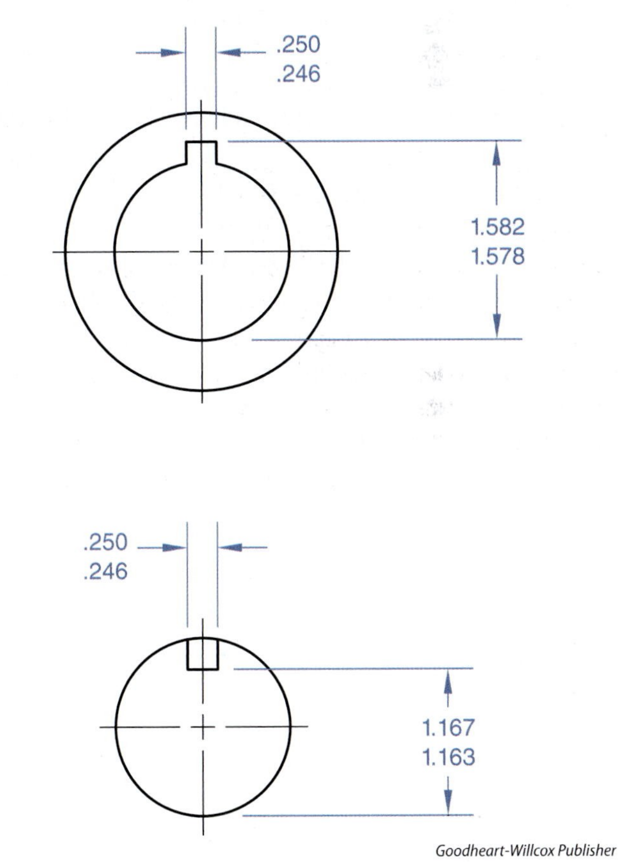

Figure 11-18. Linear dimensions are used to specify exact distances for keyways.

before knurling and a minimum acceptable diameter after knurling should be specified. Knurling standards are covered in ASME B94.6. A knurl should be specified with respect to pitch and type, and if the diameter after the knurl is critical, it may also be specified. See **Figure 11-19**. As shown, while simulating the knurl pattern in graphic form has been a common practice throughout the years, chain lines can be used with a local callout note, and the simulated graphic pattern can then be omitted.

Broaching

Broaching is done on a special machine using a machine tool called a broach. The broach progressively "punches" a shape with a series of cutting teeth. The teeth are set in such a way that each tooth is a few thousandths of an inch higher than the preceding one. Broaching can produce holes of circular, square, or irregular outline; keyways; internal gear teeth; splines; or flat external contours. Broaching is fast and accurate and produces a good-quality finish. As discussed several times in this unit, the practice of specifying the machine tool to make a particular feature is discouraged, but if a broaching operation is required, a local callout note may specify the process.

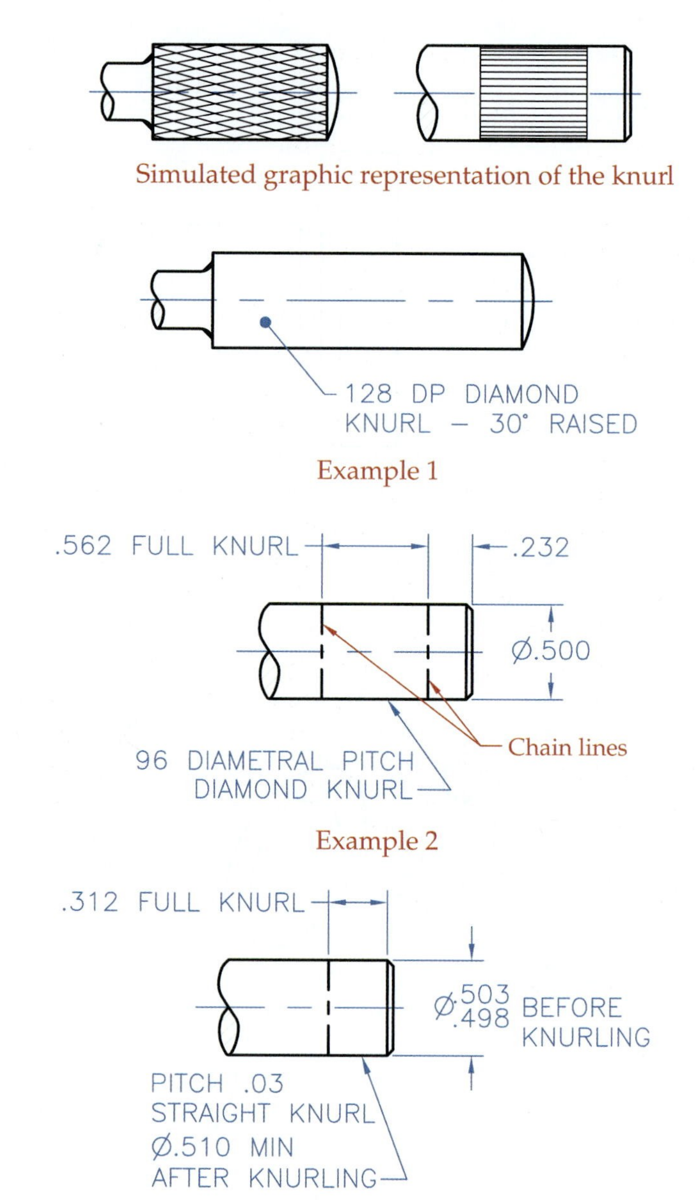

Figure 11-19. A callout for a knurl may specify several items, including pitch, type, and before and after diameters. Chain lines can be used to specify the area to which the knurl applies, and a simulated graphic representation can be omitted.

Summary

- Additional annotations to the views and dimensions are referred to as notes, specifications, or callouts.
- Notes fall into the two broad categories of local notes and general notes.
- ASME standard symbols such as counterbore, countersink, and depth help clarify local notes without using words.
- Drilled and reamed holes can be dimensioned with leader lines, but linear dimensions can also help clarify sizes in certain situations.
- A cylindrical enlargement at one end of a hole is called a counterbore or spotface, and serves a variety of purposes.
- A conical enlargement at one end of a hole is called a countersink, while a counterdrill is when a cylindrical hole is also drilled larger than, and coaxial with, a smaller hole, creating a conical transition.
- Chamfered edges are characterized as slightly beveled edges that may eliminate burrs and sharpness, or may allow for easier assembly.
- Necks and undercuts provide for retaining rings or for better transition between adjacent cylinders, and they should be dimensioned in a fashion that complements inspection.
- Keyways that accommodate a variety of standard keys can be specified with local callout notes, but particular linear dimensions can also be given that complement the inspection process.
- ASME standard knurls can be dimensioned with callouts that specify pitch, type, and finish diameter, using chain lines to indicate the area to which the knurl is applied.
- Broaching is a machining process using a tooth-based tool that progressively punches a hole of a particular shape—perhaps a square, hexagon, or keyway profile.

Unit Review

Name ______________________ **Date** ____________ **Class** ____________

Answer the following questions using the information provided in this unit.

Know and Understand

____ 1. Which of the following terms is out of place with respect to the other three?
A. Callouts
B. Linear dimensions
C. Notes
D. Specifications

____ 2. *True or False?* There are two classifications for notes—local and general.

Match the following symbol descriptions to the symbol names. Answers are used only once.

____ 3. An open-top rectangular shape similar to a horseshoe

____ 4. A term that uses three letters as an abbreviation

____ 5. A letter "T" shape with an arrow incorporated into it

____ 6. A V-shaped symbol

____ 7. Modification of another symbol incorporating a two-letter abbreviation

____ 8. A note that uses four letters as an abbreviation

A. Countersink
B. Depth
C. Spotface
D. Counterbore
E. Through
F. Typical

____ 9. Which of the following tools would best be used to create a hole?
A. Twist drill
B. Mill cutter bit
C. Laser
D. None of the above can be specified as the best method without additional information.

____ 10. *True or False?* Holes can be square, and the callout for that can incorporate a symbol for square.

____ 11. *True or False?* If a hole is to be reamed, the word REAM must appear on the drawing preceding the diameter of the reamer.

Match each of the four descriptions to a corresponding term. Answers are used only once.

____ 12. Hole with two diameters, each quite deep, with a conical transition

____ 13. Hole enlarged cylindrically at one end, perhaps to accommodate the head of a hex-head bolt

____ 14. Hole enlarged conically at one end, perhaps to accommodate the head of a flat-head screw

____ 15. Hole with a shallow cylindrical enlargement, perhaps for a flat washer

A. Counterbore
B. Spotface
C. Countersink
D. Counterdrill

____ 16. *True or False?* A callout for a countersink can also specify a radius that applies to the rim around the top of the countersink.

____ 17. *True or False?* If a second counterbore is applied to a first counterbore, the depth measurements of the through hole and both counterbores are all referenced to the same surface.

____ 18. *True or False?* There are situations with counterbores, countersinks, spotfaces, and counterdrills wherein linear dimensions may be more clear than local callout notes.

____ 19. A beveled edge is referred to as a _____.
A. chamfer
B. keyway
C. undercut
D. knurl

____ 20. *True or False?* For necks and undercuts, a local note pointing to the feature is the recommended dimensioning method.

____ 21. *True or False?* For keyways, a local note pointing to the feature and stating the standard type of key is the recommended dimensioning method.

____ 22. A type of line that helps identify the scope or extent of a knurl pattern is the _____ line.
A. phantom
B. section
C. chain
D. hidden

____ 23. *True or False?* Broaching is a process that cuts metal away by force.

Critical Thinking

1. What are some of the key advantages of using the symbols discussed in this unit instead of words?

____

____

____

____

____

____

2. What might be some of the drawbacks of using the symbols discussed in this unit, even though their use is worth it in the overall scope of industrial prints?

____

____

____

____

____

____

Apply and Analyze

Name ____________________ Date ______________ Class ______________

Review Activity 11-1

For each of the written descriptions below, sketch the ASME standard symbols, if there are any, along with the numeric values next to the leader line shoulder. An example has been done for you. Also refer to Figure 11-4 as needed, as well as Appendix D.

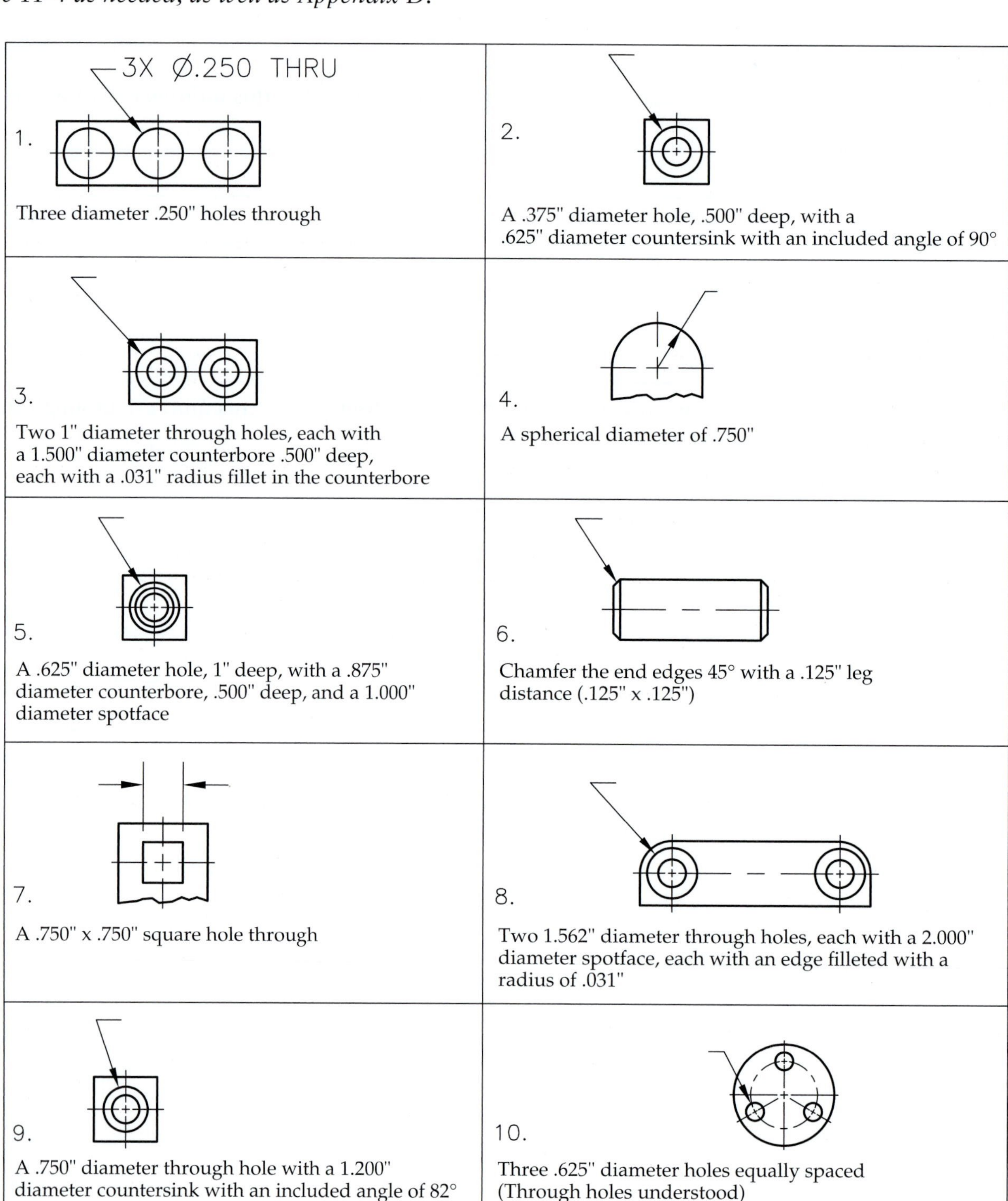

Apply and Analyze

Name ______________________ **Date** ____________ **Class** ______________

Review Activity 11-2

For each local note or callout given, use lettering to put the notes into common words and expressions that describe the note. An example has been done for you. Refer to Figure 11-4 and Appendix D as needed.

1. 2X ⌀.500 ↧.375	1. Two .500" diameter holes, .375" deep
2. ⌀.250 ↧1.000 ⌵ ⌀.375 X 82°	
3. ⌀.750 THRU ⌴ ⌀1.250 ↧.375	
4. SR.562	
5. 2X ⌀.750 ↧1.500 ⌴ ⌀1.000 ↧.250 ⌴ ⌀1.250 ↧.125	
6. .250 X 45° TYP	
7. □.250 ↧.12	
8. 4X ⌀.562 THRU SF ⌀.625 R.031	
9. ⌀.500 THRU ⌵ ⌀.750 X 90°	
10. 5X ⌀.125 EQUALLY SPACED	

Apply and Analyze

Name ______________________ Date ______________ Class ______________

Industry Print Exercise 11-1

Refer to the print PR 11-1 and answer the questions below.

1. How many counterbored holes does this part have? ______________
2. How many countersunk holes does this part have? ______________
3. What depth is specified for the counterbores? ______________
4. What process will be used to put the part number on the part?

____

5. What machining process will be used to eliminate sharp corners?

____

6. For this print, is the letter "X" used to indicate "by," or "times," or both? ______________

Review questions based on previous units:

7. How many leader lines are there on this print? ______________
8. If the view with the cutting-plane line is the front view, what is the width of the object?

____

9. How many threaded holes does this part have?______________
10. Of the threads that feature a major diameter of 5/8″, what is the pitch?______________
11. What type of section view is the right side view?

____

12. What is the name of this part?

____

13. What size paper was the original drawing for this print?______________
14. Is the left side of this part symmetrical with the right side?______________
15. What tolerance is specified for the linear dimensions of this print? ______________

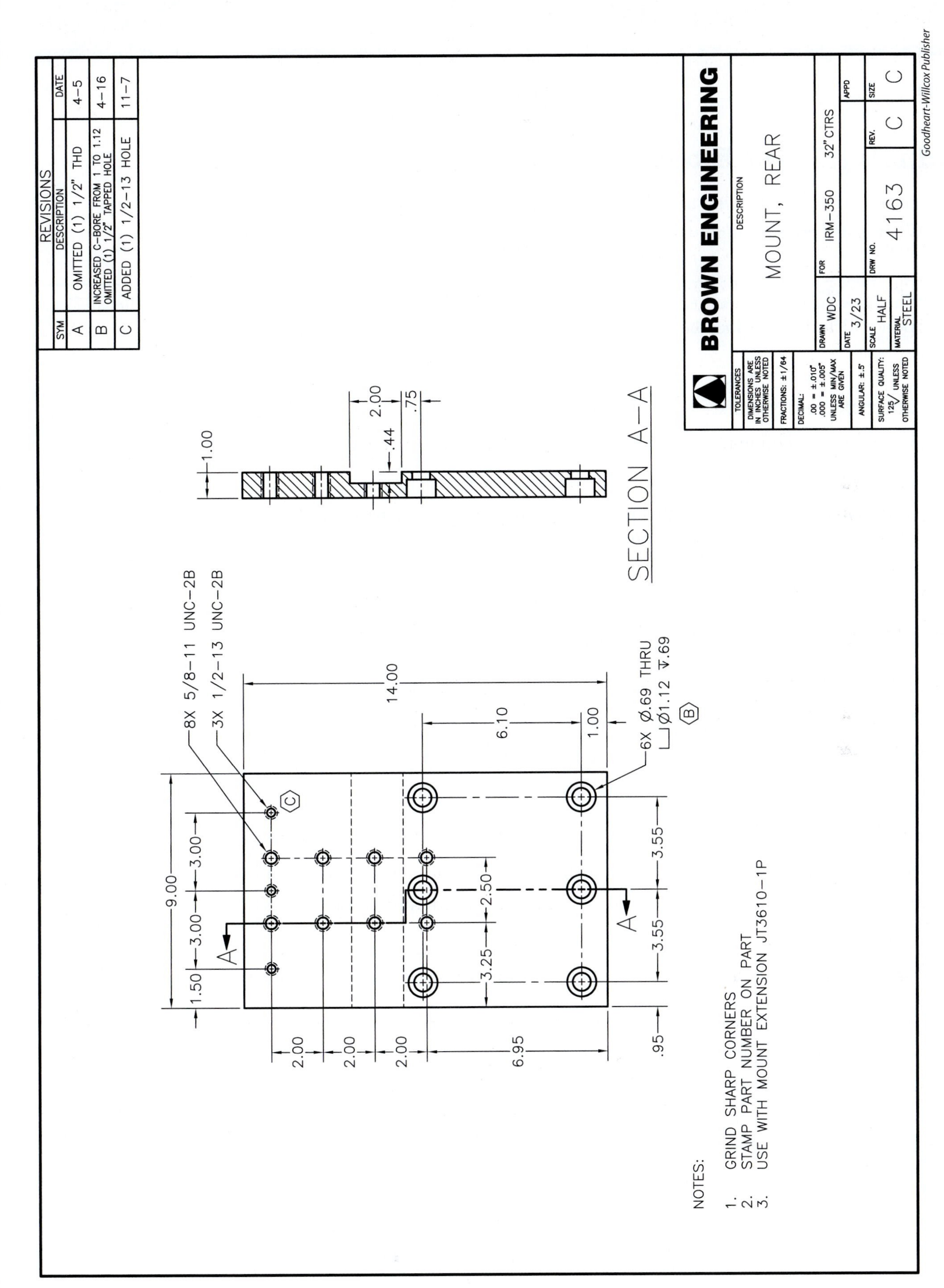

PR 11-1. Rear Mount

Apply and Analyze

Name ______________________ **Date** ______________ **Class** ______________

Industry Print Exercise 11-2

Refer to the print PR 11-2 and answer the study questions below.

1. What is the primary manufacturing process for this part?

____

2. What text height is specified for the "Made in USA" letters, which are to be raised above the surface upon which they are located by .010″?

____

3. How many local notes does this drawing have? ______________________
4. How many general notes does this drawing have, not counting those in the title block?____________
5. Is any of the lettering for this part recessed into the part? ______________________
6. One of the specifications for this part is to "wheelabrate," a process that is similar to sand blasting. What other finishing processes are covered in the same note?

____

Review questions based on previous units:

7. Who was the engineer for this drawing?

____

8. How many auxiliary views were used in this drawing? ______________________
9. How many cutting-plane lines are shown in this drawing? ______________________
10. If the section view is thought of as the top view, what would the other views be called?

____

11. Does this object have any features or surfaces that could be described as conical? ____________
12. What eight-digit specification number is assigned to the material for this part?

____

13. Are the dimension values placed on the drawing with the aligned system or the unidirectional system?

____

14. What is the largest diameter specified on the print?______________________
15. What is the smallest radius value specified on this drawing for a fillet or round?

____

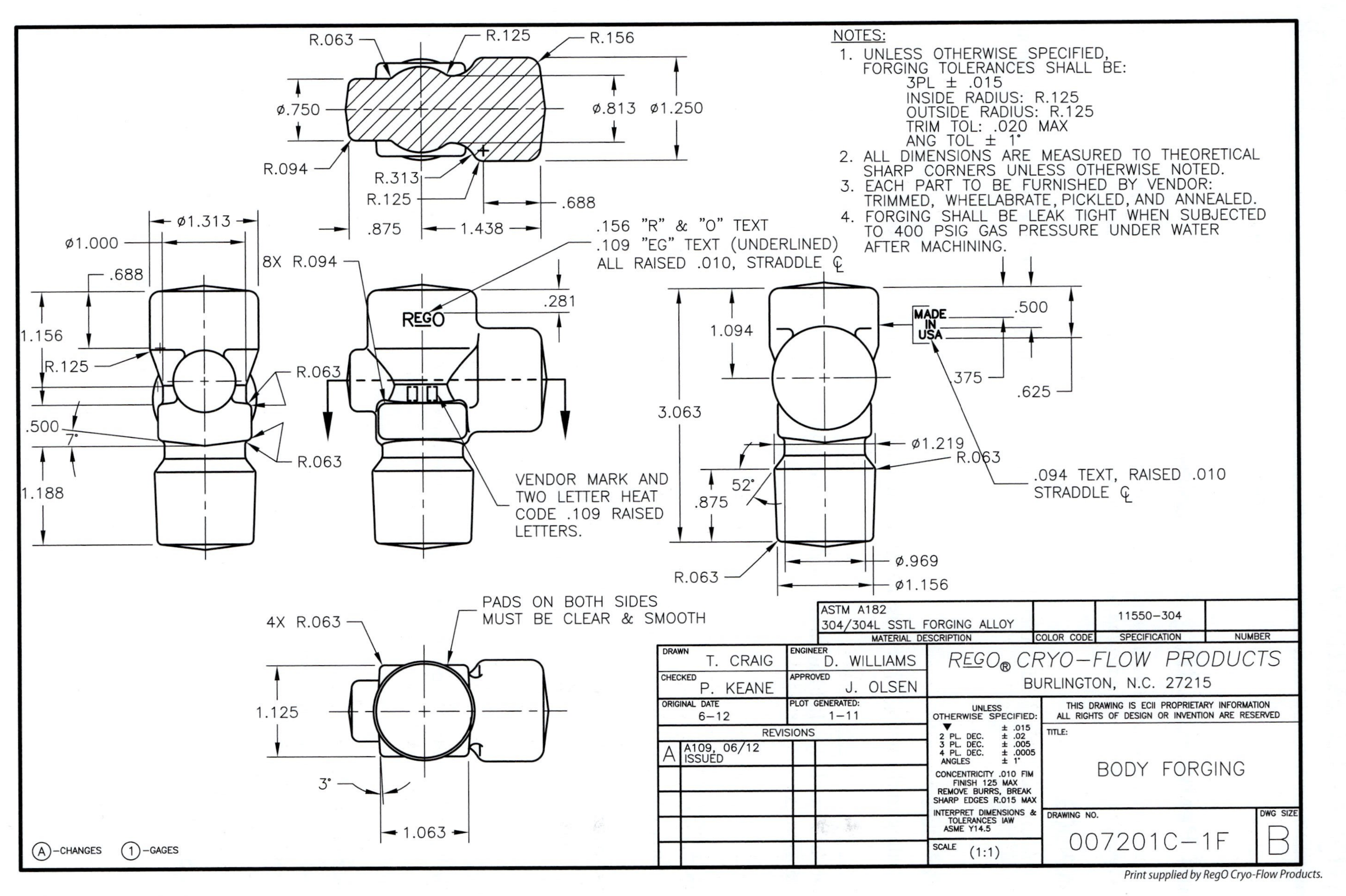

Print supplied by RegO Cryo-Flow Products.

PR 11-2. Body Forging

Apply and Analyze

Name ______________________ Date ______________ Class ______________

Bonus Print Reading Exercises

The following questions are based on various bonus prints located in the folder that accompanies this textbook. Refer to the print indicated, evaluate the print, and answer the question.

Print AP-003

1. How many local-note leader lines direct the print reader to the general notes for the specification?

____

Print AP-004

2. According to the notes, what must be performed per military standard A-8625?

____

3. There are two holes that have a counterbore on one end. What is the basic depth for those counterbores?

____

Print AP-007

4. List three conditions considered unacceptable for the valve seat surface.

____

____

____

Print AP-008

5. To help facilitate machining and inspection, what is permitted on both ends of this object?

____

Print AP-009

6. Based on the specifications given, how much pull-off force will the mounting stud need to endure without damage or distortion?

____

Print AP-010

7. What substance will be used on the set screws when they are installed?

____

Print AP-011

8. Is the part number stamp specified by a local note, general note, or both?

____

9. What is the maximum amount of break to be used on sharp corners and edges?

____

Print AP-013

10. How much torque is to be applied when installing the flange screws?

____

11. What process is specified for part #15 before assembly?

____

Print AP-014

12. Which note was added with revision B?

____

13. This assembly drawing is for a cutting tool. According to the notes, what is the maximum speed for it?

____

Print AP-016

14. There are three optional holes on this part. If used, what are they for?

____

Print AP-017

____ 15. Which of the following processes is *not* mentioned on this print?
  A. Forging
  B. Carburizing
  C. Grinding
  D. Casting

16. What term is applied to diminishing the sharp edges of the teeth?

____

Print AP-019

17. What is the basic radius for fillets and rounds, unless otherwise specified?

____

Print AP-020

18. Some of the dimensions are notated with a "DFT." What does that stand for?

____

19. Section B-B shows the countersink angle at 82°. Is the depth determined by a linear measurement, or by a diameter?

____

Print AP-022

20. Unless specified, what imperfect condition along the edges is acceptable?

____

UNIT 12 Surface Texture Symbols

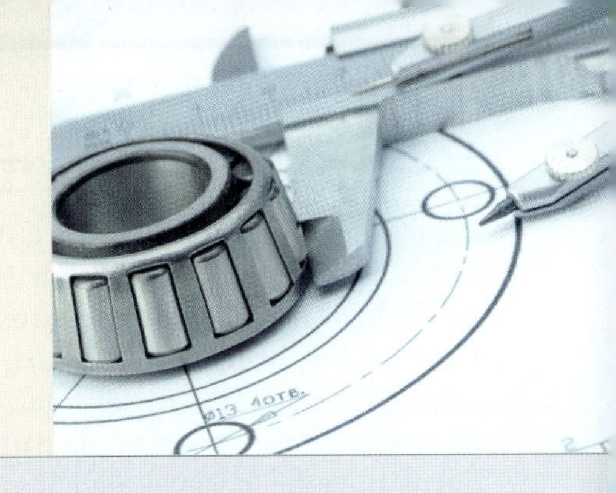

LEARNING OBJECTIVES

After completing this unit, you will be able to:

- Explain common terms related to surface quality and surface texture symbols.
- Identify and interpret the components of surface texture symbols.
- Identify lay symbols used in surface texture symbols.
- Describe the units used by various values within surface texture symbols.
- Explain standard practices for applying surface texture symbols on a print.
- Identify special designations related to surface quality that may appear on a print.

TECHNICAL TERMS

average roughness (R_a)
evaluation length
finish
flaw
honing
lapping
lay
profilometer
sampling length
surface roughness
surface texture
surface waviness
warpage

Products with metal parts may have specific requirements for how smooth the metal surfaces must be. For any particular feature of the part, the degree of roughness (or degree of smoothness) may need to be specified to ensure a proper function or finish. While some surfaces may need to be smooth to help gaskets seal against them, others may need an element of roughness so that lubricants are retained. By nature, objects cast in sand molds will have rough surfaces. Some features of the part will need to be finished by machining, and holes may need to be smoothed and made functional by drilling, boring, or reaming. ASME B46.1 sets forth the terms and definitions for surface texture. ASME Y14.36 sets forth standards for how the surface texture symbols are to be specified on the print, which is the primary focus of this unit. For this topic, there are some differences between ASME and ISO standards, so ISO 1302 may also be useful for understanding the more complex applications of surface texture symbols. Other symbols and systems for controlling the general flatness of surfaces are covered in Unit 13, *Geometric Dimensioning and Tolerancing*, but that is more concerned with overall general measurements and geometric form, and not with the finer aspects of the surface finish. Similar to the application of tolerances, engineers and designers are responsible for establishing the level of surface finish, trying to balance quality and performance with cost.

Surface Texture Terms

Prior to the development of current surface texture symbols, drafters indicated a finished surface with an italic "f" or a simple sans serif "V" symbol, **Figure 12-1**. In those cases, notes or company guidelines were available to set forth the precise standard of work. In today's manufacturing enterprise, standards have been developed that provide a thorough system for not only noting finished surfaces, but also for specifying the roughness, waviness, and lay of a surface.

In general, surface texture is measured with a ***profilometer***, a term applied to a wide variety of devices that can measure a surface's profile and report data related to the terms discussed in this unit. The original profilometer was a device that had a stylus, or needle, that moved relative to the surface profile. In addition to these contact-based profilometers, other technologies have evolved, many of which use noncontact optical methods to obtain the data required for inspection. Some of these can provide feedback dynamically as the part is being processed. **Figure 12-2** shows a profilometer. This unit does not attempt to explain the manufacturing or inspection processes required to measure surface texture, but rather prepares the print reader to interpret the symbols on the print.

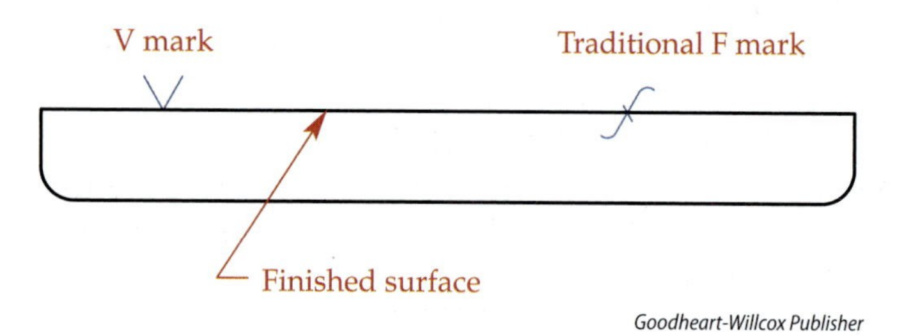

Figure 12-1. This figure shows finish marks that were in common use before surface texture symbols were standardized.

A fundamental understanding of the standard surface texture symbols requires an understanding of a few terms, as shown in **Figure 12-3**. ***Surface texture***, or ***finish***, is the overall term that refers to the roughness, waviness, lay, and flaws of a surface. The surface texture, sometimes stated as "surface finish," is the specified smoothness required on the finished surface of a part. It is usually obtained by machining, grinding, or lapping. In this context, ***lapping*** is polishing a surface to achieve smoothness. ***Honing*** is another term that indicates a high-precision finishing process using an abrasive block and rotary motion.

Surface roughness is the fine irregularities in the surface texture. It is a result of the production process used, and it varies depending on material and hardness. Included in surface roughness measurements are irregularities that result from the machine production process, such as traverse feed marks, feed rate, or tool condition. Roughness height is measured by a profilometer in microinches (millionths of an inch) or micrometers (millionths of a meter). To put these terms into decimal values, 32 microinches is equal to .000032″, and 0.8 micrometers is equal to 0.0008 millimeters. These values make

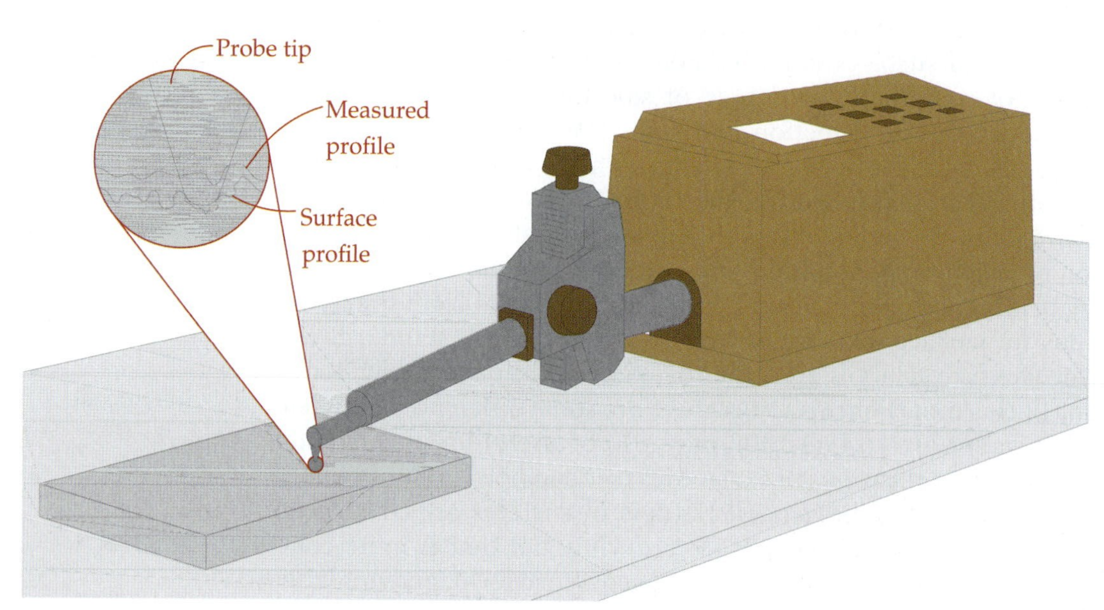

Figure 12-2. A profilometer has a small probe, or stylus, that moves across the surface and provides a measurement profile of the roughness.

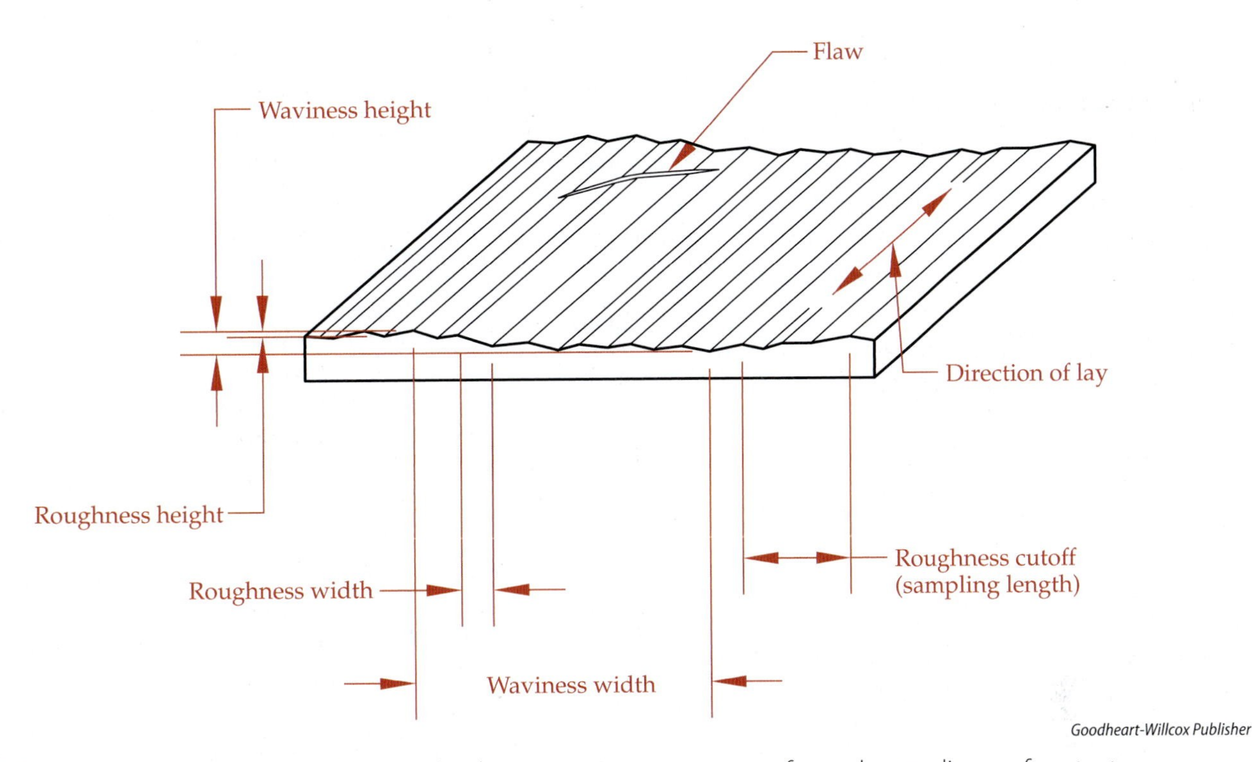

Goodheart-Willcox Publisher

Figure 12-3. Roughness, waviness, and lay are the three most important terms for understanding surface texture symbols. Flaws are also a concern if precise surface quality is required.

it clear that surface roughness measurements are working with microscopic distances. Most profilometers measure surface roughness height from 1 to 1000 microinches.

Surface waviness is the widely spaced component of surface texture due to such factors as machine chatter, vibrations, work deflections, ***warpage***, and heat treatment. Waviness is rated in inches. If a spline curve were drawn through the average roughness peaks and valleys, surface waviness would be the variation in the total curve.

Lay is the term used to describe the direction of the predominant surface pattern that results from the manufacturing process. There are seven symbols that are used to indicate the direction of the lay, if it is considered essential to a particular surface finish. Measurements should be taken at maximum surface roughness, so contact-based instruments should, in general, measure across the lay.

A ***flaw*** is an unintentional interruption in the surface, such as a crack, pit, or dent. In general, surface texture symbols do not deal with flaws, but rather just the roughness, waviness, and lay of a surface.

A surface texture specification of 8 microinches requires a roughness average of less than .000008″. In order to create this finish, some very fine honing or lapping of a surface must be done. The table in **Figure 12-4** indicates some recommended roughness height values in both micrometers and microinches, with a description of the surface and the process or processes by which the surface may be produced. In ISO standards, roughness grade numbers for each of the value settings are also available, with N8 equal to 3.2 micrometers, N7 for 1.6 micrometers, etc., with N values decreasing with the finer levels of smoothness.

Surface Texture Symbol

The surface texture symbol resembles a check mark, as shown in **Figure 12-5**. There are five basic forms the symbol can take. In **Figure 12-5A**, the basic surface texture symbol indicates the surface may be produced by any method. Often, especially with former practice, only the surface roughness average specification is used with this symbol. There are some industrial prints that may include the casting information about a part, as well as the finished size information after some material is to be removed. As shown in **Figure 12-5B**, a horizontal bar, forming a small triangle, is added to indicate material removal is required by machining. **Figure 12-5C** shows a value can be placed directly to the left of the short leg of the check mark to indicate a minimum amount of material must be removed, and that value is specified in inches or millimeters.

Some parts are manufactured with processes that do not remove material, such as powder metallurgy, injection molding, cold finishing, casting, or

Roughness Height Rating		Surface Description	Process
Micrometers	Microinches		
25.2 √	1000 √	Very rough	Saw and torch cutting, forging, or sand casting
12.5 √	500 √	Rough machining	Heavy cuts and coarse feeds in turning, milling, and boring
6.3 √	250 √	Coarse	Very coarse surface grind, rapid feeds in turning, planing, milling, boring, and filing
3.2 √	125 √	Medium	Machining operations with sharp tools, high speeds, fine feeds, and light cuts
1.6 √	63 √	Good machine finish	Sharp tools, high speeds, extra-fine feeds and cuts
0.8 √	32 √	High-grade machine finish	Extremely fine feeds and cuts on lathe, mill, and shapers required. Easily produced by centerless cylindrical and surface grinding
0.4 √	16 √	High-quality machine finish	Very smooth reaming or fine cylindrical or surface grinding or coarse hone or lapping of surface
0.2 √	8 √	Very-fine machine finish	Fine honing and lapping of surface
0.05 0.1 √	2–4 √	Extremely smooth machine finish	Extra-fine honing and lapping of surface

Goodheart-Willcox Publisher

Figure 12-4. This table indicates recommended roughness height values in microinches and micrometers.

forging. In those cases, a circle indicates that material removal is prohibited, as shown in **Figure 12-5D.** The prints may need to specify the texture for surface finish even though material will not be removed. For situations wherein more information needs to be specified, a horizontal bar is added to the long leg of the check mark to help locate parameters that define the surface texture to the print reader, **Figure 12-5E.** Above the horizontal bar, the production method can be indicated and other notes can be specified. Below the bar, the ***sampling length***, or *roughness cutoff*, should be specified, but if not, a cutoff of .03″ or 0.8 mm is common. ASME and ISO standards may vary on this, but they are good resources if company standards do not address default values. The sampling length applies to the distance increment used for checking the roughness parameter, whereas a total ***evaluation length*** is across the entire surface, which would also reveal the waviness.

A key to locating numeric values, parameters, and notes within the surface texture symbol is illustrated in **Figure 12-6.** In this diagram, "a" indicates the location for the maximum average roughness value, given in microinches or micrometers. Two values

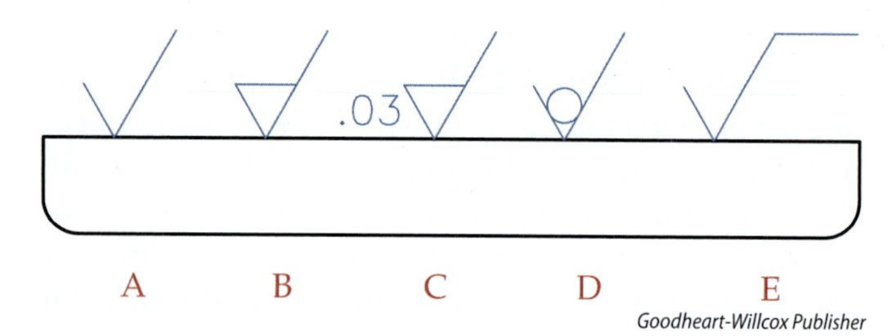

Goodheart-Willcox Publisher

Figure 12-5. These are the five basic forms of the surface texture symbol, minus numeric or symbolic annotations, as explained in the text.

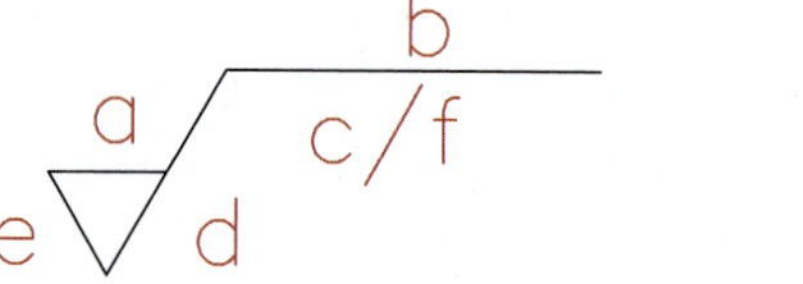

Goodheart-Willcox Publisher

Figure 12-6. The values that are applied to a surface texture symbol must be appropriately located with respect to the linework. The positions are explained in the text.

could be stacked at this location to indicate a range of roughness. The ***average roughness*** (R_a) is defined as the average deviation of the measurements taken by the profilometer. For a contact-based profilometer, the stylus tip traverses the surface, moving up and down along the tiny ridges, and the R_a simply averages the readings. If it is important to indicate the production method, such as MILL, GRIND, or LAP, this is placed in position "b." Other notes related to treatments or coatings can also be given at this location. In former practice, waviness height or width was also given at this location, but current standards call for waviness parameters to be given in position "f."

Under the bar, position "c" indicates the roughness cutoff, and this value is given in inches or millimeters. Location "d" is for a lay symbol. Lay symbols are illustrated in **Figure 12-7** and specify the acceptable direction of the lay. Location "e" is for the minimum material removal requirement amount, specified in inches or millimeters. Location "f" is for roughness parameters other than the arithmetic mean average (R_a). Some engineers or designers may specify a root

Lay Symbol	Meaning	Example Showing Tool Marks
=	Lay is somewhat parallel to the "edge view" line to which the symbol is applied.	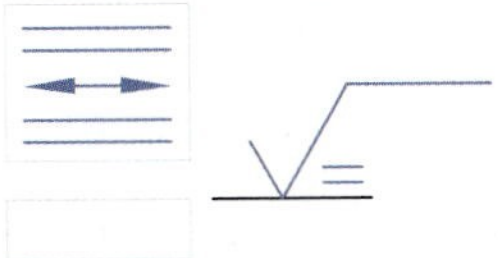 =
⊥	Lay is perpendicular to the "edge view" line to which the symbol is applied.	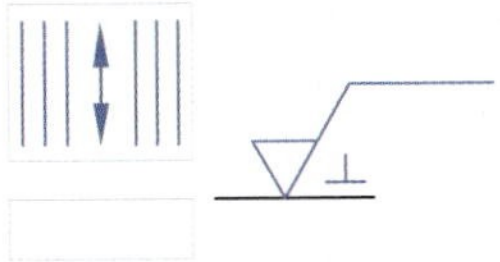 ⊥
X	Lay is angular in both directions to the "edge view" line to which the symbol is applied.	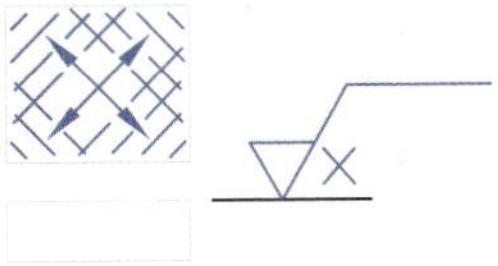 X
M	Lay is multidirectional or random.	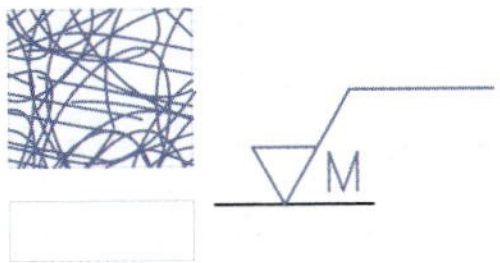 M
C	Lay is approximately circular to the center of the surface to which the symbol is applied.	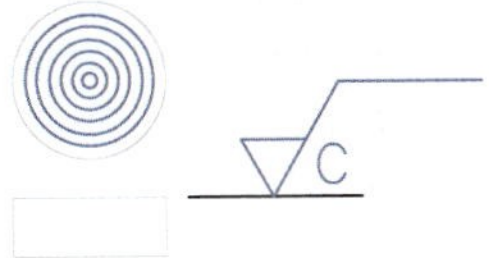 C
R	Lay is approximately radial to the center of the surface to which the symbol is applied.	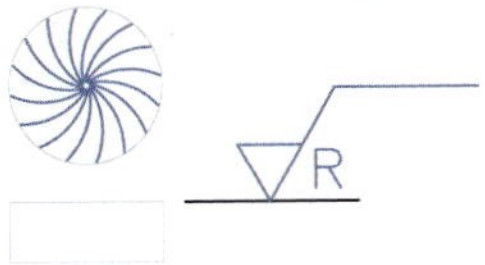 R
P	Lay is particulate, nondirectional, or protuberant (bulging).	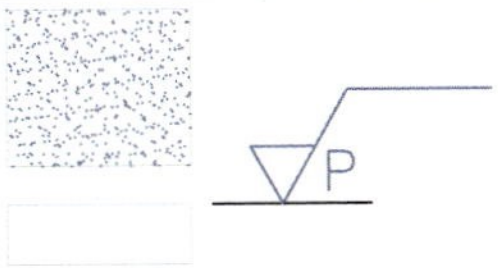 P

Goodheart-Willcox Publisher

Figure 12-7. One of the seven lay symbols illustrated here may be indicated within the surface texture symbol. The lay symbol is based on the direction or pattern of the lay.

mean square (RMS) method with an R_q parameter. Since averaging methods sometimes do not control the highs and lows, there are other parameters that specify and analyze high points, or low points, or perhaps the distance between high and low points.

Not all parameters are comparable, as they do evaluate different aspects of surface deviation. **Figure 12-8** is a basic illustration that describes how certain parameters are applied. In the example, R_v specifies a maximum profile valley depth within each sampling length, R_p specifies a maximum profile peak height within each sampling length, and R_z specifies the total distance between highs and lows within the sampling length. Any one of these parameters can be specified, based on the design requirements. Other parameters can be specified besides roughness values, and ranges of value can also be specified. Waviness average (W_a) and waviness height (W_t) can be specified if waviness control is required. Spacing parameters such as peak count (RP_c) or mean roughness spacing (S_m) can also be specified.

It is beyond the scope of this text to cover all the parameters that can be specified for surface texture. In summary, decreasing the roughness of a surface will increase the cost, and certain manufacturing methods or inspection techniques may be more costly, so surface texture symbols are not applied without a lot of careful consideration by the designer and engineer.

Application Examples for Surface Texture Symbols

There are various techniques for locating the surface texture symbol on the print. The V-point of the symbol is placed on a line that is considered the edge view of the surface being controlled. To improve readability of the drawing, the symbol can also be attached to an extension line leading out from the surface, or to the shoulder of a leader line that points to the surface. As shown in **Figure 12-9**, notes can also help alleviate drawing clutter by alphabetically flagging the surface texture symbol near the views and detailing the numeric specifications in a general note.

Since many of the standards for application of surface texture symbols are relatively new, you may not see a full use of all available designations. As a student of print reading, your value within the company setting may be increased by helping spread the word of new and more in-depth standards that help prints specify additional information to others within the manufacturing enterprise. Examples of surface texture symbol applications are given in **Figure 12-10.** The first three are shown with US customary units, and the next five are shown with SI units.

In **Figure 12-10A**, the symbol specifies a roughness average rating of 63 microinches with a sampling length of .05 inches. Since there is only one roughness value specified, the number indicates the maximum amount of roughness (63 microinches). In **Figure 12-10B**, the range of roughness average is specified as 32–63 microinches with a sampling length of .10 inches. In **Figure 12-10C**, the roughness average is specified as in **Figure 12-10A**, but this specification indicates no removal of material is allowed.

The symbols are the same for metric values, but the values are different. It is also common practice in the metric system to apply a leading zero in front of the decimal. The values specified in **Figure 12-10** are for print reading practice only. Recommended values for various scenarios may be indicated in ASME standards or perhaps company standards.

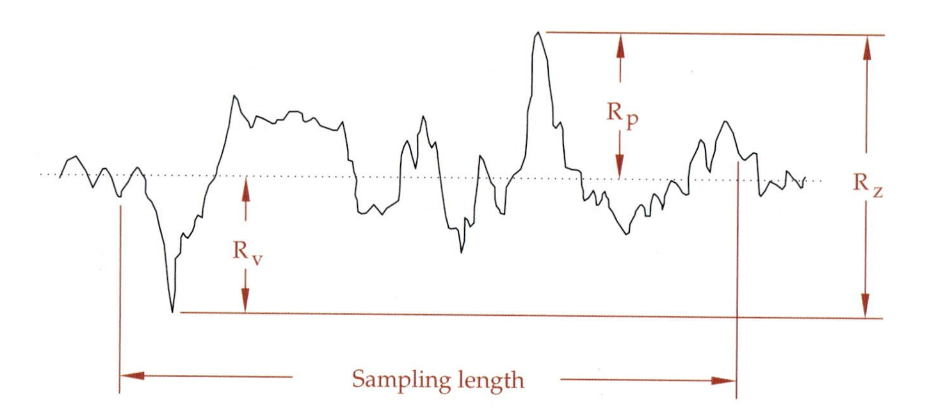

Goodheart-Willcox Publisher

Figure 12-8. There are several parameters for analyzing roughness data. In this basic example, three parameters are shown, R_v (maximum profile valley depth), R_p (maximum profile peak height), and R_z (total valley to peak within a sampling length).

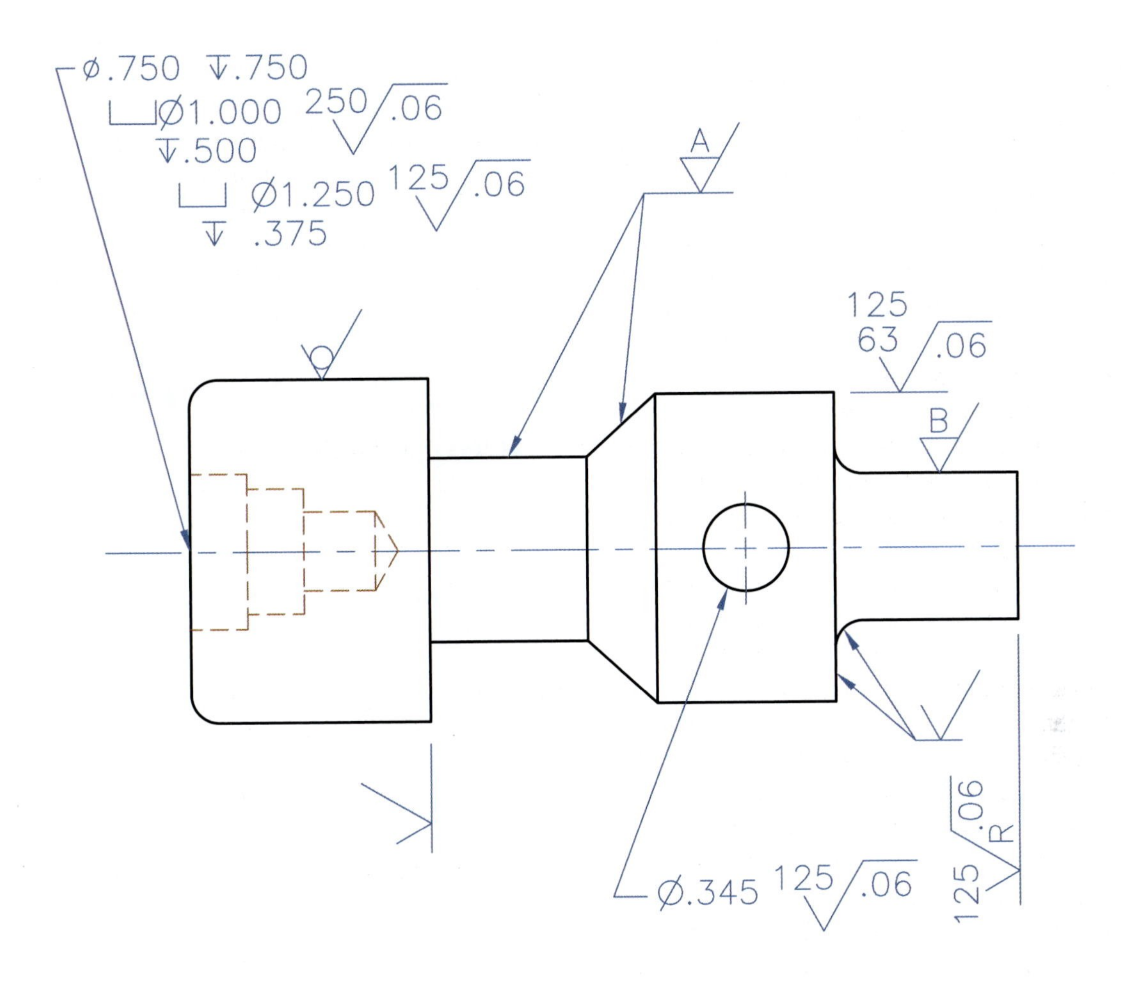

NOTES: 1. UNLESS OTHERWISE SPECIFIED ALL SURFACES 32/.03

2. √ = 63/.06   A = 125/.03

B = 16/.03   ⌀ = 63/.06

Goodheart-Willcox Publisher

Figure 12-9. The surface texture symbol should be placed on the edge view of the surface being specified, but extension lines, leader lines, and notes can be used to help keep the shape description from being too cluttered.

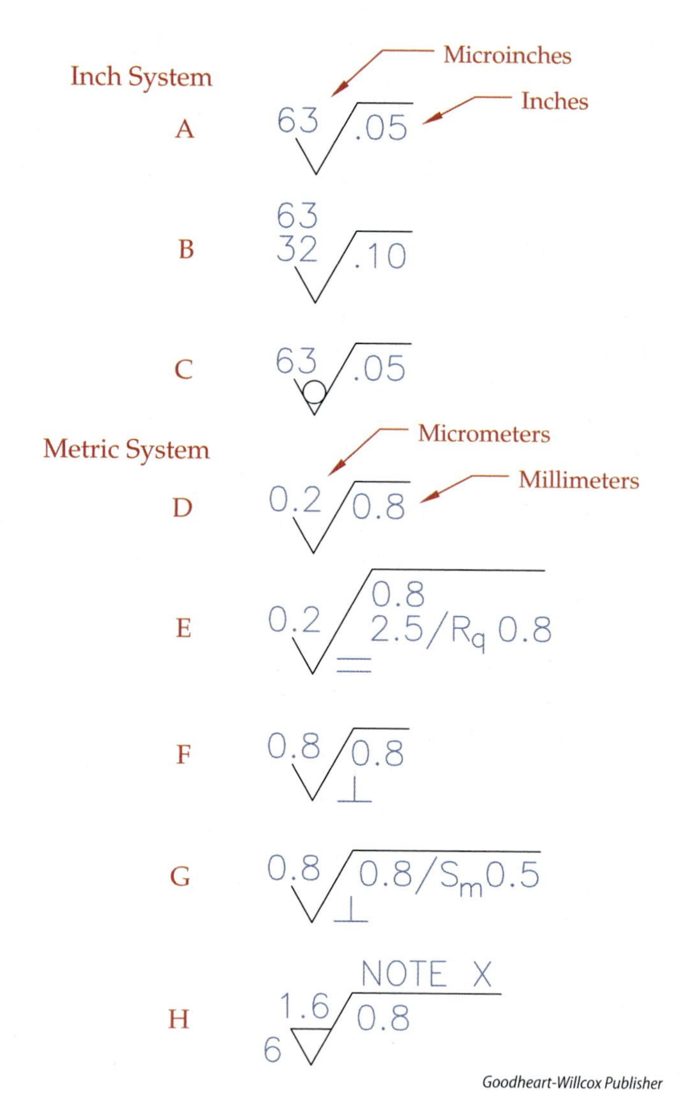

Goodheart-Willcox Publisher

Figure 12-10. Specific applications of the surface texture symbol are shown here to help you appreciate the multitude of options for applying the symbol.

In **Figure 12-10D**, the 0.8 below the bar indicates a sampling length of 0.8 millimeters, and the roughness average (R_a) is specified as 0.2 micrometers. In **Figure 12-10E**, the additional 2.5/R_q0.8 specification indicates a root mean square value should also be held at no more than 0.8 micrometers sampled over a length of 2.5 millimeters. Also shown in **Figure 12-10E** is the lay symbol for the parallel direction.

In **Figure 12-10F**, the lay designation is the perpendicular symbol. This means the lay runs perpendicular to the line to which the symbol is applied. In **Figure 12-10G**, the S_m parameter is placed after the sampling length. The average roughness is still held at 0.8 micrometers per sampling length of 0.8 millimeters, although the maximum roughness spacing is 0.5 millimeters. In **Figure 12-10H**, the symbol indicates material removal is required to produce the surface. The minimum amount of stock to be removed is 6 millimeters, the roughness average is 1.6 micrometers, and the sampling length is 0.8 millimeters. The NOTE X directs the print reader to a general note. In place of this note, a default process such as MILL or GRIND could have been specified.

Special Surface Finish Designations

As with other print reading symbols and notes, many special designations may be needed to cover special situations for surface finish. **Figure 12-11A** shows how specific processes can each be applied, with increased surface quality as each process is completed. **Figure 12-11B** shows how an area can be designated on the surface, in this case the rim of a hole that may have a special relationship with a mating part. Standards also specify that a drawing or print for a part to be coated or plated should indicate whether or not the surface texture symbol applies before, after, or both before and after coating or plating. This can be specified above the horizontal bar on the right leg of the symbol, or in a local or general note on the drawing.

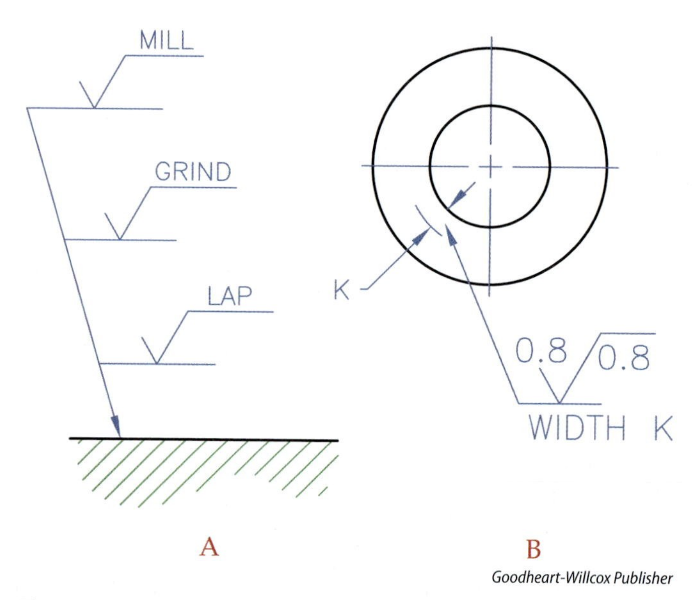

Goodheart-Willcox Publisher

Figure 12-11. Special applications of the surface texture symbol may be needed for certain situations. A—Surface quality is specified for each process step. B—A designated area is specified on the surface.

Summary

- The application of standard symbols that specify the texture or finish of a surface is a regular occurrence in industrial prints.
- Surface texture is primarily measured at a microscopic level using contact-based or noncontact optical profilometers.
- Surface texture, or finish, is the overall term that refers to the roughness, waviness, and lay of a surface, and may include other flaws that were unanticipated.
- Surface roughness is concerned with peaks and valleys measured at a microscopic level, reported as microinches or micrometers, and is sampled over short distances.
- Surface waviness is concerned with the widely spaced averages of peaks and valleys over the evaluation length of the entire surface.
- Lay is a term that describes the direction of the predominant surface pattern that results from the manufacturing process, and symbols are available that communicate the lay requirements.
- There are several variations of the surface texture symbol, with considerations for specifying the roughness average and material removal allowed, specified, or forbidden, as well as common rules for specifying processes, lay, cutoff length, and other parameters.
- A wide array of parameters can be incorporated into the surface texture symbol, and many of these parameters are chosen and specified by the designer or engineer to ensure quality and performance at the most efficient cost.
- Surface texture symbols are applied to the views of the object in a variety of ways, including the attachment of the symbol to visible edge view lines, extension lines, or leader lines.

Unit Review

Name ________________ Date ________ Class ________

Answer the following questions using the information provided in this unit.

Know and Understand

____ 1. Which of the following manufacturing processes will most likely require more specifications about surface finish?
A. Sand-cast aluminum
B. Wooden parts for cabinetry
C. Plastic injection molding
D. Sheet metal stamping

____ 2. *True or False?* There are two standards for surface texture, one that establishes the terms and definitions, and another that sets forth how symbols are to be specified on the print.

____ 3. The current standard surface texture symbol most resembles a(n) _____.
A. V
B. check mark
C. f
D. encircled S

____ 4. Which of the following surface characteristics is an unintentional interruption in the surface, such as a crack or pit, and is *not* covered by the surface texture symbol?
A. Roughness
B. Waviness
C. Lay
D. Flaw

____ 5. *True or False?* Waviness deals with the immediate ups and downs of a surface, while roughness deals with a wider width distance.

Match the following roughness/smoothness terms to the processes that may produce the finish. Answers are used only once.

____ 6. Sharp tools, high speeds, extra-fine feeds and cuts

____ 7. Extra-fine honing and lapping

____ 8. Saw and torch cutting, forging, or sand casting

____ 9. Coarse surface grind, rapid feeds in turning and milling

A. Very rough
B. Coarse
C. Good machine
D. Extremely smooth

____ 10. *True or False?* The level of precision for surface roughness is micrometers or microinches, wherein "micro" means "ten thousandths."

____ 11. A device for measuring surface texture in microinches is a _____.
A. micrometer
B. profilometer
C. dial caliper
D. digital caliper

For 12–16, match the symbol descriptions to the lettered figure labels. Answers are used only once.

.03

A B C D E

____ 12. Specifies the minimum amount of material to be removed and the final roughness

____ 13. Specifies the particular method to achieve specified roughness value

____ 14. Specifies, basically, that any production method can be used

____ 15. Specifies that material must be removed, but no amount stated

____ 16. Specifies no material can be removed

____ 17. With a more involved surface texture symbol, all of the following information can be specified within the symbol environment *except* which choice?
A. The method of obtaining the average value (arithmetic average, 10-point mean, etc.)
B. The process by which the surface is machined (mill, grind, lap, etc.)
C. The type of inspection machine to be used
D. The length of material on the surface that can be sampled

____ 18. *True or False?* The term "lay" is used to describe the direction of the predominant surface pattern caused by the machining process.

Match the following lay symbols to the descriptions. Answers are used only once.

____ 19. Somewhat parallel to the edge view where symbol is attached

____ 20. Somewhat perpendicular to the edge view where symbol is attached

____ 21. Angular in both directions to the edge view where symbol is attached

____ 22. Multidirectional or random

____ 23. Approximately circular to the center of the surface

____ 24. Approximately radial to the surface

A. R
B. X
C. C
D. ⊥
E. M
F. =

____ 25. *True or False?* It is common practice to attach the surface texture symbol to a line that is considered to be an edge view of the surface being controlled.

____ 26. *True or False?* Surface texture symbols can be attached to extension lines and leader lines, but not visible lines of the object.

____ 27. Which of the following is placed near the short stem of the surface texture symbol if a roughness average range is desired?
A. The upper number, such as 63
B. The lower number, such as 32
C. Both numbers, such as 63 over 32
D. An average number with a tolerance, such as 47±15

Critical Thinking

1. Give two or more reasons that engineers or designers would specify a precision smoothness to a surface.

____

____

____

____

____

2. Why would engineers *not* specify super smooth surfaces for all surfaces of an object?

____

____

____

____

____

3. Give a few reasons why symbols are advantageous over other forms of specifying, such as general or local notes.

____

____

____

____

____

Apply and Analyze

Name ____________________ Date ____________ Class ____________

Review Activity 12-1

For each surface texture symbol given below, fill in the maximum roughness height average, the sampling length, the lay symbol, and the minimum amount of material that must be removed. If the value is not specified, then fill in the blank with N/A. The first five are in the US customary system, and the second five are in the metric system.

	Maximum Roughness Average	Sampling Length	Lay Symbol	Minimum Material Removal
INCH 1. 125 √ .05 X				
2. 63 32 √ .10				
3. 16 √ (lay symbol ○)				
4. 250 .06 √ C				
5. 8 √ .05				
METRIC 6. 1.6 √ 0.8				
7. 0.2 √ 0.8 2.5/R_q 0.8				
8. 0.8 √ 2.5 ⊥				
9. 3.2 √ 0.8/S_m0.5 =				
10. NOTE 3 6.3 3 √ 0.4				

Apply and Analyze

Name ______________________ Date ______________ Class ______________

Review Activity 12-2

For each of the written descriptions below, sketch a surface texture symbol with the proper notation(s) required to accomplish that description. The first five are in the US customary system, and the second five are in the metric system.

INCH 1. √	Maximum roughness average is 250 microinches, with a sampling length of .05".
2. √	Roughness average range is 32 to 63 microinches, with a sampling length of .05".
3. √	No material is to be removed. Maximum roughness average is 32 microinches, with a sampling length of .10".
4. √	Maximum roughness average is 250 microinches, lay is multidirectional, and no sampling length is specified.
5. √	Minimum material removal is .06", maximum R_a is 8 microinches, sampling length is .05", and lay is circular.
METRIC 6. √	Maximum roughness average is 0.8 micrometers, with a sampling length of 0.5 millimeters.
7. √	Instead of roughness average, use the R_q parameter at 0.8, with a sampling length of 2.5 mm.
8. √	Maximum roughness average is 0.8 micrometers, with a sampling length of 2.5 mm, and lay is parallel to the indicated line.
9. √	Maximum roughness average is 3.2 micrometers, with a sampling length of 0.8 mm, and lay is radial.
10. √	Maximum roughness average is 6.3 micrometers, with a sampling length of 0.4 mm, and 6 mm or more must be removed by milling.

Apply and Analyze

Name ______________________ Date ____________ Class ____________

Industry Print Exercise 12-1

Refer to the print PR 12-1 and answer the questions below.

1. How many surface texture symbols are there on the views of this drawing? ______________
2. The two face surfaces perpendicular to the axis of this part that have surface texture specification should have a surface texture of _____ microinches or smoother. ______________
3. For the surface texture symbols that have a horizontal bar, what does the .05 indicate?

   ____
4. According to the lay symbols shown in the round view, should the lay on the two flat surfaces be in the same direction as the axis of the cylinder or perpendicular to the axis?

   ____
5. What surface texture value should be applied to all surfaces not specified on the drawing views?

   ____

Review questions based on previous units:

6. What scale are the views on the original drawing for this print? ______________
7. For the untoleranced dimensions, what tolerances should be applied?

   ____
8. What is the part name? ______________________________
9. Are there any section lines on this drawing? ______________________________
10. What type of section view is used in this drawing?

    ____
11. What is the pitch of the external thread?

    ____
12. How are the threads represented?

    ____
13. Why is the countersink hole on the right end not dimensioned?

    ____
14. There is a limit-method dimension labeled P.D. What does P.D. stand for?

    ____
15. What is the MMC of the counterbore diameter? ______________________________

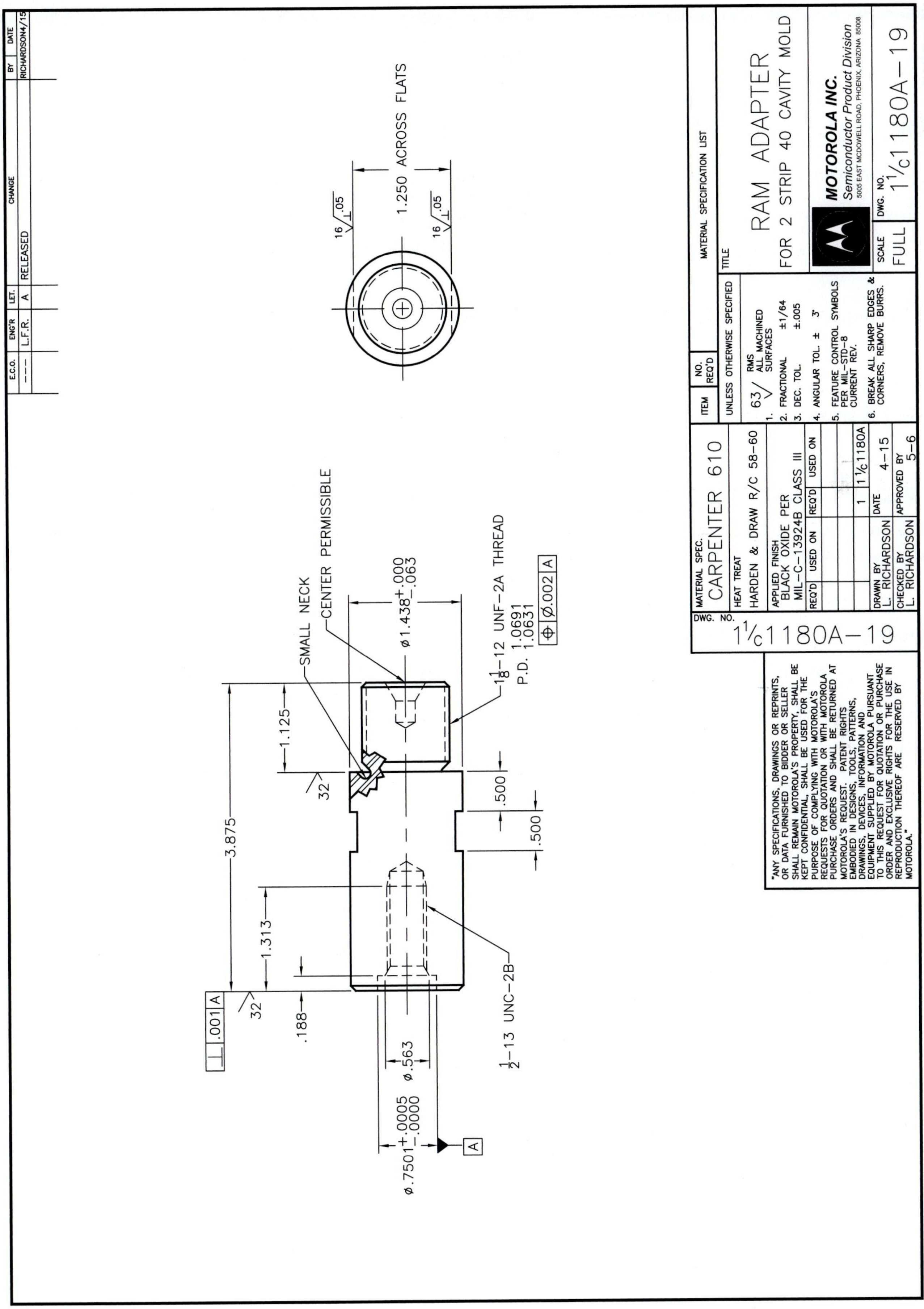

E.C.O.	ENG'R	LET.	CHANGE	BY	DATE
---	L.F.R.	A	RELEASED	RICHARDSON	4/15

⊥ .001 A
3.875
32
1.313
.188
1.125
32
SMALL NECK
CENTER PERMISSIBLE
ø.7501 +.0005 −.0000
ø.563
A
ø1.438 +.000 −.063
1/2−13 UNC−2B
.500
.500
1 1/8−12 UNF−2A THREAD
P.D. 1.0691 1.0631
⌖ Ø.002 A
16 ⊥.05
1.250 ACROSS FLATS
16 ⊥.05

"ANY SPECIFICATIONS, DRAWINGS OR REPRINTS, OR DATA FURNISHED TO BIDDER OR SELLER SHALL REMAIN MOTOROLA'S PROPERTY, SHALL BE KEPT CONFIDENTIAL, SHALL BE USED FOR THE PURPOSE OF COMPLYING WITH MOTOROLA'S REQUESTS FOR QUOTATION OR WITH MOTOROLA PURCHASE ORDERS AND SHALL BE RETURNED AT MOTOROLA'S REQUEST. PATENT RIGHTS EMBODIED IN DESIGNS, TOOLS, PATTERNS, DRAWINGS, DEVICES, INFORMATION AND EQUIPMENT SUPPLIED BY MOTOROLA PURSUANT TO THIS REQUEST FOR QUOTATION OR PURCHASE ORDER AND EXCLUSIVE RIGHTS FOR THE USE IN REPRODUCTION THEREOF ARE RESERVED BY MOTOROLA."

DWG. NO. 1 1/C 1180A−19

MATERIAL SPEC. CARPENTER 610
HEAT TREAT HARDEN & DRAW R/C 58−60
APPLIED FINISH BLACK OXIDE PER MIL−C−13924B CLASS III

REQ'D	USED ON	REQ'D	USED ON
		1	1 1/C 1180A

DRAWN BY L. RICHARDSON | DATE 4−15
CHECKED BY L. RICHARDSON | APPROVED BY 5−6

ITEM	NO. REQ'D	MATERIAL SPECIFICATION LIST

UNLESS OTHERWISE SPECIFIED

1. 63 RMS ALL MACHINED SURFACES
2. FRACTIONAL ±1/64
3. DEC. TOL. ±.005
4. ANGULAR TOL. ± 3°
5. FEATURE CONTROL SYMBOLS PER MIL−STD−8 CURRENT REV.
6. BREAK ALL SHARP EDGES & CORNERS, REMOVE BURRS.

TITLE
RAM ADAPTER
FOR 2 STRIP 40 CAVITY MOLD

MOTOROLA INC.
Semiconductor Product Division
5005 EAST MCDOWELL ROAD, PHOENIX, ARIZONA 85008

SCALE FULL | DWG. NO. 1 1/C 1180A−19

Print supplied by Motorola, Inc.

PR 12-1. Ram Adapter

Apply and Analyze

Name ______________________ Date ______________ Class ______________

Industry Print Exercise 12-2

Refer to the print PR 12-2 and answer the questions below.

1. How many surface texture symbols have been specified on the views of this drawing?______________
2. Which surface is to be smoother, the inside of the .500″ diameter hole or the 15° tapered conical surface?

   ____
3. Which machining process, if any, is specified for the surface texture symbol applied to the 1.5″ diameter face surface?

   ____
4. For the surface texture symbols indicated on this drawing, are the values in microinches or micrometers?

   ____
5. What lay symbol is indicated on one of the surface texture symbols? ______________
6. Do any of the surface texture symbols indicate the surface texture should be obtained without removing material?

   ____

Review questions on previous units:

7. Are there any cutting-plane lines shown on this drawing?______________
8. What material is specified for this part? ______________
9. What scale are the main views on the original drawing? ______________
10. How many threaded features does this part have? ______________
11. How many 45° conical surfaces are there on this part? ______________
12. Looking over the entire drawing, how many dimensions are expressed using the limit method?

    ____
13. There is a solid, equilateral triangle floating next to some dimension values. What tolerance applies to those values?

    ____
14. What is the maximum diameter of the drill spot that may exist at the bottom of the main hole in this part?

    ____
15. In the view that shows the hexagon shape, why is the dimensional value in parentheses?

    ____

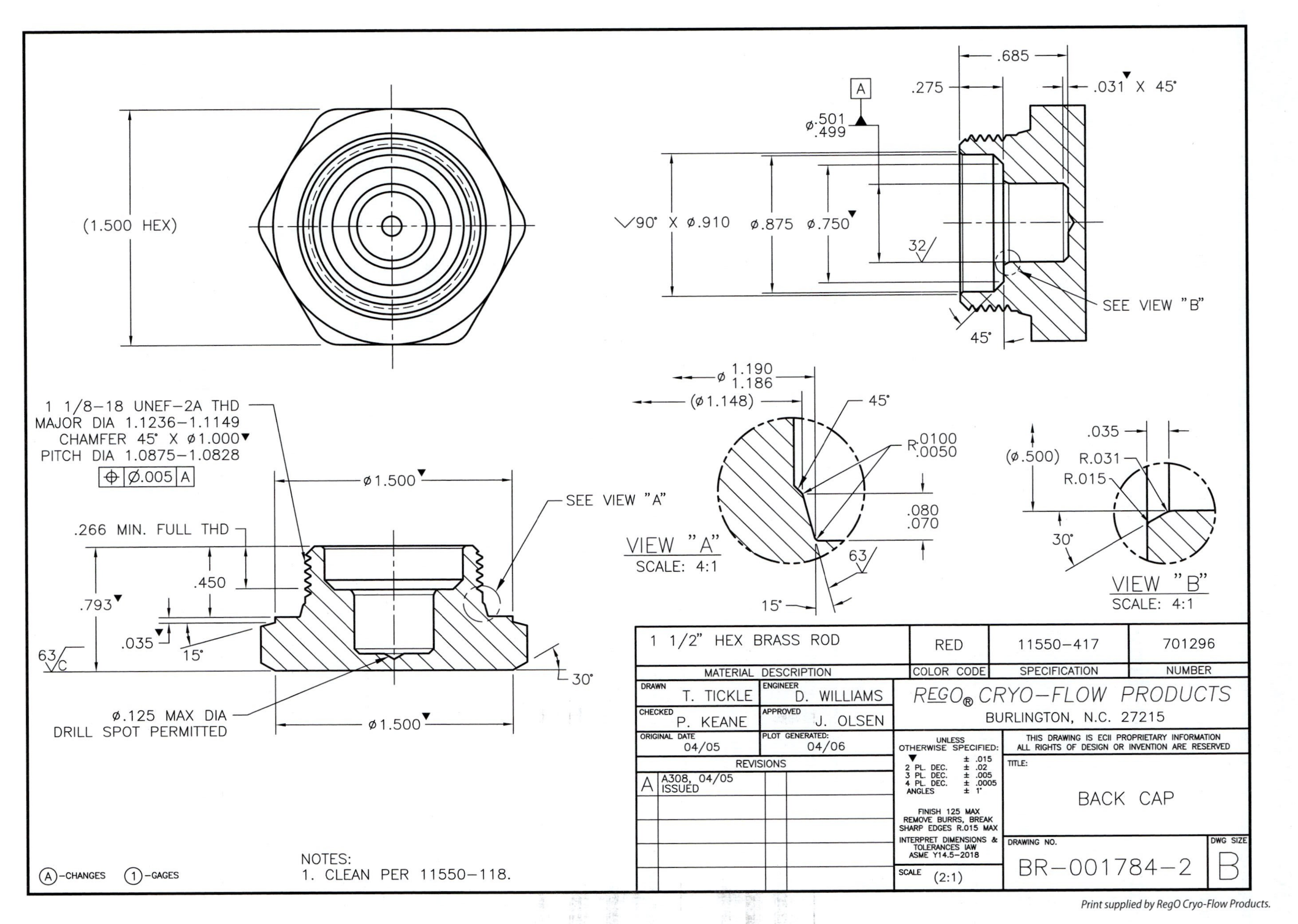

Print supplied by RegO Cryo-Flow Products.

PR 12-2. Back Cap

Apply and Analyze

Name ____________________ **Date** ____________ **Class** ____________

Bonus Print Reading Exercises

The following questions are based on various bonus prints located in the folder that accompanies this textbook. Refer to the print indicated, evaluate the print, and answer the question.

Print AP-001

1. Datum A has a surface texture of 32 microinches specified. What surface texture specification applies to the other symbols?

____

Print AP-006

2. The groove surfaces are specified to have a surface texture of 0.8 micrometers. Are any surfaces on the object specified to have a better quality than that?

____

3. What maximum surface texture (in micrometers) is specified for all surfaces?

____

Print AP-008

4. This print features international N values on the surface texture symbols. N5 is equivalent to 0.4 micrometers, N6 is equivalent to 0.8 micrometers, and N7 is equivalent to 1.6 micrometers. What value (in micrometers) is represented by N8?

____

Print AP-011

5. What is the surface texture specification for the cylindrical surface that has a diameter of 31.70–31.80?

____

6. What is the smoothest surface texture specified on this print?

____

Print AP-012

7. According to the notes, what is the UOS (unless otherwise specified) surface texture specification?

____

Print AP-016

8. What is the maximum roughness specification for machined surfaces?

____

Print AP-017

9. The surface texture symbols on this print have a horizontal bar rather than an open check mark appearance. What does that specify?

____

10. What is the surface texture specification for the cylindrical surface that will establish datum A?

____

Notes

UNIT 13
Geometric Dimensioning and Tolerancing

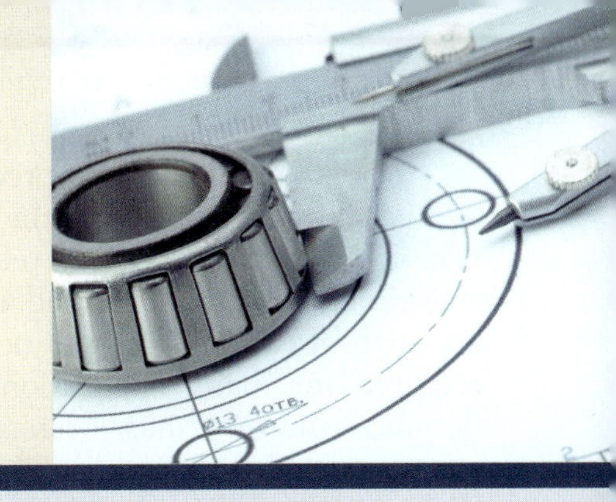

LEARNING OBJECTIVES

After completing this unit, you will be able to:

- Describe the purpose and objectives of geometric dimensioning and tolerancing (GD&T).
- Identify current and former ASME Y14.5 symbols used in GD&T.
- Define terms related to GD&T.
- Explain the purpose and function of datums.
- Identify proper datum identification techniques on a print.
- Explain basic dimensions as featured in drawings that use GD&T methods.
- Read and interpret the use of modifiers as they apply to basic GD&T applications.
- Read and interpret basic applications of feature control frames for each of the GD&T control symbols.
- Explain how composite tolerances are applied in a basic GD&T application.

TECHNICAL TERMS

angularity
circularity
circular runout
composite tolerancing
concentricity
cylindricity
datum feature
datum precedence
datum reference frame
datum target
feature
feature control frame
feature of size
flatness
form tolerance
full indicator movement (FIM)
geometric dimensioning and tolerancing (GD&T)
independency
least material boundary (LMB)
material boundary modifier
material condition modifier
maximum material boundary (MMB)
orientation tolerance
parallelism
perpendicularity
positional tolerance
profile of a line
profile of a surface
profile tolerance
regardless of feature size (RFS)
regardless of material boundary (RMB)
runout tolerance
straightness
symmetry
total runout
true position

Modern manufacturing requires more preciseness or exactness in the design and production of parts than formerly required. Simply using traditional dimensioning and tolerancing methods set forth in earlier units will not meet the precision required for many applications. For the last few decades, industry has been using an advanced system of print annotation known as ***geometric dimensioning and tolerancing (GD&T)*** to control the quality of mass-produced parts with a system of geometric control symbols, feature control frames, basic dimensions, and modifiers.

In Unit 10, you studied tolerances of size and location. However, these methods do not address the preciseness of the geometry. For example, how flat is flat? How round is round? The design team can implement GD&T methods to address these issues. With GD&T, the print can express the geometric form required for the part to function effectively.

The ASME standard for GD&T is ASME Y14.5, titled *Dimensioning and Tolerancing*. This standard is by far the most extensive of the ASME standards, including over 300 pages of text and illustrations for standard dimensioning and tolerancing practices, as well as the preferred practices for GD&T. While the material presented in this unit is based on the current standard, examples of former practice are also covered to help you read older prints. When working with a print containing GD&T symbology, the print reader should expect to find a statement on the drawing indicating which ASME standard edition applies. Information presented in this unit is based on ASME Y14.5-2018, the most current standard at the time of publication. One of the key mandates from the standard is that the drawing should specify the standard and year.

GD&T is an in-depth, intermediate-level field of knowledge that often requires intensive study for a complete understanding. It is recommended you check into some of the complete texts available solely for GD&T. For those who wish to be certified in GD&T, certification is available through ASME. The goal of this unit is to provide an overview of the basics of GD&T to get you started on the road to reading and interpreting prints containing GD&T symbology.

As GD&T has developed, many changes have made it a complicated topic for the print reader. Rules of interpretation have changed on more than one occasion, especially with respect to modifier use and datum identification. You should study GD&T beyond the material presented in this textbook and pay attention to older standards as well as the current standards. This will help you interpret the symbols and specifications used by designers and drafters on both older and more recent industrial prints.

In summary, geometric dimensioning and tolerancing is an extensive and sometimes complex system for describing the quality of geometry and geometric locations on industrial prints. You will need to keep abreast of the latest standards as they continue to evolve for this field. This will help your company maintain clear and concise dimensioning and tolerancing practices, resulting in high-quality manufactured parts that function well.

Geometric Dimensioning and Tolerancing Symbols

A simple glance at a drawing will quickly indicate whether it contains GD&T specifications. A drawing with GD&T will have square or rectangular boxes associated with local notes and around some dimensions. These symbols will often be found throughout the drawing, although they may be isolated to just one or two features. See **Figure 13-1**. The boxes are datum feature symbols, basic dimensions, or feature control frames. Each of these components is addressed in this unit.

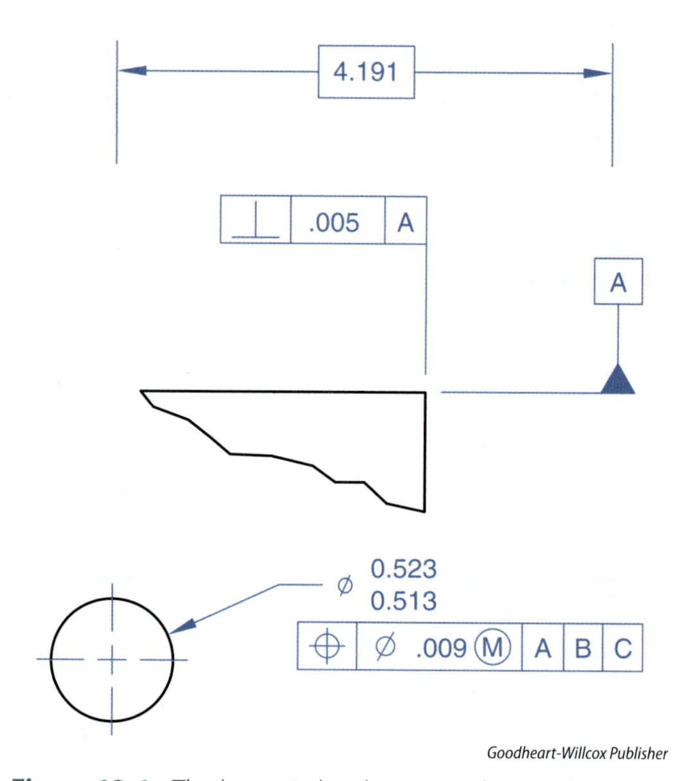

Figure 13-1. The boxes in local notes and around some dimensions quickly identify a drawing as containing GD&T specifications.

Feature control frames are long rectangular boxes with compartments, and they are attached to extension lines or leader lines, or floating near certain dimensions. Within the feature control frames are geometric characteristic symbols that specify control of the geometry for a particular feature. **Figure 13-2** shows the 12 characteristics introduced in this unit, as well as the two characteristics that have been discontinued. The characteristics have been grouped into common categories, and former practice will also be briefly discussed in this unit.

GD&T Terminology

As defined in Unit 9, a *datum* is a theoretically exact point, axis, or plane used as the origin from which the location or geometric control of features is established. A datum is identified on a drawing by a symbol that appears as a box on a leader stem terminating with a triangle that may be either filled solid or unfilled. See **Figure 13-3**. As discussed in Unit 10, this symbol is called the *datum feature symbol*. For identifying datum axes and datum center planes within 2D views, the datum feature symbol is usually aligned with the dimension line of the feature used to determine the axis or center plane, but it may also be attached to the visible lines of the feature. The datum feature symbol *cannot* be simply attached to a center line without reference to any measurements or surfaces. If you are reading a print drawn in this manner, you must seek clarification of which feature (or set of features) is to be used to establish the datum reference. As shown in **Figure 13-3**, when datum feature symbols are applied to a 3D model, they may be attached to surface areas or to available dimensions.

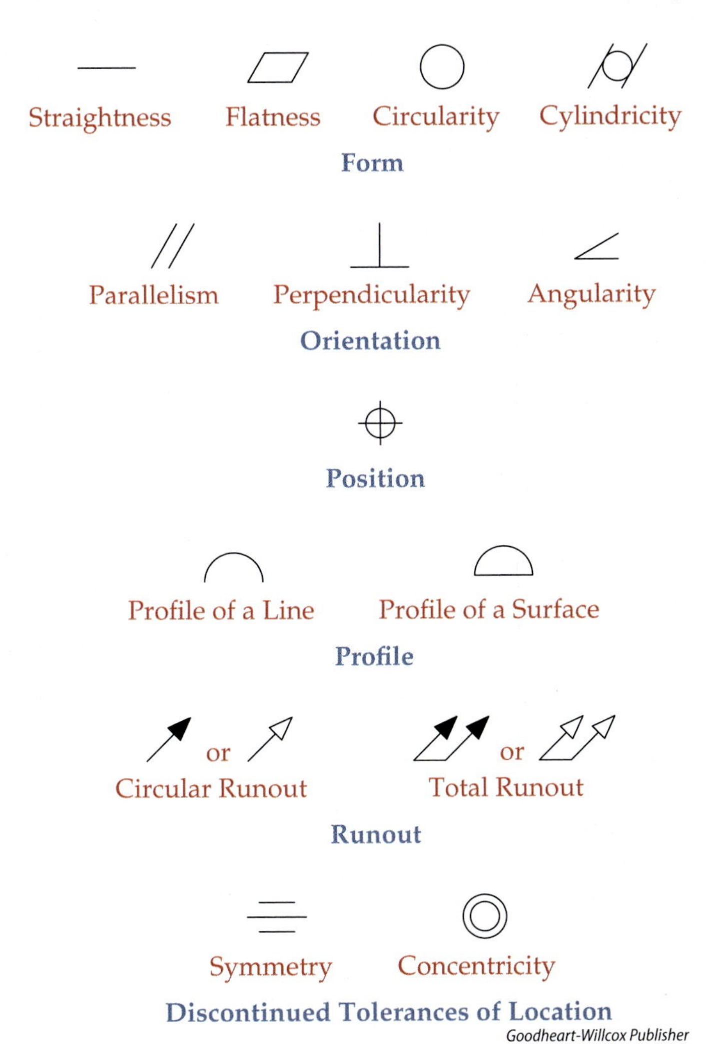

Figure 13-2. These are the 12 geometric control characteristics currently used in GD&T, with two additional controls that were recently discontinued.

In former practice, the datum feature symbol was drawn as a rectangular box, and the datum identification letter was preceded and followed by a dash. See **Figure 13-4**. The box had no stem, and it was attached to the view with an extension line or leader line. For a datum axis or center plane identification, the box was to be floating next to a measurement of a feature of size, such as the diameter specification for a hole or width dimension of a slot. As with the current symbol, there was an association of the symbol with the dimensions of the real features of the object that were to be used to establish the hypothetical datum axis or datum center plane.

As defined in Unit 10, a *basic dimension* is a numeric value used to describe the theoretically exact size, profile shape, orientation angle, or location distance for a feature or datum target. Tolerances are *not* directly applied to basic dimensions. In the case of location dimensions for holes, the location tolerance is given in the feature control frame. In the case of defining an irregular profile, the shape tolerance is also given in the feature control frame. In other cases, such as for locating datum targets, gage makers and inspection personnel use their own tolerance guidelines at a higher level of precision. The symbol for a basic dimension is a rectangle enclosing the dimension numeral, as shown in **Figure 13-5**. Again, just because there is no tolerance shown with the dimension value, it does not mean there is no size or locational tolerance. The tolerance is just applied differently through standard GD&T practices.

The term ***feature*** refers to a physical portion of a part. For example, a surface is a feature. Slots, tabs,

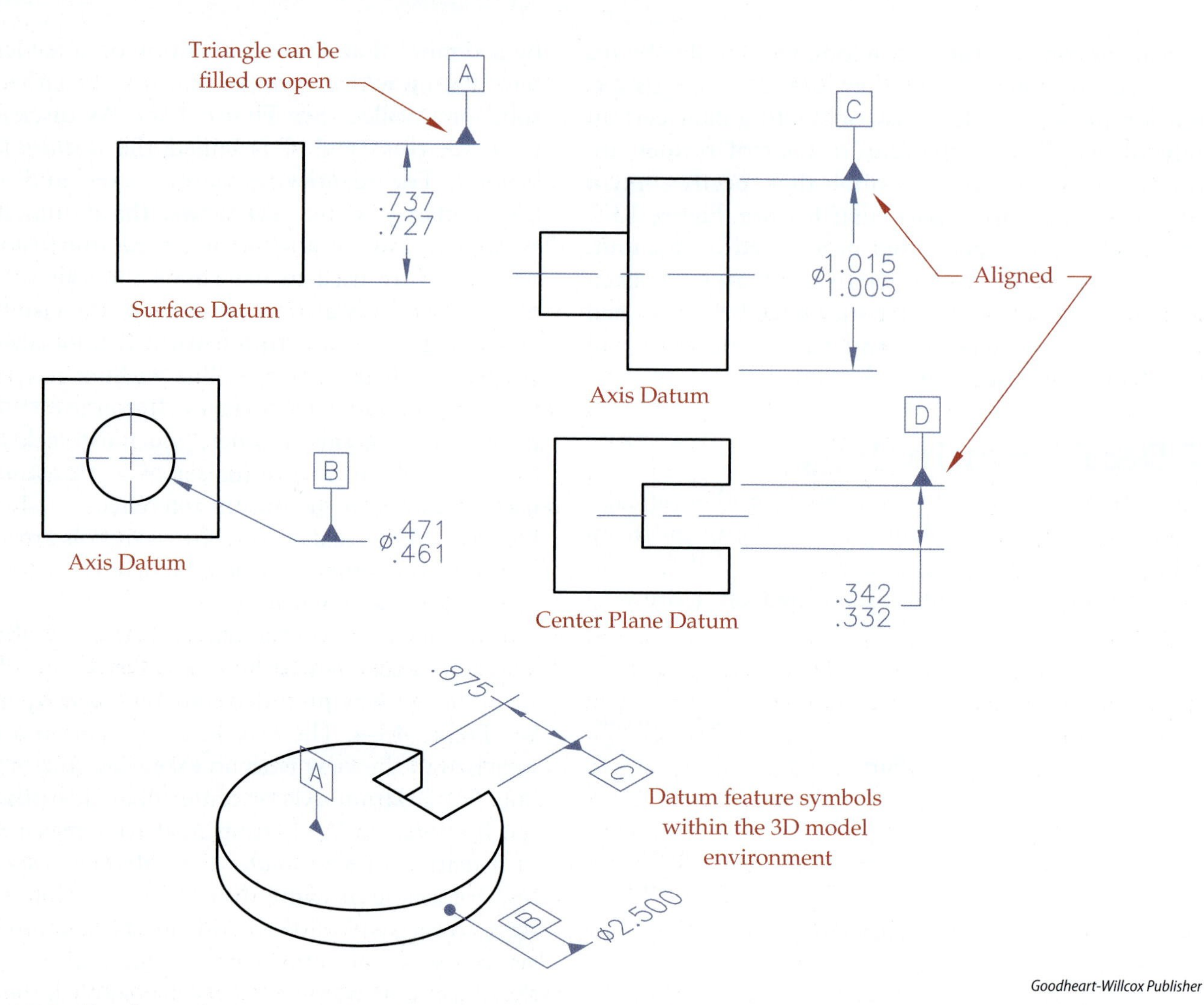

Goodheart-Willcox Publisher

Figure 13-3. The datum feature symbol is a box on a leader stem terminating with a triangle. It can be applied to orthographic views or reside in the 3D model environment.

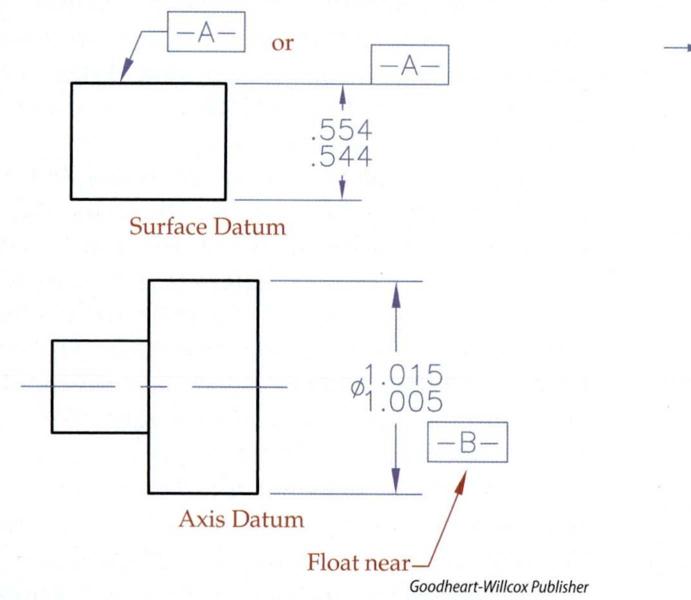

Goodheart-Willcox Publisher

Figure 13-4. In older standards, the datum feature symbol was a rectangular box with the datum letter preceded and followed by a dash.

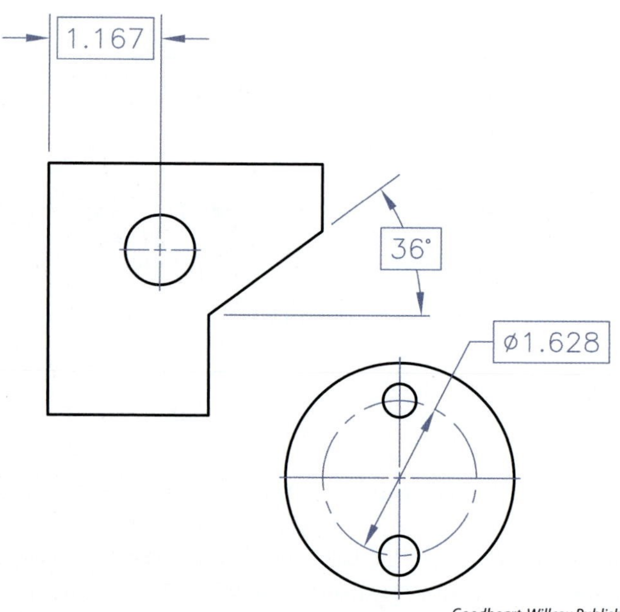

Goodheart-Willcox Publisher

Figure 13-5. A basic dimension features a rectangle enclosing the dimension value.

posts, holes, keyways, and threaded parts of an object are also called features. Most likely, an object has many features. Another important concept in GD&T is the term ***feature of size***, which is simply a feature that has a center plane or center axis. These features of size are things such as holes, slots, tabs, and posts. There are certain benefits in GD&T that can be applied to a feature of size. Features of size, if used as datums, also offer additional opportunities for flexibility and bonus tolerance while specifying geometric control.

A ***feature control frame*** is a rectangular box with compartments enclosing such items as the geometric characteristic symbol, specified geometric tolerance, and datum reference, **Figure 13-6**. The feature control frame can be very simple for some controls, but very complex for others. For a positional tolerance, three datums are often required to establish a locked-in three-dimensional position. If referenced, each datum is in a separate compartment in the feature control frame. On occasion, two features can form a combined, or simultaneous, datum. In this case, the datum reference letters are placed in a single compartment separated by dashes. Some of the earliest GD&T standards called for the datum reference to precede the amount of tolerance, but current standards place the datum references at the end.

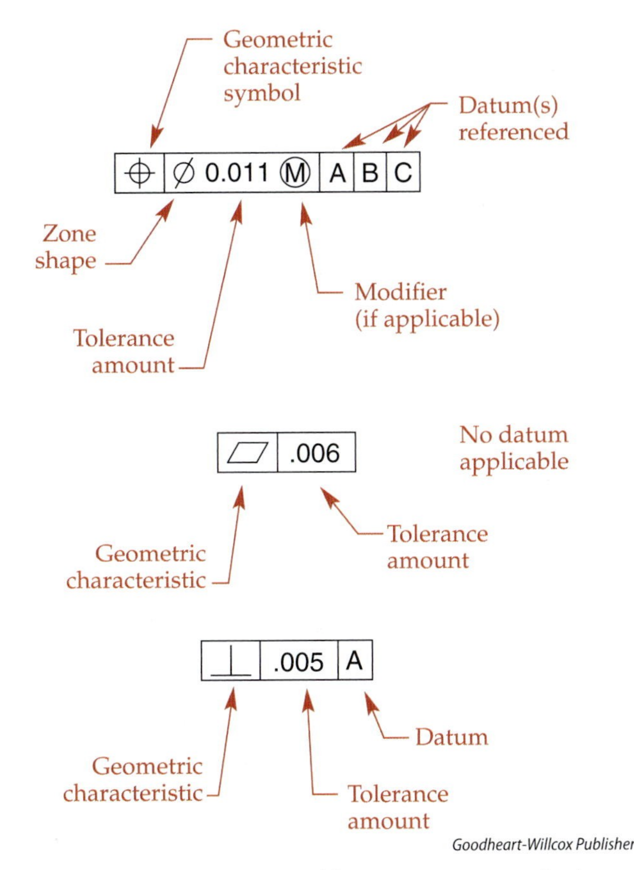

Figure 13-6. A feature control frame is a rectangular box with compartments enclosing the geometric characteristic symbol, a possible zone diameter symbol, the specified geometric tolerance, various possible modifiers, and one or more possible datum references.

As defined in Unit 3, *tolerance* is the range within which a specified dimensional value is permitted to vary. As discussed in Unit 10, the tolerance is the difference between the larger and smaller limits of a size or location value. For GD&T, tolerance should be considered as two-dimensional or three-dimensional space within which the geometry of a feature can vary. Each feature of the object can be specified to fall within an envelope of space determined by geometric tolerance specifications. For example, a straightness tolerance zone is the area between two line segments (2D zone), while a flatness tolerance zone is the area between two flat planes (3D zone). A locational tolerance zone for an axis is most often a cylindrical location zone. The tolerance zone size is indicated in a feature control frame along with the geometric characteristic.

While many of the controls truly do deal with the geometric shape and form of the features, one concept that also dominates the study of GD&T is the geometric position of features such as holes, slots, posts, and tabs. The term ***true position*** is defined as the theoretically exact location of a feature's axis or median plane. True position is established by basic dimensions, which locate a three-dimensional locational zone for the true position desired by the designer. Due to the extensive and sometimes complex nature of positional tolerancing, an entire section of the standards, titled *Tolerances of Position*, is dedicated to this aspect of GD&T.

In addition to the 12 geometric control symbols, there are several other symbols that help communicate GD&T specifications. Some of them are used within the feature control frame, and some are used in the open area surrounding the feature control frame. The symbols for "all around" and "all over" may appear at the corner of the leader shoulder. **Figure 13-7** shows the available symbols, some of which have been introduced in earlier units but have extended meaning within GD&T applications. An explanation for these symbols will be given later in this unit.

Two of the symbols are considered ***material condition modifiers*** and are used to indicate a bonus tolerance is to be applied. The bonus tolerance is based on the deviation of the actual size of the feature from either the maximum material condition (MMC) or the least material condition (LMC). Review Unit 10 for a

GD&T Symbols	
Ⓜ	Maximum Material Condition (applied to a tolerance) or Maximum Material Boundary (applied to a datum reference)
Ⓛ	Least Material Condition (applied to a tolerance) or Least Material Boundary (applied to a datum reference)
Ⓢ	Regardless of Feature Size (old standard - discontinued)
Ⓟ	Projected Tolerance Zone
Ⓕ	Free State
Ⓣ	Tangent Plane
Ⓤ	Unequally Disposed Profile
Ⓘ	Independency
⟨ST⟩	Statistical Tolerance
⟨CF⟩	Continuous Feature
←→ *	Between
	All Around
	All Over
▷	Translation (movable datum)
Δ	Dynamic Profile
→ *	From-To

*May be filled or open

Goodheart-Willcox Publisher

Figure 13-7. Various symbols are used in a variety of dimensioning and tolerancing applications.

discussion of MMC and LMC definitions and principles. To indicate this bonus tolerance modification, an encircled M, for MMC, or an encircled L, for LMC, is used in the feature control frame.

Datums

With respect to datums, the terminology should be carefully considered. For example, there is a fine distinction to be made between a datum *feature* and a datum *plane* or *axis*. One is real and one is hypothetical. ***Datum features*** are actual geometric features of the part, such as flat surfaces or cylindrical surfaces, used as a reference from which to establish the measurements and readings. For example, a surface is a real datum feature used to establish a theoretically exact datum plane. Likewise, the surface of a hole is a real datum feature used to establish a theoretically exact datum axis.

Datum features used to set up the part for inspection are identified on the drawing. The hypothetical datum planes and axes are represented by items such as precision inspection equipment, gages, gage pins, the axes of a coordinate measuring machine, or the surface of a marble inspection block. For example, the real points of an object rest on the marble block surface, and the marble surface then becomes the hypothetically perfect datum plane from which flatness or parallelism can be checked.

Another datum concept is the ***datum reference frame***, which consists of three mutually perpendicular planes that represent how the part will be oriented in space for inspecting the geometric controls. See **Figure 13-8**. These planes are simulated by positioning the part on appropriate datum features and restricting the motion of the part. See **Figure 13-9**.

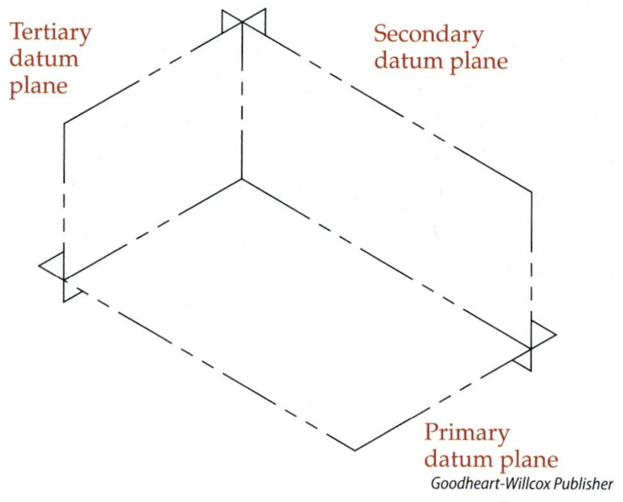

Goodheart-Willcox Publisher

Figure 13-8. A datum reference frame consists of three mutually perpendicular planes representing the features of the part that are the most important in relation to the geometric controls being applied.

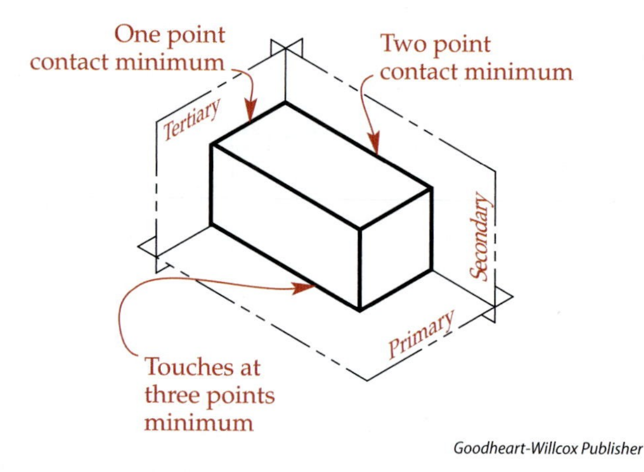

Goodheart-Willcox Publisher

Figure 13-9. A datum reference frame is necessary for checking certain geometric characteristics.

In the Y14.5 standard, this is referred to as "constraining the degrees of freedom" of a part. As mentioned previously, if datums are features of size, and modifiers are applied, the part can move around within the material boundaries in such a way as to have more parts passing inspection. This flexible datum setup is an advantage of GD&T over former practice.

Not all geometric controls require three datums to be established or identified on the print. In some cases, cylindrical datum features are represented by two theoretical planes intersecting at right angles on the datum axis, and one additional surface for perpendicularity. As a result, a cylindrical object may be set into a datum reference frame with two identified datum features. In some cases, additional datums are specified for other geometric or relational controls.

Datum precedence is the order in which an object is placed into the datum reference frame. The precedence is indicated by the order of the datum reference letters, from left to right, in the feature control frame. From the left, these datums are designated as primary, secondary, and tertiary datum features. They are selected in order of functional design importance, if applicable. The drafter usually attempts to use the letters A, B, and C in the primary-secondary-tertiary order, but that is not required, nor of great importance.

In basic theory, the primary datum feature relates the part to the datum reference frame with three points on the surface in contact with the first datum plane. The secondary datum feature relates with two points in contact with the second datum plane. The tertiary datum feature relates with one point in contact with the third datum plane.

Datum targets are specific points, lines, or areas of contact on a part that can be specified to establish the datum reference frame. Datum targets can be used to correct for the irregularities of some features, such as nonplanar, thin, or uneven surfaces, especially when it is not economically feasible to machine or stabilize them. The datum targets are basically defining the location of fixture pins that hold a part in place during manufacturing or inspection. In these cases, the datum targets are used by the gage maker to establish how the part is to be mounted. Datum targets are identified by an X on the drawing, or a target area of the desired shape is specified. See **Figure 13-10**. Recent standards also introduced a movable datum target symbol, for those situations where a datum setup might use movable target pins.

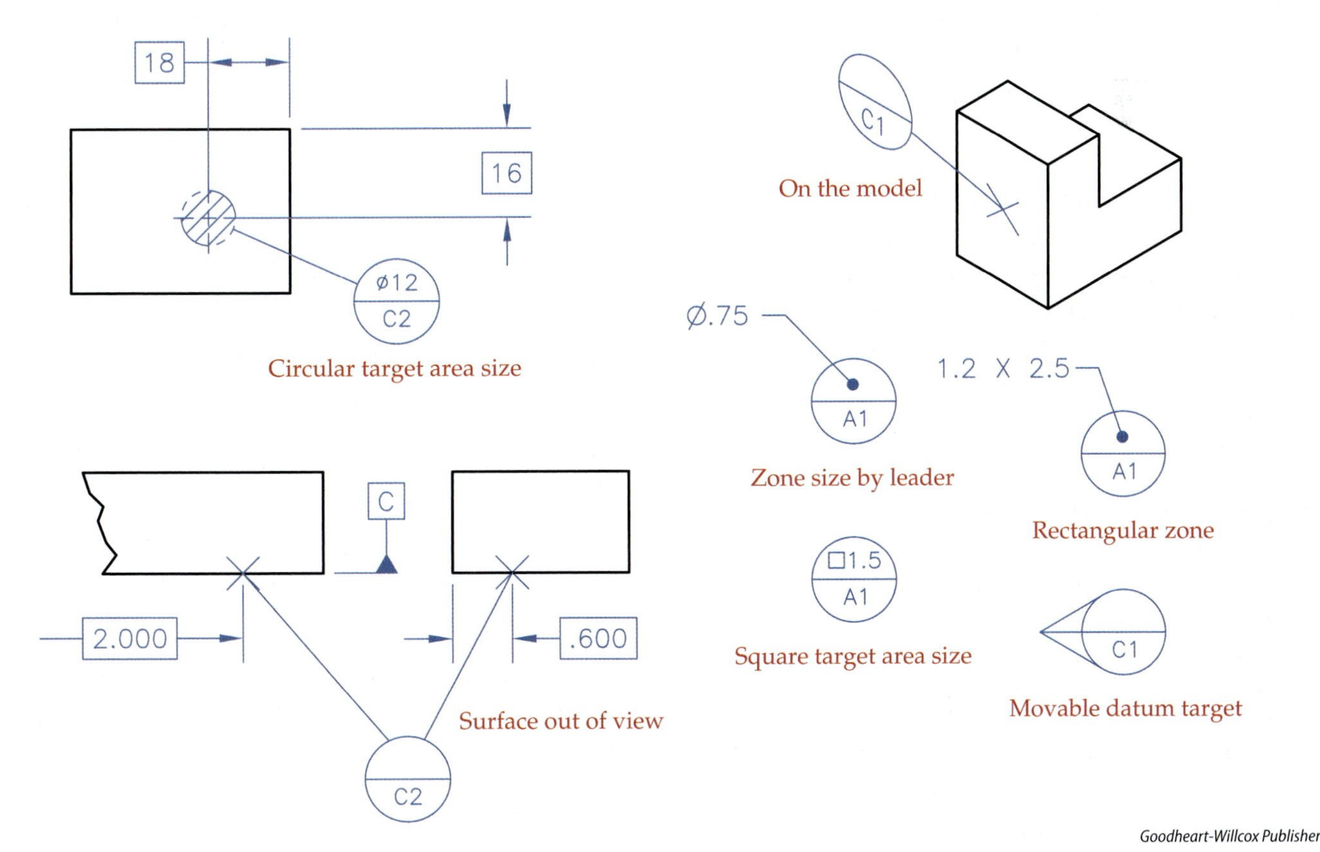

Goodheart-Willcox Publisher

Figure 13-10. Datum targets are identified by an "X" on the drawing, or a target area of the desired shape is specified. Also shown are target area specifications and the newer movable datum target.

The basic dimensions on the print are used by the gage maker, who in turn applies tolerances according to gage maker standards or guidelines.

Standardized Meaning of Size

Before applying geometric controls to an object, it is critical to standardize the meaning of "size." Unless otherwise specified, the limits of size control the geometry, and GD&T symbols are not needed. This standardized definition of size, also referred to as *Rule #1: The Envelope Principle*, can be paraphrased to read "where only a tolerance of size is specified, the geometric form of an individual feature of size is controlled by the limits of size." In other words, a part cannot geometrically deviate beyond a boundary of perfect form when produced at the maximum material conditions specified by the dimensions given on the drawing views or contained within the model.

Therefore, by definition, the size of a cylinder controls the shape of that cylinder. If a cylinder has specified limits of 1.490″ to 1.500″, and if its actual size is at the maximum material condition of 1.500″, then it must be perfectly straight. If it is .005″ undersize, some of the elements technically could deviate up to .005″. When a size tolerance amount is so large that geometric variance might become problematic, GD&T controls are used to control the geometric variance and help maintain the form of the feature.

Rule #1 also does not apply to stock materials, such as sheets, tubing, or structural shapes. Other items exempt from Rule #1 include items produced to certain government standards that prescribe certain geometric control. Where it is desirable to permit a surface or surfaces of a feature to exceed the boundary of perfect form, a note such as PERFECT FORM AT MMC NOT REQUIRED can be specified on the drawing. Also, recent standards have introduced an ***independency*** symbol, simply an encircled I, which can be placed next to a dimension to indicate that perfect form is not required for that dimension.

Material Condition and Boundary Applicability

The principles of geometric dimensioning and tolerancing allow for a flexible bonus tolerance for certain features, or a flexible datum setup, thus increasing productivity. As defined earlier, a *feature of size* is a feature with a center axis or a center plane, such as a hole, slot, or tab. As discussed in Unit 10, by definition, the *maximum material condition (MMC)* of these features is when they contain the most material allowed by the size dimension. That is, a hole's MMC is the smallest diameter allowable, because the object has more material when the hole is smaller. This forces us to think "backward" for internal features. A shaft's MMC is the largest diameter, which is easier to visualize. This principle of MMC allows us to benefit from GD&T modifiers and gain bonus tolerance as features vary from the MMC condition.

The symbol for MMC as a *material condition modifier* is an M contained within a circle. As an example, the MMC material condition modifier is often applied to the positional tolerance of a hole. In this case, the designer is specifying that as the hole gets bigger (away from MMC), it can be more out of position and still be functional. Likewise, in the case of a pin, as the pin gets smaller (away from MMC), it can be more out of position and still clear a mating hole. The mathematical calculations are easy. For each .001″ of diameter departure from MMC, the diameter of the positional zone can increase by .001″. Before GD&T, there was not a way to specify this bonus tolerance.

Material condition modifiers can also be applied to the datum setup. As with material condition modifiers used for features of size, modifiers can be added to datum features if the datums are features of size, such as holes, slots, tabs, or posts. Adding modifiers to datum features allows for more parts to pass inspection, as the datum reference frame can be adjusted for conditions that are still within the boundaries of the design intent. The most recent standards now define these datum modifiers as ***material boundary modifiers***. When the encircled M symbol is applied to a datum feature, the datum feature is to be established at the ***maximum material boundary (MMB)***. Use of the term MMB is now current practice, even though within drawings, the symbology appears the same as former standards. In summary, the encircled M symbol is used in the feature control frame to modify an individual tolerance, a datum reference, or both. See **Figure 13-11**.

As defined in Unit 10, the *least material condition (LMC)* occurs when the object contains the least amount of material allowed by a size dimension. That is, a hole's LMC is the largest diameter, while a shaft's LMC is the smallest diameter. The symbol for LMC, an encircled L, is used in the feature control frame to modify an individual tolerance. See **Figure 13-12**.

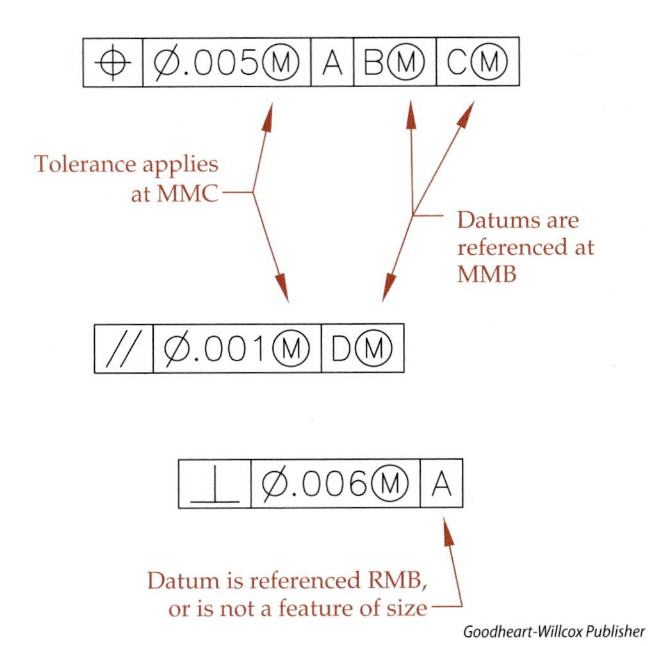

Figure 13-11. The symbol for MMC and MMB is an M contained within a circle. It indicates bonus tolerance is given as the feature departs from MMC or allows a datum feature to be established in a more flexible manner.

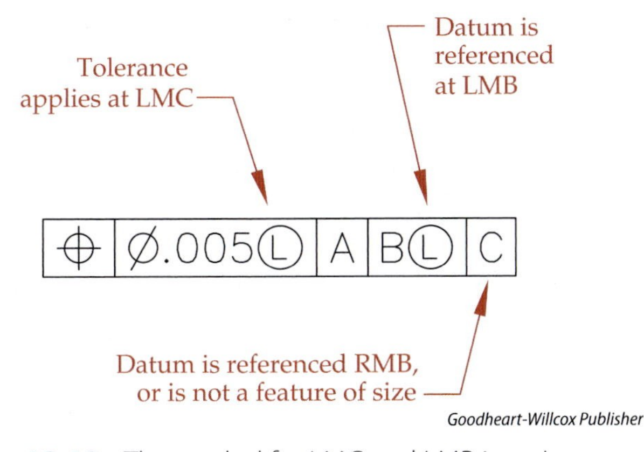

Figure 13-12. The symbol for LMC and LMB is an L contained within a circle. It indicates bonus tolerance is given as the feature departs from LMC or allows a datum feature to be established in a more flexible manner.

When the encircled L symbol is applied to a datum feature, the datum feature is to be established at the ***least material boundary (LMB)***. Using LMC and LMB as modifiers is not nearly as common as using MMC or MMB. They are often applied in situations where a designer wants to guarantee a minimum edge distance between two features. For example, in the case of a hole near the edge of the part, as the hole gets smaller (away from LMC), it can be more out of position without endangering the minimum thickness of the material around the hole.

For years, another modifier symbol known as ***regardless of feature size (RFS)*** was used to specify those situations wherein an MMC or LMC modifier could be applied, but the desire was to keep the tolerance the same regardless of the feature size. In some cases, perhaps there was no relationship between the controlled feature and any mating parts or clearance conditions, so bonus tolerance was of no concern or benefit. In this case, the specification was represented by an S contained in a circle. See **Figure 13-13**. Under current standards, if an MMC or LMC modifier is not present, RFS is automatically assumed. The RFS modifier is not to be used. As with the other datum modifiers, terminology has been clarified. If a datum feature of size is not modified by MMB or LMB, it is assumed to be established ***regardless of material boundary (RMB)***. If GD&T requirements are checked with a fixture gage that establishes the datum reference frame, a pin, collet, or gage block that represents the datum feature will need to be adjustable to accommodate the RMB condition, whereas an MMB specification allows the fixture to be sized at the datum's MMC, and the part can move around within the fixture—within those limits.

Historical Summary

A summary of GD&T symbol history may be helpful to those who deal with older prints. At one point in time, MMC was so common for positional tolerancing that the rule was that if no modifier symbol

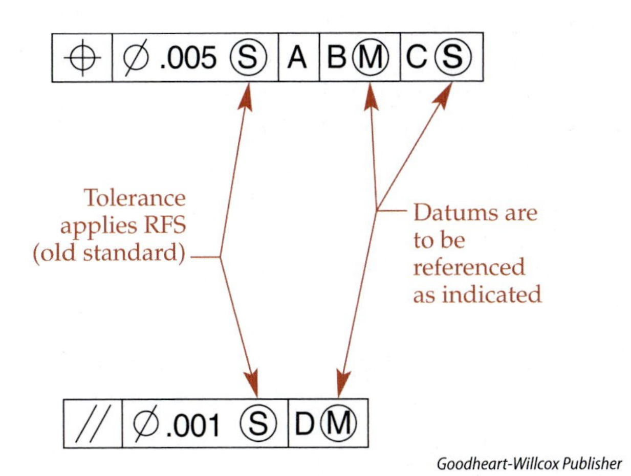

Figure 13-13. In former practice, the symbol for "regardless of feature size" was an S contained in a circle. It was formerly a required specification for positional tolerance and for a datum that was a feature of size, unless MMC or LMC was desired. Now, RFS is assumed and the symbol is no longer needed.

appeared in a positional tolerance specification, MMC was *assumed*. This apparently caused some confusion, so for another period in time, the standard rule was that a symbol for MMC, LMC, or RFS was *always* required for the positional tolerance specification, but not for the other characteristics.

Under current standards, MMC or LMC for geometric position and MMB or LMB for datums must be specified, if applicable and desired, and RFS or RMB are always to be assumed if no modifiers are present. This is known as Rule #2, which simply states "RFS is the default condition for geometric tolerance values" and "RMB is the default condition for datum feature references." If you encounter a print that seems contrary to this, or if there are written notes and specifications that do not match this rule, consult with the proper personnel to be sure of the correct interpretation. The concept of material condition modification takes time to understand. You need to carefully study the geometric characteristics to see how the modifiers have an impact on interpreting a print.

Form Tolerances

With all the foundational topics covered, it is now time to discuss the individual geometric tolerances. As illustrated in **Figure 13-2**, these are grouped into categories, the first of which is the tolerances that apply to the individual geometric form of features. ***Form tolerances*** are used to control straightness, flatness, circularity (roundness), and cylindricity. They are perhaps the easiest GD&T controls to understand, although aspects of these geometric controls also reside within the size tolerance or within the other geometric controls. Form tolerances are *not* referenced or related to datums but are applicable to single (individual) features or elements of single features. Form tolerances are not required on the drawing if the size tolerance is sufficient to control the geometry.

Straightness is a measure of an element of a surface or a derived median line existing in a straight line. Straightness tolerance specifies a tolerance zone within which the considered element or derived median line must lie. See **Figure 13-14**. Straightness

Straightness tolerance - Straightness of surface line elements

On orthographic views

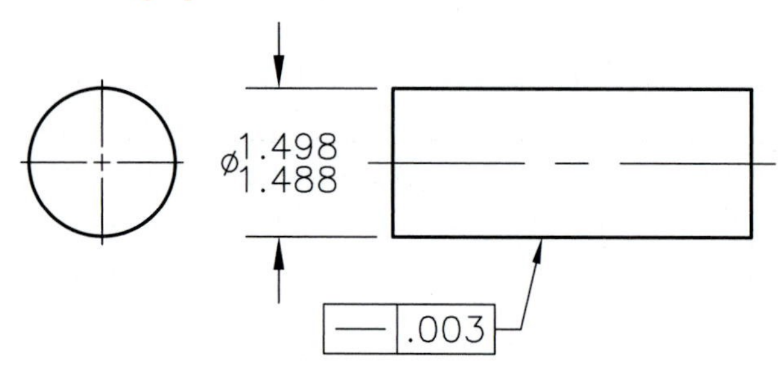

Each longitudinal element of the surface must lie between two parallel lines. The feature must be within the specified limits of size and cannot exceed perfect form at MMC. Waisting or barreling must not exceed the limits of size of the feature.

On the 3D model

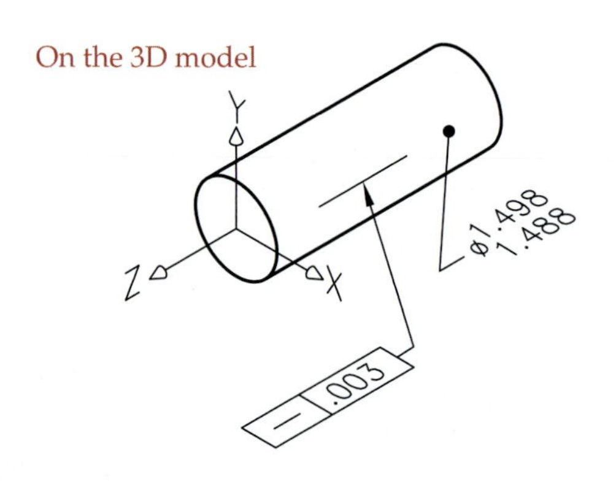

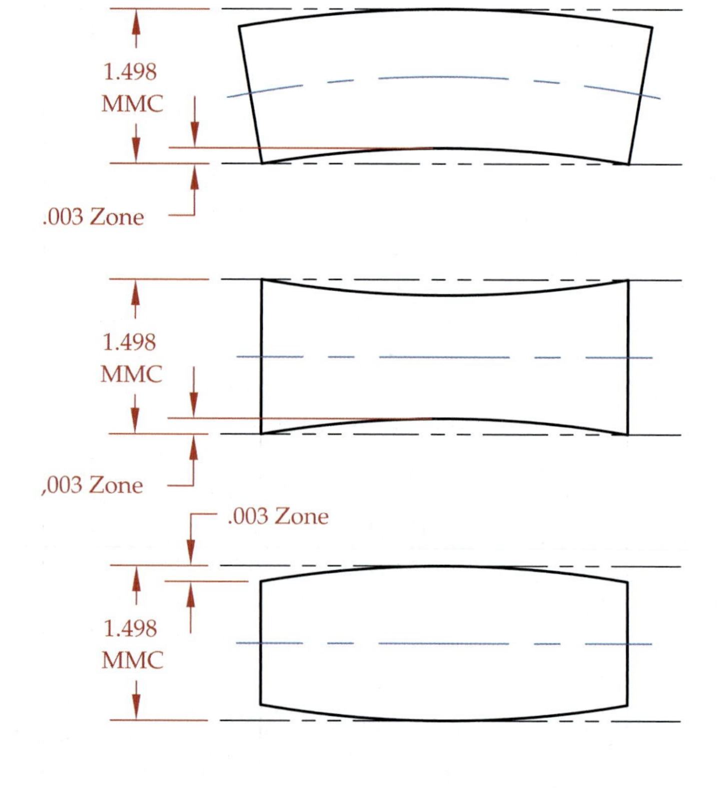

Goodheart-Willcox Publisher

Figure 13-14. Straightness can control elements of a surface as the feature departs from MMC, especially in cases where the size tolerance would allow for too much error in straightness.

is one control that has two possible applications for cylindrical features. In former standards, the terminology expressed these two options as *element straightness* and *axis straightness*. For element straightness, Rule #1 applies, so at MMC all elements must be straight. Only as local measurements show that the size is less than the MMC will there be potential for the elements to be bent or crooked, and the geometric tolerance ensures straightness is controlled.

The straightness of all elements as a composite is now expressed as the straightness of the derived median line. For this application of straightness, Rule #1 (perfect form at MMC) does not apply, and the MMC modifier can be applied, which allows the part to be bowed or bent beyond the MMC. See **Figure 13-15**. In situations like this, the MMC and the geometric tolerance are combined to form a virtual condition, a term that describes the worst-case scenario of size and form. For these applications, a "go" or "no-go" gage can be used to check the geometric form, although that is not required. Straightness can also be applied to noncylindrical features, as well as on a unit basis.

Flatness is the condition of a surface having all surface elements, or the elements of a derived median plane, within one plane. The flatness tolerance zone is defined by two parallel planes within which the surface must lie. See **Figure 13-16**. Flatness can be thought of as a three-dimensional version of straightness. Flatness is often used to qualify a datum surface. In these cases, the datum feature symbol is often attached to the feature control frame, but the flatness control is not referencing the datum. See **Figure 13-17**. Flatness is to be determined first and then, if the feature passes inspection, the surface qualifies to be used as a datum feature.

Similar to straightness, flatness can also be applied to a median plane, in which case the feature control frame will be placed adjacent to the dimension value for the two surfaces whose points will determine the derived median plane. For median plane flatness, Rule #1 (perfect form at MMC) does not apply, and the MMC modifier can be applied, which allows the feature to extend beyond the MMC boundary envelope. Also, in similar fashion to straightness, flatness can be applied on a unit basis rather than across the entire length of the surface. In these cases, it is

Straightness tolerance - Straightness of a derived median line at MMC

On orthographic views

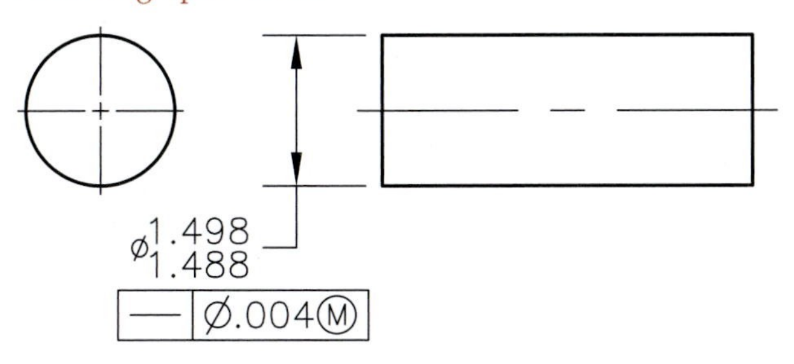

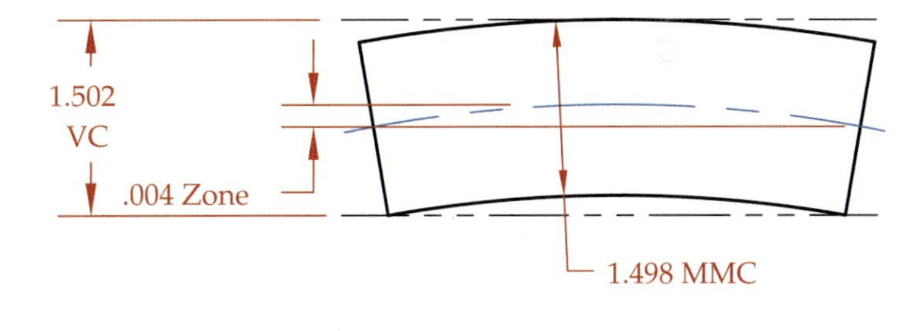

The derived median line of the feature must lie within a cylindrical tolerance zone with a diameter of .004 at MMC, which creates a virtual condition (VC) of 1.502. As each local size departs from MMC, the tolerance zone increases by an amount equal to the departure. Each circular element is still required to be within the specified size limits.

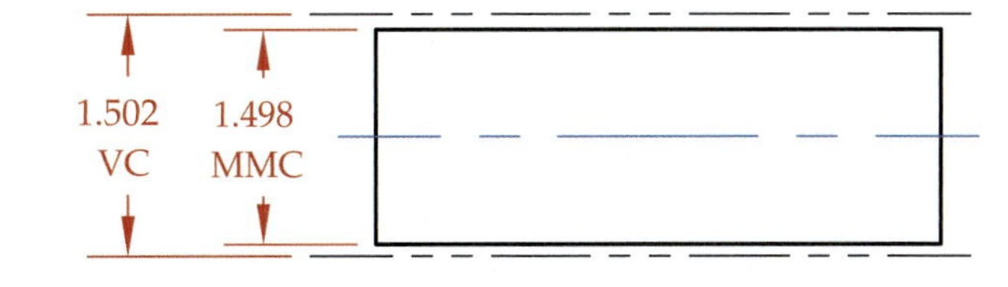

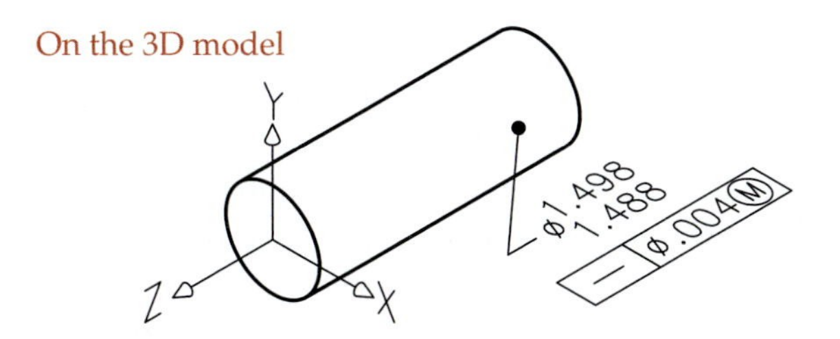

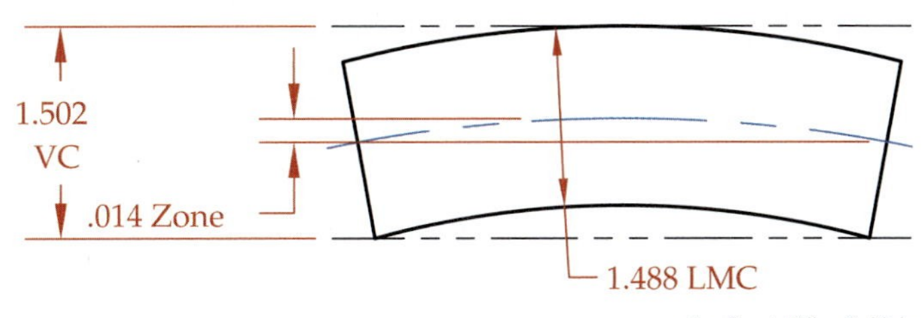

Goodheart-Willcox Publisher

Figure 13-15. Straightness can also control the composite of all surface elements in the form of axis control, in which case perfect form at MMC is not required, and a material modifier can be applied.

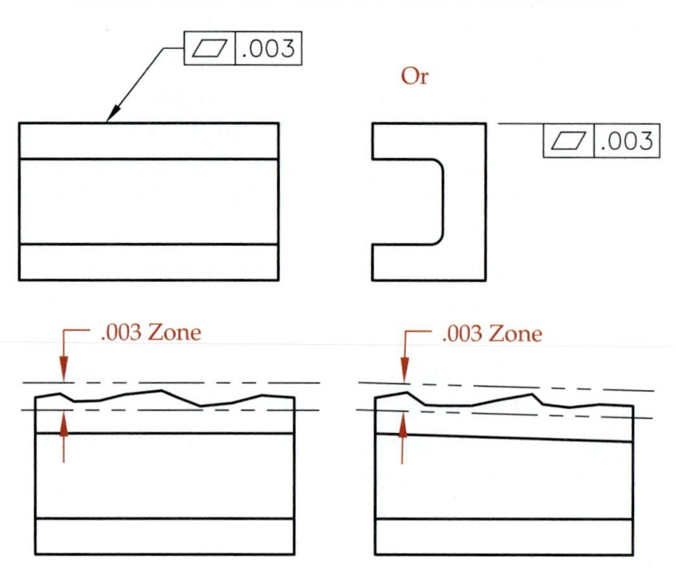

Goodheart-Willcox Publisher

Figure 13-16. Flatness is the condition of a surface having all elements in one plane.

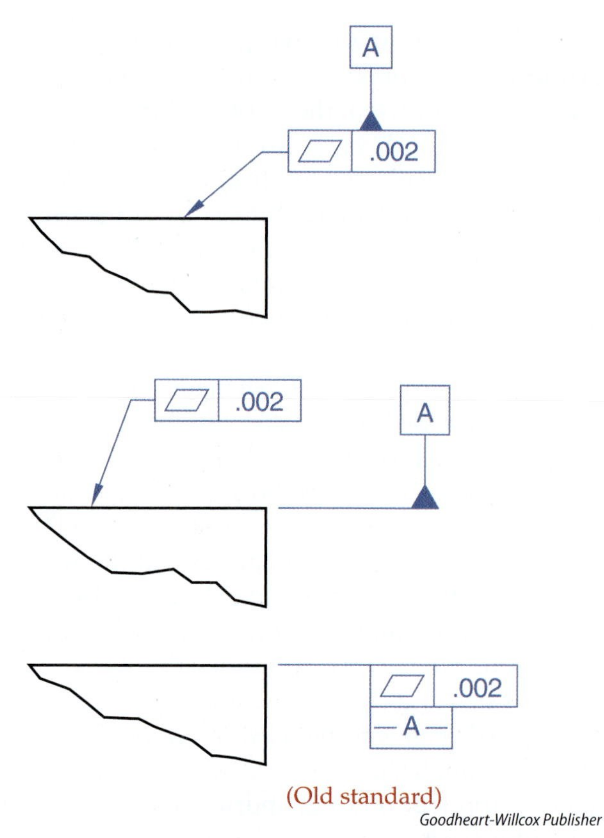

Goodheart-Willcox Publisher

Figure 13-17. A datum feature symbol attached to a surface or feature control frame indicates the feature controlled by the geometric tolerance is also to be used as a datum, assuming it passes inspection.

recommended that an overall flatness is still specified, in which case a double-layer feature control frame is used, and the overall flatness is specified above, with a units per area specified below, **Figure 13-18.**

There may be situations wherein flatness is applied to a surface but an exception to Rule #1 is desired. For median plane flatness the exception is understood, but for an individual surface that is not a feature of size, the standard method for indicating an override of Rule #1 is to use the independency symbol, placing it next to the dimensional value. This extends the outer boundary of the thickness of this part beyond the MMC by the geometric tolerance amount, as shown in **Figure 13-19.**

Circularity or *roundness* for a cylinder or cone is a condition where all points of the surface intersected by any plane perpendicular to a common axis are equidistant from that axis. See **Figure 13-20.** For a sphere, all points of the surface intersected by any plane passing through a common center must be equidistant from that center. Simply put, roundness asks the question, "How round is any given cross section of a round feature, regardless of a datum axis?" Circularity tolerance specifies a radius value for a tolerance zone bounded by two concentric circles within which each circular element of the surface must lie. As with straightness and flatness, circularity control is not required to be specified if the limits of size control the geometric form sufficiently. Only as local measurements indicate the size is less than the MMC will there be potential for the elements to become crooked, and the geometric tolerance ensures roundness is controlled.

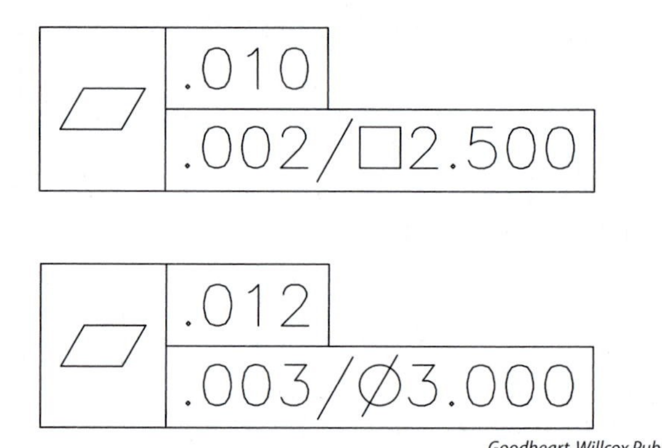

Goodheart-Willcox Publisher

Figure 13-18. Flatness can also be applied in a per unit manner, as shown. An overall flatness is still recommended to prevent unacceptable waviness.

Flatness tolerance - Flatness with independency specified

On orthographic views

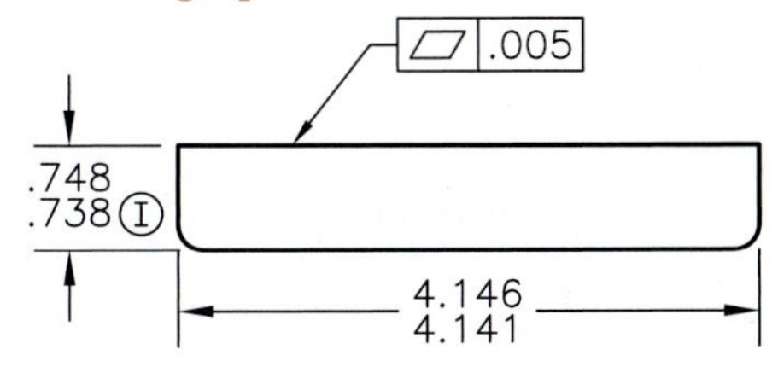

The surface must lie between two parallel planes .005" apart, but perfect form at MMC is not required, as specified by the independency symbol. Each local size shall still be within the size limits.

On the 3D model

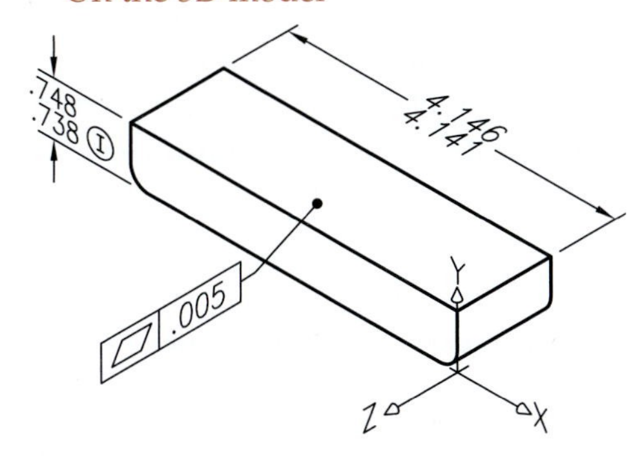

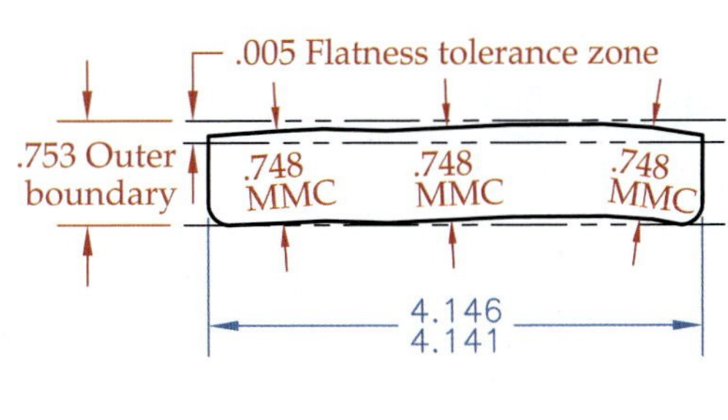

Goodheart-Willcox Publisher

Figure 13-19. For situations wherein flatness control is applied to a surface but perfect form at MMC is not desired, an independency symbol can be applied.

Circularity tolerance - Circularity of surface elements

On orthographic views

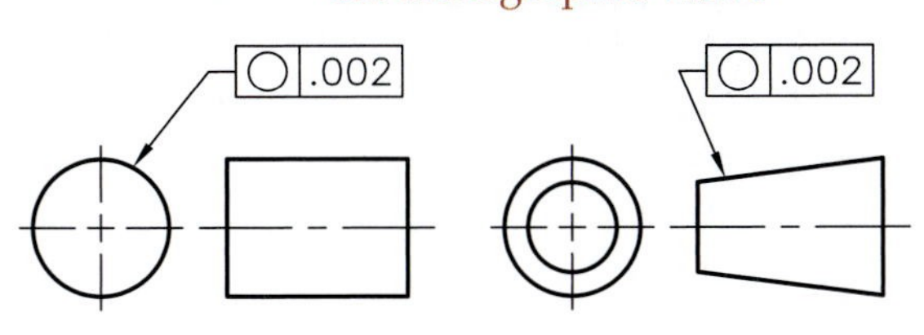

On the 3D model

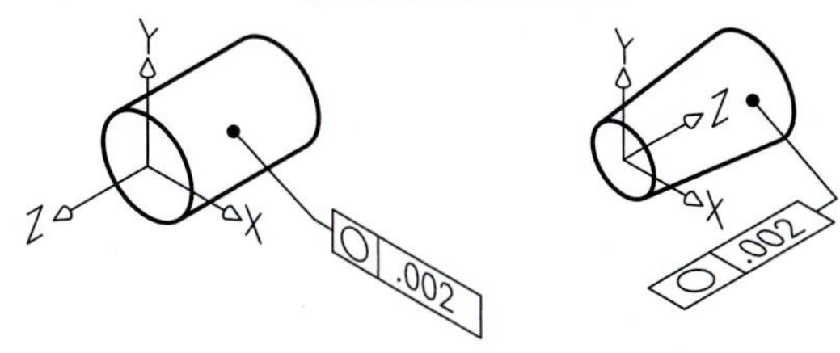

.002 Circularity tolerance zone

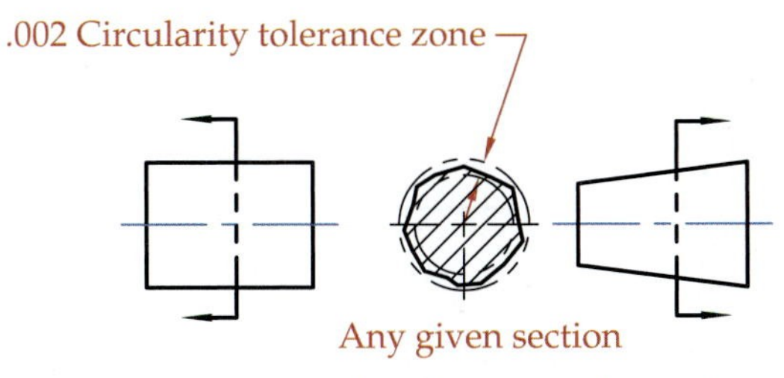

Any given section

Any given circular element on the surface must lie between two concentric circles .002" apart. The feature cannot exceed perfect form at MMC.

Goodheart-Willcox Publisher

Figure 13-20. Circularity or roundness for a cylinder or cone is a condition where all points of the surface intersected by any plane perpendicular to a common axis are equidistant from that axis.

Cylindricity is a condition of a surface of revolution in which all points on the surface are equidistant from a common axis. See **Figure 13-21**. Cylindricity tolerance specifies a radius value for a tolerance zone bounded by two concentric cylinders within which the controlled surface must lie. Simply put, cylindricity asks the question, "How close to a true cylinder is the feature?" Cylindricity combines roundness and straightness but also controls taper. Cylindricity is not easily checked, as there are no datums to be referenced, but precision equipment is available to determine cylindricity error. See **Figure 13-22**.

Orientation Tolerances

Orientation tolerances control angularity, parallelism, and perpendicularity. Since these tolerances control the orientation of features to one another, they always refer the controlled feature to at least one datum. As with form tolerances, orientation tolerances are specified when size tolerances do not adequately control the orientation desired. The designer or engineer who specifies orientation tolerances will also be aware that other geometric tolerance controls, such as runout, profile, and position, may or may not control the orientation sufficiently. For example, a geometric control such as position may have a cylindrical tolerance zone that is generous enough to a degree that perpendicularity error could be unacceptable. In a case like this,

Cylindricity tolerance -
Cylindricity of surface elements

On orthographic views

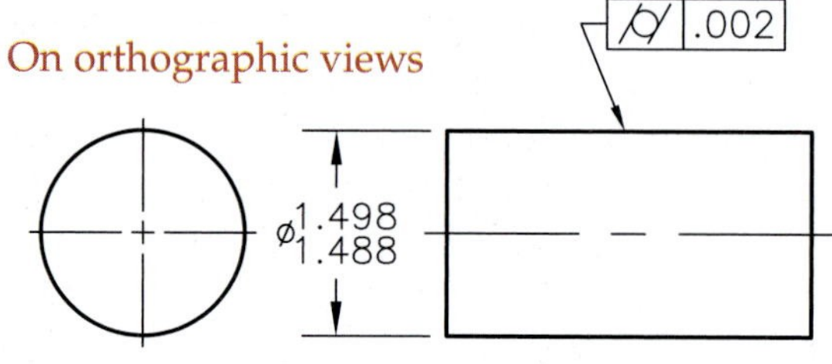

On the 3D model

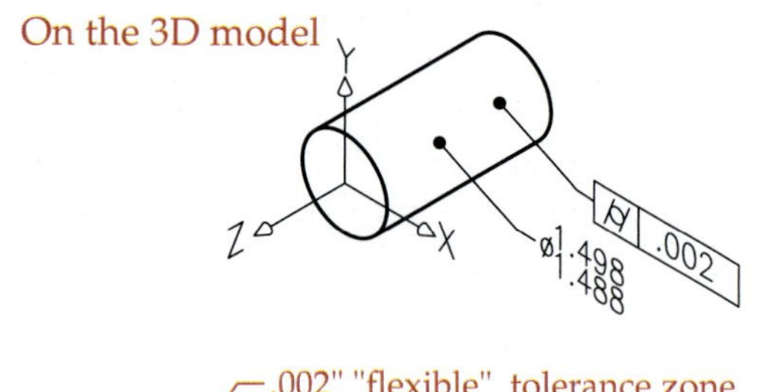

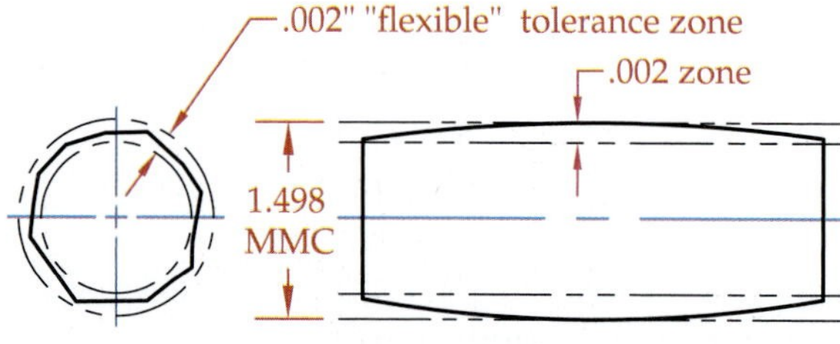

The cylindrical surface must lie between two coaxial cylinders, one .002" larger than the other. The feature cannot exceed perfect form at MMC. This controls straightness, circularity, and taper.

Goodheart-Willcox Publisher

Figure 13-21. Cylindricity is a condition of a surface of revolution in which all points on the surface are equidistant from a common axis.

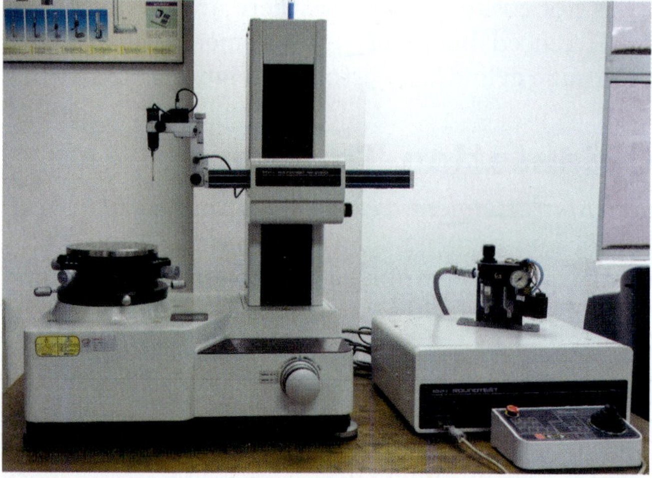

mofaez/Shutterstock.com

Figure 13-22. Cylindricity is not easily checked, as there are no datums to be referenced, but precision equipment is available, such as the device shown here.

a double-layer feature control frame can be used to secondarily control orientation more precisely than by the position control.

Parallelism is the condition of a surface or axis being equidistant at all points from a datum plane or datum axis. A parallelism tolerance zone for a surface feature is defined by two planes or lines parallel to a datum plane or axis within which the line elements of the feature must lie. See **Figure 13-23.** As shown, the size tolerance provides a parallelism control of .010, but as the size of the object varies from MMC, the potential for parallelism error increases. The two parallel planes forming the tolerance zone "float" between the limits of size. Material modifiers

Parallelism tolerance - Parallelism of surface elements

On orthographic views

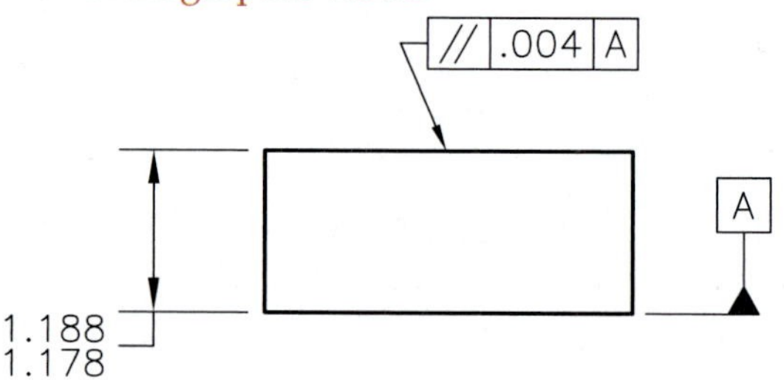

On the 3D model

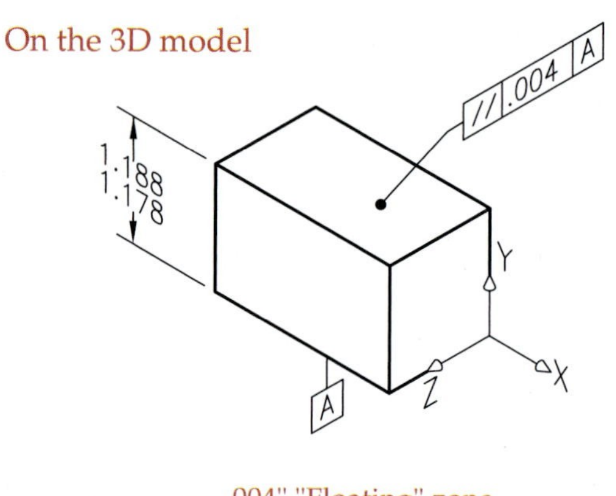

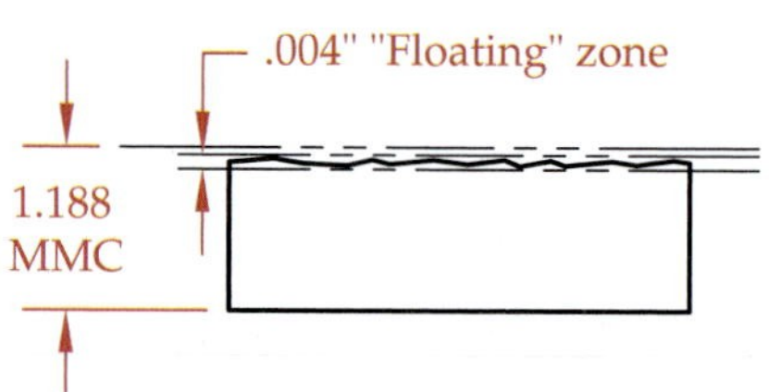

The surface must lie between two planes that are .004" apart and parallel to datum A. The feature cannot exceed perfect form at MMC.

Goodheart-Willcox Publisher

Figure 13-23. Parallelism is the condition of a surface or axis being equidistant at all points from a datum plane.

do not apply because the surface being controlled is not a feature of size.

A parallelism tolerance zone for a cylindrical feature is defined by a cylindrical zone whose axis is parallel to a plane or datum axis. The axis of the feature must lie within this zone, **Figure 13-24**. In the case of a hole, the positional tolerance may be specified and controlled within a datum reference frame that requires two or more datums to be identified. If this position zone is larger than the desired parallelism control, a second feature control frame can be attached to the bottom of the feature control frame, and a parallelism control to one datum is specified, as shown in **Figure 13-24**. Also notice that as parallelism is applied to a feature of size, an MMC or LMC modifier can be applied. If the datum is a feature of size, an MMB or LMB modifier can also be applied to the datum. In the absence of modifier symbols, the tolerance is determined regardless of feature size, and the datum is established regardless of material boundary.

Perpendicularity is the condition of a surface, median plane, or axis existing at a right angle to a datum plane or axis. A perpendicularity tolerance zone for a surface is defined by two parallel planes perpendicular to a datum plane or axis between which the surface or axis of the feature must lie. See **Figure 13-25**. A perpendicularity tolerance zone for

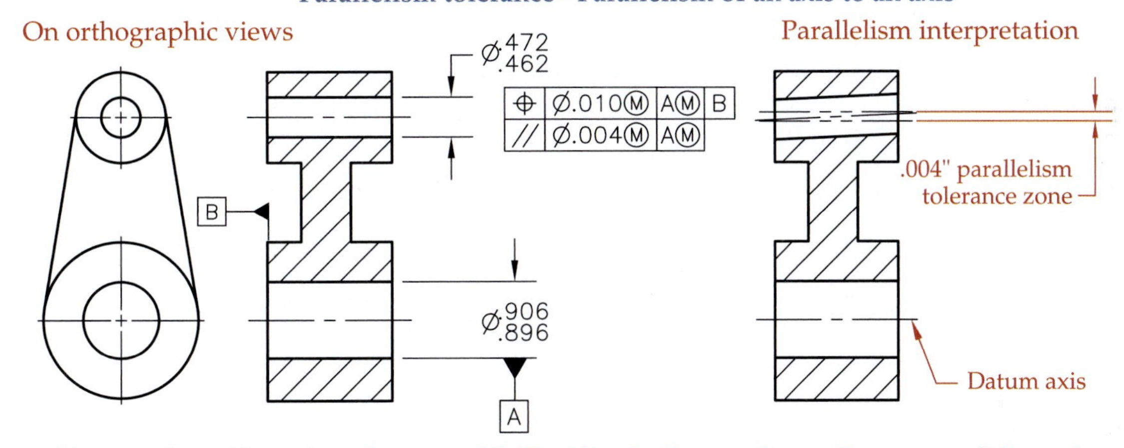

Goodheart-Willcox Publisher

Figure 13-24. A parallelism tolerance zone for a cylindrical feature is defined by a cylindrical tolerance zone whose axis is parallel to a datum plane or datum axis.

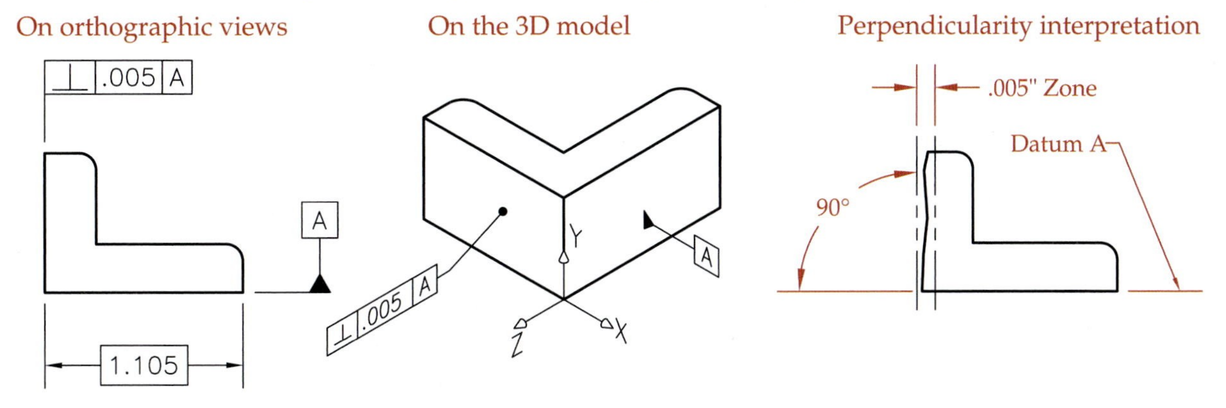

Goodheart-Willcox Publisher

Figure 13-25. Perpendicularity is the condition of a surface, median plane, or axis existing at a right angle to a datum plane or axis.

a cylindrical feature is a cylindrical zone perpendicular to a datum plane within which the axis of the feature must lie. See **Figure 13-26**. A diameter symbol appears in the feature control frame. Like parallelism of an axis, perpendicularity control of an axis is often a refinement of a positional tolerance, and a modifier can be applied.

Angularity is the condition of a surface or axis existing at a specified angle to a datum plane or axis, wherein the angle is not 90°. An angularity tolerance zone for a surface is defined by two parallel planes at the specified basic angle from a datum plane or axis between which the surface or axis of the feature must lie. See **Figure 13-27**. The GD&T angularity control creates a uniform tolerance zone, whereas traditional plus-and-minus degree tolerances always create a fan-shaped zone. Additional applications of angularity for specific situations are covered in the standards but are beyond the scope of this unit.

Profile Tolerances

Profile tolerance is a control that most often describes the amount of tolerance to be maintained for an irregular outline of an object or feature, but it is not limited to that application. The elements of a profile can be straight lines, arcs, and other curved lines, and are defined by basic dimensions or the 3D computer model within the data set. The profile tolerance specifies a uniform boundary along the true profile, within which the elements of the surface must lie. See **Figure 13-28**. The profile can be referenced to a

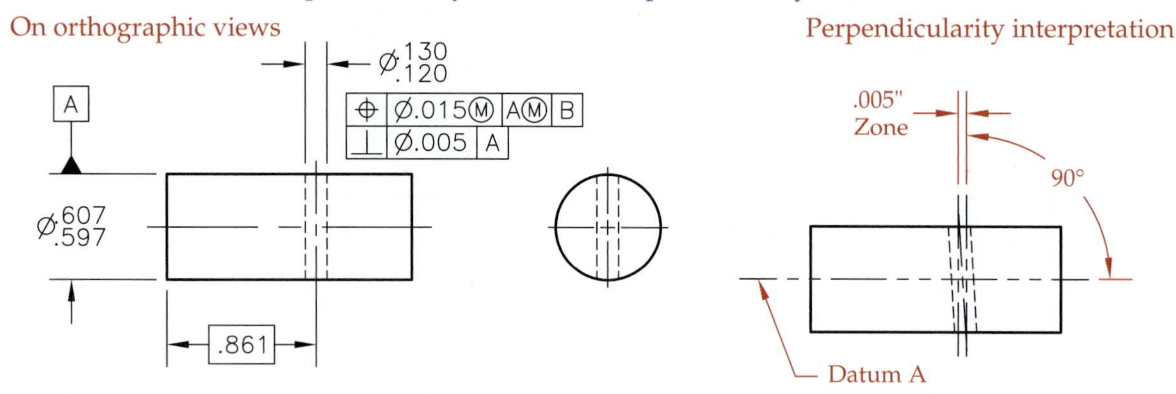

Goodheart-Willcox Publisher

Figure 13-26. A perpendicularity tolerance for a cylindrical feature is a cylindrical tolerance zone perpendicular to a datum plane within which the axis of the feature must lie.

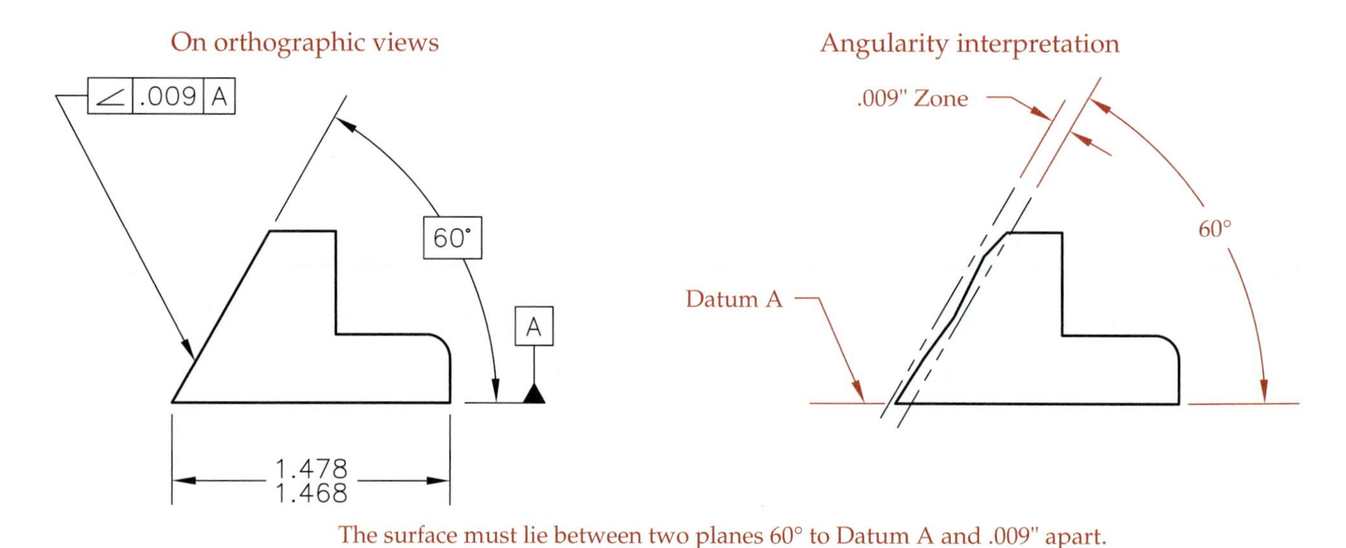

Goodheart-Willcox Publisher

Figure 13-27. Angularity is the condition of a surface or axis existing at a specified angle to a datum plane or axis.

datum in a fashion similar to the parallelism control. In some applications, profile is specified between two points, such as BETWEEN X & Y. The symbol for between is shown in **Figure 13-28.**

Profile is divided into two categories—profile of a line and profile of a surface. See **Figure 13-29.** ***Profile of a line*** is a two-dimensional control of the elements of an irregular surface. It can be thought of in the same way as straightness or roundness, but for shapes defined by basic dimensions. ***Profile of a surface*** is a three-dimensional control that extends throughout the irregular surface, so at least one datum is usually established to help locate the tolerance zone profile or to control the surface element's perpendicularity to a datum.

Both profile tolerances distribute the tolerance zone bilaterally and equilaterally each side of the true profile. If a unilateral or unequal bilateral application is desired, the unequally disposed profile symbol, an encircled U, is placed in the feature control

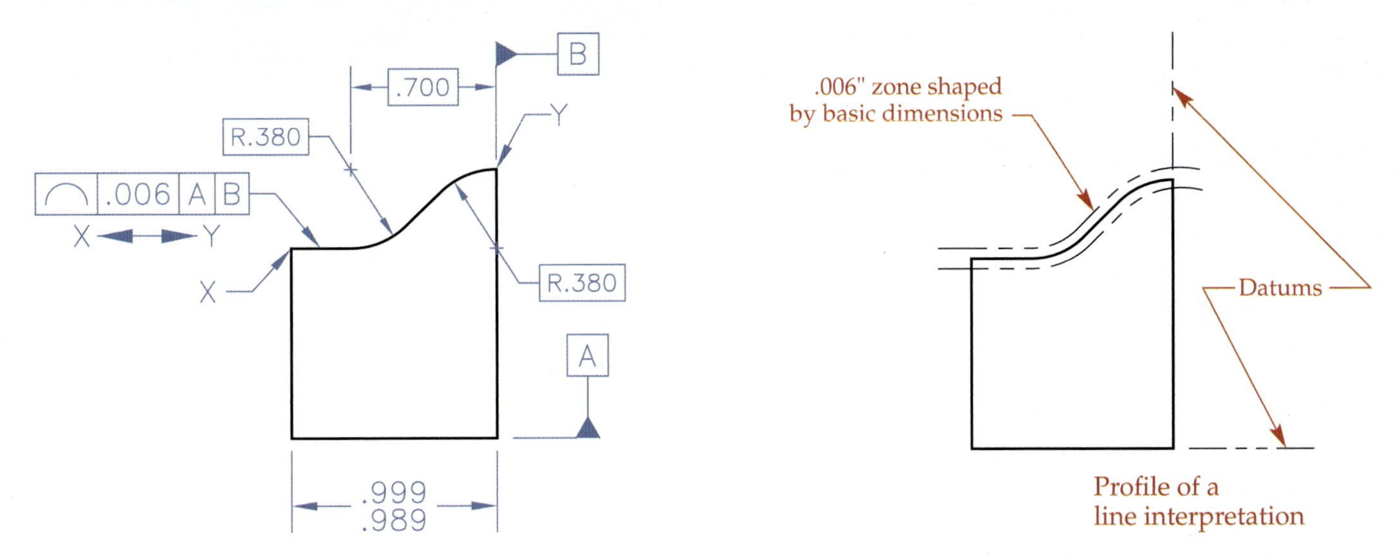

Goodheart-Willcox Publisher

Figure 13-28. A profile of a line tolerance specifies a uniform boundary along the true profile within which the elements of the surface must lie, but each element is individually controlled.

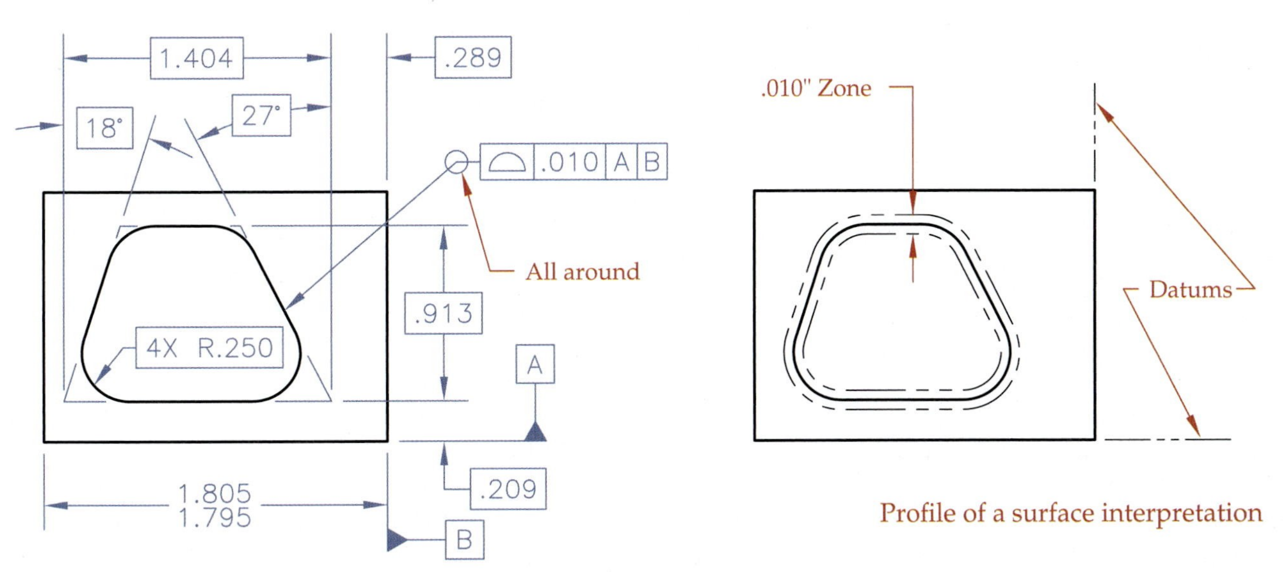

Goodheart-Willcox Publisher

Figure 13-29. A profile of a surface tolerance zone is often defined by basic dimensions and considers all elements of the surface as a whole.

frame following the tolerance amount. As shown in **Figure 13-30**, the value specified after the unequally disposed symbol indicates the offset of the true profile in the direction that adds material. If the value is zero, then the profile tolerance zone is unidirectional within the true profile.

Profile of a surface is also used in some special circumstances, such as (1) controlling the angle of a surface to an axis (other than 90°), (2) controlling the coplanar relationship between two surfaces, (3) controlling a basic offset of one or more parallel planar surfaces to be used as a datum reference, or (4) controlling the trueness of a cone-shaped surface.

Runout Tolerances

Runout tolerance is a composite tolerance used to control the functional relationship of one or more features of a part to a datum axis. This includes those surfaces constructed around a datum axis and those constructed at right angles to a datum axis. The two types of runout control are circular runout and total runout. Runout is often measured with a dial indicator and is measured regardless of feature size. A dial indicator provides a needle reading as the part is rotated. See **Figure 13-31**. ***Full indicator movement (FIM)*** is the total movement of the dial needle in measuring the variance of a surface. FIM has the same meaning as the older terms *full indicator reading (FIR)* and *total indicator reading (TIR)*. Other methods of measuring with lasers, optics, or even noncontact sensors are available, with the same principle—that runout describes surface error measured while the feature is being rotated about a datum axis.

alterfalter/Shutterstock.com

Figure 13-31. Full indicator movement is the total movement of a dial indicator when measuring the variance of a surface as it rotates about a datum axis.

Unequally Disposed Profile Tolerance Zone

.002" lies beyond the true profile boundary, and .008" lies within the true profile boundary

Unequally Disposed Profile Tolerance Zone

Unilateral tolerance with all .015" within the true profile boundary

Goodheart-Willcox Publisher

Figure 13-30. For situations wherein profile control is not to be applied equilaterally and bilaterally, an unequally disposed symbol can be applied within the feature control frame.

Circular runout controls circular elements of a surface. For surfaces constructed around a datum axis, circular runout controls circularity and concentricity with respect to the datum. See **Figure 13-32**. Bonus tolerance can never be specified for runout, as the dial indicator always makes contact with the surface as it rotates about the datum axis, regardless of the feature size.

Total runout provides composite control of all surface elements. See **Figure 13-33**. For surfaces constructed around a datum axis, total runout is used to control cumulative variations of circularity, straightness, coaxiality, angularity, taper, and profile of a surface. It is important to note that the features themselves could be perfectly cylindrical, but if they are not *coaxial*, then runout can occur. It is also important to note that the features could be perfectly coaxial, but if surface irregularities exist, then runout can occur.

A runout tolerance can be applied to a specific portion of a surface by using a chain line drawn adjacent to the surface profile on one side of the datum axis for the desired length, as specified with a basic dimension.

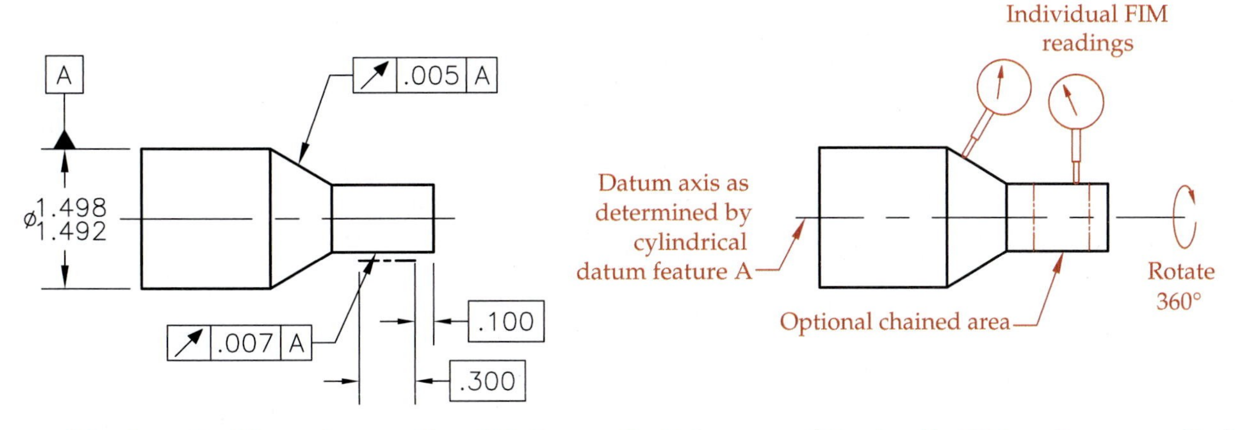

Figure 13-32. Circular runout controls individual circular elements of a surface as the part is rotated about a datum axis. The dial indicator is positioned perpendicular to the surface elements as the surface is rotated.

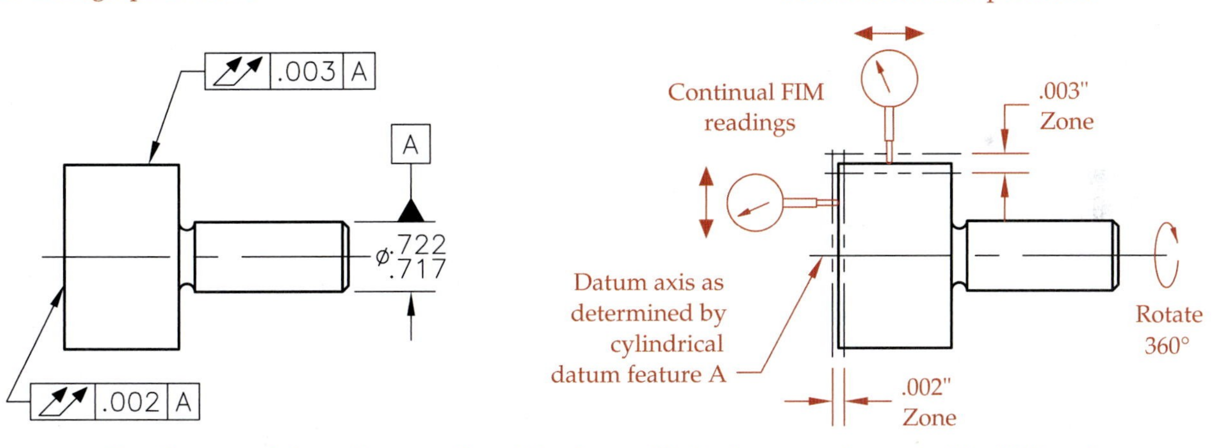

Figure 13-33. Total runout provides composite control of all elements of a surface as the part is rotated about a datum axis. Total runout can be applied to an "end" surface that is perpendicular to the axis.

Tolerances of Location and Position

Throughout the history of GD&T, three geometric control symbols grouped together as tolerances of location were position, symmetry, and concentricity. As GD&T has evolved, there have continued to be adjustments as to which symbol can best be used to control the location of surface points. For a period of time, symmetry was discontinued in ASME standards but continued to be used in ISO standards, but it was later reinstated in ASME for specialized situations wherein true symmetry on an RFS basis was required. Concentricity was also a geometric control used to control coaxiality on an RFS basis when true balance was required. With ASME Y14.5-2018, symmetry and concentricity were discontinued, and the chapter of the standards dedicated to location is now titled *Tolerances of Position*, with the position symbol being the only symbol remaining in that chapter. In summary, there are many applications of the position control, including the center distance between features of size, the location of a group of features as a pattern, coaxial relationships of features of size, and symmetrical relationships of features of size.

The tolerance of position is often referred to as ***positional tolerance***. It defines a zone within which the center axis or center plane of a feature of size is permitted to vary from the theoretically exact position. See **Figure 13-34**. The true position is established by basic dimensions from specified datum features and between interrelated features. The tolerance is three-dimensional in nature, as it is oriented perpendicular to one plane of the datum reference frame. A positional tolerance is specified by the position symbol, a tolerance, and a modifier, and then the appropriate datum reference, all in a feature control frame. The tolerance zone is cylindrical for a hole, or it is the distance between two planes for a slot or tab. This GD&T control alone is one of the most beneficial controls available. Several pages of application examples and explanations are given in ASME Y14.5.

For historical purposes, and for those who are in transition to the newest standards, a brief discussion about symmetry and concentricity is included in this section. ***Symmetry*** was a form of positional tolerance wherein a feature was controlled identically about a center plane of a datum feature, as shown in **Figure 13-35**. In the early years of GD&T, symmetry was used for most center plane applications. However, the position symbol was later adopted for all center plane applications, especially in the case of bonus tolerance applications at MMC. When symmetry returned to the standards, it was intended only for situations applied regardless of feature size wherein true balance was the desired result, regardless of the feature's shape on each side of the center plane.

In similar fashion, concentricity was a control that was used early in GD&T for coaxial applications, but as GD&T evolved, the position symbol was more appropriately applied when MMC bonus tolerance

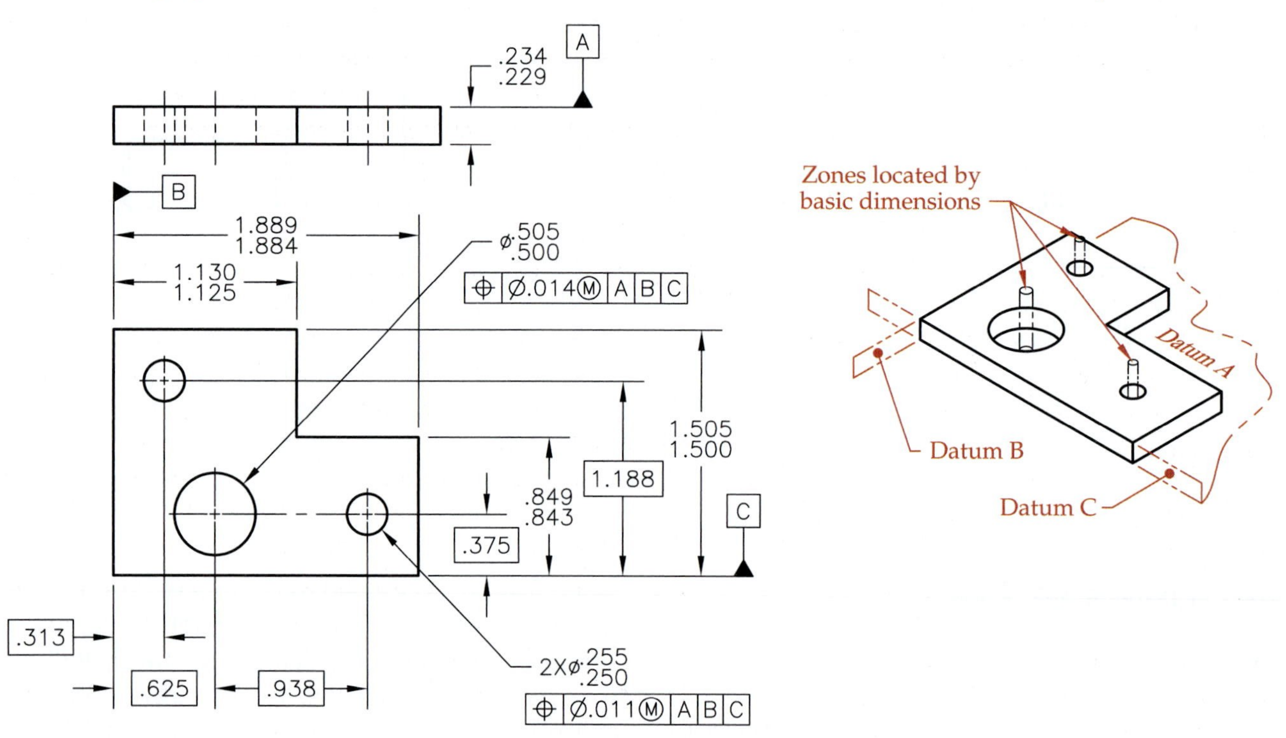

Figure 13-34. A positional tolerance defines a zone within which the center axis or center plane of a feature of size is permitted to vary from the theoretically exact position. In most cases, a bonus tolerance can be applied.

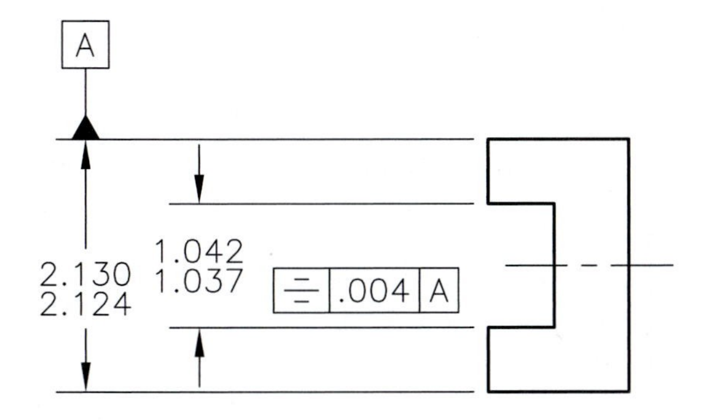

Goodheart-Willcox Publisher

Figure 13-35. In former practice, symmetry was a form of locational tolerance used to control surfaces identically on each side of a center plane established by a datum feature, regardless of feature size.

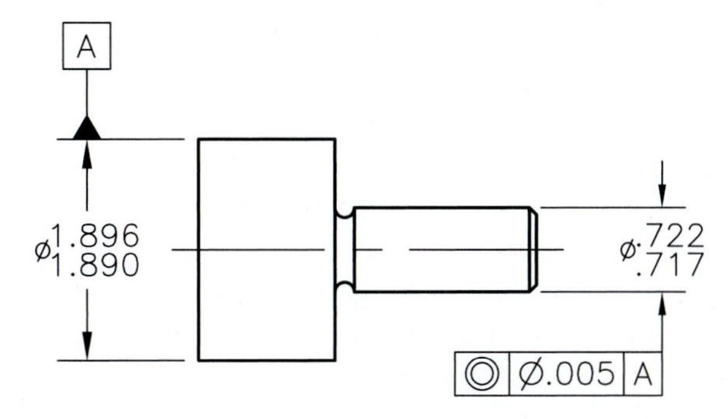

Goodheart-Willcox Publisher

Figure 13-36. In former practice, concentricity was used to control an axis, established by cross-sectional elements of a surface, to be coaxial with the axis of a datum feature and was only to be used regardless of feature size. It examined how well the feature was balanced along the datum axis, regardless of the surface shape.

could be given, and true balance was not a functional requirement. In summary, ***concentricity*** as a geometric control was only used to control the condition wherein axes of all cross-sectional elements of a surface of revolution were to be common to the axis of a datum feature. See **Figure 13-36.** Like symmetry, this tolerance was only to be applied regardless of feature size, and only when needed to control the true balance with respect to the center axis, regardless of the shape of the part. To specify two parts as generally being coaxial, a positional tolerance or runout was usually a more appropriate specification.

The changes to the standards that have been made in recent decades can make it challenging to read older prints. This also points out the importance of studying the standards under which the print was developed, as well as the importance of continuing education in an industrial setting.

Composite Tolerancing

Sometimes a feature control frame is a double-layer frame with a single entry of the geometric characteristic in the first section but two individual layers that specify different tolerance amounts or datum references. See **Figure 13-37.** This is referred to as ***composite tolerancing.*** When this happens for a positional tolerance and a pattern of holes, the top layer indicates the larger positional tolerance for locating the pattern of holes as a group, with respect to the specified datums. This datum framework is known as a pattern-locating tolerance zone framework. The bottom layer applies a smaller tolerance to each individual feature within the pattern, in reference to other members of the pattern. This datum framework is known as a feature-locating tolerance zone framework. For each feature within the pattern, the smaller-diameter zones "float" within the other zones, which prevents each feature within the pattern from drifting too far apart. Notice Datum A is still specified in the lower layer, which establishes the perpendicular orientation for the holes.

Profile of a surface may also be specified in a composite manner. For example, the tolerance zone for the irregular shape of a vent opening on a dashboard may have a very generous tolerance with respect to the datum reference frame, but the surface profile of the shape must be maintained more tightly.

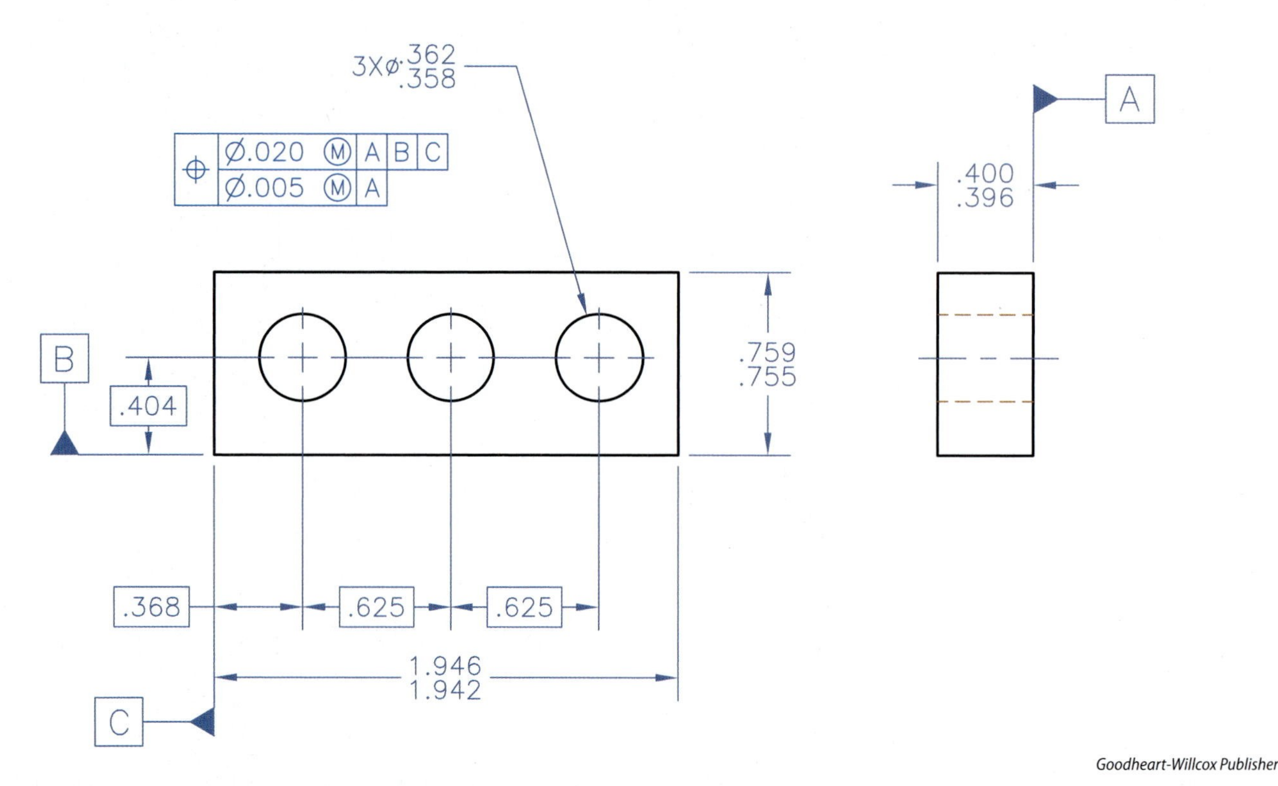

Goodheart-Willcox Publisher

Figure 13-37. The upper row of a composite tolerance feature control frame applies to the location of the pattern of holes with respect to the datum reference frame, while the lower row applies to the individual features with respect to each other within the pattern.

Other Specifications

There are numerous additional GD&T examples and applications that are beyond the scope of this text. In **Figure 13-38**, a few additional GD&T specifications are shown.

In **Figure 13-38A**, a projected tolerance zone is shown. This specification is especially useful for threaded features. It removes the inspection from the threaded part and projects it to an area above the hole that is usually an amount that coincides with the thickness of a mating part, such as the hole in a cover that must correspond and mate with a fastener that will be screwed into the threaded hole.

In **Figure 13-38B**, the profile of a surface symbol is being used to indicate a coplanar relationship between two flat surfaces. In **Figure 13-38C**, the free state symbol is being used with circularity (roundness). This indicates the check should be made with the part in a free, or unconstrained, state. In **Figure 13-38D**, the tangent plane symbol indicates the parallelism control is to be applied with respect to a plane that is tangent with all the high points of the surface. This modification specifies a different checking procedure and is less concerned with overall flatness.

In **Figure 13-38E**, the dynamic profile tolerance modifier is applied in the lower row of a profile composite feature control frame. It allows the profile shape to dynamically increase or decrease within the larger tolerance zone specified in the upper row, as long as the profile remains true to the design measurements within the smaller tolerance zone specified. This focuses on the profile shape more than location.

In **Figure 13-38F**, the datum translation symbol is used in conjunction with datum C to indicate datum C translates, or moves, to establish the datum reference frame. In **Figure 13-38G**, the all over symbol allows the profile tolerance to be applied all over the three-dimensional profile of the model. As an alternative, the term ALL OVER can also be applied under the feature control frame.

In **Figure 13-38H**, the from-to symbol is used to indicate a profile tolerance zone that uniformly decreases from point R to point S, with the range specified within the feature control frame. In **Figure 13-38I**, the statistical tolerance symbol, if applied to the geometric tolerance, is to be placed inside the feature control following the tolerance value. In **Figure 13-38J**, the continuous feature symbol, if applicable, is placed outside the feature control frame rather than within the frame.

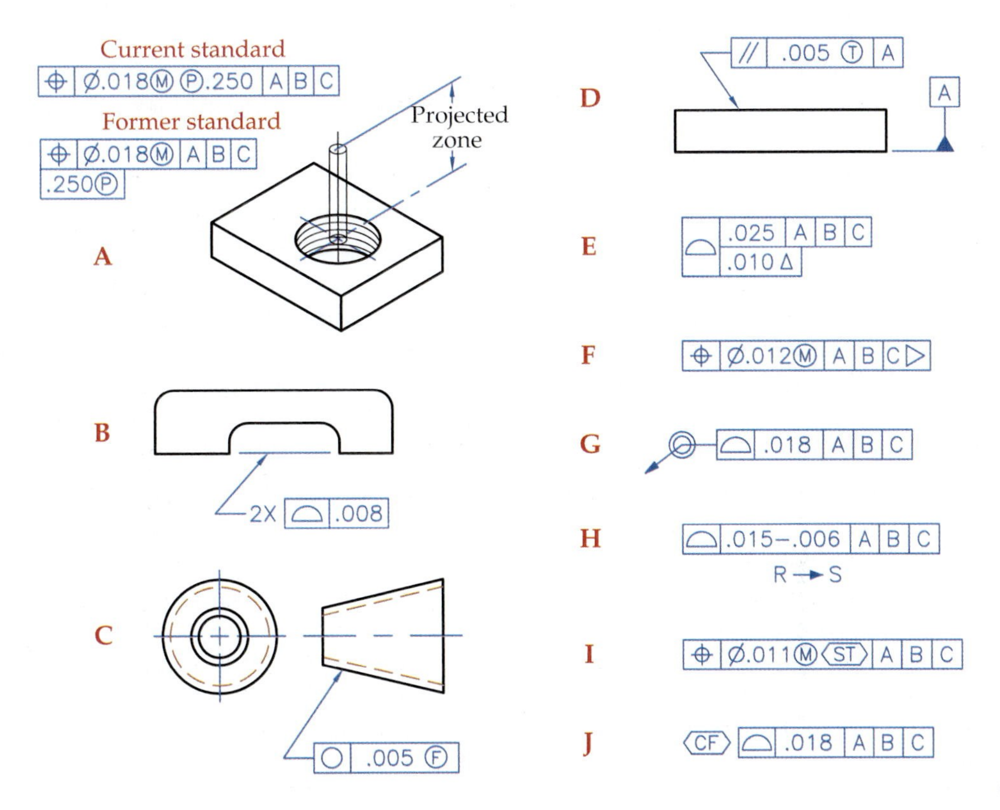

Goodheart-Willcox Publisher

Figure 13-38. Additional GD&T specifications include: (A) a projected tolerance zone, (B) profile of a surface for coplanar relationships, (C) free state symbol, (D) the tangent plane symbol, (E) the dynamic profile symbol, (F) the datum translation symbol, (G) the all over symbol, (H) the from-to symbol, (I) the statistical control symbol, and (J) the continuous feature symbol.

In summary, as geometric dimensioning and tolerancing continues to be used on industrial prints and models, you may need to consult additional resources and continue your education periodically. While it does require a more educated workforce, the goals of GD&T are worthy. They include increased productivity, more parts passing inspection, clarified specifications for the precise conditions under which parts can best function, improved quality, and a finer product.

Summary

- Geometric Dimensioning and Tolerancing (GD&T) is an advanced system of print notation that controls the geometry of mass-produced parts with a system of geometric control symbols, feature control frames, datum establishment symbols, basic dimensions, and modifier symbols.
- The most recent ASME Y14.5-2018 standard, titled *Dimensioning and Tolerancing*, is an extensive and thorough guide to the rules and procedures for implementing GD&T.
- There are currently 12 geometric control symbols that control geometric characteristics such as form, orientation, position, profile, and runout.
- Datums are theoretically exact points, axes, or planes established by real features of the object, as identified on the 3D model or print with datum feature symbols and datum target symbols.
- Feature control frames are rectangular frames that contain a geometric tolerance value and a variety of symbols, including, if applicable, the geometric control characteristic, the zone shape, the tolerance zone size, material condition modifiers, and datum references.
- Basic dimensions in GD&T, wherein values are enclosed in rectangular boxes, are used to establish theoretically exact dimensions for such characteristics as the location of

hole axes, geometric profiles, datum target locations, and inclined surface angularity.
- Tolerances for basic dimensions, depending on the application, may be found within feature control frames or may be determined by quality control inspection methods.
- In addition to the many symbols used in general dimensioning and tolerancing, GD&T methods also require an understanding of several more symbols, some of which are considered modifier symbols.
- Features of size are defined as those that have a center plane or center axis, and these features qualify for a bonus tolerance, which can be specified with an MMC modifier or an LMC modifier.
- Datum features are actual geometric features of the object that establish datum planes and axes, represented most often by precision equipment used in the manufacturing or inspection process.
- Some controls, such as position and surface profile, require that the object be constrained within a three-dimensional datum reference frame comprised of a primary, secondary, and tertiary datum plane.
- The standardized meaning of size, referred to as *Rule #1: The Envelope Principle*, states that an object cannot exceed its perfect form at maximum material condition, which in turn controls geometric form to some degree.
- Material condition modifiers for maximum material condition (MMC, shown as an encircled M) and least material condition (LMC, shown as an encircled L) can be applied to the control of features of size to allow for a bonus tolerance.
- In similar fashion, the encircled M and L symbols can also be used to allow for a flexible datum establishment known as maximum material boundary (MMB) or least material boundary (LMB).
- The four form tolerances for individual features are straightness, flatness, roundness, and cylindricity, none of which reference datums.
- Straightness and flatness have special applications that allow them to be applied on a unit basis or to a feature of size.
- The three orientation tolerances that relate one feature to another are parallelism, perpendicularity, and angularity, each of which must reference another datum.
- When applied to a feature of size, orientation controls are sometimes used to further refine position, which controls orientation to some degree.
- The two profile tolerances are profile of a line (2D) and profile of a surface (3D), both of which have a wide range of uses but primarily were first designed to help control irregular or composite profiles defined with basic dimensions.
- The two runout tolerances are circular runout (2D) and total runout (3D), both of which require a datum axis and are applied regardless of feature size, as the inspection process involves the full indicator movement (FIM) of a dial indicator making contact with a surface while it revolves 360°.
- Positional tolerancing is a geometric control primarily involved with the location of features of size, such as center planes or axes, or patterns of these features, relative to the true position specified.
- Tolerances known as symmetry and concentricity, formerly approved for use as tolerances of location, have been discontinued from the Y14.5 standard.
- Composite tolerancing incorporates a two-layer feature control frame with the same geometric control and features a larger tolerance on the upper row that locates a pattern or profile and a smaller tolerance on the lower row that internally controls the pattern or profile with a smaller tolerance zone that "floats" within the larger tolerance zone.
- Additional specifications for GD&T are numerous and varied and continue to evolve with the field of GD&T.

Unit Review

Name ______________________ Date ______________ Class ______________

Answer the following questions using the information provided in this unit.

Know and Understand

____ 1. *True or False?* The ASME standard for dimensioning and tolerancing that includes GD&T is ASME Y15.4, last revised in 2015.

____ 2. One way to identify that a drawing is using GD&T is the presence of _____ associated with local notes, or around dimension values, or floating next to size dimensions.
A. circles
B. arrows
C. boxes
D. triangles

____ 3. *True or False?* The latest ASME standard for GD&T provides for 12 different geometric control symbols, including straightness and parallelism.

____ 4. *True or False?* A basic dimension means there is no tolerance associated with that dimension.

____ 5. Which of the following has a circular shape?
A. Datum target symbol
B. Feature control frame
C. Basic dimension
D. Datum feature symbol

Match the following descriptions of symbols to the symbol names. Answers are used only once.

____ 6. Dynamic profile
____ 7. Tangent plane
____ 8. Translation datum
____ 9. Unequally disposed
____ 10. Between
____ 11. Free state
____ 12. Maximum material boundary

A. Encircled U
B. Encircled M
C. Encircled T
D. Encircled F
E. Two arrows
F. Triangle pointing right
G. Triangle pointing up

____ 13. *True or False?* Some geometric controls may need the object to be constrained within a 3D datum reference frame.

____ 14. Which of the following is established by *Rule #1: The Envelope Principle*?
A. Size tolerances do not control geometry.
B. A feature of size is best represented by a flat plane surface.
C. Perfect form is required at MMC.
D. Regardless of feature size applies unless MMC or LMC are specified.

____ 15. What is the main reason a material modifier is specified?
A. To allow bonus tolerance
B. To more tightly control the number of parts passing inspection
C. To identify the target size for a local dimension
D. To identify the feature as a feature of size

____ 16. *True or False?* Tolerances of form can reference a datum if they are controlling a center plane or center axis.

____ 17. Which of the following is *not* in the category of orientation control, even though it does control orientation to some extent?
A. Parallelism
B. Perpendicularity
C. Angularity
D. Position

____ 18. With respect to the two profile tolerances, which statement is *false*?
A. Profile of a line is a 2D control, while profile of a surface is a 3D control.
B. Both profile controls can be specified between two points.
C. Profile of a surface must be specified "all over."
D. Both profile controls are assumed to be bilateral and equilateral unless modified.

____ 19. *True or False?* Runout is often checked with a dial indicator that renders an FIM (full indicator movement) value.

____ 20. *True or False?* If a hole is located with a positional tolerance, the tolerance zone is cylindrical and should be specified with a diameter symbol in the feature control frame.

____ 21. A _____ tolerance is specified in a double-layer feature control frame.
A. profile
B. composite
C. position
D. runout

Critical Thinking

1. Give two or more reasons that engineers or designers would use geometric dimensioning and tolerancing.

____

____

____

____

____

____

2. Are there any drawbacks or difficulties that may cause a company to avoid GD&T? List a couple of ideas, but also give your reason that a company should overcome that difficulty.

____

____

____

____

____

____

Apply and Analyze

Name ______________________ Date ______________ Class ______________

Review Activity 13-1

For each feature control frame given below, write out the words that would be used to read the feature control frame.

1. | // | .003 | A |

____

____

2. | ∠ | .012 | B |

____

____

3. | ⌒ | .010 | A | B |

X ◄—► Y

____

____

4. | ⌖ | Ø.009Ⓜ | A | BⓂ |

____

____

5. | ⌭ | .008 |

____

____

6. | ▱ | .0003 |

____

____

7. | ⌰ | .004 | A |

____

____

8. | ⊥ | Ø.005Ⓜ | A |

____

____

9. | — | .003 |

____

____

10. | ⌖ | Ø.020Ⓜ | A | B | C |

| // | Ø.003 | B |

____

____

Apply and Analyze

Name ______________________ **Date** ______________ **Class** ______________

Review Activity 13-2

For each of the written descriptions below, sketch a feature control frame that matches the description.

1.	This feature must be flat within .003".
2.	The cross-sectional shape of this feature must be round within a radial tolerance zone of .002".
3.	FIM reading of this surface should be within .005", measured at any location along the axis, with respect to datum A.
4.	This hole should be located within a cylindrical zone of .005" at MMC with respect to datums A, B, and C.
5.	The axis of this feature should be parallel within .006" to the datum axis, with the datum established RMB.
6.	The profile of this surface should be within a .010" zone with respect to datums A, B, and C, between X and Y.
7.	The surface elements of this cylinder must be straight within .002".
8.	This surface must be perpendicular to datum C within .005".
9.	The runout of this cylindrical surface should be .003" or less, measured totally along the surface, to datum A.
10.	The axis of this post should be located within a cylindrical zone of .015" at MMC, with respect to datums A, B, and C, but the perpendicularity to datum A must also be within .002" at MMC.

Notes

Apply and Analyze

Name ______________________ **Date** ______________ **Class** ______________

Industry Print Exercise 13-1

Refer to the print PR 13-1 and answer the questions below.

1. How many datums are identified? ______________________
2. Before datum A can be used as a datum, what qualification must it meet?

   ____
3. What feature is used to establish datum B?

   ____
4. Why do the coordinate location dimensions have boxes around them?

   ____
5. How many feature control frames specify a bonus tolerance? ______________________
6. The depth (front-to-back distance) of the part has a tolerance specified by the title block as .02″ (±.01″). Nevertheless, the front surface must still be parallel to the back by what amount?

   ____
7. What is the MMC of datum feature B, which has a title block tolerance specification of .010″ (±.005″)?

   ____

Review questions based on previous units:

8. What is the part number of this print? ______________________
9. How many threaded holes are there on this part? ______________________
10. What is the radius of the four corners of this part? ______________________
11. What type of metal should be used for this part?

    ____
12. What is the size tolerance on the four largest counterbore diameters?

    ____
13. What is the overall height, width, and depth of this part?

    ____
14. What type of section view is the right side view?

    ____
15. What type of representation is used to show the threads in the section view?

    ____

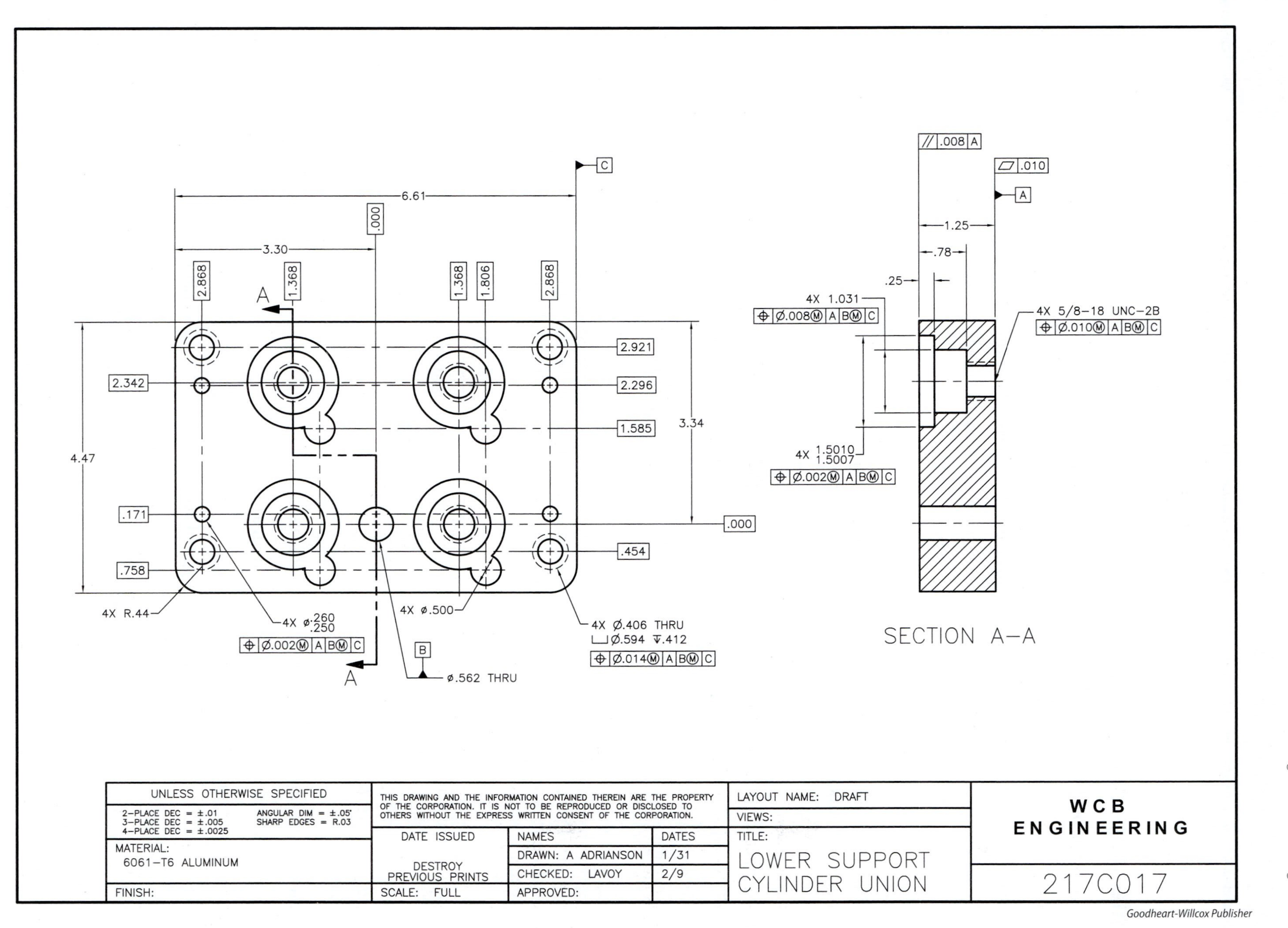

PR 13-1. Cylinder Union Lower Support

Apply and Analyze

Name ______________________ **Date** ______________ **Class** ______________

Industry Print Exercise 13-2

Refer to the print PR 13-2 and answer the questions below.

1. Is datum feature A an axis, center plane, or flat surface? ______________________
2. What is the MMC of the cylindrical surface that establishes datum B? ______________________
3. What qualification is required before datum feature B can be used?

____

____

4. The counterbore diameter (3.754/3.752) is located with a positional tolerance of .002″. How much bonus tolerance could possibly be added to the .002″?

____

5. For that same feature, is datum B referenced at MMB, LMB, or RMB? ______________________
6. For the threaded holes, how tall is the projected tolerance zone specified? ______________________
7. According to the title block, the tolerance for the .88″ depth measurement is ±.010″, as it is a two-place decimal. For the surface opposite datum feature A, how much could that surface be "unparallel" if the parallelism control was not specified?

____

8. For the eight counterbored holes, the positional tolerance for the through hole is the same as the position for the counterbore. How much bonus tolerance could possibly be added on to the .005″ tolerance specified?

____

9. One of the feature control frames has a CF symbol preceding it. What does this mean?

____

Review questions based on previous units:

10. For what size paper was this drawing created? ______________________
11. How many threaded holes are there on this part? ______________________
12. Are the specified chamfers more or less than 1/32″? ______________________
13. What is the pitch of the threads? ______________________
14. Based on the cutting-plane line, what type of section view is the right side view?

____

15. What type of representation is used to show the threads in the section view?

____

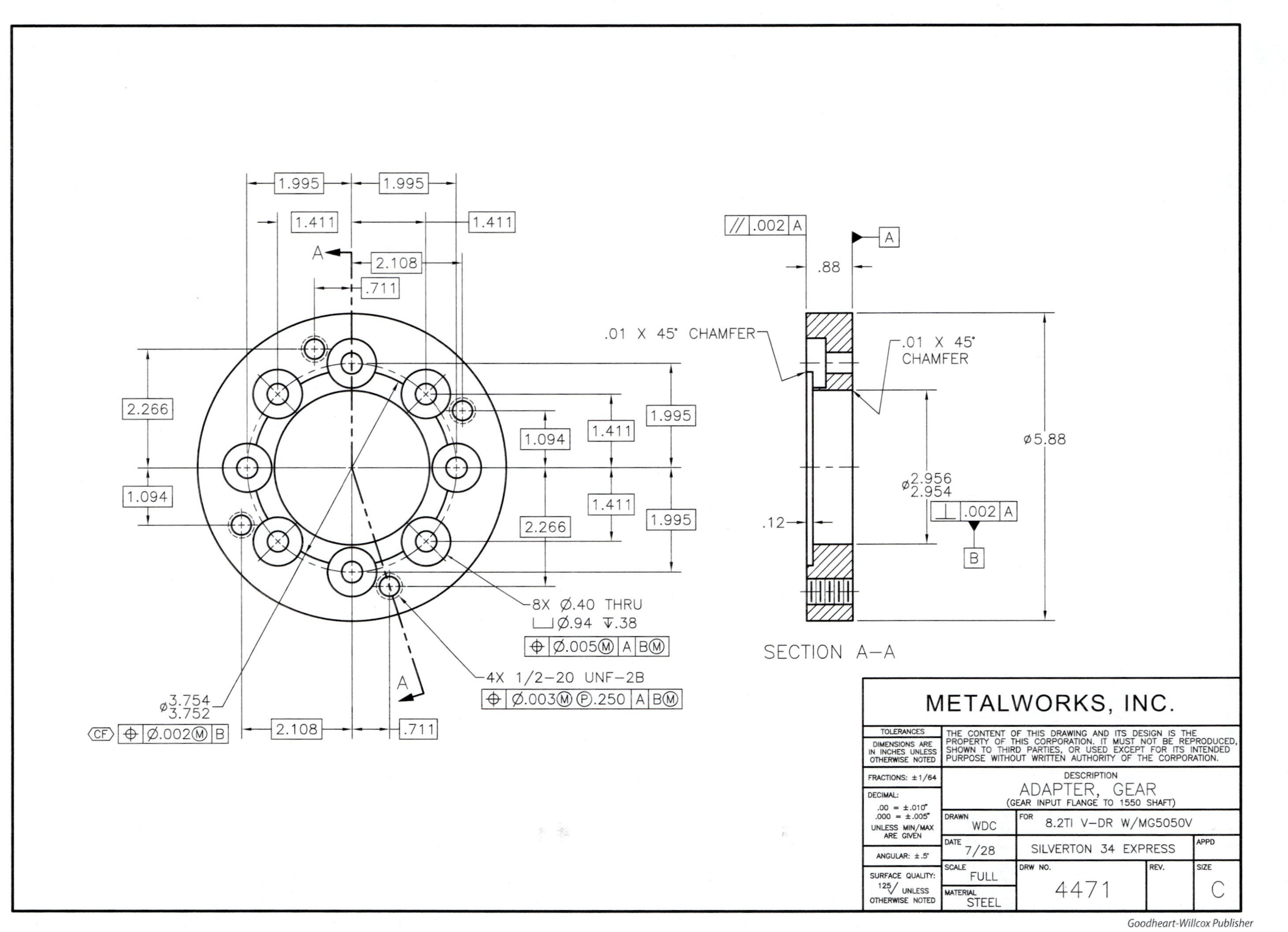

Goodheart-Willcox Publisher

PR 13-2. Gear Adapter

Apply and Analyze

Name ______________ Date ______________ Class ______________

Bonus Print Reading Exercises

The following questions are based on various bonus prints located in the folder that accompanies this textbook. Refer to the print indicated, evaluate the print, and answer the question.

Print AP-001

1. Is the feature that will be used to establish datum B a threaded feature?

____

2. What term applies to how the positional tolerance is to be applied—MMC, RFS, or LMC?

____

Print AP-003

3. Is datum A established as a plane, a center plane, or an axis?

____

4. For the .066 slot, is datum B to be established MMB, RMB, or LMB?

____

Print AP-004

5. How many datums are identified on this assembly drawing?

____

6. What geometric control is used to specify a coplanar relationship between two surfaces that also establish datum C?

____

7. Most of the basic dimensions on this print are linear dimensions for true position. Give another example of a basic dimension on this print that is not a linear dimension.

____

8. What is the mean diameter of the holes that have a positional tolerance of .020 modified at LMC?

____

Print AP-005

9. What geometric tolerance is applied to control the individual form of the larger part of this subassembly?

____

10. What geometric tolerance is applied to the orientation of the two subparts to each other within this subassembly?

____

Print AP-006

11. For this part, is datum A established by a planar surface or by a cylindrical surface?

____

12. Calculate the MMC of the hole that is to be coaxial within a diameter zone of 0.5 mm with respect to datum B.

____

13. What geometric qualification is required for datum D?

____

Print AP-008

14. What additional geometric control is specified for the surfaces with an N5 surface texture specification?

____

15. What geometric control qualifies datum feature A to be acceptable?

____

Print AP-010

16. What geometric control is specified for this assembly?

____

17. What two three-letter acronyms apply to the modifiers found in the feature control frames?

____

Print AP-011

18. List the four geometric tolerancing controls specified within feature control frames on this print.

____

19. What is the MMC of the feature that establishes datum D?

____

Print AP-017

20. What geometric control is applied on this drawing?

____

Notes

UNIT 14
Drawing Revision Systems

LEARNING OBJECTIVES

After completing this unit, you will be able to:

- Describe drawing practices related to drawing revisions.
- Identify revision information on an industrial print.
- Explain the information contained within a revision history block.
- Explain the information found in a revision status of sheets block for multisheet drawings.

TECHNICAL TERMS

all sheets same
by drawing method
by sheet method
drawing deviation (DD)
drawing revision
engineering change order (ECO)
not-to-scale method
now condition
revision authorization document
was condition

As products and parts are designed, the activities that surround the process are documented in a variety of ways. As design ideas are developed, prototypes made, and drawings created, the approval process for the design activity is usually structured in such a way that all parties in the manufacturing enterprise sign off at various stages. In the case of drawings and prints, those with authority to do so add signatures to the prints. These signatures often include initials or names for DRAWN BY, CHECKED BY, and APPROVED BY. Within the CAD system, signatures are issued in electronic form. This unit addresses the system for approving changes to the original design idea of a part or assembly. Standards for revision systems are covered by ASME Y14.35M, titled *Revision of Engineering Drawings and Associated Documents*.

The term ***drawing revision*** refers to any change made to a drawing after final approval has been given to the drawing. This term does not apply to changes made *prior* to final approval of the drawing. Revisions may occur for a variety of reasons. Changes may relate to an improvement in the product's function or reliability, a new method of manufacture, cost-reducing implementations, alterations to quality control standards, or the correction of errors in the original drawing. In some cases, customers may request a change to incorporate a new or improved design into the product. In addition to changing the drawing, a ***revision authorization document*** is usually generated as part of the revision system within the company. Whatever the reason for a revision, an understanding of a company's drawing revision system is required in order to correctly interpret revised prints.

This unit explains some of the drawing practices that are impacted by revisions. Considerable variation exists among industries in how print revisions are processed and recorded. The information presented in this unit will help the print reader develop an understanding of the revision process in general. More importantly, the print reader must understand how changes are indicated on a print, and how those changes impact the current status of the part or assembly. You should take additional steps to become thoroughly familiar with the details of the revision system used by your company.

Revision Process

Revisions made to a drawing vary in the degree of change. Revisions are usually permanent changes in the design or manufacturing process. However, a revision may be a temporary change to accommodate a special situation, with another revision required to reverse the temporary change. With electronic documents and computer systems, the process is quite different than former practice when originals and prints were still output to paper and manually filed.

Revision authorization documents may take on a variety of terms, including ***engineering change order (ECO)***, which will be used in this unit for the sake of clarity. Other names for revision authorization documents include *alteration notice (AN)*, *change in design (CID)*, *engineering notice (EN)*, *engineering change notice (ECN)*, and *advance drawing change notice (ADCN)*. **Figure 14-1** shows an ECO form example similar to what many companies might use. This document could be in the form of a paper

ENGINEERING CHANGE ORDER

Title:	SEAL DIMENSIONS UPDATE	Part:	SEAL DIMENSIONS UPDATE
Change Category:	MFG-PR	Primary Assembly:	MFG-PR
ECO Number	9-21414	Group number:	021414
Request Code	153	Critical Code:	153

Reason for change *(Describe change including both scope and impact below)*

Seal requirements indicate unnecessary cost involved in current storage and size.

Description of change *(Describe solution below)*

Change A - Zone B5 - Note 9 removed. Was : STORAGE AND USE LIMITATIONS PER ATC-STD-001 AND 002

Change B - Zone C4 - 63.060 / 63.000 Was: 63.035 / 63.000

Originator:	Alan Thurston	Originator's Signature:	
Date:	2/17/XX		

APPROVAL

Due Date: 3/17/XX

Requirements:
- Is the ECO filled out completely? ☒ yes ☐ no
- Is the Original Drawing attached? ☒ yes ☐ no
- Is the Updated Drawing attached? ☒ yes ☐ no
- Is the Due Date understood by the team? ☒ yes ☐ no

Execution:
- Modify existing part? ☒
- Create new parts? ☐
- Scrap existing parts? ☐

Name & Dept:		Signature:	Date:
Frank Williams	Mfg Engineering		
Kevin Newton	Pdctn Engineering		
Sarah McLaughlin	Engineering Mgmnt		

Goodheart-Willcox Publisher

Figure 14-1. This engineering change order (ECO) form is just one example of a document that helps track a change through the system.

sheet or an electronic file with protected fields to be filled in by authorized individuals as the document is routed through the system.

In addition to being used for making parts in-house, drawings are documents representing a binding contract between a supplier and the company. Therefore, one person seldom has the right to individually change the drawing once it has been adopted and approved. Most companies periodically have engineering team meetings, and proposals for changes can be brought up for discussion at these meetings. Smaller companies may have individuals who have the authority to initiate a change without a committee meeting.

In some companies, initiating and approving a change to a part, therefore a change to the print, requires the ECO document to be generated first. An ECO document might be a form filled out and registered by a clerk, secretary, or administrative assistant. A number is usually assigned to the ECO, and then a drafter or technician is assigned to make the drawing or computer model changes indicated by the designer or engineer. In the case of CAD files, the revision letter is usually built into the file name. A file should exist for each revision of the part, with the existing CAD files being used until the ECO is approved.

In the case of prints, if a new plot or drawing is created, copies of the new "original" may be circulated to several departments, whether affected by the change or not. The purchasing department may need to know if new materials or different quantities of materials must be ordered. The sales department may need to be notified if the appearance of a part changes, so literature can be updated. Manufacturing and process control may need to be notified if the processes used to make the parts are affected. The product design department needs to be notified in case the current parts are used in new products in development.

Either by paper or electronic mail, the ECO revision document circulates through all appropriate departments. Approval signatures are gathered as needed, either on paper or electronically. While the revision approval is in process, something such as ECO PENDING might be incorporated into current prints issued to personnel from the engineering or print-control group, as shown in **Figure 14-2.** In a situation where the documents are electronic, the company will have a system for flagging parts or prints that are progressing through the revision process.

The paperwork or electronic files being circulated may contain both the old and the new conditions. The old condition is called the ***was condition***. The new condition is called the ***now condition***. Since the design revision may include more than one part—and therefore more than one print—all prints may be circulated with the same ECO document. A sequence number can also be assigned to each element of the revision, and the changes can be described on the drawing in addition to the ECO document.

Noting the Revision

Once the ECO has been prepared, the drawing is ready to be changed, with the changes recorded in the revision block on the drawing. Items of the drawing can be changed by deleting, crossing out, or changing dimensional values and views. Removing information, such as a detail or dimension, should make room for the new items. However, it may be desirable to leave the information on the drawing but crossed through, so that it can still be read while also informing the print reader that information has been replaced or

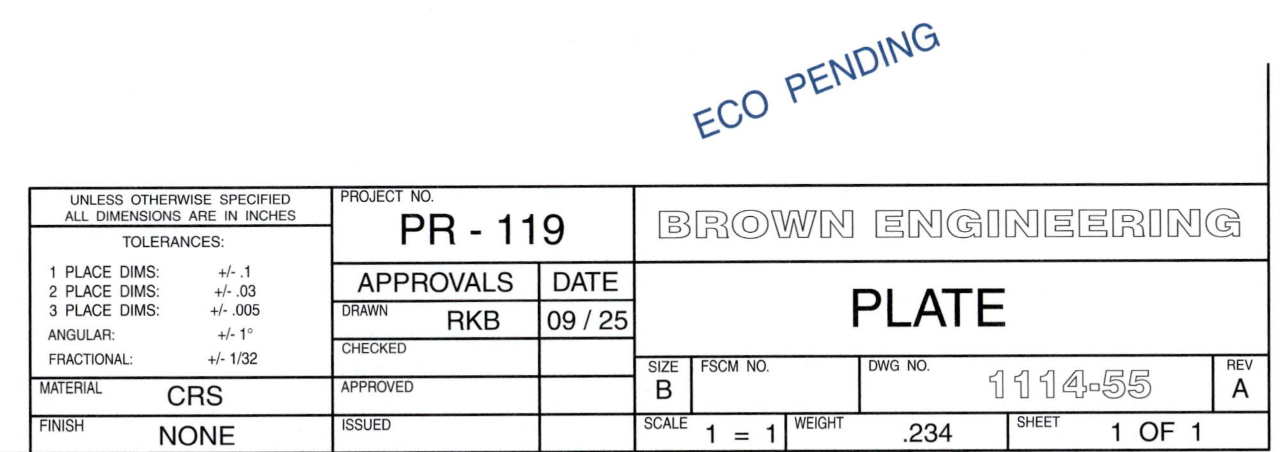

Figure 14-2. When "ECO PENDING" is added near the title block, an ECO is in process for the drawing.

deleted. Standards recommend the replaced information, such as a note, be crossed through with one or two lines or with a hatching pattern.

After changes have been made, revision letters are applied next to the specific areas of the drawing wherein the change occurred. As recommended by standards, uppercase lettering should be used for revisions, beginning with the letter A and progressing through the alphabet. However, the letters I, O, Q, S, X, and Z are not used. These letters may be confused for numbers or drawing elements (for example, the letter "I" looks like a one and the letter "O" looks like a zero). If the revision of a part progresses beyond the letter Y, dual letters AA, AB, AC, etc., are next in sequence.

The current standards recommend the revision letter be enclosed in a circle. Some companies may elect to use other geometric shapes. As shown in **Figure 14-3**, leaders may be used to indicate a specific position or a multitude of positions. This figure also illustrates the use of revision item numbers. If a single revision contains several changes, the changes can be numerically itemized. The item number can then be combined with the revision letter (for example, B1 and B2).

Sometimes, the revision letter may simply float in an empty space where a detail once existed but is now gone. In cases where a drawing has been redrawn without change, the revision letters can be omitted from the revised drawing, although the revision letter of the new drawing does not need to be changed. If revision letters might be confused with other symbols on the drawing, they may also be omitted. In cases where a drawing has been redrawn with change, the revision letters for the current change should be implemented. In both of these cases, former revisions can still be shown and listed. Be aware that each print may have its own unique traits, and many factors are taken into consideration by the person who approves the print and revisions.

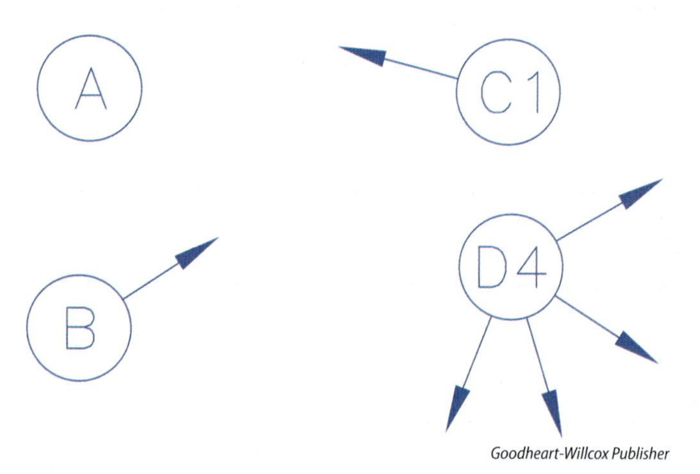

Figure 14-3. Revision letters are usually enclosed within a geometric shape. Leaders can be used for clarity. Multiple changes in one revision can be itemized with numeric values.

Some companies prefer to apply the letter A to the initial release of a drawing, so the print begins with revision A as a new part. It is also permissible to begin the release of a new part or print with a dash as the revision indication. If a drawing contains a dash in the revision box, this indicates the part has never been revised.

Digital data have provided a new set of rules to the revision system. According to ASME standards, revisions of digital data files and copies of digital data files are to be considered redraws, except wherein the document is first converted from a manual drawing to a digital drawing file. It is also acceptable practice to indicate the revision letter within the electronic part file number. For example, the file name 236-2345-B or the file name 236-2345-REV-B could be used to help track the revision letter in the computerized archive system.

Revision History Block

As explained in Unit 3, the *revision history block* should be located in the upper-right corner for most sheet layouts. This is shown in **Figure 3-11** in that unit. The revision history block often includes information such as the zone, or area, in the drawing in which the change is located, the alphanumeric character applied to the change, an itemized description of the change or changes, a reference to the ECO number, the date, and the approval initials as set forth by the company standards. Additional columns can also be added as necessary. For multisheet drawings, there may be a *revision status of sheets block* (first defined in Unit 3) attached near the revision history block.

Figure 14-4 shows a typical revision history block. A horizontal dividing line should appear between revisions. The columns are used in this fashion:

- The first column (A) identifies the zone on the drawing where the change is found. If the revision covers several zones, the zone locations can be listed in the description.
- The second column (B) identifies the revision letter. Nothing else should be entered into this column. The revision may have several changes with several zones or itemized descriptions and perhaps even multiple sheets, but the revision letter is listed once in the revision history block.

REVISION HISTORY				
ZONE	REV	DESCRIPTION	DATE	APPROVED
	A	REDRAWN WITHOUT CHANGE	05/09	DCW
B3	B	SEE ECO 8746	06/10	DCW
C4	C	REMOVED HOLE AND KEYWAY	02/11	RKB
Ⓐ	Ⓑ	Ⓒ	Ⓓ	Ⓔ

Goodheart-Willcox Publisher

Figure 14-4. The revision history block provides a guide to each revision that has been made to the original drawing.

- A description column (C) is used to provide a concise explanation of the change, if possible. For example, when a note is added to the drawing, the type or number of the note can be referred to in the description block, such as ADDED NOTE 1. When a dimension is changed, the old value can be given, such as WAS .999–1.003. If the description is too lengthy, the ECO number can be listed here. Some companies prefer to always list the ECO number and may even have a column for that.
- A date column (D) is for the date the change was made to the drawing.
- A column for the initials of the approving engineer (E) is also recommended.

Figure 14-5 illustrates how the first line of the revision history block may indicate that the drawing was redrawn without change, as in A, or redrawn with change, as in B. The revision history block may also include information when a drawing supersedes or is superseded by another drawing. The drawing that supersedes, or replaces, another drawing may have a note such as REPLACES WITH CHANGE DRAWING 236-2345-REV B in the description area. An additional description of how this new drawing is different may also be included. If the description is too lengthy or cumbersome to place on the drawing, a reference may be made to the ECO number instead. The ECO then describes the differences.

REVISION HISTORY				
ZONE	REV	DESCRIPTION	DATE	APPROVED
	C	REDRAWN WITHOUT CHANGE (ECO 24359)	05/09	DCW

A

REVISION HISTORY				
ZONE	REV	DESCRIPTION	DATE	APPROVED
F4	A	ADDED VENDOR SPECIFICATION – NOTE 4	01/08	RKB
D3	B	REDRAWN WITH CHANGE (ECO 24360)	05/09	DCW

B

Goodheart-Willcox Publisher

Figure 14-5. A redraw should always be indicated in the revision history block. A—The drawing was redrawn without change. B—The drawing was redrawn with change.

With CAD systems, most drawings have associative dimensions that are automatically maintained by the software. These dimensions are always real size within the CAD database and, if a drawing is properly plotted, the views will always be to scale. In the case of manually created drawings, it is also desirable that the views of a drawing always be created to scale, with the scale specified on the drawing. However, there are occasions when a change to the views of the drawing is too extensive and the drawing or views can be revised by simply changing the dimension value. In these cases, if the views are still clear, the ***not-to-scale method*** is used to indicate that the dimension values as shown are *not* to the drawing scale stated. In this method, a heavy line is drawn below the dimension value of the revised feature. This indicates the dimension is changed but the feature is not redrawn to scale. See **Figure 14-6.**

In summary, the revision authorization document should contain a complete record of the changes that have been made to the original drawing. Throughout the life of a drawing and as it is redrawn, some of the revision history is lost from the original. The ECO files should be maintained throughout the life of a part to assist in tracking down older dimensions or other design features that may be associated with old inventory.

Multisheet Revision History

If a part has multiple sheets, the revision history can become more complex. Standards set forth guidelines for three approaches to multisheet revision history: by drawing, by sheet, or all sheets the same. In the ***by drawing method***, the revision history block is on sheet 1 and applies to the whole drawing set. The revision history block on sheet 1 can reference all sheets, as illustrated in **Figure 14-7**, either separated with subheadings, as shown in A, or with zone markings that indicate page, as shown in B.

In the ***by sheet method***, each sheet is treated independently. A revision status of sheets block is added to the left of the revision history block. Each sheet also has its own revision history block. For example, sheet 1 may be identified as revision B, sheet 2 may be identified as revision A, and sheet 3 may be identified with a dash to indicate no revisions have ever been made to that sheet. If the only sheet that changes is sheet 1 and a sheet 4 is added, sheet 1 becomes revision C, sheet 2 remains revision A, sheet 3 remains a dash, and sheet 4 begins with a dash in the revision block. **Figure 14-8** illustrates the STATUS OF SHEETS table for a WAS and NOW of a multisheet drawing.

A slightly different method is referred to as the ***all sheets same*** approach. When this approach is used, a revision will cause all sheets that have any change to be changed to the same letter. **Figure 14-9** illustrates a WAS and NOW for a multisheet drawing that had four sheets until a revision added a fifth sheet and changes on sheets 1, 2, and 4. As a result of the all sheets same approach, sheets 1, 2, 4, and 5 all became revision C, and sheet 3, which was unchanged, remained at revision A status.

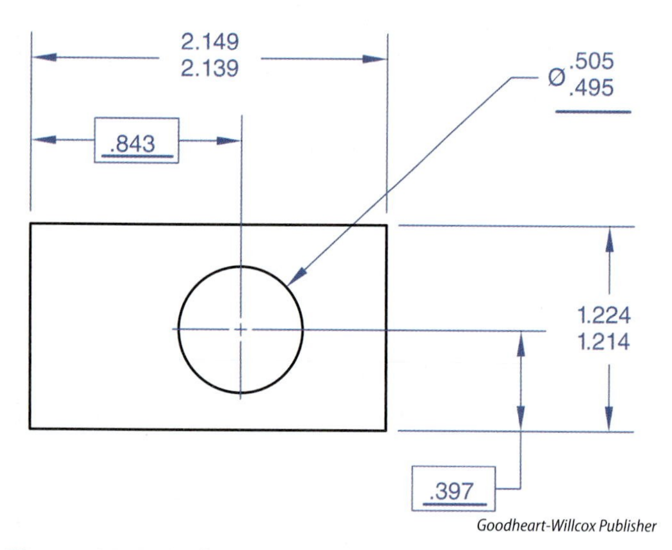

Figure 14-6. In the not-to-scale method, a heavy line drawn below a dimension value indicates the value has been changed but the feature was not redrawn to scale.

Other Revision System Features

Although not expressly covered by ASME standards, there are many variations that have been used for revisions throughout the years. As a member of the manufacturing enterprise, you need to be aware of standards that are unique to the company or perhaps used by suppliers and subcontractors. The next sections cover a few examples of additional revision system features.

Revision History Block Columns

Other columns may be included in a revision history block to clarify the revision history of a drawing. Sometimes this is to further document the changes made in the original drawing. Some other listings that may appear include:

- An authority or ECO column that can be used to record the engineering change order number.

REVISION HISTORY				
ZONE	REV	DESCRIPTION	DATE	APPROVED
	A	SHEET 1	01/08	RKB
A7		CHANGED .500 TO .510		
D3		ADDED DATUM D		
		SHEET 2		
C4		CHANGED .495 TO .490		
C4		ADDED WELDING NOTE		
		SHEET 3		
		NO CHANGE		
	B		04/09	RKB

A

REVISION HISTORY				
ZONE	REV	DESCRIPTION	DATE	APPROVED
	A		01/08	RKB
A7–1		CHANGED .500 TO .510		
D3–1		ADDED DATUM D		
C4–2		CHANGED .495 TO .490		
C4–2		ADDED WELDING NOTE		
	B		04/09	RKB

B

Goodheart-Willcox Publisher

Figure 14-7. For a multisheet drawing, the revision history block on sheet 1 can be used to note the revision status for the whole drawing set.

STATUS OF SHEETS

WAS

REV STATUS	
SH	REV
1	B
2	A
3	–

NOW

REV STATUS	
SH	REV
1	C
2	A
3	–
4	–

Goodheart-Willcox Publisher

Figure 14-8. For a multisheet drawing, a revision status of sheets block is added to the left of the revision history block. The revision status of sheets block treats all sheets independently. Each sheet progresses to the next letter if something on that sheet is changed.

- An "effective on" column that can be used to give the serial number or ship number of the machine, assembly, or part on which the change becomes effective. Sometimes this is a separate block. The change may also be indicated as effective on a certain date.

Drawing Deviation

Another possible term used by some companies is ***drawing deviation (DD)***. This is basically a formal description of a temporary revision that authorizes a deviation from the print. The deviation is allowed only as described on a DD sheet or form. These DDs can be issued for revisions such as:

- A temporary change in the standard part
- A temporary substitution in the method of manufacture
- A design change that requires rework of existing units

SHEET 1 WAS

REV STATUS		REVISION HISTORY				
SH	REV	ZONE	REV	DESCRIPTION	DATE	APPROVED
1	B	D3	B	ADDED FLATNESS TOLERANCE (ECO 29037)	11/10	DCW
2	A					
3	A					
4	–					

SHEET 1 NOW

REV STATUS		REVISION HISTORY				
SH	REV	ZONE	REV	DESCRIPTION	DATE	APPROVED
1	C	D3	B	ADDED FLATNESS TOLERANCE (ECO 29037)	11/10	DCW
2	C	E4	C	ADDED SURFACE TEXTURE (ECO 29124)	02/12	DCW
3	A					
4	C					
5	C					

Goodheart-Willcox Publisher

Figure 14-9. This revision status of sheets block uses the ***all sheets same*** approach. Each sheet is independently revised, but when a revision affects a sheet, the same letter is used for all other sheets affected by the same revision.

A DD is prepared on the same type of sheet as an ECO, but a colored sheet may be used to call attention to the fact that the revision is temporary. A sequence number is assigned to the DD and may be recorded in the revision history. The DD remains in effect until altered or voided by another DD or a revision in the drawing.

Summary

- Changes to a part or a product may occur for a variety of reasons, including quality improvement, error correction, or new manufacturing methods, resulting in a change to the prints that document the shape and specifications for the parts.
- Companies have a procedure and revision authorization plan that includes documents, tracking methods, and standard practices for indicating revisions on the print.
- An Engineering Change Order (or similar term) will document the *was condition* and the *now condition* of a part, communicate the revisions to all parties impacted by the change, and document the revisions for historical purposes.
- On the print, revisions are noted by letters of the alphabet placed near individual view, dimension, or note changes, and an entry in the revision history block documents the change or group of changes with reference to the ECO document number.
- Revision history blocks, usually located in one of the corners of the title block, will contain information such as revision letter, drawing zone, description, date, and signatures.
- Revisions for parts involving multiple sheets can be more involved and are recorded in one of three ways: by drawing, by sheet, or all sheets the same.
- Drawing deviations (DDs) are a systematic way of creating a print for alternate parts with differences or revisions for a period of time before reverting back to the original print.

Unit Review

Name ______________________ **Date** ______________ **Class** ______________

Answer the following questions using the information provided in this unit.

Know and Understand

____ 1. *True or False?* There are no ASME standards that companies can refer to for implementing revisions to drawings.

____ 2. Which of the following is *least* likely to be the reason why a print would undergo a revision process?
A. An improvement in the part's function is made by resizing one dimension.
B. The engineer who designed and approved the part has left the company.
C. The cost of inspection is reduced by increasing the tolerance on a noncritical dimension.
D. A review of the geometric position tolerance shows that bonus tolerance can be implemented in a situation wherein it was not.

____ 3. *True or False?* All companies refer to their drawing revisions as engineering change orders (ECO).

____ 4. *True or False?* An ECO form, whether a paper document or electronic document, helps track and document change through the system.

____ 5. *True or False?* When dimensions or notes are changed on the drawing, the former information will no longer be present once the revision process is complete.

____ 6. Some letters, such as I, can be confused with a numeral, so they are not recommended for use as revision letters. Which of the following letters of the alphabet is also *not* recommended for use as a revision letter?
A. K
B. M
C. E
D. O

____ 7. *True or False?* The first release of a drawing is often identified as revision A.

____ 8. Which of the following is *least* likely to be found in a revision history block?
A. Scale
B. Drawing zone
C. Date
D. ECO number

____ 9. Which of the following is *not* an accepted method for maintaining revisions of multisheet drawings?
A. Revisions are not shown on any sheets but are kept in separate files.
B. One revision letter is used for all sheets, and one revision history block is found on sheet 1.
C. Each sheet maintains its own revision letter and revision block, and each sheet is totally independent.
D. Each sheet maintains its own revision letter and revision block, but all sheets that are affected by a revision will then feature the same revision letter as a result of the revision.

____ 10. *True or False?* A temporary revision that authorizes a change for reworking some existing units is often called an Engineering Change Notice (ECN).

Critical Thinking

1. List three departments within a company setting that may need to be informed of revisions and changes, and give a brief rationale for your choices.

2. Give a rationale for each of the following to be included in the revision history block: drawing zone(s), ECO number, description, date, and approved by.

Notes

Apply and Analyze

Name ______________________ **Date** ______________ **Class** ______________

Industry Print Exercise 14-1

Refer to the print PR 14-1 and answer the questions below.

1. How many columns are featured in the revision history block of this drawing?

   ____

2. What is the current revision letter assigned to this part? ______________________________

3. What revision specified the surface texture of the tapered portion?

   ____

4. What are the initials of the person who implemented or approved ECO 68-557?

   ____

5. Which revision specified the vendor part number of the bushing?

   ____

6. Revision F changed the maximum axial deflection to .500 at 2000 pounds. What had it been before this revision?

   ____

7. Instead of revisions D, E, and F being indicated by three letters, what other method of marking could have been implemented for these three changes?

   ____

Review questions based on previous units:

8. What is the thread form specified for the internal threaded holes?

   ____

9. What method was used to show external threads on this part?

   ____

10. Is this a drawing of multiple parts assembled? ______________________________

11. Does this drawing feature any cutting-plane lines? ______________________________

12. What is the radius of the undercut indicated by a local note? ______________________________

13. What is the maximum material condition diameter of the largest cylindrical feature of this part?

    ____

14. How deep is the blind drilled hole?

    ____

15. What dimensional value is given as the rubber wall thickness?

    ____

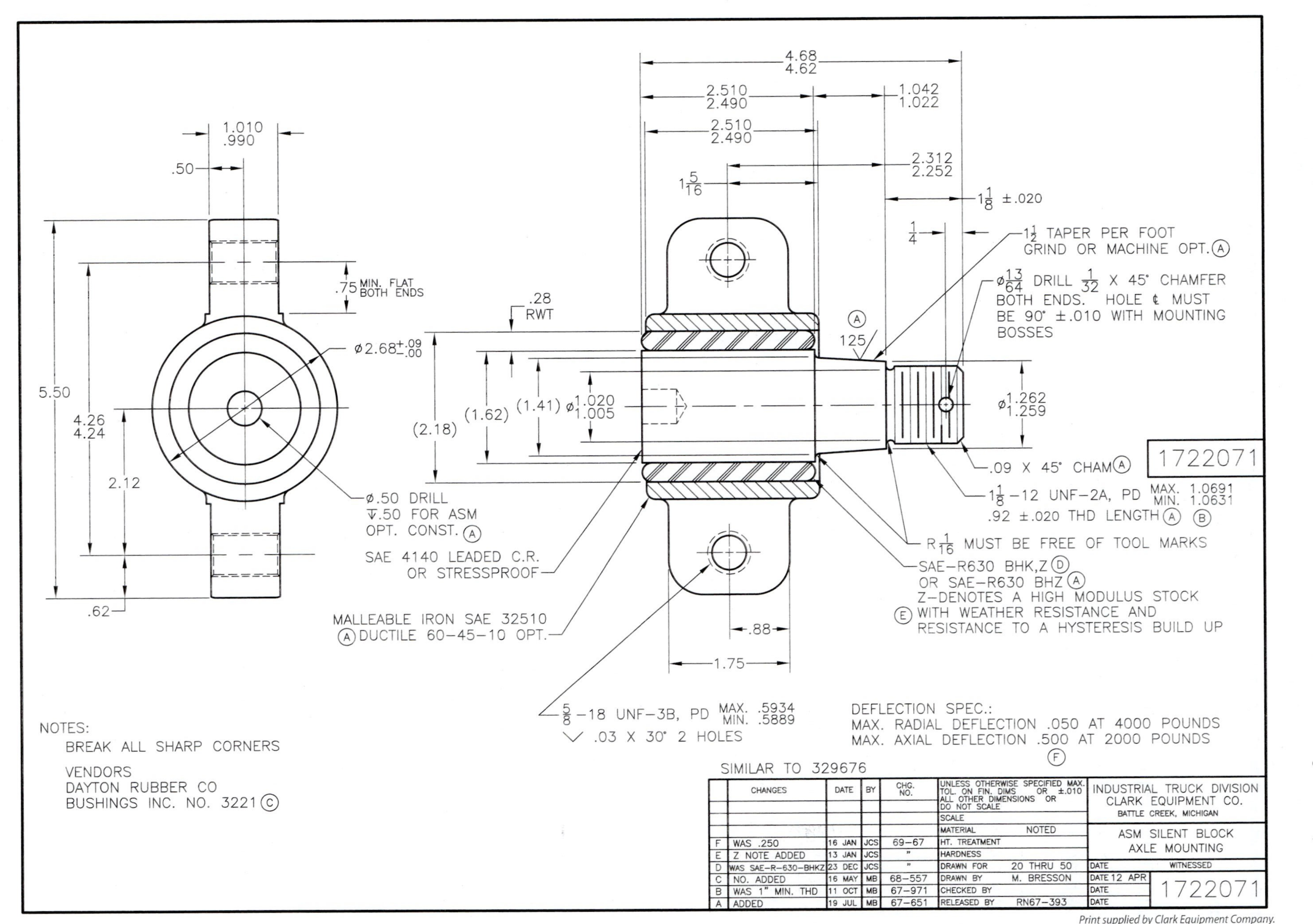

Print supplied by Clark Equipment Company.

PR 14-1. ASM Silent Block Axle Mounting

Apply and Analyze

Name ______________________ **Date** ____________ **Class** ____________

Industry Print Exercise 14-2

Refer to the print PR 14-2 and answer the questions below.

1. What information is given in the three columns of the revision history block of this drawing?

____

2. What revision letter/item number is assigned to the addition of datum W?

____

3. Revision A9 is floating next to an empty space where _____ used to be.

____

4. Revision A5 added a triangle and a feature control frame. What does the triangle mean?

____

5. What revision letter/item number is assigned to the material specification?

____

6. In what zone is revision A2 located?

____

7. What type of pin is referred to in revision A1?

____

8. What change was made in revision A8?

____

Review questions based on previous units:

9. What is the name of this part? ____
10. How many general notes are shown on this print?

____

11. If not specified, what is the tolerance for a three-place decimal value? ____________
12. Are there any surface texture symbols shown? ______________________
13. What is the maximum material condition diameter of the main central hole?

____

14. How much larger is the enlarged view than the principal views?

____

15. What four geometric tolerances are invoked on this drawing?

____

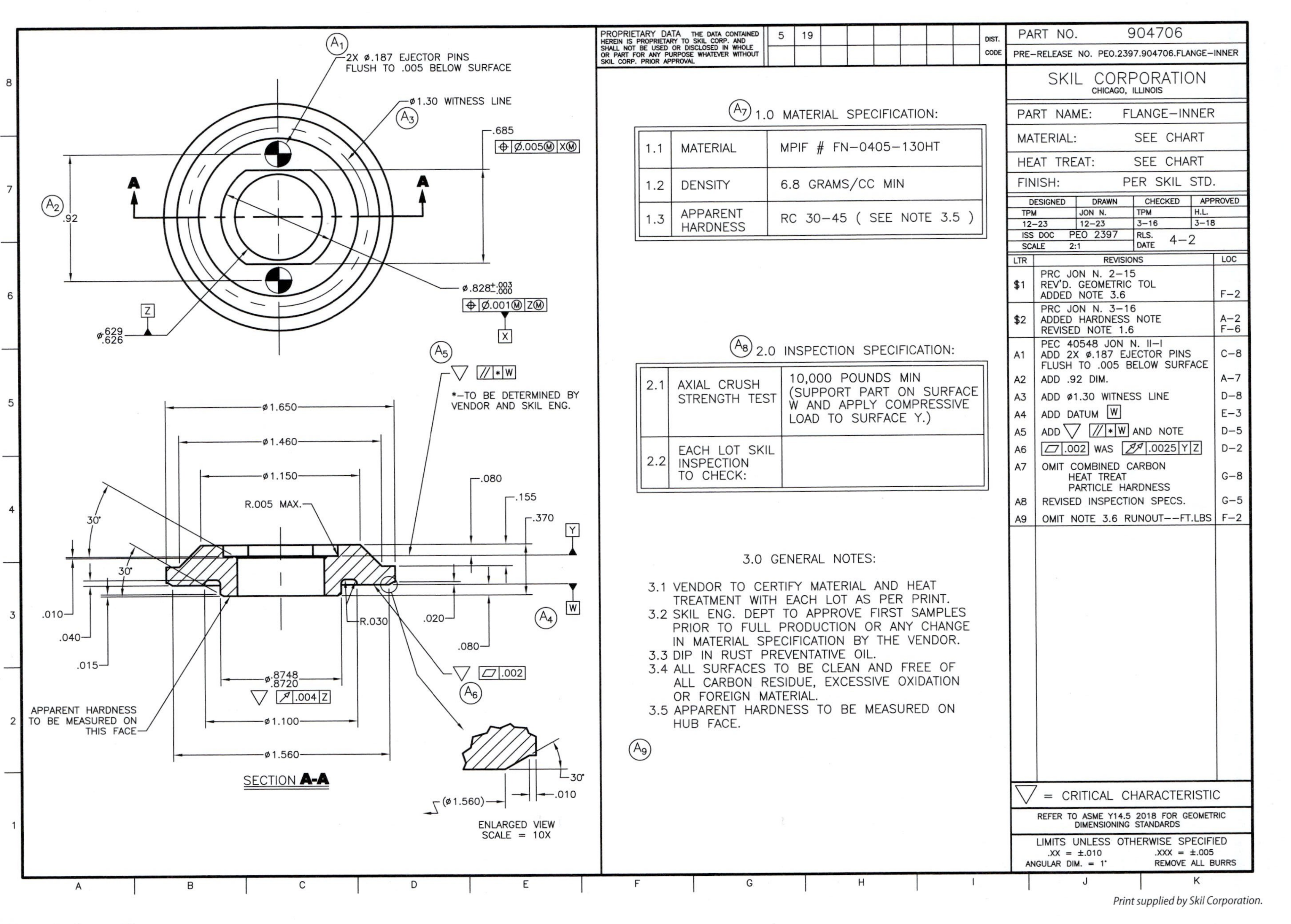

Print supplied by Skil Corporation.

PR 14-2. Inner Flange

Apply and Analyze

Name ______________________ **Date** ______________ **Class** ______________

Bonus Print Reading Exercises

The following questions are based on various bonus prints located in the folder that accompanies this textbook. Refer to the print indicated, evaluate the print, and answer the question.

Print AP-003

1. According to the revision block, revision C added a radius dimension. Is that radius dimension found in Section A-A, Section B, or one of the other views?

____

Print AP-004

2. What letter represents the revision status of this print?

____

Print AP-005

3. Does this company refer to an ECO document number for its drawing revisions?

____

4. What was revision C primarily about, and how many flags help the print reader find revision C items on the print?

____

____

Print AP-006

5. In the title block, there is a B in the lower-right corner. Is that the sheet size or the revision status?

____

6. What ECN (Engineering Change Notice) number was assigned to the latest revision of this print?

____

Print AP-008

7. To how many areas on the part did revision B apply?

____

Print AP-009

8. During which revision was Detail B added?

____

Print AP-010

9. What engineering control number was assigned to revision A for this print?

Print AP-011

10. How many individual changes are shown in the revision history for change number 510?

Print AP-014

11. What geometric shape is used to help identify revisions on the drawing?

12. List the three categories indicated by the initials in the revision block.

13. How many changes occurred during revision 05?

Print AP-017

14. How many changes occurred during revision C?

15. In general terms, what was changed as a result of revision D?

Print AP-018

16. How many revision C changes are flagged on the print?

17. In the revision history block, what does ECO most likely stand for, as expressed in this unit?

Print AP-020

18. During which of the revisions was a finishing process changed, and what was changed?

19. In the revision block, four items are noted for revision B. Outside the revision block, how many dimensions and/or notes are flagged by revision symbols for revision B?

Print AP-022

20. What drawing revision document number was assigned to revision A for this drawing?

SECTION 4

Industrial Drawing Types

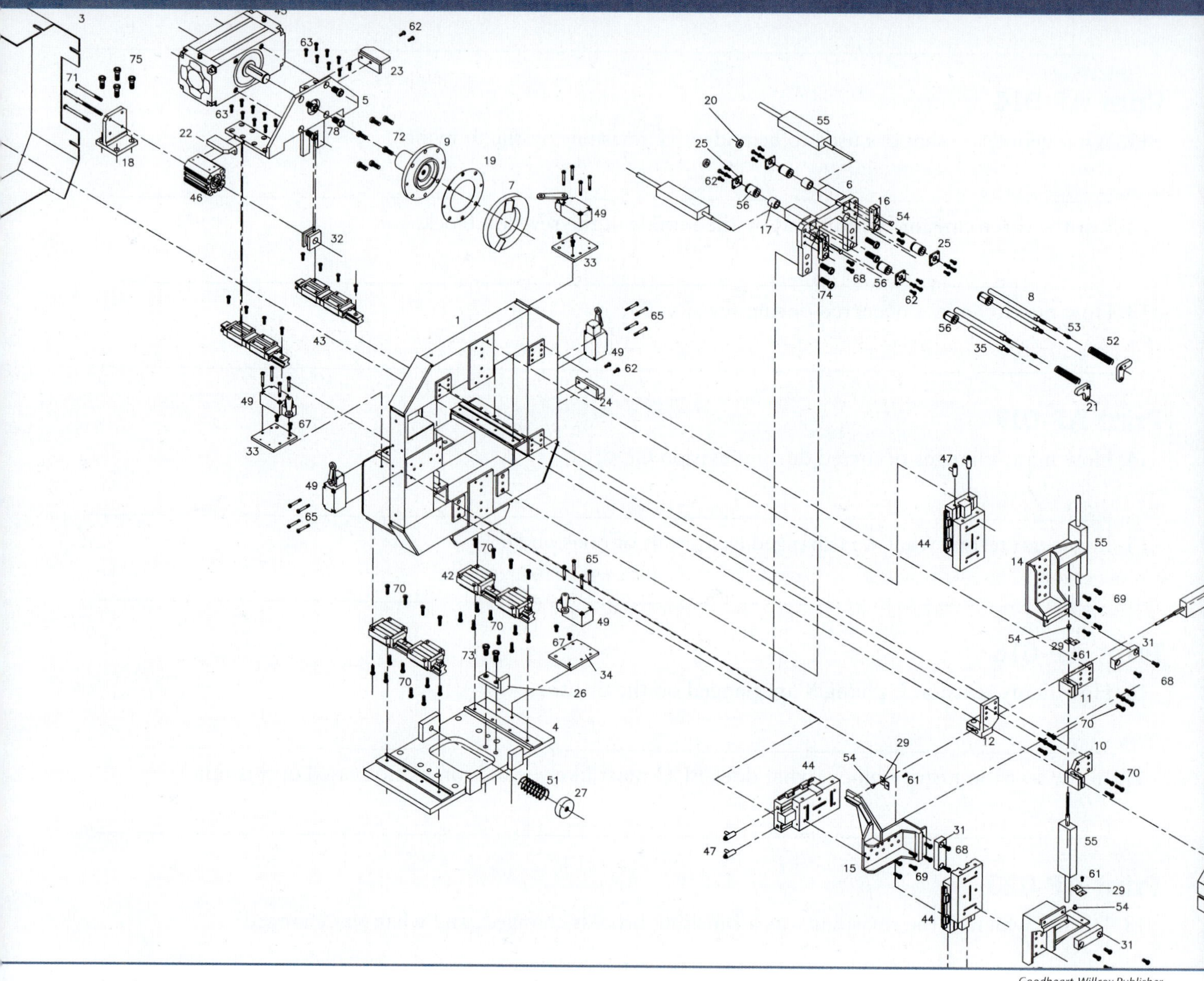

Goodheart-Willcox Publisher

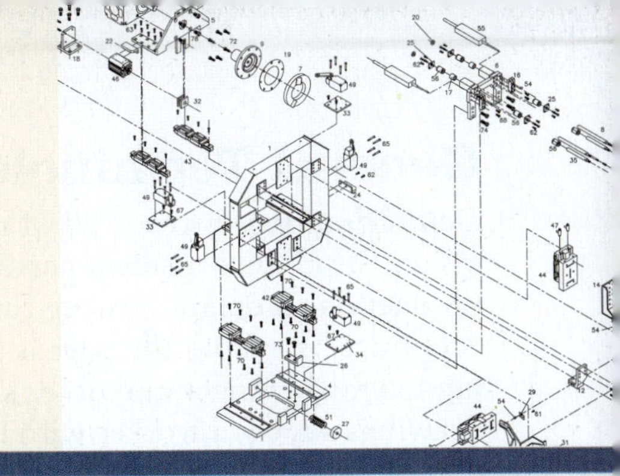

UNIT 15 Detail Drawings

LEARNING OBJECTIVES

After completing this unit, you will be able to:

- Explain terms and standards related to various types of drawings used in industry.
- Describe how detail drawings are defined and categorized in industry.
- Identify the unique characteristics of casting and forging drawings.
- Identify the unique characteristics of purchased part control drawings.
- Identify the unique characteristics of modifying drawings.
- Describe the unique characteristics of pattern development and welding drawings.
- List and describe other specialized types of drawings used in industry.

TECHNICAL TERMS

altered item drawing
casting drawing
control drawing
detail drawing
machining drawing
modification drawing
modifying drawing
monodetail drawing
multidetail drawing
pattern development drawing
welding drawing
working drawings

To produce a part, assembly, or structure, a set of drawings is necessary to provide details for the production of each component and correct assembly of all parts. The terminology for drawing types is covered in ASME Y14.24, titled *Types and Applications of Engineering Drawings*. This standard covers a wide array of drawing types, each classified according to its purpose or unique role in manufacturing. In drafting texts, the types of drawings are often simply referred to as ***working drawings*** and consist of two basic types—detail drawings and assembly drawings. While detail drawings provide information on how to produce the parts, assembly drawings provide information on how each unit or subunit is put together. In this unit, you will examine the nature and purpose of detail drawings. Assembly drawings are discussed in Unit 16.

General Terminology

Detail drawings provide all of the information necessary to produce a single part. Every part of every product has a detail drawing that is, in essence, the "contract" for how the part is to be made, including tolerance ranges and other specifications. Detail drawings supply a worker with the following:

- The name of the part
- A shape description of the part
- The dimensional size of the part and the part's features
- Notes detailing such things as material, special machining, surface texture, and heat treatment

Detail drawings can be very complex or very simple. Usually only one part is placed on a detail drawing, as shown in **Figure 15-1**. The standard term for this is a ***monodetail drawing***, but in practice it is often simply called a detail drawing. Very complex products, such as automobiles or airplanes, may require several hundred detail drawings to fully explain all parts and subassemblies. In some cases, one detail drawing of a large part may even require more than one sheet. As covered in earlier units, print readers need to begin by orienting themselves to the print through the title block, part name, and page information.

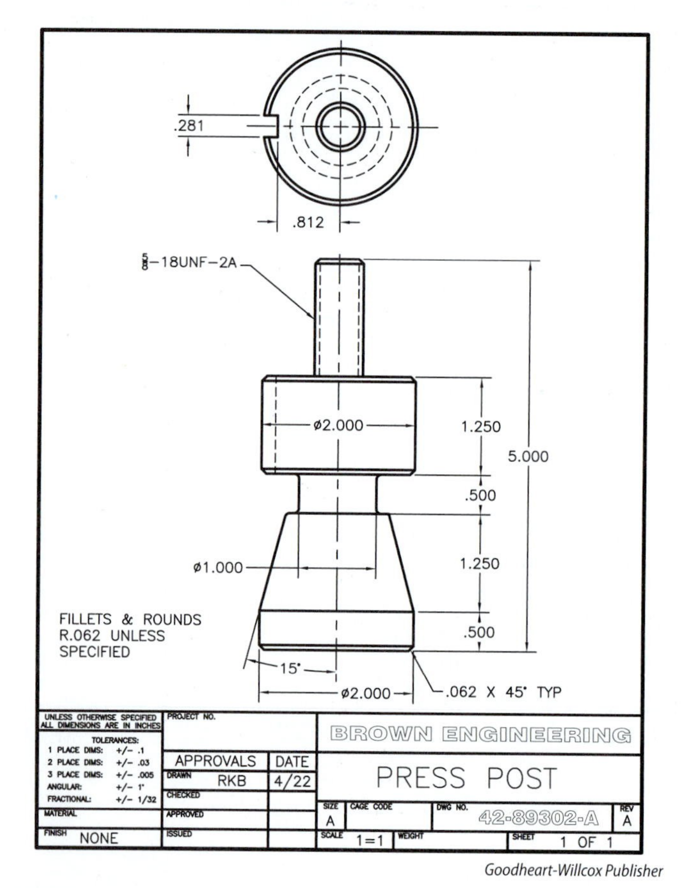

Goodheart-Willcox Publisher

Figure 15-1. A detail drawing can be a very simple drawing that shows the shape description, dimensions, specifications, material, and other contractual descriptions needed to manufacture the part.

Although a detail drawing is considered to be one drawing of one part, a series of parts can be shown on one sheet if the only difference between parts is a few dimensions. For example, a tabulated drawing would be useful for a series of threaded rods that differ only in length. In this case, several parts in a family of products can be included on one sheet with a table, as shown in **Figure 15-2**.

A ***multidetail drawing*** shows more than one multiview drawing on one sheet. While the standards caution against using multidetail drawings, there may be situations where two or more unique parts are integral to each other and it is deemed important to show both parts on one sheet. If this is done, one revision letter should apply to both parts on the sheet and records of both parts should be updated anytime the drawing is updated. In educational settings, drafting exercises and problems often incorporate many details on one large sheet to streamline the learning process, but, in industry, combining more than one part within one detail drawing can be problematic.

Casting and Forging Drawings

Parts that are forged or cast create a unique scenario for the designer and drafter. Cast or forged components often have two sets of requirements: (1) the dimensional size of the mold and (2) the dimensional size of the finished product after machining. These parts often use a type of drawing called a ***casting drawing*** to show the dimensions necessary to create the mold. See **Figure 15-3**. Additional mold-related information is often provided, including information about the mold parting line. The material to be cast is called out in the title block or as a general note on the drawing.

Machining drawings are the detail drawings that show the dimensions needed to convert a cast part into a finished part. A machining drawing describes the location and sizes of machined surfaces and features, such as threaded holes or datum reference surfaces. The drawing in **Figure 15-4** is an example of a machining drawing for the cast part shown in **Figure 15-3**. For any one casting drawing, there may be several different machining drawings. Each different drawing may show a different way of machining the same casting to create a family of parts. In this

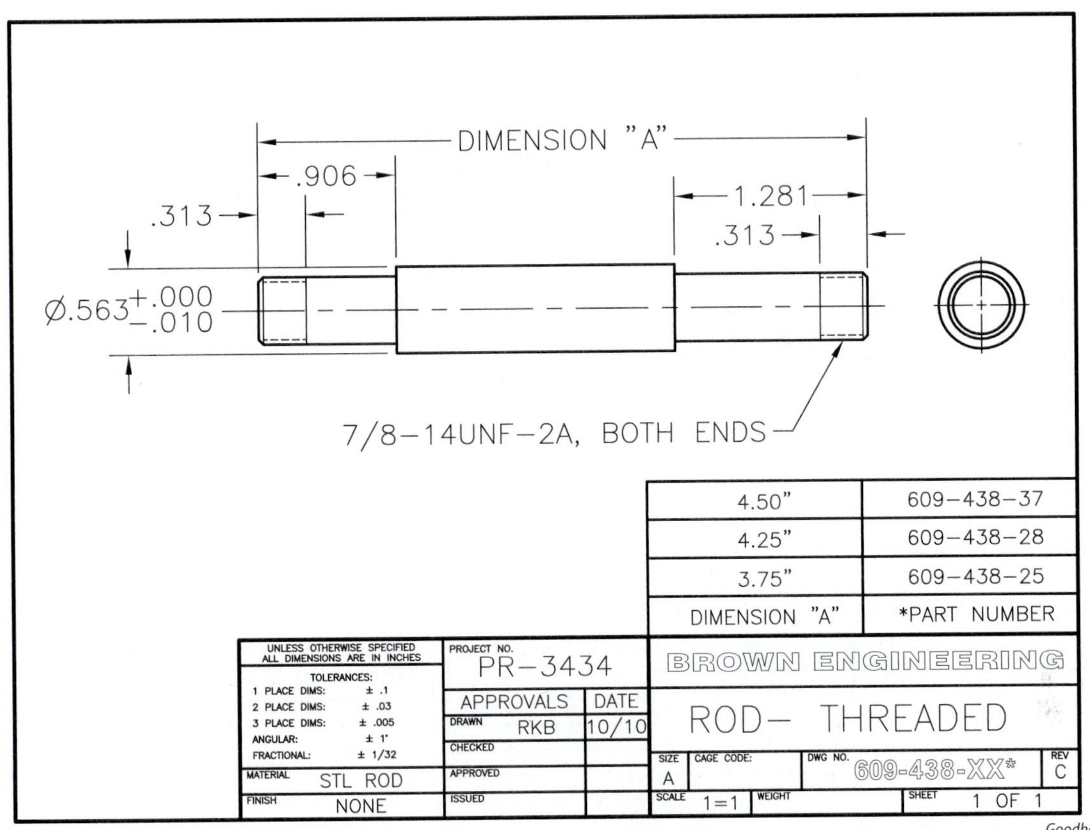

Goodheart-Willcox Publisher

Figure 15-2. The monodetail drawing allows more than one part to be covered with one flexible detail drawing featuring a tabulated dimension.

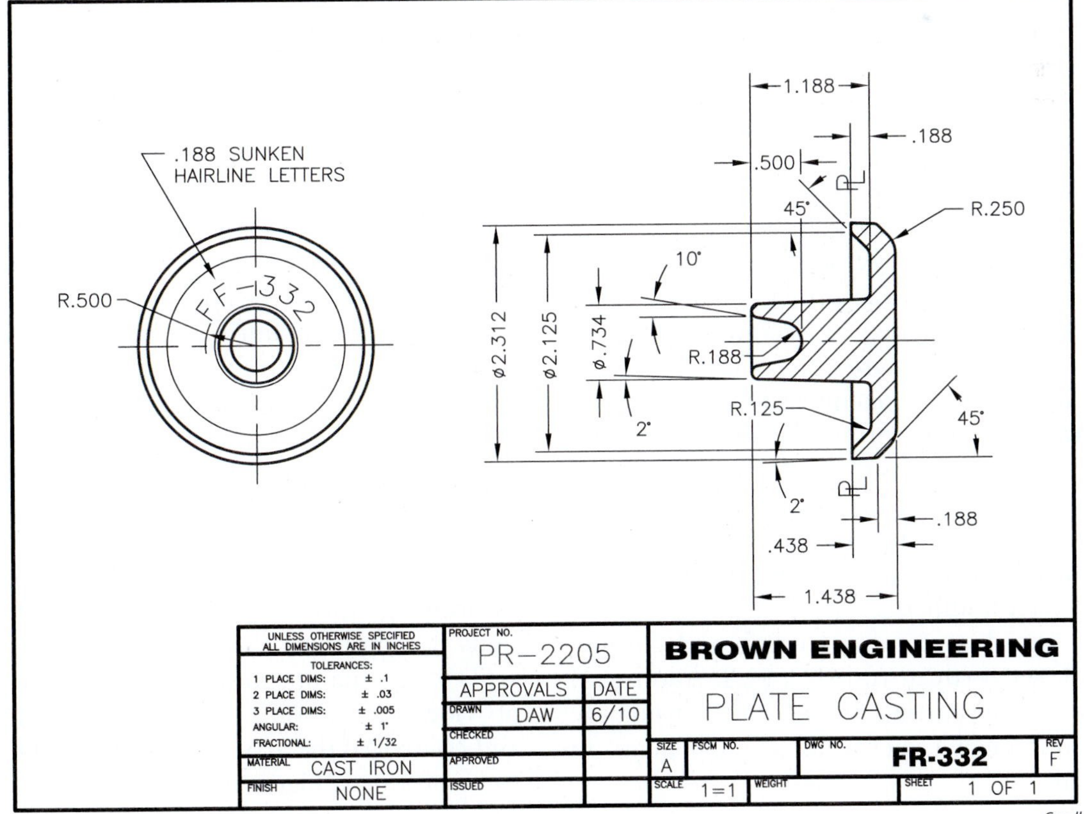

Goodheart-Willcox Publisher

Figure 15-3. A casting drawing shows the dimensions necessary to create a mold.

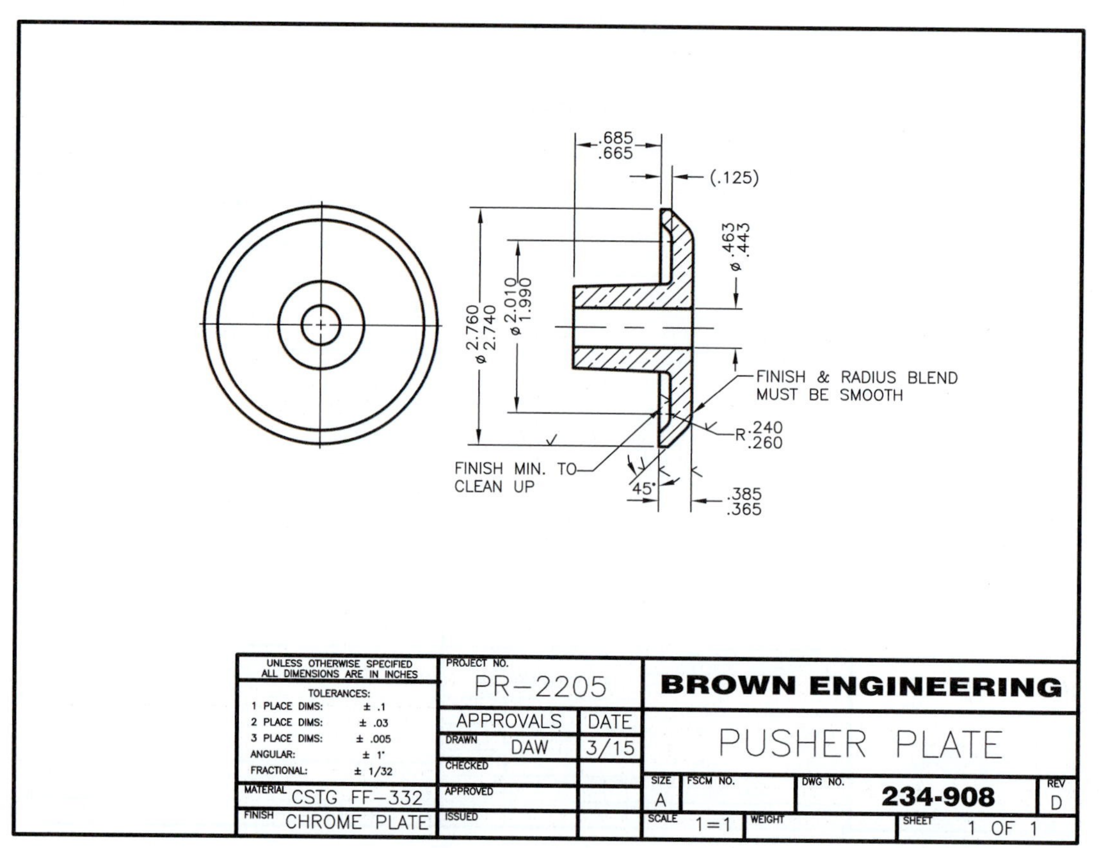

Goodheart-Willcox Publisher

Figure 15-4. A machining drawing shows the dimensions needed to convert a cast part into a finished part.

example, the part is made by machining away a large section of the center post, which may be needed for a different part created from the casting. The material is specified as cast iron for the casting drawing, **Figure 15-3**, but the material specified for the machining drawing, **Figure 15-4**, is casting FF-332.

Depending on the complexity of the part, the casting drawing and machining drawing can be combined on one print. In this case, the finished size is represented by the visible lines, and the outline of the casting shape is represented by phantom lines. See **Figure 15-5**. With this scenario, it is crucial that both the mold maker and the machinist be able to work from the one combination print. Also in this scenario, the cast object in its unfinished form does not have a separate part number, but it may be assigned a process control number for inventory. Once the part is machined to the finished dimensions, it can then be stored by its part number.

Standards for documenting casting and forging drawings are set forth in ASME Y14.8. Additional information about parting lines, draft, mismatch, and datum referencing is covered in more detail in that standard.

Purchased Part Control Drawings

Components and parts that are purchased, such as a switch or a timer, often do not require detail drawings. However, ***control drawings*** are often created for these items to document various engineering requirements to assure the interchangeability of items each time they are purchased. These requirements may include the appearance; the function, size envelope, and mounting and mating dimensions; and such things as interface or performance characteristics. More importantly, a purchased part control drawing serves as a document from which bids can be obtained from suppliers. **Figure 15-6** shows an example of a purchased part control drawing.

ASME Y14.31, *Undimensioned Drawings*, sets forth guidelines for undimensioned drawings, such as drawings that document artwork or graphic designs. Examples of these types of drawings include art layout or paint configuration drawings that could provide registration marks to aid in the inspection and control of the artwork that might be included in some products, **Figure 15-7**.

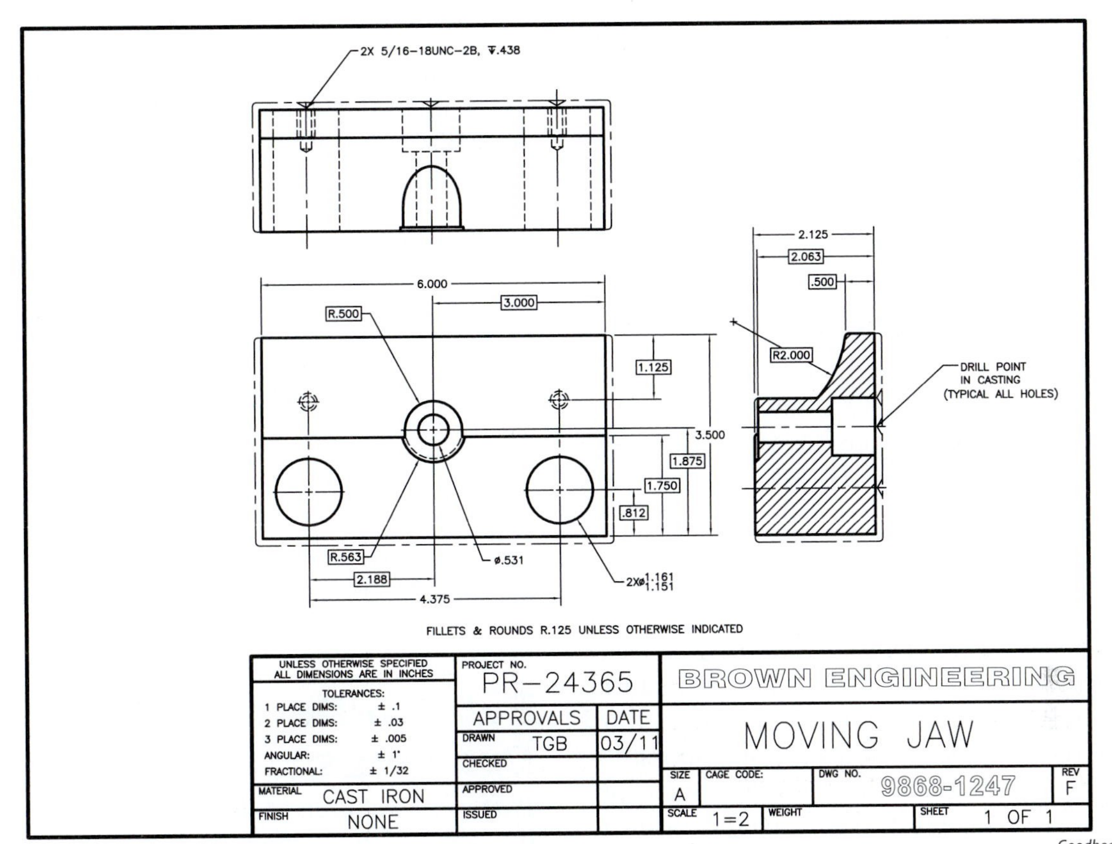

Goodheart-Willcox Publisher

Figure 15-5. A machining drawing may contain lines, notes, or dimensions that provide the mold maker with required information, making a separate casting drawing unnecessary.

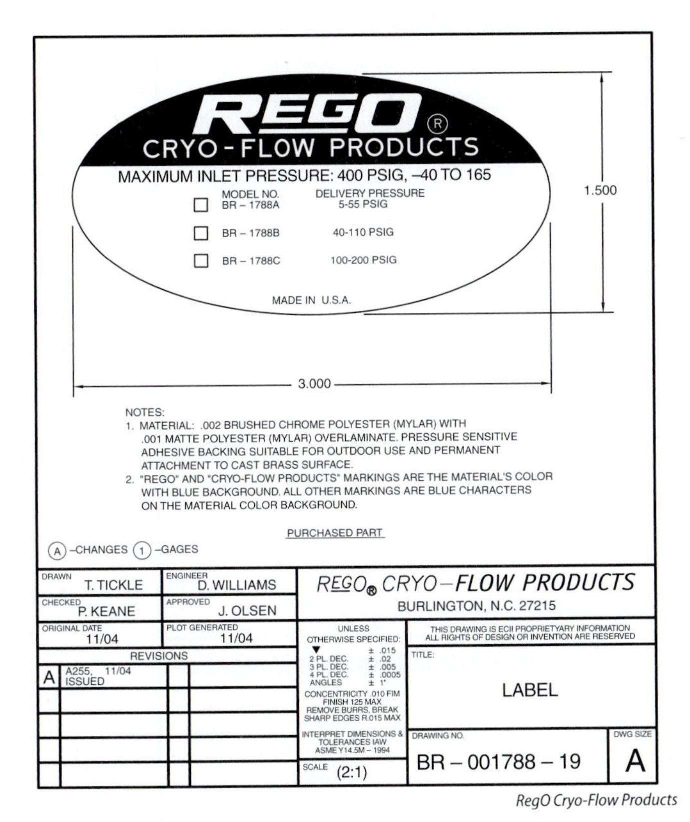

RegO Cryo-Flow Products

Figure 15-6. Purchased part control drawings can be used to record the measurements of a purchased part or product.

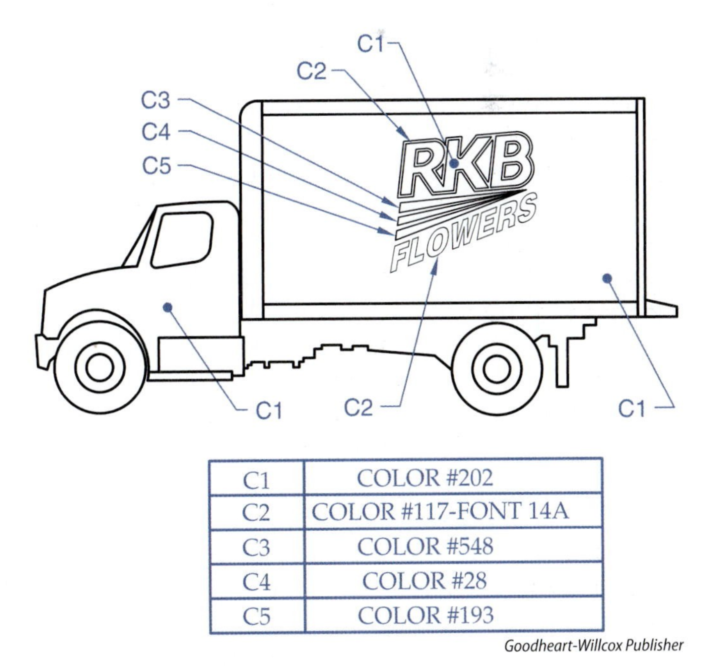

C1	COLOR #202
C2	COLOR #117-FONT 14A
C3	COLOR #548
C4	COLOR #28
C5	COLOR #193

Goodheart-Willcox Publisher

Figure 15-7. Undimensioned drawings document artwork, such as the graphic logo paint scheme for a truck.

Standard industry items, such as nuts, bolts, and fasteners, usually do not require control drawings. These items are often well-documented in standard manuals and handbooks with regard to size, shape, strength, tolerances, etc. Libraries of CAD drawings and models also make these parts readily available to the design team.

Modifying Drawings

By definition, some drawings are referred to as ***modifying drawings***. These drawings describe objects that are not made from raw or bulk materials, but rather from other parts. Types of modifying drawings include altered item drawings and modification drawings.

Altered item drawings are drawings that describe an alteration procedure that transforms one part into another part. This alteration establishes a new part number for the altered part and can be performed by the original manufacturer or a third party. Only the dimensions necessary in making the transformation are required. Some of the dimensions of the original part may be given as reference dimensions. Altered item drawings are similar in nature to a machining drawing that describes only the changes to be made to a casting. The material is listed as the part being altered, not the raw material of the original part. ASME standards recommend the notation ALTERED ITEM DRAWING be given adjacent to the title block. The drawing in **Figure 15-8** is an altered item drawing that serves as the drawing for making a new part from the part shown in **Figure 15-4**.

Modification drawings are created to describe changes to items after they have been delivered. By standard definition, modification drawings are often prepared to satisfy a sudden change in requirements and may involve adding to, removing from, or reworking items to satisfy a given situation. In this scenario, it is desirable to revise the original detail drawing specifications to accommodate future parts. ASME standards recommend the notation MODIFICATION DRAWING be given adjacent to the title block.

Pattern Development Drawings

Pattern development drawings apply dimensions to the shape of the part while it is a flat piece of metal or other stock. The detail drawing may show the final

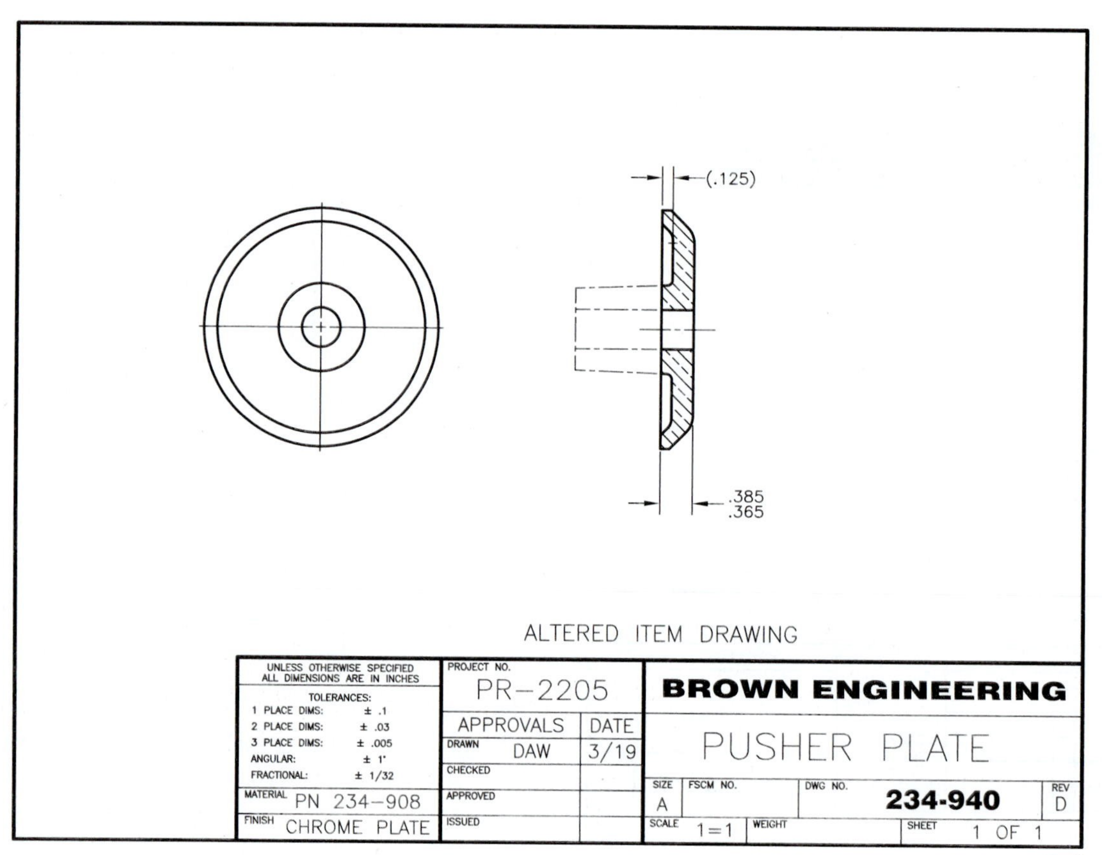

Goodheart-Willcox Publisher

Figure 15-8. Altered item drawings describe the procedure used to transform a part into another part.

shape description of the part and include phantom lines to show the size of the object before bending. Sometimes the part is shown flat, or "unfolded," and a pictorial view is shown to indicate the final form. See **Figure 15-9**. This type of drawing is discussed in more detail in Unit 21, *Precision Sheet Metal Parts*.

Flat pattern developments are also covered in AMSE Y14.31, *Undimensioned Drawings*, as some products created in this fashion could easily be described, manufactured, and inspected with a true geometry view or with CAD electronic data. Not all developments are sheet metal. Composite materials and cloth products are also candidates for drawings that are created in this fashion.

Welding Drawings

Welding drawings are detail drawings that show all components of a welded part assembled and include welding specifications. The welding drawing takes on the appearance of an assembly drawing due to the multiple parts that are assembled in position to be welded. The drawing in **Figure 15-10** is a welding drawing. Welding drawings and welding symbols are discussed in detail in Unit 22, *Welding Prints*.

Other types of inseparable assembly drawings are examined in Unit 16, *Assembly Drawings*.

Other Types of Drawings

There are many other types of drawings classified by ASME Y14.24. Most of these drawings incorporate the same print reading skills covered in this text, so a thorough coverage of the drawing types is unnecessary. Here are a few other types of drawings that may help you understand the purpose and scope of other industrial prints:

- An installation drawing provides information about how to position or install parts in relation to another part or assembly.
- A selected item drawing is a modifying drawing that specifies criteria for selecting a part from inventory that may have a further restriction or specification beyond the other parts. Perhaps the selected items survive a test or tolerance requirement. ASME standards recommend the notation SELECTED ITEM DRAWING be given adjacent to the title block.
- A tube bend drawing is a drawing that may use pictorial representation or tables to describe the bends necessary to make a final product.

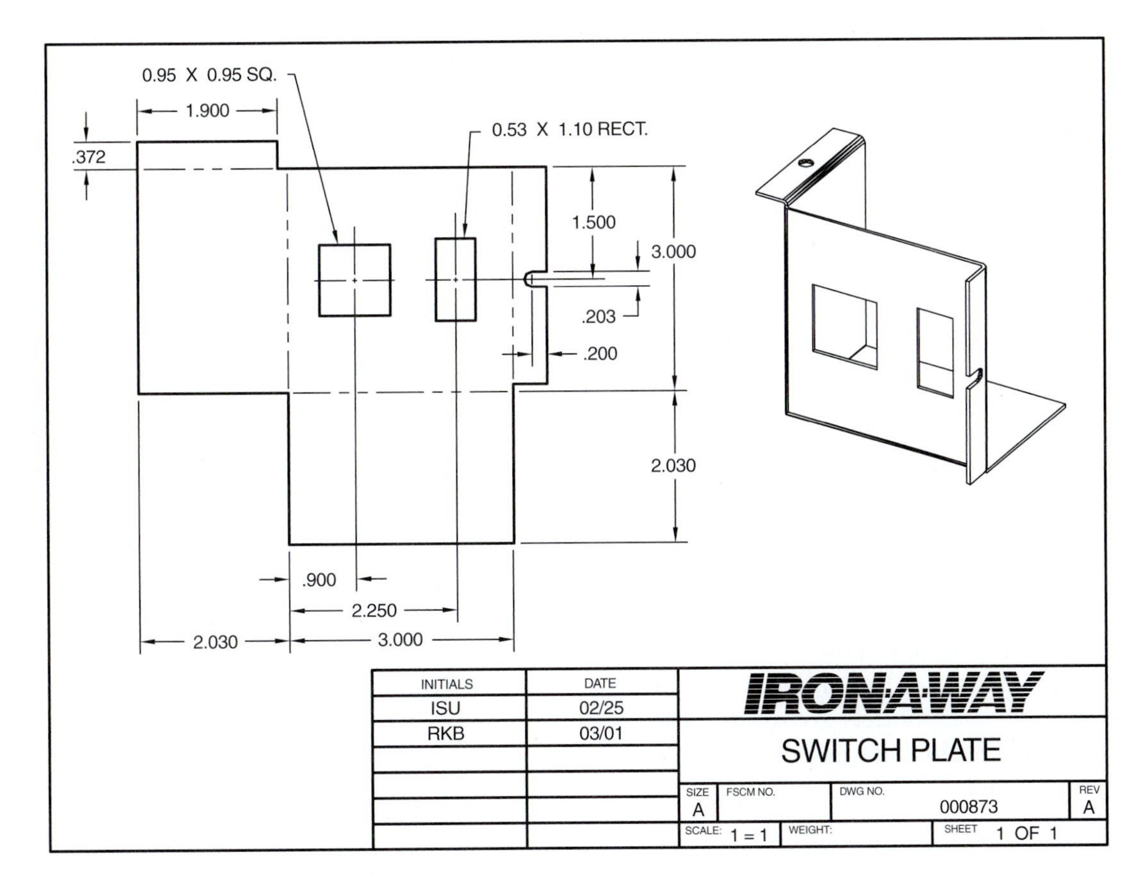

Iron-A-Way

Figure 15-9. Pattern development drawings apply dimensions to the shape of the part while it is a flat piece.

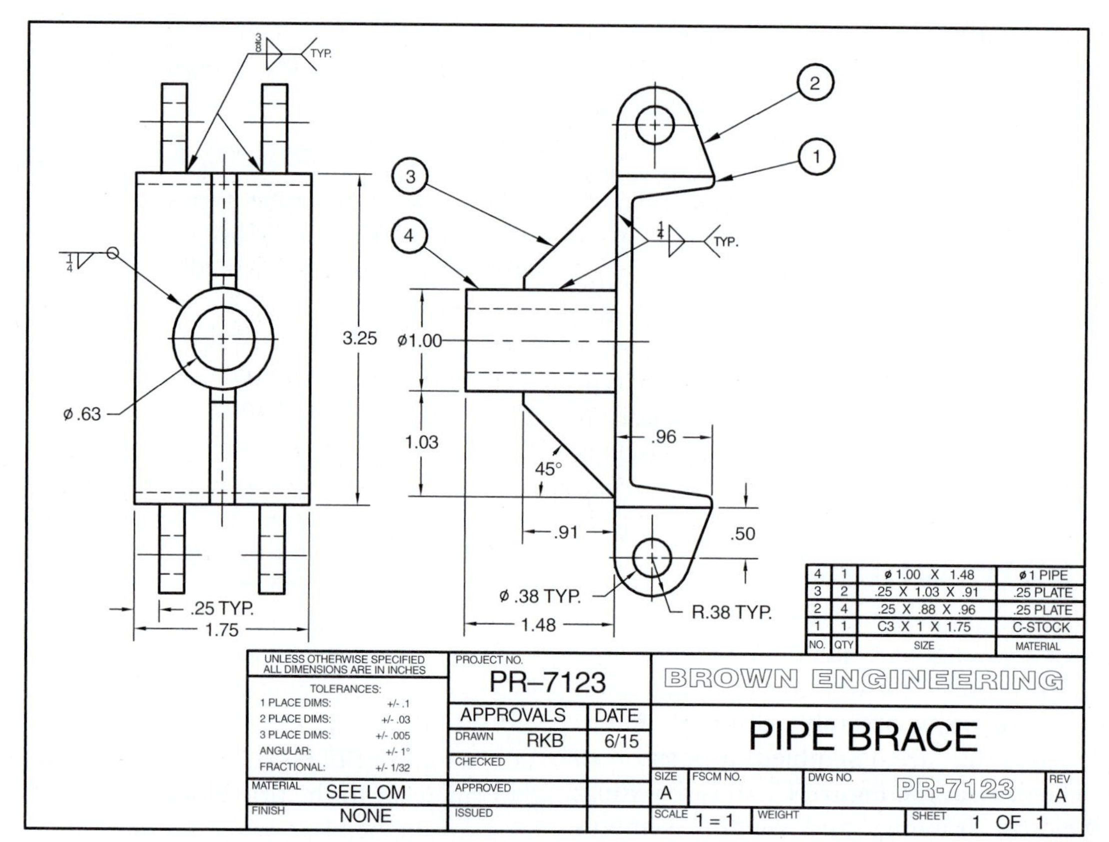

Goodheart-Willcox Publisher

Figure 15-10. Welding drawings show all components of a welded part assembled with welding specifications.

- A matched set drawing describes parts that are uniquely required to be produced to function within a matching relationship. A matched set drawing specifies the matching requirements by dimensions or other specifications. The dimensions of the individual parts not related to the matching features can be supplied on the matched set drawing or on other drawings. ASME standards recommend the notation FURNISH ONLY AS A MATCHED SET be given adjacent to the title block.
- A contour definition drawing contains the mathematical, numeric, or graphic definition required to define or build a contoured or sculptured surface. This type of drawing is needed for parts such as car fenders and airplane bodies. The definition of the contoured surface may require a series of graphic sections, a table of coordinates, or mathematical equations. In computer-integrated manufacturing, CAD models are usually the best way to control the data associated with contour definitions.

Additional types of drawings listed in ASME Y14.31, *Undimensioned Drawings*, include the following:

- A printed circuit drawing is often an undimensioned drawing that defines the circuitry pattern that is to be etched or screened onto the base material. The drawing, if printed to paper or plastic film, would most likely be created at an enlarged scale that could serve as an optical comparator diagram. See **Figure 15-11**.
- A wire harness drawing is often an undimensioned drawing that shows the configuration, and perhaps the identification, of items attached to the harness assembly. This drawing can serve as a layout guide or fixture that helps with the inspection of wire length between connections. See **Figure 15-12**.

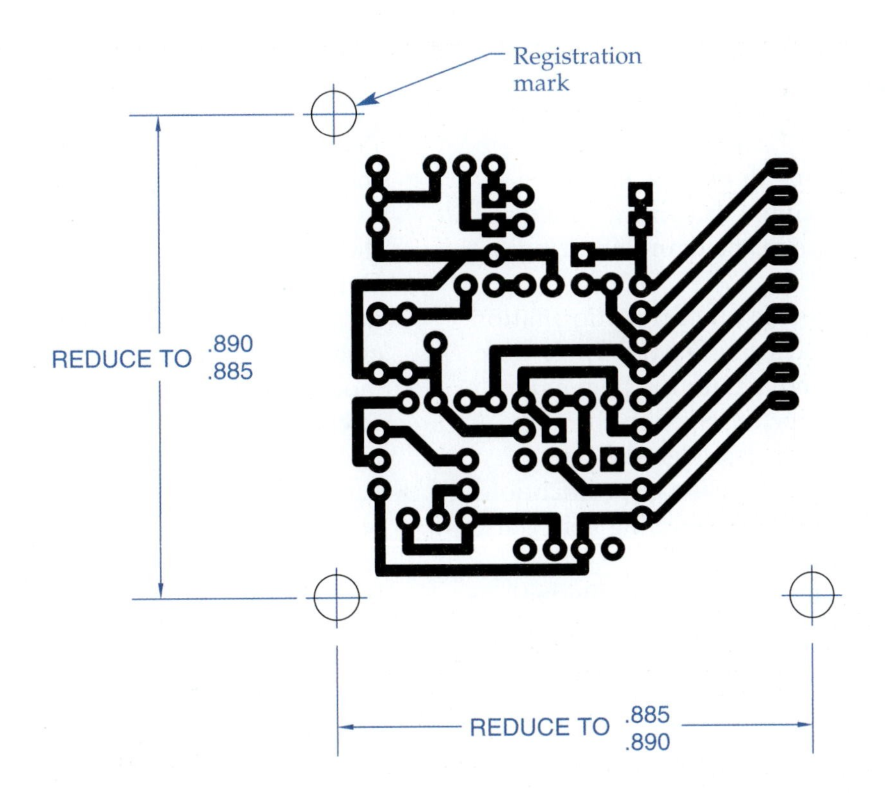

Goodheart-Willcox Publisher

Figure 15-11. Printed circuit drawings can be undimensioned drawings that serve as an inspection aid or reference.

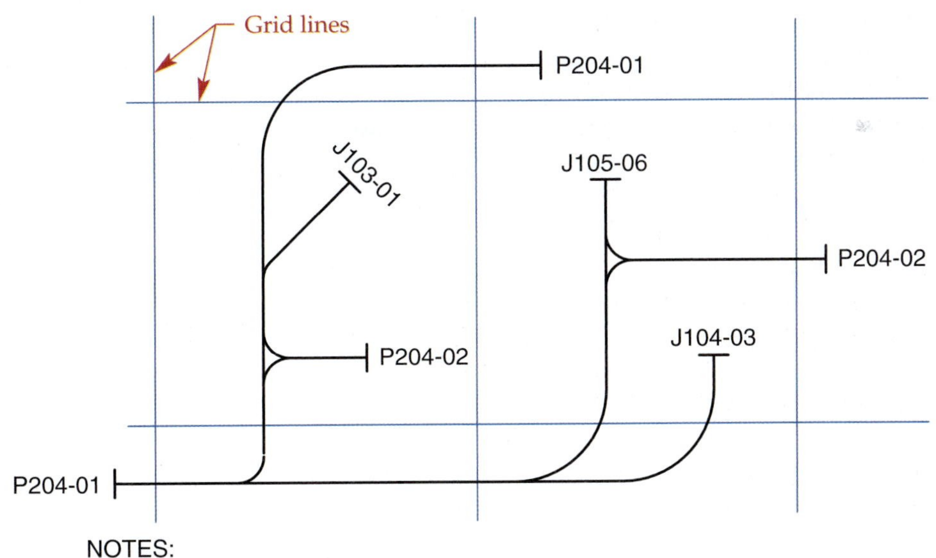

Goodheart-Willcox Publisher

Figure 15-12. Wire harness drawings can be undimensioned drawings that serve as an inspection aid or reference.

Summary

- Drawings and prints have generally been referred to as working drawings, including both detail and assembly drawings, which each serve a purpose in the manufacturing enterprise.
- Detail drawings provide all of the information necessary to produce a single part. Their wide array of information includes the part name and number, the shape description through views and/or a computer model, the size description through dimensions, specifications, and notes, including notes detailing materials, finishes, and treatments.
- A detail drawing created for one individual part is more specifically referred to as a monodetail drawing, versus a multidetail drawing of multiple parts for one print.
- Detail drawings referred to as casting or forging drawings provide all the information for the mold or formed part but do not provide all the information for the finished part.
- Detail drawings called machining drawings provide the dimensions necessary to process a cast or forged part into a finished product by machining various surfaces to a smoother finish, or creating such features as holes, threads, and keyways.
- Purchased part control drawings document the contractual specifications and appearance of purchased parts, which aids in controlling the consistency of those parts from the vendor.
- Detail drawings referred to as modifying drawings include altered item drawings, which create new parts by altering existing parts, and modification drawings, which may be needed to clarify modifications needed to parts that have been delivered.
- Pattern development drawings often deal with flat-pattern developments of parts such as sheet metal parts that are then formed by bending, or perhaps fabric or canvas patterns that may then be formed in some fashion into a free-form product.
- Welding drawings are considered detail drawings even though they feature one or more parts that will be welded together into an inseparable assembly.
- Various other types of drawings are classified by ASME standards, some of which fall into an additional category of undimensioned drawings, including installation drawings, selected item drawings, tube bend drawings, matched set drawings, contour definition drawings, and printed circuit board drawings.

Unit Review

Name ______________________ Date ______________ Class ______________

Answer the following questions using the information provided in this unit.

Know and Understand

____ 1. *True or False?* There are no ASME standards that companies can refer to for types and applications of engineering drawings.

____ 2. What are the two basic categories of working drawings?
A. Casting and Machining
B. Monodetail and Multidetail
C. Altered Item and Modification
D. Detail and Assembly

____ 3. *True or False?* Most detail drawings are monodetail—one multiview drawing of one part on one sheet.

____ 4. Which of the following drawings is usually of an unfinished product?
A. Purchased part control drawing
B. Machining drawing
C. Casting drawing
D. Altered item drawing

____ 5. *True or False?* A machining drawing and a casting drawing can be combined on one sheet.

____ 6. Of the following drawings, which is *least likely* to need a purchased part control drawing?
A. Standard nuts and bolts
B. A special elliptical metal label with the company name imprinted on it
C. A custom-made spring designed just for the company
D. Artwork for the company logo that will be placed on the sides of panel trucks

____ 7. *True or False?* A modification drawing transforms one part into another part, establishing a new part number, while an altered item drawing is usually created to document a short-term process that needs to be applied to a group of parts that have already come in the door.

____ 8. *True or False?* Another name for a flat pattern drawing is a development.

____ 9. Which of the following drawing types is technically an assembly drawing?
A. Metal
B. Development
C. Welding
D. Installation

____ 10. Refer to the ASME drawing types listed in the unit. Which one of the following is *not* an ASME drawing type?
A. Tube bend
B. Contour definition
C. Matched set
D. Skeleton

Critical Thinking

1. Give two reasons that support the practice of using monodetail drawings, with one part represented on each print, versus using multidetail drawings, with more than one part represented on each print.

2. Of the items listed below, identify a type of drawing from the unit that may be used in the manufacture of that item, and explain your answer if you feel it is helpful.
   A metal toolbox tray
   A video graphics card for a desktop computer
   A brake-line tube for a riding mower
   A custom-made spring for a pipe valve

Notes

Apply and Analyze

Name ______________________ Date ______________ Class ______________

Industry Print Exercise 15-1

Refer to the print PR 15-1 and answer the questions below.

1. Is this a detail drawing of just one part?
2. What material is specified for this part?
3. Is this drawing a casting drawing?
4. What are the overall dimensions of this part (diameter × length)?
5. Several of the dimensions must be checked with gages. How does the inspector know which gage to use for which dimension?

Review questions based on previous units:

6. What process is to be used to create the hexagon-shaped cavity?
7. What is the name of this part?
8. What surface texture (in microinches) is specified for the cylindrical surface with a maximum material condition of 1.250″?
9. How many threaded holes are there on this part?
10. Why is the 1.50″ diameter a reference dimension?
11. How many dimensions are expressed using the limit method?
12. How many times is the countersink symbol used in this drawing?
13. Is the sectional view a full section or a half section?
14. What is the primary description of revision A?
15. What tolerance applies to the 20° angle specified in the detail view?

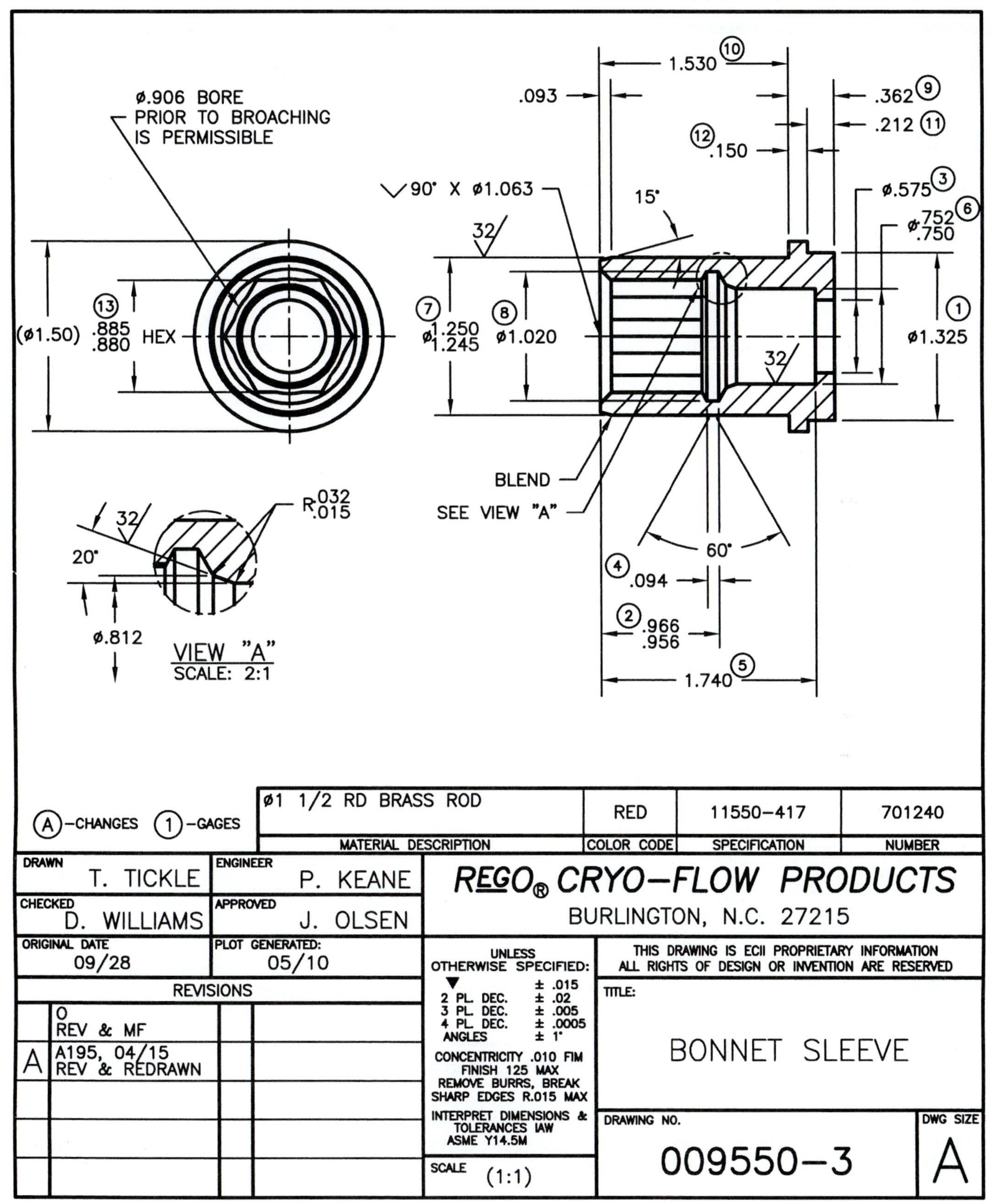

Print supplied by RegO Cryo-Flow Products.

PR 15-1. Bonnet Sleeve

Apply and Analyze

Name ______________________ **Date** ______________ **Class** ______________

Industry Print Exercise 15-2

Refer to the print PR 15-2 and answer the questions below.

1. What company specification number is given to the material for this part?

   ____

2. What revision letter (issue) is the current revision status for this print?

   ____

3. Is this a monodetail or a multidetail drawing?

   ____

4. How many critical characteristics are specified in the dimensions and notes?

   ____

5. Is this a machining drawing? ____

Review questions based on previous units:

6. At what scale are the regular views on the original print? ______________________
7. What GD&T control(s) is/are specified for this part?

   ____

8. What type of section view is Section A-A? ______________________
9. What method of tolerance expression is used for the overall depth of this object (as shown in the rectangular view)?

   ____

10. Are there any auxiliary views shown? ______________________
11. In the revision history block, what does ECN most likely stand for?

    ____

12. What document number would explain the current revision status of this print?

    ____

13. What is the maximum material condition of the feature that will establish datum A?

    ____

14. This part is a 16-position rotor. The contour of the outer surface determines the positions. How many degrees are there between each rotor position?

    ____

15. Approximately how tall should the recessed lettering be that marks the cavity I.D. on this part?

    ____

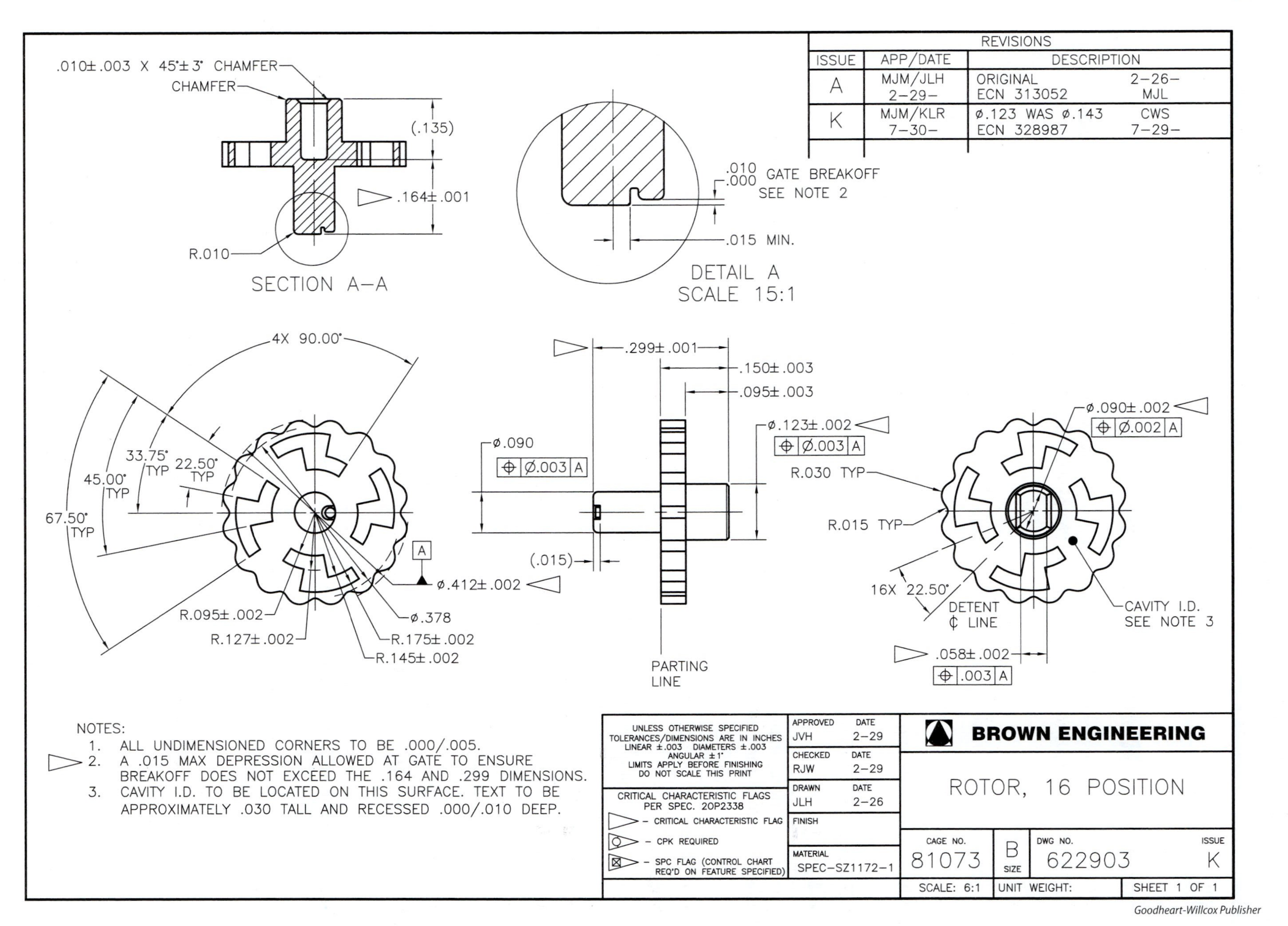

Goodheart-Willcox Publisher

PR 15-2. Sixteen-Position Rotor

Apply and Analyze

Name ______________________ Date ______________ Class ______________

Bonus Print Reading Exercises

The following questions are based on various bonus prints located in the folder that accompanies this textbook. Refer to the print indicated, evaluate the print, and answer the question.

Print AP-001

1. Besides being a monodetail drawing, what other term applies to this type of drawing?

____

2. What is the name and number of the specified material for this drawing?

____

Print AP-003

3. The general notes indicate this is a cast part. Is this detail drawing showing all dimensions for the finished part or the casting dimensions for an unfinished part?

____

Print AP-006

4. Is this print a monodetail drawing, a subassembly drawing, or both?

____

5. Is there any indication that this detail drawing is for a casting that will later be machined to a finished part?

____

Print AP-007

6. Is this print a detail drawing or an assembly drawing?

____

7. This print features a pictorial drawing at a scale of 4:1. Is that a requirement for a detail drawing?

____

Print AP-011

8. What is the name and number of the specified material for this part?

____

9. Is datum A specified as a machined surface or a cast surface?

____

10. In Section A-A, what do the hidden lines represent?

____

Print AP-012

11. If this part is still in the rough, unfinished state, what is its part number?

____

Print AP-016

12. Is this detail drawing for one part, or does it have a flexible table that enables it to represent multiple parts?

____

Print AP-020

13. What material is specified for this part?

____

14. Is this part to be machined or cast?

____

Print AP-022

15. Based on categories covered in this unit, what type of detail drawing is this print?

____

Notes

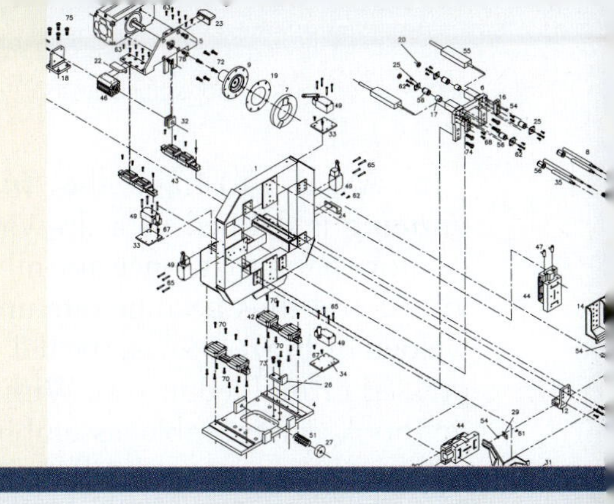

UNIT 16 Assembly Drawings

LEARNING OBJECTIVES

After completing this unit, you will be able to:

- Explain the nature and role of assembly drawings in the industrial setting.
- Explain different ways of creating pictorial and multiview assembly drawings used in industry.
- Discuss the role of the subassembly drawing.
- Identify and read information about the components within an assembly drawing.
- Identify sectioning techniques used to delineate component parts in an assembly drawing.
- Identify and read parts list information about assembly drawings that are drawn for multiple variations.

TECHNICAL TERMS

assembly drawing
detail assembly drawing
diagram assembly drawing
exploded assembly drawing
inseparable assembly drawing
installation assembly drawing
subassembly
subassembly drawing

Unit 15 discussed the nature of detail drawings in industry. These drawings are necessary to manufacture every piece or part of a product. This unit examines assembly drawings. These drawings are helpful by showing how each unit or subunit is put together. It is important to note that the terminology and definitions related to assembly drawings can vary greatly depending on the industry. Terms are not clearly defined and textbooks do not all agree, although standards such as ASME Y14.24, *Types and Applications of Engineering Drawings*, are working toward that end.

Drawing types can also be combined. For example, an assembly drawing may describe precise manufacturing processes that need to be applied after the assembly of two or more parts, but it may also contain detailed information about a part. In these cases, some drawings may function both as detail drawings and assembly drawings. It is common to express several specifications for the entire assembly using general notes.

In ASME Y14.24, the ***inseparable assembly drawing*** is defined as a drawing composed of two or more parts that, once assembled, are permanently joined and thus become one unit. Examples include welded or brazed parts, riveted parts, and parts with pressed or molded inserts. Within the context of this textbook, inseparable assembly drawings serve in the same capacity as detail drawings of single parts. Specific examples, such as welding drawings, are covered in other units.

Assembly Drawings

In general, ***assembly drawings*** are those drawings that show the working relationship of the various parts of a machine, structure, or product as they fit and function together. Usually, each part in the assembly has its own unique part number. Within the assembly drawing, the part number is indexed and listed in a parts list on the drawing, as described in Unit 3. The assembly drawing usually provides the following:

- Name of the subassembly or assembly mechanism, as well as a part number that can be used in higher assemblies and inventory
- Visual relationship of one part to another in order to correctly assemble components
- Parts list or bill of materials
- Overall size and location dimensions, when necessary to provide critical information
- Information that cannot be determined from the separate part drawings

Assembly

An assembly drawing often serves as the detail drawing of the entire product, machine, or device as a whole. See **Figure 16-1**. This drawing is often a multiview drawing and may be sectioned in one or more views, although one view may be sufficient to describe the assembly. The drawing contains only the hidden lines critical to explaining how the parts are assembled. This usually means few, if any, hidden lines appear on the assembly drawing. Center lines are still appropriate. Dimensions that express the overall measurements or the range of movement for a particular part may also be applied.

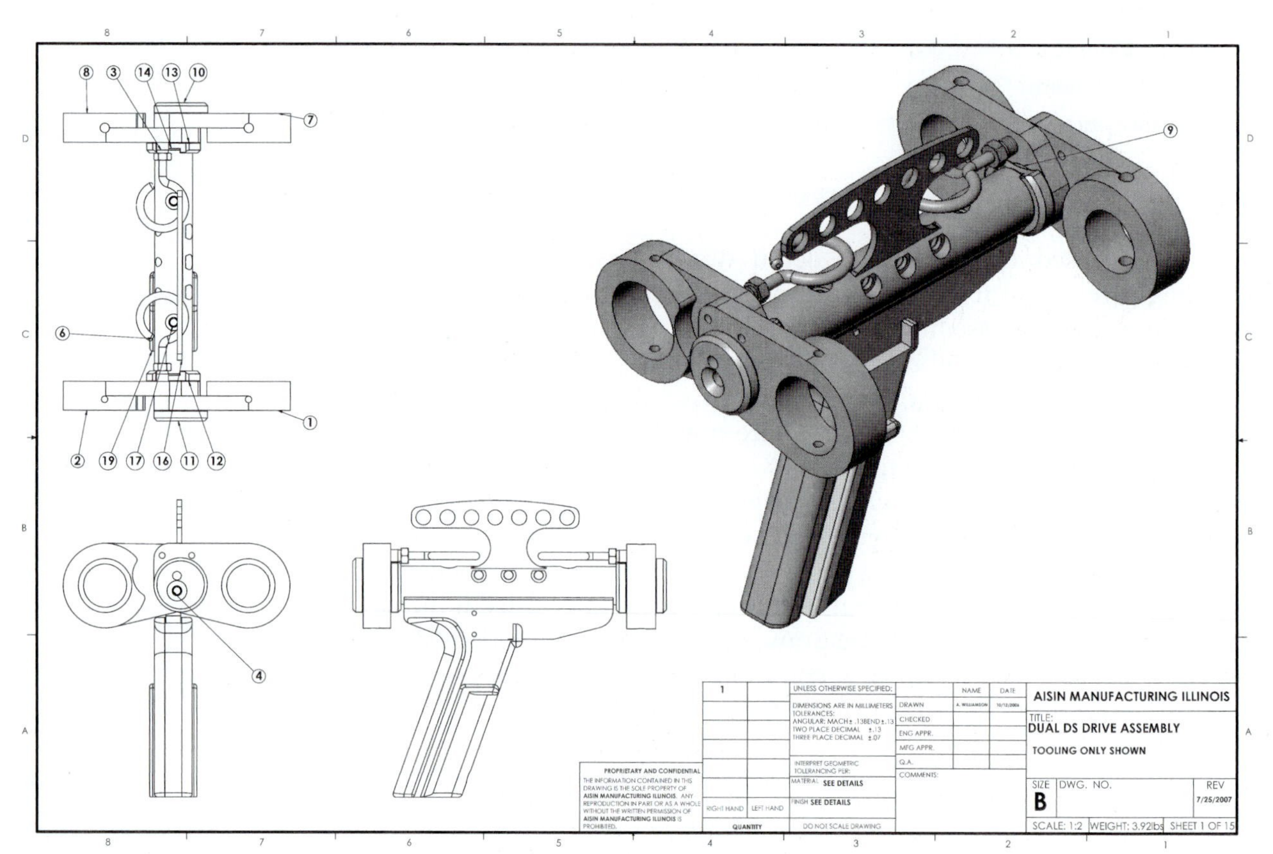

AISIN Manufacturing Illinois

Figure 16-1. Assembly drawings serve as the detail drawing of the entire product, machine, or device.

Subassembly

While an assembly drawing may be made for a complete mechanism or machine, more complex machines and structures may require several subassembly drawings for subunits. ***Subassembly*** is a term for any assembly that fits within a larger assembly. ***Subassembly drawings*** are similar in nature to other assembly drawings. They include a group of related parts composing a subunit of a larger mechanism, such as a drill press spindle assembly, automatic transmission assembly, or drive sprocket assembly.

The drawing in **Figure 16-2** is an example of a subassembly drawing. In this figure, the subassembly is a multiview projection that provides information about a part composed of other parts. In essence, this subassembly drawing is like a detail drawing, in that it provides identification about the group of parts as a new unit. The subassembly part number is used for inventory and process control. As with any assembly drawing, dimensions are not usually required because all the materials are already manufactured and inspected, although the final dimensions of the subassembly can be included, as well as any other dimensions that may need to be inspected or controlled.

Detail Assembly

In ASME Y14.24, the term "detail assembly" is applied to an assembly drawing that also contains all dimensions for all features of the individual parts within the assembly. These ***detail assembly drawings*** are useful for very simple assembly mechanisms. The drawing in **Figure 16-3** is a detail assembly drawing. The company saved documentation time and expense by showing the dimensions of all parts in this one drawing. However, this practice can be problematic with respect to maintaining one print for individual components as well as the assembly. Unless the pieces of this item are all manufactured and assembled before arriving in inventory, tracking can be difficult.

Diagram Assembly

Diagram assembly drawings use conventional symbols joined with single lines to show piping flow and wiring assemblies. The drawing explains to the reader how to assemble parts, but the parts are shown as symbols only. An example of a diagram assembly drawing is shown in **Figure 16-4**.

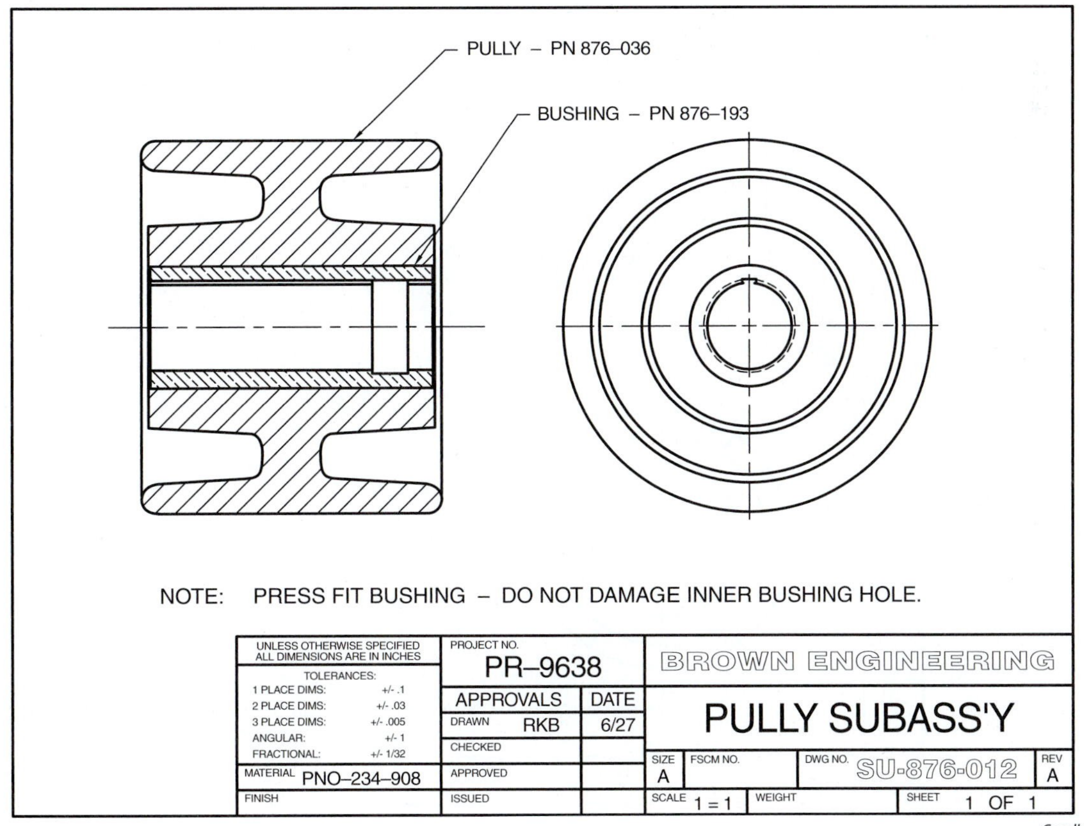

Goodheart-Willcox Publisher

Figure 16-2. Subassembly drawings include a related group of parts composing a subunit of a larger mechanism.

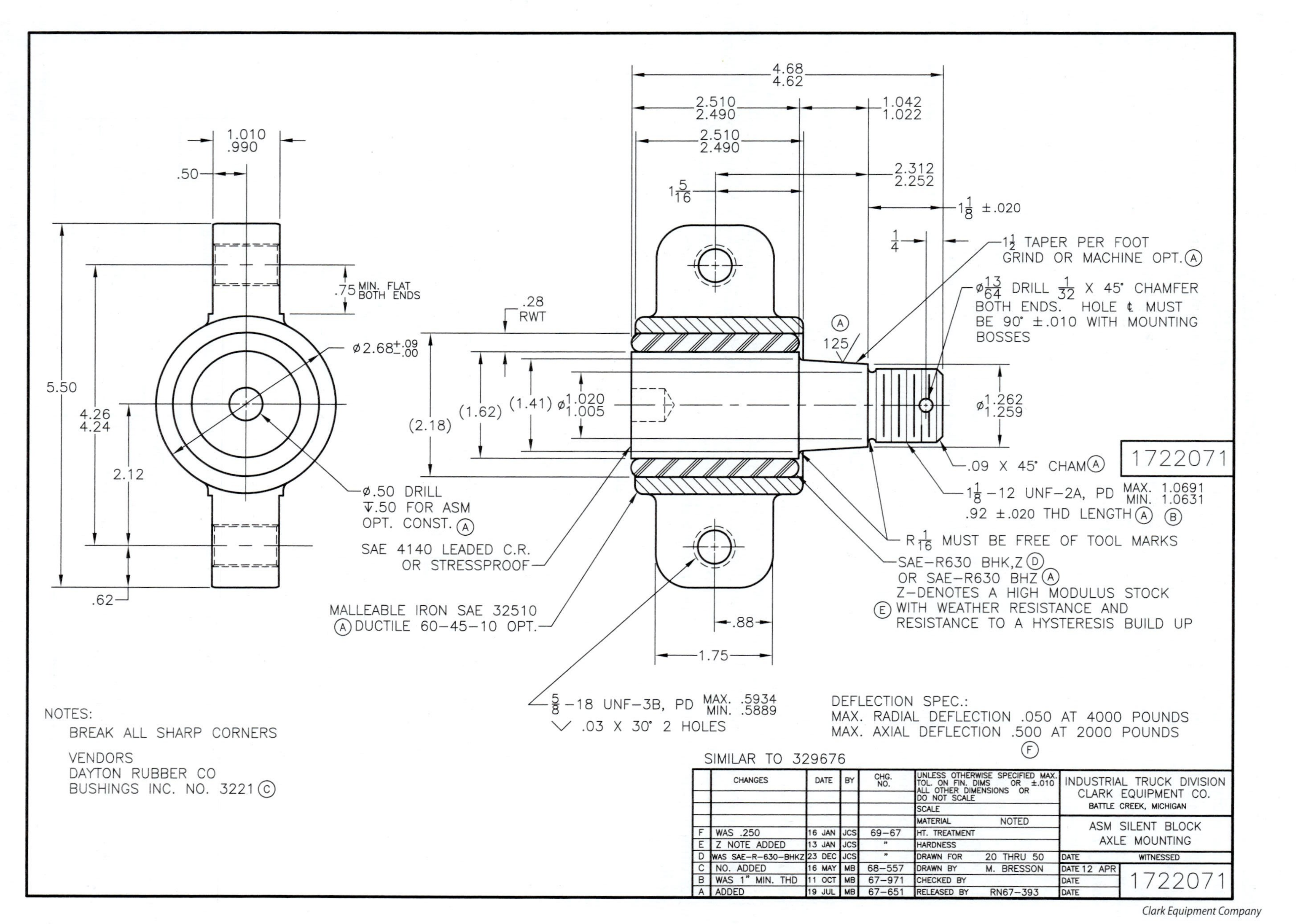

Figure 16-3. Detail assembly drawings combine the features of detail drawings and assembly drawings.

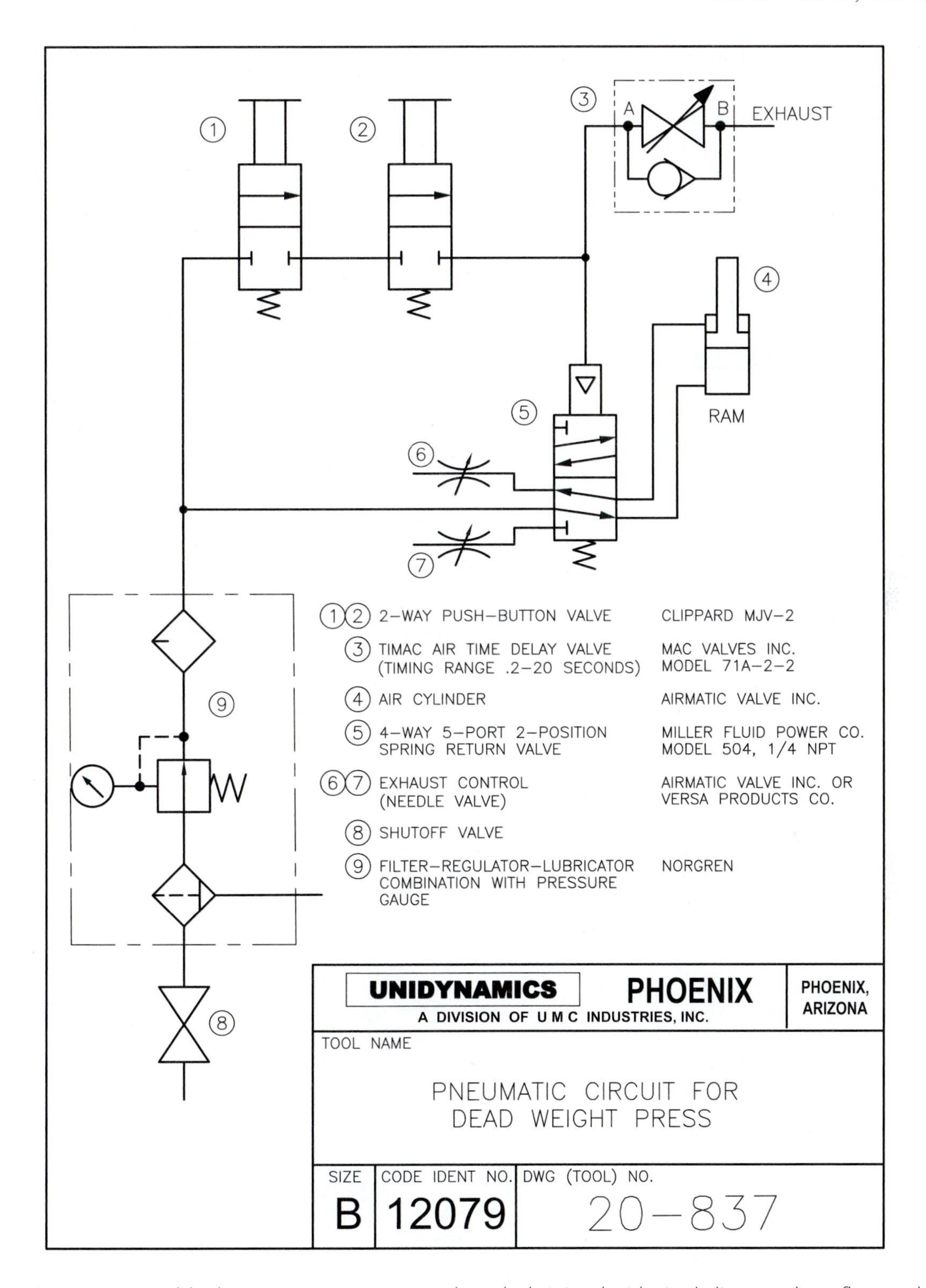

Unidynamics

Figure 16-4. Diagram assembly drawings use conventional symbols joined with single lines to show flow and wiring.

Installation Assembly

Installation assembly drawings provide the necessary information to install or erect a piece of equipment. For example, a dust collector system or other large piece of equipment may be shipped knocked down and then erected at the customer's plant or field location. An installation assembly drawing provides the instructions for correctly installing the equipment. Installation assembly drawings are often shown in pictorial form.

Exploded Assemblies

Exploded assembly drawings show some or all of the parts separated from each other to help show the correct order or method of assembly, **Figure 16-5**. Exploded assembly drawings are really not a separate type of assembly drawing. Both multiview and pictorial assemblies can be drawn in an exploded manner. Exploded assemblies are easy to read and understand when analyzing a mechanism or product. Thus, they are most often used for service manuals or part request diagrams intended for the general public.

Part Identification

A system is necessary for identifying the parts of the assembly. Most assembly or subassembly drawings use leaders to identify the parts. Many assembly drawings use balloons neatly arranged on the drawing to identify the parts. As defined in Unit 3, *balloons* are circles containing a number or letter and are usually connected to leader lines, **Figure 16-6**.

Sections in Assembly Drawings

Section lines for different parts of an assembly drawing are turned at different angles, or the scale and spacing of the section lines may be different. This is important to keep in mind when reading an assembly drawing. As covered in Unit 6, *Section Views*, there are a few conventional practices related to assembly sections. All section lines assigned to any one part should all be the same angle and spacing. Adjacent parts should feature section lines at different angles or spacing, with spacing getting closer on smaller parts. See **Figure 16-7**. To clarify an assembly drawing with many section lines, it is also helpful to use the section line patterns representing the part material, rather than general-purpose section lines. Thin materials such as gaskets may have the sectioned area filled in solid. For parts such as shafts, bolts, and other fasteners, it is common practice to leave those parts uncut by the central cutting plane.

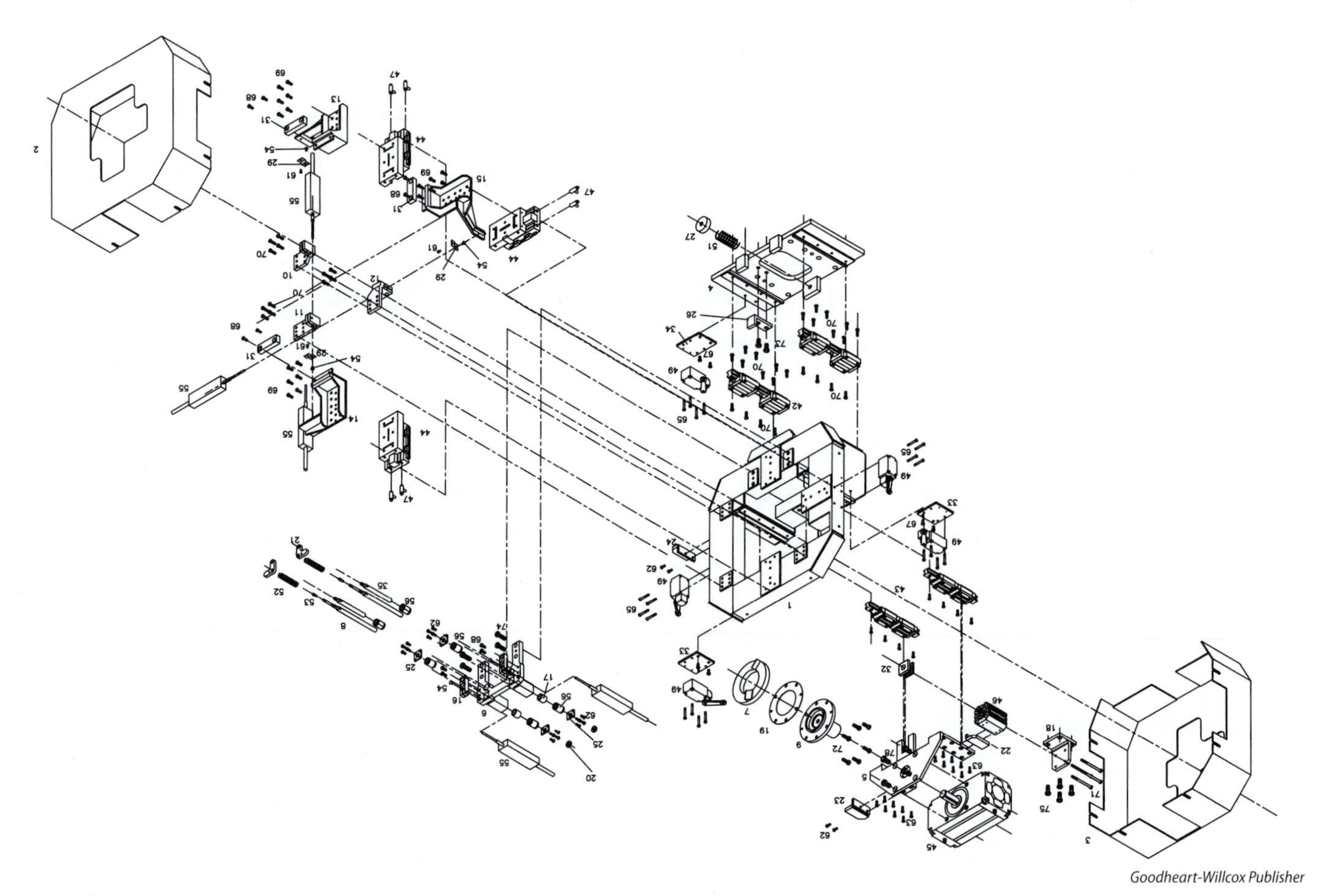

Goodheart-Willcox Publisher

Figure 16-5. Exploded assembly drawings represent parts separated from each other to help show the correct order or method of assembly. This drawing also serves as a maintenance document.

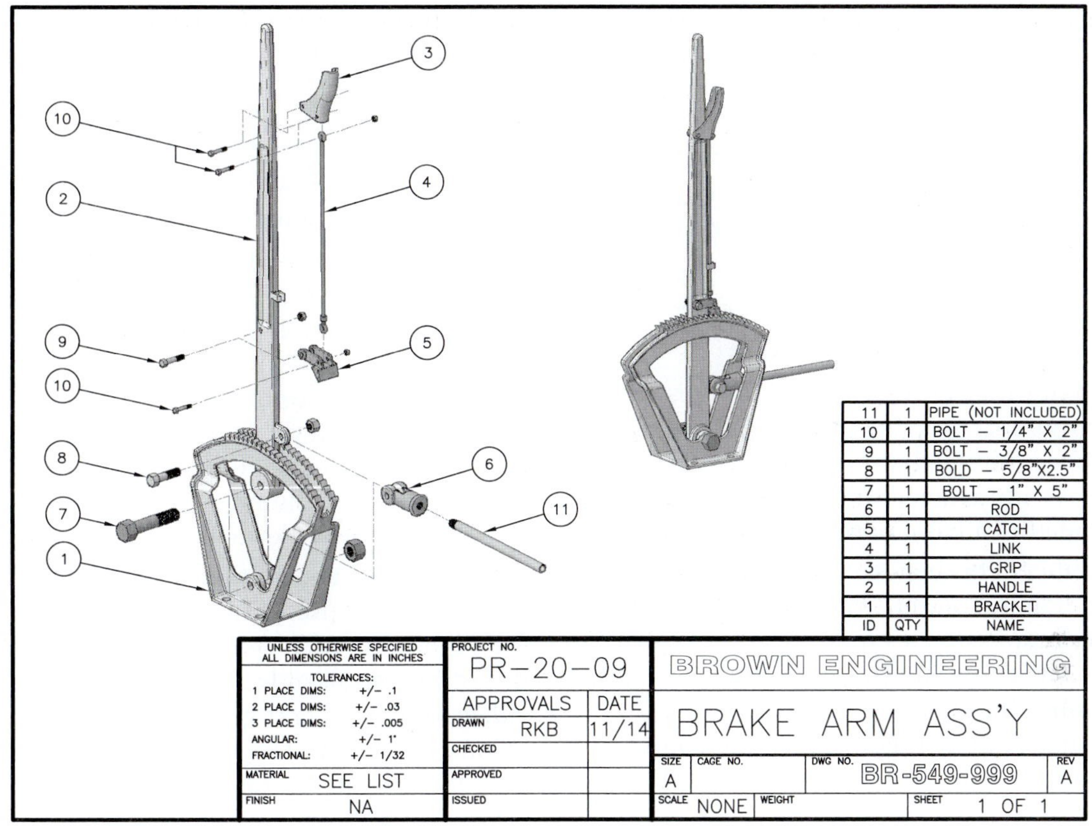

Goodheart-Willcox Publisher

Figure 16-6. Balloons contain the find numbers for the parts list.

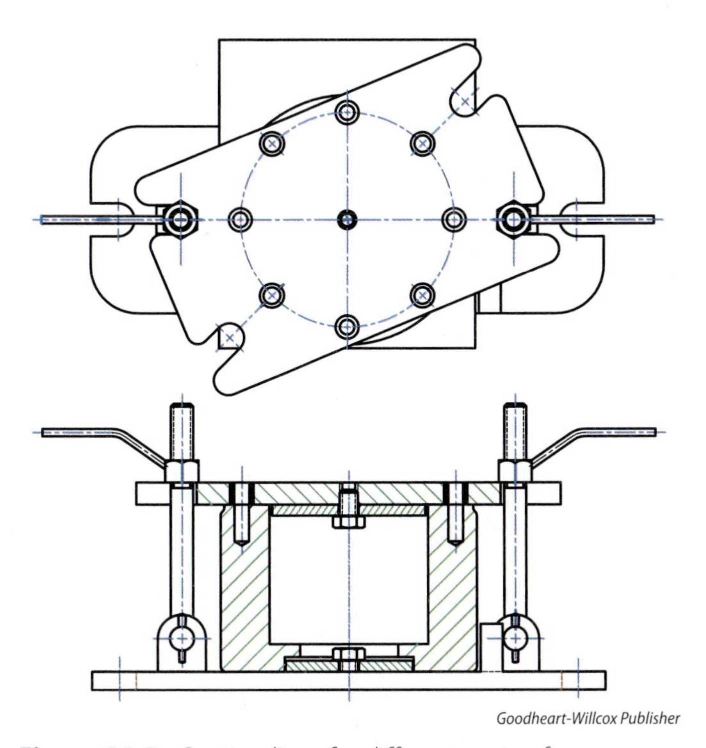

Goodheart-Willcox Publisher

Figure 16-7. Section lines for different parts of an assembly drawing are turned at different angles. Different spacing is also used to help delineate parts.

Application Blocks

In some cases, an application block is located adjacent to the title block. As mentioned in Unit 3, an *application block* provides the opportunity to indicate the larger unit, subassembly, or assembly for a particular part or subassembly. As illustrated in **Figure 16-8**, the application block usually has a column for the part number or assembly number titled

BR–548X–4	548Z
BR–549X–4	549Z
NEXT ASSEMBLY	USED ON
APPLICATION	

Goodheart-Willcox Publisher

Figure 16-8. Application blocks are used to indicate what assembly is next in line for the part or subassembly. They may include a number for the model or main system in which the assembly is used. Application blocks are sometimes incorporated into the parts list.

NEXT ASSEMBLY. Another column titled USED ON indicates a higher level system number, such as the model number of the product in which the part or assembly is used. Additional quantity columns can be added to indicate the number of subassemblies required at the next assembly or the total quantity used in the product. The application block can also be incorporated into the parts list.

Parts List Information

As discussed in Unit 3, the *parts list* is a chart listing the parts within an assembly drawing, **Figure 16-9.** It usually appears immediately above the title block on a print. As illustrated, the following items are usually included in a basic parts list:

- **Find number.** This matches the number found within a balloon on the drawing.
- **Quantity required.** This indicates the number of the item needed within the assembly.
- **Part number (PIN).** This indicates the identification number assigned by the company to the part.
- **Nomenclature or description.** This provides an opportunity to name or describe the part.

Additional Listings

Additional columns are often found in the parts list for assembly drawings. These columns may include information about materials, code numbers, or even application block information. See **Figure 16-10** for an expanded parts list that incorporates the following columns:

- Cage code column
- Material column
- Procurement specifications column
- Unit weight column
- Notes or remarks column

The CAGE code column appears on assembly drawings that use parts with these supplier codes. See **Figure 16-10A.** As mentioned in Unit 3, *CAGE codes* are applicable to activities and parts provided for the federal government. CAGE stands for "Commercial and Government Entity." This code is only listed if the part is purchased from a registered

5	4	725–P45	¼" LOCK WASHER
4	4	3452–M12	¼" X 1" MACHINE SCREW
3	2	658–4512	SLIDE BAR
2	2	985–4567	BASE GROOVE GUIDE
1	1	985–1345	BASE
FIND NO.	QTY REQD	PART NO.	NOMENCLATURE OR DESCRIPTION
PARTS LIST			

Goodheart-Willcox Publisher

Figure 16-9. At minimum, a parts list should contain a find number (if applicable), the quantity required, the part number, and the description of that part.

5	4		725–P45	¼" LOCK WASHER		ALLIED FSNRS	.05	
4	4		3452–M12	¼" X 1" MACHINE SCREW		ALLIED FSNRS	.04	SLOT HEAD ONLY
3	2	81349	658–4512	SLIDE BAR	AL–6061		.57	
2	2	81349	985–4567	BASE GROOVE GUIDE	STL		.3	
1	1	81349	985–1345	BASE	CI		2.5	
FIND NO.	QTY REQD	CAGE CODE	PART NO.	NOMENCLATURE OR DESCRIPTION	MATERIAL	PROCUREMENT SPECIFICATIONS	UNIT WEIGHT	NOTES OR REMARKS
PARTS LIST								
		A			B	C	D	E

Goodheart-Willcox Publisher

Figure 16-10. Parts lists can also include columns for (A) CAGE codes, (B) material, (C) procurement specifications, (D) unit weight, and (E) notes or remarks.

vendor. At one time, this was referred to as the Federal Supply Code for Manufacturers (FSCM) number.

The material column lists the commercial name of the material used in making the part. See **Figure 16-10B**. This may be a trademark name, such as Lexan, or an industry abbreviation, such as CRS (cold rolled steel). The procurement specification column furnishes the commercial specification and stock size of the part material. This column is sometimes labeled material specification, sizes, notes, or suppliers. See **Figure 16-10C**. The names and addresses of manufacturers of purchased parts are sometimes included in this column.

The unit weight column gives the actual weight of the part. See **Figure 16-10D**. This information is typically supplied only when it is required by contract. An additional column for notes or remarks can be added to provide information related to a vendor source, general notes on the drawing, or specific conditional information about the part. See **Figure 16-10E**.

Notes concerning interchangeability and replaceability are sometimes included in a column. These notes indicate parts that are interchangeable with parts in other assemblies. These notes are also used to indicate parts that are likely to wear rapidly and thus need replacement. Replacement parts is another designation for replaceability. When these notes appear on a drawing, they often indicate that additional parts must be stocked. A zone column is used for a zone designation code. Detail parts on larger prints may be more easily located within the prints using the zone designation code.

Parts lists can also be created as a separate document. In these cases, the note SEE SEPARATE PARTS LIST should be placed on the drawing above the title block.

Versions in Assembly Drawings

It is sometimes practical to create an assembly drawing that illustrates more than one version of the assembly. In these cases, each version is identified by a dash number following the assembly number, such as 65302-1, 65302-3, and 65302-5. When dash-numbered assembly drawings are created, columns and rows are added to accommodate all versions of the assembly.

Figure 16-11 illustrates an assembly drawing parts list configured in such a way as to define two different assembly options. The first column is for assembly number 557-2100-5, while the second column is for 557-2100-3. The third column is for shared parts, in this case the frame and slide. In this example, the -3 assembly has a base, while the -5 assembly has a channel, two glides, and two setscrews. Drawings with options and versions can quickly become rather confusing, especially to those who are viewing the drawing for the first time. The print reader may have to study the print carefully and be patient until all the information is digested.

2			AR5–13	6	GLIDE		ACETAL RSN		
2			G84–3S	5	SETSCREW			ALLIED FSN	
1			557–1845	4	CHANNEL		ALY6061–0	5–2X15	
	1		557–1843	3	BASE		ALY6061–0	8–14X20	
		2	557–3547	2	SLIDE		STL–4392		
		1	557–1754	1	FRAME		CI		
–5	–3	X	557–2100–X		SLIDE ASS'Y				
NO. RQD. ASS'Y	NO. RQD. ASS'Y	NO. RQD. ASS'Y	PART OR IDENTIFYING NO.	ITEM	NOMENCLATURE OR DESCRIPTION	CODE IDENT.	MATERIAL	MATERIAL SPECIFICATION SIZES, NOTES, SUPPLIERS	UNIT WT.
PARTS LIST									

Goodheart-Willcox Publisher

Figure 16-11. One assembly drawing can sometimes be configured to cover more than one version of the assembly, each designated with a dash number.

Summary

- Assembly drawings are those drawings that show the working relationship of the various parts of a machine, structure, or product as they fit and function together.
- An assembly drawing serves as the detail drawing for a group of parts that have already been detailed, so some of the hidden lines and dimensions that are not critical to the assembly may not be present.
- Subassembly drawings, often given a part name and number, are similar to assembly drawings and serve to define a subset of parts that fall within the total machine or product.
- Detail assembly drawings are sometimes created that contain all the dimensions and information required for all of the parts within the assembly.
- Diagram assembly drawings use conventional symbols and linework to show how parts are assembled, and they are commonly applied to wiring diagrams or piping flow drawings such as those used to describe hydraulic and pneumatic mechanisms.
- As the name implies, installation assembly drawings are those that help explain how parts are to be assembled or installed, and they are often shown in an exploded pictorial format.
- A system for identifying the parts and listing them on the drawing often involves identification balloons and a parts list chart near the title block.
- When sectional views are used within an assembly drawing, section line patterns are systematically spaced and angled to help distinguish the individual parts.
- An application block, in the form of a small table near the title block, can also be implemented to assist with identifying the larger assembly in which a part or subassembly will be used.
- A wide variety of information can be incorporated into the parts list, including CAGE codes, weight, procurement specs, and other notes or remarks.
- Parts lists are sometimes created to facilitate multiple assemblies within one assembly drawing.

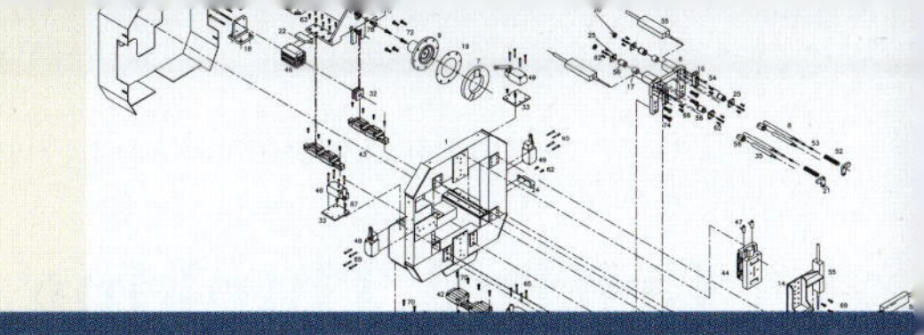

Unit Review

Name ________________ Date ________ Class ________

Answer the following questions using the information provided in this unit.

Know and Understand

____ 1. *True or False?* ASME standards define drawings of multiple parts that will be permanently assembled (for example, by welding) as inseparable assembly drawings.

____ 2. Which of the following is least characteristic of, or common to, an assembly drawing?
A. A parts list table
B. A unique part name and part number
C. Critical overall size and location dimensions
D. Surface texture symbols for each part

Match the following symbol descriptions to the symbol names. Answers are used only once.

____ 3. An assembly that still includes the dimensions specific to the individual parts

____ 4. A term that could apply to many assemblies, as it describes how the parts are positioned

____ 5. An assembly, usually pictorial, that describes the sequence and process for assembling the parts

____ 6. An assembly drawing of a group of parts that will function as one part of a larger unit

____ 7. An assembly drawing that most likely is comprised of lines and symbols

A. Subassembly
B. Detail assembly
C. Diagram assembly
D. Installation assembly
E. Exploded assembly

____ 8. Which term was used in this unit to describe the circles and numerals, often connected to parts with a leader line, that identify the parts for the parts list?
A. Bubbles
B. Balloons
C. Tags
D. Flags

____ 9. Which of the following statements about an assembly section view is *false*?
A. The section lines for each part should be created at a different spacing or a different angle.
B. Features such as shafts and bolts may be shown as whole rather than cut.
C. Thin materials such as gaskets may be filled solid rather than hatched with section lines.
D. Section lines on one side of the center line should be a mirror image of the other side.

____ 10. Which term was used in this unit to describe the table near the title block that serves to provide information about the next assembly or assemblies in which a part or subassembly is used?
A. Parts list
B. Revision block
C. Application block
D. Assembly status

Critical Thinking

1. What are some assembly drawings that are common to everyday life?

2. What might be some of the factors involved in deciding whether or not to do an exploded assembly and whether or not the views should be pictorial or orthographic?

Notes

Apply and Analyze

Name ______________________ Date ______________ Class ______________

Industry Print Exercise 16-1

Refer to the print PR 16-1 and answer the questions below.

1. How many assemblies are described by this one assembly drawing? ______________________
2. Each assembly described by this drawing uses one of two subassemblies, either part number _____ or part number _____.

   ____
3. What part number (not item number) is common to all of the assemblies described by this drawing?

   ____
4. How many different BODY options are there with respect to this drawing?

   ____
5. How will the date code be applied to the assembly?

   ____
6. What is the find (item) number of the part that is adjusted to set the spring compression?

   ____
7. What is the basic height of this assembly as shown in the section view?

   ____
8. What process is to be performed just prior to assembly?

   ____
9. Is the main spring shown as a section?

   ____
10. Part PRV250-3 is a threaded part whose top surface must be assembled to a distance of _____ below the top surface of the body.

    ____

Review questions based on previous units:

11. What paper size is the original version of this print? ______________________
12. Who checked this drawing? ______________________
13. What does it mean that the .875 HEX dimension value is in parentheses?

    ____
14. What is the scale of the views on the original drawing?

    ____
15. Are there any cutting-plane lines shown in this drawing? ______________________

NOTES:
1. CAUTION: SPRING MATERIAL MUST BE STAINLESS STEEL.
2. DRAWING SHOWS SPRING GUIDE ASSEMBLY PRV250–4.
3. THIS DISTANCE IS REQUIRED FOR SPRING COMPRESSION, AND SEAT'S COMPRESSION SET REMOVAL.
4. DRAWING SHOWS BODY PRV375–1, WITH WEEP HOLE.
5. CLEAN ITEMS 1–4 PER B–11550–400 JUST PRIOR TO ASSEMBLY.
6. STORE CLEANED ITEMS IN CLEAN, SEALED CONTAINERS IN THE CLEAN ROOM.

COL 1	COL 2	COL 3	COL 4	COL 5
ASSEMBLY	RANGE (PSIG)	SPRING	SPRING GUIDE ASSEMBLY	BODY
PRV9433F–A	10–39	BX250–025	PRV250–10	PRV375–1
PRV9433F–B	40–89	BX250–065	PRV250–10	PRV375–1
PRV9433F–C	90–139	BX250–115	PRV250–10	PRV375–1
PRV9433T–D	140–199	BX250–180	PRV250–4	PRV375–1
PRV9433T–E	200–299	BX250–260	PRV250–4	PRV375–1
PRV9433T–F	300–379	BX250–340	PRV250–4	PRV375–1
PRV9433T–G	380–459	BX250–420	PRV250–4	PRV375–1
PRV9433T–H	460–550	BX250–500	PRV250–4	PRV375–1
PRV9433FP–A	10–39	BX250–025	PRV250–10	PRV375P–1
PRV9433FP–B	40–89	BX250–065	PRV250–10	PRV375P–1
PRV9433FP–C	90–139	BX250–115	PRV250–10	PRV375P–1
PRV9433TP–D	140–199	BX250–180	PRV250–4	PRV375P–1
PRV9433TP–E	200–299	BX250–260	PRV250–4	PRV375P–1
PRV9433TP–F	300–379	BX250–340	PRV250–4	PRV375P–1
PRV9433TP–G	380–459	BX250–420	PRV250–4	PRV375P–1
PRV9433TP–H	460–550	BX250–500	PRV250–4	PRV375P–1

ITEM	QTY.	DESCRIPTION	NUMBER
4	1	ADJUSTING SCREW	PRV250–3
3	1	SPRING	SEE COL 3
2	1	SPRING GUIDE ASSEMBLY	SEE COL 4
1	1	BODY	SEE COL 5

DRAWN T. TICKLE | ENGINEER J. OLSEN
CHECKED D. WILLIAMS | APPROVED J. OLSEN
ORIGINAL DATE 05/25 | PLOT GENERATED: 08/02

REVISIONS
A A206, 05/25 ISSUED

REGO® CRYO–FLOW PRODUCTS
BURLINGTON, N.C. 27215

UNLESS OTHERWISE SPECIFIED:
2 PL. DEC. ± .015
3 PL. DEC. ± .02
4 PL. DEC. ± .005
ANGLES ± .0005
± 1°
CONCENTRICITY .010 FIM
FINISH 125 MAX
REMOVE BURRS, BREAK SHARP EDGES R.015 MAX
INTERPRET DIMENSIONS & TOLERANCES IAW ASME Y14.5M

THIS DRAWING IS ECII PROPRIETARY INFORMATION ALL RIGHTS OF DESIGN OR INVENTION ARE RESERVED

TITLE: PRV BLANK

DRAWING NO. PRV009433XX–X

DWG SIZE B

SCALE (2:1)

Print supplied by RegO Cryo-Flow Products.

PR 16-1. PRV Blank

Apply and Analyze

Name ______________________ **Date** ____________ **Class** ______________

Industry Print Exercise 16-2

Refer to the print PR 16-2 and answer the questions below.

1. How many assemblies are described by this one assembly drawing? ____________________
2. The catalog numbers for the assemblies described by this drawing all begin with KM50-MCLN. List the unique numbers that finish the catalog numbers for each assembly.

   ____
3. Some of the assemblies described by this drawing are left-hand and some are right-hand. Which of the two is the way the drawing is represented?

   ____
4. Is STEEL BODY 224292-01 a left-hand or right-hand body?

   ____
5. List the find (detail) numbers of the parts that are common to all assemblies.

   ____
6. In inches, what is the largest overall dimension value specified in this assembly drawing?

   ____
7. What part number and description are associated with find number 6?

   ____
8. In general, what is the name of this series of assemblies? ____________________
9. Give a one-word description of the items that must be purchased separately.

   ____
10. List the three angular measurements that are important to this assembly.

    ____

Review questions based on previous units:

11. What do the numerical values in brackets indicate?

    ____
12. What does the comma represent in the dimension that reads "50,000 OVER N.R. GAGE INSERT"?

    ____
13. What former value was specified for the 7° angular measurement?

    ____
14. What are the initials of the person who approved adding the KM logo? ____________________
15. Is the drawing first-angle projection or third-angle projection? ____________________

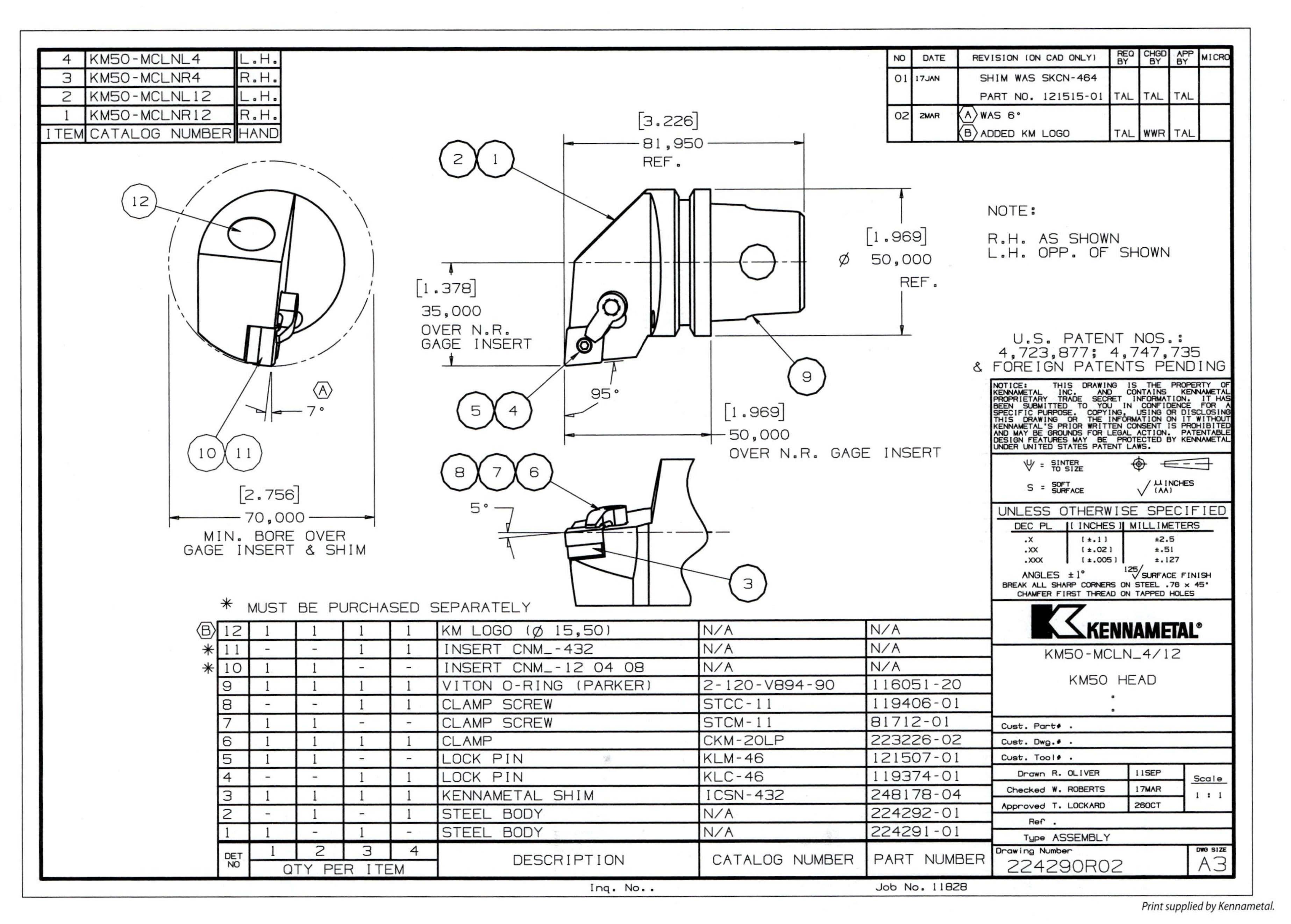

Print supplied by Kennametal.

PR 16-2. KM50 Head

Apply and Analyze

Name ______________________ Date ______________ Class ______________

Bonus Print Reading Exercises

The following questions are based on various bonus prints located in the folder that accompanies this textbook. Refer to the print indicated, evaluate the print, and answer the question.

Print AP-004

1. Of the types of assembly drawings discussed in the text, which type best describes this drawing?

____

2. Name the three subparts indicated on this drawing.

____

Print AP-009

3. If this assembly drawing were changed to identify the parts within a parts list, how many part rows would be needed?

____

4. Is any part of this assembly drawing considered "exploded"?

____

Print AP-010

5. Carefully examine this assembly drawing. Within each identification bubble, there is an upper number and a lower number. What does each number stand for?

____

6. Does the 109.41 dimension on this drawing apply before assembly or after?

____

7. On this print, the assembly is sectioned. Are there any parts that have *not* been sectioned?

____

Print AP-013

8. Not counting the two closures (not shown), give the total number of pieces required for this assembly.

____

9. How many products or models are covered by this one assembly drawing?

____

10. What must be done to the label before it is attached?

11. Give the part names of the two parts that are different depending on the model assembly number.

12. How is item 5 to be positioned in relation to item 13?

13. Calculate the total height of the assembly, as shown on the drawing.

Print AP-014

14. How many products are covered by this one assembly drawing?

15. How do we know that item 5 is not shown or identified in the views of the drawing?

16. Are items 1 and 2 both visible in the top view?

17. Is item 4 visibly shown in either view?

18. If you think of the top view as a clock, there is one insert shown at 6 o'clock. How many inserts are there on the actual manufactured milling cutter?

Print AP-015

19. Analyze the assembly and describe the basic purpose for the assembly.

20. How will item 5 be attached to item 9.1?

SECTION 5

Specialized Parts and Prints

Artie85/Shutterstock.com

UNIT **17**

Springs and Fasteners in Industrial Prints

LEARNING OBJECTIVES

After completing this unit, you will be able to:

- Define terms related to springs used in industrial applications.
- Define terms related to fasteners used in industrial applications.
- Identify types of threaded fasteners shown on industrial prints.
- Identify types of non-threaded fasteners shown on industrial prints.
- Read specifications related to springs and fasteners on a print.

TECHNICAL TERMS

bolt
cap screw
clevis pin
closed end
compression spring
cotter pin
dowel pin
extension spring
garter spring
machine screw
nut
open end
retaining ring
rivet
set screw
spring
spring washer
stud
torsion spring
washer

Within the realm of product design, designers have a vast supply of standard parts from which to choose when assembling the components of a product. These objects include springs, pins, keys, and other fasteners such as nuts, screws, and bolts. Although springs are items often purchased to complete an assembly of parts, sometimes the designer needs to custom design a spring for a particular application. In this unit, springs and fasteners are presented as objects that often do not need a detail drawing. However, these items are visually drawn and represented in an assembly drawing, so the well-trained print reader should be familiar with them to better read the drawings and specifications for a product.

ASME standards for springs and fasteners focus on the complete dimensional and general data associated with these parts, including strength and accuracy specifications. They do not focus on the drawing of the parts, although thread representation is covered in ASME Y14.6, as presented in Unit 8. In general, drawings of the parts are often simplified to represent the fastener or spring and sometimes do not represent the accurate production measurements. Since these parts are for sale, most of them are available for CAD systems in symbol libraries and can be accessed on the internet from fastener vendors or CAD forums.

Multiple ASME standards have been established regarding springs and fasteners. Some common ASME standards that define the size and characteristics of fasteners include the following:

- B17.1, *Keys and Keyseats*
- B17.2, *Woodruff Keys and Keyseats*
- B18.2.1, *Square and Hex Bolts and Screws*
- B18.3, *Socket Cap, Shoulder, and Set Screws, Hex and Spline Keys*
- B18.5, *Round Head Bolts*
- B18.8.1, *Clevis Pins and Cotter Pins*
- B18.8.2, *Taper Pins, Dowel Pins, Straight Pins, Grooved Pins, and Spring Pins*
- B18.25, *Square and Rectangular Keys and Keyways*
- B18.27, *Tapered and Reduced Cross Section Retaining Rings*

The B18 series of standards includes separate documents for lock washers, round-head bolts, rivets, square-neck bolts, hex nuts, and machine screws. ASME B18.12 is titled *Glossary of Terms for Mechanical Fasteners*. ASME Y14.13, *Mechanical Spring Representation,* was discontinued, but many companies still refer to that standard as good practice for drawing springs, and it is still available as a historical reference.

Springs

A ***spring*** is a device that is designed to either compress or expand due to a force and, in doing so, counteract that force with an equivalent force. Springs are made from wire material. Companies that produce springs often also create other wire forms for various applications. See **Figure 17-1**.

Types of Springs

A common example of a spring is a helical wound coil of wire that, if pulled, will stretch to become longer and will then return to its original, shorter shape if the pulling stops. This type of spring is called an ***extension spring***, **Figure 17-2**. Extension springs are characterized by coils that touch each other in the free position.

Master Spring and Wire Form Co.

Figure 17-2. Extension springs are used to help pull parts toward each other.

Master Spring and Wire Form Co.

Figure 17-1. Springs are used to help maintain or control certain relationships between other parts. This figure also shows other parts that can be made from wire or rods.

A ***compression spring*** is a coil of wire designed to hold its shape with the coils separated from each other in the free position, as if it is already stretched out. See **Figure 17-3**. A force is required to compress the coils, and the spring exerts a force that attempts to return the coil to the longer length.

Extension springs and compression springs are designed to have force applied in the same direction as the axis of the spring coil. ***Torsion springs*** are designed to be used in situations where the force is at a right angle to the axis of the coil of wire. Torsion springs are wound either in a helical or spiral form and are commonly used on parts that rotate about an axis. These springs can be used to apply force to the rotation of one part about an axle or pivot pin. See **Figure 17-4**.

Other springs found in industrial assemblies come in the form of garter springs and spring washers. ***Garter springs*** are long extension springs designed to encompass a circular feature in a belt-like manner. See **Figure 17-5**. ***Spring washers*** are flat springs fabricated from thin metals that can hold a bowed shape. A spring washer is designed to exert a force if another part, such as a threaded fastener, attempts to flatten it. Common styles of spring washers are shown in **Figure 17-6**.

Study **Figure 17-7** as you examine how each of the following terms related to springs may be applied:

- **Active coils.** For a compression or extension spring, these are the windings in the helical shape that are subject to move when the spring is under force to stretch or compress. Some coils on the end of an extension spring are not active, but rather form the end of the spring for connecting to other objects.

Master Spring and Wire Form Co.

Figure 17-3. Compression springs are used to push parts away from each other.

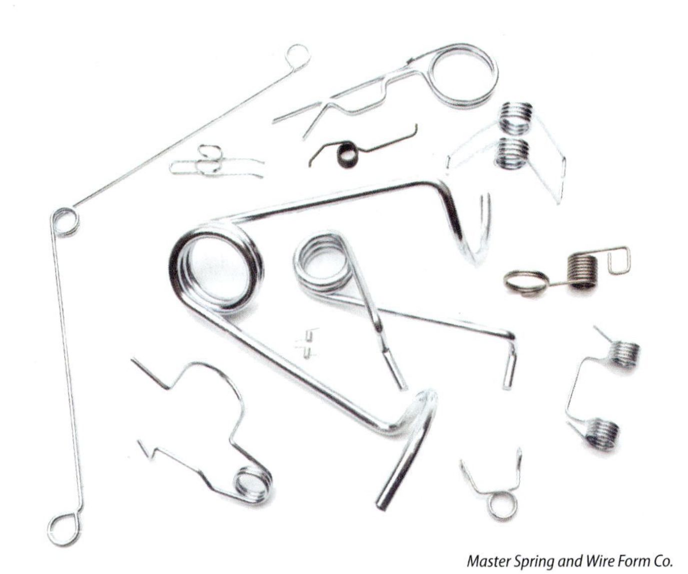

Master Spring and Wire Form Co.

Figure 17-4. Torsion springs can be used to provide tension or force around an axis.

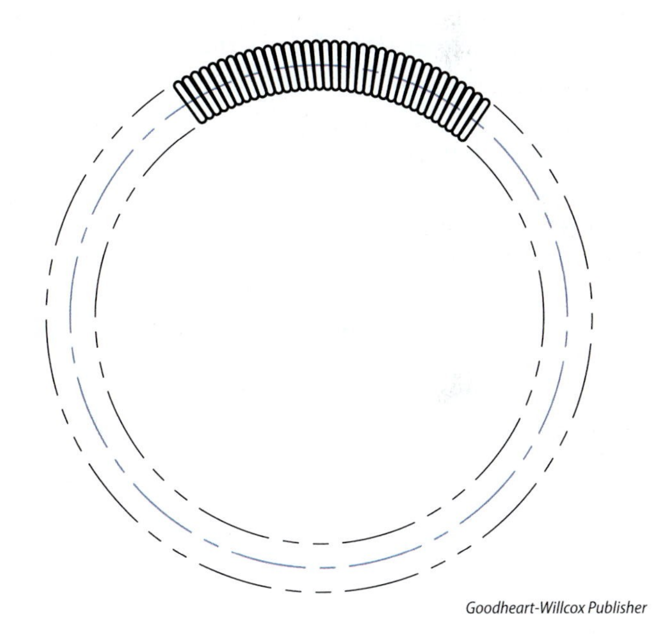

Goodheart-Willcox Publisher

Figure 17-5. Garter springs are long extension springs designed to encompass a circular feature in a belt-like manner.

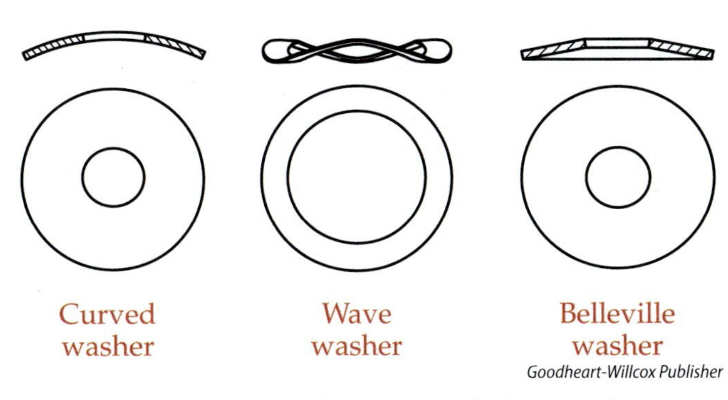

Goodheart-Willcox Publisher

Figure 17-6. Spring washer types include curved, wave, and Belleville.

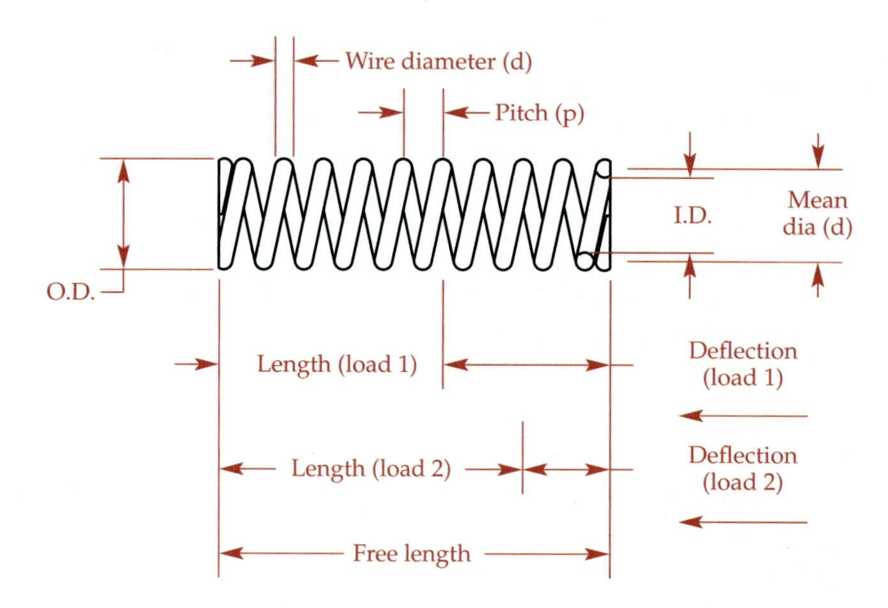

Goodheart-Willcox Publisher

Figure 17-7. Terms related to springs are illustrated in this compression spring example.

- **Inside diameter (ID).** This is the diameter on the inside of the coil of a compression or expansion spring.
- **Outside diameter (OD).** This is the diameter on the outside of the coil of a compression or expansion spring.
- **Average diameter.** This is the diameter average between the ID and OD, usually the diameter measured from the center axis of the wire across the axis of the spring from side to side.
- **Free length.** This is the overall distance from one end of the spring to the other when no force is being applied.
- **Handed.** This describes the helical winding direction, either left-handed (LH) or right-handed (RH). This may not have an impact on the design of the spring, unless other parts are inserted into the spring.
- **Load.** This is the force applied to the spring in a particular application.
- **Pitch.** This is the distance from the center of one wire to the center of the next wire in the active coils. For an extension spring, the pitch is usually equal to the wire diameter. Specifying the number of coils is more common than specifying the pitch.

Drawing Representation

Before the advent of CAD systems, it was tedious to draw every coil of a spring. Therefore, it became common to only draw each end of the spring, including one or two coils. The ends of the spring are then connected with phantom lines, as shown in **Figure 17-8**. Another common technique is to draw the spring as a schematic single-line representation, still showing the number of coils (also shown in **Figure 17-8**).

Most CAD systems can easily duplicate entities. This has made the phantom-line method of drawing springs unnecessary, but it can still be used to keep the number of lines to a minimum. Garter springs use circular phantom lines to assist the drafter in showing the spring without complex details.

End Styles

When drawing compression springs, the type of end finish is usually shown. **Figure 17-9** shows the common end styles. The ***open end***, also called plain end, style simply features the wire ending after a specified

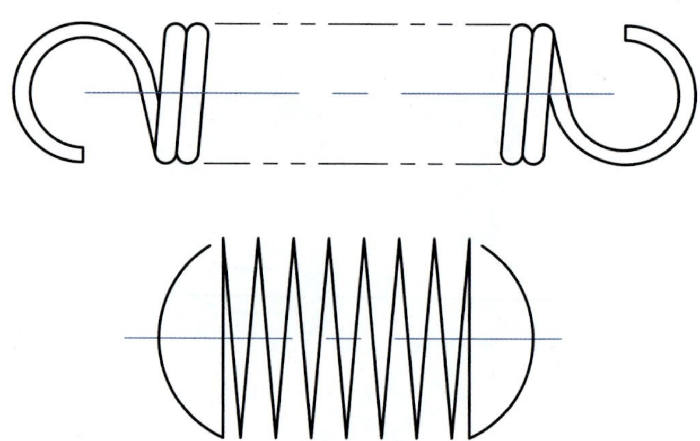

Goodheart-Willcox Publisher

Figure 17-8. In the past, phantom lines were commonly used to show repetitive detail like that found in spring drawings, or springs were drawn in a schematic format. With CAD systems, these techniques are not as common.

number of coils, spaced evenly based on the pitch. The ***closed end***, also called square end, style changes the pitch on the last coil in such a way that the end of the spring is somewhat perpendicular to the axis without grinding. The other options for end styles are based on whether or not a grinding process is applied to create a more perpendicular surface for the spring. Closed and ground and open and ground are two additional end styles that can be specified when ordering springs from most spring manufacturers.

For extension springs, the hook type is often drawn at each end. As shown in **Figure 17-10**, hook types include machine hook, full round, full loop, double loop, loop on side, and extended hook. Garter springs may be connected end-to-end with a tapered end, a connection, or loops.

Typical Specifications

Depending on the application, the typically specified spring data may include the wire material, wire diameter, outside diameter, inside diameter, number of coils, and free length. For extension springs, the free length is often clarified to be "inside hooks." The gap between the hook tip and the body length can also be specified. See **Figure 17-11**. For compression springs, a specification similar to "works inside a ____ diameter hole" or "works outside a ____ diameter shaft" can be indicated. For expansion springs, an initial tension force can be specified to indicate the conditions under which the spring will be at rest. For extension and compression springs, it is common to specify load lengths, such as "length at load 1 = ____" or "length at load 2 = ____," for example. For compression springs, a solid height can be specified. Sometimes special requirements are specified related to temperature, chemical exposure, or finish.

Torsion springs can be drawn to show various specifications. The drawing usually includes the free state leg angle between the two legs, the installed deflection, and the maximum deflection. See **Figure 17-12**.

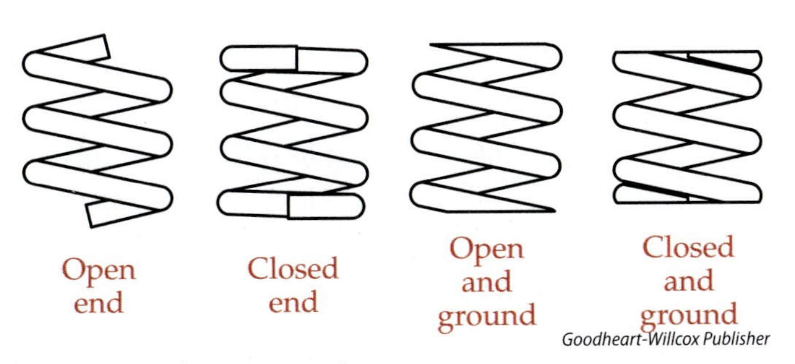

Figure 17-9. Compression spring end styles include either open (plain) or closed (square), each either ground or not ground.

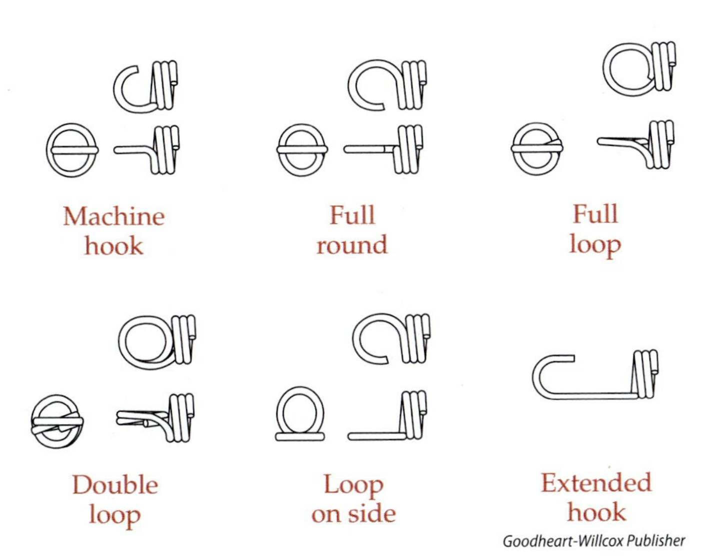

Figure 17-10. Extension springs can be made with a variety of ends to attach to other parts.

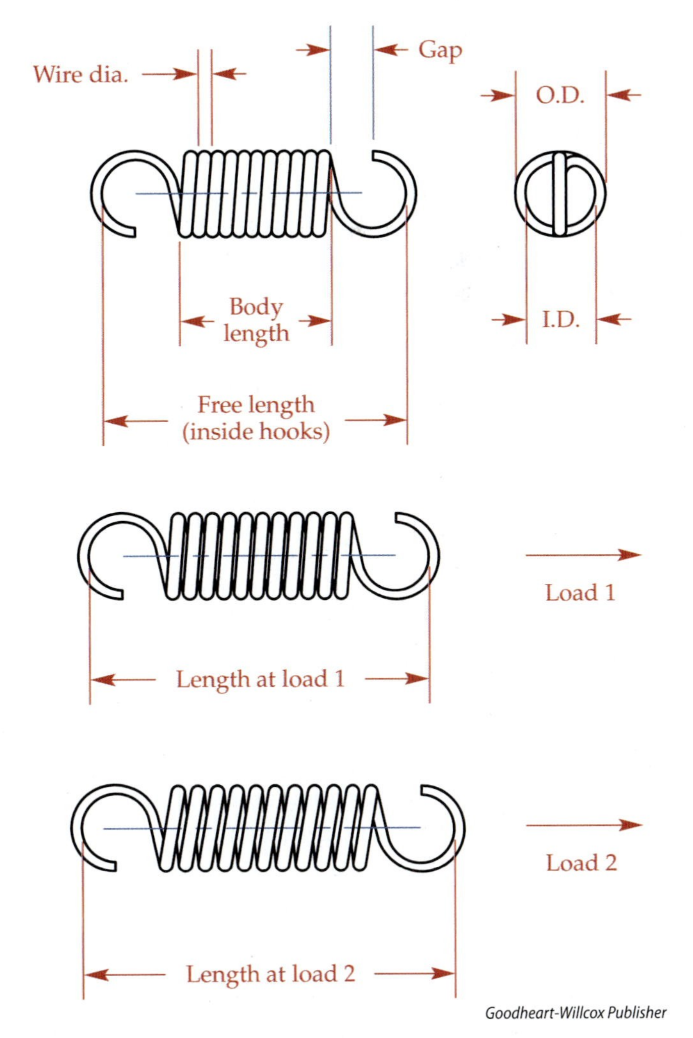

Figure 17-11. Terms related to extension springs are illustrated in this example.

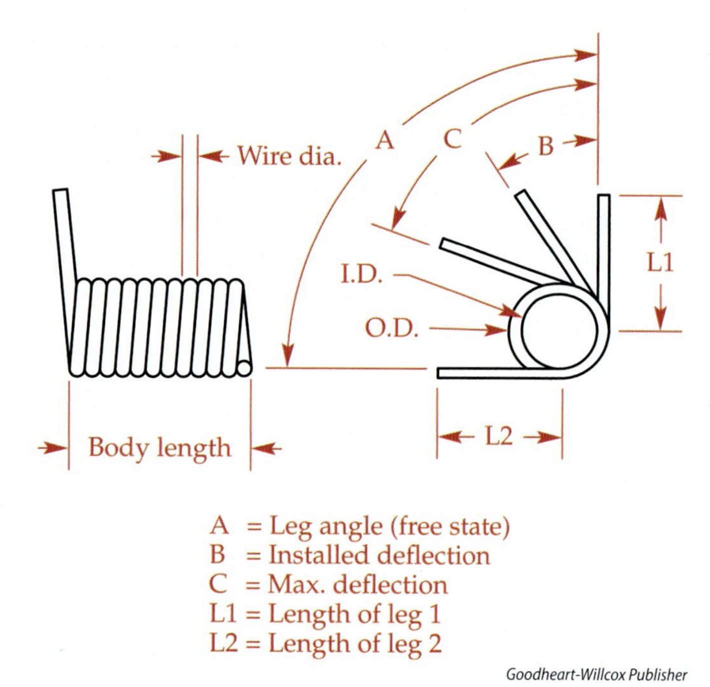

Figure 17-12. Terms related to torsion springs are illustrated in this example.

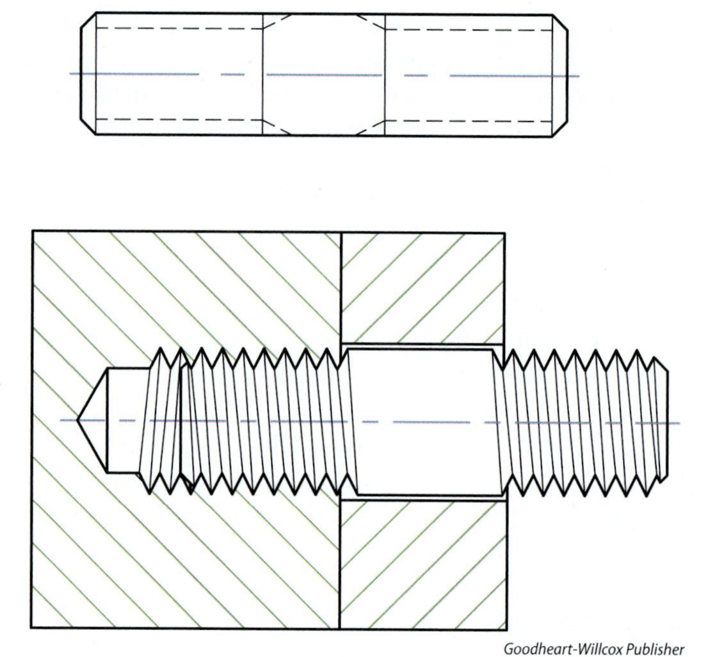

Figure 17-13. A stud is a cylindrical fastener with threads on both ends, shown here in the simplified (upper) and detailed (lower) methods. In the detailed view, the stud is shown assembled in a blind tapped hole with a cover plate and clearance hole.

Threaded Fasteners

The most common way of fastening two or more parts together is with threaded fasteners, such as nuts and bolts. This section discusses the types of threaded fasteners and the terminology associated with threaded fasteners.

General Terminology

Terminology for threaded fasteners is not as standardized across the industry as you would expect. There are arguments for and against many definitions. Within the scope of this textbook, the term ***bolt*** is used as a general term for a uniform-diameter cylinder that has external threads on one end and a head on the other. A ***stud*** is a rod that has external threads on both ends, **Figure 17-13**. The threads may or may not be the same on each end. Typically, one end of a stud may be screwed into a threaded hole, leaving a threaded post sticking up that accepts a nut. In general, a ***nut*** is a device with internal threads designed to fit on the threaded end of a bolt or stud.

A universal distinction between a bolt and a screw does not exist. One of the common terms used throughout the years is ***cap screw***, which usually applies to bolts that are finished. Cap screws may have hexagonal heads with built-in washer surfaces, or they may have other head types, as discussed in the next section. At one time, the term cap screw indicated the body was completely threaded from the head to the tip. The term ***machine screw*** is similar in definition to a cap screw, but the term is usually applied to smaller fasteners of 1/4″ diameter or less.

Both cap screws and machine screws can accept nuts on the end, so as to hold two parts together in a "floating fastener" manner. They can also be used through a clearance hole in one part and into a threaded hole in the second part, in a "fixed fastener" application. **Figure 17-14** illustrates (from left to right) a bolt and nut; a cap screw in a blind, tapped hole; and a machine screw in a tapped through hole.

Head Types

There are a wide variety of head styles for cap screws and machine screws, **Figure 17-15**. The most common style is the hexagon head, which may or may not include a washer surface underneath the head. Flat-head and round-head fasteners with a slot are normally shown looking through the slot. If a round view or top view is drawn, the slot is rotated 45°, even though the other view is looking through the slot. This is a common conventional practice that ignores true projection for the sake of clarity. Other common head styles include the hex flange, button, fillister, hex socket, and oval. Print readers should be aware that abbreviations such as FHMS (flat-head machine screw) or BHCS (button-head cap screw)

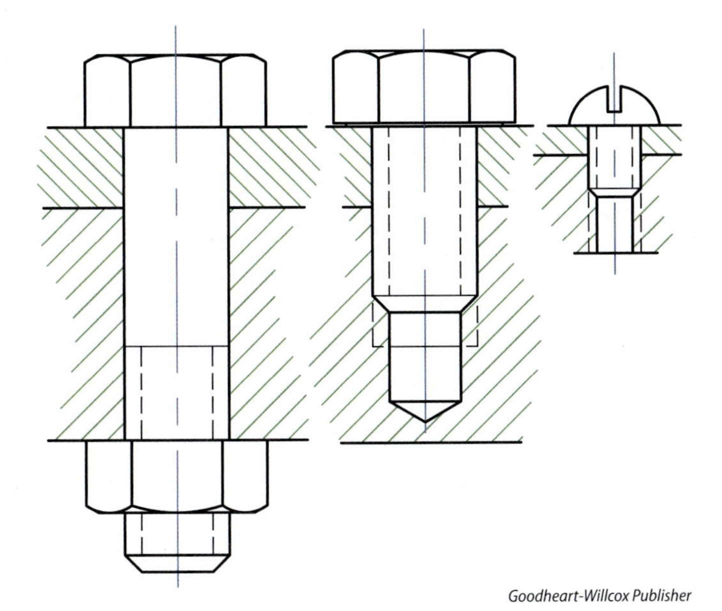

Figure 17-14. Bolts typically have a head on one end. Often, a nut is threaded onto the other end. Cap screws are bolts that can also be screwed into a threaded hole. Machine screws are smaller screws that may or may not use a nut.

may be included on the print, so a little research may be necessary from time to time if you do not deal with fasteners on a regular basis.

Nut Types

The common hexagon nut is the item that mounts on the end of a bolt to provide the "second head." By tightening the nut, the fastener remains in place. Nuts are identified by the bolt body size, not the distance across the flats of the square or hexagon head. If hexagon nuts are finished, there is a slight chamfer on one or both sides. A heavy nut is thicker than a standard nut.

In addition to the hexagon nut, there are several variations designed to function in particular ways. As illustrated in **Figure 17-16**, these include the flange, jam, slotted, and square nut. The flange nut has a built-in washer surface that provides the same function as a washer. Jam nuts can be used in situations where a regular nut is too thick, but also as a second nut that tightens against the other nut to lock it in place. Slotted nuts have slots that accommodate cotter pins that can be placed through a hole in the bolt. Square nuts are not as common as hexagon nuts but are available in various sizes. The square nut requires an open-end wrench but provides more bearing surface for the wrench than the hexagon nut. Hex coupling nuts and sleeve nuts are designed to connect two pieces of threaded material. The length of a coupling nut is approximately three times the nominal size.

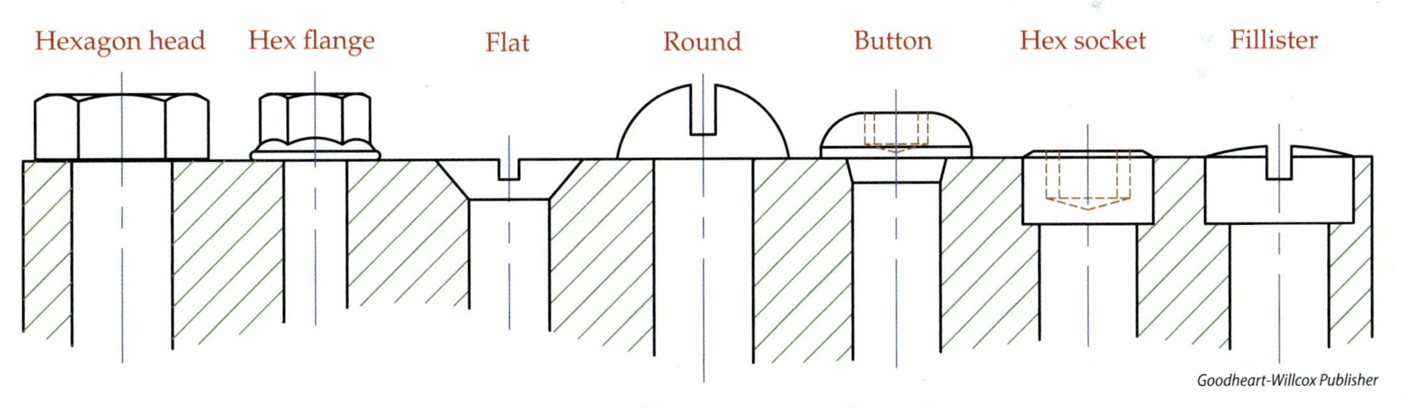

Figure 17-15. Machine screws and cap screws are available in a variety of head shapes.

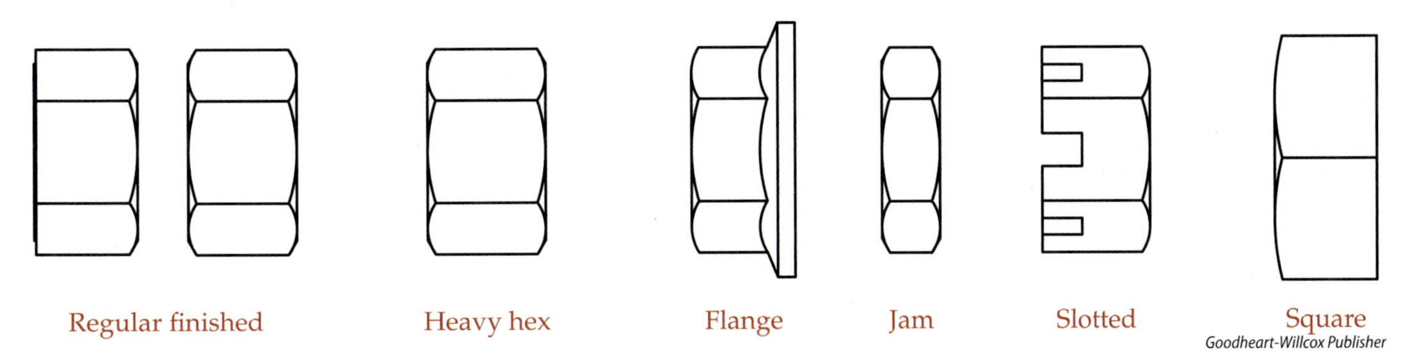

Figure 17-16. Nuts are available in a variety of shapes and for a variety of purposes.

Screws

Set screws are designed to help hold parts in place. For example, a set screw can be used to hold the hub of a pulley onto a shaft by tightening the set screw into a flat area on the axle. **Figure 17-17** shows an example of a set screw used as a stop pin. While set screws are often headless, as with the common socket head or slotted head, a set screw can also have a square head. **Figure 17-18** shows four common types of head for set screws and six common point types. Set screw points can be categorized as flat, cup, oval, cone, half dog, and full dog.

There is some disagreement about using the word "screw" for a non-tapered, threaded body. In some standards, the word "screw" is proposed as a definition for a tapered body that is threaded but also designed to cut its own threads. These screws are sometimes referred to as tapping screws, but this description also includes such fasteners as wood screws or lag screws, both of which might require a pilot hole.

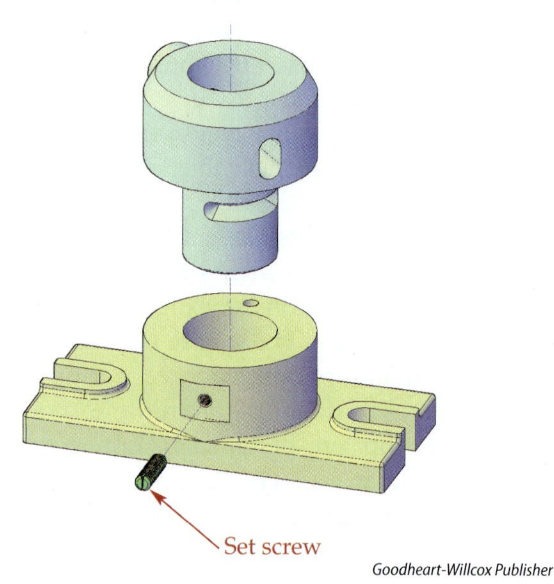

Figure 17-17. Set screws can be used to help hold one part to another. In this illustration, the set screw serves as a stop pin that allows a pivoting part with a mating groove to rotate between limits.

Non-threaded Fasteners

The previous sections discussed threaded fasteners. This section discusses non-threaded fasteners. Included in this category are washers, rivets, pins, keys, and retaining rings.

Washers

Washers are thin, flat cylinders that have a clearance hole for a fastener. They are usually made of metal and are designed to work in conjunction with bolts, nuts, pins, and other fasteners. See **Figure 17-19.** Washers help distribute the pressure of the tightened fastener across more surface area and also may be used to fill in space.

Flat washers are simply rings of metal with a clearance hole for the fastener. Fender washers have an outside diameter larger than a typical flat washer to help spread the tightening forces even more. Spring lock washers are designed with a split in the washer and a slight helical warp. This design feature places

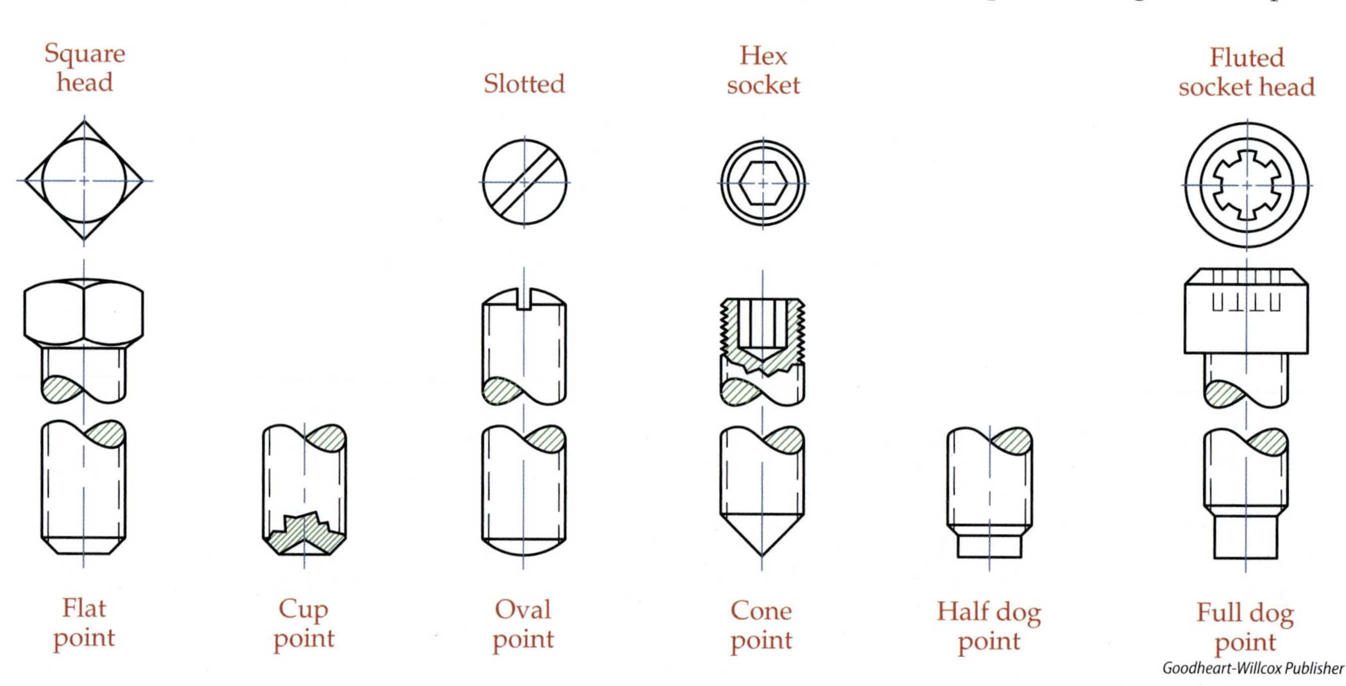

Figure 17-18. Set screws are available in a variety of head and tip options.

tension against the nut as the nut is tightened, which helps prevent it from loosening on its own. Some tooth lock washers are designed with external or internal teeth that provide locking tension.

Rivets

A ***rivet*** is a mechanical fastener designed to be permanent after installation. A solid rivet typically has one head and a smooth, cylindrical body that is malleable. After the rivet is inserted through the materials being connected, the opposite end of the rivet can be formed to also have a head. The new head of the solid rivet can be formed with a hammer or compression tool. **Figure 17-20** shows various solid rivet heads that are available and the standard print expression and line techniques for indicating how rivet heads are to be finished.

Blind rivets, or pop rivets, are designed to fasten thin materials when the installer can only access one side of the riveted hole. A special tool is used to pull

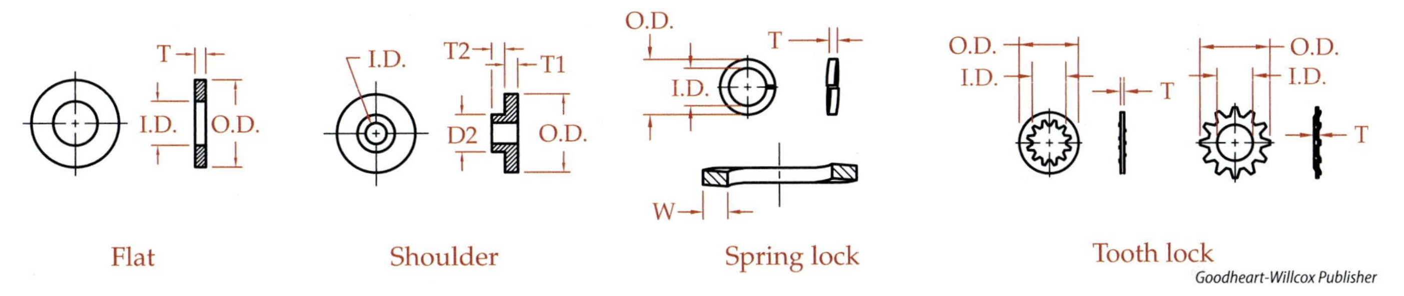

Goodheart-Willcox Publisher

Figure 17-19. Washers are available in a variety of sizes and shapes, and for a variety of purposes.

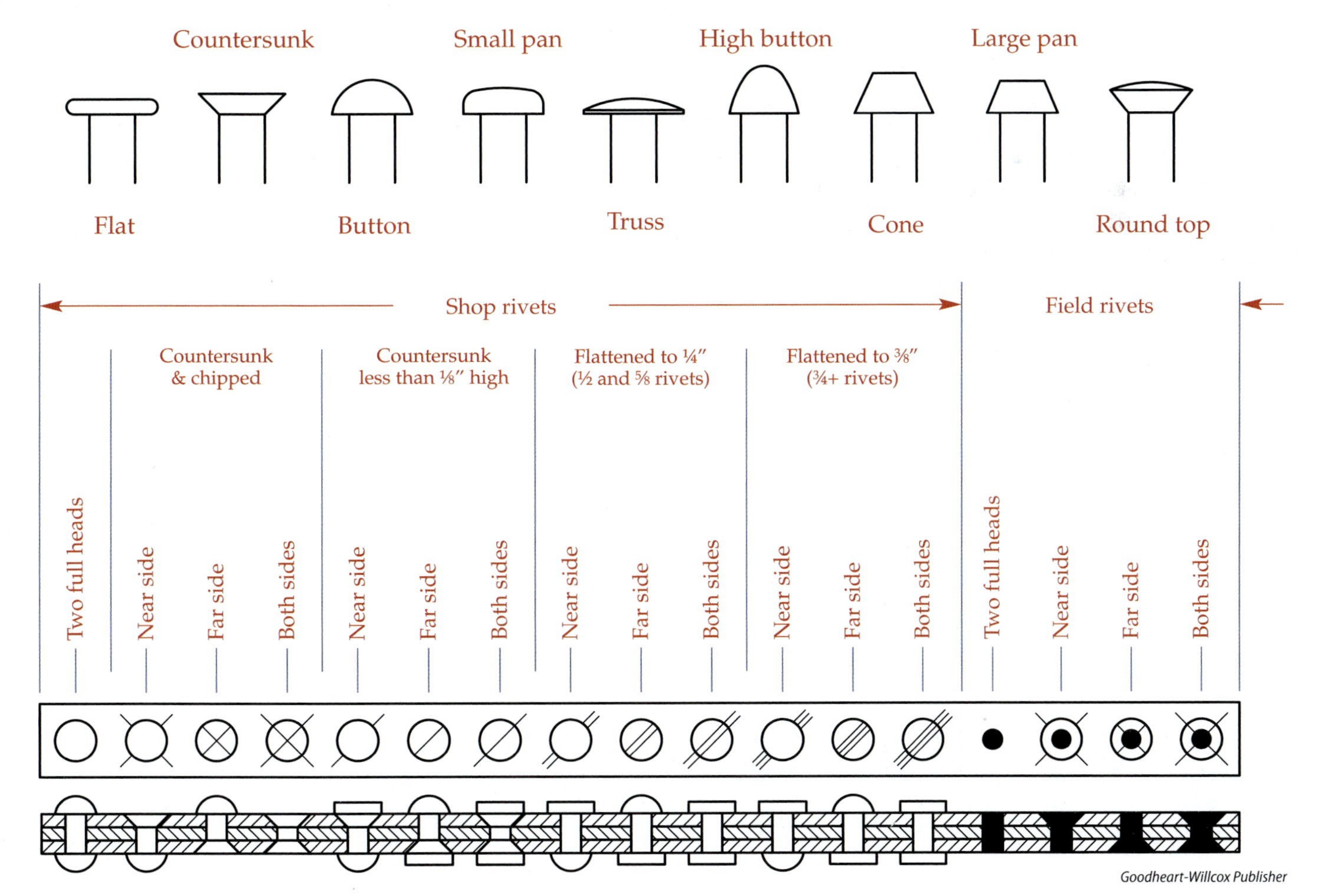

Goodheart-Willcox Publisher

Figure 17-20. Solid rivets are available in a variety of head shapes. On a print, 45° lines are used in a variety of combinations to indicate various finish techniques.

the shaft of the rivet through a hollow center in the rivet to form the second head. Blind rivets are not as strong as solid rivets.

Pins

Some parts are assembled and then fastened with pins. Pins are classified in various ways, including tapered, straight, spring, grooved, clevis, and cotter. See **Figure 17-21**. Tapered pins taper 1/4″ per foot (1:48) on the diameter and are classified by number. A #0 taper pin is the smallest and a #14 pin is the largest. The large end of the taper pin is constant, as defined by the pin number, but the pin length determines the diameter of the small end. Drilling the hole in steps or using a tapered reaming tool is necessary to get a good fit.

Straight pins are more difficult to assemble. Tight tolerances are required to ensure a good fit, so the pin does not fall out. Spring pins are rolled and have a slot that allows for expansion and contraction, thus assuring a more secure fit. In similar fashion, grooved pins have grooves pressed into their cylindrical or tapered bodies that are designed to help the pin remain securely in place. Vendors of grooved pins usually have a wide selection of groove types, each with certain advantages. Many pins, including grooved pins, are designed to require a press fit.

Clevis pins are designed for quick assembly or disassembly. The clevis pin has a head on one end and a hole designed to accept a cotter pin on the other end. See **Figure 17-22**. Clevis pins come in a range of diameters and lengths. The important dimensions of a clevis pin include the main shank diameter, the distance from the head to the cotter pin hole, and the diameter of the cotter pin hole.

Cotter pins are made of a folded strip of metal, which provides a split end to the pin that can be flared by separation. The folded end of the pin bulges enough to form a head. The tips of the split end are offset enough to be able to grab the ends and bend them with pliers or other tools. Cotter pins are used in conjunction with clevis pins and slotted nuts. They are also available in shapes that do not require flaring, **Figure 17-23**.

Dowel pins are another type of pin common in industry. These heat-treated and precision-made pins are used more in subassemblies than as fasteners. Dowel pins are designed to be press fit into holes to create a boss or guide pin that helps parts line up for fasteners. For example, a base or container may have two dowel pins press fit into the protruding edges

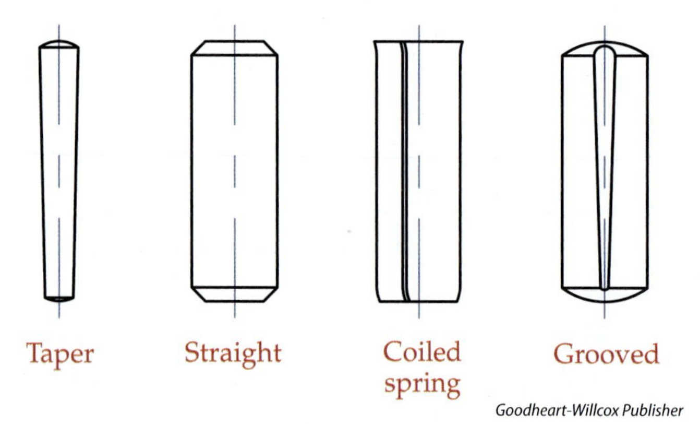

Figure 17-21. Pins are available in a variety of sizes and shapes, and for a variety of purposes.

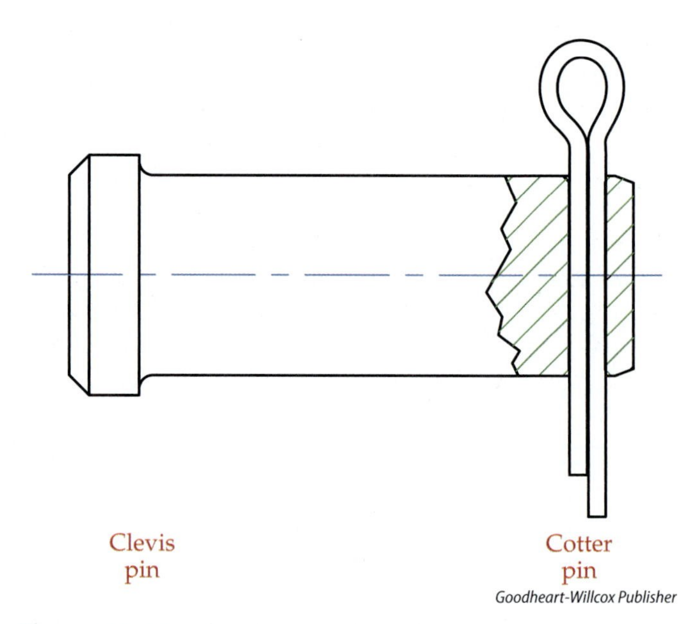

Figure 17-22. The clevis pin has a head on one end and a hole for a cotter pin on the other end of the shaft.

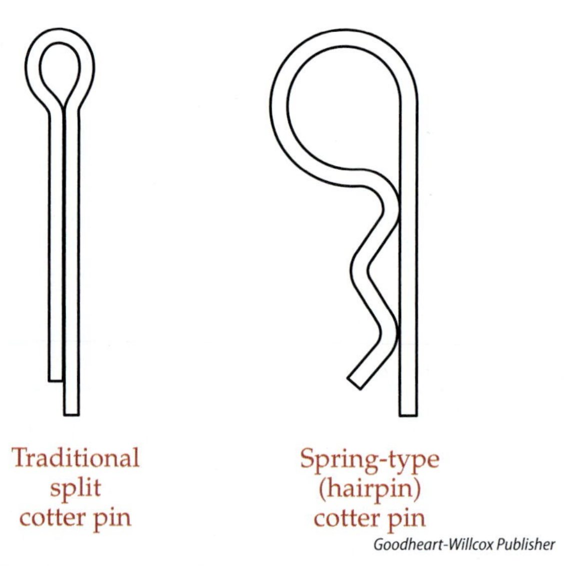

Figure 17-23. A traditional cotter pin features a split end that can be flared by separation. Another type of cotter pin is the hairpin, or spring, type that can be snapped into place.

to help align the cover, with clearance holes that fit over the pins, as well as additional clearance holes for threaded fasteners. The dowel pins are usually about the same diameter as the fasteners used in the other holes and are designed to have twice the length of the diameter embedded into each part. Dowel pins usually have a small chamfer on each end, although some are designed with spherical ends.

Keys

Keys come in a variety of shapes and forms, **Figure 17-24**. As discussed in Unit 11, *keys* are primarily used to keep a part from rotating about the shaft to which it is mounted. For example, a wheel or pulley is often mounted to a shaft, and a key is inserted into grooves in the hub of the wheel and the shaft. The grooves are designed to accommodate the particular key that is used. A square or rectangular key can be placed into a groove in the shaft that is long enough to accommodate the key. The slot in the hub is then aligned to slide over the key.

Woodruff keys are somewhat semicircular, so the keyseat is cut into the shaft with a Woodruff cutter. The tapered key and gib-head key are both designed to be inserted into the keyway after the hub has been mounted over the shaft. Both of these keys taper at 1/8″ per foot. The hub length is designed to engage in the mating parts in such a way that the key can be grabbed and removed. Other key shapes are available through various vendors. These may be named with trademark names or with general descriptive names.

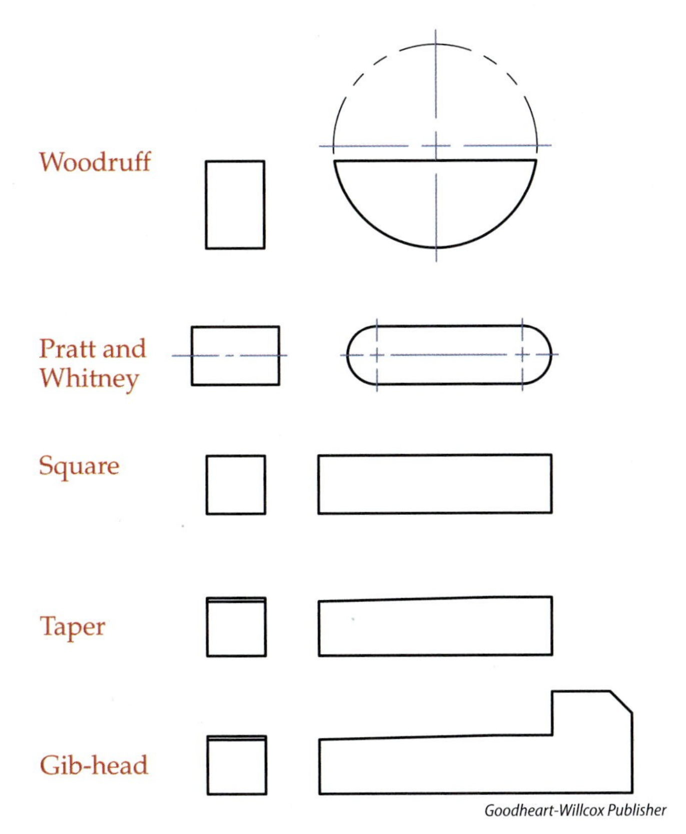

Figure 17-24. Keys are available in a variety of sizes and shapes, each with different characteristics. Tapered and gib-head keys are designed to be inserted into the key groove after the hub is mounted on the shaft.

Retaining Rings

Retaining rings are used in assemblies to help parts stay together. They are somewhat permanent, although they are designed in such a way that they can be removed with special tools. Retaining rings typically are inserted into grooves. The groove may be external, with the retaining ring fitting into a groove near the end of a shaft. The groove may also be internal, with the retaining ring snapping into a groove near the end of a hole. See **Figure 17-25**.

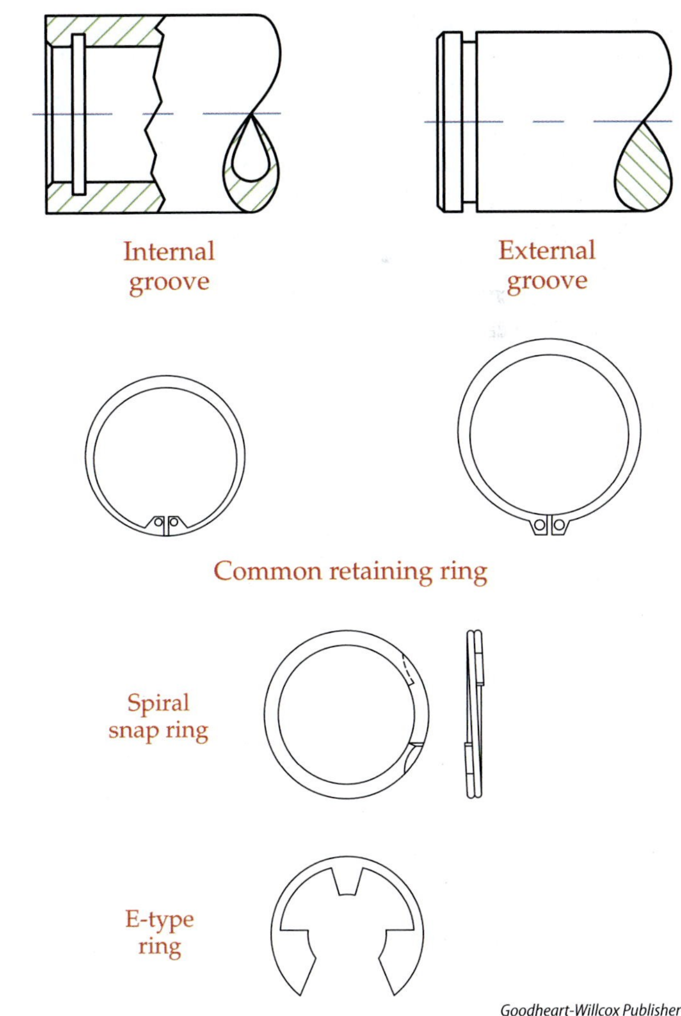

Figure 17-25. Retaining rings commonly have holes for a tool, but they are also available in styles that snap in and out with other techniques.

The most common retaining rings contain tiny holes on the two ends to allow special tools to expand or compress the rings. Other retaining rings include snap-style rings designed to function in a similar fashion, but without the ears of traditional retaining rings. E-style rings are also available and can be found in a variety of sizes.

Summary

- There are a wide array of standard parts used in assembling the components of a product, including springs, pins, keys, and other fasteners, such as nuts, screws, and bolts.
- There are many ASME standards that set forth guidelines for the manufacture of keys, bolts, pins, and rings.
- Springs are devices made from wire that exert forces in such a way as to control parts in assembly, falling into categories such as extension, compression, and torsion.
- Terms related to springs are similar to terms related to threads, as a portion of the spring is based on a helical geometric form.
- Spring representation can be simplified, with phantom lines representing the repetitive features, but it can also be schematic in nature.
- Terms and conditions related to springs include types of spring ends, lengths at free state versus loads, wire diameter, and coil diameter.
- Threaded fasteners in industrial applications include bolts, studs, nuts, cap screws, machine screws, and set screws.
- There are a variety of head types for cap screws, machine screws, and set screws.
- Nuts are available in a variety of shapes and for a variety of purposes.
- Set screws are available in a variety of head styles as well as tip styles.
- Non-threaded fasteners in industrial applications include washers, rivets, pins, and retainer rings, each available in a variety of shapes and sizes for a variety of purposes.

Unit Review

Name ______________________ Date ____________ Class ____________

Answer the following questions using the information provided in this unit.

Know and Understand

____ 1. *True or False?* There is one ASME standard that covers all springs and fasteners.

____ 2. Which of the following is *not* a type of spring?
A. Expansion
B. Compression
C. Torsion
D. Garter

____ 3. Which of the following terms is *not* a real term related to springs?
A. Active coils
B. Outside diameter
C. Lead
D. Free length

____ 4. Which of the following is *not* a type of spring washer?
A. Curved
B. Fender
C. Wave
D. Belleville

____ 5. *True or False?* As defined in this unit, "bolt" is a general term that can be applied to both cap screws and machine screws.

____ 6. Which of the following is *not* a type of cap screw head?
A. Jam
B. Round
C. Fillister
D. Hex socket

____ 7. *True or False?* While nuts come in a variety of sizes and thicknesses, and some have washer-like flange surfaces or slots for cotter pins, all of these are based on a hexagon shape.

____ 8. Which of the following is *not* a type of set screw tip?
A. Half dog
B. Cup
C. Oval
D. Shoulder

____ 9. *True or False?* Tooth lock washers have little "teeth" that help provide locking tension, and come in both internal and external versions.

____ 10. *True or False?* Rivets come in three forms: solid, blind, and threaded.

____ 11. Which of the following is *not* a type of pin fastener discussed in this unit?
A. Taper
B. Straight
C. Gib-head
D. Coiled spring

____ 12. *True or False?* Clevis pins have a head on one end and a hole on the other end designed to accommodate a cotter pin.

____ 13. *True or False?* Rather than serve as fasteners, dowel pins are commonly used to help align parts within the assembly.

____ 14. Which of the following is *not* a type of key discussed in this unit?
A. Woodruff
B. Hairpin
C. Taper
D. Pratt & Whitney

____ 15. *True or False?* Retainer rings are designed to snap into place in external grooves, or be removed from the grooves, using a Phillips screwdriver.

Critical Thinking

1. What are some devices common to everyday life around the house that incorporate springs and/or fasteners?

____

____

____

____

____

2. What benefits are there to using a CAD system to draw springs and fasteners compared to the drafting techniques that were used before CAD systems?

____

____

____

____

____

Notes

Apply and Analyze

Name ______________________ Date ______________ Class ______________

Industry Print Exercise 17-1

Refer to the print PR 17-1 and answer the questions below.

1. What type of head does this fastener have?

____

2. Based on the thread specification, what is the nominal diameter of this fastener?

____

3. In general, is the across-the-flats measurement of the head 1.5 times the nominal body diameter (same as the thread major diameter), which is a common proportion for bolt heads?

____

4. What is the minimum diameter of the bearing surface of the flange on the head?

____

5. What is the total maximum length of this fastener (from head tip to body tip)?

____

6. What is to be applied to a portion of the threads of this fastener?

____

7. In the noncircular view, is the head of the bolt shown across the flats or across the corners?

____

Review questions based on previous units:

8. At what scale was the original drawing created?

____

9. What is the part number that is replaced by this part?

____

10. What paper size is the original version of this print?

____

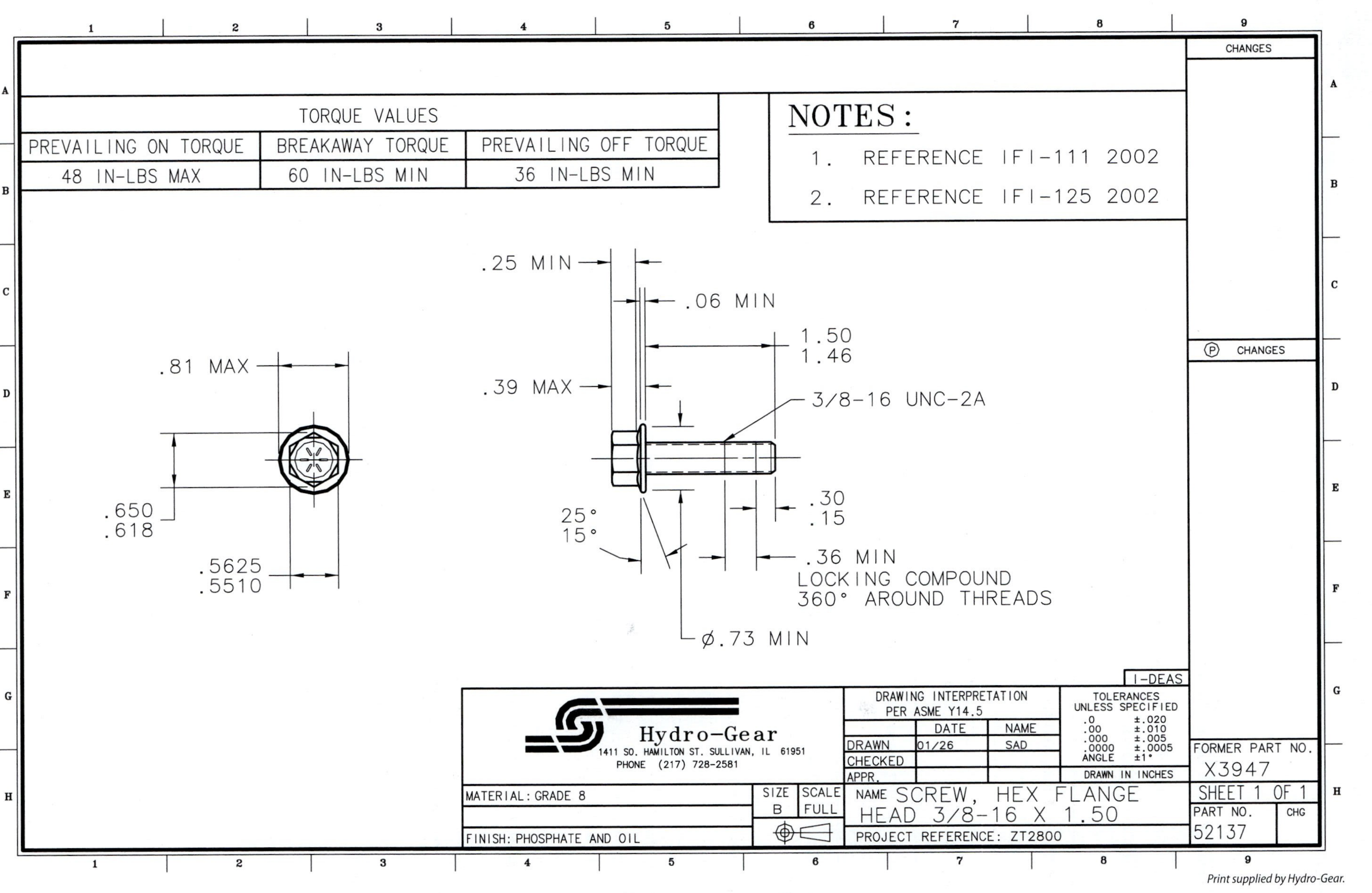

Print supplied by Hydro-Gear.

PR 17-1. Hex Flange Screw

Apply and Analyze

Name ______________________ Date ______________ Class ______________

Industry Print Exercise 17-2

Refer to the print PR 17-2 and answer the questions below.

1. Is the locknut shown sectioned or not sectioned?

____

2. How many total washers are used in this assembly?

____

3. What type of spring is used in this assembly: compression, extension, or torsion?

____

4. Does the spring used in this assembly have a ground end?

____

5. Does the part associated with find/item number 13 fit into the part associated with find/item number 14?

____

6. What must be done to the handwheel before adding the washer and locknut?

____

Review questions based on previous units:

7. What term describes the manner in which the parts associated with find/item numbers 9–14 are shown in the detail, as compared to the way they are shown in the main section view?

____

8. How many pressure seal rings are needed?

____

9. What are the part numbers of the two subassemblies (*not* find/item numbers)?

____

10. What is applied to parts 7509-9 and 7509-42 prior to assembly?

____

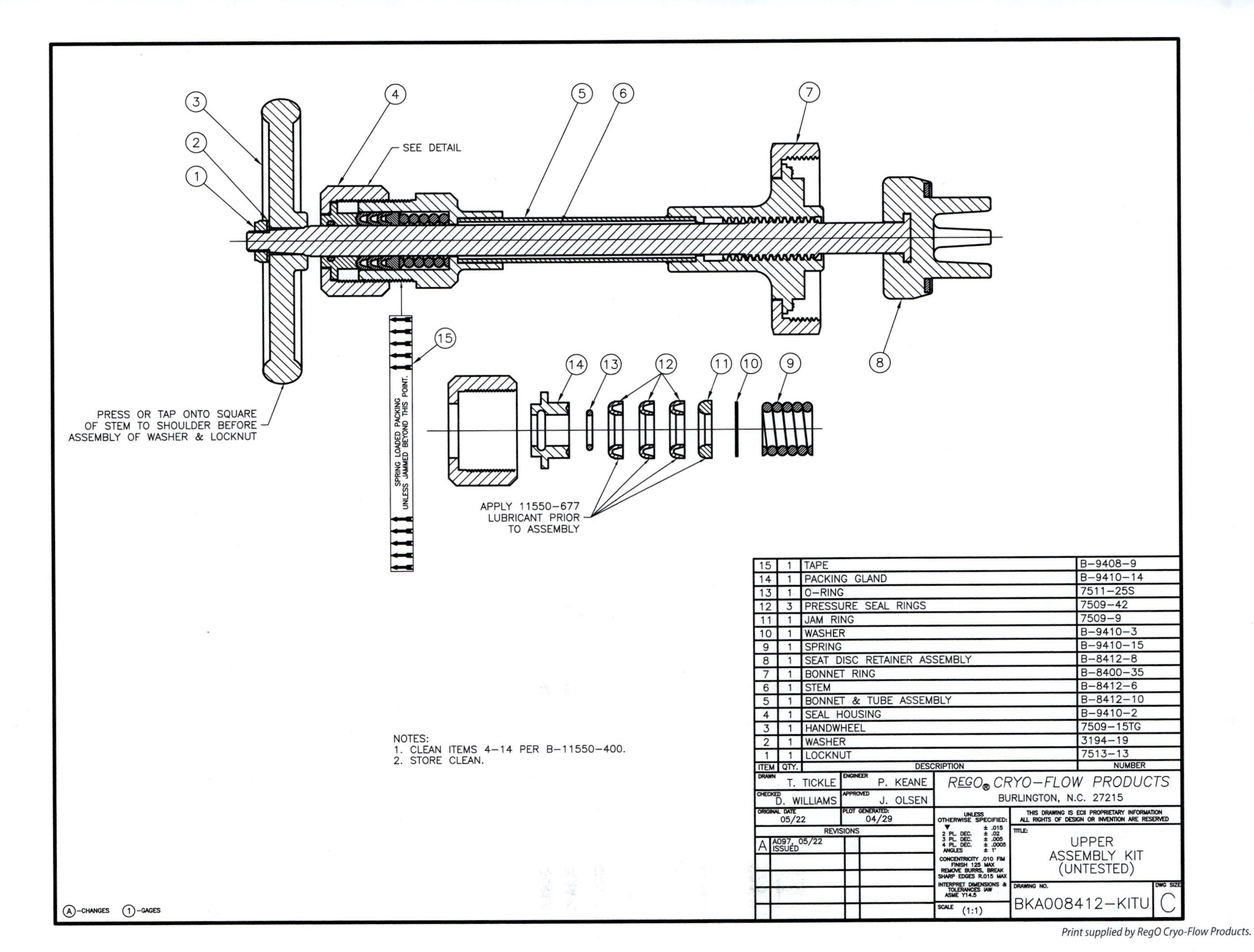

ITEM	QTY.	DESCRIPTION	NUMBER
15	1	TAPE	B-9408-9
14	1	PACKING GLAND	B-9410-14
13	1	O-RING	7511-25S
12	3	PRESSURE SEAL RINGS	7509-42
11	1	JAM RING	7509-9
10	1	WASHER	B-9410-3
9	1	SPRING	B-9410-15
8	1	SEAT DISC RETAINER ASSEMBLY	B-8412-8
7	1	BONNET RING	B-8400-35
6	1	STEM	B-8412-6
5	1	BONNET & TUBE ASSEMBLY	B-8412-10
4	1	SEAL HOUSING	B-9410-2
3	1	HANDWHEEL	7509-15TG
2	1	WASHER	3194-19
1	1	LOCKNUT	7513-13

Print supplied by RegO Cryo-Flow Products.

PR 17-2. Upper Assembly Kit

Apply and Analyze

Name ______________________ Date ______________ Class ______________

Bonus Print Reading Exercises

The following questions are based on various bonus prints located in the folder that accompanies this textbook. Refer to the print indicated, evaluate the print, and answer the question.

Print AP-002

1. Analyze the assembly and determine whether the spring featured in Section A-A is an extension spring, compression spring, or torsion spring.

____

Print AP-004

2. According to the parts list, what type of fastener is part #3, and what thread specification is given for it?

____

Print AP-007

3. This part has a groove that must be free of parting line flash. What item will be placed within that groove?

____

Print AP-009

4. Analyze the assembly, and determine whether the spring is an extension spring, compression spring, or torsion spring.

____

Print AP-010

5. Give a specific and complete description for item 3.

____

6. By visual identification, what type of head does item 5 feature?

____

7. What instructions are given about the 1/8 × 1/4 roll pins?

____

8. By visual identification, what type of tip does item 6 feature?

____

Print AP-013

9. While this drawing covers more than one assembly number, how many springs are required for any one of the assemblies?

____

10. What type of spring ends are shown for item 7?

____

11. What type of head is illustrated for the fasteners that hold the bonnet to the body?

____

12. What type of head is illustrated for the adjusting screw?

____

Print AP-015

13. Fully describe the fasteners (including the washers) that hold the turntable subassembly to the turntable.

____

14. Fully describe the fasteners (including the nuts) that hold the flag pole base to the base plate.

____

Print AP-022

15. This part will have three fasteners pressed into place. Give the specific, complete description for these fasteners.

____

Notes

UNIT 18

Gears, Splines, and Serrations

LEARNING OBJECTIVES

After completing this unit, you will be able to:

- Describe various classifications and types of gears common in industrial applications.
- Identify and discuss various industrial standards pertinent to gears and splines.
- Identify the most common terms and characteristics of gears.
- Identify spur gears and their representations and specifications.
- Identify terms and characteristics for metric gears as compared with inch-based gears.
- Describe materials and manufacturing processes relevant to gears.
- Identify bevel gears and their representations and specifications.
- Identify worm gears and their representations and specifications.
- Define splines and serrations and identify their representations and specifications.

TECHNICAL TERMS

addendum (a)
addendum circle diameter
backlash
base circle
bevel gear
center distance
chordal addendum
chordal tooth thickness
circular pitch (P_c)
circular tooth thickness
clearance
dedendum (b)
dedendum circle diameter
diametral pitch (P_d)
face gear
gear blank
helical gear
herringbone gear
hobbing machine
hypoid gear
internal diameter
internal gear
lead angle
linear pitch
miter gears
module (m)
outside diameter
pin diameter
pinion
pin measurement
pitch circle
pressure angle
rack
screw gears (crossed helical)
serrations
spur gear
throat diameter
tooth face
tooth flank
tooth space
whole depth
working depth
worm
worm gear

The purpose of this unit is to explore gears, splines, and serrations, including the associated terminology and representations on prints. While the primary focus will be gears of various types, splines and serrations have similar characteristics and manufacturing processes, so it is logical to include an overview of them in this unit. In the simplest of terms, a gear is a wheel or cylinder that features teeth, allowing the wheel to perform in union with another gear. The purpose of gears as a set may be to transmit power, to change rotation direction or speed of various axles, or to allow for precise rotation of parts, such as the gears within a stepper motor. The wide array of gear types and applications continues to expand with new technologies, and new manufacturing methods have led to many innovations in this field.

Overview of Gears in Industry

There are many types and sizes of gears used in industry today, with some of the more common gears illustrated in **Figure 18-1**. As shown in the figure, gears are usually categorized by how the axes of the gears are oriented to each other. Gears with axes parallel to each other include the spur gear, helical gear, double helical gear, and internal gear. The ***spur gear*** has teeth that are parallel with the axis, while the ***helical gear*** has teeth that are at an angle to the axis of revolution, so they are curved in a helical manner. A helical gear set requires a left-hand and right-hand pair of gears. Helical gears work more smoothly than straight spur gears, but they are more difficult to manufacture. A ***herringbone gear***,

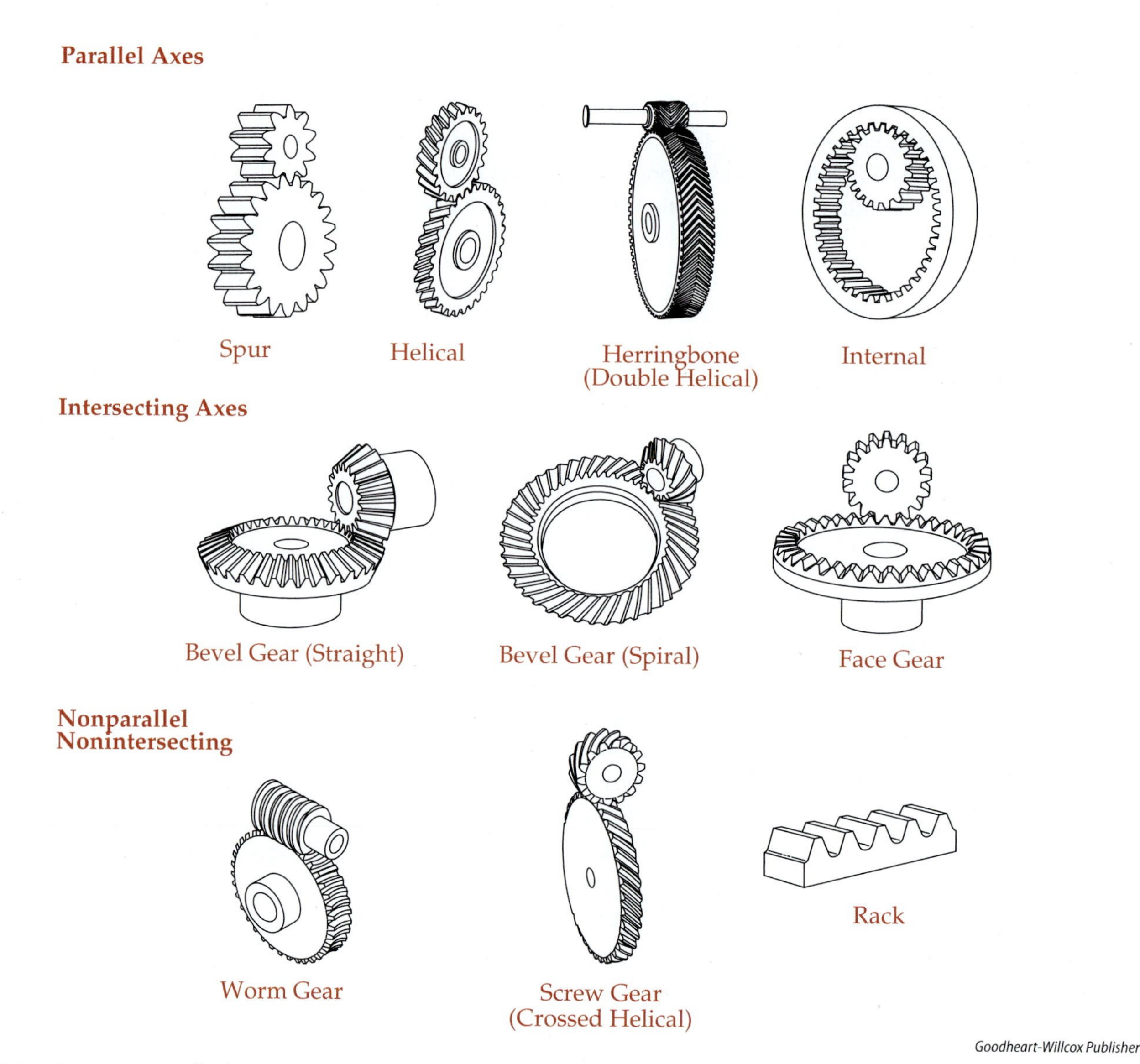

Figure 18-1. Gears are usually divided into three categories: parallel axes, intersecting axes, and nonparallel/nonintersecting axes. The gears illustrated here are some of the more common types of gears.

also known as a *double helical gear*, combines two helical gear teeth of opposing directions, featuring a V-shape appearance. The herringbone gear operates without the axial load that a helical gear incurs. An ***internal gear*** features teeth cut on the inside of a cylinder so that a spur gear can mate and function within the internal gear. A spur gear can work in conjunction with a ***rack***, which is a straight bar with gear teeth. The driving spur gear, often smaller and referred to as the ***pinion***, causes the rack to move in a linear fashion. One of the more common rack-and-pinion devices is incorporated into the steering system of a vehicle. In theory, the rack is a spur gear with an infinite diameter.

Another category of gears includes those gears whose axes intersect at some angle between 0° and 180°. The gears are based on cone-shaped surfaces that intersect and interact with each other. In this category, the ***bevel gear*** transfers motion from one axis to another intersecting axis. For bevel gears, if the axes are 90° to each other and the gears are the same size and number of teeth, they are called ***miter gears***, which simply transfer rotation from one axis to another without a change in speed. Bevel gear teeth can be straight, or they can be helical and curved in a spiral fashion, allowing for smoother and quieter action. The ***face gear*** is another variation that works along the same principles as bevel gears in a 90° relationship. The face gear allows for a spur gear to drive a circular disc with a ring of teeth cut in its side face, allowing it to be manufactured without some of the equipment needed to manufacture bevel gears.

A third category of gears is used for nonparallel and nonintersecting axes. Gears in this category include the ***worm gear***, which is driven by a cylindrical ***worm*** with characteristics resembling Acme screw threads. The worm gear itself is a version of a spur gear designed to mesh with the worm. Also within this third category are the ***screw gear***, also referred to as a *crossed helical gear*, and the ***hypoid gear***, a complex variation of the spiral bevel gear that allows for axes to be offset. Face gears with spiral teeth that can be driven by a worm gear also fall into this category.

In summary, there are many ways to implement gear design in a wide array of mechanisms. With respect to industrial prints, this unit will explore basic applications of spur gears, bevel gears, and worm gears, and expose you to the types of information to expect.

Standards Related to Gears and Splines

Under the umbrella of ANSI, the American Gear Manufacturing Association (AGMA) publishes a wide array of standard publications related to gears. In general, these standards address such topics as nomenclature, design criteria, selection, lubrication methods, testing and load ratings, and more. The Society of Automotive Engineers (SAE) also publishes gear standards, some of which address specific gear applications in the automotive industry, such as design issues related to rack-and-pinion steering, starting motor pinions, or planetary gear systems. Numerous international (ISO) standards for gear design and manufacture are also available, covering the terminology applied to metric applications.

With respect to general gear representation in drawings, there are two ASME standards, ASME Y14.7.1, titled *Gear Drawing Standards-Part 1: For Spur, Helical, Double Helical and Rack*, and ASME Y14.7.2, titled *Gear and Spline Drawing Standards-Part 2: Bevel and Hypoid Gears*.

Spur Gears

As defined earlier, a spur gear resembles a wheel with a number of equally spaced teeth cut parallel to the axis. See **Figure 18-2**. It is the most common type of gear. Spur gears are used for drives on mechanisms such as machine lathes and mills where the axes of the gears are parallel and the gears are in the same plane with each other. The reverse gear in a manual automotive transmission is a spur gear, as are most of the gears in a watch or clock.

Spur Gear Terminology

There are many terms associated with spur gears. The terms explained in this section are commonly used and should be understood. A few of the terms and abbreviations will be different in the metric system, although the principles are the same.

For inch-based gears, ***diametral pitch*** (P_d) is the number of teeth in a gear per inch of pitch diameter. For example, a gear having 48 teeth and a pitch diameter of 3″ has a diametral pitch of 16 (48 ÷ 3 = 16). Mating gears must have identical diametral pitches. In industrial prints, it is common to identify gears by the diametral pitch, with just the word "pitch." For example, a set of gears in which each gear has a diametral pitch of 48 would be referred to as 48 PITCH GEARS.

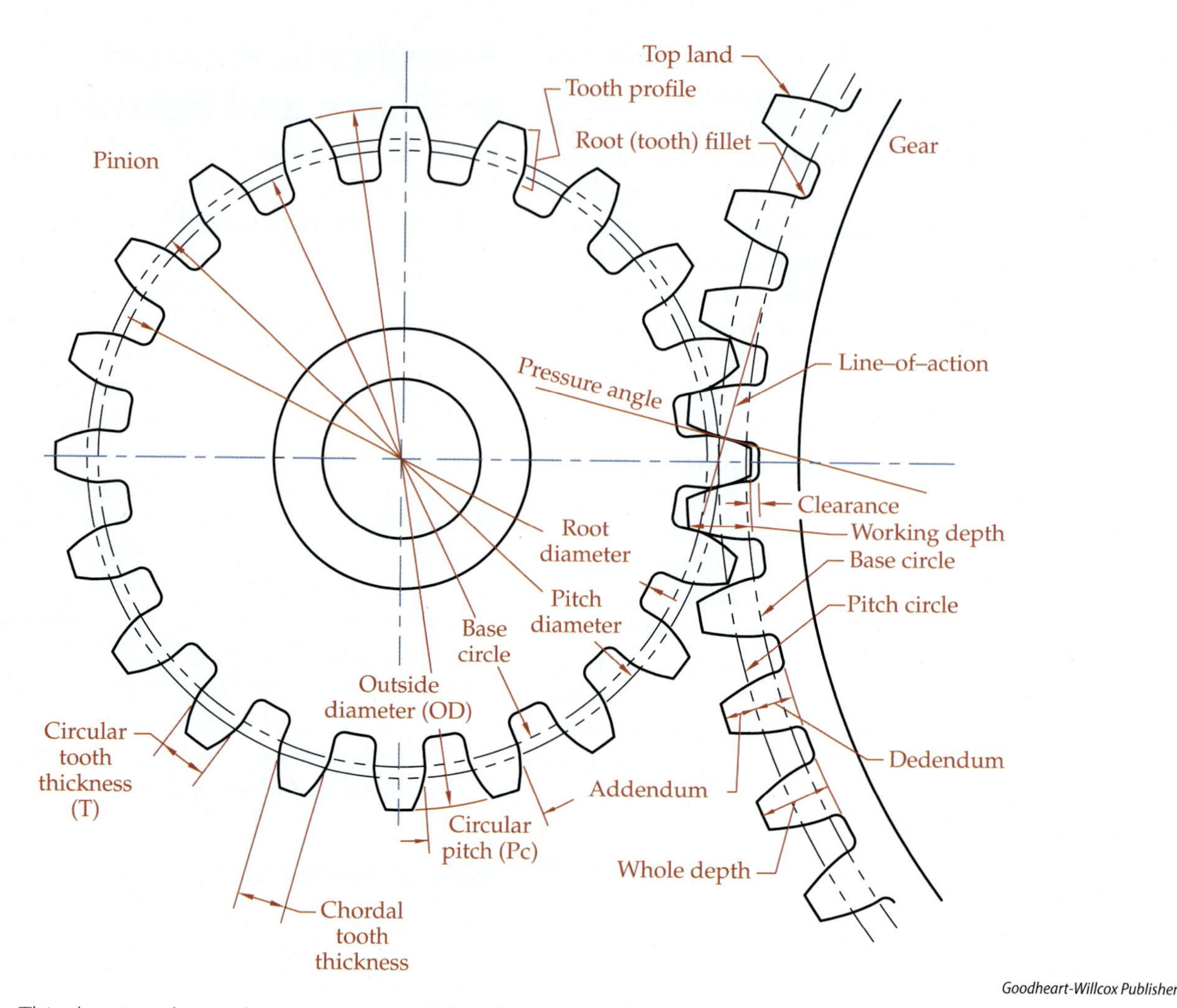

Figure 18-2. This drawing shows the geometry and identification of spur gear parts.

The ***pitch circle*** is an imaginary circle located approximately halfway between the roots and tops of the gear teeth. A gear's pitch circle is tangent to the pitch circle of any mating gear. See **Figure 18-3**. The *pitch diameter (D)* is the diameter of the pitch circle. Pitch diameters of mating gears are compared to determine the gear ratio. As illustrated in **Figure 18-3**, the relationship of one pitch diameter to another is identical to the relationship of two friction rollers.

Circular pitch (P_c) is the length of the arc along the pitch circle from the midpoint of one gear tooth to the midpoint of the next tooth. This value can be calculated by dividing the circumference of the pitch circle (πD) by the number of teeth on the gear. Nominally, this value is two times the circular tooth thickness, since each circular pitch distance contains one tooth and one matching tooth space. A gear with a pitch diameter of 6″ and 48 teeth would have a circular pitch of .3925″ and a circular tooth thickness of .1963″.

The ***addendum (a)*** is the radial distance between the pitch circle and the top of the gear tooth. This distance is usually $1/P_d$ (one divided by the diametral pitch). For example, the addendum for a gear with a diametral pitch of 4 is typically 1/4″. The ***addendum circle diameter*** is equal to the pitch circle diameter plus twice the addendum (D + 2a). The addendum circle diameter is equivalent to the outside diameter of the gear.

The ***dedendum (b)*** is the radial distance between the pitch circle and the bottom of the gear tooth. This distance is usually $1.25/P_d$ (1.25 divided by the diametral pitch). For example, the dedendum for a gear with a diametral pitch of 4 is typically 5/16″. The dedendum must be larger than the addendum to allow for clearance between the mating gear teeth. The ***dedendum circle diameter*** is equal to the pitch circle minus twice the dedendum (D – 2b). The dedendum circle diameter is equivalent to the root diameter of the gear.

Clearance is the radial distance between the top of a tooth on one gear and the bottom of the tooth space on the mating gear. See **Figure 18-2**. Clearance can be expressed as the dedendum minus the addendum (b – a).

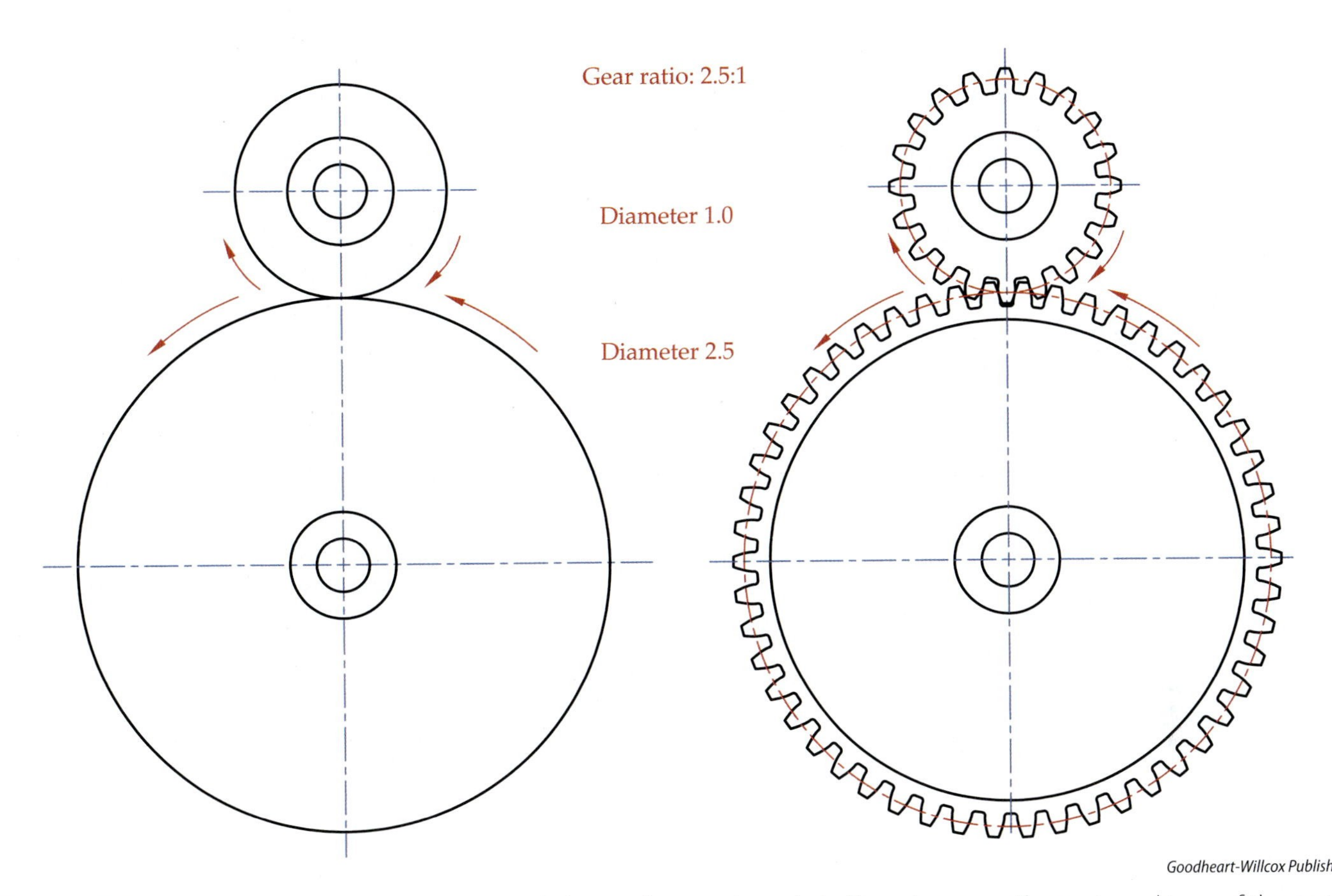

Figure 18-3. The pitch circle is an imaginary circle located approximately halfway between the roots and tops of the gear teeth.

The ***tooth face*** is the curved surface of the gear tooth that lies outside of the pitch circle. The ***tooth flank*** is the curved surface of a gear tooth that lies inside of the pitch circle. ***Tooth space*** is the distance at the pitch circle between two adjacent gear teeth. In general, the tooth space is the circular pitch divided by two ($P_c \div 2$). However, a calculated amount is added to allow for clearance between the gear tooth and the tooth space, so tooth space is slightly larger than the corresponding tooth.

The ***circular tooth thickness*** is the length of the *arc* along the pitch circle between the two sides of a gear tooth. This thickness is equivalent to the circular pitch divided by two ($P_c \div 2$). The ***chordal tooth thickness*** is the length of the *chord* between the intersection of the pitch circle and the two sides of the gear tooth. The chordal tooth thickness is the distance measured when a gear-tooth caliper is used to measure the tooth thickness at the pitch circle. See **Figure 18-4**. The ***chordal addendum*** is the distance from the top of the gear tooth to the chord at the pitch circle. It is the height dimension used in setting gear-tooth calipers to measure tooth thickness. Chordal thickness and addendum values are available in table form. The table values are often given for a diametral pitch of one, but specific distances can be found by dividing the table value by the diametral pitch for a particular gear.

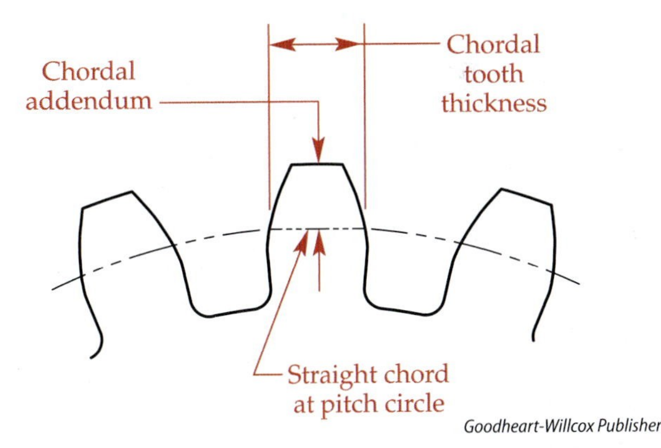

Figure 18-4. The chordal tooth thickness is the tooth thickness at the pitch circle, as measured by a gear-tooth caliper.

The ***working depth*** of two mating gears is the sum of their addendums (a + a). The ***whole depth*** is the total depth of a tooth space, or addendum plus dedendum (a + b).

The ***pressure angle*** is the angle between the tooth profile and a radial line at the intersection of mating gear teeth on the pitch circle. The pressure angle can also be illustrated by drawing a line through the points of contact between engaged gear teeth. Standard values for pressure angles throughout the years have been 14.5°, 20°, and 25°. The pressure angles of mating gears must be the same. In general, the 20° pressure angle is the most common and versatile.

Backlash is the amount by which the width of a tooth space exceeds the thickness of the engaging tooth on the pitch circles. While the basic calculations for a gear would create a tooth space width identical to a tooth width, some backlash must be allowed to compensate for manufacturing error, expansion, and lubrication.

The ***outside diameter*** is the diameter of a circle coinciding with the tops of the teeth of an external gear. This is the equivalent to the addendum circle. For an internal gear, the ***internal diameter*** is the diameter of a circle coinciding with the tops of the gear teeth.

The ***center distance*** is the distance between the center axes of mating gears. The ***base circle*** is the diameter from which an involute tooth curve is generated or developed. **Figure 4-12** in Unit 4 illustrates how the involute shape is generated. The base circle diameter is calculated by multiplying the pitch circle diameter by the cosine of the pressure angle. As a result, gear teeth with a 20° pressure angle have more involute surface length than gear teeth with a 14° pressure angle.

Pin measurement is sometimes used to specify a condition or size for gears or splines. This measurement is made over pins for an external gear and between pins for an internal gear. ***Pin diameter*** is the diameter of the measuring pin or ball used between gear teeth.

Spur Gear Representation and Specification

Rather than draw all tooth profiles of a gear, spur gears can be represented by showing the addendum circle (outside diameter) and the dedendum circle (root diameter) as phantom lines or long-dash lines. This practice was developed primarily to save pencil drafters from tedious work tracing involute shapes of intricate gear teeth. While CAD systems can easily create all teeth on a gear, phantom lines are still often used as the conventional standard practice. This conventional practice can also avoid confusion. For example, if the gear teeth are not straight but are instead helical, multiview drawings showing all tooth profiles can be rather confusing, as the gear teeth features are not normal to the viewing direction. For drawings that do not show all teeth, a circular center line is also drawn to show the pitch diameter.

The specifications required to machine a gear are usually given in a data block or table located on the drawing. This eliminates the need for calculations to be done in the shop, thus reducing possible errors. However, it may be necessary at times to calculate gear data. When the diametral pitch and number of teeth are known, most remaining data can be calculated with the aid of formulas common in most machinist's handbooks or gear supplier catalogs.

AGMA has established quality class standards for gears. First, gears can be classified as coarse pitch or fine pitch. For coarse-pitch gears, the quality class can range from 3 (least precise) to 15 (most precise). For fine-pitch gears, the quality class can range from 5 (least precise) to 16 (most precise).

Gear Manufacturing

A variety of processes and materials can be used to manufacture gears. Gears often begin with the creation of a cylinder without the gear teeth, known as a ***gear blank***, **Figure 18-5**. The gear blank is often the starting point for manufacturing gears, and it may include a hub, a central core, and a set screw hole for mounting the gear after it has been finished. The gear blank can be created by many common processes, including casting, forging, milling, machining, or additive manufacturing. Several vendors supply a full gear blank product line in a variety of materials, including brass, steel, or plastic.

Many gears are manufactured by subtracting material to form the gear teeth. Processes for forming the teeth includes milling or broaching. For either of these processes, cutting or broaching tools are designed to cut or punch the precise shape of the space between teeth. The gear blank is often rotated to perform the operation for each individual space, although a broaching tool could be designed for multiple teeth, especially for a small internal gear.

A specialized milling process for cutting gears, splines, and sprockets is known as hobbing. A ***hobbing machine***, or ***hobber***, progressively cuts gear teeth with a helical cutting tool known as the hob, **Figure 18-6**. This involves the rotary motion of the hob, as the gear blank rotates while the space between the teeth is being hobbed. Multiple blanks can also be hobbed simultaneously, making this one of the more popular processes for many types of gears. Depending on the accuracy and surface finish required for a gear, several additional operations may be required, such as grinding, honing, and lapping.

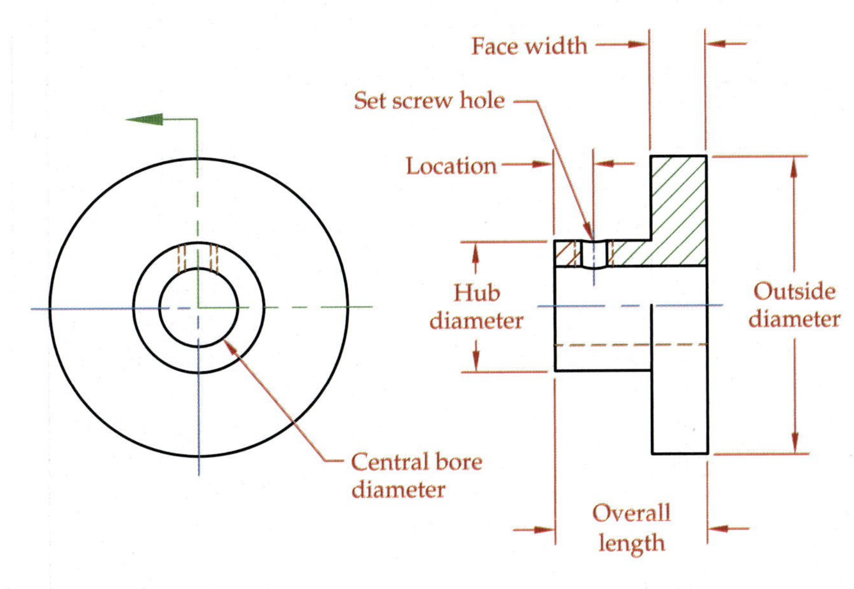

Goodheart-Willcox Publisher

Figure 18-5. Gear blanks are available with flat face cylinders, center bored holes, and optional hubs or set screw holes. Only the teeth remain to be formed.

Dmitry Kalinovsky/Shutterstock.com

Figure 18-6. A hobbing machine is a very effective machine that uses a hob for milling out the space between gear teeth.

For plastic gears, injection molding is a process that produces gears very efficiently. Although plastic gears are not as strong, they operate more quietly and have a self-lubricating nature in some cases. Some plastic gears may also incorporate a metal core through the hub.

Metric Terminology

While metric gear design is based on the same principles and proportions as inch gears, some of the terminology is different, mostly due to the difference in size between the millimeter and the inch. For example, a gear with a diameter of 6″ and 48 teeth has a diametral pitch of 8 (48 ÷ 6). If the gear is converted to a metric size of 152.4 mm, the diametral pitch translates into an unusual value that is not common to our way of thinking. Diametral pitch, as used in the inch system, provides a convenient integer value that helps the design and manufacturing team compare gears. The metric system requires a slightly different way of thinking, and a similar way of discussing the size of a gear or set of gears.

Instead of diametral pitch (P_d), metric gear designs use the term ***module (m)***, a ratio defined as the pitch diameter (d) divided by the number of teeth (z). Module serves the same function as diametral pitch, but notice it is the inverse proportion, with diameter divided by the number of teeth, rather than the number of teeth divided by the diameter. For the gear just mentioned, the module value would therefore be 152.4 mm ÷ 48 teeth = 3.175. Like the inch-based system, mating gears would also need to have a module value of 3.175. For metric gear applications, it is easier to think in terms of millimeter values to begin with rather than translating inches, but if manufacturers have equipment that is inch-based, working (or at least thinking) in both systems may be necessary. For those who are used to thinking of integer values with diametral pitch, the odd module value of 3.175 may seem strange. If the designer can design and manufacture in millimeters, it would be more logical to design the gear as a 150 mm gear with 50 teeth, providing an integer module value of 3.

In the metric system, the other terms for gear design are still used, but the formulas are adjusted to accommodate a module value. The abbreviations used for some of the terms are also different, or perhaps the abbreviation is lowercase instead of uppercase.

Formulas impacted by a *module (m)* value (instead of *diametral pitch*) are shown below, using the gear just discussed as an example.

For a module (m) value of 3:

Addendum (h_a) = m (h_a = 3 mm)
Dedendum (h_f) = 1.25 × m (h_f = 3.75 mm)
Whole depth (h) = 2.25 × m (h = 6.75 mm)
Working depth (h_w) = 2 × m (h_w = 6 mm)
Top clearance (c) = 0.25 × m (c = 0.75 mm)
Circular pitch (p) = π × m (p = 9.45 mm)

Similar in nature to the manufacture and use of threads and threaded fasteners, gear design is another area wherein the United States is slow to convert to metric units and systems. There are vast costs involved with changing over to metric machinery processes. Even if the change is made, a transition period would likely be required, during which equipment and stock for both systems of measurement and tooling would need to be maintained. It is encouraging to realize that many resources and computer-based solutions are available to assist the print reader in a world of mixed units and associated terms.

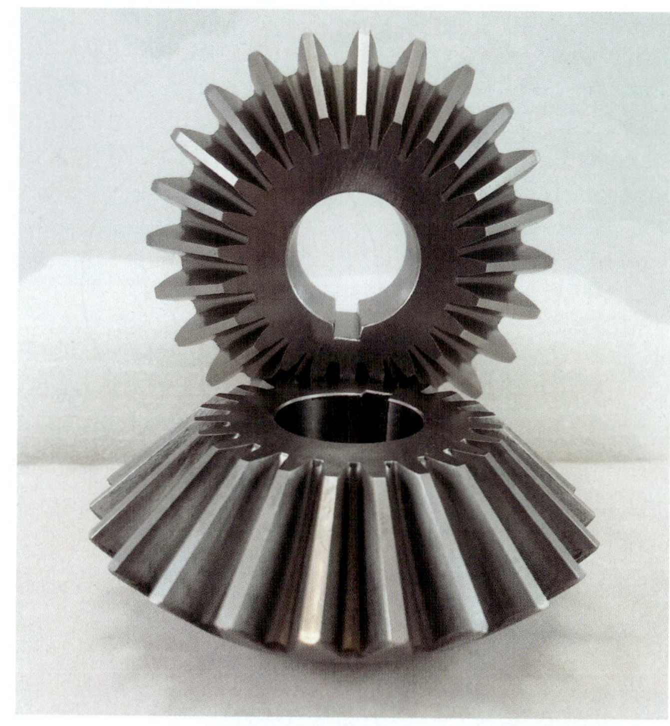

Bill Haag/Shutterstock.com

Figure 18-7. A bevel gear resembles a cone with teeth on its conical side.

Bevel Gears

As defined earlier, a *bevel gear* resembles a cone with teeth on its conical side. See **Figure 18-7**. Like all gears, bevel gears transmit motion and power to a mating gear. Bevel gears are commonly used in industry. Most mating bevel gears have their shafts at right angles. However, the shaft angle can be other than 90°. The geometry and identification of bevel gear parts are shown in **Figure 18-8**.

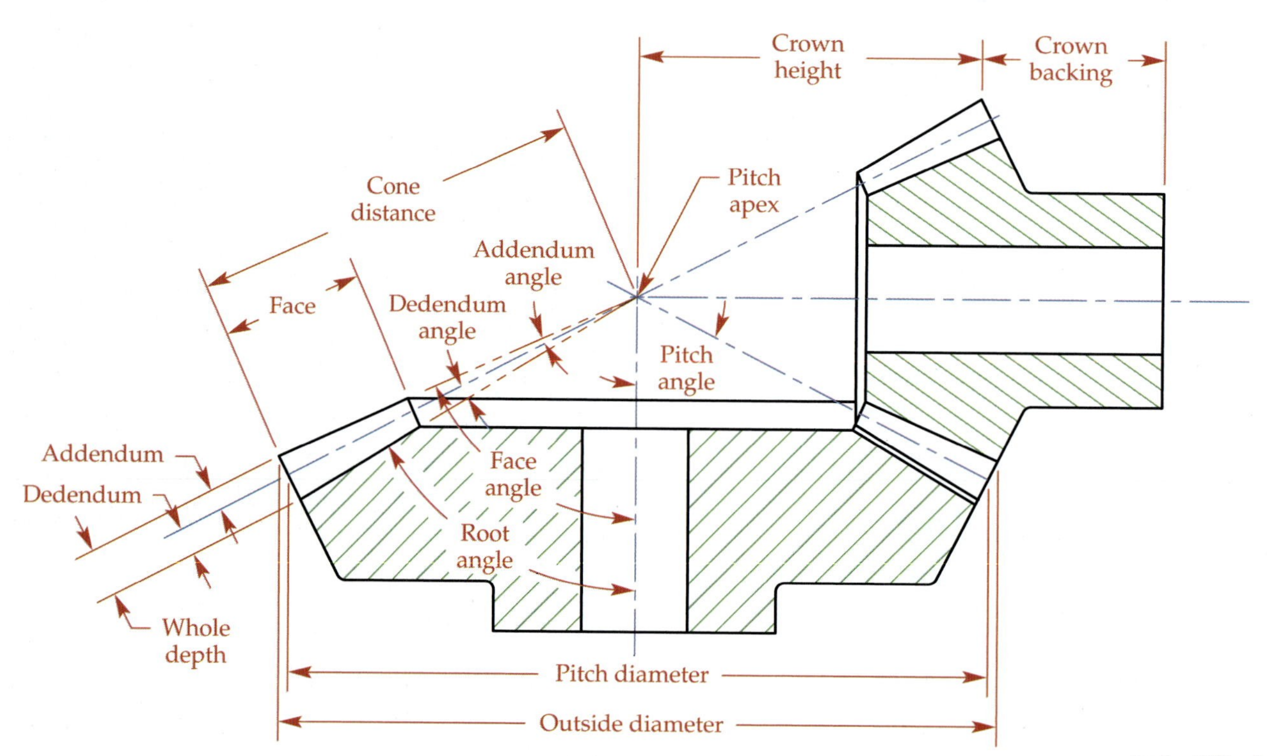

Goodheart-Willcox Publisher

Figure 18-8. This drawing shows the geometry and identification of bevel gear parts.

The specifications necessary for cutting a bevel gear are usually given in the gear data block on the drawing. Due to the cone-shaped nature of a bevel gear, it is more complicated to manufacture than the spur gear, as the tooth thickness changes along the cone-shaped surface. For bevel gears with helical teeth, the complexity increases. Leading the way in gear design and manufacture are a few companies, such as Gleason Works and the Klingelnberg Group, each having patented processes, specialized equipment, and trademarked gears for complex machinery. While bevel gears in their various forms can have intimidating design features, the basic gear data and terms on the print are much the same as with spur gears. Resources for inspecting and evaluating the prints include the machinist's handbook and the many resources that gear companies provide to the industrial team that implements their products or equipment.

Worm Gears

As defined earlier, a *worm gear* is a cylindrical gear, similar in form to a spur gear, with teeth cut on an angle. The worm gear is driven by a *worm*, which is similar in form to a screw. See **Figure 18-9**. Usually, worms and worm gears are used for transmitting motion and power at a 90° angle. The worm and worm gear are also used for speed reduction. This is because one revolution of a single-thread worm only advances one tooth, or one pitch, on the worm gear.

To increase the length of action, the worm gear can be made in a throated, or concave, shape, which allows the worm gear to partially wrap around the worm. Another technique is to use a "double-enveloping," or globoid, worm, which has an hourglass shape. While these specialized worms increase efficiency or load capacity, they are often characterized by higher manufacturing costs and fewer supplier choices. The specifications for cutting a worm gear and worm are usually given in the drawing data block. If additional data is needed, check the required formula in a machinist's handbook.

ra3rn/Shutterstock.com

Figure 18-9. A worm gear is a cylindrical gear, similar to a spur gear, with teeth cut on an angle and driven by a worm.

The terminology for worm gearing is much the same as for spur and bevel gearing. See **Figure 18-10**. However, there are some additional terms associated with worms and worm gears. These terms are discussed below.

- ***Linear pitch*** is the distance from a given point on one worm thread to the next. This distance is equal to the circular pitch of the worm gear.
- In screw thread applications, *lead* is the distance a thread advances in one revolution. For a single-threaded worm, the linear pitch and the lead are the same. For a double-threaded worm, the lead is twice the linear pitch; for a four-threaded worm, the lead is four times the linear pitch; and so on.
- ***Lead angle*** is the angle the lead makes with a perpendicular line to the worm axis.
- ***Throat diameter*** is the diameter of a circle coinciding with the tops of the worm gear teeth at their center plane.

Splines and Serrations

Splines and ***serrations*** are like multiple keys or grooves on a shaft that prevent rotation between the shaft and a related member. By definition, splines are externally raised features and serrations are internal notches. As applied in industrial parts, splines and serrations often appear similar to wide, small-diameter spur gear teeth. Traditionally, splines have teeth with parallel sides. However, splines with involute sides are increasing in popularity. See **Figure 18-11**. Involute splines are produced with the same technique and equipment as is used for gears, including hobbing machines.

As notches, serrations are primarily used for parts permanently fitted together, perhaps by wedging. Serrations have different tooth proportions and higher pressure angles than splines. Serrations are well adapted for use on thin-wall tubing. Serrations are also common on hand tools like pliers as a type of gripping surface.

Terminology associated with involute splines and serrations is often the same as for spur gears. Usually, data for producing splines or serrations is given on the print. When a formula is needed, check a machinist's handbook.

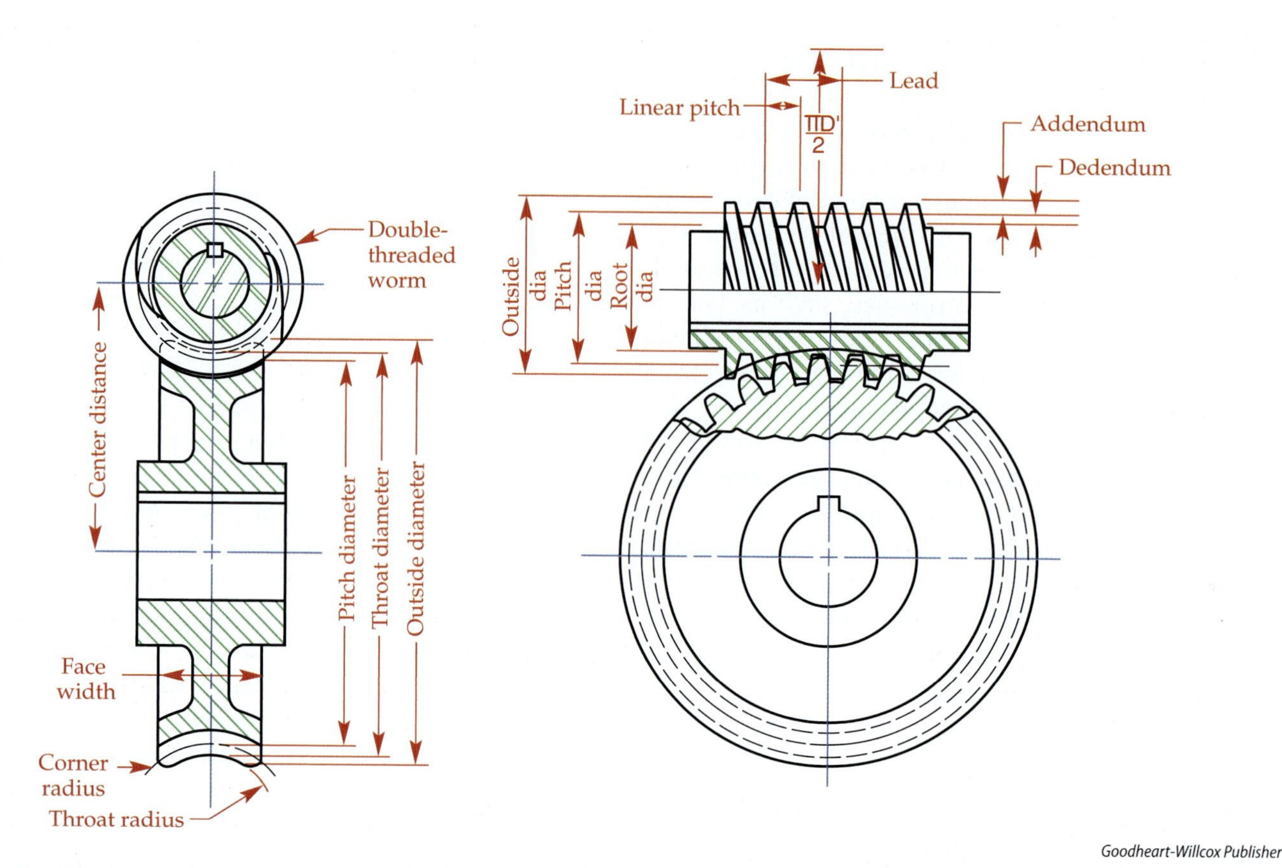

Goodheart-Willcox Publisher

Figure 18-10. The terminology for worm gearing is much the same as for spur and bevel gearing.

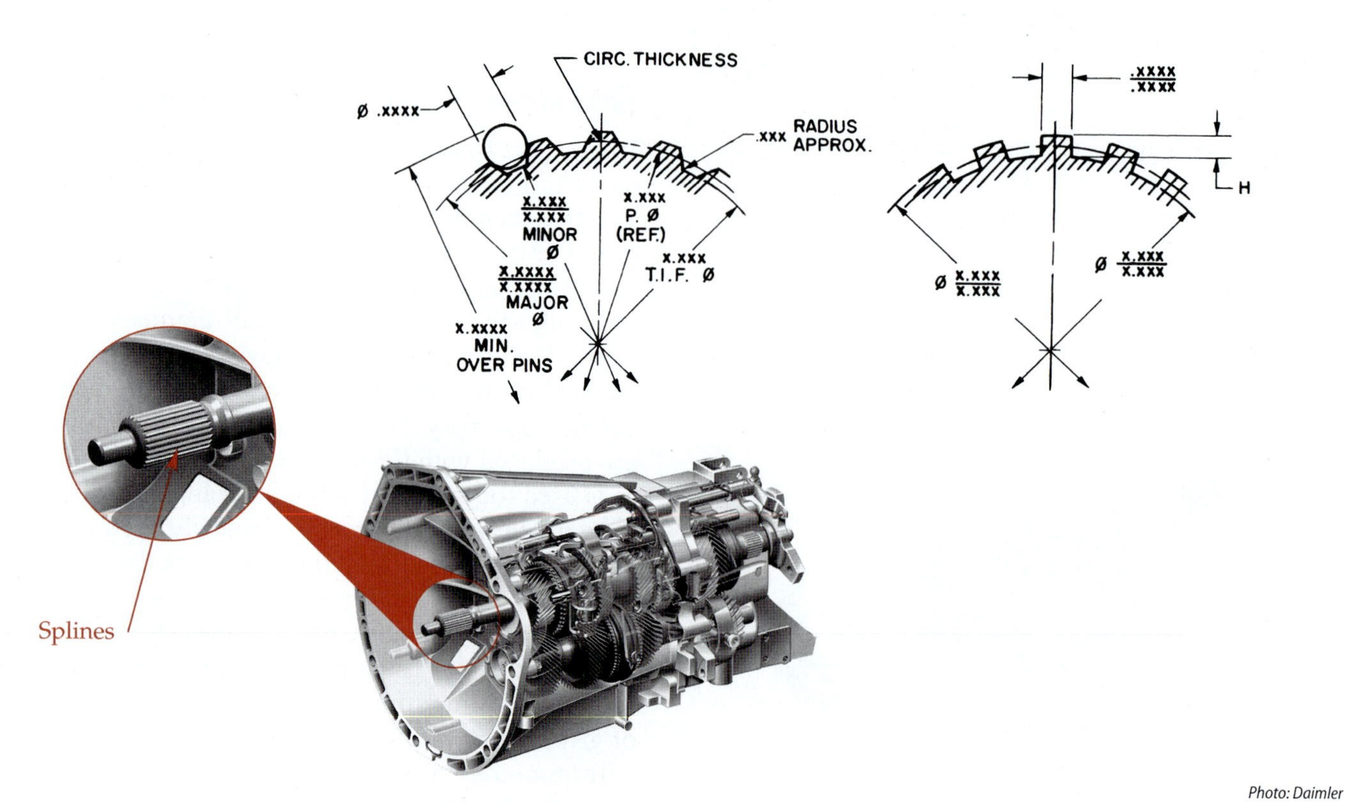

Photo: Daimler

Figure 18-11. Splines are like multiple keys on a shaft that prevent rotation between the shaft and its related member. The input shaft on an automotive transmission has splines to mate it to the torque converter or clutch plate.

Summary

- Gears are designed in a variety of forms to work in conjunction with each other to perform mechanical tasks such as transferring motion or changing speeds from one axle to another or adjusting the rotational position of various pieces within the assembly.
- Gears can be categorized by how the axes of gears within a set are oriented to one another, either parallel, at an angle, or nonintersecting and nonparallel.
- Parallel axis gears include the spur gear, helical gear, herringbone gear, and internal gear.
- Intersecting-axis gears include the bevel gear, both straight and spiral, and the face gear.
- Nonintersecting and on-parallel gears include the worm gear, screw gear, and hypoid gear.
- Standards are published for gear design and manufacturing by the American Gear Manufacturing Association, the Society of Automotive Engineers, and the American Society of Mechanical Engineers, with the ASME standards primarily related to the representation of gears within engineering drawings.
- Spur gear terminology is quite extensive, with many terms and parameters that are interrelated and determined by such features as the pitch diameter, number of teeth, and pressure angle.
- Through conventional practices involving phantom lines and center lines, spur gears can be represented on an industrial print without constructing every tooth profile or true projection views of complex helical shapes and forms.
- Spur gear drawings will typically contain extensive gear data tables that define the size and specifications for the gear teeth.
- Gears can be manufactured with a variety of materials and processes, including forging, casting, broaching, milling, additive manufacturing such as 3D printing, and a specialized milling process known as hobbing.
- The terminology for metric gears is very similar to inch-based gears but involves pairing gears based on equal module values instead of diametrical pitch values.
- Bevel gears, commonly designed as cone-shaped gears that transfer power and motion at intersecting angles, are more complex to manufacture, but many of the spur gear terms are still directly applied and are found in the gear data tables on the print.
- Worms and worm gears transmit motion and power in ways that are slightly different from other gear sets, and a few additional terms such as linear pitch, lead, and lead angle help specify the gear information to the print reader.
- The design of shaft assemblies sometimes incorporates external raised features known as splines and internal grooved features known as serrations, which are similar in shape to gear teeth and created with similar processes and equipment.

Unit Review

Name ______________________ Date ______________ Class ______________

Answer the following questions using the information provided in this unit.

Know and Understand

____ 1. *True or False?* There are approximately five categories under which gears can be classified.

____ 2. For gear sets that have a parallel axis, which of the following does *not* belong in this category?
A. Internal
B. Worm
C. Spur
D. Helical

____ 3. For gear sets that have intersecting axes, which of the following does *not* belong in this category?
A. Face
B. Miter
C. Herringbone
D. Bevel

____ 4. Which of the following organizations does *not* publish standards related to gear design?
A. Society of Manufacturing Engineers
B. American Gear Manufacturing Association
C. American Society for Mechanical Engineers
D. Society of Automotive Engineers

____ 5. *True or False?* The diametral pitch describes a ratio between the number of teeth on a gear and the diameter of that gear.

For questions 6–12, match the terms with the abbreviated clues. Answers are used only once.

____ 6. Top of tooth to bottom of tooth space on mating gear

____ 7. Pitch of circle plus two addendums

____ 8. Radial distance—bottom of a tooth to pitch circle ($1.25/P_d$)

____ 9. Radial distance—top of a tooth to pitch circle ($1/P_d$)

____ 10. Addendum + addendum

____ 11. Half of the circular pitch

____ 12. Circumference of pitch circle divided by number of teeth

A. Circular pitch
B. Addendum
C. Dedendum
D. Clearance
E. Circular tooth thickness
F. Working depth
G. Outside diameter

____ 13. What term can be illustrated by drawing a line through the points of contact between engaged teeth?
A. Base angle
B. Pressure angle
C. Tangent angle
D. Pitch angle

____ 14. *True or False?* The chordal thickness of a tooth is different from circular tooth thickness and can be checked with a gear-tooth caliper.

____ 15. *True or False?* Backlash occurs when the gear tooth is slightly larger along the pitch circle than the tooth space on the corresponding mating gear's pitch diameter.

____ 16. Which of the following would you least expect to find on a gear blank?
A. Set screw hole
B. Hub
C. Hole through the center
D. Helical teeth

____ 17. Which of the following processes is similar to hobbing, although hobbing is a more specialized form of that process?
A. Forging
B. Casting
C. Milling
D. Grinding

____ 18. *True or False?* Mating metric gears must have identical module values, just as inch-based gears must have identical diametral pitch values.

____ 19. *True or False?* If bevel gears have a 90° relationship, they are called miter gears and must feature an identical number of teeth.

____ 20. Which of the following terms is *not* exclusive to a worm gear set?
A. Circular pitch
B. Throat diameter
C. Lead
D. Linear pitch

____ 21. *True or False?* Since splines and serrations always feature involute sides, they are also manufactured by hobbing.

Critical Thinking

1. What are some devices common to everyday life around the house that incorporate gears?

____

____

____

____

____

____

2. What benefits are there to using a CAD system to draw gears compared to the drafting techniques that were used before CAD systems?

____

____

____

____

____

____

Apply and Analyze

Name ______________________ Date ____________ Class ______________

Industry Print Exercise 18-1

Refer to the print PR 18-1 and answer the questions below.

1. What is the diametral pitch (P_d) of this gear?

   ____

2. Is the addendum $1/P_d$?

   ____

3. What is the pitch diameter of this gear?

   ____

4. A mating gear for this gear must have a pressure angle of _____.

   ____

5. Is the part number to be marked on both sides of the gear? ______________________

6. What size pin is specified for measuring the diameter over pins? ______________________

7. After hobbing, the circular tooth thickness should be _____, and then reduced to _____ by grinding.

   ____

8. Is the AGMA gear class specified by the print very precise or not very precise?

   ____

9. What type of line is used to indicate the outside diameter of the gear? ______________________

10. What is the range of backlash specified for this gear?

    ____

Review questions based on previous units:

11. What is the 1/2″ notch on the central hole called?

    ____

12. Are the section lines indicative of the material?

    ____

13. What material is used for this gear?

    ____

14. What geometric tolerances are specified by a general note instead of a feature control frame?

    ____

15. Are any dimensions specified using the limit method of tolerancing?

    ____

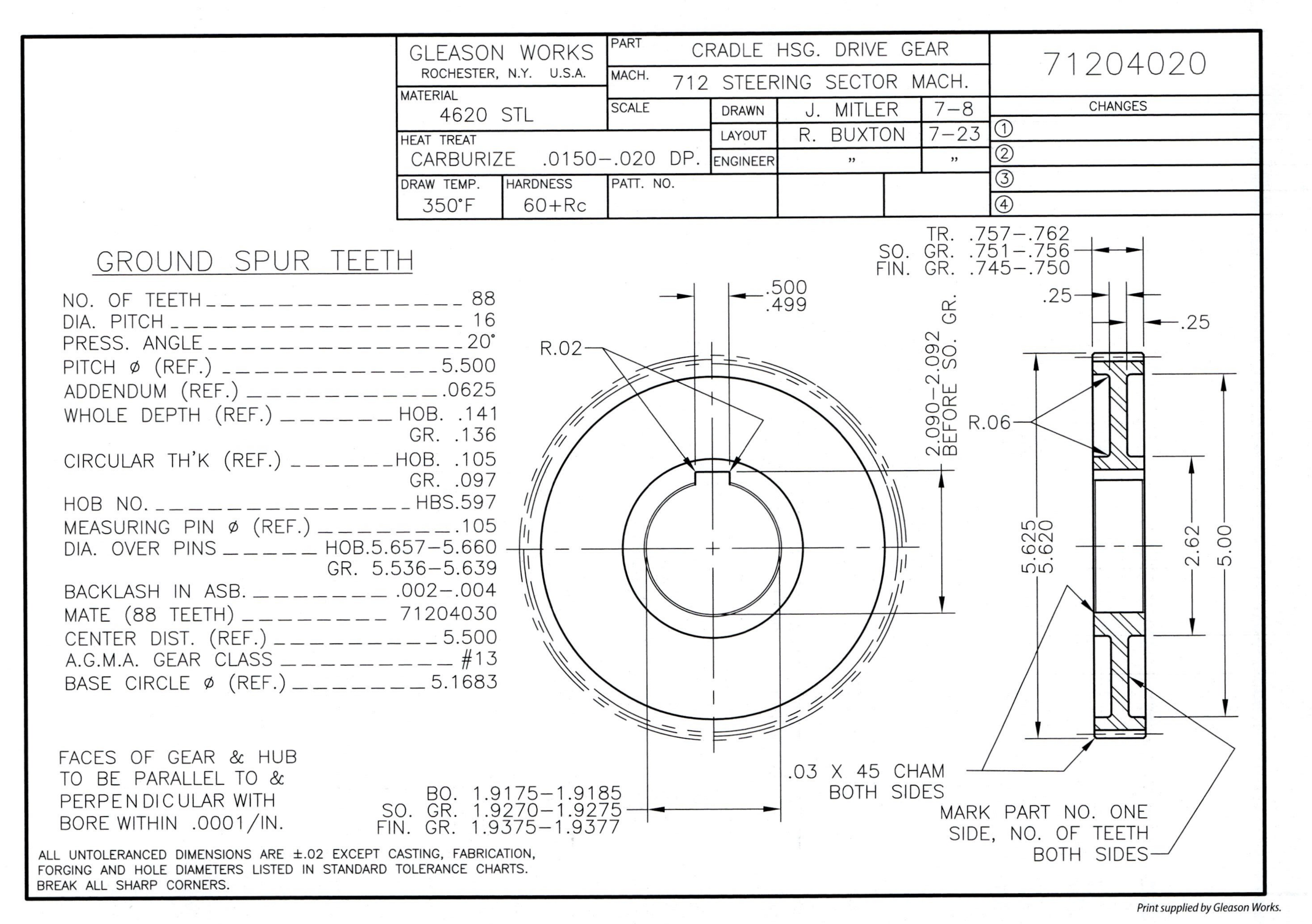

PR 18-1. Cradle Housing Drive Gear

Apply and Analyze

Name ______________________ **Date** ______________ **Class** ______________

Industry Print Exercise 18-2

Refer to the print PR 18-2 and answer the questions below.

1. What is the total number of gears specified by this one print?

____

2. What is the diametral pitch (P_d) of the gears specified by this one print?

____

3. How many options are there for the basic size of diameter C?

____

4. What is the pressure angle specified for the gears of this print?

____

5. All of the parts covered by this one print have a depth of .460″, but the face width of the worm wheel is specified as a fractional distance of _____.

____

6. What hole size is specified for part W5-15S? ______________________

7. What three part numbers each have 90 teeth?

____

8. What is the smallest pitch diameter specified? ______________________

9. What material is specified for these gears? ______________________

10. What worm is specified as the mate for this worm wheel? ______________________

Review questions based on previous units:

11. Is Section E-E shown in alignment with other views or as a removed view?

____

12. What scale is specified for this print?

____

13. How many feature control frames are used on this print?

____

14. What is the current revision for this print?

____

15. What surface finish is specified for the parts?

____

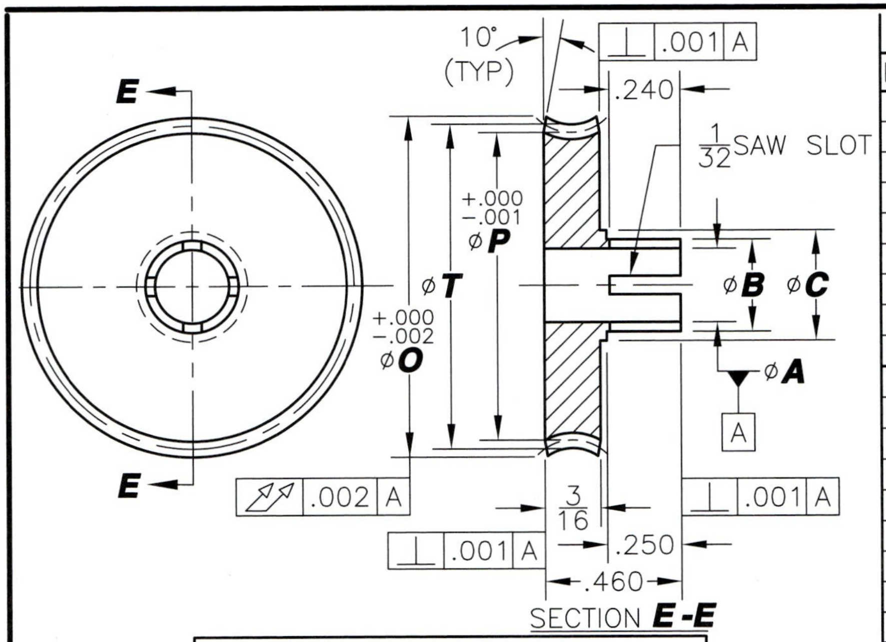

10° (TYP)

⊥ .001 A

.240

$\frac{1}{32}$ SAW SLOT

+.000 −.001 ⌀P

⌀T

+.000 −.002 ⌀O

⌀B ⌀C

⌀A

A

.002 A

$\frac{3}{16}$

⊥ .001 A

⊥ .001 A

.250

.460

SECTION E-E

E

E

WORM WHEEL DATA	
PITCH (DIAM.)	48
THREAD	FOUR (R.H.)
LEAD OF WORM	.2618
PRESSURE ANGLE	25°
LEAD ANGLE	14°−2'
WHOLE DEPTH	.0407
A.G.M.A.	PREC 1
TESTING PRESSURE	20 OZ.
TOOTH FORM	INVOLUTE

WORM WHEEL DATA (MATES WITH W11)

PART NO.	NO. OF TEETH	⌀P PITCH	⌀T THROAT	⌀O	⌀A	⌀B	⌀C
W5−1S	30	.6250	.6624	.6832			
−2S	40	.8333	.8707	.8915			
−3S	50	1.0417	1.0791	1.0999			
−4S	60	1.2500	1.2874	1.3082			
−5S	70	1.4583	1.4957	1.5165	+.0005 −.0000 ⌀.1248	+.000 −.003 ⌀.1875	±.005 ⌀.250
−6S	80	1.6667	1.7041	1.7249			
−7S	90	1.8750	1.9124	1.9332			
−8S	100	2.0833	2.1208	2.1415			
W5−9S	120	2.5000	2.5374	2.5582			
W5−10S	30	.6250	.6624	.6832			
−11S	40	.8333	.8707	.8919			
−12S	50	1.0417	1.0791	1.0999			
−13S	60	1.2500	1.2874	1.3082			
−14S	70	1.4583	1.4957	1.5165	+.0005 −.0000 ⌀.1873	+.000 −.003 ⌀.2500	±.005 ⌀.312
−15S	80	1.6667	1.7041	1.7249			
−16S	90	1.8750	1.9124	1.9332			
−17S	100	2.0833	2.1208	2.1415			
W5−18S	120	2.5000	2.5374	2.5582			
W5−19S	30	.6250	.6624	.6832			
−20S	40	.8333	.8707	.8919			
−21S	50	1.0417	1.0791	1.0999			
−22S	60	1.2500	1.2874	1.3082			
−23S	70	1.4583	1.4957	1.5165	+.0005 −.0000 ⌀.2498	+.000 −.003 ⌀.3125	±.005 ⌀.375
−24S	80	1.6667	1.7041	1.7249			
−25S	90	1.8750	1.9124	1.9332			
−26S	100	2.0833	2.1208	2.1415			
−27S	120	2.5000	2.5374	2.5582			
W5−28S	180	3.7500	3.7874	3.8082			

REV	DATE	APP'D	CHANGE
F	12/4	GUY	ECO1321−2314
			CONC. .002 WAS .0005
E	1/11	MU	ECO1321−2119
D	10/4	AEH	ECO1321−959
C	9/18	AEH	ECO1321−941

ITEM NO.	NO. REQ'D	USED ON	REV	DATE	APP'D	CHANGE

DO NOT SCALE THIS DRAWING

REMOVE ALL BURRS AND SHARP EDGES UNLESS OTHERWISE SPEC

SURFACE FINISH 125 UNLESS OTHERWISE SPECIFIED — WEIGHT

TOLERANCES UNLESS SPECIFIED — FRACT DIM ± 1/64 — DECIMAL DIM ± .005 — ANGULAR DIM ± −− — SCALE NONE

DRAWN BY A.E.H. — APPROVED BY

CHECKED BY A.J.C. — ENG'R

CLASS NO. — JOB NO.

HEAT TREAT −−−

FINISH −−−

MATERIAL BRONZE AS PER QQ−B−637, COMP. 1

STERLING PRECISION CORPORATION INSTRUMENT DIVISION NEW YORK

TITLE WORM WHEEL (CLAMP TYPE) 48 PITCH R.H. 3/16 FACE

NEXT ASSEMBLY N/A — DRAWING NUMBER W5−1S TO W5−28S — REV F

Print supplied by Sterling Precision Corporation.

PR 18-2. Worm Wheel (Clamp Type)

Apply and Analyze

Name ______________ Date ______________ Class ______________

Bonus Print Reading Exercises

The following questions are based on various bonus prints located in the folder that accompanies this textbook. Refer to the print indicated, evaluate the print, and answer the question.

Print AP-008

1. What mean dimension represents the diameter of the cylinders that feature involute splines?

   ____

2. Considering the splines on the right end as a continuous feature, and ignoring the relief groove, calculate the basic length of the splines.

   ____

3. In inches, what diameter is specified for the mean diameter of the splines as measured over .060″ diameter pins?

   ____

4. Are any surfaces besides the splines and grooves to be heat treated?

   ____

5. What pressure angle is specified for the splines?

   ____

Print AP-016

6. For the spur gear on this part, give the 1) number of teeth, 2) pitch diameter, 3) addendum, and 4) maximum circular tooth thickness.

   ____

   ____

7. For the worm gear on this part, give the 1) lead, 2) lead angle, 3) working depth of thread, and 4) theoretical diameter when measuring over wires.

   ____

   ____

8. What is the mating gear part number for the spur gear, how many teeth does it have, and what is the gear ratio between it and the spur gear?

   ____

   ____

9. Does either of the two gears on this part have backlash specified?

   ____

10. What pressure angle is specified for the two gears of this part?

Print AP-017

11. How many teeth are specified for this helical gear?

12. What pressure angle is specified for this helical gear?

13. What is the pitch diameter of this gear?

14. At the standard pitch diameter, what maximum circular tooth thickness is specified?

15. Some measurements for gears and splines can better be performed over balls or pins. What diameter balls are specified for checking this gear?

16. What is the maximum diameter across this gear, as shown in the sectional view?

Notes

UNIT 19
Cam Diagrams and Prints

LEARNING OBJECTIVES

After completing this unit, you will be able to:

- Explain the principles of cams as applied to common objects and products.
- Identify types of cams used in industrial applications.
- Explain terms related to cams and followers.
- Interpret displacement diagrams with respect to the rise, fall, and dwell of a cam follower during a cycle.
- Identify different methods for calculating motion transition in displacement diagrams.

TECHNICAL TERMS

base circle
cam
cam profile
conjugate cam
cylindrical cam
displacement
displacement diagram
dwell
endface cam
face cam
follower
follower constraint
globoidal cam
linear cam
radial cam

Within the realm of mechanism design are devices known as ***cams***, which are specially shaped parts that cause other parts to move in a controlled fashion. While some mechanisms use linkages and bars to connect parts, cams often transfer motion by surface-to-surface contact of an irregular shape with another surface or roller wheel. Many cams transmit the rotary motion of an irregularly shaped wheel into a linear displacement, or into an angular pivoting motion. The camshaft of an automotive engine is a common example of cams working within a mechanical device. As the camshaft rotates, there are several lobes causing other valves to open and close in a precise manner. See **Figure 19-1**. However, the history of the cam and camshaft is hundreds of years older than the internal combustion engine.

D Balamut/Shutterstock.com

Figure 19-1. The camshaft is one of the most common examples of the cam.

There are other examples of cams in products around the home or shop. Before computer-controlled solutions became available, sewing machines provided the homemaker with a variety of sewing stitch patterns through a set of cams that could be inserted into the sewing machine by the user. Within an adjustable-length shower curtain rod, there is usually a cam that serves to friction-lock two metal tubes together. In the shop, clamps and tooling fixtures sometimes incorporate cam surfaces to produce quick-release handles. **Figure 19-2** features a double-cam handle used to help lock tooling into a secure position. These cam surfaces are often a simple spiral design.

Goodheart-Willcox Publisher

Figure 19-2. Cam surfaces can also be incorporated into clamps and tool fixtures to produce quick release handles.

The principles of cam design can also be applied to small mechanisms within items such as door latching or locking hardware, wherein a cam-shaped piece works in conjunction with a mating part to position it by surface-to-surface contact. **Figure 19-3** shows a cam-shaped hub rotating within a mating hole of a sliding part. In this example, the cam-shaped hub can be rotated 90° either way to accomplish the same linear movement of the sliding part. In examples such as this, a spring can be incorporated into the mechanism to keep the mating surfaces together or to cause one of the parts to move back toward a home position.

While the principles of cam design are incorporated into many products around the home and shop, this unit focuses on cams that are used in industrial mechanisms to help transfer motion in a systematic manner.

Cam Types and Terminology

In general, cams change uniform rotary motion into oscillating up-and-down or back-and-forth linear motion. There are a variety of design types for cams, including radial or plate cams, cylindrical or barrel cams, endface cams, and groove or face cams. See **Figure 19-4**. Even a wedge or linear slot can serve as a cam, as illustrated in **Figure 19-5**.

There is most likely a ***follower*** device that moves as a result of contact with the cam, but not all cams have followers. The follower may have a piston-type or a pivoting-type motion. The contact surface of the follower may be a knife edge, flat face, or roller. **Figure 19-6** shows examples of followers. The roller type provides the most flexibility. Precision-roller followers are available in various degrees of quality and types of bearing.

Radial cams or plate cams can be manufactured primarily as a wheel with an irregular shape, perhaps a spline shape, a pear shape, or with a spiral outer profile. This type of cam performs primarily as a rotating disk with a profile surface that contacts the follower.

Goodheart-Willcox Publisher

Figure 19-3. Principles of cams can be incorporated into devices wherein a specially shaped part works in conjunction with a mating part to position it by surface-to-surface contact.

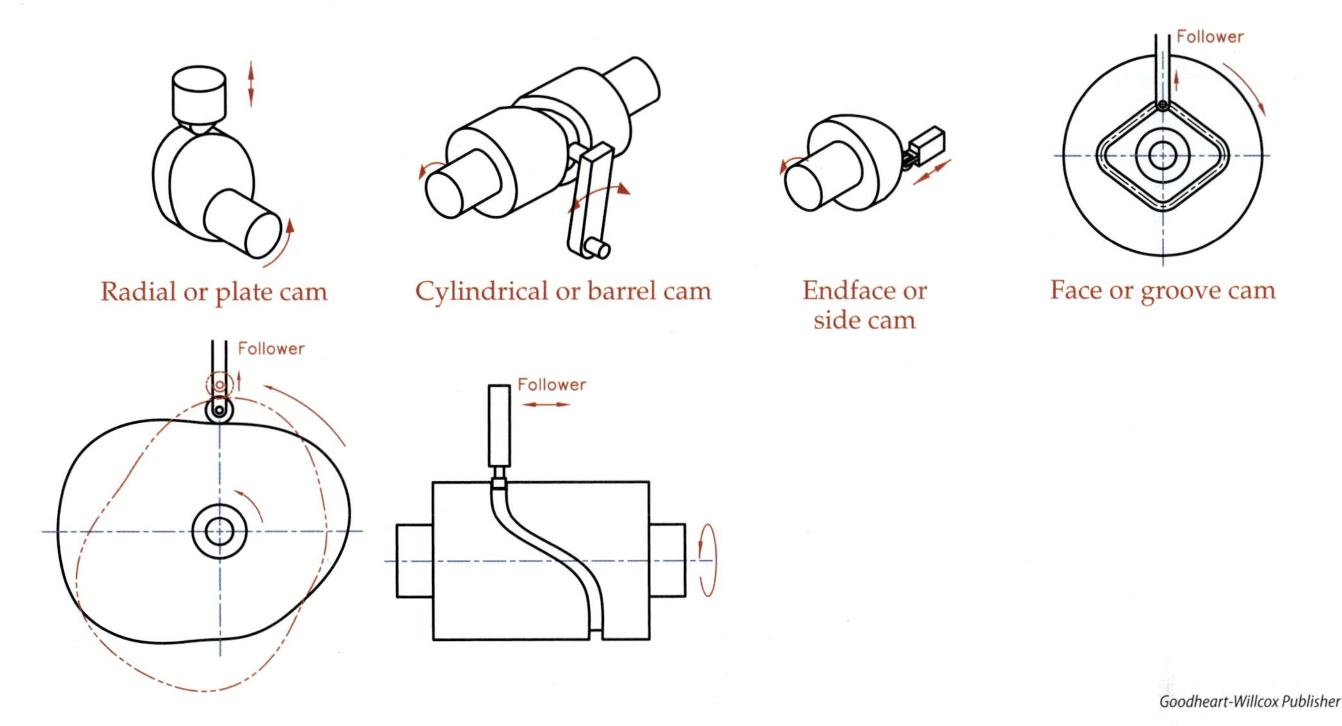

Goodheart-Willcox Publisher

Figure 19-4. Types of cams include radial or plate cams, cylindrical or barrel cams, endface or side cams, and face or groove cams.

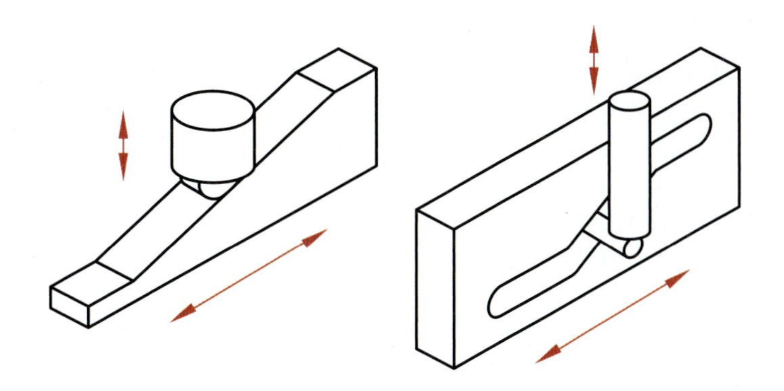

Goodheart-Willcox Publisher

Figure 19-5. Even a wedge or an angled slot can serve as a cam, causing a follower to move or oscillate.

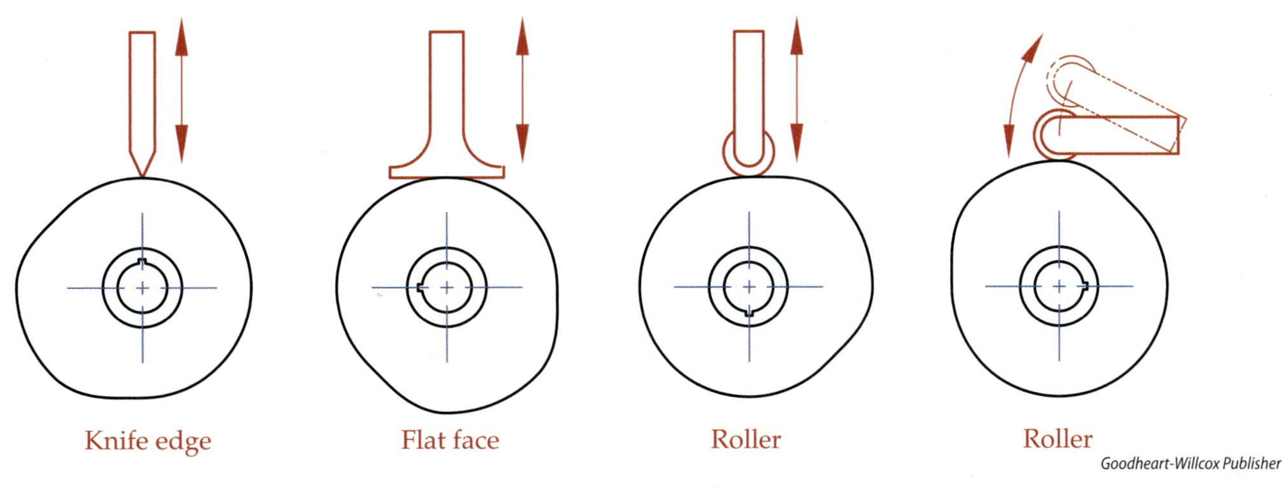

Goodheart-Willcox Publisher

Figure 19-6. Types of followers include knife edge, flat face, and roller. Roller followers in various degrees of quality and bearing types are available from vendors.

Cylindrical cams, also referred to as *barrel cams*, feature a follower groove cut within the cylindrical surface of a cylinder. These types of cams are often positioned with the axis of rotation vertical. Then, the rotating cam causes a follower to oscillate. This type of cam may be found in certain clock mechanisms or in sewing machines. A barrel cam may be referred to as an ***endface cam*** or *side cam* if the cylinder has only one cam surface, allowing the follower to ride on the rim of the cylinder.

Face cams are manufactured with a follower groove in the flat surface of a disk. The principle is the same as that of the radial cam, but with this cam the follower is nested in a groove. Some references refer to these as *groove cams*.

A ***linear cam*** gets its name from the manner in which the cam moves, in a straight line. For example, as a linear cam moves a certain horizontal distance, the follower nested within the groove is designed to move up or down as the slot dictates. These cams offer a nonelectrical solution in mechanisms that repeat motion over and over.

Follower constraint is a term that applies to how the follower is controlled. For example, one form of constraint could be gravity, wherein the weight of the follower is the force that keeps the follower in constant contact with the cam. Another form of constraint could be a spring designed to exert the proper force upon the follower to keep it in constant contact with the cam. For cams that incorporate a groove, such as the cylindrical cam, the follower is constrained by the groove.

Newer developments in cam design include the ***conjugate cam***, wherein two cams are affixed together, rotating together with two profiles controlling two rollers mounted to the same oscillating follower device. This design creates a mechanical constraint for the follower wherein one cam may push the follower in one direction, while the other cam causes the follower to return in the opposing direction, **Figure 19-7**. Another newer and more complex style of cam is the ***globoidal cam***, which has more than one design option. One form of the globoidal cam is similar in form to the barrel cam, but with a convex or concave cylinder, as shown in **Figure 19-8**. This form results in an oscillating follower. Another form of globoidal cam is also referred to as a roller gear cam. This form features a twisting "blade" or thread that resembles the worm that drives a worm gear. In the case of the globoidal cam, the cam drives a turret with rollers, as shown in **Figure 19-9**, and the follower movement is an intermittent rotation of the wheel.

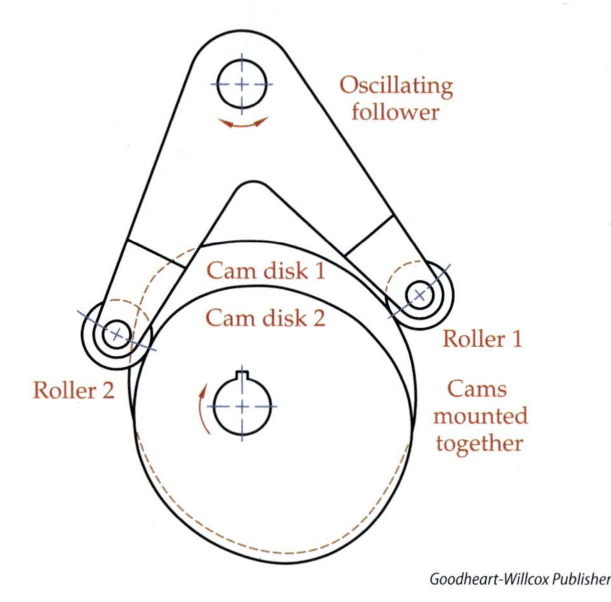

Goodheart-Willcox Publisher

Figure 19-7. A conjugate cam features two plate cams mounted together that as a unit drive two rollers, both mounted on one follower.

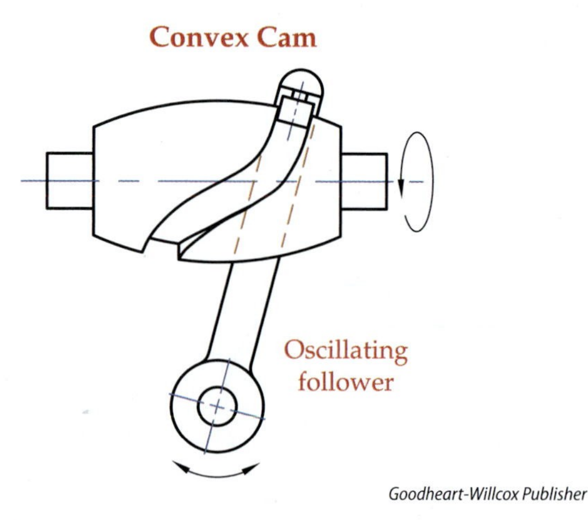

Goodheart-Willcox Publisher

Figure 19-8. One form of globoidal cam is similar in appearance to a barrel cam, but with concave or convex shape and grooves that drive an oscillating follower constrained by a roller wheel within the groove.

The following terms are also used in discussions relating to cams. A basic understanding of these terms will help you when reading prints containing cams and related components. **Figure 19-10** should also be examined as you study these terms.

- **Cam.** A machine element that transmits or delivers motion through rolling or sliding contact with a follower.
- **Cam profile.** For most cams, the ***cam profile*** is the shape of the surface on which the follower has contact. For a face or barrel cam, this is the shape of the groove.
- **Follower.** The machine or mechanism that moves with reciprocating movement by following the cam as it rotates or oscillates.
- **Oscillate.** To move back and forth in a repeating fashion. A follower may oscillate in a linear fashion (straight up and down or in and out) or may rock back and forth about a pivot point.
- **Displacement.** The ***displacement*** is the distance that a follower rises or falls during a portion of the cycle.

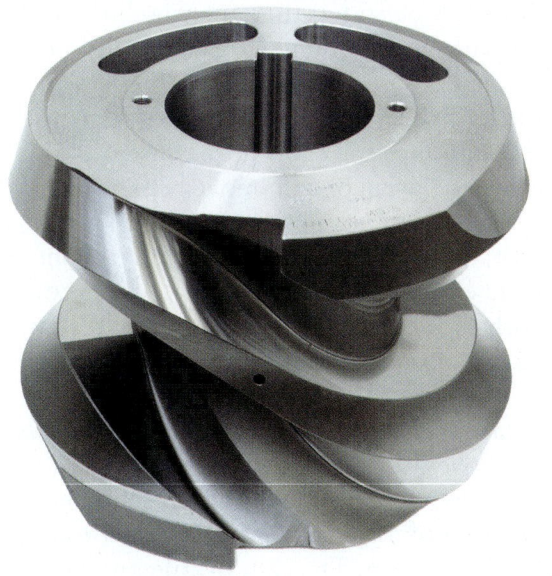

Colombo Filippetti

Figure 19-9. One form of globoidal cam features a twisting "blade" or thread that resembles the worm that drives a worm gear, driving a turret with rollers, resulting in intermittent rotation.

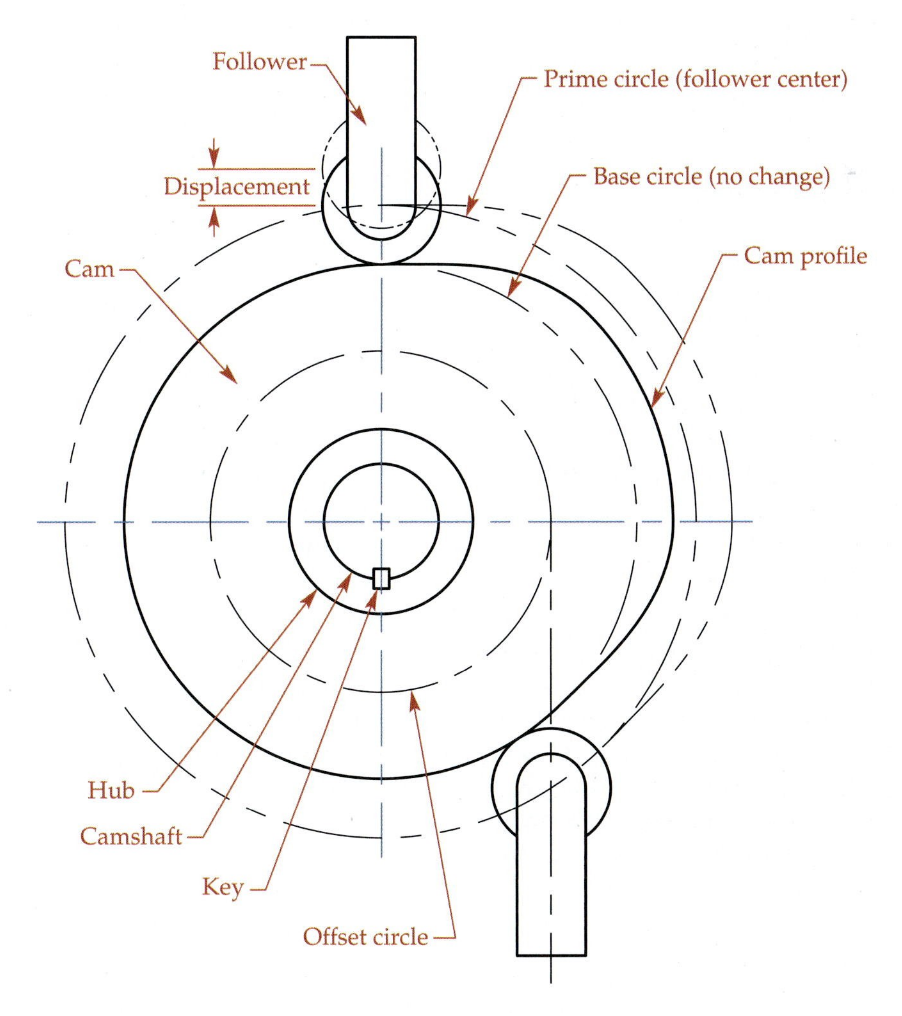

Goodheart-Willcox Publisher

Figure 19-10. This illustration features some of the terms common to radial cams.

- **Dwell.** A period of time for which the follower does not move or change position is known as *dwell*.
- **Cycle.** In the case of round cams, one 360° revolution of the cam. In the case of a linear cam, one cycle of cam movement from a home position to other position(s) and then back. The displacement diagram usually charts one cycle.
- **Base circle.** The *base circle* is a circle concentric with the center axis of a radial (plate) cam having a radius equivalent to the follower edge at rest or in the home position. If the cam had a base circle as its overall shape, there would be no movement of the follower.
- **Prime circle.** A circle that represents the position of a roller follower's center axis at the home position. The cam profile must take into consideration the roller wheel diameter at the prime circle position to accommodate surface-to-surface tangencies.
- **Offset circle.** A circle that represents the offset amount for a roller follower's center axis. For a roller follower offset from the center axis of the cam, the cam profile must take into consideration the offset position to accommodate surface-to-surface tangencies.

Cam Displacement Diagrams

Cams are usually designed with the assistance of ***displacement diagrams***. Displacement amounts can be charted over the course of one cycle and then transferred into the cam profile, which is often a spline shape not easily dimensioned. Reading a print that describes a cam may require an understanding of the displacement diagrams that are used to graphically represent the movement of the follower. **Figure 19-11** shows a displacement diagram representing the rises and fall for one complete revolution of a cam.

The displacement diagram not only represents 360° of revolution, but also usually represents a certain period of time. For example, the diagram illustrated in **Figure 19-11** indicates the follower is to rise a given amount in one second, dwell for .5 seconds, rise an additional amount in 1.5 seconds, dwell for .5 seconds, fall back to the beginning level in two seconds, and then dwell for .5 seconds before starting the cycle again. The total cycle is, therefore, six seconds.

Depending on the accuracy of the desired cam and the level of accuracy required, the data from the displacement diagram can be transferred to the cam drawing in increments of 10°, 15°, or 30°. In terms of two-dimensional drafting, CAD systems can calculate spline curves through the points and provide useful data for the CNC machine or other manufacturing equipment.

The displacement information may also be calculated using computer programs and spreadsheets. In these cases, the data can be placed on the print in table form, as well as stored electronically. Cam design software is also available for most cam applications, allowing designers to use additional tools to automate the design process.

Follower Motion

There are different methods of designing the rise and fall of the follower. Some methods cause more stress on the follower as the cam forces it through various transitions. The designer must consider velocity, acceleration, and even surges or jerks that may be created by moving parts. To prevent the components from being stressed too much by the changing motion, some motion transitions are desirable. The designer understands this information and, therefore, it is seldom a concern for the print reader. In **Figure 19-12**, methods of creating the fall and rise displacements in the diagram include:

- **Uniform motion.** In **Figure 19-12A**, the rise and fall lines are simply straight line segments. This motion is characterized by a constant rate but may subject the follower to more sudden change.
- **Modified uniform motion.** In **Figure 19-12B**, a radius is added at the beginning and end of a straight line to smooth the transition.
- **Harmonic motion.** In **Figure 19-12C**, the shape is determined by projecting from an equally divided semicircle.
- **Parabolic motion.** In **Figure 19-12D**, the displacement curve is created with parabolic construction.
- **Cycloidal motion.** In **Figure 19-12E**, the displacement curve is generated from a cycloid.

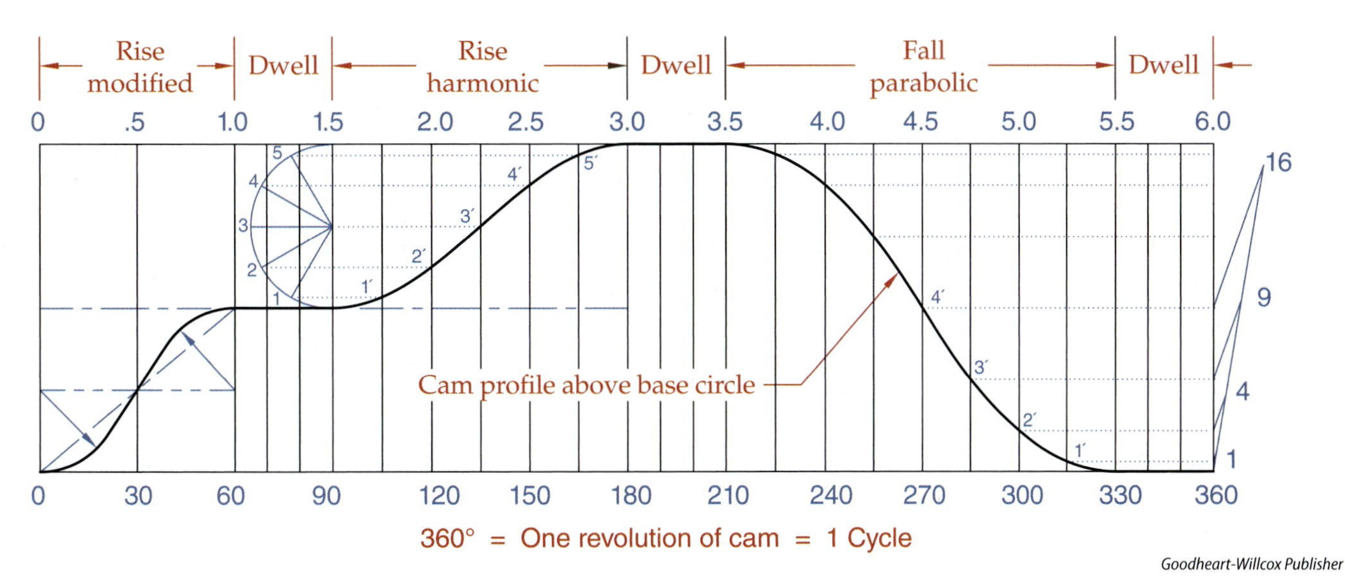

Goodheart-Willcox Publisher

Figure 19-11. This displacement diagram represents a 360° cycle that occurs in six seconds. It features a rise during the first second, a dwell of .5 seconds, an additional rise equal to the first but over the next 1.5 seconds, another dwell of .5 seconds, and then a fall over the next 2.5 seconds.

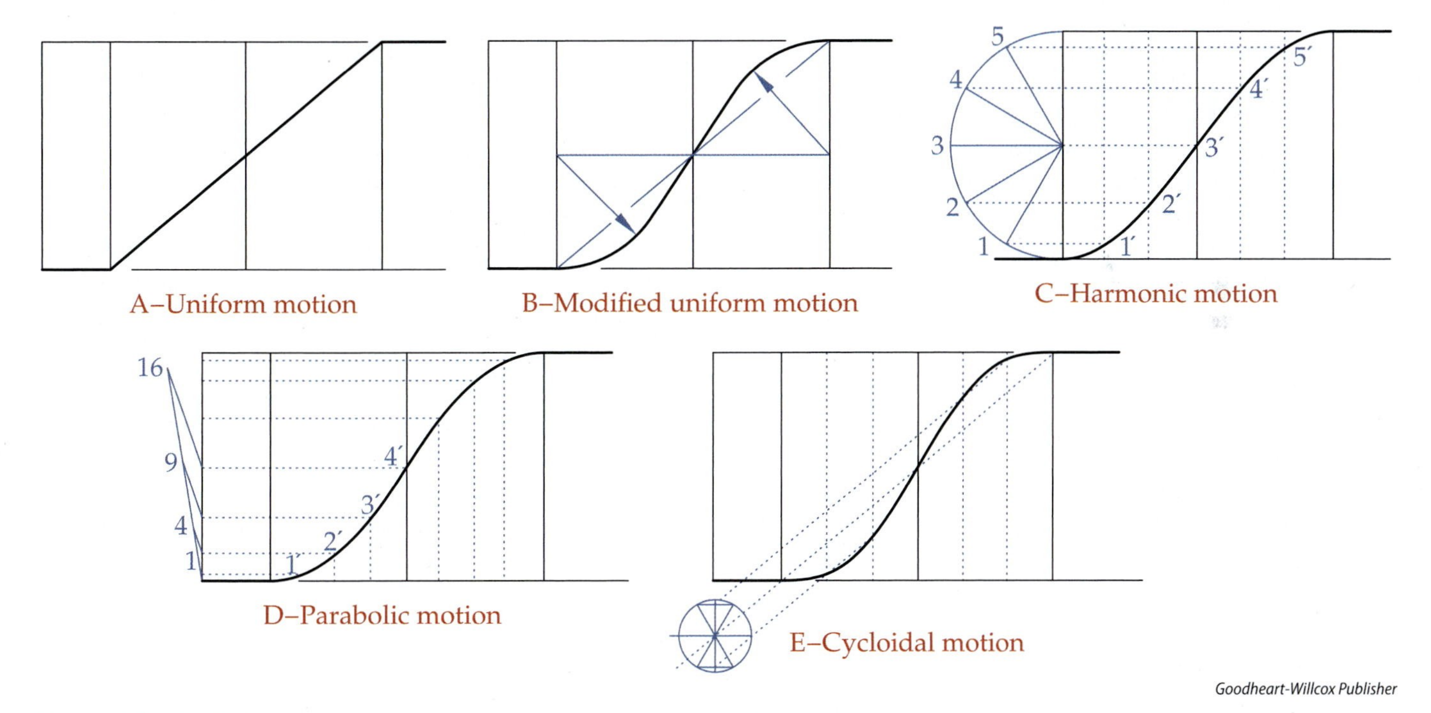

Goodheart-Willcox Publisher

Figure 19-12. Different techniques are available for designing the displacement motion as transitions occur.

Reading Cam Prints

Drawings for cam devices follow standard practices. They are often created as multiview drawings, but with the cam profile dimensioned in a table or displacement diagram. However, angular dimensions, such as HARMONIC RISE 45° or DWELL 90°, may be given. See **Figure 19-13**. For plate cams, CAD-based splines can also be transferred to the manufacturing and inspection equipment. In recent years, the ability to describe a part and share information and contract specifications with a 3D CAD model has been of great benefit. With respect to the tolerance of the geometry, profile of a surface is a common geometric control for the cam profile.

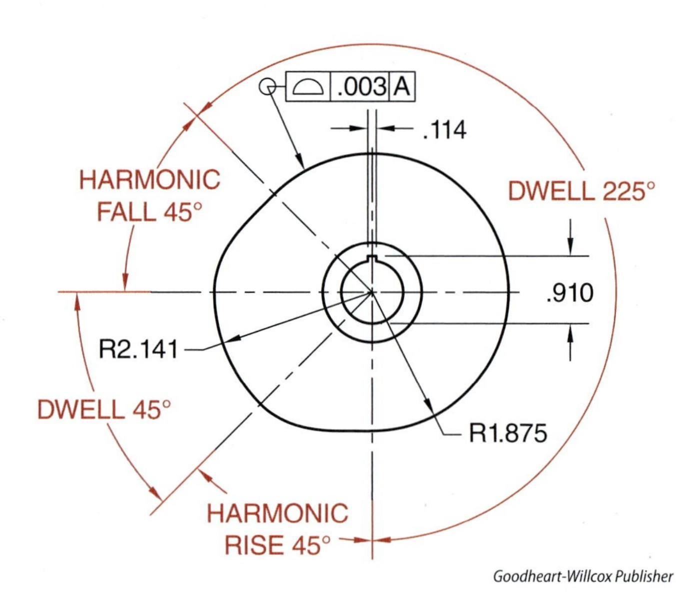

Goodheart-Willcox Publisher

Figure 19-13. While cam drawings usually follow standard practices, additional terms or techniques may be incorporated to describe the irregular profile.

In summary, cam design can be complex, with a lot of design consideration given to velocity, acceleration, follower constraints, and unique spline-based contours. Fortunately, computerized equipment has made it easier to design, analyze, and describe cam shapes, with computer-controlled manufacturing and inspection equipment doing much of the work. Yet, the importance of accuracy in the manufacturing processes and inspection methods cannot be overemphasized. The function of the cam can be compromised by surface variations that may not even be obvious to the casual observer, creating issues of stress, vibration, and noise when the parts begin working together.

Summary

- Cams are specially shaped parts that cause other parts to move in a controlled fashion.
- Cams are found in many objects around the house and shop, including sewing machines, shower rods, and hold-down clamps.
- The principles of cam design can also be applied to knobs, latches, and other devices that move slides and plates into position to perform tasks.
- Many cams change uniform rotary motion into oscillating linear or angular motion.
- Cams are made in a variety of shapes, including cylindrical plates and grooved cylinders, most of which are designed to move a follower device in a specified manner.
- Followers come in a variety of forms, including knife edge, flat face, and roller.
- Followers are constrained to the cam through various means, including gravity, springs, grooves, or mechanical means.
- New and more complex forms of cams are still being developed, including conjugate cams combining multiple plates and followers and helical screw-like globoidal cams that incorporate screw-shaped cams and roller gears.
- Cam design requires a basic understanding of terms such as cam profile, oscillation, displacement, dwell, and cycle.
- Cam design is enhanced by diagrams that plan out the rise and fall displacement, as well as the dwell, of the follower movement over a full 360° revolution of a cam rotation.
- Cam design requires an understanding of harmonic, parabolic, and cycloidal motion with respect to the implications of each on the acceleration and velocity of the cam.
- Describing cam profiles on industrial prints can be accomplished in a variety of ways, including tables, displacement descriptions, CAD-based spline definitions, or 3D geometry model files that can be transferred to manufacturing and inspection equipment.

Unit Review

Name ______________________ Date ______________ Class ______________

Answer the following questions using the information provided in this unit.

Know and Understand

____ 1. *True or False?* Cams were invented around 1950 and have been incorporated into different machine devices ever since.

____ 2. *True or False?* The principles of cam design have been incorporated into several products around the home and shop.

____ 3. Which of the following terms does *not* apply to follower types?
A. Knife edge
B. Flat face
C. Roller
D. Valve

For questions 4–9, match the types of cams with the descriptions. Answers are used only once.

____ 4. Cam moves in a straight line

____ 5. Roller gear cam—may resemble a worm gear

____ 6. Barrel cam—groove cut in it for follower

____ 7. Two cams mounted together—two rollers on one follower

____ 8. Plate cam—irregular-shape wheel

____ 9. Disc with a groove cut into the face of the cam

A. Radial
B. Cylindrical
C. Face
D. Linear
E. Conjugate
F. Globoidal

____ 10. *True or False?* One example of follower constraint is a spring exerting the force necessary to keep the follower engaged with the cam.

____ 11. Which of the following terms describes the shape of the cam surface or groove that engages with the follower?
A. Cam profile
B. Follower path
C. Base circle
D. Offset shape

____ 12. *True or False?* During a cycle, the period of time that a follower does not move, but maintains its position, is called the *freeze*.

____ 13. *True or False?* The term *oscillate* refers to the rocking back and forth of a follower on a pivot point, and the term *vacillate* refers to the back and forth linear movement of a sliding follower.

____ 14. Which of the following terms is used to describe the 360° turn of a round cam on a displacement diagram?
A. Rhythm
B. Cycle
C. Phase
D. Sequence

____ 15. All of the following are usually shown in a displacement diagram *except* _____.
A. time
B. fall
C. rise
D. base circle diameter

____ 16. *True or False?* If a cam had the shape of the *base circle*, the follower would not move at all.

____ 17. Which of the following is a circle that represents the position of a roller follower's center axis at the home position?
A. Home circle
B. Prime circle
C. Base circle
D. Roller circle

____ 18. *True or False?* Forces caused by the velocity and acceleration of the follower as it oscillates are of concern to the cam designer more than they are to the print reader.

____ 19. Which of the following is *not* a displacement motion term?
A. Cycloidal
B. Parabolic
C. Hyperbolic
D. Harmonic

____ 20. *True or False?* Cam prints may specify the dimensions for a cam shape with a table.

Critical Thinking

1. Compare the knife-edge follower with a roller follower, and discuss some of the advantages or disadvantages of each.

2. What benefits are there to using a CAD system to design cams compared to the drafting techniques that were used before CAD systems?

Apply and Analyze

Name ____________________ Date ____________ Class ______________

Review Activity 19-1

Analyze the following displacement diagrams for one cycle of the follower displacement. Then, answer the questions for each diagram.

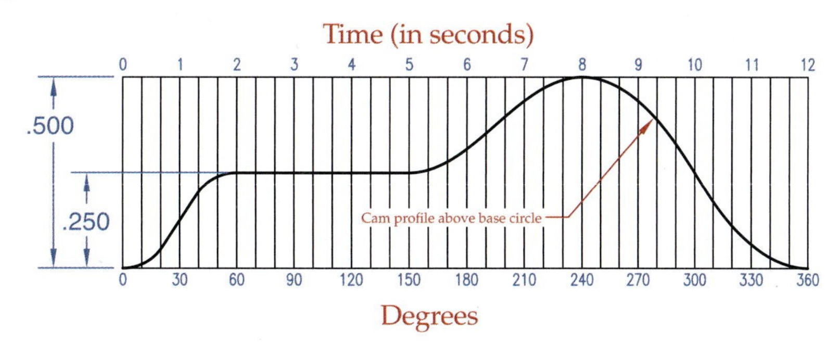

A. Time for one cycle: ____________________

B. Total dwell time: ____________________

C. Total rise after two seconds: ____________________

D. Total rise after eight seconds: ____________________

E. Total time to fall: ____________________

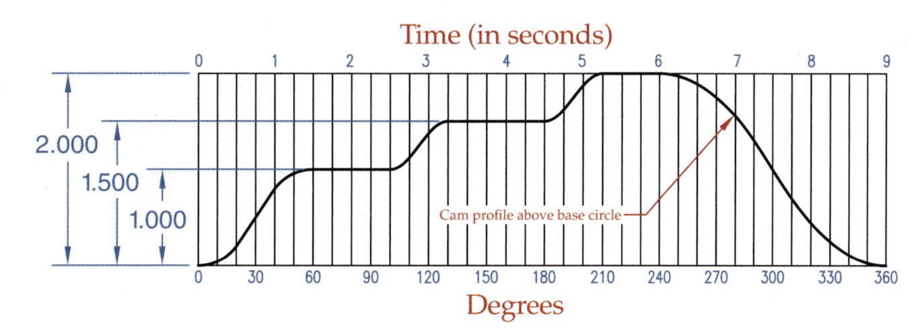

A. Degrees in one cycle: ____________________

B. Total dwell time: ____________________

C. Degrees in first rise: ____________________

D. Degrees in second rise: ____________________

E. Degrees in fall: ____________________

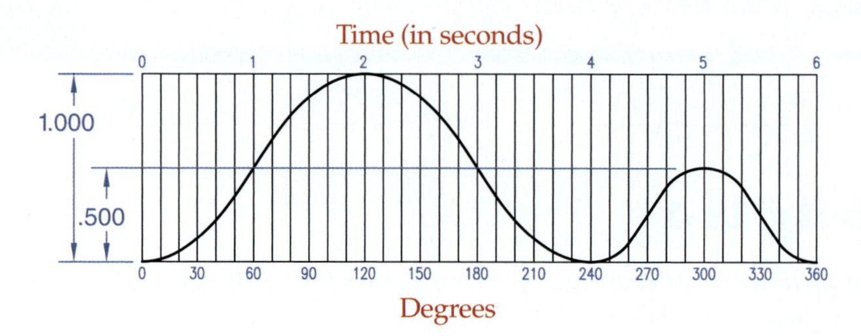

A. Time for one cycle: ______________________

B. Total dwell time: ______________________

C. Time to make first rise: ______________________

D. Total displacement in second rise: ______________________

E. Total time spent falling: ______________________

Notes

Apply and Analyze

Name ______________________ Date ______________ Class ______________

Industry Print Exercise 19-1

Refer to the print PR 19-1 and answer the questions below.

1. What view of the drawing details the size measurements of the cam profile?

____

2. Which cam type describes the cam featured on this part?

____

3. Is there a displacement diagram featured on this print? ______________________
4. As drawn, is the cam profile symmetrical above and below the horizontal center line?

____

5. As drawn, is the cam profile symmetrical with respect to the vertical center line?

____

6. Calculate the displacement of the follower surface if the base circle has a radius of .173″.

____

7. How many *sets* of displacement measurements are given to describe the profile of the cam surface between the 9 o'clock position and the 3 o'clock position?

____

8. How much geometric tolerance (profile of a surface tolerance) is applied to the cam profile?

____

9. What surface hardness is specified for the cam surface?

____

10. How "wide" is the cam surface, as specified in the main view?

____

Review questions based on previous units:

11. At what scale was the original drawing of this part created? ______________________
12. At what scale is Detail B created? ______________________
13. What name is given to the ridges formed on the left end of this part?

____

14. How deep is the center drill in the right end of this part? ______________________
15. What surface quality is specified for the .465″ diameter surface in the .142″ wide groove just to the left of datum feature A?

____

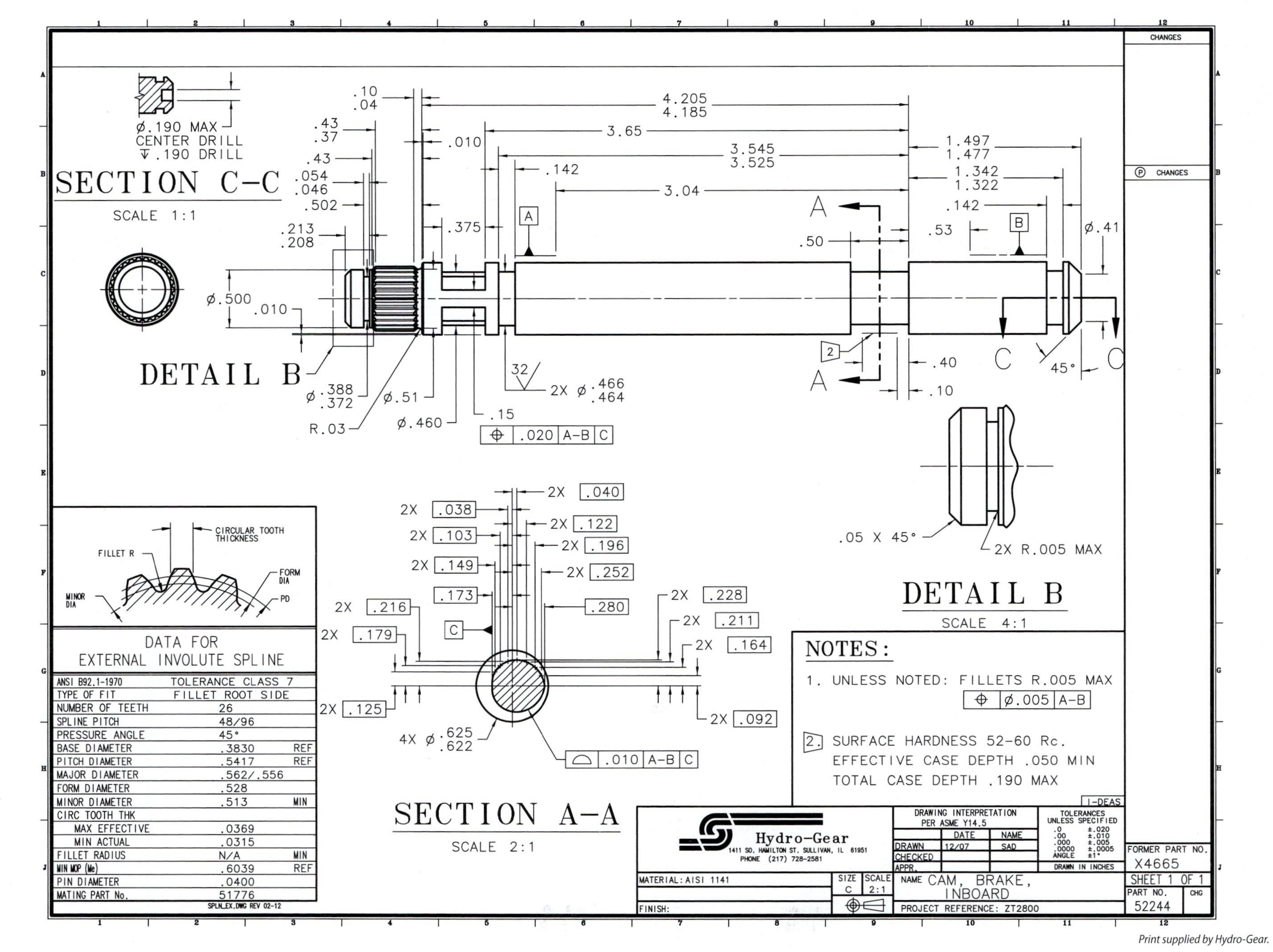

Print supplied by Hydro-Gear.

PR 19-1. Inboard Brake Cam

Apply and Analyze

Name ______________________ Date ____________ Class ____________

Industry Print Exercise 19-2

Refer to the print PR 19-2 and answer the questions below.

1. Given the principal three views, top, front, and bottom, which view best details the shape of the profile that forms the cam?

   ____

2. Calculate the height of the cam surface profile.

   ____

3. Calculate the distance from the center axis of the hub to the peak of the R.094 curve.

   ____

Refer to Figures 19-3 and PR 19-2 to answer the next four questions.

4. Is one direction, clockwise or counterclockwise, the more effective rotation for this cam hub to move the mating part, or are both directions equally effective?

   ____

5. What follower constraint will be incorporated into this design?

   ____

6. When the cam hub is rotated 45°, as shown in the second position in Figure 19-3, what is the radius of the curved surface that is "pushing" the sliding part?

   ____

7. Study the position illustrated at the far right in Figure 19-3. Which dimension on the print (PR 19-2) best approximates the distance from the axis of the hub to the peak of the curve on the slider that mates into the R.094 depression in the position illustrated at the far left?

   ____

8. Subtract answer 3 from answer 7 to determine the approximate movement of the sliding part.

   ____

Review questions based on previous units:

9. At most, how deep is the cavity on one end of the cam hub? ______________
10. Are the external cylinders dimensioned using the "cylinder rule"? ______________
11. What term applies to the rim around the edge of the cavity? ______________
12. What surface quality is specified for the surfaces of this part? ______________

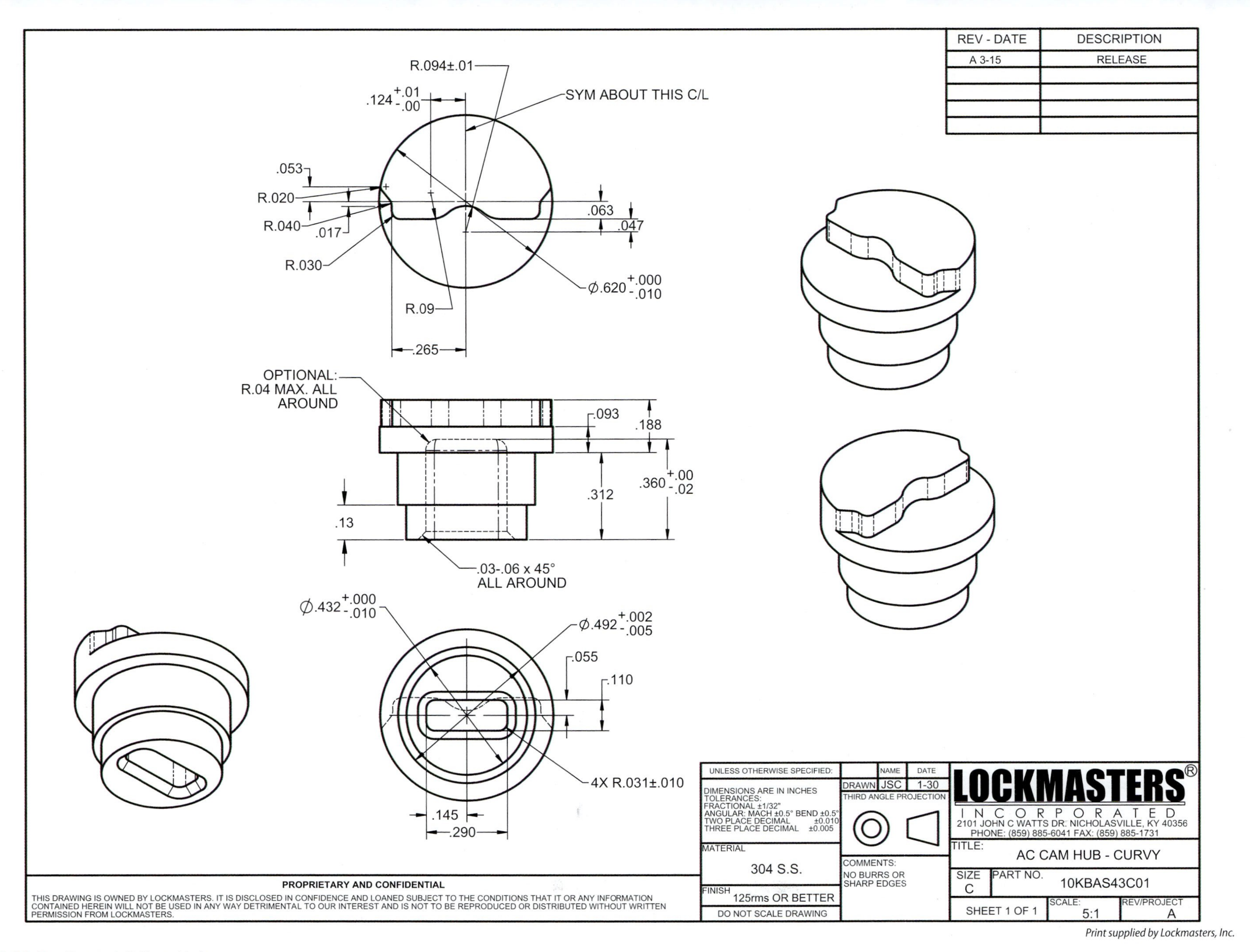

Print supplied by Lockmasters, Inc.

PR 19-2. Curvy AC Cam Hub

Notes

UNIT 20
Plastic Parts

LEARNING OBJECTIVES

After completing this unit, you will be able to:

- List common names and abbreviations used in the manufacture of plastic parts.
- Explain the difference between thermoplastic and thermoset plastics.
- Explain various processes used in the plastics industry.
- Describe methods used to fasten and join plastics.
- Read and interpret prints from the plastics industry.

TECHNICAL TERMS

boss
casting (plastic)
cavity
core
draft
ejector pin
elastomer
extrusion (plastic)
flash
gate
gate vestige
gusset
insert
knockout
mold
molding
mold shrinkage
monomer
overmold
parting line
plastic
polymer
polymerization
resin
rotational molding
runner
sink mark
sprue
thermoforming
thermoplastics
thermoset plastic
warpage

The term *plastic* refers to a product of polymer chemistry—a synthetic material formed by chemical reaction rather than a substance of natural origin. Early forms of plastics had limited use as toys, kitchen utensils, and other products that do not require great strength, durability, or resistance to heat and chemicals. Now, plastics are an important engineering material and have a wide variety of applications. Some of these applications include electrical parts, electrical insulation, gears, bearings, packaging, automobile bodies, motor housings, and architectural products. By the latter part of the twentieth century, the plastics industry represented a major segment of manufacturing, and it continues to do so today.

Materials in the Plastics Industry

Plastics are broadly classified as either thermoplastics or thermoset plastics. The division is based on the effect temperature has on their properties. The next sections cover the two types of plastics.

Thermoplastics

Thermoplastics soften when heated and become solid when cooled. Thermoplastics are classified further into categories based on solvent resistance, degree of volumetric change during processing, and optical properties. Amorphous thermoplastics are typically transparent, exhibit relatively little shrinkage during processing, and have poor solvent resistance. Some common amorphous thermoplastics are polystyrene (PS), polyvinyl chloride (PVC), and polycarbonate (PC). Semicrystalline thermoplastics are typically translucent, exhibit more shrinkage than amorphous plastics, and have much higher solvent resistance than amorphous plastics. Some common semicrystalline thermoplastics include variations of polyethylene (PE), polypropylene (PP), and polyamide (PA), commonly known as nylon. Plastics are made opaque by adding pigments. A listing of common plastics, their abbreviations, their classification, and some typical applications is given in **Figure 20-1**.

Thermoplastics are solid at room temperature but become liquid when heated beyond their melting points. This melting/freezing cycle is the principle used to manufacture the plastic into finished parts. The initial form of the thermoplastics is either pellets or powders (solid forms). Some fabrication techniques using melted thermoplastics include injection molding, extrusion, blow molding, and rotational molding. Several plastic products made with these processes are shown in **Figure 20-2**. When a sheet of thermoplastic is softened to just below its melting point, it can be fabricated by thermoforming using vacuum or pressure. In their solid state, thermoplastics can be processed by standard metal-cutting techniques, such as drilling and milling.

Thermoplastics		
Abbreviation	**Name**	**Typical Applications**
Amorphous		
ABS	Acrylonitrile-butadiene-styrene	Appliance housings, refrigerator liners, 3D printer filament
PC	Polycarbonate	Helmets, headlights, compact discs, 3D printer filament
PMMA	Polymethyl methacrylate	Skylights, stoplights
PS	Polystyrene	Cutlery, packaging, insulation
PVC	Polyvinyl chloride	Pipes, window frames, siding, hoses
Semicrystalline		
PA	Polyamide (Nylon®)	Bearings, gears, housings for tools
PE	Polyethylene	Bottles, bags, tubing
PET	Polyethylene terephthalate	Microwavable or freezable packaging, drink bottles, plastic film (Recycle 1)
HDPE	High-density polyethylene	Oil bottles, milk jugs, detergent & bleach bottles (Recycle 2)
LDPE	Low-density polyethylene	Thin, flexible products, plastic bags, shower curtains (Recycle 4)
POM	Polyoxymethylene (Acetal)	Gears, plumbing fixtures
PP	Polypropylene	Suitcases, Tupperware®
Special (Amorphous or Semicrystalline)		
PLA	Polylactic acid	Biodegradable substitute for PET, DIY 3D printing

Thermoset Plastics		
Abbreviation	**Name**	**Typical Applications**
EP	Epoxy	Adhesive, aerospace panels
MF	Melamine-formaldehyde	Heat-resistant surfaces, dishes
PF	Phenol-formaldehyde	Cooking pan handles, breaker boxes
UP	Unsaturated polyester	Tubs and showers, car panels, boats

Goodheart-Willcox Publisher

Figure 20-1. This table shows common plastics, their abbreviations, and typical applications.

StanislauV/Shutterstock.com

Figure 20-2. Examples of thermoplastics made using various processes can be found in everyday life. The thermoplastic items pictured here began as solid pellets.

Thermoset Plastics

Thermoset plastics, unlike thermoplastics, cannot be made liquid again by heat or solvent. Thermoset materials are chemically altered through a curing process known as *cross-linking polymerization*, which occurs during the molding process. The chemical makeup of thermoset materials causes extensive bonding between molecules, which results in the formation of a rigid mass. Curing can be initiated by the addition of an initiator chemical (catalysis), the application of heat, ultraviolet light, or by the simple mixing of two components, depending on the type of thermoset.

The most common thermoset plastics are phenolics, unsaturated polyesters, epoxies, and melamine. Some processes used to make products from thermoset materials include compression molding, transfer molding, and fiberglass lay-up. The initial form of thermoset materials may be granules, pastes, or liquids. To obtain the better product qualities, such as higher strength and durability, molding compounds made from these resins always contain additional fillers and reinforcing agents.

Processes in the Plastics Industry

There are many processes applied in industry to form plastics into useful products. Those most widely used are discussed here to define terms that might appear on prints.

Molding is the process of forming a plastic object by forcing molten plastic material into a hollow mold using pressure. Examples of molding processes are injection molding and blow molding. Injection molding is the most widely used process to create products of great complexity with the lowest cost, **Figure 20-3**, so many of the prints that involve plastic parts are for injection molding. Blow molding is typically used to make plastic bottles, **Figure 20-4**.

Thermoforming is the process of forming a plastic object by forcing a softened sheet of thermoplastic onto a mold using vacuum or positive pressure. **Figure 20-5** illustrates a part that was formed about a mold. Many large sheet-type parts, such as motorcycle windscreens, are made by thermoforming.

Extrusion is a continuous process of melting thermoplastics, then pumping the molten plastic through a heated die. The shape of the die exit forms the product. Once the molten plastic passes through the die, it is cooled and finally sized by downstream

Engel, Inc.

Figure 20-3. This machine is a typical injection molding machine.

Uniloy

Figure 20-4. These plastic bottles are made using a blow molding machine.

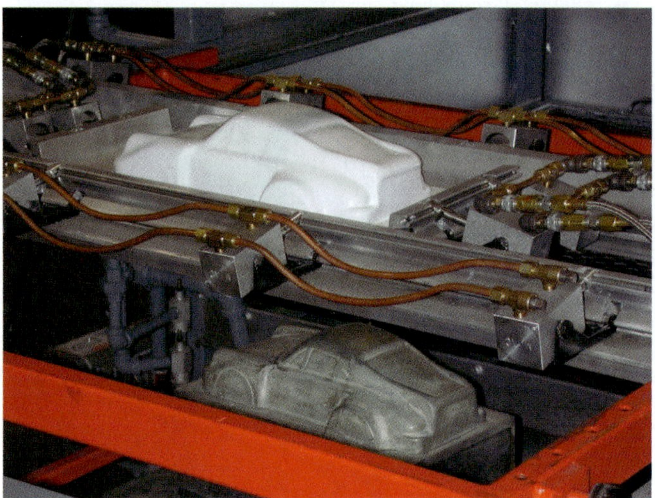

MAAC Machinery

Figure 20-5. This plastic car model was created by thermoforming.

equipment. The types of products made from extrusion depend on the die. Some products made from extrusion include:

- Films and sheets used in single or multiple layers or as coatings for cloth and paper
- Profile shapes such as rod, pipe, or channels, **Figure 20-6**
- Coatings around wire and cable

Casting is the process of forming a plastic object by pouring molten material into an open mold. No pressure is used in casting. In the case of a thermoset, the cast material polymerizes, or cross-links, in the mold. ***Rotational molding*** is a casting process that converts powdered thermoplastics into hollow objects by rotating the powder in a closed, heated mold until the powder melts and fuses into a solid mass.

Machining, finishing, decorating, and assembly for plastic products are the same as in other fabricating industries. Therefore, the drawings and prints providing the design and manufacturing specifications of plastic parts and assemblies are also similar.

Fastening and Joining Plastics

There are four principal techniques to join plastics to each other and to other materials:

- **Mechanical fasteners.** These applicable fasteners include thread-cutting screws, rivets, and press-in inserts.
- **Mechanical fits.** Press-in and snap-on fits can be used to join plastics.
- **Welding.** Plastics can be joined using ultrasonic welding, heat welding, and spin welding. An ultrasonic welding machine is shown in **Figure 20-7**.
- **Adhesives.** Elastomers, epoxies, and solvents can also be used to join plastics.

Plastics Terminology

The terminology associated with plastics technology differs somewhat from that of metalworking. You must understand plastics terms and features in order to read prints for the plastics industry. Many of the terms discussed in this section frequently appear on prints related to manufacturing with plastics. Other general terms are also discussed.

A ***monomer*** is a molecule of low molecular weight capable of reacting with identical or different monomers to form a polymer. A ***polymer*** is the basic molecular structure of a plastic material. It is the product of polymerization. ***Polymerization*** is the joining of two or more molecules to form a new and more complex molecule whose physical properties are different. ***Resin*** is the basic organic material

4level/Shutterstock.com; Bocos Benedict/Shutterstock.com

Figure 20-6. Plastic pipes and tubing are created using the extrusion process. The bottom picture shows plastic pipe coming out of an extruder, where it has been heated and shaped by a die.

from which plastics are formed. The term is also used to describe plastic not yet converted into a product. An ***elastomer*** is a material having similar characteristics to those of rubber. Elastomers stretch to at least twice their normal length. In addition, they rapidly return to their original length. Thermoplastic elastomers developed in recent years have some advantages over rubber.

As previously explained in this unit, the molding process involves the placing of a plastic material, in a heated or liquid state, into a hollow mold under pressure to form a plastic part. The ***mold*** is a hollow form into which a plastic material is placed to give the material its final shape. A ***cavity*** is the depression, or set of matching depressions, in a mold used for plastic forming. The cavity shapes the surfaces of the case or molded article. This term also refers to the stationary part of a mold used for injection molding. The ***parting line*** is the edge of a mold cavity where mold halves come together.

Dukane

Figure 20-7. This ultrasonic welding machine is used to join plastics.

An ***ejector pin*** is a pin designed to push a plastic product out of a mold. The ejector pin is typically made of hardened steel, and its location is often indicated on the print for the associated part. A ***knockout*** is any part of the mechanism of a mold used to eject the molded articles. In some cases, an acceptable location for a knockout is specified on the print.

The term ***core*** can have three meanings. First, a core is part of a mold that hollows out a section of a part. In this usage, a core is also called a core pin. Second, a core is the movable portion of a mold used for injection molding through which the ejector pins typically pass. Third, a core is a channel in a mold for circulation of heat-transfer media.

A ***draft*** is a slight outward taper in a mold wall designed to facilitate removal of the molded object. A back draft is a taper in the opposite direction. This tends to impede removal of the object or hold it in place.

In injection molding and transfer molding, the ***sprue*** is the main feed channel that connects the mold orifice to the runners leading to each cavity gate. The term is also given to the piece of plastic material formed from this channel. A ***runner*** is the usually circular channel that connects the sprue with the gate of the mold cavity. The term is also given to the piece of plastic material formed from this channel. A ***gate*** is the channel connecting the runner to the part cavity. The gate may have the same diameter or cross section as the runner but is often restricted to a diameter of 1/8″ or less. ***Gate vestige*** is the residual material that remains after a gate is torn or cut away from a molded part. See **Figure 20-8**.

Mold shrinkage is the immediate shrinkage of the plastic part after it has been removed from the mold. Both the mold and molded part are measured at normal room temperature. ***Warpage*** is the distortion of the plastic part caused by nonuniform shrinkage. A ***sink mark***, or *sink*, is a shallow depression or dimple on the surface of an injection-molded article caused by a short shot or local internal shrinkage after the gate seals.

Flash is the thin web of excess material that is forced into the crevices between mating mold surfaces. Flash remains attached to the part when removed from the mold. Often, flash is removed in a later finishing step.

A ***boss*** is a circular protrusion from the plastic body. A boss is used to anchor a threaded fastener into the plastic. A ***gusset*** is an angular piece of material added to strengthen two adjoining walls. An ***insert*** is a material placed into the mold cavity prior to molding that is then surrounded by the molding material. ***Overmold*** is the process of injection molding a plastic over an existing plastic part. This term also refers to the material that forms the overmolding.

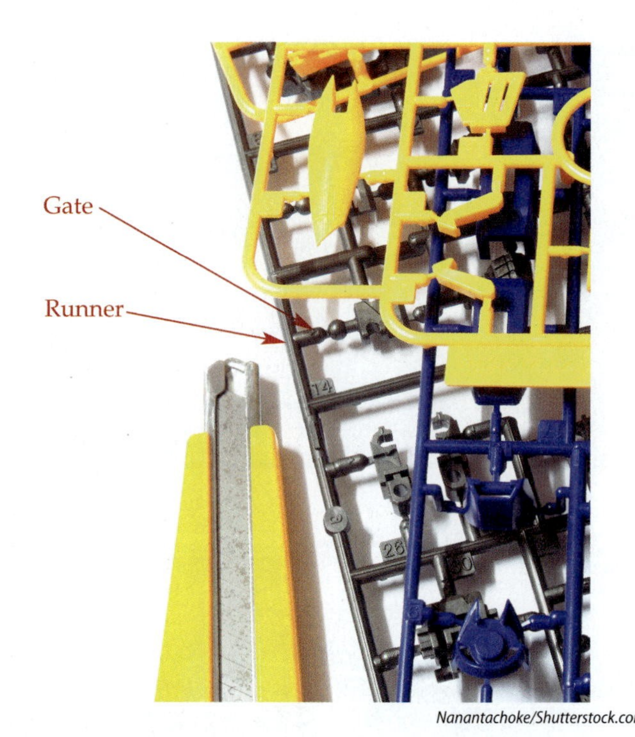

Nanantachoke/Shutterstock.com

Figure 20-8. The parts for this model kit are made by an injection molding process. The runners and gates are clearly visible and need to be cut away before the model can be assembled.

Reading Prints from the Plastics Industry

Once you are familiar with plastic materials, processes, and terminology, reading prints for plastic parts and assemblies is much like reading prints for other manufacturing industries. The drawing layout, title block, notes, and dimensioning procedures are the same.

Summary

- Synthetic materials known as plastics have become an important part of modern industrial design applications.
- Plastics can broadly be defined as thermoplastics or thermoset plastics, but are further defined by properties such as solvency, shrinkage factors, and freeze/thaw melting points.
- Unlike thermoplastics, thermoset plastics cannot be made liquid again by heat or solvents, but they still avail themselves to many applications that require rigid and strong parts.
- Processes within the plastics industry include various forms of molding, thermoforming, and extrusion, as well as the fabrication of plastic parts with fasteners, fits, adhesives, and welding.
- An extensive glossary for plastic part prints will include many molding terms, such as *cavity*, *ejector pin*, *core*, *knockout*, *draft*, *sprue*, *runner*, *gate*, *mold shrinkage*, and *flash*.
- Other than the terms related to plastic processes, materials, and molding, reading plastic part prints is similar in nature to the process of reading most other industrial prints.

Unit Review

Name ______________________ Date ______________ Class ______________

Know and Understand

Answer the following questions using the information provided in this unit.

____ 1. *Plastic* is a general term that indicates a _____ material formed by chemical reaction, rather than a substance of natural origin.
A. hard
B. rubber-like
C. synthetic
D. soft

____ 2. *True or False?* In the industry, some use the term *thermoplastics* and some use the term *thermoset plastics*, but they both mean the same thing.

____ 3. *True or False?* Amorphous thermoplastics are typically transparent, and semicrystalline thermoplastics are typically translucent, but there are other general differences between the two types.

____ 4. The very first, or *initial*, form of thermoplastics is often _____ or powder.
A. liquid
B. pellets
C. sheets
D. wire

____ 5. Which of the following is *not* a common use of ABS plastic?
A. Appliance housings
B. Refrigerator liners
C. 3D printer filament
D. Grocery bags

____ 6. *True or False?* Thermoset plastics cannot be melted once they are formed.

____ 7. *True or False?* Thermoforming, just like the name implies, is the process of forcing a thin film of thermoplastic that has been warmed and softened onto a molded shape.

____ 8. Which of the following is *not* used to join plastic materials together?
A. Soldering
B. Adhesive
C. Fasteners
D. Welding

____ 9. Which of the following terms specifically applies to a plastic material that has rubber-like characteristics?
A. Elastomer
B. Polymer
C. Monomer
D. Resin

____ 10. Which of the following is a common plastic process wherein liquid plastic is forced into a hollow mold with great pressure?
A. Rotational molding
B. Blow molding
C. Injection molding
D. Casting

____ 11. On a print, what term is used to indicate the location on a molded object where the two halves of the mold meet?
A. Point of demarcation
B. Core datum
C. Ejection edge
D. Parting line

____ 12. *True or False?* The *sink mark* is the location spot where a mechanical device, such as a pin, assists the part out of the mold.

____ 13. For parts that are molded, the print reader should expect to see information about the taper of a surface that helps the part come out of the mold more easily. This is called _____.
A. allowance
B. shrink
C. draft
D. sprue

____ 14. *True or False?* The thin web of excess material that may be on a molded plastic part that just came out of the mold is called the *flash*.

____ 15. *True or False? Overmolding* is a term applied to the placement of a metal insert into the plastic mold before the plastic part is formed.

Critical Thinking

1. For the four plastic products listed below, think of a material used in the past, or think of a substitute material that could be used instead of plastic, and briefly discuss why that material may be more or less desirable, costly, or difficult to produce.

   Milk jug

   Plastic knife and fork set

   TV remote

   Food storage container/bowl with sealing lid

2. Why might plastic parts be more difficult to design than other parts made from other materials?

Apply and Analyze

Name ______________________ Date ______________ Class ______________

Review Activity 20-1

Match the following plastics to the corresponding acronym.

____ 1. Polystyrene
____ 2. Polyvinyl chloride
____ 3. Polycarbonate
____ 4. Polyethylene terephthalate
____ 5. Polypropylene
____ 6. Nylon
____ 7. Polymethyl methacrylate
____ 8. Phenol-formaldehyde
____ 9. Unsaturated polyester
____ 10. Acetal
____ 11. High-density polyethylene
____ 12. Polylactic acid

A. PA
B. PC
C. PET
D. PF
E. PLA
F. PMMA
G. POM
H. PP
I. HDPE
J. PS
K. PVC
L. UP

Apply and Analyze

Name ________________ Date ________________ Class ________________

Industry Print Exercise 20-1

Refer to the print PR 20-1 and answer the questions below.

1. How much draft, if any, is specified for this part?

____

2. The material specified for this part is a common plastic. What is its trade name?

____

3. What is the maximum allowable depth specified for a knockout pin mark?

____

4. What datum surface is referred to with respect to flash and gate breakoff?

____

5. To prevent sharp edges on the final molded part, what radius is allowed on all edges unless otherwise specified?

____

Review questions based on previous units:

6. What is the cage number for this part? ________________

7. What scale are the views on the original print? ________________

8. Using the mean values, calculate the diameter of the small circular flat surface at the top of the three posts.

____

9. What is the diameter of the feature that is used to establish datum axis B?

____

10. What type of dimensions are those that have boxes around the values?

____

11. What scale are the isometric pictorial views? ________________

12. What MMC is specified for the diameter of each of the three posts?

____

13. Taking tolerances into consideration, what is the maximum amount the three posts stick up above the main body of the part?

____

14. What revision issue is this drawing? ________________

15. Issue A of this drawing was in conjunction with ECN 328945. What does ECN most likely stand for?

____

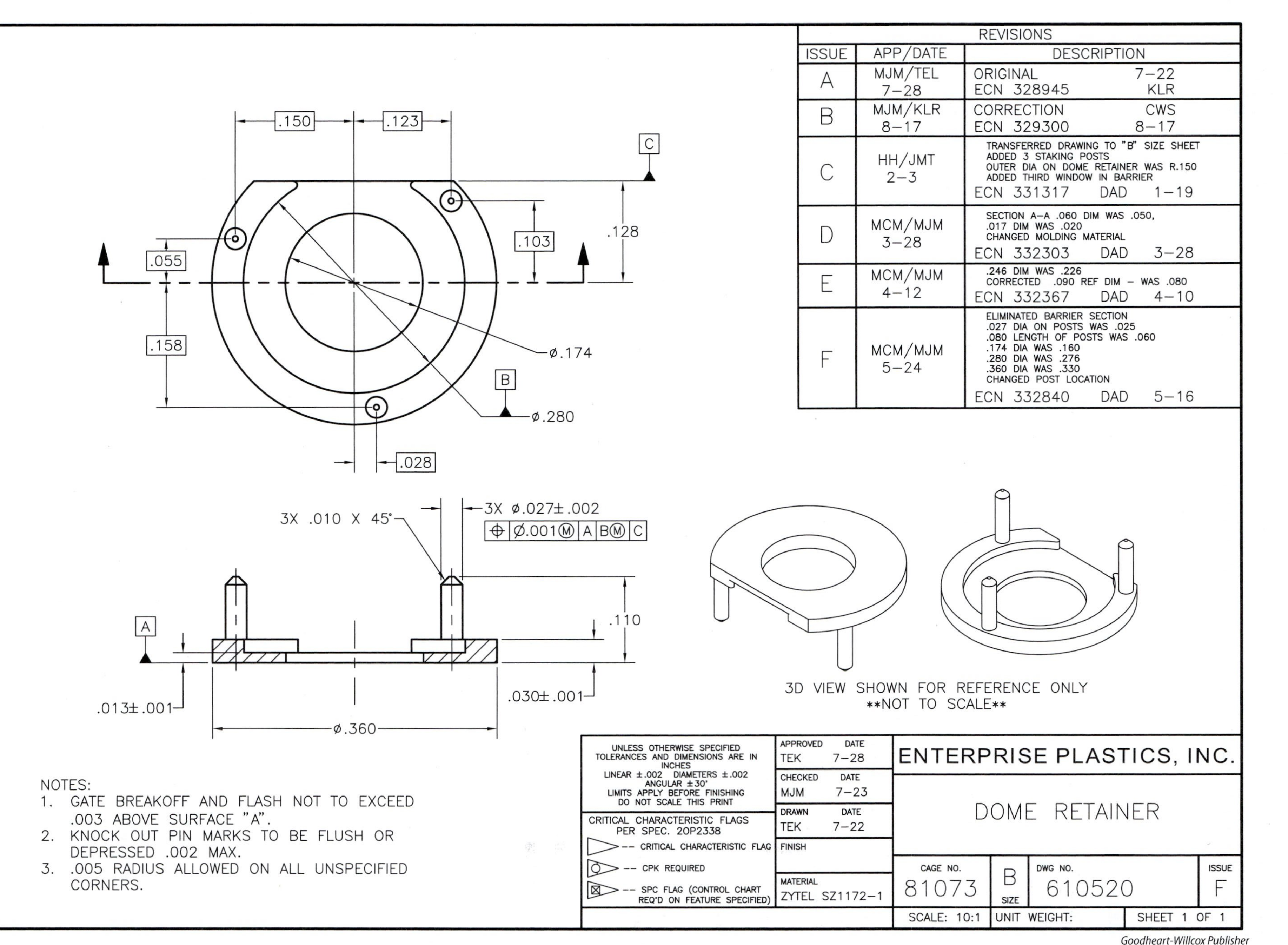

REVISIONS		
ISSUE	APP/DATE	DESCRIPTION
A	MJM/TEL 7-28	ORIGINAL 7-22 ECN 328945 KLR
B	MJM/KLR 8-17	CORRECTION CWS ECN 329300 8-17
C	HH/JMT 2-3	TRANSFERRED DRAWING TO "B" SIZE SHEET ADDED 3 STAKING POSTS OUTER DIA ON DOME RETAINER WAS R.150 ADDED THIRD WINDOW IN BARRIER ECN 331317 DAD 1-19
D	MCM/MJM 3-28	SECTION A-A .060 DIM WAS .050, .017 DIM WAS .020 CHANGED MOLDING MATERIAL ECN 332303 DAD 3-28
E	MCM/MJM 4-12	.246 DIM WAS .226 CORRECTED .090 REF DIM - WAS .080 ECN 332367 DAD 4-10
F	MCM/MJM 5-24	ELIMINATED BARRIER SECTION .027 DIA ON POSTS WAS .025 .080 LENGTH OF POSTS WAS .060 .174 DIA WAS .160 .280 DIA WAS .276 .360 DIA WAS .330 CHANGED POST LOCATION ECN 332840 DAD 5-16

Goodheart-Willcox Publisher

PR 20-1. Dome Retainer

Apply and Analyze

Name ______________________ Date ______________ Class ______________

Industry Print Exercise 20-2

Refer to the print PR 20-2 and answer the questions below.

1. What draft is applied to the part feature with the largest diameter of .470″?

____

2. In the section view, what draft is applied to the six ridges of the knob that are 60° apart?

____

3. The small central cavity with a diameter of .165″ simply indicates +DFT. According to drawing notes, what draft is allowed for that feature?

____

4. The material used for this part is a blended plastic developed by GE. What is its trade name?

____

5. For the angular dimensions that are not draft angles, what angular tolerance is specified?

____

6. As measured perpendicular to the axis, how thick is the material of the feature at the right-hand end of the part, as shown in the section view?

____

Review questions based on previous units:

7. What company document specifies the finish for the outside surfaces of this part?

____

8. Are there any leader lines shown on this print? ______________________

9. At what scale are these views created? ______________________

10. Who checked this drawing? ______________________

11. Does this object have any dimensions that use the limit method of tolerance expression?

____

12. How many cutting planes are shown? ______________________

13. Unless otherwise specified, what tolerance is to be applied to linear measurements?

____

14. For some of the dimensions, there is a TYP note. What does that stand for?

____

15. How many changes were made as part of revision A? ______________________

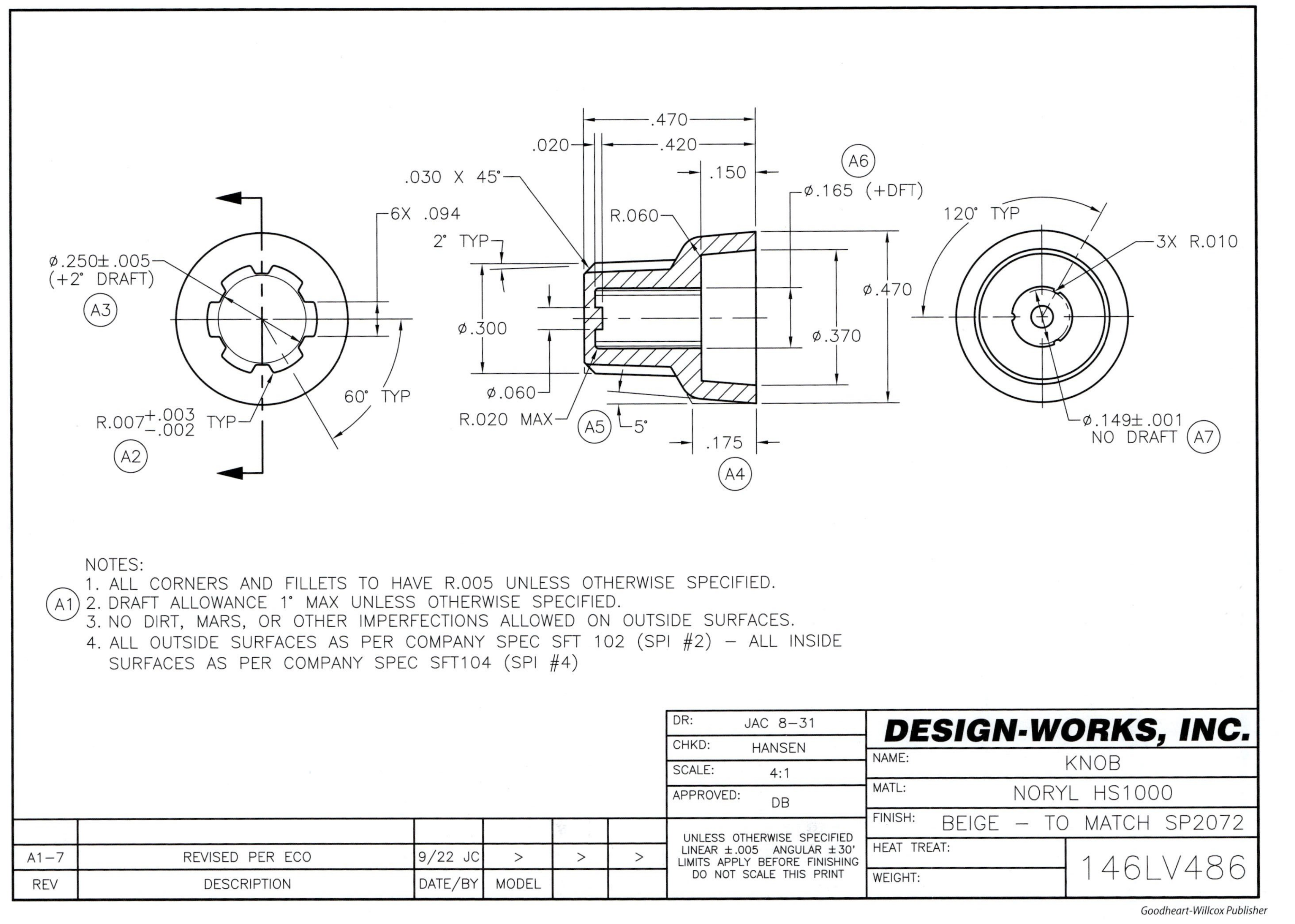

Goodheart-Willcox Publisher

PR 20-2. Knob

Apply and Analyze

Name ______________________ Date ______________ Class ______________

Bonus Print Reading Exercises

The following questions are based on various bonus prints located in the folder that accompanies this textbook. Refer to the print indicated, evaluate the print, and answer the question.

Print AP-003

1. What material is specified for this part?

____

2. What is the maximum amount of flash that is allowed on the threads?

____

3. In how many views are the knock-out pin relief areas shown?

____

4. Will the gate for this molded part be attached to a planar surface or a cylindrical surface?

____

Print AP-007

5. Based on the section line pattern, what material is indicated for this part? (Refer to Appendix D.)

____

6. List two features of this part that must be free of flash.

____

7. List two mold defects that are specified on the print as unacceptable.

____

Print AP-018

8. What material is specified for this part?

____

9. In which view is the parting line specified?

____

10. Will the cavity number appear inside the part or outside?

____

11. What type of mold release *cannot* be used?

____

12. In the view where the cutting-plane line is shown, what do the circles represent?

____

13. Give the maximum amount of flash allowed at the parting line.

____

14. What is the typical draft on the outer sides of the part, with the one side that features the hole being an exception?

____

Print AP-019

15. What document establishes the color for this part, and in general, what color is it?

____

16. What type of plastic is specified for this part?

____

17. How much flashing or mismatch is permitted at the parting line?

____

18. In Section A-A, a .352″ diameter hole is specified. If the other end of that hole can be larger or smaller, which would it be?

____

19. What is the maximum amount of draft that can be applied to any feature of this part?

____

Print AP-020

20. This print is for a cast aluminum part, but list three terms found on this print that are the same as what would be found on a print for a molded plastic part, as discussed in this unit.

____

Notes

UNIT **21**

Precision Sheet Metal Parts

LEARNING OBJECTIVES

After completing this unit, you will be able to:

- Explain the design implications of bending sheet metal to form a part.
- Discuss characteristics of sheet metal thickness and gage numbers.
- Explain the basic characteristics of a tool and die maker's role in sheet metal fabrication.
- Define and apply various terms related to sheet metal bends.
- Define and interpret various terms related to sheet metal drawing notations.
- Use a setback chart to calculate bend radii for precision sheet metal.
- Use proper procedures and formula to calculate bend allowance for precision sheet metal.
- Read and interpret precision sheet metal prints.

TECHNICAL TERMS

bend
bend allowance
bend angle
bend axis
bend deduction (setback)
bend instruction
bend radius
bend tangent line
blank length
brake
brake process
center line of bend (CLB)
developed length
die
flange
form block line (FBL)
gage
hydroforming
inside mold line (IML)
joggle
K-factor
leg
neutral arc bend radius
neutral plane
outside mold line (OML)
overbending
setback
springback
thickness
tool and die maker

Growth in the instrumentation and electronic industries in the latter part of the twentieth century brought about the need for a special type of sheet metal work. Precision sheet metal parts are those manufactured from thin metals to machine shop tolerances. The metal is machined in the flat position and then folded or assembled. The calculations and layout must be precise so the relationship and location of the resulting planes and features of the folded part are within specified tolerances. See **Figure 21-1**. Geometric dimensioning and tolerancing controls may be applied to the flat pattern as well as the final folded part. In some cases, these drawings can be created as undimensioned drawings, as covered by ASME Y14.31, but they may require some special notations with respect to the bends.

The focus of this unit is primarily on forming parts from sheet metal with cutting, bending, punching, and stamping technologies, with attention to both dimensioned and undimensioned drawings. Most often, information from 2D or 3D CAD files is converted into the machine code that drives computer numerical controlled (CNC) machines that precisely cut or stamp sheet metal shapes. New technologies in laser cutting, waterjet cutting, and air bending are continuing to increase the quality of sheet metal products. Forming processes may incorporate the use of bending machines known as ***brakes***, although sometimes the principles of a brake are incorporated into other machines, such as presses. Heavy-duty presses use custom-made tools known as ***dies*** to punch holes of any shape, or cut the metal, while other tools are used to force or press the sheet metal into shape.

As these newer technologies continue to evolve, some allow for sheet metals to be formed into more complex shapes. One such technology is known as ***hydroforming***, which can be applied to both sheet metal and metal tubing. Hydroforming is a special die process that uses pressurized hydraulic fluid in a flexible bladder or diaphragm to force thin metal sheets or tubes into a die shape. This process is similar in nature to plastic thermoforming, although heating the material is not required. The engineer's choice of manufacturing process is usually based on the productivity advantages of the various technologies. Other fluid-forming technologies that do not require flexible diaphragms, and some that use high-pressure water instead of hydraulic oil fluids, are continuing to evolve. As a print reader, be ready to

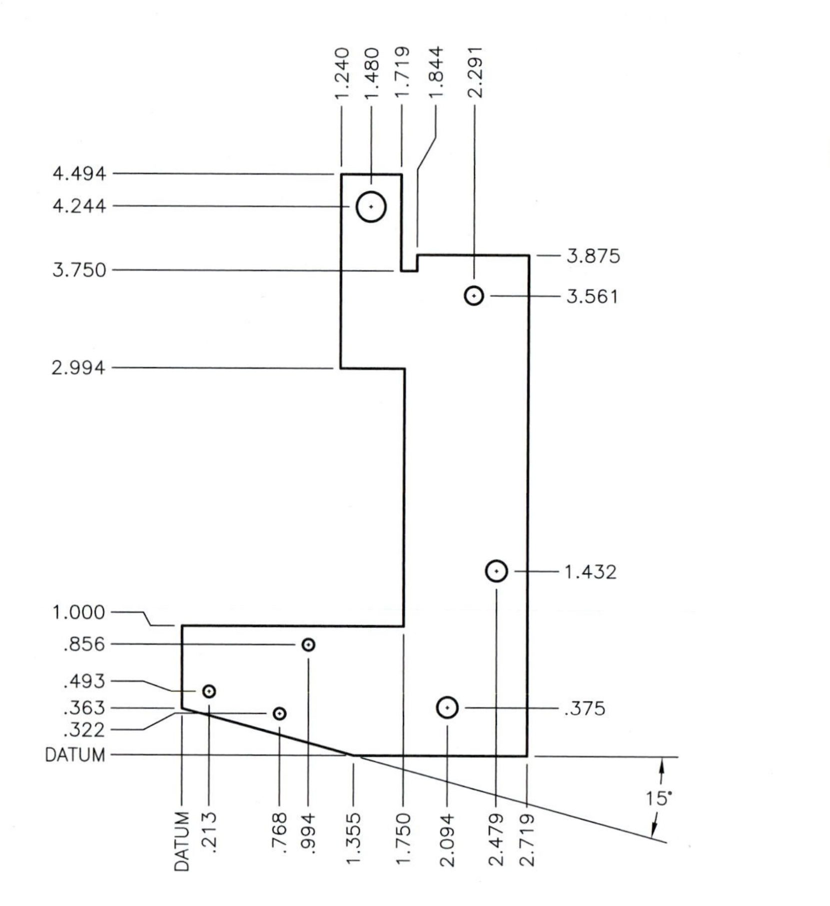

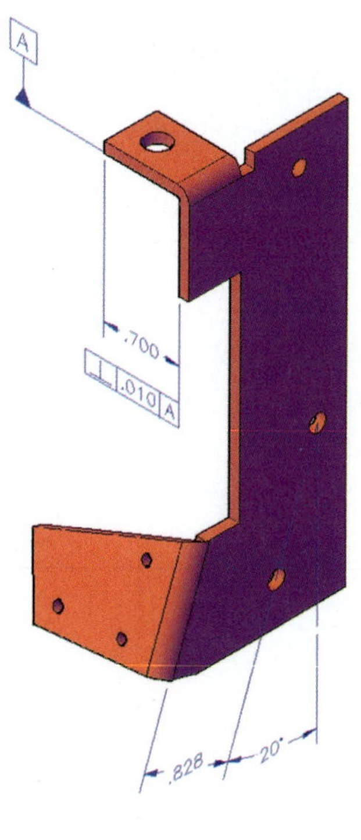

Goodheart-Willcox Publisher

Figure 21-1. A folded or "bent" part begins as a flat blank of stock material, as shown on the left.

research terms and new notes that may find their way into prints. These newer technologies that allow for parts that are free-form in nature diminish the need for traditional prints with dimensions, as the 3D model data can be transferred directly to the manufacturing processes and checking equipment.

Another field of study beyond the scope of this text is the work performed by a ***tool and die maker***, which includes the design and manufacture of tools and dies primarily used to form metal and plastic parts. Tool and die makers often use models, drawings, or prints that are unique to the nature of some manufacturing process, involving tool and die drawings that are also different in nature from detail drawings or contract documents. While tool and die makers are involved in precision machining for the tools and dies used in sheet metal processes, they also make jigs, fixtures, molds, and other important equipment for the manufacturing enterprise.

Sheet Metal Thickness

In the US customary units system, various materials such as wire, rods, and sheet metal are described with ***gage*** (abbreviated GA) numbers that describe a dimensional aspect of the material. Sometimes the spelling *gauge* is used. For wire, the gage number refers to the diameter or cross-sectional area of the material, whereas for sheet metal, the gage number refers to the thickness of the material. One confusing aspect of this term is that the higher the gage, the thinner the material, because the standardized term is based on the weight of the material. For example, 16 gage cold rolled steel (CRS) is 2.5 pounds per square foot, while 18 gage is 2.0 pounds per square foot and 20 gage is 1.5 pounds per square foot. Accordingly, not all materials have the same thickness value per gage value. For example, 15 gage standard steel is .0673″ thick, while 15 gage aluminum is .0571″ thick. Gage numbers for steel may be as thick as .2391″ (3 GA) or as thin as .006″ (38 GA). Typically, sheet metal thinner than .006″ is referred to as foil, while sheet metal .250″ or thicker is considered metal plate.

Gage charts or tables provide a handy resource for designers, engineers, and print readers. ASME standards recommend that material thickness be specified in inches or millimeters, with the gage number shown in parentheses after the linear value, if desired.

Bending Terms

While terms associated with sheet metal bending are not necessarily standardized, there are some common terms that will help the print reader develop an understanding of how flat patterns are developed. **Figure 21-2** illustrates some of the terms.

- ***Bend***: A bend is a uniformly curved section of material that serves as the edge between two sides or legs.
- ***Flange***: A flange is a portion of an item being bent, such as the side or leg of a sheet metal strip.
- ***Bend allowance***: The bend allowance is the amount of metal required to create the desired bend, based on a given bend angle, material thickness, and bend radius.
- ***Bend angle***: The bend angle is the number of degrees between flat surfaces on each side of a bend. This angle can be described as acute (<90°), square (=90°), or obtuse (>90°). Acute angles may also be referred to as *closed angles*, while obtuse angles may be referred to as *open angles*.

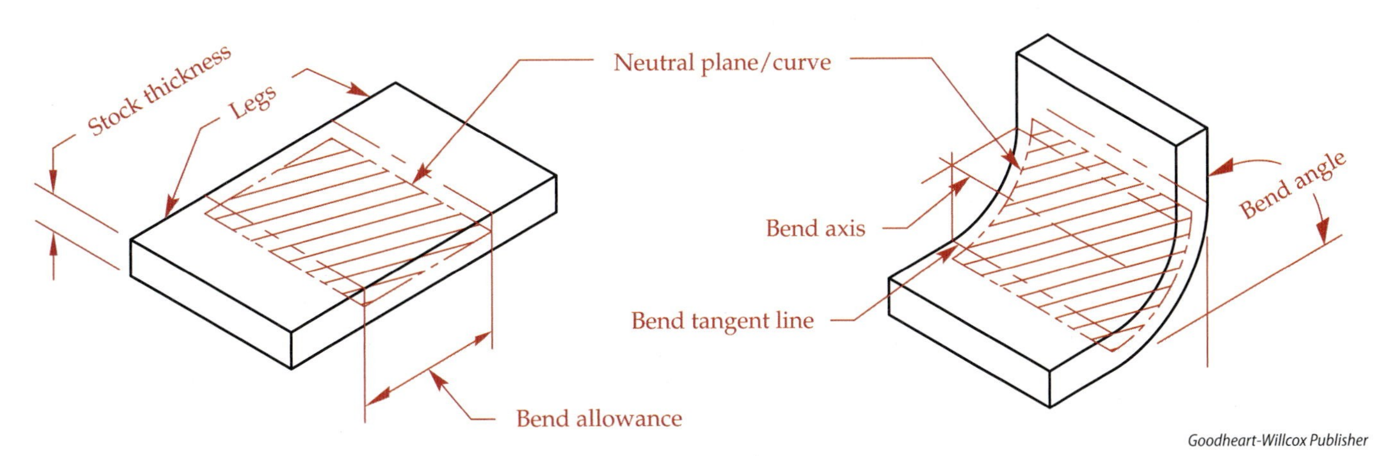

Figure 21-2. Bend terminology common to most handbooks and references is illustrated here.

- ***Bend axis*:** The bend axis is the theoretical center axis of the bend.
- ***Bend deduction (setback)*:** The bend deduction, or setback, is the amount of material to deduct from the blank length to accommodate a bend, rather than using a sharp corner, between two legs. The setback method of calculating is often used as an alternative method of calculating the total stock length, rather than calculating the leg lengths plus bend allowance.
- ***Bend tangent line*:** The bend tangent line is the line of transition between the flat surface leg and the curved portion of the bend.
- ***Bend radius*:** The bend radius is the desired amount of the curvature measured on the inside of the bend, usually identified as R.
- ***Thickness*:** Thickness refers to the stock thickness of the sheet metal, usually identified as T.
- ***Leg*:** A leg is the straight segment of stock on either side of a bend.
- ***Neutral plane*:** The neutral plane is the area within a flat blank that will neither compress nor stretch as bending occurs. This plane is not necessarily in the exact central portion of the material. After bending, this plane is characterized by a neutral radius curve. By definition, bend allowance can be determined by calculating the circumference of the neutral plane.
- ***K-factor*:** The K-factor is a factor used in formulas to allow for the neutral plane offset. The K-factor varies and may be impacted by bend angle sharpness, bend tools, and material properties. It is reasonable to assume that the K-factor will be between .25 and .50, which means the neutral plane falls between 1/4T and 1/2T on the axis side of the center plane of the bend. See **Figure 21-3**.
- ***Neutral arc bend radius*:** The neutral arc bend radius is the bend radius plus a percentage (K-factor) of the metal thickness.
- ***Blank length*:** The blank length is the total distance across all legs and the bend allowance.
- ***Springback*:** Springback is the opening of a bend after forming due to the elasticity of the material, usually specified in degrees or a percentage.
- ***Overbending*:** Overbending is bending an extra increment to compensate for springback.

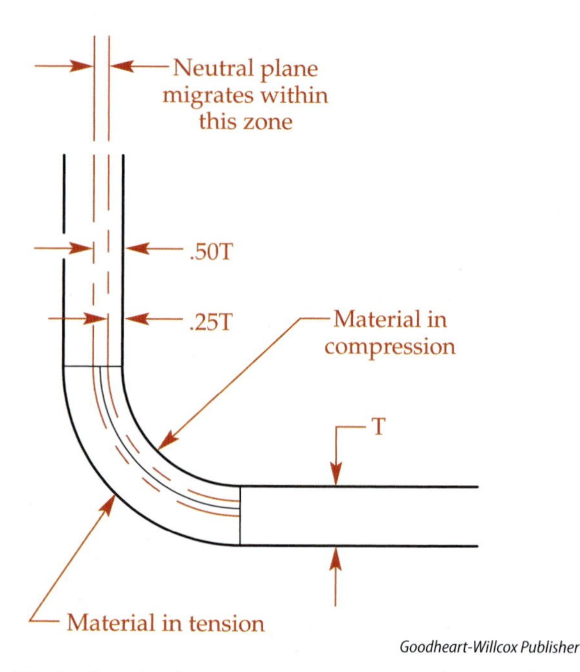

Figure 21-3. Bend calculations require an understanding of the K-factor, which approximates the offset position of the neutral plane.

Sheet Metal Drawing Terminology

Prints for precision sheet metal parts are read in the same manner as other prints in the metal machining and fabrication industries. Since sheet metal drawings are usually created as flat pattern developments, the ordinate, or arrowless, method of dimensioning is common, as featured in the **Figure 21-1** print, and the values can be easily incorporated into the CNC machine program. Flat pattern developments can also be created as undimensioned drawings, using template checking methods as set forth in ASME Y14.31. Within this standard, the unit on flat pattern developments standardizes some of the notations and linework that can be used to clarify and describe the bending processes. While prints are often created and dimensioned in such a way that manufacturing processes are not specified, sometimes sheet metal drawings may include notes or specifications that favor specific methods likely to be used to manufacture the part. Some of the definitions and terms set forth in ASME Y14.31 that are applicable to sheet metal drawings include:

- ***Outside mold line (OML)* and *inside mold line (IML)*:** The outside mold line (OML) represents the intersection of outside surfaces if projected to a sharp edge, and the inside mold line (IML) represents the intersection of inside surfaces if projected to a sharp edge. See **Figure 21-4**.

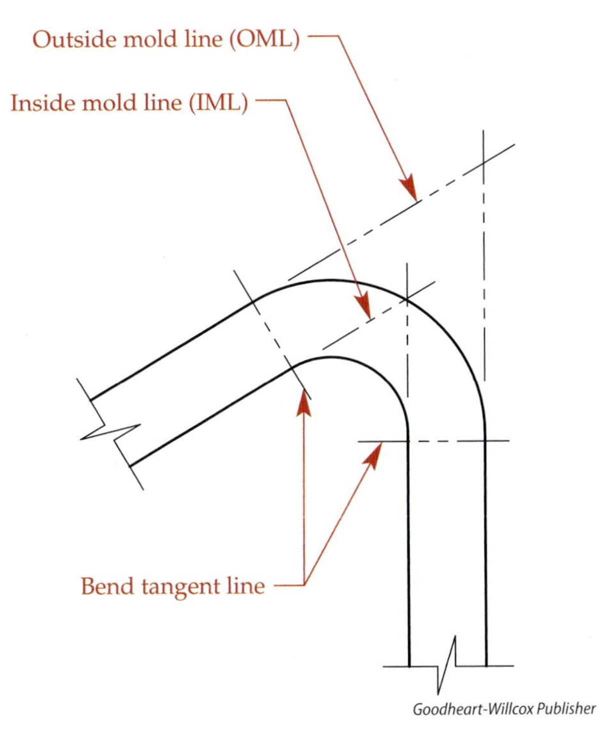

Figure 21-4. Terms related to bends on sheet metal drawings are outside mold line (OML), inside mold line (IML), and bend tangent line.

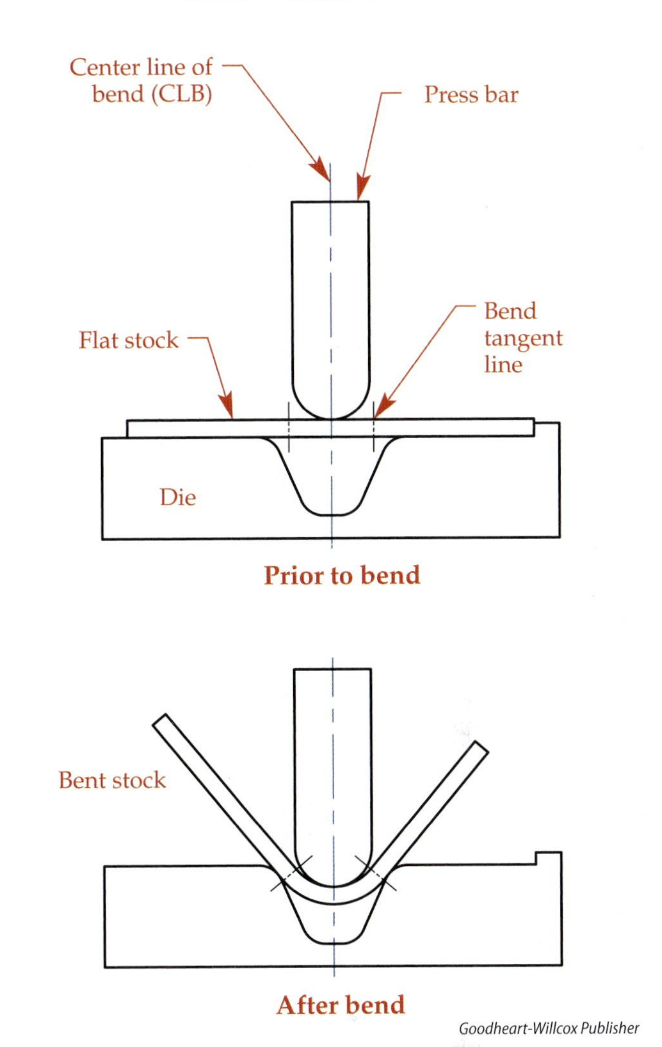

Figure 21-5. A brake press bar and forming die can be used to effectively bend sheet metal to a precise angle.

- ***Brake process***: The brake process is a method of forming using a punch or press bar with a die to bend the surface up and around the shape of the tool. See **Figure 21-5**. By the nature of the process, bend instructions are often bend up (BUP), and the lines shown on the drawing will be the center line of bend (CLB) and outside mold lines (OML). With the bending processes being shown on the drawing, there are some implications for linework, as illustrated in **Figure 21-6**. Notice that CLB lines are shown as center lines. OML lines are shown as solid lines, about .5″ (13 mm) in length at each end of the bend. If present, IML and FBL lines are also shown as center lines. If the flanges are to be trimmed after forming, the trim lines should also be shown as center lines.
- **Hydroforming:** As described earlier in the unit, hydroforming processes can bend the material over a form block using high-pressure fluid applied through a flexible bladder, rubber sheet, or diaphragm. By the nature of the process, bend instructions are often bend down (BDN), and the bend lines are referred to as the inside mold line (IML) and form block line (FBL). See **Figure 21-7**.
- ***Center line of bend (CLB)***: The center line of bend (CLB) is a line midway between bend tangent lines that indicates a bend contact line for the punch in a brake process.
- ***Form block line (FBL)***: The form block line (FBL) is a line representing the surface of a forming tool about which the metal will be bent.
- ***Bend instruction***: A bend instruction is a notation on the drawing giving the number of degrees and direction for forming a flange relative to the surface to which the instruction note is associated. The bend instruction should be attached to a CLB, IML, or FBL with a leader line.

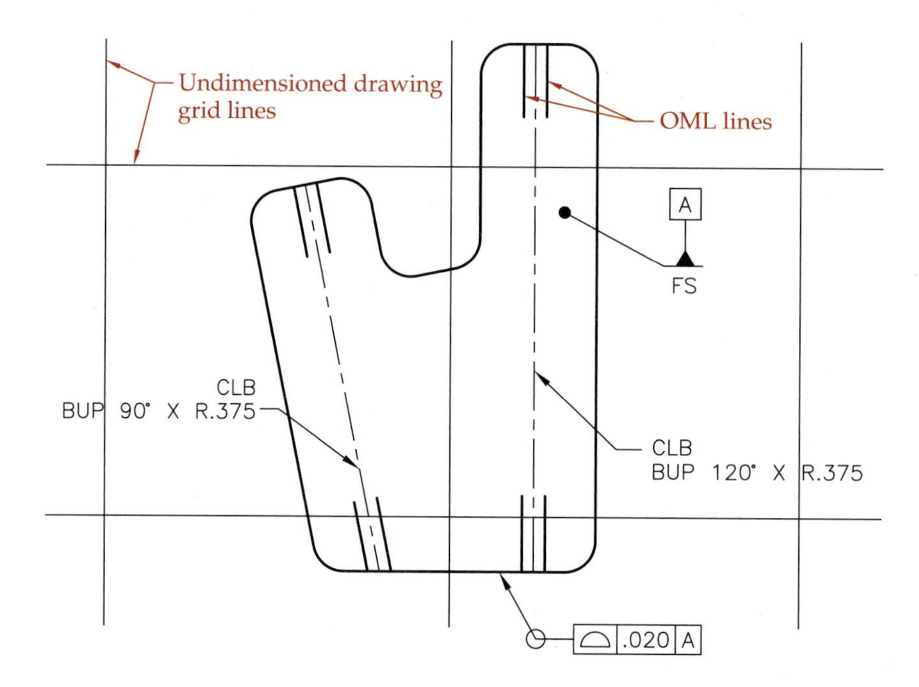

Goodheart-Willcox Publisher

Figure 21-6. Sheet metal drawings are sometimes created as undimensioned drawings. In these cases, bend instructions are attached to CLB lines, and OML lines are indicated with short solid lines. The shape of the unfolded part may be inspected with respect to the grid lines using a stable overlay comparator drawing.

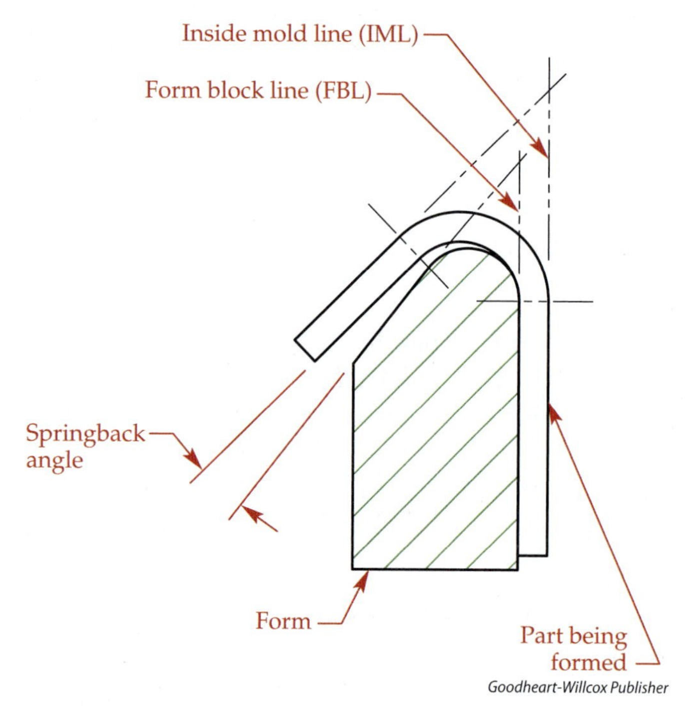

Goodheart-Willcox Publisher

Figure 21-7. A form block can also be used to assist with a bend. In this case, the IML represents the surface of the molding tools, while the other surface is pressed against the form block at the form block line (FBL). The form block must also allow for springback.

- ***Joggle*:** A joggle is an offset displacement of material from its original plane. For example, the last inch of a sheet metal surface may need to have a joggle to accommodate for the thickness of a mating sheet metal flange for welding overlap. Joggle instructions can be placed on the drawing with a JOG UP or JOG DOWN notation. A couple of basic 90° bends with joggles are shown in **Figure 21-8**, although hydro processes efficiently allow for more complicated forms at various angles. Notice that the joggle lines are shown as hidden lines in the detail drawing.

Calculations and Layouts

Prints are often dimensioned with the part shown flat, and a pictorial view is also often available on the print. With CAD systems, the part can be modeled in the 3D folded condition, and the software calculates the bends and creates the flat pattern. This unit will also examine those situations wherein manual calculations are made and is designed to assist the print reader in understanding the transition from a folded 3D part to a flat pattern.

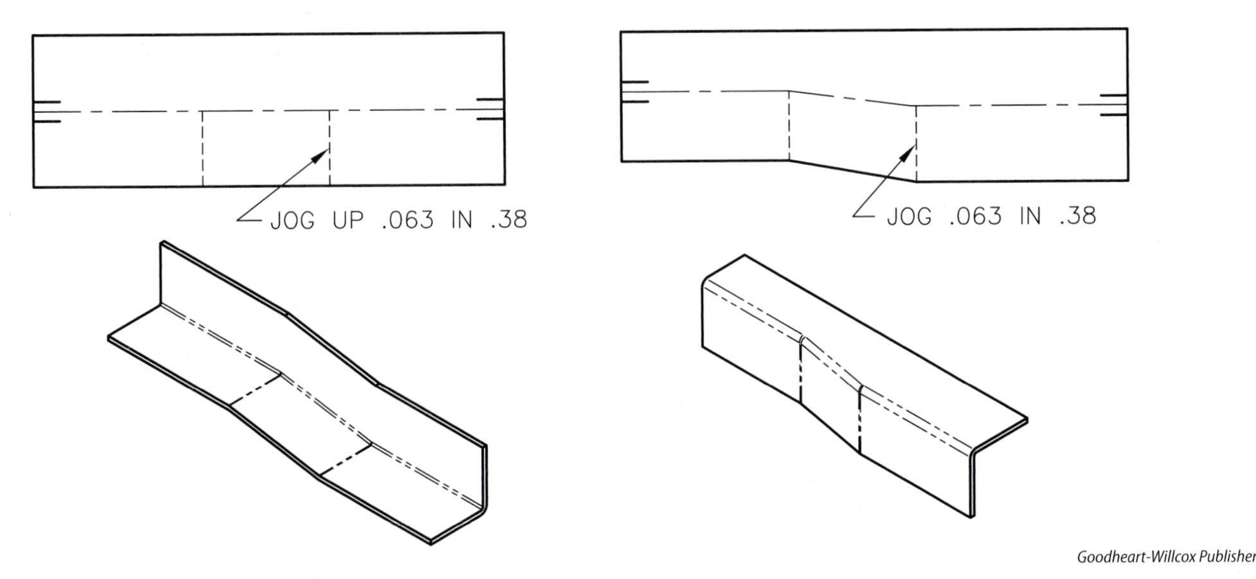

Figure 21-8. Joggle lines are shown as hidden lines, and directions in the local note give the offset depth and length of the joggle.

The ***developed length*** represents the dimensions of the flat part that result in the correct dimensions for the folded part. There are two primary methods of calculating the developed length of a part. One is to use a setback chart, and the other is to use bend allowance calculations. Various bend allowance values can also be placed in a chart for quick reference. While these calculations are seldom the responsibility of the print reader, it is useful to know how the dimensions are calculated to double-check accuracy in the shop or field. Folding procedures and notes are sometimes added to the print, either in local note or chart form.

Setback Charts

One method of calculating bends is to use a bend deduction value as found in a machinist's handbook chart or table known as a setback chart or bend deduction chart. ***Setback*** is the amount of distance subtracted from the total length of two sides that would otherwise form a sharp square corner. In other words, setback is the amount deducted from the distance of the square corner when cutting across the corner in a curved manner. These charts are available in most precision sheet metal industries, saving time and eliminating calculation errors. In some handbooks, more detailed charts are available that allow for a wider range of material thickness, bend angles, and bend radii. Designers should be aware that as various new bending technologies evolve, chart values may need to be slightly adjusted for a company's particular methods and materials.

To illustrate the principles discussed above, an example of a small setback chart for 90° bends is shown in **Figure 21-9**. The leftmost column indicates the bend radius, and the top row indicates the material thickness. As illustrated by the red entries in **Figure 21-9**, a 1/8″ bend radius with a metal thickness of .040″ requires a setback, or bend deduction, of .106″.

Precision Sheet Metal Setback Chart

90° Bend Radius	Material Thickness						
	.016	.020	.025	.032	.040	.051	.064
1/32	.034	.039	.046	.055	.065	.081	.097
3/64	.041	.046	.053	.062	.072	.086	.104
1/16	.048	.053	.059	.068	.079	.093	.110
5/64	.054	.060	.066	.075	.086	.100	.117
3/32	.061	.066	.073	.082	.092	.107	.124
7/64	.068	.073	.080	.089	.099	.113	.130
1/8	.075	.080	.086	.095	.106	.120	.137
9/64	.081	.087	.093	.102	.113	.127	.144
5/32	.088	.093	.100	.109	.119	.134	.150
11/64	.095	.100	.107	.116	.126	.140	.157
3/16	.102	.107	.113	.122	.133	.147	.164

Goodheart-Willcox Publisher

Figure 21-9. A bend radius of 1/8″ on metal with a thickness of .040″ requires a setback of .106″, as shown here in red.

Figure 21-10 and the formula that follows this paragraph illustrate the application of the setback example just given followed through to calculate the developed length of a precision sheet metal part. The formula for calculating the developed length of a precision sheet metal part is:

Developed length = X + Y – Z

In this formula, X represents the outside overall distance of one side (as if the corner is square), Y represents the outside overall distance of the other side, and Z represents the setback deduction for the bend, as indicated by the chart.

Developed length = X + Y – Z

Developed length = 1.00 + .750 – .106

Developed length = 1.750 – .106

Developed length = 1.644

Bend Allowance Charts

Other bend allowance charts are available that express calculations in a different manner from the bend setback method discussed earlier. These charts are *additive* in nature. **Figure 21-11** shows a portion of a bend allowance chart that provides the user with the amount of material to add to the straight sides (legs) of a part. In this manner, the distances that are straight on the finished product are added together. The distance provided on the chart is added to the

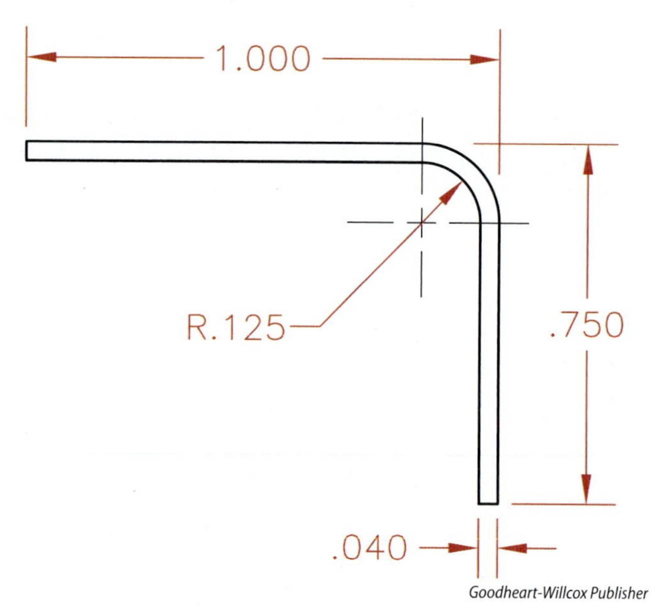

Figure 21-10. When calculating the developed length with a setback chart, use the measurements of the sides as if there was a sharp corner, and then subtract the amount provided in the setback chart.

Bend Allowance for 90° Bends

Radius → / Thickness ↓	.031	.063	.094	.125
.016	.060	.110	.159	.208
.020	.062	.113	.161	.210
.025	.066	.116	.165	.214
.032	.071	.121	.170	.218
.040	.077	.127	.176	.224
.051		.134	.183	.232
.064		.144	.183	.241

Goodheart-Willcox Publisher

Figure 21-11. This chart is an example of a bend allowance chart for 90° bends for various material thicknesses and bend radii.

flat pattern development. In summary, this chart requires the designer to calculate the leg distances by subtracting R + T from the side length. For example, with the part illustrated in **Figure 21-10**, the finished leg lengths are calculated as follows:

Leg one = X – (R + T) = 1.000 – (.125 + .040) = .835

Leg two = Y – (R + T) = .750 – (.125 + .040) = .585

The bend allowance from the table indicates .224 should be added for the bend to determine the total developed length:

.835 + .585 + .224 = 1.644

Formulas

Without charts, the bend allowance can be calculated using a formula. Calculating the bend allowance is the equivalent of determining the circumference, or lineal length, of the bend. The circumference of a complete circle is πD, so the circumference of a 90° arc is $1/4\pi D$ or $1/2\pi R$. However, for bend allowance, the radius value should be based on the bend radius of the neutral arc. This can be determined by adding the K-factor distance to the inside bend radius.

The various means of determining the K-factor are beyond the scope of this text. There are several factors involved, including material properties, bending tools, and the ratio of material thickness to bend radius. Some references recommend reverse engineering the K-factor using a sample piece of material bent with the intended bending tool.

In the following examples, two different K-factors are given for calculating the bend allowance. Many companies have developed their own formulas or guidelines for determining the K-factor. These two examples are representative of the types of rules a company may use. Refer to **Figure 21-12** to assist

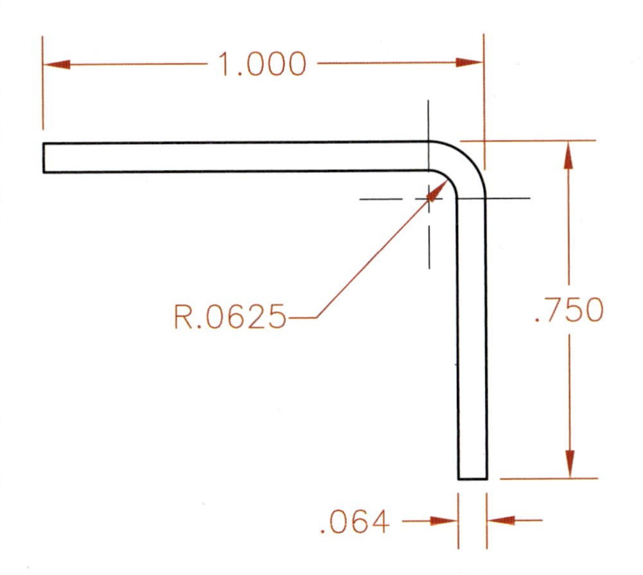

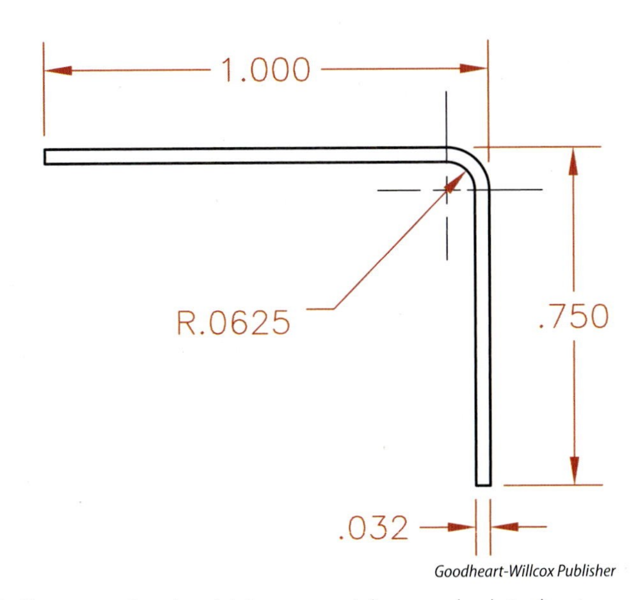

Goodheart-Willcox Publisher

Figure 21-12. These examples are explained in the text. The bend allowance for the thicker material was calculated using a K-factor of .4. The bend allowance for the thinner material was calculated using a K-factor of .5.

you in the examples below. In these examples, each formula is calculating the bend allowance of a 90° bend. If the angle of bend is not 90°, a different fractional amount will need to replace 1/2 in the formula.

Example Rule A

When the inside radius is less than twice the material thickness and the bend is 90°, use this formula, which incorporates a K-factor of .4:

$$A = 1/2\pi\ (R + .4T)$$

In the formula above, A = the bend allowance, R = the inside radius, and T = the material thickness.

As an example, to find the lineal length of a 90° bend when the stock thickness is .064″ and the inside radius is 1/16″ (.0625″), apply the formula as follows:

$$A = 1/2\pi\ (R + .4T)$$

$$A = 1/2 \times 3.1416 \times (.0625 + .4 \times .064)$$

$$A = 1.5708 \times (.0625 + .0256)$$

$$A = 1.5708 \times .0881$$

$$A = .1384$$

The bend allowance for this example is .1384″.

Example Rule B

When the inside radius is *more* than twice the material thickness and the bend is 90°, use this formula, which incorporates a K-factor of .5:

$$A = 1/2\pi\ (R + .5T)$$

As an example, to find the lineal length of a 90° bend when the stock thickness is .032″ and the inside radius is 1/16″ (.0625″), apply the formula as follows:

$$A = 1/2\pi\ (R + .5T)$$

$$A = 1/2 \times 3.1416 \times (.0625 + .5 \times .032)$$

$$A = 1.5708 \times (.0625 + .016)$$

$$A = 1.5708 \times .0785$$

$$A = .1233$$

The bend allowance for this example is .1233″.

Developed Length

After the bend allowance (A) has been found, the developed length of the entire part can be calculated. In the earlier example, each leg length was calculated by subtracting R + T. The following formula incorporates the subtraction of R + T from both legs:

$$\text{Developed length} = X + Y + A - (2R + 2T)$$

In the formula above, X = the outside distance of one side (as if the corner was square), Y = the outside distance of the other side, A = the bend allowance, R = the inside radius, and T = the stock thickness.

The .064″ thick material in the earlier example using Rule A results in a developed length of 1.6354″:

$$\text{Developed length} = 1.00 + .75 + .1384 - (.125 + .128)$$

$$\text{Developed length} = 1.8884 - .253$$

$$\text{Developed length} = 1.6354$$

The .032″ thick material in the earlier example using Rule B resulted in a developed length of 1.6843″:

Developed length = 1.00 + .75 + .1233 – (.125 + .064)

Developed length = 1.8733 – .189

Developed length = 1.6843

Layout and Fold Procedure

The bend allowance formulas and setback charts for precision sheet metal parts provide approximate data. The coefficient of expansion for various materials can vary. Therefore, it is a good idea to have a duplicate flat blank for reference when making adjustments. Once all overall measurements and feature locations meet specified tolerances, the duplicate blank can be discarded or used. The procedure to lay out and fold precision sheet metal parts is:

1. Calculate and lay out the developed length and feature locations.
2. Make two identical flat blanks to the overall dimensions.
3. With the forming die, form one piece and check measurements.
4. If the folded part is not within tolerances, make machine adjustments as needed.
5. Make two more identical blanks.
6. With the forming die, form one piece and check measurements.
7. If the folded part is not within tolerances, make machine adjustments as needed.

Repeat steps 5 and 6 until the folded part is within tolerances.

8. Once the folded part is within tolerances, the identical blank is the pattern to be used for building all blanks.

Reading Precision Sheet Metal Prints

While prints for folded sheet metal parts are not much different from prints of other parts, the material thickness is one element unique to sheet metal parts. For a flat pattern development, often the key description is a single-view drawing of the pattern with the depth, or material thickness, not shown in a top or side view. Often, the print will include the thickness of the material as a reference value, usually in the material block.

As discussed earlier, with 3D parametric solid modeling programs, designers can build the final folded model of the sheet metal part in the CAD system. A variety of modeling tools allow the designer to automatically fold at a line, add a flange, cut an opening, or even add a hem. Parameters such as thickness and radius can be changed, and the model updates without rebuilding it from scratch. When the completed part is unfolded to develop a pattern, the CAD system uses programmed settings that calculate the bend allowance, incorporate the K-factor, and assign true material properties to the part. **Figure 21-13** shows a simple part created in a 3D CAD program. While the software allows the drafter to hide or show the bend tangent lines, the drafter may have to use manual methods to show partial outside mold lines. As with any drawing, dimensions and notes can be added to clarify the final product.

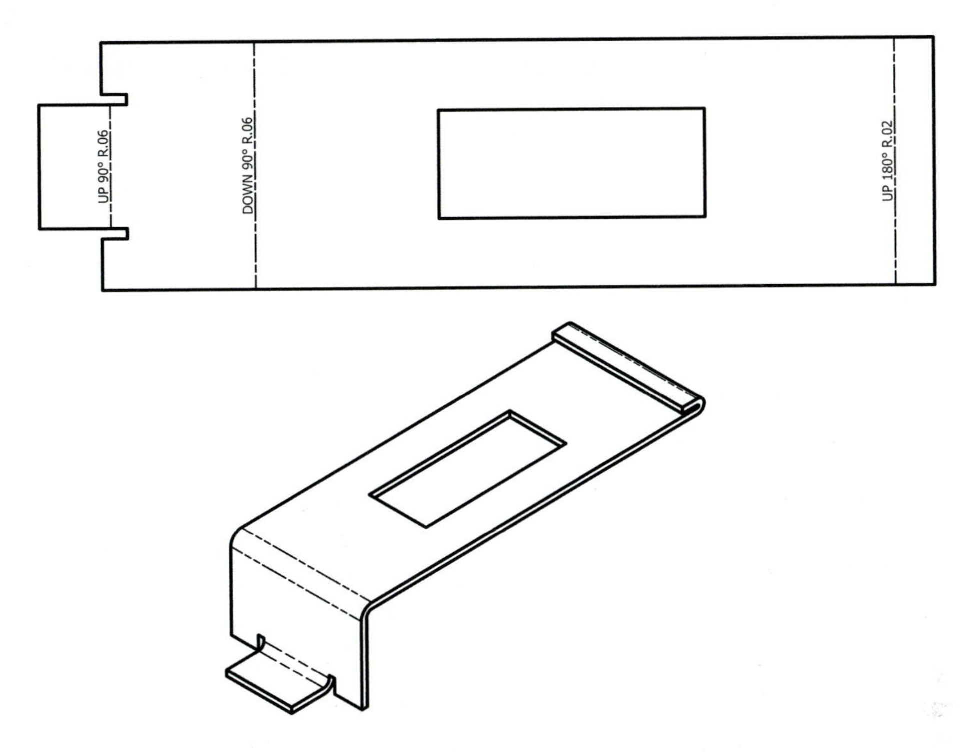

Goodheart-Willcox Publisher

Figure 21-13. Three-dimensional parametric solid modeling CAD programs can easily develop the flat pattern from the completed part, and the user can determine how the linework is displayed. By default, this flat pattern drawing shows center lines for the center line of bend (CLB) lines, but bend tangent lines were turned off. Bend notes are added automatically by the software, but they can also be edited. (Note: Dimensions and other notes are not shown in this example.)

Summary

- Precision sheet metal parts are those manufactured from thin metals in the flat position and then folded into a final form.
- While sheet metal thickness should be specified by inches or millimeters on the print, the US customary system also categorizes metal thickness by a gage value, which varies according to the type of metal.
- The print reader must be familiar with several terms associated with sheet metal bending processes, including bend allowance, bend angle, flange, leg, and neutral plane.
- ASME Y14.31, titled *Undimensioned Drawings*, sets forth guidelines for line appearance based on terms associated with bending processes, such as outside mold line (OML), inside mold line (IML), center line of bend (CLB), and form block line (FBL).
- Flat pattern layouts for parts that will be bent and formed require bend allowance calculations, taking into consideration the material thickness, the bend angle, and the bend radius.
- While CAD systems are programmed to calculate bend allowance, other manual methods can be of value, including charts or tables that provide a bend setback value, which is subtractive, or a bend allowance value, which is additive.
- Developed lengths can be calculated for any bend angle, taking into consideration the K-factor for the material thickness and material properties and the circumference of the bend arc.
- Sheet metal prints are similar in nature to prints representing other fabrication methods and often incorporate standardized arrowless dimensioning or undimensioned drawing practices.

Unit Review

Name ____________________ Date ____________ Class ____________

Know and Understand

Answer the following questions using the information provided in this unit.

____ 1. *True or False?* Precision sheet metal prints are characterized by a flat pattern view that, if manufactured according to the dimensions given, can be folded into a functional part or product.

____ 2. *True or False?* Sheet metal is referred to by a gage value that indicates not only the thickness, but also the weight.

____ 3. The amount of material required to form the rounded edge on the part is called the bend _____.
A. radius
B. angle
C. setback
D. allowance

____ 4. The line of transition between a flat surface leg and the curved portion of the bend is called the bend _____ line.
A. radius
B. axis
C. tangent
D. neutral

____ 5. *True or False?* Bend radius, as specified on a print, usually applies to the radius of the bend at the exact central plane of the material thickness.

____ 6. *True or False?* As sheet metal is folded or bent, there is a *neutral plane* within the material, describing the area of metal that neither stretches nor compresses.

____ 7. Which of the following terms helps describe the *neutral plane* position?
A. K-factor
B. Leg
C. N-thickness
D. Flange

____ 8. *True or False?* Bending an extra increment to compensate for *setback* is referred to as *overbending*.

____ 9. If the outside surfaces of a folded part are projected to a sharp edge, which of the following abbreviations could be applied?
A. FBL
B. OML
C. CLB
D. IML

____ 10. If the last inch of a sheet metal surface needs to accommodate for the thickness of a mating sheet metal flange for welding overlap, what feature is incorporated into the design?
A. Trim
B. Fold
C. Joggle
D. Hem

____ 11. *True or False?* The difference between the sheet metal material needed to do a perfectly square corner with sharp edges versus a rounded corner with an inside radius is the *bend setback*.

____ 12. What do R and T represent when calculating the length of a leg with this formula: Leg one = X – (R + T)?
A. Bend radius and material thickness
B. R-factor and T-factor
C. Neutral plane radius and T-factor
D. Outer radius and setback value

____ 13. *True or False?* When doing manual calculations for a bend, a K-factor of .45 is appropriate for a 45° bend, while a K-factor of .60 is appropriate for a 60° bend.

____ 14. *True or False?* Since the circumference of a circle is πD, the circumference of a 90° bend can be calculated as $1/2\pi R$.

____ 15. *True or False?* For sheet metal parts, CAD designers usually model a part in its final folded form, and the CAD system automatically develops the flat pattern.

Critical Thinking

1. List three devices or products common to everyday life around the house or shop that incorporate precision sheet metal fabrication.

2. Briefly discuss why products such as those you listed in question 1 are still being manufactured with sheet metal versus other materials such as plastic or wood.

Apply and Analyze

Name ____________________ Date ____________ Class ____________

Industry Print Exercise 21-1

Refer to the print PR 21-1 and answer the questions below.

1. How many bends does this part have? ____________________
2. Calculate the distance between the two bend center lines. ____________________
3. What gage is specified for the sheet metal, and how thick is that? ____________________
4. What is the bend radius for this part? ____________________
5. There are some short solid lines perpendicular to the main object lines. What term is applied to those lines? ____________________
6. In the note, what does BUP stand for? ____________________
7. Based on the discussion in this unit, what type of process, brake or hydro, is logical for this part? ____________________

Review questions based on previous units:

8. What geometric tolerance is applied to the size of this blank? ____________________
9. What process will be used to eliminate burrs? ____________________
10. Calculate the size of the rectangular slot. ____________________
11. Taking into consideration the geometric tolerance, calculate the MMC for the overall blank measurements. ____________________
12. How many holes feature a diameter of .136″? ____________________

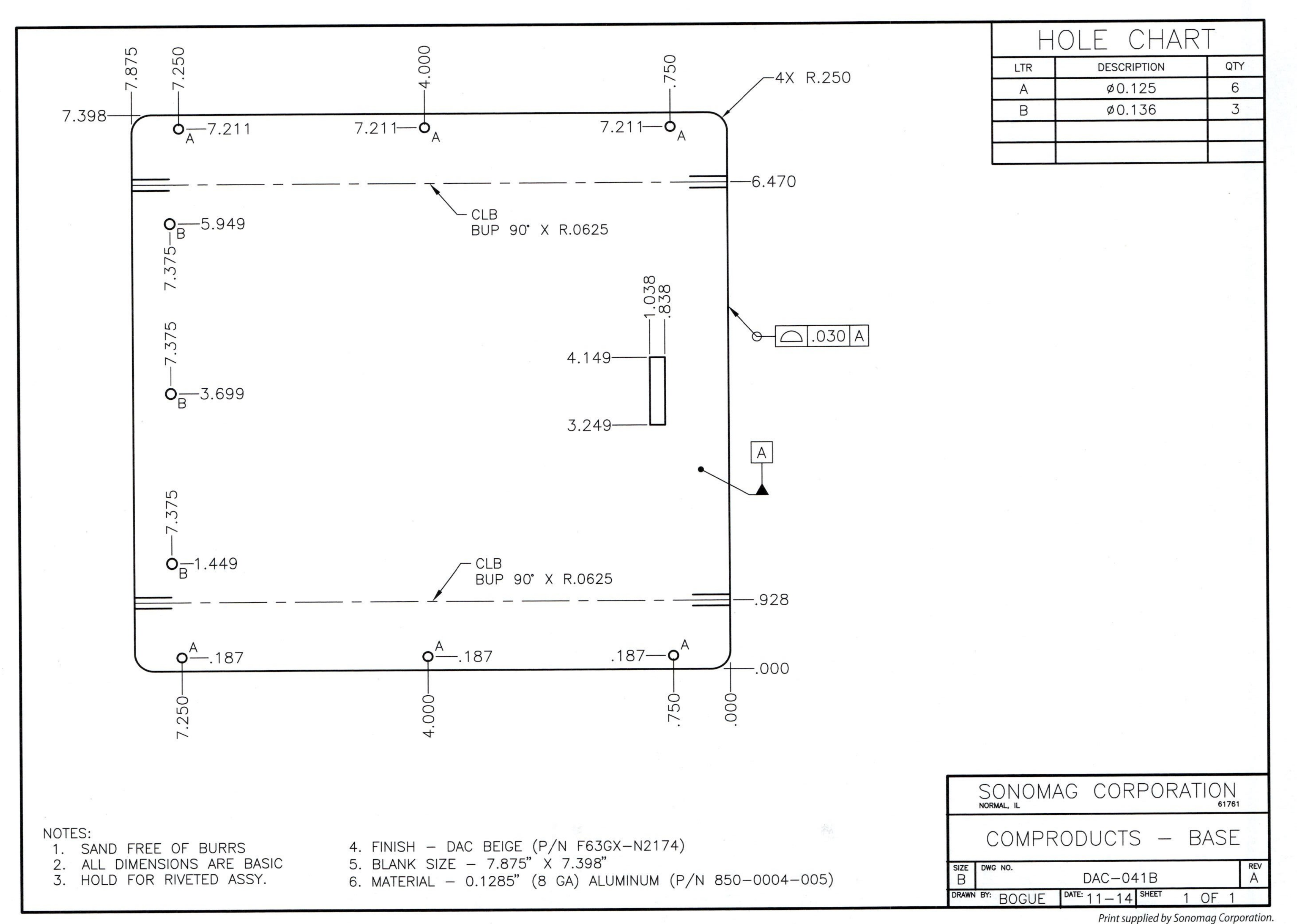

Print supplied by Sonomag Corporation.

PR 21-1. Base – Comproducts

Apply and Analyze

Name ______________________ Date ______________ Class ______________

Industry Print Exercise 21-2

Refer to the print PR 21-2 and answer the questions below.

1. How many bends does this part have? ______________________
2. What is the basic thickness of the metal for this part? ______________________
3. What bend deduction (setback) was used in calculating the flat pattern? ______________________
4. What is the bend radius for this part? ______________________
5. In the formed view, is the 10.574″ dimension to the inside or outside of the bends?
   ____
6. What size are the slotted holes of this part? ______________________
7. Even though the finished part is 5.5″ tall, what is the developed length of the flat pattern in that direction? ______________________
8. How wide are the clearance notches on each side of the metal that is folded along the top, as shown in the formed view? ______________________
9. What is the radius for breaking sharp edges? ______________________
10. What amount of tolerance is permitted for the dimensions shown in the formed view?
    ____
11. Analyze the formed object in the pictorial view. What term can be applied to the features that contain the holes, including two dimensioned as .750 BOTH ENDS and one dimensioned as a .750 wide strip along the top edge? ______________________
12. What is the overall size of the blank required for this part? ______________________

Review questions based on previous units:

13. Calculate the offset for hole C with respect to its alignment with holes A and B.
    ____
14. What part number will this part be used on? ______________________
15. How many holes on this part are square holes? ______________________

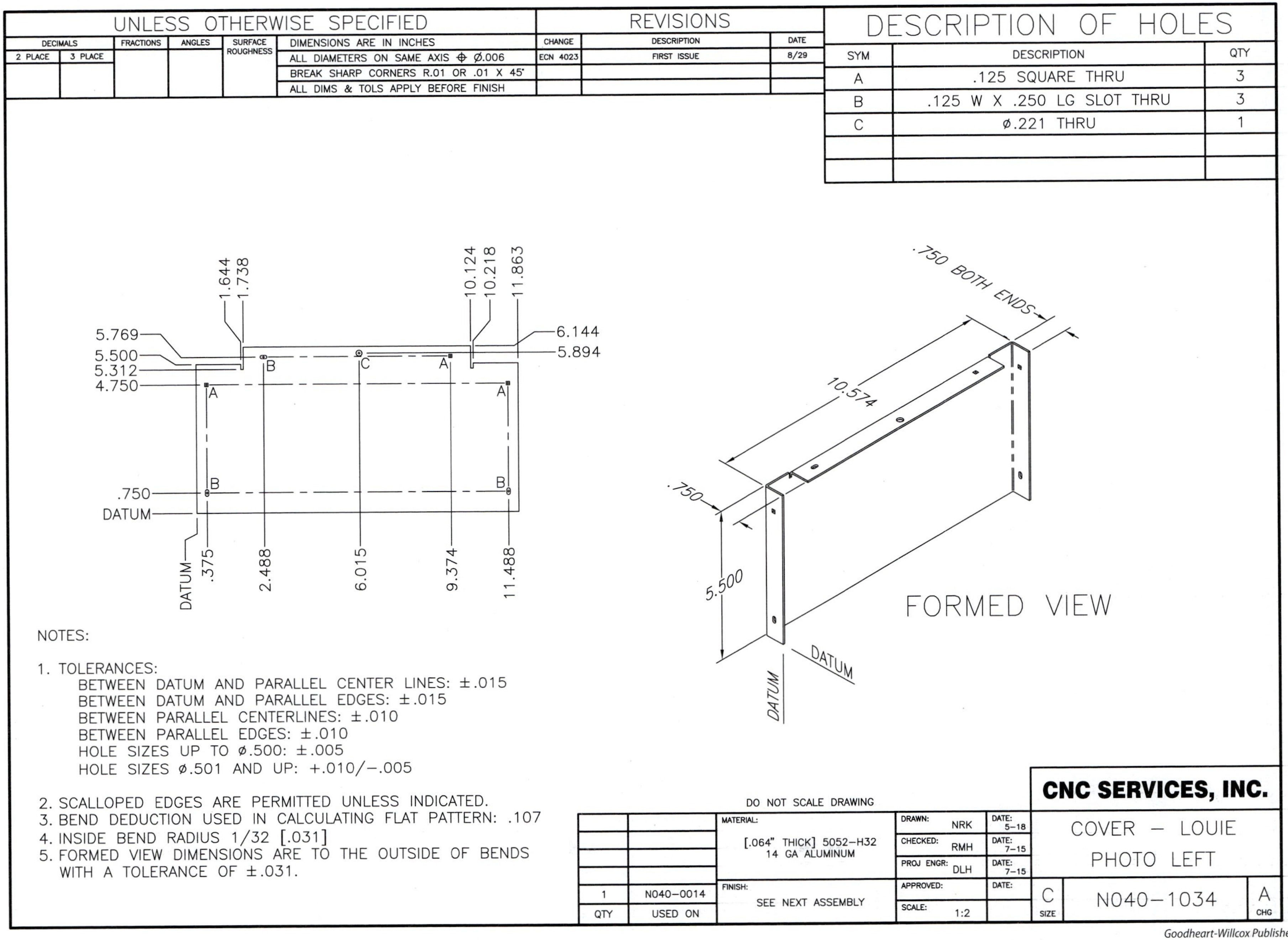

UNLESS OTHERWISE SPECIFIED					
DECIMALS		FRACTIONS	ANGLES	SURFACE ROUGHNESS	DIMENSIONS ARE IN INCHES
2 PLACE	3 PLACE				ALL DIAMETERS ON SAME AXIS ⊕ ∅.006
					BREAK SHARP CORNERS R.01 OR .01 X 45°
					ALL DIMS & TOLS APPLY BEFORE FINISH

REVISIONS		
CHANGE	DESCRIPTION	DATE
ECN 4023	FIRST ISSUE	8/29

DESCRIPTION OF HOLES

SYM	DESCRIPTION	QTY
A	.125 SQUARE THRU	3
B	.125 W X .250 LG SLOT THRU	3
C	∅.221 THRU	1

5.769
5.500
5.312
4.750
.750
DATUM
1.644
1.738
10.124
10.218
11.863
6.144
5.894
A B C A A B B
DATUM
.375
2.488
6.015
9.374
11.488

.750 BOTH ENDS
10.574
.750
5.500
DATUM
DATUM

FORMED VIEW

NOTES:

1. TOLERANCES:
   BETWEEN DATUM AND PARALLEL CENTER LINES: ±.015
   BETWEEN DATUM AND PARALLEL EDGES: ±.015
   BETWEEN PARALLEL CENTERLINES: ±.010
   BETWEEN PARALLEL EDGES: ±.010
   HOLE SIZES UP TO ∅.500: ±.005
   HOLE SIZES ∅.501 AND UP: +.010/−.005

2. SCALLOPED EDGES ARE PERMITTED UNLESS INDICATED.
3. BEND DEDUCTION USED IN CALCULATING FLAT PATTERN: .107
4. INSIDE BEND RADIUS 1/32 [.031]
5. FORMED VIEW DIMENSIONS ARE TO THE OUTSIDE OF BENDS WITH A TOLERANCE OF ±.031.

DO NOT SCALE DRAWING

		MATERIAL: [.064" THICK] 5052–H32 14 GA ALUMINUM	DRAWN: NRK	DATE: 5–18	CNC SERVICES, INC.
			CHECKED: RMH	DATE: 7–15	COVER – LOUIE PHOTO LEFT
			PROJ ENGR: DLH	DATE: 7–15	
1	N040–0014	FINISH: SEE NEXT ASSEMBLY	APPROVED:	DATE:	SIZE C · N040–1034 · CHG A
QTY	USED ON		SCALE: 1:2		

Goodheart-Willcox Publisher

PR 21-2. Louie Photo Left – Cover

Apply and Analyze

Name ______________________ Date ____________ Class ______________

Bonus Print Reading Exercises

The following questions are based on the various bonus prints located in the folder that accompanies this textbook. Refer to the print indicated, evaluate the print, and answer the question.

Print AP-021

1. What thickness of sheet metal is used for this part, in gage and inches?

____

2. In its final form, what are the three overall dimensions of this part?

____

3. What part number (indicated in parentheses) protrudes up from the sheet metal and is located 7.260″ from the left?

____

4. How many 3″ long slots are featured on this part?

____

5. How many holes are identified as hole "A"?

____

Print AP-022

6. What term is used on this print to refer to the pictorial illustration of the part in its final shape?

____

7. How many 4.2 mm x 18.2 mm notches are featured on this part?

____

8. What is the overall mean size of the flat pattern for this part?

____

9. What inside bend radius is specified for this part?

____

10. What bend reduction was applied in calculating the flat pattern dimensions?

____

UNIT 22
Welding Prints

LEARNING OBJECTIVES

After completing this unit, you will be able to:

- Discuss the nature of industrial prints that feature welding processes.
- Identify the standards that set forth methods of detailing welding specifications on an industrial print.
- Identify the basic types of joints that are common to welded assemblies.
- Explain basic welding processes that may be identified or specified in an industrial print.
- Define and explain various welding terms important to describing the formation of the weld bead.
- Identify the standard elements of a welding symbol, including the various numerical values and notations that can be contained within it.
- Identify the basic *weld* symbols used in *welding* symbols, including the various edge shapes commonly used in groove welds.
- Identify the numeric notations that join with the weld symbol to express size, depth, length, quantity, and spacing.
- Identify supplementary symbols that are added to the welding symbol to clarify special instructions such as field welding, welding all around, the addition of inserts and spacers, or the contour of the weld face.
- Interpret the complete meaning of various welding symbol examples using appropriate resources.

TECHNICAL TERMS

arc welding
back gouging
backing symbol
backing weld
back weld
brazing
broken arrow
consumable insert
depth of fusion
electron beam welding (EBW)
face
faying surfaces
filler
fillet weld
fusion face
fusion welding
gas welding
groove weld
heat affected zone (HAZ)
laser beam welding
leg
plug weld
resistance welding
root
root opening
slot weld
soldering
solid-state welding
spot weld
stud welding
supplementary symbol
surfacing weld
throat
toe
weld
weld bead
welding
welding symbol
weld joint
weld symbol

Reading welding drawings is similar in technique to reading other drawings, with a main difference being the symbols involved. There are also a wide variety of techniques and processes that create an extensive glossary of terms specific to this unit. The American Welding Society (AWS) has developed and adopted standard procedures for using symbols to indicate the exact location, size, strength, geometry, and specifications needed to describe the weld required. The symbols in this unit are based on AWS A2.4, titled *Standard Symbols for Welding, Brazing, and Nondestructive Examination.* As noted by the title, this standard also provides a means of specifying brazing operations as well as nondestructive testing methods, including the frequency, extent, and types of examination methods. This unit deals with welding and the symbols used in that process. Additional information that may provide insight to the print reader can be found in AWS A3.0, titled *Standard Welding Terms and Definitions.* In total, there are over 40 standards published by the AWS, covering a wide array of welding processes and technologies.

Welding Defined

Welding is a fabrication process that usually involves joining metal pieces by melting them along a joint, with a filler material being added in some cases. ***Filler*** is the material that is added to the weld, perhaps in the form of consumable electrode welding rods, inserts, or spacers. As the pieces cool, the junction becomes a strong joint and is referred to as a ***weld***. One aspect of learning how to read welding prints is understanding the different joint types and material shapes along the joint, and another aspect is understanding the characteristics of the weld itself.

Different energy sources can be used for welding, but these are not necessarily indicated on the print. Welding can be accomplished with gas flame, electric arc, lasers, and ultrasound power supplies. ***Brazing*** and ***soldering*** are similar in nature to welding but are fabrication processes that melt a joint material (filler) at a lower melting point than the materials being joined, so the pieces themselves are not actually melted. Brazing and soldering notations can be incorporated into welding symbols.

Welding Drawings

As discussed in Units 15 and 16, welding drawings are considered detail drawings, although as inseparable assemblies they share characteristics of assembly drawings. The print usually serves the same purpose as a detail drawing for one part, even though the object is comprised of individual pieces shown in their assembled positions. A parts list is often included. The joints or seams between pieces to be welded together are shown on the drawing, although various beveled edges, grooves, or holes in the pieces may not be shown in detail. It is the welding symbols that indicate the type and size of the weld to be applied, so the print reader must be familiar with the resources and standards that explain how to interpret those symbols.

Weld Joint Types

The location where two pieces of metal come together to be welded is referred to as the ***weld joint***. Weld joints often dictate the type of weld. The selection of a weld is determined by three basic factors: 1) the thickness of the metals to be joined, 2) the depth of penetration of the weld required to meet the strength specifications, and 3) the type of metal or metals being joined. Designers and engineers must take into consideration these factors and choose a weld that will handle the stress and forces that may act on the joint after welding, as well as allow for the proper finish and surface quality. Codes and specifications will help determine and specify the level of welding quality required.

Five basic weld joints are shown in **Figure 22-1**. Butt joints and T joints are very common, but corner joints, lap joints, and edge joints are also used,

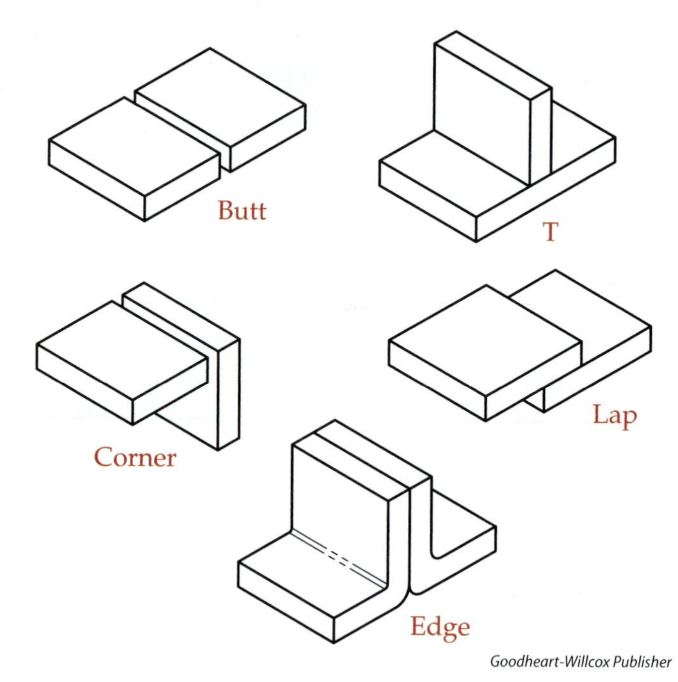

Figure 22-1. Five basic weld joints are the butt, T, corner, lap, and edge.

depending on the application. The metal edges of the pieces may also be beveled or chamfered with shapes that allow for the form of the weld to be created with different strength characteristics or finishing options. Some weld symbols derive their name from the contour shapes featured along the joints, as illustrated in **Figure 22-2**. Since these geometric shapes are not necessarily constructed on the drawing, learning to read and interpret the welding symbols is very important.

Welding Processes

As described earlier, a weld is often created by a process that melts the metal material along a joint. The welding process creates a deposit along the joint known as a ***weld bead***. While there are many process technologies, the weld bead along the joint is the goal, and the object of inspection. In manufacturing, the term ***faying surfaces*** is applied to surfaces that contact each other for the purpose of bonding by adhesive method, bolts, rivets, or, in this case, weldment. The weld bead must penetrate the faying surfaces but may also be built up with filler material during the process.

A main category of welding processes is ***fusion welding***, in which the pieces are joined together using a process that melts the metals of each piece as well as any filler material. In scientific terms, a phase change occurs where the joint surfaces convert from solid to liquid and then return to a solid. In general terms, fusion welding technologies can also be grouped by the heat source, be it electricity or fuel. ***Arc welding*** methods use an electric current via electrodes to generate the necessary heat to melt the metal. ***Gas welding*** methods, also referred to as *oxyfuel welding*, use a gas fuel mixed with pure oxygen to create intense heat with a welding torch. Brazing and soldering can be accomplished without an oxygen component.

Other more specialized forms of fusion welding include ***resistance welding*** methods used in spot welding and seam welding, wherein phase change occurs due to resistance to electric current channeled through two pieces of metal held together under pressure for a defined amount of time, causing the materials to join. Another fusion technology category is generally known as *intense energy beam welding*, wherein sophisticated equipment causes molecular phase change through technologies such as ***electron beam welding (EBW)*** and ***laser beam welding (LBW)***. *Electron beam welding* bombards the workpiece with a focused stream of electrons traveling at high speed within a vacuum, which results in enough heat to create fusion. As the name implies, *laser beam welding* uses a laser beam optically focused on the workpiece surface to be welded, generating an energy beam that brings the solid metal to a liquid phase change. These technologies have specialized benefits with some materials that could not easily be welded using the more traditional methods.

Sometimes specific welding methods are specified within the tail of the welding symbol. You may need to consult company standards or other resources to determine additional information about the welding method indicated. There are a multitude of specifications, including those shown in **Figure 22-3**.

In general, welding accomplished without fusion is referred to as ***solid-state welding***. In this method, joining of the metals is accomplished without phase change. Pieces to be welded are heated to an extent but not to the point of melting. Application of pressure is usually needed, and types of welding in this category include roll welding, diffusion welding, and friction welding.

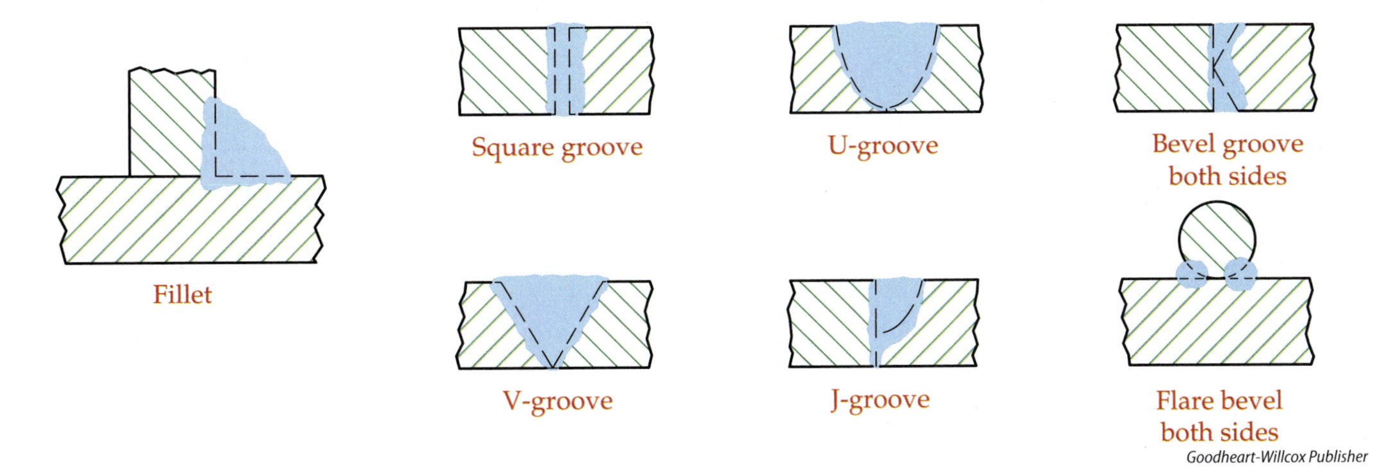

Figure 22-2. A weld can take many forms, including various contoured edges.

Welding Methods and Abbreviations	
Abbreviation	**Welding Method**
EBW	Electron Beam Welding
FCAW	Flux Cored Wire
GMAW	Gas Metal Arc Welding (MIG/MAG)
GTAW	Tungsten Inert Gas Welding (TIG)
LBW	Laser Beam Welding
OAW	Oxy-Acetylene Welding
OFW	Oxyfuel Gas Welding
PGW	Pressure Gas Welding
PAW	Plasma Arc Welding
RSEW	Resistance Seam Welding
RSW	Resistance Spot Welding
SMAW	Shielded Metal Arc Welding
SAW	Submerged Arc Welding

Goodheart-Willcox Publisher

Figure 22-3. Various welding methods can be specified on the drawing or within the welding symbol tail.

Welding Terms

As with many topics within this text, there are several terms unique to welding manufacturing methods or processes on prints. Some terms are familiar but have unique meanings within the context of welding applications. Some terms are related to the fusion process, and some are related to the spacing of pieces before welding or the size and spacing of the welds. **Figure 22-4** exhibits a few terms related to welding as applied to fillet welds, groove welds, spot welds, and surface welds. A ***fillet weld*** is created in a triangular fashion, often within a right-angle corner joint or on both sides of a T-joint. A ***groove weld*** describes most welds, including a weld that fills between two surfaces wherein an opening is provided, perhaps with beveled or shaped edges providing additional space for the weld. Other welds can be applied to materials that are lapped. Study the figure with respect to the following terms:

- ***Face***: the surface of the weld bead that is exposed and can be contoured as flat, concave, or convex

Actual throat
Theoretical throat
Effective throat
Toe
Weld interface
Face
Fusion face
Toe
Leg weld size
Depth of fusion
Root opening
Butting member
Non-butting member
Heat affected zone (HAZ)
Leg weld size

Weld terms as applied to a fillet weld

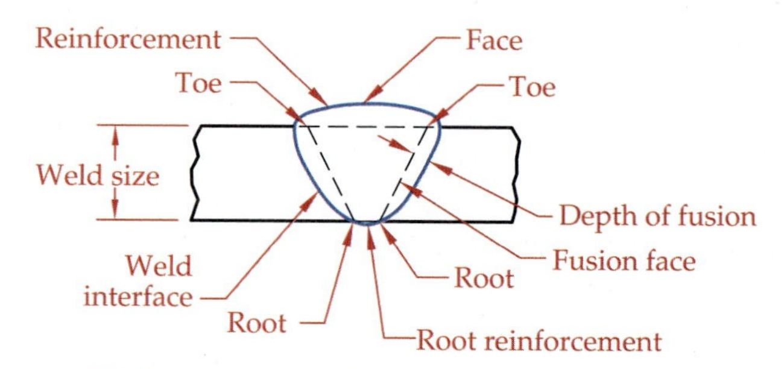

Weld terms as applied to a groove weld

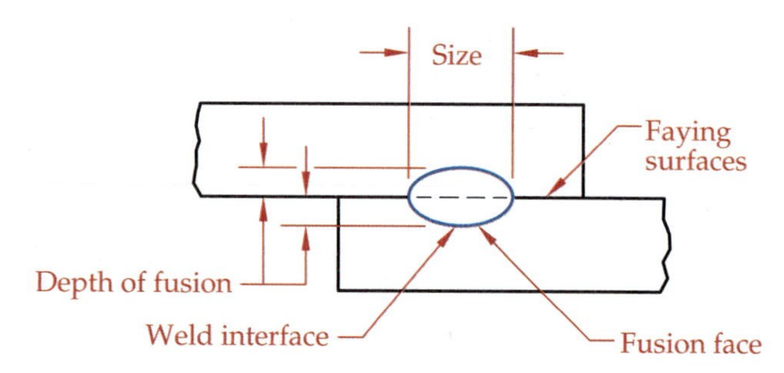

Weld terms as applied to a spot or seam weld

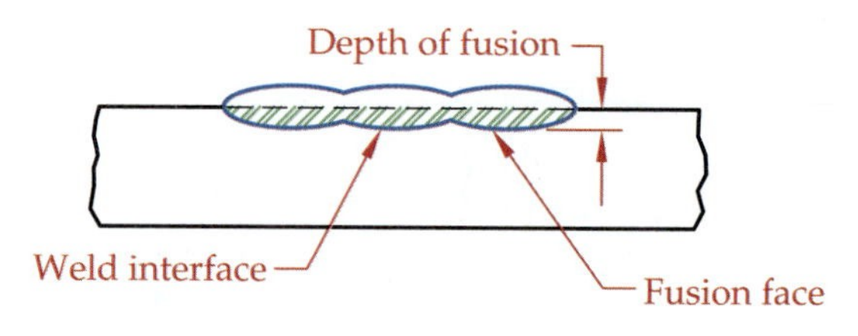

Weld terms as applied to a surface weld

Goodheart-Willcox Publisher

Figure 22-4. Welding terms are shown here as applied to various weld types.

- *Fusion face*: the portion of surface along the joint that will experience a phase change, joining the parent metal to the weld bead
- *Depth of fusion*: the distance the fusion face of the weld penetrates the parent piece
- *Heat affected zone (HAZ)*: the area around the weld bead that has been impacted by the welding process and may be subject to change in properties or weakness
- *Throat*: the depth of the weld bead from the face to the root, which for a fillet weld is measured in actual, theoretical, or effective terms
- *Leg*: the size of the weld, which for a fillet weld is expressed with two values for each side of the weld bead triangle
- *Root*: the bottom portion of a weld bead, which for a fillet weld is the corner
- *Root opening*: the distance pieces are separated to leave some open space in which the weld bead will form
- *Toe*: the top portion of a weld bead as measured along the pieces, which for a fillet weld is the leg distance away from the root

Welding Symbol

A ***welding symbol*** is the symbolic notation used to indicate the type and size of weld to be applied at a seam, joint, or location on the assembled pieces. The symbol consists of a leader line with an arrow, a reference line, and a tail section. See **Figure 22-5.**

The reference line is the horizontal line portion of the welding symbol. Attached to one end is a leader line with an arrow. Attached to the other end is an optional tail section. The reference line may appear vertically on some prints if that assists with getting the symbols on the print. The reference line will have weld symbols attached to it, with several numeric values positioned around the symbol if needed.

The leader line with an arrow is used to connect the welding symbol reference line to one side of the joint. This side of the joint is considered the *arrow side*. The side opposite of the arrow is considered the *other side* of the joint. The point of the arrow is usually attached to a seam or joint between two pieces to be welded. For spot welds, it may point to the precise location or center line intersection.

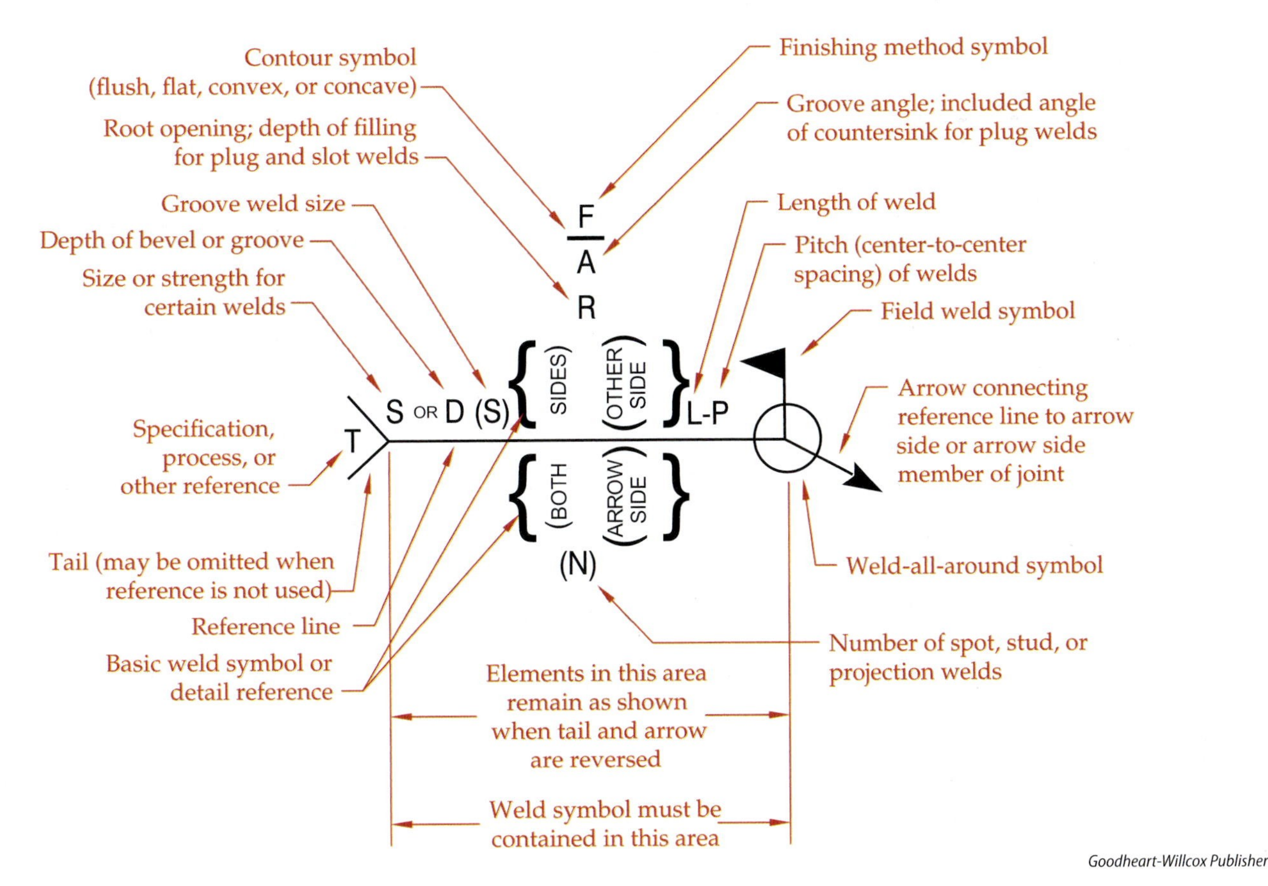

Goodheart-Willcox Publisher

Figure 22-5. The welding symbol provides much information to the reader.

The tail section, shown in **Figure 22-6**, is used for designating a wide array of information, including the welding specification, the welding process, or another reference such as an industry specification. Notes such as SEE NOTE A or SEE DETAIL A can be added to the tail if a drawing view helps explain the weld. If a reference is not given, the tail lines can be omitted from the reference line.

The dimensional units specified in conjunction with the welding symbol should be given in US customary units or SI units. No attempt to provide both units of measurement within the symbol should be made. If conversion values are desired, a separate table can be implemented. If tolerance values are desired for the dimensional values, they can be placed within the tail of the symbol or addressed by general notes. If a tolerance is specified, it should directly reference the dimensional feature of the weld being toleranced.

When welding characteristics are presented within this unit of study, many examples of the welding symbol will be discussed or illustrated. You may wish to refer back to **Figure 22-5** as a reminder of how information is specified and located with respect to the reference line.

Weld Symbols

The main component along the reference line is the weld symbol. A ***weld symbol*** indicates the type of weld only. Basic weld symbols for various types of welds are shown in **Figure 22-7**, with several of the types being groove welds. In cases where the weld can be applied on both sides of a joint, it is appropriate to show the weld symbol on both sides of the reference line. In other situations, such as a spot weld or seam weld, it may even be appropriate to show the symbol centered with the reference line.

When the symbol is placed on the side of the reference line nearest to the reader (below the reference line), the weld is to be on the *arrow side* of the joint. See **Figure 22-8**. When the symbol is on the side of the reference line away from the reader (above the reference line), the weld is to be on the *other side* of the joint. If the weld symbol appears on both sides of the reference line, the weld is to be on *both sides* of the joint.

Weld dimensions such as depth, weld size, length, and pitch are shown on the same side of the reference

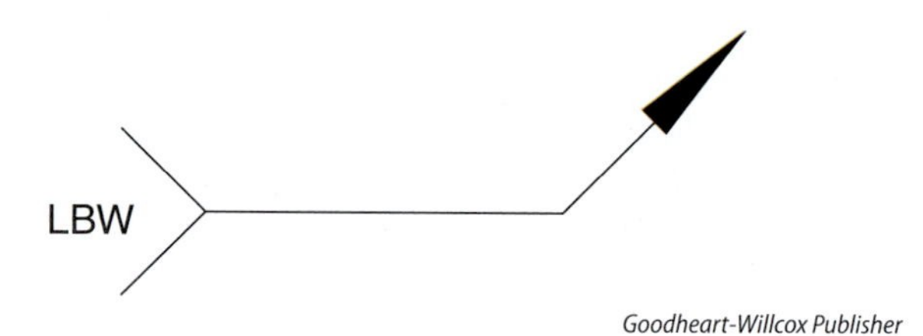

Goodheart-Willcox Publisher

Figure 22-6. The tail section of the welding symbol can designate many instructions, including the welding specification, method, sizes, or other reference.

Groove							
Square	Scarf	V	Bevel	U	J	Flare-V	Flare-bevel

Fillet	Plug	Slot	Stud	Spot or projection	Seam	Back or backing	Surfacing	Edge
	Ø Ø							

The American Welding Society

Figure 22-7. Basic weld symbols for various types of welds are shown here.

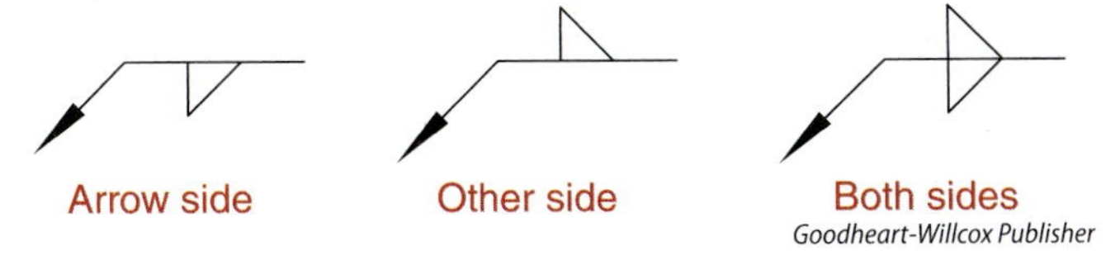

Figure 22-8. The weld location is indicated by where the weld symbol appears in relation to the reference line.

line as the weld symbol. **Figure 22-9** shows how numeric values are applied with the fillet weld symbol. Sometimes dimensions can be covered by a general note, such as ALL FILLET WELDS 3/8″ IN SIZE UNLESS OTHERWISE NOTED, in which case the welding symbol does not need to contain the dimensions, but the weld symbol is still required, **Figure 22-9A**. For fillet welds where both legs are the same size, the weld size can be specified in front of the weld symbol, **Figure 22-9B**. When both welds on two sides have the same dimensions, either or both welds may be dimensioned, as in **Figure 22-9C**, which also shows the length of the weld following the weld symbol. The pitch, or distance between intermittent welds, is shown following a dash after the length value, **Figure 22-9D**. In this example, the welds on the other side are staggered, so the weld symbol is likewise shifted.

Groove welds are often used if larger pieces of metal are to be welded. Various specifications about the size of the grooves can be incorporated into the welding symbol. The depth indicates the bevel distance, whereas the weld size, in parentheses, indicates the weld penetration. While there are too many scenarios to cover within the scope of this unit, **Figure 22-10** shows some examples of groove angle specifications. With all welding symbols, additional numeric values that describe size or spacing should be shown on the same side of the reference line as the weld symbol. Refer to **Figure 22-5** to review how the numeric values are placed in relationship to the weld symbol.

In **Figure 22-10A**, the depth of the groove weld is specified as 3/8″ with a weld size of 1/2″, and the angle of the groove is 45°, with the weld to be performed on the *other side*, not the *arrow side*. The angle specification is the included angle between the surfaces of both pieces as positioned for assembly, not the individual slope of either part alone. If the thickness of the material is greater than 1/2″, a *partial joint penetration* is being specified. For groove welds that do not specify depth and size, a *complete joint*

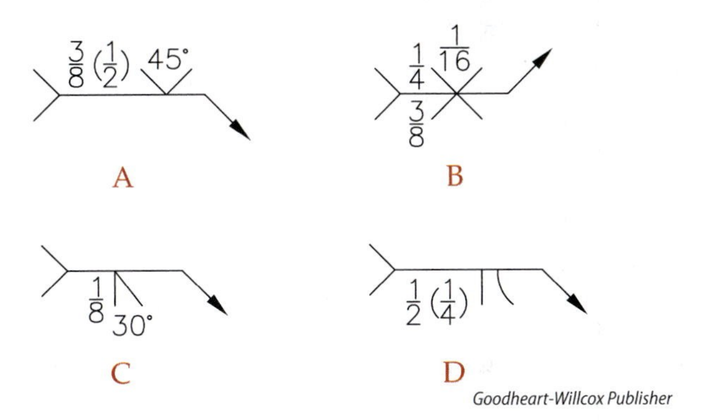

Figure 22-10. Groove welds are specified in a variety of ways.

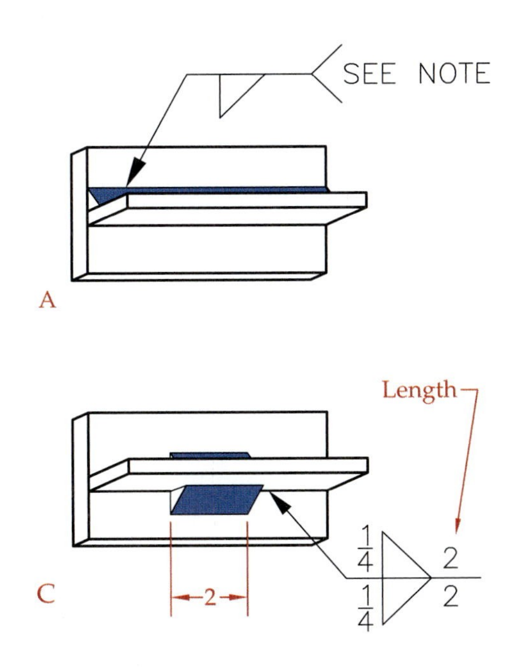

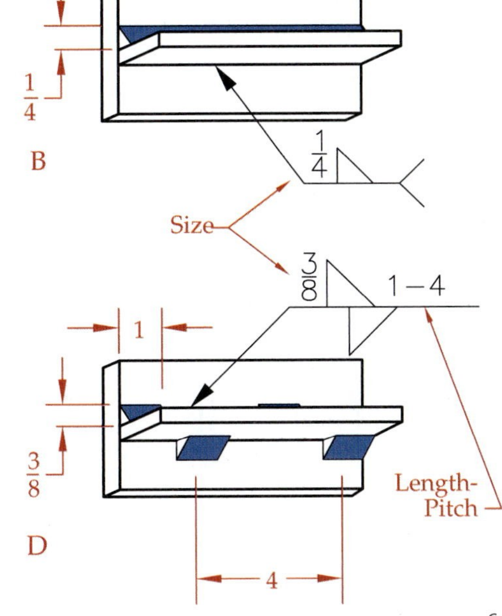

Figure 22-9. Fillet welds are specified in a variety of ways.

penetration is required. This can be further clarified by placing CJP in the tail of the symbol.

As illustrated in **Figure 22-10B**, the depth on the arrow side is different than the depth on the other side. Since the angle is not specified within the welding symbol, perhaps it is covered by a general note. The root opening size of the groove weld is specified as 1/16″ by placing the value inside of the weld symbol. If there is a weld symbol on both sides of the reference line, the root opening only needs to be specified on one side.

In **Figure 22-10C**, a 30° bevel groove is specified, with the depth at 1/8″, so the weld size is understood to be at least that amount. With a straight arrow pointing to the joint, either piece can feature the beveled edge. However, if the designer or engineer wishes to specify which piece features the bevel, a ***broken arrow*** should be used. **Figure 22-11** applies this standard technique to a double bevel groove. The AWS standard uses the term "broken arrow" for an arrow that has a bend between the arrow and the reference line. The broken arrow application only applies to bevel groove and J-groove welds, as they allow for only one piece to have the contoured edge. As shown, without the broken arrow, there are four options on how to arrange the contour edges.

In **Figure 22-10D**, a flare groove is specified with a depth of 1/2″, but the weld size is only 1/4″. Flare groove welds are often applied to a cylindrical or curved surface mating with a flat surface or other curved surface. For flare groove welds, the depth is measured from the weld bead face to the element of tangency. In these cases, the weld size often does not need to extend to the element of tangency, **Figure 22-12**, so the depth value is larger than the weld size.

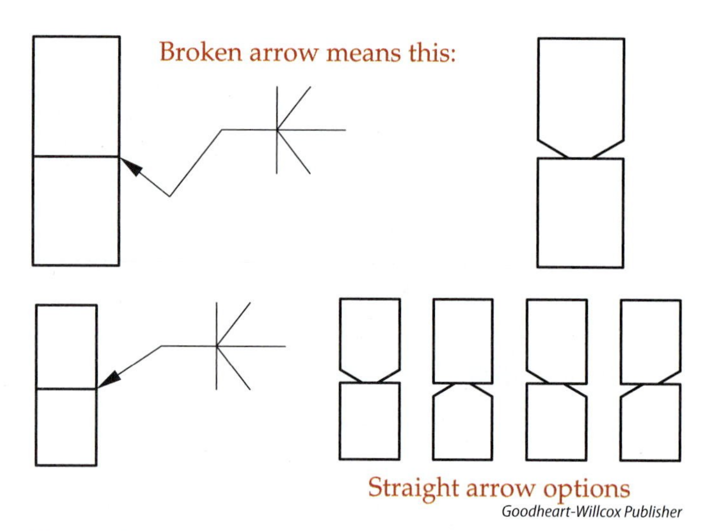

Figure 22-11. For some groove joints, the broken arrow is used to specify which piece features the contoured edge.

Additional Applications

In addition to fillet welds and groove welds, other welding processes and techniques are also prevalent in industrial applications. The spot weld symbol is used to specify a ***spot weld***, commonly used to weld lapping sheet metal. It may specify the spot weld diameter (**Figure 22-13A**), strength in pounds (**Figure 22-13B**), center-to-center pitch (**Figure 22-13C**), and number of spot welds (**Figure 22-13D**). In **Figure 22-13C**, the spot weld symbol is centered with the reference line, indicating a "weld through."

Slot welds and ***plug welds*** are welds implemented through a slot (*slot weld*) or hole (*plug weld*) in one member to join it to an adjacent member piece. For slot welds, the width across the slot is specified on the left side of the weld symbol, and the length is specified on the right side of the symbol. A plug weld is round, as indicated by the diameter symbol and value given in front of the rectangular symbol. Other values, such as the depth of the weld, can be given within the rectangular symbol, while beveled surfaces or countersinks can be specified outside the rectangular symbol. If a depth is not specified, the weld depth

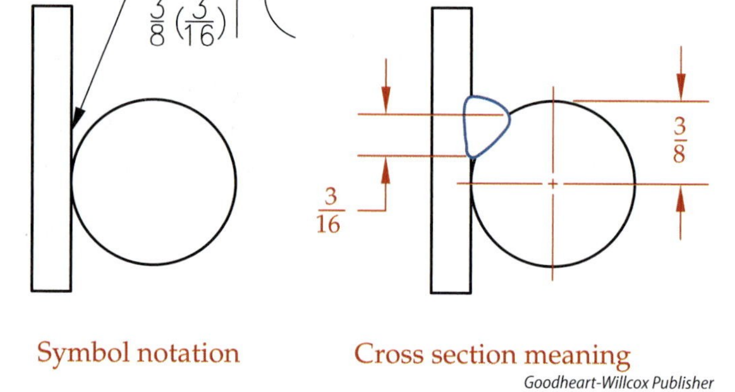

Figure 22-12. For flare groove joints, the depth is specified to an element of tangency.

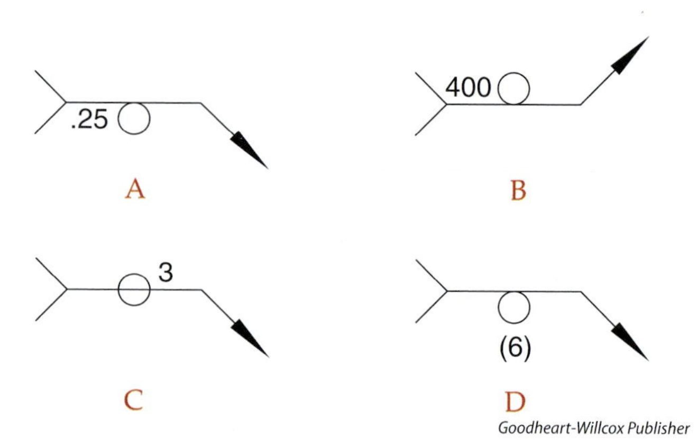

Figure 22-13. The spot weld symbol can be annotated in many ways.

is the total thickness of the plug or slot. A larger round hole with a fillet weld all around is *not* considered to be a plug weld. Slot welds and plug wells are also commonly specified in quantity, with pitch distance indicated to the right of the symbol, and quantity indicated in parentheses above or below the weld symbol. For slot orientation, additional notes may be required that can be called out in the tail.

Stud welding is the process of welding fasteners in the form of studs or nuts onto a base metal using equipment and fasteners specially designed for that purpose. The studs or nuts may or may not be threaded. The stud weld symbol is only used on the arrow side of the reference line, and other values such as size, quantity, and spacing are arranged around the symbol.

The ***back weld*** symbol or ***backing weld*** symbol can be placed opposite a groove weld symbol to indicate a weld is to be run along the back side of that groove weld. A back weld is created *after* the groove weld, while a backing weld is created *before* the groove weld. The symbol is shown on the opposite side of the reference line from the groove weld symbol. As both procedures use the same symbol, a note indicating BACK WELD or BACKING WELD should be added in the tail. Another option is to use multiple reference lines, as discussed later in this unit. Multiple reference lines can show the sequence by using the back/backing weld symbol on one reference line and the remaining information on the other reference line.

Back welds or backing welds are not to be confused with another process known as ***back gouging***, wherein a root opening needs to be opened or cleaned on the other side of a partially welded joint. This process can be accomplished with welding torch equipment or a grinder. BACK GOUGE may also appear as a special instruction in the tail.

The ***surfacing weld*** symbol should be shown arrow side only and indicates a weld buildup is to be created on a flat or cylindrical surface for the purpose of changing the dimensions or properties of that surface. The size of the buildup is usually stated. There are many options for surfacing welds, including the direction of the weld beads that may be applied. Directional terms can be given in the welding symbol tail, but sometimes a detail view is appropriate to clarify how the surfacing weld is to be formed.

Multiple specifications are also permissible in a variety of ways. One reference line can address more than one joint or location with multiple leader lines with arrows. Also, there are cases in which it is desirable to have a combination of weld symbols. **Figure 22-14** is an example of a T-joint with a bevel weld specified in combination with a fillet weld. The bevel weld symbol is closer to the reference line, indicating the bevel weld will be created before the fillet weld. In addition to multiple leaders and weld symbols, multiple reference lines can also be used to specify a sequence of welding processes. The reference line closest to the arrow will be the first operation to be performed.

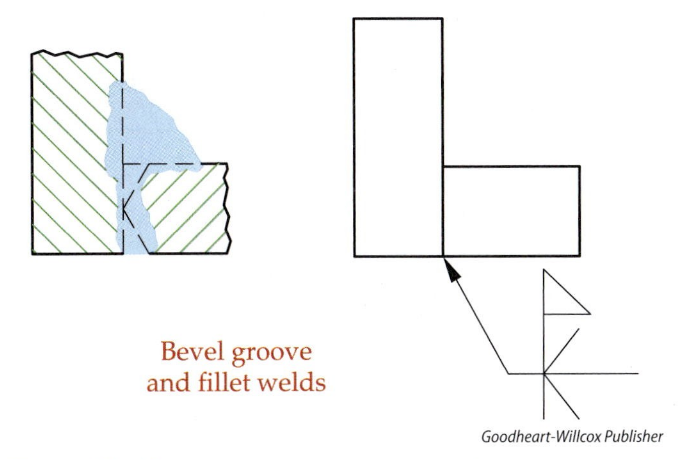

Figure 22-14. Weld symbols can be combined in cases where two different weld types are required.

It is beyond the scope of this text to cover all the specialized welding processes and scenarios that may appear on prints with welding symbols. For a full interpretation of the welding symbols specified on more complex industrial prints, additional references, such as the AWS A2.4 standard, should be consulted.

Supplementary Symbols

Supplementary symbols are used with the welding symbol to further specify characteristics of the weld. Supplementary symbols include the weld-all-around, field weld, melt-through, insert, backing, spacer, and contour symbols, **Figure 22-15.**

The weld-all-around symbol indicates the weld extends completely around a joint, similar in fashion to how the circular symbol in the elbow of the leader line works in other drawing specifications.

The field weld symbol indicates a weld is not to be made in the shop or place of initial construction. For example, the components of a large storage tank may be manufactured at a machine shop. The components must then be shipped to the job site and welded together. The assembly drawings for the tank would make use of the field weld symbol for all welds that would be done "in the field" instead of in the shop. The field weld symbol resembles a small pennant flag and can be positioned to the left or right, with the stem either up or down.

Weld-All-Around	Field Weld	Melt-Through	Consumable Insert (Square)	Backing (Rectangle)	Spacer (Rectangle)	Contour		
						Flat or Flush	Convex	Concave

Goodheart-Willcox Publisher

Figure 22-15. Supplementary symbols can be added to the welding symbol to provide additional information symbolically.

The melt-through symbol indicates welds where complete joint penetration plus root reinforcement is required of a weld made from one side. When melt-through welds are to have a certain finished shape, a contour symbol is added. For edge welds that require complete joint penetration, the melt-through symbol is required.

Consumable inserts are specially designed metal fillers placed between the metal parts to be welded. A common example is metal rings placed between butt-jointed pipes. These inserts are available in a variety of metal materials and cross-sectional shapes, as classified by the AWS A5.30 standard, titled *Specification for Consumable Inserts*. The consumable insert symbol is a square opposite the weld symbol. In addition, the AWS classification number value is specified in the tail of the welding symbol. These five AWS class values define the cross-sectional shape of the insert: inverted T (Class 1), J (Class 2), solid ring (Class 3), Y (Class 4), or rectangular (Class 5).

In some cases, backing material is desired. This can take the form of a metal plate temporarily tacked to the back of the pieces to be welded. The ***backing symbol*** is in the form of a rectangle and is placed opposite the weld symbol. If the backing material is to be removed after the weld, then an R is placed within the symbol. The removal may specify a process such as milling or grinding. The size of the backing material and other specifications are to be shown in the welding symbol tail.

On a double groove weld, a spacer can also be incorporated into the weld. The symbol for this is a rectangle centered with the reference line. An unusual aspect of this symbol is the way in which the weld symbol is divided into two parts, with one part on each end of the rectangle, **Figure 22-16.** The size and material of the spacer is indicated in the welding symbol tail or may be included in a detail view.

The contour symbol is a supplementary symbol that can be shown next to the weld symbol to indicate if the weld surface is to be 1) flush or flat-faced, 2) convex, or 3) concave. In addition to the three contour symbols, a finishing designator can be added if an additional process is required beyond the weld process itself, **Figure 22-17.** The finishing designators indicate the method of finishing, not the surface texture. The letter C = chipping, G = grinding, M = machining, P = planishing, R = rolling, and H = hammering, or a U can be added to specify the finishing method as unspecified but required.

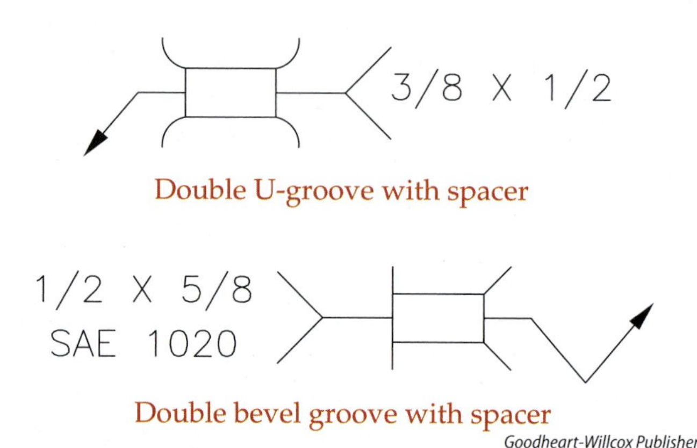

Goodheart-Willcox Publisher

Figure 22-16. The supplementary symbol for a spacer is unique in the way that it straddles the reference line and splits the weld symbol into two parts.

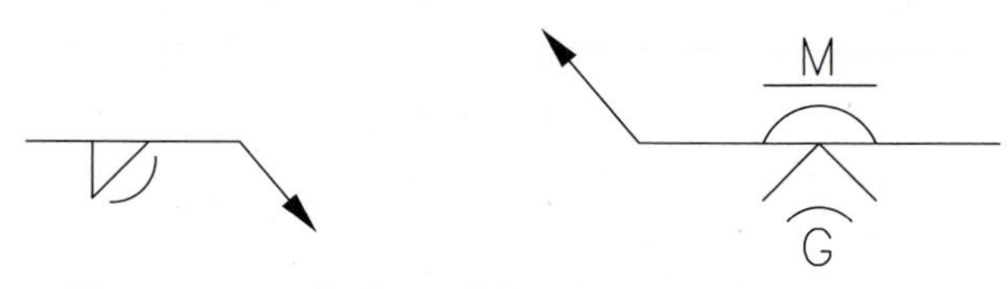

Goodheart-Willcox Publisher

Figure 22-17. The contour symbol can be added to specify the form of the weld face. Finishing designators can be added to specify a process.

In a few cases, the print may include hatching to help the print reader identify the area to be welded. This is especially helpful when dimensions give an indication of weld length, or when intermittent spacing is dimensioned on a view itself, **Figure 22-18.** Hatching can also help clarify the surfaces affected by edge joint welds.

While the preceding paragraphs have attempted to cover many of the ways in which symbol components work together to specify the welding processes, there are other aspects of welding still to be explored. As a student of print reading, rest assured that resources in the form of posters, charts, tables, and online references exist so that the information presented in this unit does not need to be committed to memory. Exercises for this unit will assist you with putting together all the various pieces of the welding symbol, including a wide array of weld symbols, numeric values, and supplementary symbols. As technologies evolve, be prepared to learn more about the new and exciting processes that impact the prints important to the manufacturing enterprise.

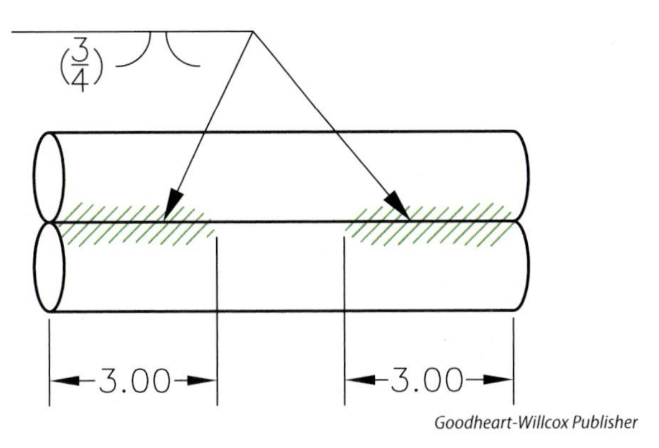

Figure 22-18. Hatching can be incorporated into the views on the print to help clarify the area of weld.

Summary

- Welding standards are published and maintained by the American Welding Society (AWS), including A2.4, *Standard Symbols for Welding, Brazing, and Nondestructive Examination*, and A3.0, *Standard Welding Definitions and Terms.*
- Welding is a fabrication process that joins metal pieces together, usually involving a phase change, or melting, of the pieces along the joint to form a weld, although there are other means of creating the weld.
- Welding drawings are considered detail drawings, even though they feature characteristics of an assembly drawing.
- Understanding the five different ways in which metal pieces are positioned to form weld joints, and the different contour shapes that can be incorporated into the edges of the metal pieces, is important to the print reader.
- Welding processes are various in nature and include fusion welding methods such as arc welding, gas welding, resistance welding, and intense beam welding, as well as solid-state welding methods that can create welds without phase change.
- Several welding terms describe the physical characteristics of the weld and the various dimensions that specify the size.
- The welding symbol—comprised of a leader with arrow, a reference line, and a tail—is essential to providing a concise and standard specification of the welding process on a print.
- The weld symbol, a critical element of the welding symbol as a whole, is located on the reference line and is often accompanied by various numeric notations and values, effectively describing characteristics of the joint, as well as the size, length, and spacing of the weld.
- In addition to common fillet and groove welds, print readers need to be familiar with a wide array of other welds, such as slot, plug, stud, back/backing, edge, and surfacing welds.
- Important supplementary symbols help specify additional weld processes or characteristics, including welding all around; welding in the field; melting through; adding inserts, backing material, and spacers; and contouring the weld bead.
- Due to the vast nature of welding processes and materials, the student of print reading will need to rely on a multitude of resources, references, and standards and be prepared to study new welding technologies as they evolve.

Unit Review

Name ______________________ Date ______________ Class ______________

Know and Understand

Answer the following questions using the information provided in this unit.

____ 1. *True or False?* Welding symbols are standardized by the American Society for Welding Engineers (ASWE).

____ 2. *True or False?* Welding, soldering, and brazing are processes that involve melting two or more metal pieces along a joint, forming a strong bond known as a weld.

____ 3. *True or False?* Welding drawings are considered detail drawings, even though they may have characteristics of assembly drawings, such as a parts list.

____ 4. Which of the following is *not* one of the five types of joints (between metal pieces) listed in the text?
A. Butt
B. Angle
C. Corner
D. Lap

____ 5. Which of the following welding processes is the category name for the other three?
A. Resistance
B. Gas
C. Fusion
D. Arc

____ 6. *True or False?* In general, welding processes that do not cause melting (i.e., phase change) are categorized as *solid-state welding*.

____ 7. Which of the following terms is exhibited only once on a *fillet* weld, while the other terms are exhibited twice?
A. Fusion face
B. Leg
C. Face
D. Toe

____ 8. Which term refers to the space between the two pieces being welded?
A. Depth of fusion
B. Heat affected zone
C. Throat
D. Root opening

____ 9. *True or False?* Welding symbols are comprised of an arrow, a reference line, and a tail section, but if no information is given in the tail, the lines can be omitted.

____ 10. *True or False?* Weld symbols are located either above or below the reference line, depending on the preference of the drafter creating the print.

____ 11. Which of the following is *not* a groove weld symbol?
A. Slot
B. U
C. V
D. J

____ 12. *True or False?* A plug weld is basically a round hole in one of two flat pieces of metal that allows for a fillet weld all around the bottom edge, thereby joining the two pieces.

____ 13. Which of the following terms refers to a process that cleans out or creates a root opening?
A. Backing symbol
B. Back weld
C. Back gouge
D. Backing weld

____ 14. Which of the following welding supplementary symbols is also used outside the realm of welding?
A. Field flag
B. Convex contour
C. Insert
D. All around

____ 15. Which of the following finishing designator processes is incorrect?
A. C = chipping
B. P = peening
C. R = rolling
D. G = grinding

Critical Thinking

1. Briefly discuss three benefits of having a standardized welding symbol system for industrial prints.

2. As the number of welding scenarios can seem overwhelming, briefly discuss three to five resources that a print reader can have available for interpreting welding prints.

Apply and Analyze

Name ____________________ Date ____________ Class ____________

Review Activity 22-1

Match the correct weld symbol to the following types of welds.

____ 1. Bevel
____ 2. Plug
____ 3. Scarf
____ 4. V
____ 5. Edge
____ 6. Stud
____ 7. Back or backing
____ 8. Flare-V
____ 9. Spot or projection
____ 10. Fillet
____ 11. Square
____ 12. Slot
____ 13. Seam
____ 14. U
____ 15. Surfacing

A. 
B. 
C. 
D. 
E. 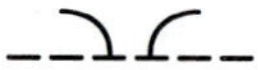
F. 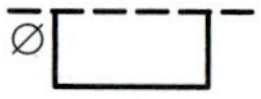 ⌀
G. 
H. 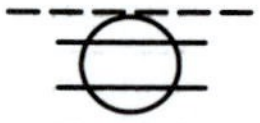
I.
J. 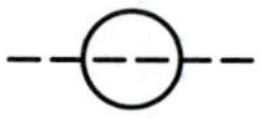
K. 
L. 
M. 
N. 
O.

Apply and Analyze

Name ______________________ Date ______________ Class ______________

Review Activity 22-2

For each of the written descriptions below, sketch a welding symbol that could replace the words, placing the straight arrow or broken arrow on the left end of the reference line for questions 1–5, and on the right end for questions 6–10.

1.	A 3/8" fillet weld on the arrow side with a convex finish by grinding.
2.	On the other side, 5/16" fillet welds, 2" long and spaced 5" apart. On the arrow side, 1/2" fillet welds with the same length and pitch.
3.	A 1/2" 90° V-groove weld with a depth of 3/8" on the arrow side, with a root opening of 1/8".
4.	A double bevel groove weld that is 1/2" on the other side, but 3/4" on the arrow side, with a flush contour on both sides, and the arrow pointing to the member that is to be beveled.
5.	A 3/16" square groove weld on the arrow side with a root opening of 1/8" and a back weld with a note.
6.	An edge weld on the arrow side with a melt-through, with no size required.
7.	A combination weld with a double bevel weld and then on the other side, a fillet weld featuring a concave face finished by machining, with sizes given in NOTE A.
8.	A 3/4" plug weld on the arrow side with a depth of 3/8" and the hole tapered with an included angle of 20°.
9.	Five [5] 1/8" spot welds that are "weld through" every 3", with the resistance welding process specified.
10.	Seven [7] stud welds for 1/2" studs to be welded every 6".

Apply and Analyze

Name ______________________ Date ______________ Class ______________

Review Activity 22-3

For each of the welding symbols below, write out the words that fully describe the specification given.

1. 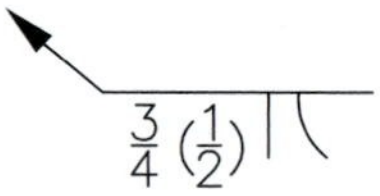 3/4 (1/2)	
2. R (5/8) 60° 1 X 1 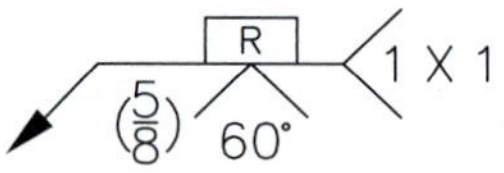	
3. 3/4 (6) 2–3 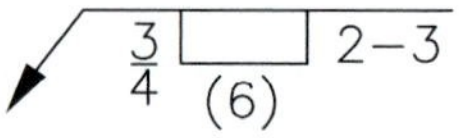	
4. 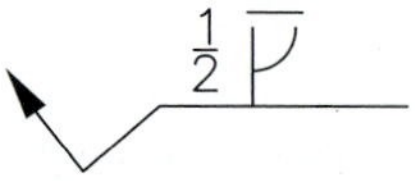 1/2	
5. 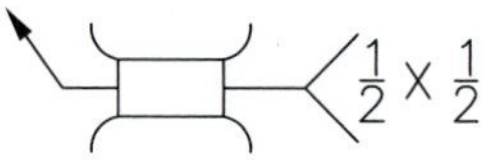 1/2 X 1/2	
6. 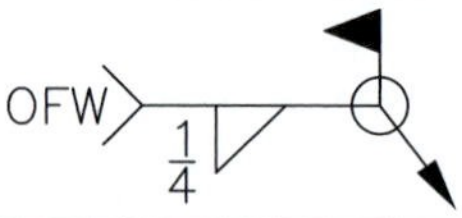 OFW 1/4	
7. 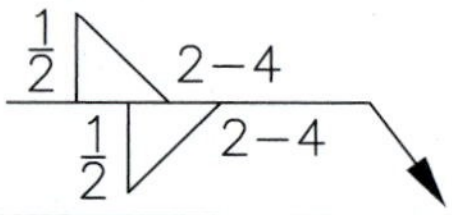 1/2 2–4 1/2 2–4	
8. SEE VIEW A–A 1/8 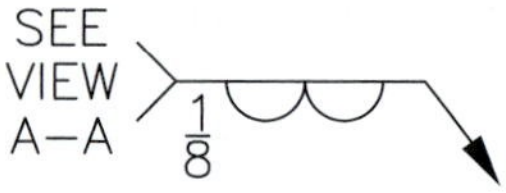	
9. 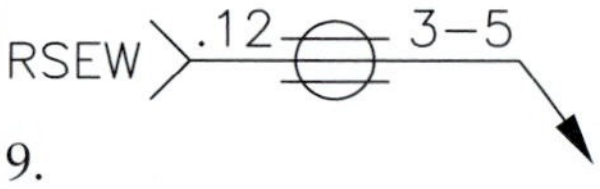 RSEW .12 3–5	
10. 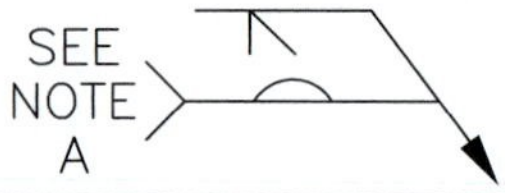 SEE NOTE A	

Notes

Apply and Analyze

Name ____________________ Date ____________ Class ____________

Industry Print Exercise 22-1

Refer to the print PR 22-1 and answer the questions below.

1. What units are used on this drawing, US customary units or SI units?
2. What is the total number of parts required to make this welded assembly?
3. How many welding symbols are on the print?
4. How many of the welding symbols apply arrow side only?
5. How many different types of welds are specified on the print and what are they?
6. For the weld between parts T150618 and T150620, what type of weld is required?
7. What company standard should be used to interpret the weld symbols on this print?
8. What note is found in the tail of the welding symbols and why is it needed?
9. For the fillet weld with a size specified at 6 mm, what length is specified for it using a linear dimension and hatching?
10. Are the welds specified on this print *fusion* welds or *solid-state* welds, and what method of welding is specified?
11. Are there any supplementary symbols applied, and if so, what are they?
12. What size is the V-groove weld at the top of piece T150619?

Review questions based on previous units:

13. What geometric tolerancing control is used to keep the two T150620 base parts aligned?
14. What geometric tolerancing control is used to control the orientation of the T150619 piece with the T150618 piece?
15. What piece number has a datum feature that is specified as Datum A?

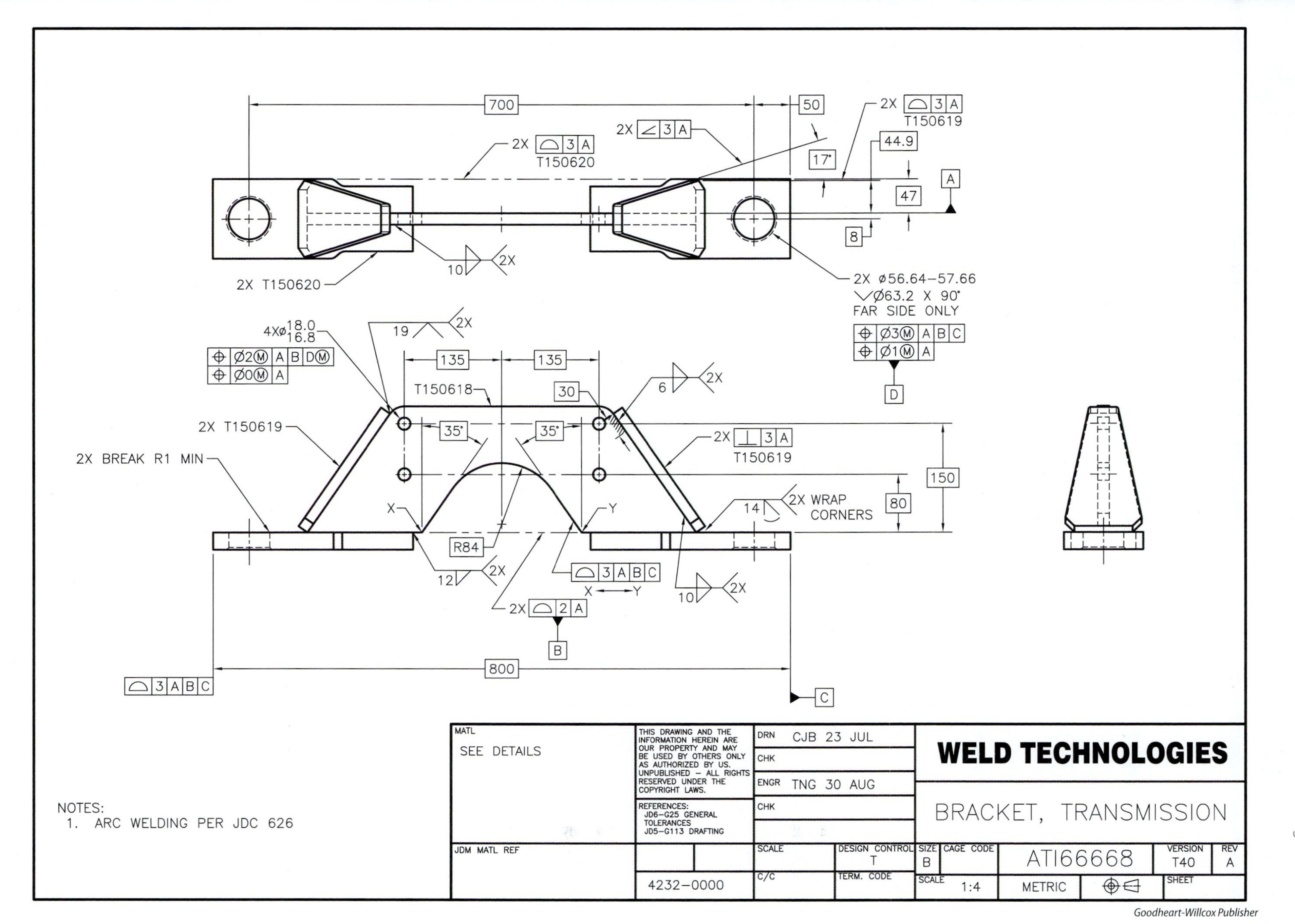

PR 22-1. Transmission Bracket

Apply and Analyze

Name ______________________ Date ______________ Class ______________

Industry Print Exercise 22-2

Refer to the print PR 22-2 and answer the questions below.

1. Counting temporary spacers, what is the total number of parts (not part numbers) required for this welded assembly? ______________________
2. What is the name and number of the part identified as item 3? ______________________
3. According to the general notes, what section of the AWS D1.1 standard should be referenced for the ultrasonic test? ______________________
4. What two weld symbols are shown in combination for the joints between the plates and the inner roll-up, and which is performed first?
5. What type of weld should be used to "tack" the parts associated with item 4 in place?
6. What machining process is specified by the finishing designators? ______________________
7. For the finishing process in question 5, does the welding symbol indicate flat, convex, or concave?
8. A couple of the welding symbols have a circle on the symbol elbow. What does that mean?
9. What contour is specified on the arrow side of the joint between pieces 1 and 2?
10. Are there complete dimensions on the drawing detailing all measurements for the individual parts?
11. Unless specified, what class are all welds to be and what AWS standard is indicated on the print as a reference?
12. On at least one occasion, what weld procedure number is specified in the welding symbol tail?
13. What is the part number and description of the item not shown on the drawing but that is used as a backing material when welding item 2?
14. How much does this part weigh once welded together? ______________________
15. What is the length of each of the six spacers? ______________________

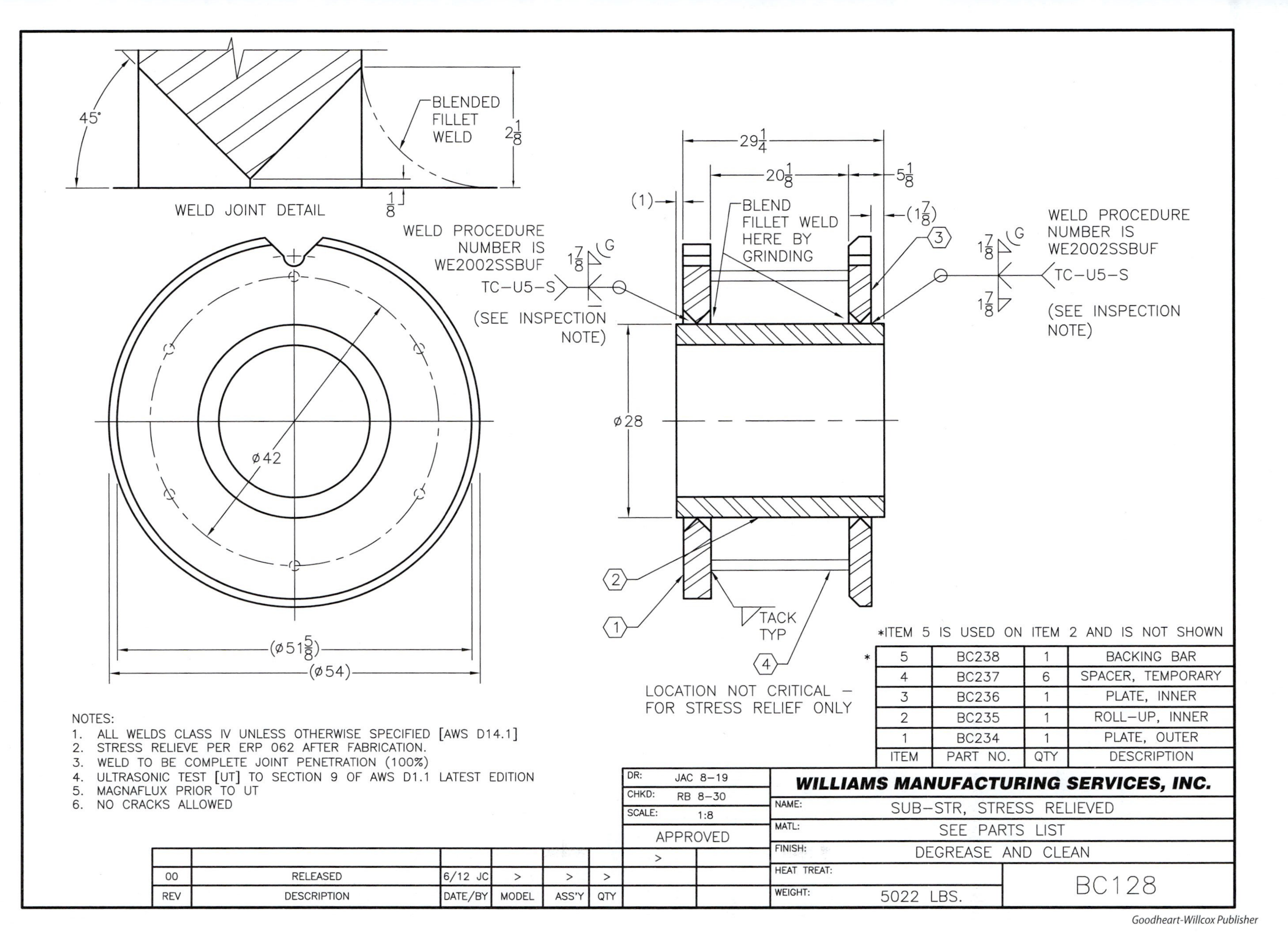

ITEM	PART NO.	QTY	DESCRIPTION
*5	BC238	1	BACKING BAR
4	BC237	6	SPACER, TEMPORARY
3	BC236	1	PLATE, INNER
2	BC235	1	ROLL–UP, INNER
1	BC234	1	PLATE, OUTER

Goodheart-Willcox Publisher

PR 22-2. Stress Relieved Sub-Str

Apply and Analyze

Name ______________________ Date ______________ Class ______________

Bonus Print Reading Exercises

The following questions are based on various bonus prints located in the folder that accompanies this textbook. Refer to the print indicated, evaluate the print, and answer the question.

Print AP-005

1. What are the names of the two parts being assembled in this print?

____

2. The engineer for this print elected to show these two parts could be assembled through brazing or soldering. How many locations are identified for this, with one being optional?

____

3. Is discoloration due to brazing or soldering acceptable?

____

4. Based on the solid black shapes that represent the brazing or soldering, what weld symbol best represents the "weld"?

____

Print AP-010

5. What type(s) of weld is(are) specified on this print?

____

6. How much applied pressure, in PSI, must the specified welds withstand?

____

7. The welding symbol for this print has a circle in the elbow of the leader line portion. What does that represent?

____

8. In the tail of the welding symbol, what does the "B" stand for in the EB or LB weld note?

____

Print AP-022

9. What type(s) of weld is(are) specified on this print?

____

10. How many total welds are required for this part?

____

UNIT **23**

Fluid Power Prints

LEARNING OBJECTIVES

After completing this unit, you will be able to:

- Define terms related to fluid power systems in industrial applications.
- Identify organizations that have a role in the standardization of fluid power systems and their associated prints.
- Describe the types of fluid power diagrams and the unique role of graphic diagrams.
- Describe the various categories of standardized fluid power symbols.
- Identify symbols used on fluid power diagrams and prints.
- Read and interpret a sequence of operations and component list as applied to a circuit diagram.
- Read and interpret a fluid power graphic diagram.

TECHNICAL TERMS

accumulator
actuate
actuator
block diagram
circuit
component
component list
conduit
control
cutaway diagram
cylinder
enclosure
envelope
fluid
fluid conditioner
fluid power
graphic diagram
hydraulic
mechatronics
pictorial diagram
pneumatic
port
reservoir
restriction
sequence of operations
solenoid

In industrial equipment, fluid power systems perform their mechanical tasks through the use of pressurized fluid within an enclosed circuit. Although we often use the term fluid as synonymous with liquid, the term applies to gases as well, as gas does not have a shape and, under pressure, can be moved through various channels. ***Fluid power*** can be defined as power that is generated, controlled, and transmitted with the use of fluids under pressure. Fluid power can be subdivided into two categories, *hydraulics* and *pneumatics*. ***Hydraulic*** systems use a liquid such as mineral oil or water pressurized with a pump device. The hydraulic fluid is channeled through tubing or pipes, controlled by valves, ultimately causing other devices such as cylinders or pistons to do the work. Hydraulic systems are closed, meaning that the fluid is always contained within the system, constantly recycling. ***Pneumatic*** systems use air or other gases that are compressed to create and control the energy source, causing devices to do the work. Pneumatic systems are open, so air is exhausted into the atmosphere after use, with the air compressor constantly taking in new air as needed. Pneumatic systems typically work with lighter tasks, with system pressure in the 80–100 psi (pounds per square inch) range, while hydraulic systems typically work with pressures in the 1000–5000 psi range.

Both systems control similar devices, incorporating valves and cylinders to create linear or rotary motion that positions other parts to accomplish goals, be it raising and lifting, such as a hydraulic jack or car lift, or powering hand tools such as nail guns or hammer drills. For many industrial applications, diagrams and drawings serve as a road map for how the fluid power components are connected. The goal of this unit is to introduce the basic symbols and methods of connecting the symbols to create fluid power circuit drawings and diagrams.

Fluid Power Standards

The National Fluid Power Association (NFPA) is a strong proponent for standards and is a resource for over 40 standards within the field of fluid power systems. Some of the standards deal with aspects of design, such as the physical characteristics of products that allow for interchanging parts. Others focus on the performance standards, including pressure ratings and strength or volume testing. Within this text, the main interest is with standards for communication that define terms and symbols, in which the NFPA is providing oversight to the ANSI and ISO organizations through technical committees. ISO standard 1219, *Fluid power systems and components — Graphical symbols and circuit diagrams* is a current resource for fluid power diagrams.

In the past, standards for fluid power drawings and diagrams were within the domain of ASME standards for engineering graphics. ASME Y32.10-1967, titled *Graphic Symbols for Fluid Power Diagrams*, last reaffirmed in 1999, is still available through ASME as a historical reference, although it has been *withdrawn*, meaning it is no longer considered an ASME-approved standard.

The entire field of fluid power could encompass many chapters, including such subtopics as system design, component specifications, implementation, and coordination of fluid power systems with electrical systems and programmable logic controllers (PLCs). ***Mechatronics*** is a newer term applied to a field of study that combines a wide array of mechanical and electrical applications in high-tech systems, such as robotics and automation. These systems often incorporate many of the technologies introduced in this unit. As a student of print reading, you are encouraged to explore the technological world beyond the print. In summary, this text limits its scope to reading prints that include hydraulic and pneumatic mechanisms, primarily expressed through schematic diagrams.

Diagram Types

Several types of diagrams or drawings are used to show how fluid power systems work. These include block, pictorial, cutaway, and graphic diagrams. The emphasis of this unit is on ***graphic diagrams***, which can also be referred to as schematics, **Figure 23-1**. Graphic diagrams use standard graphic symbols composed of various geometric shapes, including circles, squares, rectangles, arcs, and arrows. Various line dash patterns are also incorporated into the diagrams to communicate meaning. After learning the basic symbols and sequence of operations used in fluid power circuits, you will have little difficulty reading any of the types of diagrams.

Before discussing graphic diagrams in more detail, a brief description of the other types of diagrams is appropriate. A ***block diagram*** essentially consists of basic boxes that represent the components of a circuit, joined by lines that show how the components are connected, somewhat like a flowchart, **Figure 23-2**. These diagrams provide an easy method of emphasizing functions of the circuit and its components. A ***pictorial diagram*** shows components in a more picture-like manner and with piping or tubing represented by single or double lines, **Figure 23-3**.

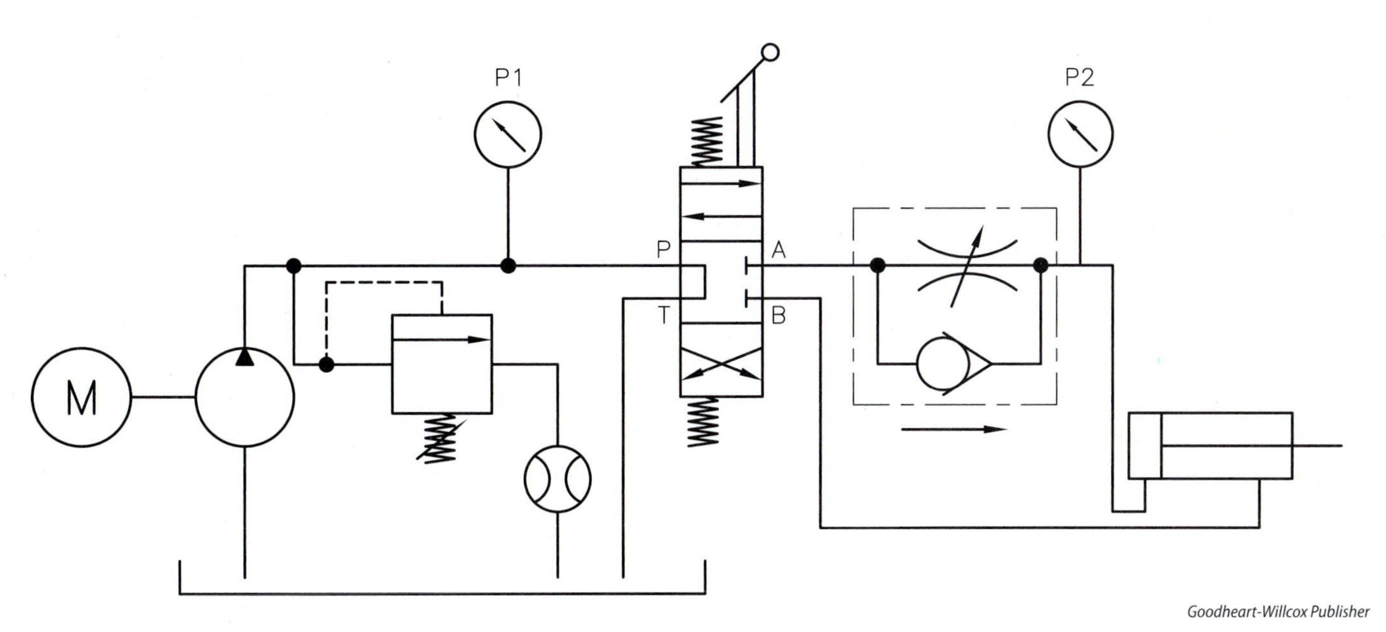

Goodheart-Willcox Publisher

Figure 23-1. A graphic diagram shows components and piping as schematic symbols.

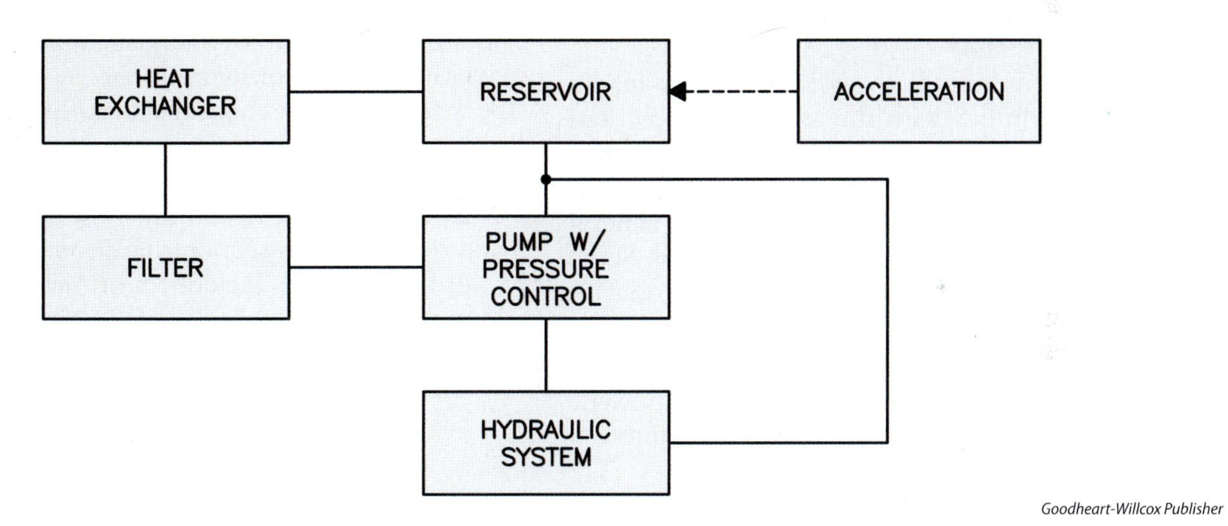

Goodheart-Willcox Publisher

Figure 23-2. A block diagram shows components or concepts with boxes connected by lines that help describe the system.

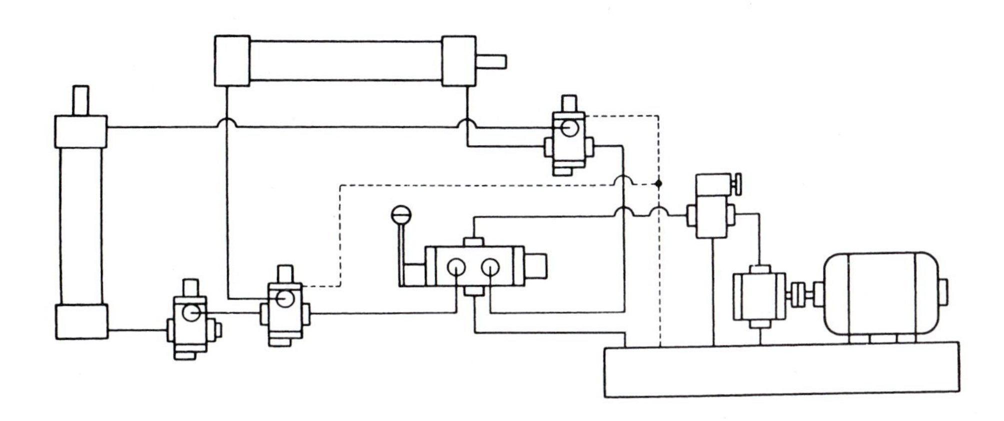

Goodheart-Willcox Publisher

Figure 23-3. A pictorial diagram shows components in a more picture-like form and piping as single or double lines.

Pictorial diagrams represent external shapes of components and may reflect scaled sizes that aid in the recognition or identification of the components. CAD systems have made it easier to show pipes or conduits more realistically, even including some of the flanges or connections that would have been omitted in former times. One of the drawbacks of pictorial diagrams is the difficulty in standardizing the functional explanation for those components. A ***cutaway diagram*** consists of cutaway symbols of components and piping that emphasize component function, **Figure 23-4**. Cutaway diagrams may use colors or hatching patterns to represent different conditions, but the function of the component is not always evident. In summary, the type of diagram used is the one that best suits or describes the purpose of the diagram.

Fluid Power Diagram Terminology

One goal of this unit is to assist the print reader in becoming familiar with the terms used for fluid power graphic diagrams. The following terms may help you understand the characteristics of the components featured within the symbols or the general categories into which symbol can be grouped. Some of the terms may appear on the prints to help clarify various aspects of the graphic diagram.

- An ***accumulator*** is a container in which fluid is stored under pressure as a source of fluid power.
- To ***actuate*** is to put devices and circuits into action.
- An ***actuator*** is a device for converting hydraulic or pneumatic energy into mechanical energy.
- A ***circuit*** is the complete path of flow in a fluid power system, including the flow-generating device, such as a pump or compressor.
- A ***component*** is a single hydraulic or pneumatic unit that performs a unique function.
- A ***conduit*** is the tubing or piping within the fluid power system via which the fluid is transmitted between components.
- A ***control*** is a device used to regulate the function of a component or system.
- A ***cylinder*** is a device that features a piston rod that moves from one position to another due to applied fluid pressure.
- An ***enclosure*** is a housing for components. A rectangle drawn with a center line dash pattern around components indicates the limits, or enclosure, of an assembly.
- A ***fluid*** is a liquid or gas.
- A ***port*** is the end of a passage in a component, either internal or external.
- A ***reservoir*** is a container for the storage of liquid in a fluid power system. It is not typically under pressure.
- A ***restriction*** is an area in a line or passage with a reduced cross section that produces a pressure drop. Examples include an orifice plate or a venturi tube.
- A ***solenoid*** is an electrical switch incorporating a metal rod that can be extended or retracted based on the application of electricity.

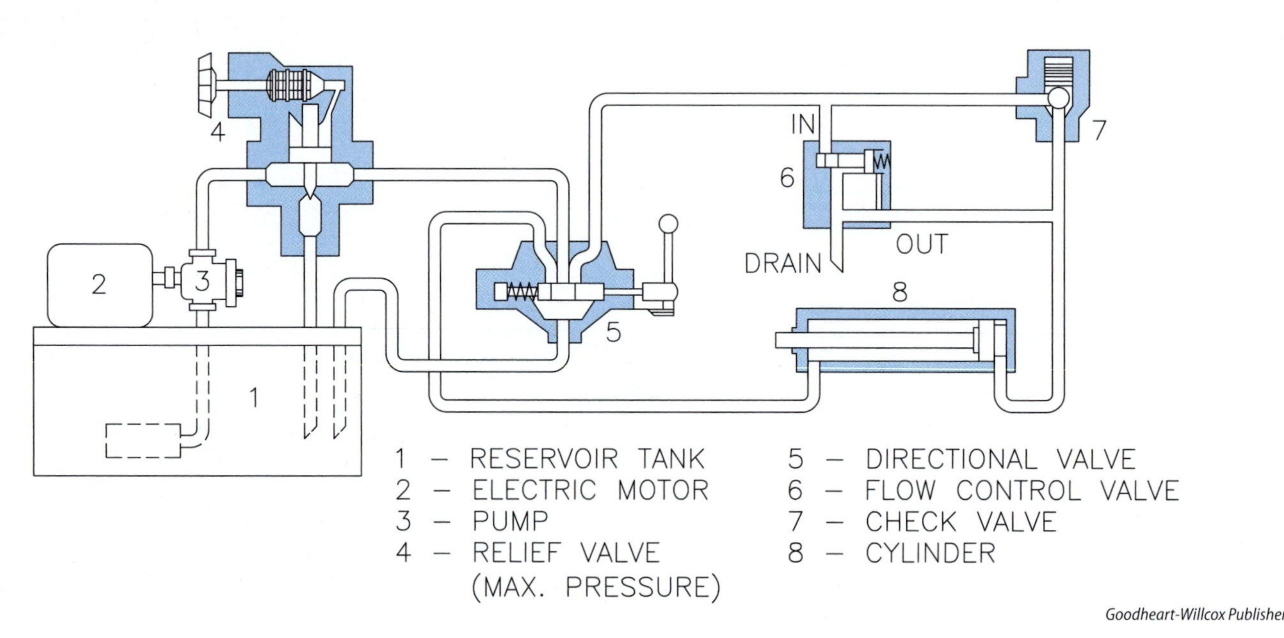

Figure 23-4. A cutaway diagram emphasizes the function of components by showing them as cutaways.

Standard Fluid Power Symbols

Standard symbols for fluid power diagrams have existed since the 1950s. Standards are reviewed and discussed by expert panels periodically to keep them current and relevant. The symbols in this text represent many of the symbols commonly found on graphic diagrams for fluid power circuits, but there are many combinations that can be arranged to address various scenarios. Once you are familiar with the symbols covered here, you will be able to read many of the diagrams, but you still may need to refer to other resources for more complex fluid power systems.

Lines and Arrow Technique

In schematic diagrams, lines and arrows are drawn or connected in ways that help describe the function of various parts within the fluid power system, as illustrated in **Figure 23-5**. In general, lines represent the conduit lines that connect components, with visible lines being the main working conduit, and dashed or hidden lines representing pilot lines or drain lines. Arcs can help distinguish where lines do not intersect, and dots can help confirm that lines connect. Hydraulic lines may feature filled arrows, while pneumatic lines may feature hollow arrows. An arrow and leader at a 45° angle can overlay symbols to represent a variable setting, meaning the component can be adjusted or varied. Curved lines with arrows can be used to represent rotation direction. Boxes drawn with a center line dash pattern can be drawn around several symbols to represent an assembly of components that serve a special function.

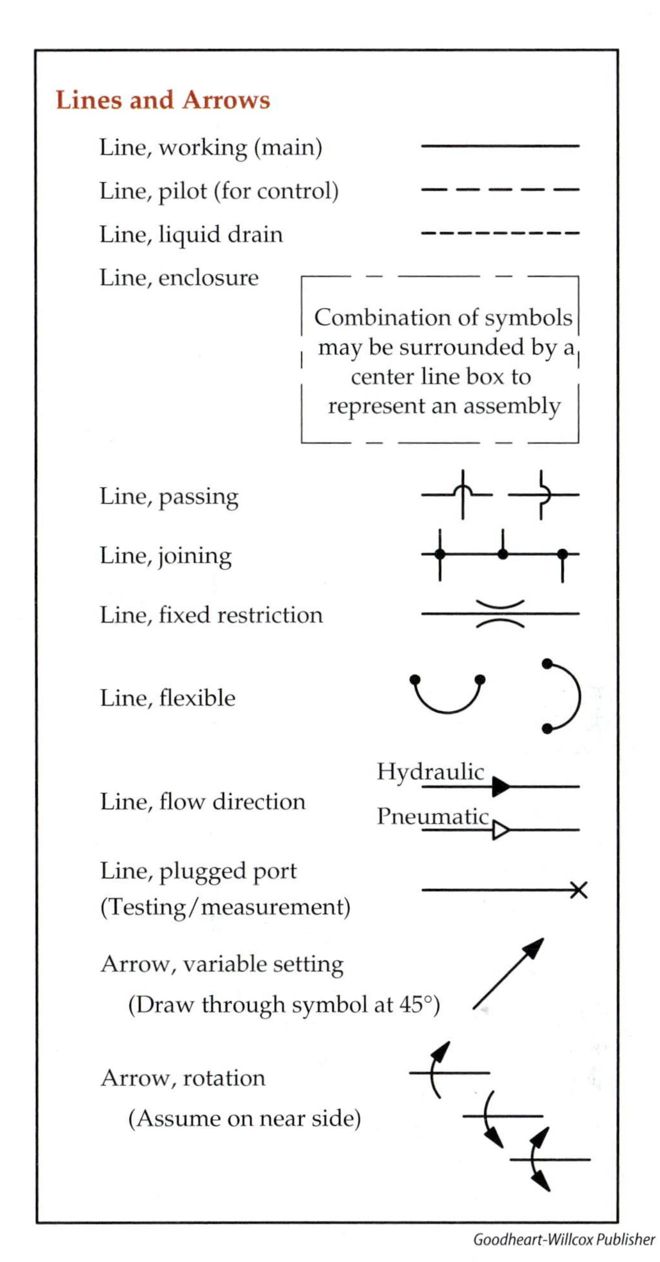

Figure 23-5. Fluid power schematic line and arrow representation helps communicate purpose, variability, and direction for various aspects of the fluid power circuit.

Fluid Storage

In general, the components related to fluid storage are primarily hydraulic reservoirs and accumulators. As shown in **Figure 23-6**, reservoirs are represented by three or four lines, depending on whether the reservoir is vented. Small reservoir symbols can be used in various locations, even though the system features only one tank. The connecting conduit line should reflect whether the line is above the fluid level or the end is submerged into the fluid, and it should always be drawn in the vertical orientation. Likewise, accumulator symbols should always be shown in an upright position. Three options for applying pressure to the stored fluid include a spring, gas or air pressure, or a weight.

Fluid Conditioners

Fluid conditioners include components that have an impact on the fluid, be it hydraulic oil or compressed gas or air. As shown in **Figure 23-7**, conditioners include filters, separators, lubricators, heaters, and coolers. Individually, a fluid conditioner symbol is in the shape of a square rotated 45°. **Figure 23-7** also shows an example of a simplified rectangular symbol combining a filter and lubricator symbol with a variable pressure regulator, but this causes the shape to change. This shows the need for a good reference featuring standardized symbols for more advanced scenarios.

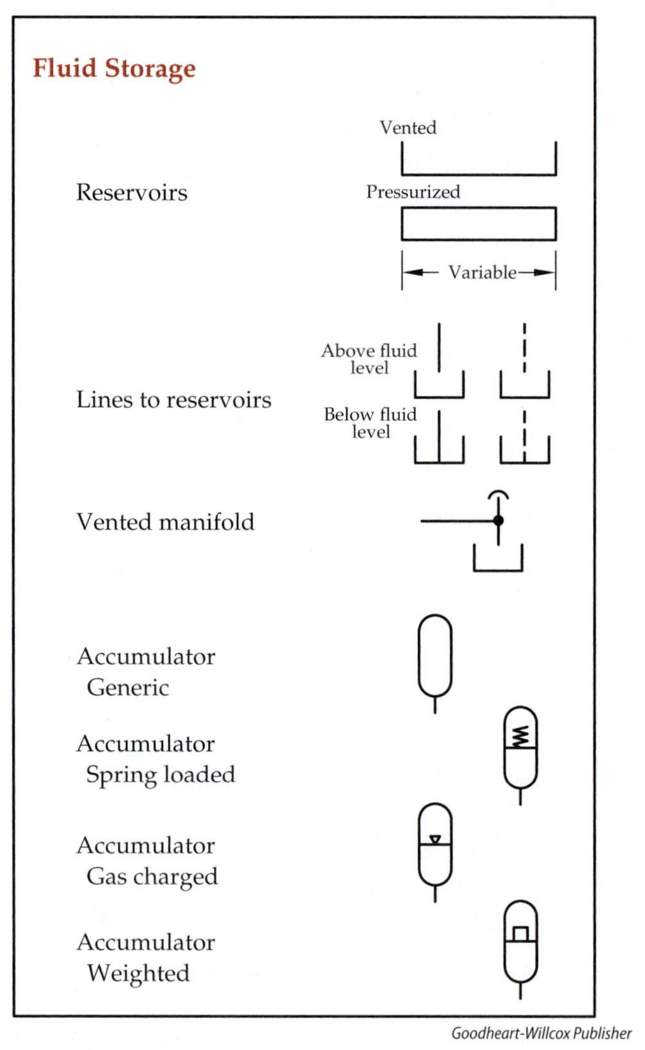

Goodheart-Willcox Publisher

Figure 23-6. Fluid storage components representing reservoirs and accumulators must be oriented logically within the graphic diagram.

Linear Devices

Devices that perform various work tasks are often in the form of linear devices known as cylinders, wherein fluid power forces a piston rod to move from one position to another. **Figure 23-8** shows common symbols for various cylinders. Single-acting cylinders move a piston with pressure arriving through one port. Perhaps a spring causes the piston to return home when pressure is removed. Double-acting cylinders have two ports, allowing pressure from two sources, causing movement in both directions. A double-end piston rod can perform work on both ends. Servo positioners can move the piston in infinite increments, rather than just the extended or retracted positions.

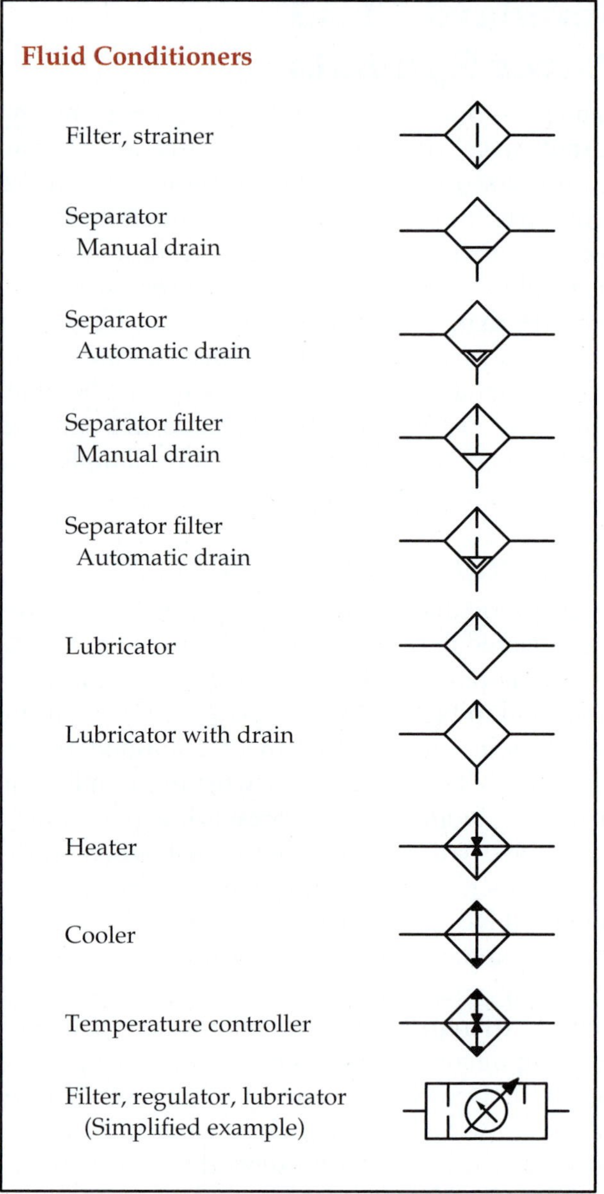

Goodheart-Willcox Publisher

Figure 23-7. Fluid conditioner components are often symbolized within rotated square symbols but may be incorporated into other shapes.

Actuation and Control Devices

Various devices are necessary to put work tasks into action, such as push buttons, pull levers, or foot pedals. Some of the devices send electric signals to solenoids, which may move valves into position to allow pressure to move along through the conduit to the cylinders. As shown in **Figure 23-9**, these controls may also take on the form of an electric motor that causes a pump to move, or a mechanical limit switch that actuates another component. Most of

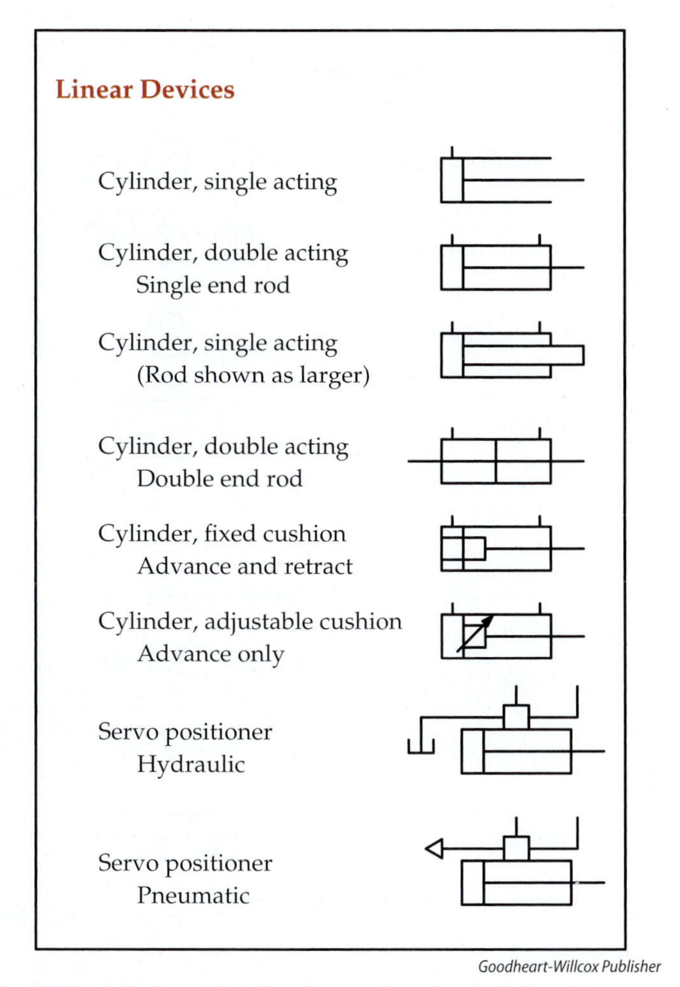

Goodheart-Willcox Publisher

Figure 23-8. Symbols for cylinder components feature port connections and lines representing the pistons.

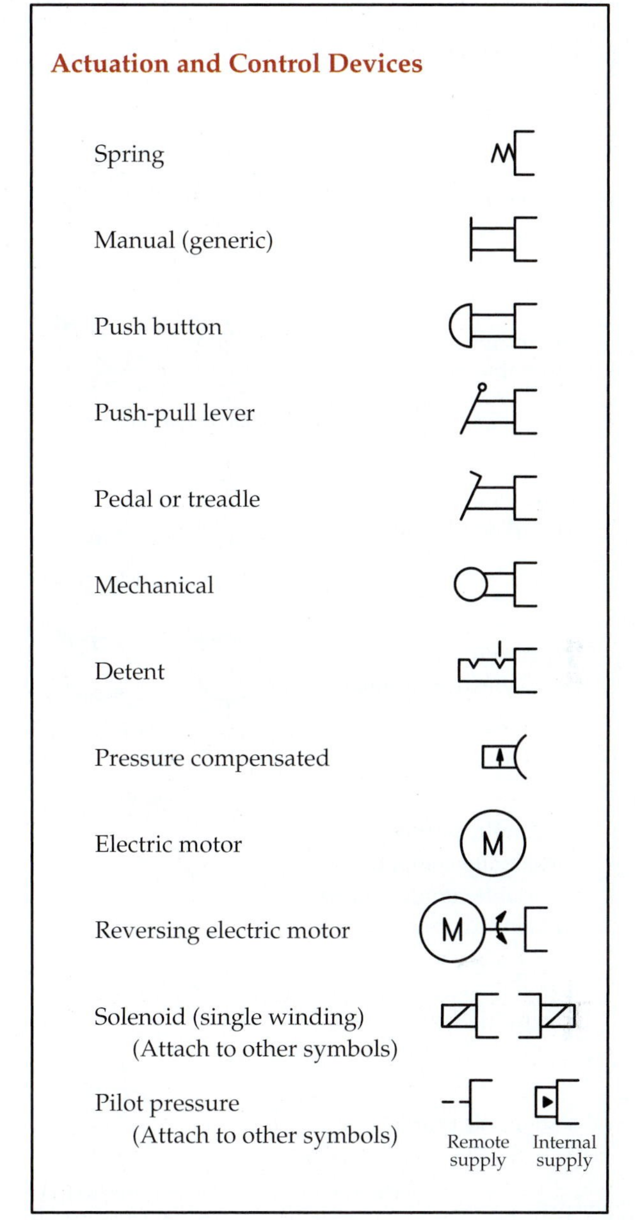

Goodheart-Willcox Publisher

Figure 23-9. Actuation and control symbols are often attached to other symbols to represent how those components may be actuated.

these actuator or control symbols are attached to other symbols, and some of them will be used in combination with each other, such as a solenoid that causes a valve to move in one direction, but a spring that causes it to move back to a home position.

Rotary Devices

In addition to the linear cylinder devices, rotary devices such as pumps and motors also perform various work tasks. **Figure 23-10** shows common symbols for rotary devices. Notice that the pump symbol is circular with an arrow pointing out at the port from which the fluid leaves the pump. As with other symbols, variable speed options are specified with the 45° leader arrow. Bidirectional pumps have two arrows within the circle. An electric motor symbol may be incorporated into the system to show the source of energy for the pump. Hydraulic and pneumatic motors that are driven by the fluid power have circular symbols with the arrows pointing in. As with pumps, these motors can be unidirectional, bidirectional, and variable speed. In addition to pumps and motors, which fully rotate, an oscillator device rotates a fixed amount, such as 180°, providing another option for hydraulic or pneumatic positioning devices.

The symbols for pumps and motors shown in the figure are simplified in the sense that all the actuators, drains, and rotary motion arrows are omitted, although complete symbology is also permitted. A complete symbol example is shown at the bottom of the figure.

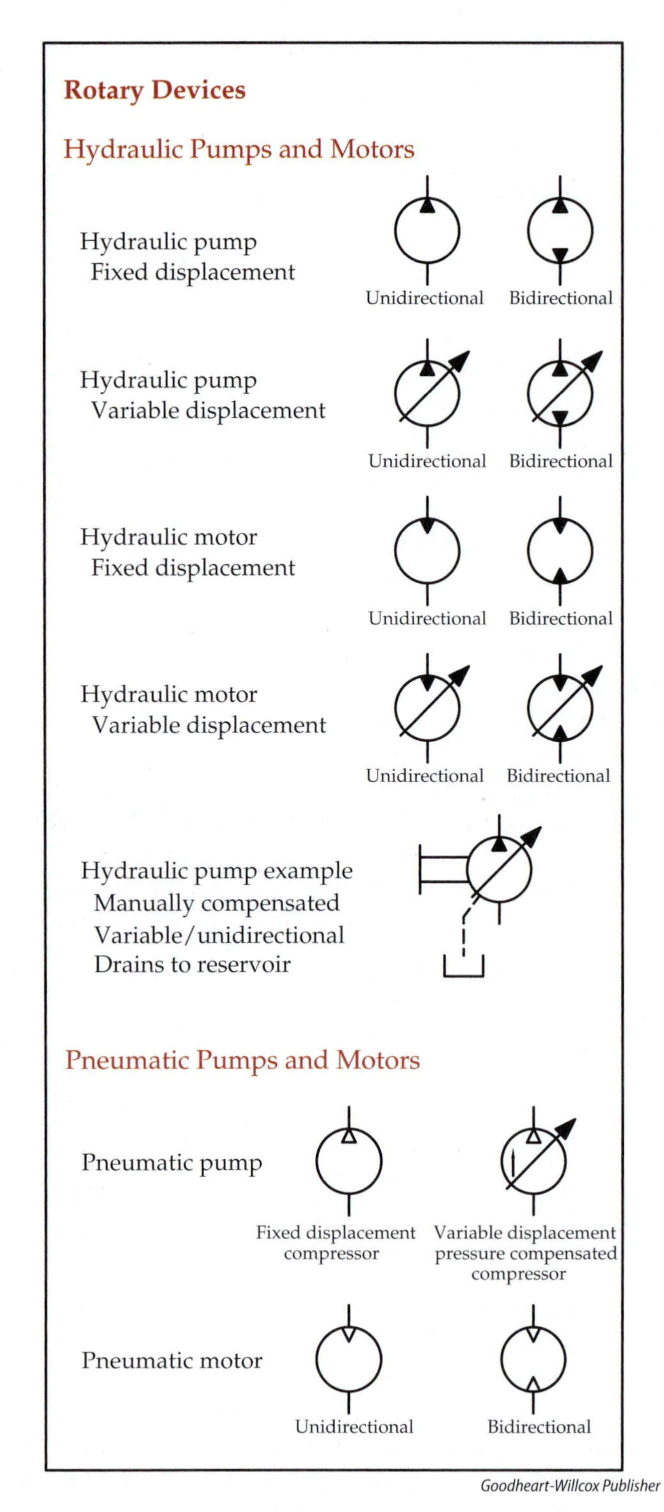

Goodheart-Willcox Publisher

Figure 23-10. Rotary devices such as pumps and motors are symbolized within circles and may be unidirectional, bidirectional, or variable in nature.

Instruments and Accessories

A few symbols are available for the various instruments found in fluid power systems, such as pressure and temperature gauges, as shown in **Figure 23-11**. The temperature symbol (a vertical line with a bulb-shaped bottom) can also be used within other

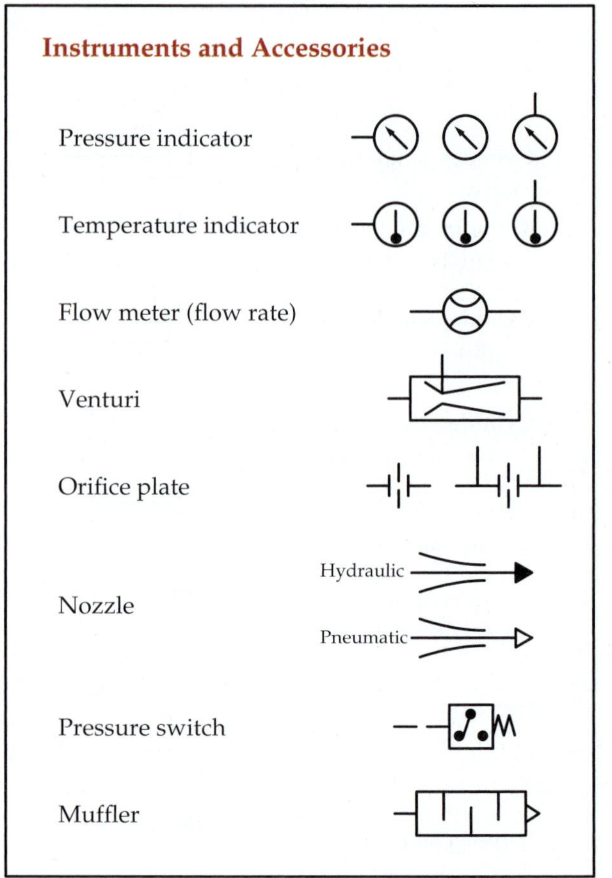

Goodheart-Willcox Publisher

Figure 23-11. Various instruments and other accessory devices require symbols that graphically represent their role in the circuit.

symbols when temperature controls are incorporated. Orifice plates and venturi tubes are devices that restrict the flow through the conduit for various purposes, including pressure readings. Notice the leads for pressure indicators. Other example symbols in this category are nozzles for both hydraulic and pneumatic applications, a pressure switch, and a muffler, which is helpful in pneumatic exhaust applications.

Valves

Valve symbols can be more complex than other symbols to learn, but with careful study, the symbols are quite useful as compared to other methods of representation. As shown in the upper section of **Figure 23-12**, a few of the symbols, such as check valves and needle valves, are simply added into the conduit line. A basic on-off valve can also be simplified in similar fashion. In many cases, though, valves are best symbolized with a rectangular ***envelope***, which may be a single square or multiple squares in combination. For multiple square envelopes, each

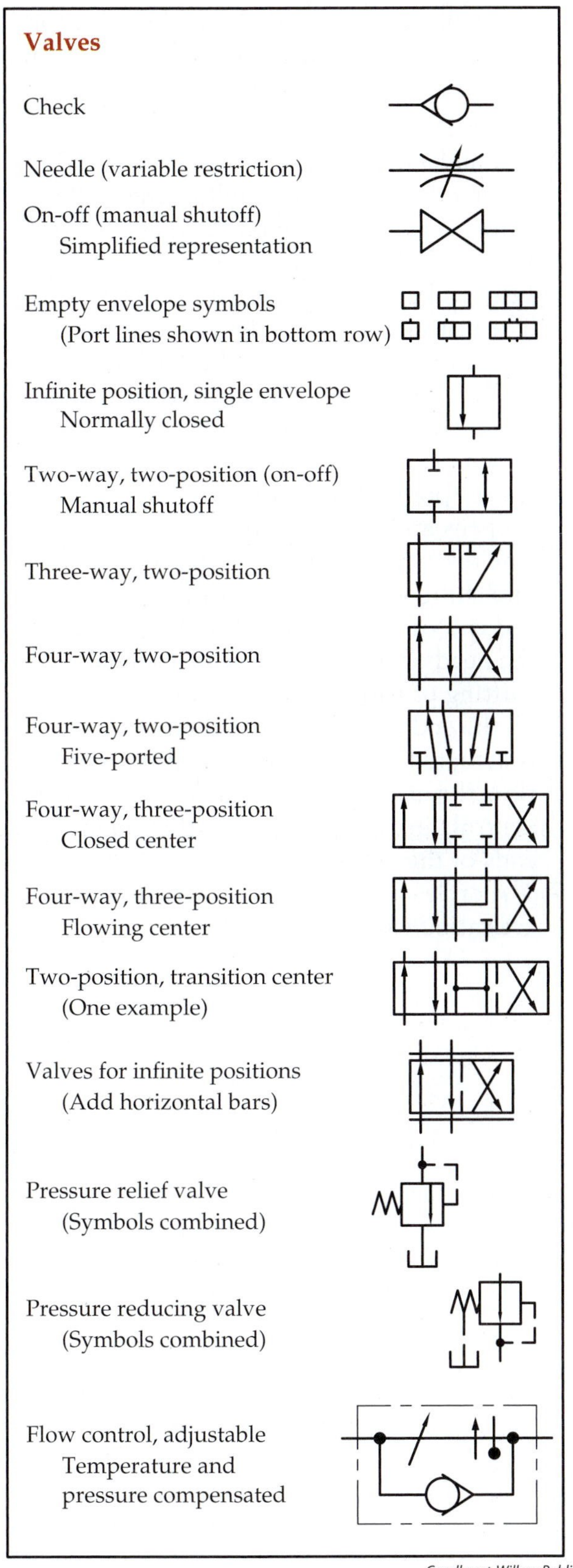

Figure 23-12. Fluid power valve symbols may be represented in a simplified manner but are often drawn within envelopes that diagram the fluid flow options for various position and port scenarios.

square represents a finite position of the valve. Ports are drawn for one of the squares, representing the home position. To read the symbol, imagine that the envelope moves (or slides) left to right, and the ports remain fixed. Valves are usually described by their number of *ways* and then number of positions. The word "way" refers to the possible fluid paths through the valve. For example, a valve may be described as a three-way, two-position valve, or simply 3/2 valve.

The single square envelope symbol shown in **Figure 23-12** represents a valve that has infinite positions. In conjunction with other symbols, such as a spring on one side or a pressurized line on the opposing side, the envelope represents a valve that could shift incrementally until the flow arrow aligns with the ports to provide complete pass-through.

The double-square envelope symbols each represent a valve that has two finite positions, but with different port configurations for different purposes. The most basic on-off symbol has no flow in one position or complete flow in the other position. For these envelope symbols, actuator symbols will be attached to the ends of the envelopes to indicate the forces that cause the valve to change from one position to the other. Additional examples show some of the options for three-port, four-port, and five-port configurations. Analyze each of the two-position valves in **Figure 23-12** to imagine how fluid can flow, or not flow, for each of the two positions.

The three-position envelope symbols represent valves that can be actuated in one direction to create a flow pattern or actuated in the other direction to create a different flow pattern. Several central position options are available, providing options for the fluid to be blocked or continue to flow through the ports. An example is also shown in **Figure 23-12** of a two-position valve that has a central transition where flow is temporarily stopped between positions. This unique application features dashed lines separating the central envelope from the ends. Valves with infinite positioning are symbolized by adding horizontal lines above and below the envelopes.

Supplementary Information

Once you are familiar with the symbols, you will be able to read and understand graphic diagrams. In addition to the graphic diagram, prints usually include a listing of the sequence of operations and a component list. These additional items help explain the function and purpose of the entire circuit and provide information related to its components.

The ***sequence of operations*** is an explanation of the various functions of the circuit in the order of occurrence. Each phase of the operation is numbered or lettered. A brief description is given of the initiating and resulting action. The sequence of operations is usually located in the upper part of the sheet or on an attached sheet.

A ***component list***, sometimes called a parts list or bill of materials, may include the name, model number, quantity, and supplier or manufacturer of components in the circuit. Each component in the diagram is numbered and keyed to the component list. For the most part, this is similar in fashion to other parts lists found in assembly drawings. In diagrams where two or more identical components are used, the same key number is used, followed by a letter to help identify each instance. However, some companies prefer that each duplicate component have a new number. The component list may appear in the upper-right corner of the print.

Reading a Graphic Diagram

To get the general idea of a graphic diagram, familiarize yourself with the pneumatic diagram shown in **Figure 23-13**. Match the numbers in the component list with their graphic symbols in the diagram, also represented by numerals in parentheses. Refer to the list of graphic symbols in this unit for further clarification. Then, try reading the pneumatic graphic circuit diagram by following the sequence of operations:

1. Manual valve 5 is depressed to start the cycle.
2. After valve 5 is shifted, follow the airflow through valve 7A and on to where it shifts valve 4. Valve 4, in turn, directs air to the head end of the cylinder (10).
3. The piston in the cylinder (10) extends to contact the workpiece and perform the riveting operation. The cylinder also depresses valve 7B.
4. When the tonnage control valve (6) reaches the set point of 80 pounds per square inch, it opens and reverses valve 5.
5. The reversing of valve 5 shifts the airflow and shifts valve 4 to direct the airflow to the rod end of the cylinder (10). Air pressure on the head of the cylinder is also vented by the shifting of valve 4.
6. At the end of the piston rod return, valve 7A is actuated. This permits the air pressure to flow through valve 7A and centers valve 4, neutralizing the pressure on the rod and head ends of the cylinder (10).
7. The cycle can begin again by manually activating valve 5.

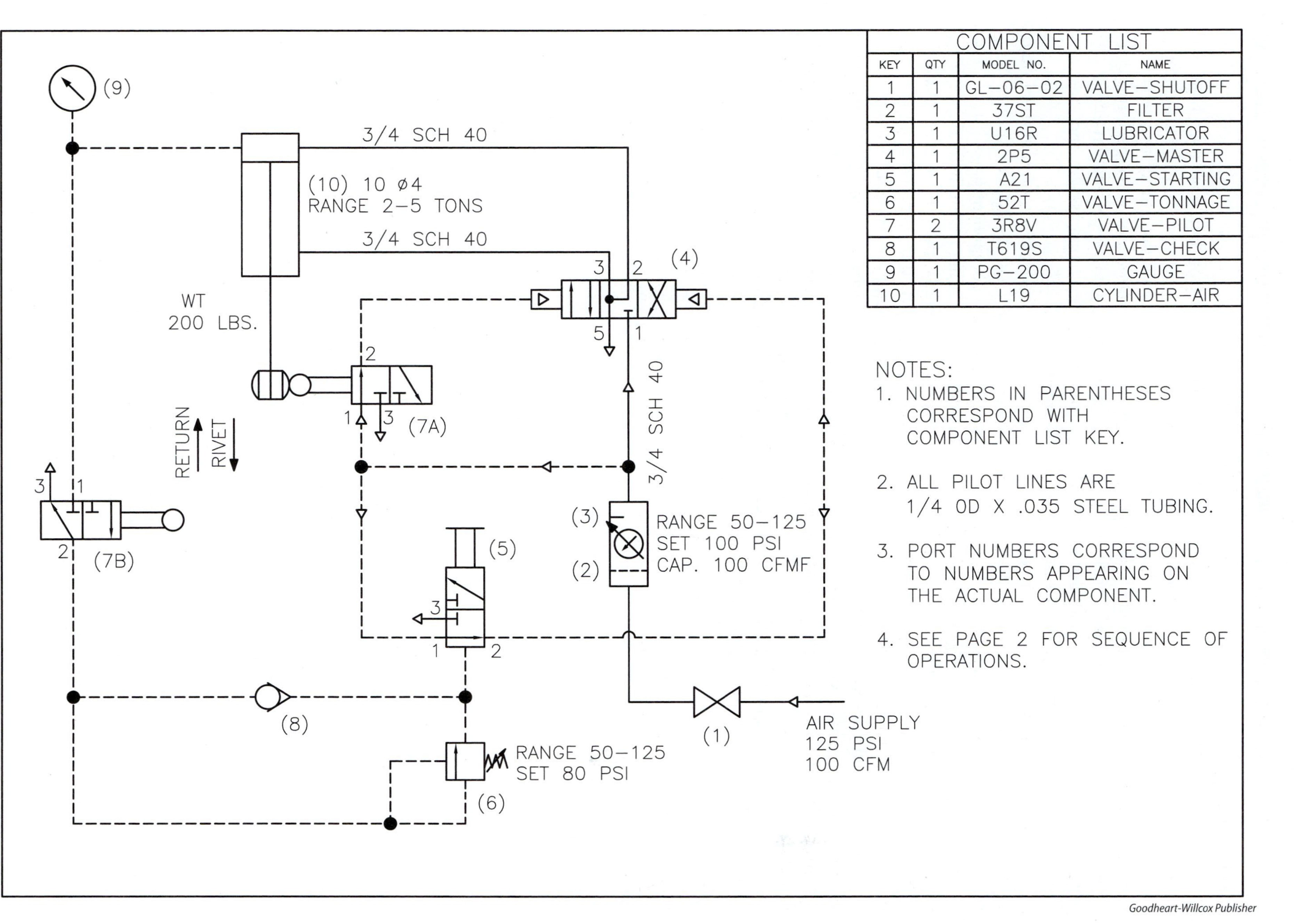

KEY	QTY	MODEL NO.	NAME
1	1	GL-06-02	VALVE-SHUTOFF
2	1	37ST	FILTER
3	1	U16R	LUBRICATOR
4	1	2P5	VALVE-MASTER
5	1	A21	VALVE-STARTING
6	1	52T	VALVE-TONNAGE
7	2	3R8V	VALVE-PILOT
8	1	T619S	VALVE-CHECK
9	1	PG-200	GAUGE
10	1	L19	CYLINDER-AIR

Figure 23-13. Refer to the text for a step-by-step sequence of operations of this circuit.

Summary

- Fluid power systems are either hydraulic or pneumatic, using pressurized fluid within a circuit to perform mechanical work.
- Organizations such as the National Fluid Power Association, ANSI, and ISO work together to provide standards for all aspects of fluid power systems, while the ASME provides a historical reference for fluid power graphic diagram symbols.
- Diagrams for fluid power system representation are most often graphic diagrams (schematics), but they can also be in the form of block, pictorial, or cutaway diagrams.
- Understanding fluid power components, symbols, diagrams, and systems requires an understanding of many terms specific to the field of study.
- Fluid power symbols are essential to providing clear and concise information about how the system functions.
- An understanding of fluid power symbols must encompass a wide range, including line and arrow representation, fluid storage and conditioning components, work-performing linear and rotary devices, actuation and control devices, instrumentation and accessories, and valves.
- Supplementary information on industrial fluid power prints will also include a sequence of operations and a component parts list.
- In addition to reading the individual component symbols, the print reader must be able to analyze and assess the complete hydraulic or pneumatic system.
- Due to the vast nature of fluid power systems, the student of print reading will need to rely on a multitude of resources, references, and standards, and be prepared to study new fluid power technologies as they evolve.

Unit Review

Name ______________________ Date ____________ Class ______________

Know and Understand

Answer the following questions using the information provided in this unit.

____ 1. *True or False?* The field of study known as fluid power encompasses both hydraulic and pneumatic technologies.

____ 2. *True or False?* There have never been ASME standards for fluid power symbols, as they are developed and published by the National Fluid Power Association (NFPA).

____ 3. Which of the four types of diagrams listed is also referred to as a schematic and is the primary standardized diagram for industrial prints?
A. Block
B. Graphic
C. Pictorial
D. Cutaway

____ 4. Which of the following four terms describes a device for converting hydraulic or pneumatic energy into mechanical energy?
A. Reservoir
B. Accumulator
C. Enclosure
D. Actuator

____ 5. *True or False?* The term that means "putting a device into action" is *actuate*.

____ 6. A _____ is the complete path of flow for an entire fluid power system, including the pump.
A. circuit
B. conduit
C. component
D. cylinder

____ 7. A center line box representing several components that are acting in assembly is referred to as a(n) _____.
A. enclosure
B. system
C. restriction
D. conditioner

Match the following component symbols to the category under which that symbol is found in the text. Answers are used only once.

____ 8. Lines and arrows
____ 9. Fluid storage
____ 10. Fluid conditioners
____ 11. Linear devices
____ 12. Actuation and control devices
____ 13. Rotary devices
____ 14. Instruments and accessories
____ 15. Valves

A. Flow meter
B. Two-position, five-connection
C. Push button
D. Bidirectional hydraulic motor
E. Cooler
F. Double-acting cylinder
G. Spring-loaded accumulator
H. Line flow direction arrow

____ 16. *True or False?* An explanation of the various functions of the circuit in the order of occurrence is often referred to as the *sequence of operations*.

____ 17. *True or False?* Within the *component list* found on fluid power prints, each component must have its own unique number or letter, even if there are additional components with the exact same model number.

Critical Thinking

1. Briefly discuss three benefits of having a standardized graphic diagram system for industrial prints as compared to the other diagram methods listed in the text. Answers may be given in terms of the disadvantages of the other diagram methods.

2. As the number of fluid power scenarios can seem overwhelming, briefly discuss 3–5 resources that a print reader can have available for interpreting fluid power prints.

Apply and Analyze

Name ______________________ Date ______________ Class ______________

Industry Print Exercise 23-1

Refer to print PR 23-1. The sequence of operations for this print is given below. Answer the questions following the sequence of operations.

Sequence of operations

1. The operator actuates push-button valves 1 and 2 until the ram (represented by the cylinder piston) starts to descend.
2. Air passes through valves 1 and 2 to the time delay valve (3) and the pilot of valve 5.
3. The two-position valve (5) is shifted, venting the head end of cylinder 4 and pressurizing the upper end.
4. The weight of the ram descends.
5. The speed of descent is controlled by the exhaust control (7).
6. The time delay valve (3) continues to time until the preset interval has passed.
7. The time delay valve (3) vents air pressure from the line between valve 2 and the pilot of valve 5.
8. With the pilot vented, the spring return of valve 5 shifts the valve back up to its normal position.
9. The rod end of cylinder 4 is then vented and the head end is pressurized.
10. The weight ascends.
11. The speed of ascent is controlled by the exhaust control (6).

Questions

1. What is the name of the circuit? ______________________
2. Give the drawing number. ______________________
3. How many different component parts are listed? ______________________
4. What starts the sequence of operation? ______________________
5. What causes valve 5 to shift? ______________________

   ____
6. When valve 5 shifts, what happens to cylinder 4? ______________________

   ____
7. What part controls the speed of the cylinder descent? ______________________

   ____
8. What causes cylinder 4 to reverse? ______________________

   ____

   ____

   ____

9. What does the center line around valve 3 indicate? What component(s) is/are included? ____________

____

____

____

10. Why is it *not* necessary to hold valves 1 and 2 down during the entire sequence?

____

____

____

____

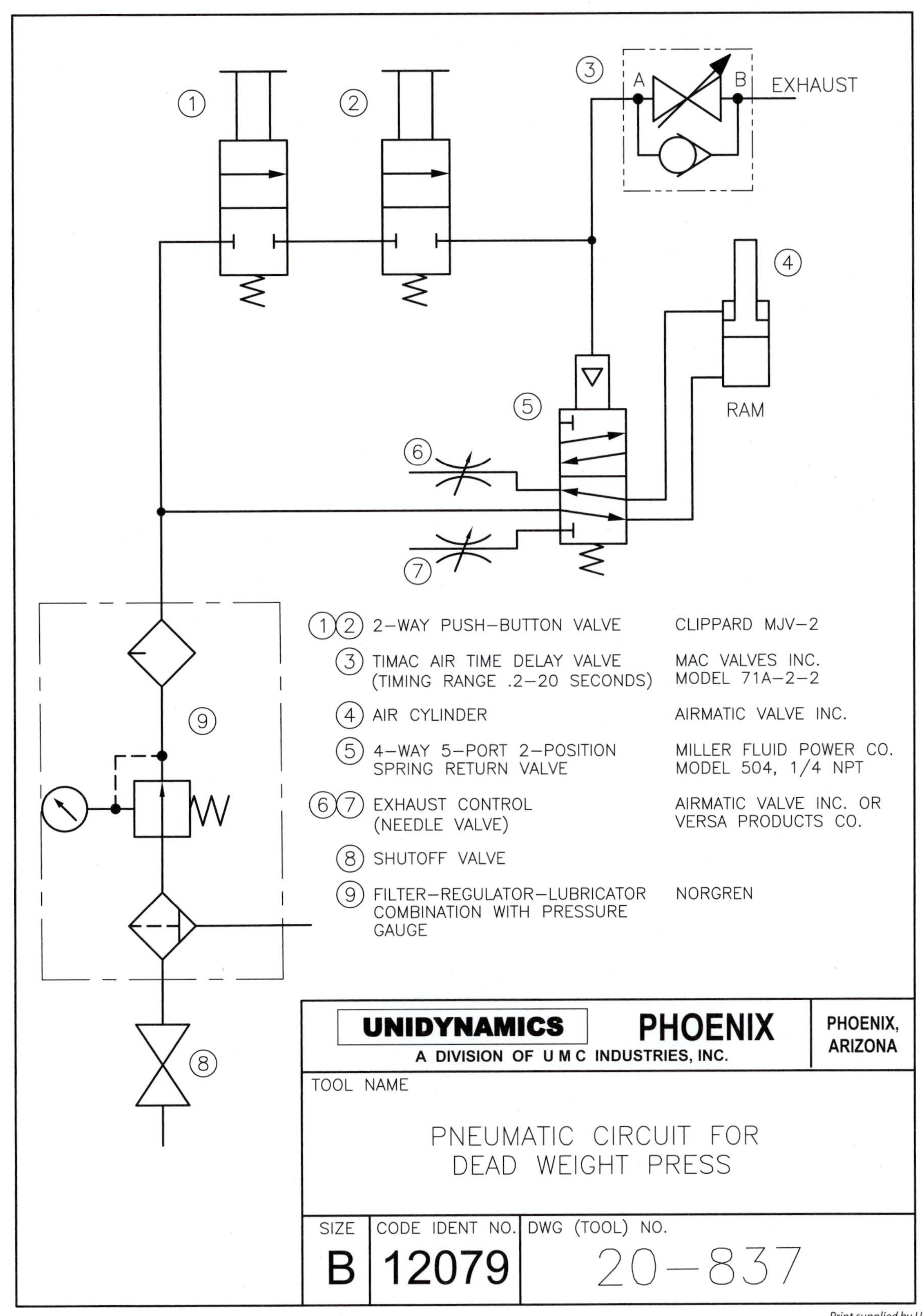

Print supplied by Unidynamics.

PR 23-1. Pneumatic Circuit for Dead Weight Press

Notes

Apply and Analyze

Name ______________________ Date ______________ Class ______________

Industry Print Exercise 23-2

Refer to print PR 23-2. The sequence of operations for this print is given below. Answer the questions following the sequence of operations.

Sequence of Operations

Description: *This is a system to control a 90-ton hydraulic, explosive-compaction press. The press ram descends at a controlled rate, slowly compacting the explosive charge to avoid detonation due to a sudden shock.*

1. The cycle is started by energizing solenoid A, which shifts control valve 3 and supplies pressure to the head end of the hydraulic cylinder (7). This operates the ram.
2. The cylinder (7) extends at a slow rate controlled by the fluid in the rod end of the cylinder, which is vented through the flow control valve (5).
3. When the cylinder is fully extended and the pressing pressure is reached, the pressure switch (6) de-energizes solenoid A and energizes solenoid B. This shifts control valve 3 to retract the cylinder (7). Control valve 5 has an integral check valve that permits the free flow of fluid in the opposite direction, causing the cylinder (7) to rapidly retract.
4. The retraction of the cylinder (7) actuates the limit switch (8), which de-energizes solenoid B. Control valve 3 returns to the neutral position. This completes the cycle. This circuit has safety features. On electrical power outage, both solenoid A and solenoid B are de-energized. This allows valve 3 to return to the center position, thus stopping ram travel. Also, valve 4 provides low-pressure relief on the ram upstroke, and valve 2 provides high-pressure relief on the ram downstroke.

Questions

1. What is the name of the circuit? ______________________
2. How many reservoir symbols are shown, and how many reservoirs are there? ______________________
3. What starts the sequence of operations? ______________________
4. When valve 3 shifts on due to the energizing of solenoid A, to which end of the cylinder (7) is pressure supplied? ______________________
5. What causes the ram to extend at a slow, controlled rate? ______________________

   ____
6. What causes the ram to retract? ______________________

   ____

   ____
7. Why does the cylinder (7) retract rapidly? ______________________

   ____

   ____

8. What completes the cycle? ______________________________

____

____

9. What happens in the event of electrical power outage? ______________________________

____

____

10. Give the function of valves 2 and 4. ______________________________

____

____

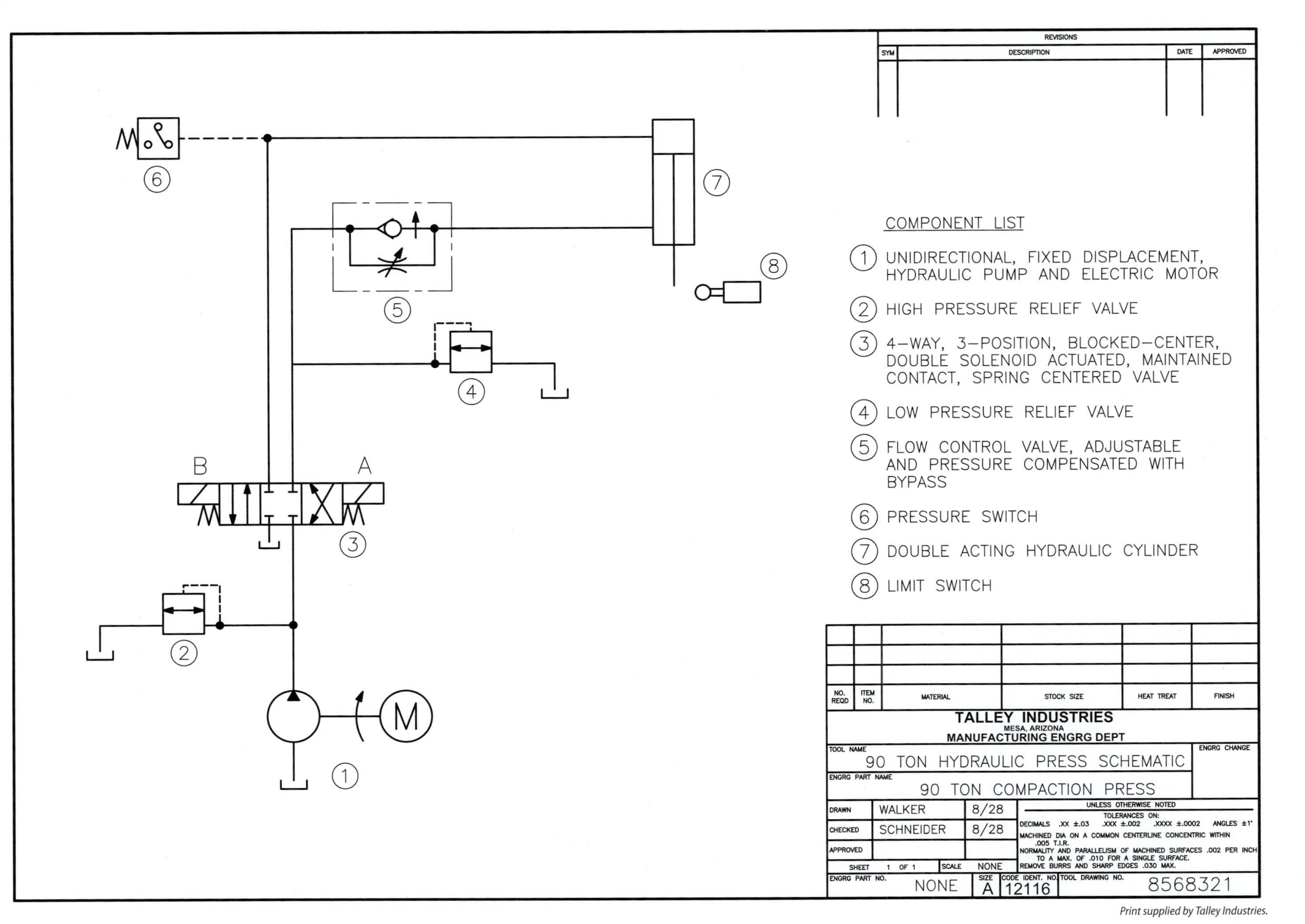

Print supplied by Talley Industries.

PR 23-2. 90 Ton Hydraulic Press Schematic

Notes

Apply and Analyze

Name ______________________ Date ______________ Class ______________

Industry Print Exercise 23-3

Refer to print PR 23-3. The sequence of operations for this print is given below. Answer the questions following the sequence of operations.

Sequence of Operations: Case Assembly Machine Hydraulic Schematic

Description: *This machine uses a fluid power system to perform two dimpling operations on a cylindrical assembly. The part is loaded into a fixture, dimpled, rotated 180°, and dimpled again. This is accomplished by the following sequence:*

1. The workpiece is loaded into the fixture.
2. When the workpiece is in position, the operator depresses valve 7. This extends cylinder 8 and clamps the piece into position.
3. The operator depresses and holds the push buttons on valves 2 and 3. These valves allow hydraulic fluid to flow to the pilot of valve 5, thus shifting the valve. Cylinder 6 extends and dimples the workpiece. *(Note: This step cannot be accomplished unless the operator depresses both palm buttons. This safety feature is to ensure that the operator is clear of the mechanism.)*
4. By releasing the push button on either valve 2 or 3, the fluid is exhausted from the pilot of valve 5, thus shifting the valve. Cylinder 6 retracts.
5. Once cylinder 6 is retracted, the operator manually actuates valve 10. Fluid pressure rotates the rotary actuator (11) by 180°.
6. When the workpiece is rotated, steps 3 and 4 are repeated to make another dimple 180° from the first.
7. On completion of step 6, the operator releases valve 10. The rotary actuator (11) returns to its original position.
8. When the part has returned to its original position, the operator releases valve 7. Cylinder 8 retracts to unclamp the workpiece from the fixture. The cycle is complete.

Note: This circuit is a dual pressure circuit. Cylinders 6 and 8 extend under high-pressure fluid. This is accomplished by pressure-relief valve 1. Retraction of these cylinders is done under low pressure, accomplished by pressure-relief valve 7.

Questions

1. What is the name of the drawing? ______________________________

2. Briefly describe the purpose of the machine. ______________________________

____

3. What actuates the workpiece clamping cylinder? ______________________________

____

4. When the push button on valve 2 is depressed, is valve 5 actuated? Why or why not?

____

____

5. What happens when cylinder 6 is extended and valve 3 is released? ______________________

____

____

6. What causes the workpiece to rotate 180° for the second part of the operation?

____

7. Describe valve 10. ______________________

8. Is cylinder 6 extended or retracted by high-pressure fluid? ______________________

9. What is the normal position (as shown) for cylinder 8? ______________________

10. Does the rotary actuator (11) operate on high-pressure or low-pressure fluid? ______________________

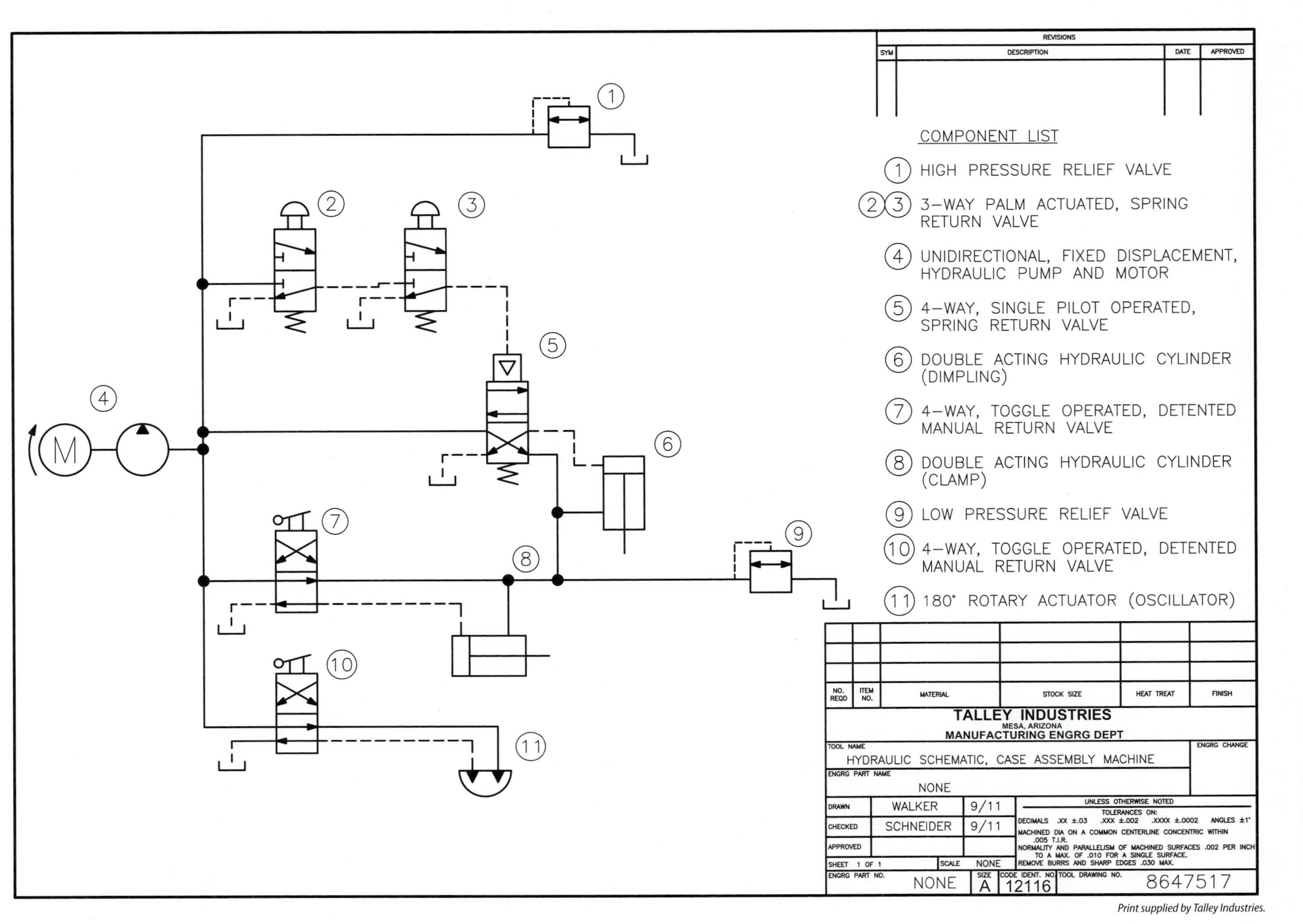

Print supplied by Talley Industries.

PR 23-3. Case Assembly Machine Hydraulic Schematic

Apply and Analyze

Name ______________________ **Date** ____________ **Class** ______________

Industry Print Exercise 23-4

Refer to print PR 23-4. The sequence of operations for this print is given below. Answer the questions following the sequence of operations.

Sequence of Operations, Part 25-086

Description: *This machine drills four 1/8″ diameter holes 90° apart through a cylindrical shell. The shell is approximately 3″ in diameter and 12″ long. After drilling, the shell is transferred vertically, where four 1/8″ diameter pins are inserted into the holes. The machine then resets itself for unloading, reloading, and the next cycle.*

1. The machine is loaded and the shell is clamped in place.
2. The automatic cycle is initiated by manually depressing the button on valve 1.
3. A pulse of line pressure opens valves 2 and 3, assuring a rest position.
4. When valve 2 shifts, transfer cylinders (4) are pressurized to lower the shell into the drilling (down) position.
5. When the shell is fully in the down position, it mechanically actuates valve 5.
6. Line pressure through valve 5 starts Timac timer 6. Line pressure also passes through valve 7 to start the four drill motors.
7. The drill motors have their own internal pneumatic controls to start and stop their rotation and to retract on reaching a predetermined depth.
8. Timer 6 times out during the drill cycle. This resets valve 7 and shifts valve 3 to allow pressure from the four valves (8) to pass when available.
9. When all drills are retracted, they mechanically shift the four valves (8), which allows pressure to pass to valve 3.
10. Pressure passes through valve 3, passes through check-speed control valve T-1, and shifts valve 2. This allows line pressure to move the transfer cylinders (4) to the up position.
11. At the full-up position of the transfer cylinders (4), valve 9 is mechanically actuated, allowing pressure to start timer 10. Pressure also passes through valve 11 and valve 18 to shift the two valves (12). This extends the four pin cylinders.
12. After timer 10 times out, pressure passes through the timer to shift valve 11. This vents the pressure on the two valve (12) controls.
13. Upon full extension of pin cylinders seating pins to a predetermined depth, microswitches (13) actuate solenoid valve 14. This allows pressure to shift the two valves (12). Shifting the valves (12) allows pressure to retract the pin cylinders.
14. Valve 15 is used as an emergency interrupt for the transfer cylinders (4). When manually actuated, the transfer cylinders retract.
15. Valves 16 and 17 are manual overrides for operating the pin cylinders. Valve 16 manually extends the pin cylinders. Valve 18 is used as a selector valve only when valve 16 is manually actuated. Valve 17 manually retracts the pin cylinders.

Bill of Materials – Part 25-086

1. Air valve, 3-port, push button, spring return
2. Air valve, 4-port, pilot operated, both ways
3. Air valve, 3-port, pilot operated, both ways
4. Air cylinder, double action, 2-1/2″ diameter × 4″ stroke
5. Air valve, 3-port, lever operated, spring return
6. Air timer, Timac, .2–20 sec. adjustable
7. Air valve, 3-port, pilot operated, spring return
8. Air valve, 3-port, lever operated, spring return
9. Air valve, 3-port, lever operated, spring return
10. Air valve, Timac, .2–20 sec. adjustable
11. Air valve, 3-port, pilot operated, spring return
12. Air valve, 4-port, pilot operated, both ways
13. Microswitch, V3-26
14. Air valve, 2-port, solenoid operated, spring return, skimmer V53DB2-150
15. Air valve, 3-port, manually operated
16. Air valve, 3-port, manually operated, spring return
17. Air valve, 3-port, manually operated
18. Air valve, 3-port, pilot operated, both ways
19. Air drills, 2800 rpm, Rockwell 1/8″ dashpot
20. Air cylinders, double action, double rod, 2-1/2″ diameter × 4″ stroke

Questions

1. What is the name of the print? ____________________
2. What is the drawing number? ____________________
3. Briefly describe the machine's function. ____________________

____

4. How is the automatic cycle started? ____________________

____

5. When the transfer cylinders (4) are in their down position, what starts the drill motors?

____

____

6. While the drilling cycle is in process, what happens to valve 7 and valve 3?

____

____

7. What activates the four valves (8)? ____________________

____

8. When the four valves (8) are shifted, what takes place? ______________________________

____

____

9. What happens when the transfer cylinders (4) are in the up position? ________________

____

____

10. What causes the pin cylinders to automatically retract? ______________________________

____

____

11. When is valve 15 used? ____

____

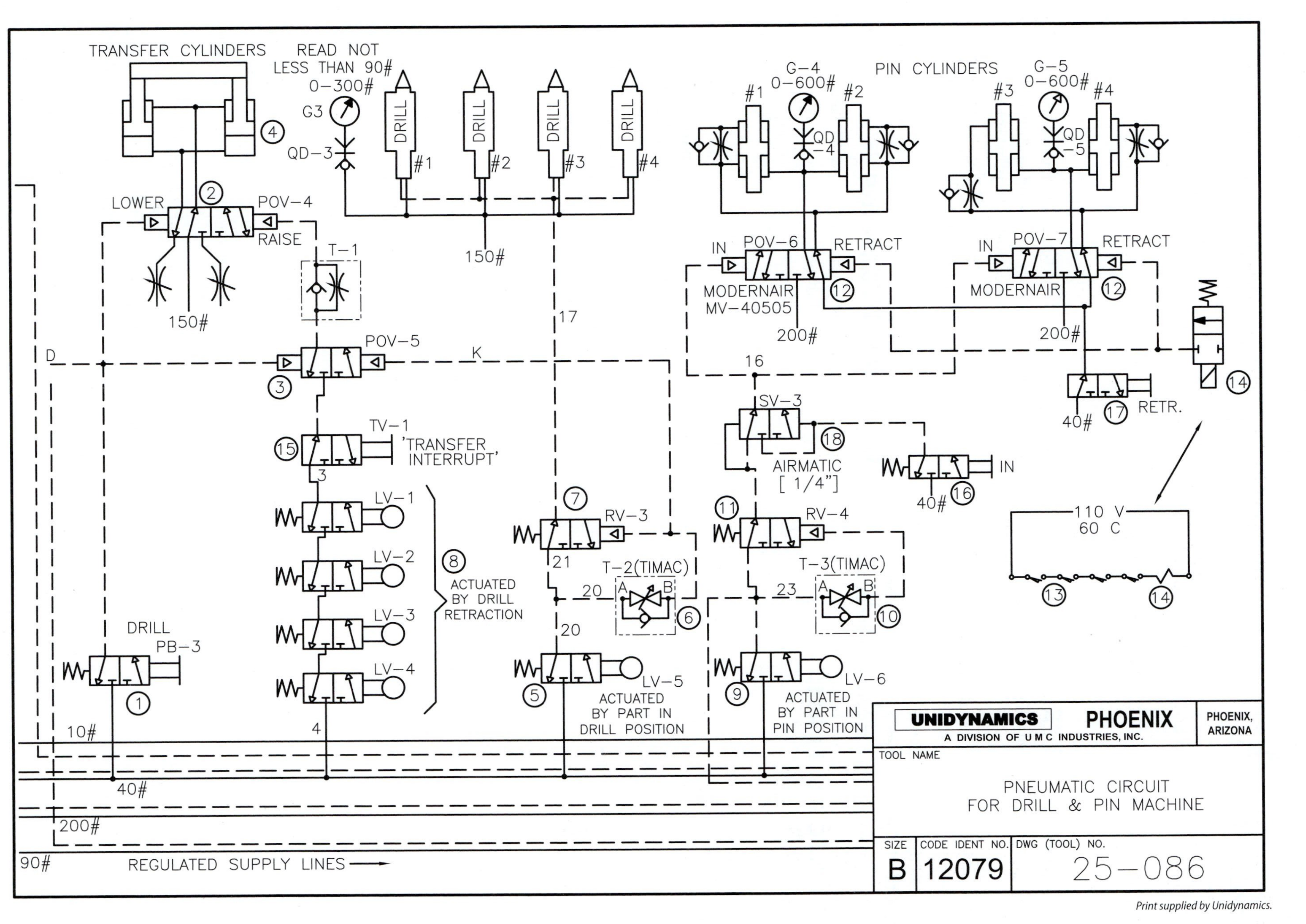

Print supplied by Unidynamics.

PR 23-4. Pneumatic Circuit for Drill and Pin Machine

APPENDIX **A**

Applied Mathematics

LEARNING OBJECTIVES

After completing this unit, you will be able to:

- Identify common fractions.
- Add, subtract, multiply, and divide common fractions.
- Identify decimal fractions.
- Add, subtract, multiply, and divide decimal fractions.
- Identify the metric system and metric numbers.
- Add, subtract, multiply, and divide metric numbers.

TECHNICAL TERMS

denominator
improper fraction
lowest common denominator (LCD)
mixed number
numerator
proper fraction

Engineers and technicians frequently need to make calculations involving common fractions and decimals in connection with print reading. If you are in need of a quick review of the fundamentals of this area of mathematics, this unit will assist you in your study.

Common Fractions

Common fractions are written with one number over the other, as follows:

$$\frac{11}{16}$$

The number on the bottom is the ***denominator***. It indicates the number of equal parts into which a unit is divided. In the above example, the unit is divided into 16 equal parts. The number on top is the ***numerator***. It indicates how many equal parts are represented by the fraction. In the above example, the fraction represents 11 of 16 equal parts. See **Figure A-1**.

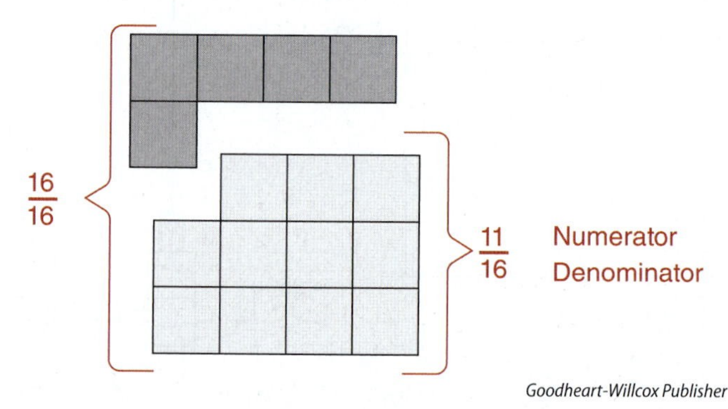

Figure A-1. A unit is divided into 16 equal parts. The denominator is 16. The numerator is the number of parts selected, in this case 11.

A ***proper fraction*** is a fraction where the numerator is less than the denominator. For example, the following are proper fractions:

$$\frac{7}{16} \text{ and } \frac{3}{4}$$

An ***improper fraction*** is a fraction where the numerator is greater than the denominator. For example, the following are improper fractions:

$$\frac{5}{4} \text{ and } \frac{19}{16}$$

A ***mixed number*** is a number consisting of a whole number and a proper fraction. For example, the following are mixed numbers:

$$2\frac{3}{4} \text{ and } 5\frac{1}{8}$$

Fundamental Steps in the Use of Fractions

There are several fundamental principles for working with fractions. These are described in the steps that follow.

1. Whole numbers can be changed to fractions by multiplying the numerator and denominator by the same number. For example, to change the whole number 6 into a fraction with 4 as the denominator, place the whole number over 1 and multiply the numerator and denominator by 4, as follows:

$$\frac{6}{1} \times \frac{4}{4} = \frac{24}{4}$$

Each whole unit contains 4 fourths. Thus, 6 units contain 6 × 4 fourths, or 24 fourths. The value of the number has not changed, because:

$$\frac{24}{4} = 6$$

2. Mixed numbers can be changed to fractions by changing the whole number to a fraction and adding the two fractions. The two fractions must have the same denominator. For example, to change 3 5/8 to a fraction, place the whole number over 1, multiply that numerator and denominator by 8, and add the resulting fraction to 5/8, as follows:

$$3\frac{5}{8} = \frac{3}{1} \times \frac{8}{8} + \frac{5}{8}$$

$$= \frac{24}{8} + \frac{5}{8}$$

$$= \frac{29}{8}$$

Each whole unit contains 8 eighths. Thus, 3 units contain 3 × 8 eighths, or 24 eighths. Adding the 5/8 part of the mixed number provides the final answer.

3. Improper fractions can be reduced to a whole or mixed number by dividing the numerator by the denominator. For example, to reduce 17/4, divide 17 by 4, as follows:

$$\frac{17}{4} = 17 \div 4 = 4\frac{1}{4}$$

The number 4 goes into 17 four times (16), with a remainder of 1. Place the 1 over 4 to get 1/4. Combining the 4 and the 1/4 provides the mixed number 4 1/4.

4. Fractions can be reduced to the lowest form by dividing the numerator and denominator by the same number. For example, to reduce 6/8 to the lowest form, divide the numerator and denominator by 2, as follows:

$$\frac{6}{8} = \frac{6 \div 2}{8 \div 2}$$

$$= \frac{3}{4}$$

Since both the numerator and denominator are divided by the same number (2), the value of the fraction is not changed.

In print reading, it can be helpful in analyzing a fractional measurement to divide by 2, even if the numerator is an odd number! For example, 15/32 can be thought of as 7 1/2 sixteenths. Sometimes it is hard to envision what 32nds of an inch are since they are so small, but converting to sixteenths may help the reader visualize the distance.

5. Fractions can be changed to a higher form by multiplying the numerator and denominator by the same number. For example, to change 5/8 to a higher form (16), multiply the numerator and denominator by the same number (2), as follows:

$$\frac{5}{8} = \frac{5 \times 2}{8 \times 2}$$

$$= \frac{10}{16}$$

The value of the fraction is not changed by multiplying both 5 and 8 by 2. This is because multiplying by 2/2 is the same as multiplying by 1 ($2 \div 2 = 1$).

Addition of Fractions

To add common fractions, the denominators must all be the same. The numerators are then simply added together. For example, to add 5/16, 3/8, and 11/32, the *lowest common denominator (LCD)* must be found. In this example, 32 is the LCD. All fractions must be changed to have 32 as the denominator. Change fractions to a higher form (32) using fundamental step number 5, as follows:

$$\frac{5 \times 2}{16 \times 2} = \frac{10}{32}$$

All fractions now have a common denominator of 32 and can be added, as follows:

$$\frac{10}{32} + \frac{12}{32} + \frac{11}{32} = \frac{33}{32}$$

The improper fraction 33/32 can be reduced using fundamental step number 3, as follows:

$$\frac{33}{32} = 33 \div 32$$

Dividing 33 by 32 results in 1 with a remainder of 1. Place the 1 over 32, to get:

$$\frac{1}{32}$$

Combining the 1 and the 1/32 results in the mixed number:

$$1\frac{1}{32}$$

Subtraction of Fractions

To subtract common fractions, the denominators must all be the same. The numerators can then simply be subtracted. For example, to subtract 5/16 from 3/4, the lowest common denominator (LCD) must be found. In this example, 16 is the LCD. All fractions must be changed to have 16 as the denominator. Change fractions to a higher form (16) using fundamental step number 5, as follows:

$$\frac{3 \times 4}{4 \times 4} = \frac{12}{16}$$

The fractions can be subtracted by subtracting the numerators, as follows:

$$\frac{12}{16} - \frac{5}{16} = \frac{7}{16}$$

Multiplication of Fractions

To multiply common fractions, first change all mixed numbers to improper fractions. Then, multiply all numerators and multiply all denominators. Reduce the resulting fraction to the lowest form. For example, to multiply 1/2, 3 1/8, and 4, the mixed number must be changed to an improper fraction. In this example, 3 1/8 must be changed to an improper fraction and then multiplied with the other fractions, as follows:

$$\frac{1}{2} \times 3\frac{1}{8} \times 4 = ?$$

$$\frac{1}{2} \times \frac{25}{8} \times \frac{4}{1} = \frac{100}{16}$$

The improper fraction 100/16 can be reduced using fundamental step number 3, as follows:

$$\frac{100}{16} = 100 \div 16$$

$$= 6\frac{4}{16}$$

Dividing 100 by 16 results in 6 with a remainder of 4. Place the 4 over 16 to get:

$$\frac{4}{16}$$

The fraction 4/16 can be further reduced using fundamental step number 4, as follows:

$$\frac{4 \div 4}{16 \div 4} = \frac{1}{4}$$

Combining the 6 and the 1/4 provides the mixed number of:

$$6\frac{1}{4}$$

The improper fraction of 100/16 reduced to the lowest form is 6 1/4.

Division of Fractions

To divide common fractions, change all mixed numbers to improper fractions and invert the divisor.

Inverting the divisor is reversing the numerator and denominator of the divisor. The divisor is the number by which you are dividing. Then, multiply all numerators and multiply all denominators. Reduce the resulting fraction to the lowest form. For example, to divide 5 1/4 by 1 1/2, the mixed numbers must be changed to improper fractions. The divisor is inverted and multiplied with the other fraction, as follows:

$$5\frac{1}{4} \div 1\frac{1}{2} = ?$$

$$\frac{21}{4} \div \frac{3}{2} =$$

$$\frac{21}{4} \times \frac{2}{3} = \frac{42}{12}$$

The improper fraction 42/12 can be reduced using fundamental step number 3, as follows:

$$\frac{42}{12} = 42 \div 12$$

$$= 3\frac{6}{12}$$

Dividing 42 by 12 results in 3 with a remainder of 6. Place the 6 over 12 to get 6/12. The fraction 6/12 can be further reduced using fundamental step number 4, as follows:

$$\frac{6 \div 6}{12 \div 6} = \frac{1}{2}$$

Combining the 3 and the 1/2 gives us the mixed number of:

$$3\frac{1}{2}$$

The improper fraction 42/12 reduced to the lowest form is 3 1/2.

Decimal Fractions

The denominator in decimal fractions is 10 or a multiple of 10. When writing decimal fractions, the denominator is omitted and a decimal point is placed in front of the numerator. For example, 1/10 is written as .1 (one-tenth) and 1/100 as .01 (one-hundredth). Other examples are:

- 3/10 is written as .3 (three-tenths).
- 87/100 is written as .87 (eighty-seven hundredths).
- 375/1000 is written as .375 (three hundred seventy-five thousandths).
- 4375/10000 is written as .4375 (four thousand three hundred seventy-five ten thousandths).

Whole numbers are written to the left of the decimal point, and fractional parts are to the right. For example:

- 1 1/10 is written as 1.1 (one and one-tenth).
- 4 35/100 is written as 4.35 (four and thirty-five hundredths).
- 5 253/1000 is written as 5.253 (five and two hundred fifty-three thousandths).

Addition and Subtraction of Decimals

Decimal fractions are added and subtracted in the same manner as whole numbers. In adding and subtracting decimals, the figures must be written so the decimal points vertically align. For example, when adding 7.3125, 1.25, .625, and 3.375, the numbers are added as follows:

$$\begin{array}{r} 7.3125 \\ 1.25 \\ .625 \\ +\ 3.375 \\ \hline 12.5625 \end{array}$$

Additional zeros can be placed to the right so each decimal fraction has the same number of digits to the right of the decimal, but this is not required. This does not change the value since 1/10 = 10/100 = 100/1000. For example:

$$\begin{array}{r} 7.3125 \\ 1.2500 \\ .6250 \\ +\ 3.3750 \\ \hline 12.5625 \end{array}$$

When subtracting decimal fractions, such as 2.25 from 8.625, the decimal points must vertically align in the same manner as addition, as shown here:

$$\begin{array}{r} 8.625 \\ -\ 2.25 \\ \hline 6.375 \end{array}$$

This can also be written as:

$$\begin{array}{r} 8.625 \\ -\ 2.250 \\ \hline 6.375 \end{array}$$

Notice the decimal point in the answer for each example is directly below the decimal points in the problem.

Multiplication of Decimals

Decimal fractions are multiplied in the same manner as whole numbers. Vertical alignment of decimal points is not necessary. To find the position of the decimal point in the answer, count the number of places to the right of the decimal point in each number being multiplied. Add them together to get the total number of places. In the answer, place the decimal point by counting off the total number of places from the right. For example, multiply 6.25 by 1.5 as follows:

```
    6.25 (two decimal places)
×    1.5 (one decimal place)
--------
    3125
    6250
   9.375 (three decimal places)
```

Division of Decimals

Decimal fractions are divided in a manner similar to whole numbers. The answer is called the quotient. However, the number of decimal places must be counted similar to multiplication of decimal fractions. To place the decimal point in the quotient (answer), count the number of places to the right of the decimal point in the divisor. Then, count this number of places to the right of the decimal point in the dividend (the number being divided). Place the decimal point directly above in the quotient. For example, to divide 36.5032 (dividend) by 4.12 (divisor), the decimal point is moved two places to the right and division is performed as follows:

```
            8.86
4.12^ | 36.50^32
        32 96
        ------
         3 543
         3 296
         ------
           2472
           2472
           ----
              0
```

Metric System

The metric system of numbers works in the same manner as the decimal-inch system. Both systems are base-ten number systems. Only the size of the units and terms differ. This makes it easy to shift from one multiple or submultiple to another.

Addition and Subtraction in the Metric System

Numbers in the metric system are added and subtracted in the same manner as they are for decimal fractions. Vertically align the decimal points, and then add or subtract as needed. For example, 38.35 millimeters, 20.666 millimeters, and 116.59 millimeters are added as follows:

```
   38.35
   20.666
+ 116.59
--------
  175.606
```

As another example, to subtract 107.902 from 118.06, align the decimal points and subtract the numbers as follows:

```
  118.06
- 107.902
---------
   10.158
```

Notice the decimal point in the answer for each example is directly below the decimal points in the problem.

Multiplication and Division in the Metric System

Numbers are multiplied and divided in the metric system in the same manner as for decimal fractions in the decimal-inch system. For example, multiply 81.6 millimeters by 3 as follows:

```
  81.6
×    3
------
 244.8
```

To divide 103.42 millimeters into four equal parts, proceed as follows:

```
     25.850
4 | 103.420
    8
    --
    23
    20
    --
     34
     32
     --
      22
      20
      --
       20
       20
       --
        0
```

Appendix Review

Name ______________________ Date ______________ Class ______________

Answer the following questions using the information provided in this appendix.

Know and Understand

____ 1. *True or False?* Fractions are written with a numerator value above a denominator value.

____ 2. What term applies to a fraction wherein the numerator is larger than the denominator?
A. Proper
B. Mixed
C. Improper
D. Common

____ 3. *True or False?* A fraction with a numerator value of 25 and a denominator value of 16 is equivalent to 1 9/16.

____ 4. *True or False?* A fraction can be reduced to the lowest form by dividing the denominator by the numerator.

____ 5. Which of the following fractions in an equation features the *lowest common denominator*?
A. 3/8
B. 1/2
C. 9/16
D. 5/32

____ 6. *True or False?* To add fractions, the denominators must all be the same.

____ 7. *True or False?* To subtract fractions, the numerators must all be the same.

____ 8. Which of the following mathematical processes first requires the fraction to be inverted?
A. Addition
B. Subtraction
C. Multiplication
D. Division

____ 9. *True or False?* When adding decimal number values, adding zeros to the end of decimal fractions will alter the final answer.

____ 10. *True or False?* Numbers in the metric system are multiplied and divided in a different manner than in the decimal-inch system.

Apply and Analyze

Name ______________________ Date ____________ Class ______________

Review Activity A-1

Adding Fractions

Solve the following problems. Reduce answers to the lowest form. Show all work in the space provided.

1. $\frac{3}{4} + \frac{1}{8} + \frac{1}{2} =$

2. $\frac{7}{8} + \frac{3}{16} =$

3. $\frac{5}{12} + \frac{3}{8} + \frac{3}{4} =$

4. $\frac{3}{10} + \frac{9}{10} + \frac{1}{4} =$

5. $\frac{7}{16} + \frac{3}{32} + \frac{1}{4} =$

6. $1\frac{3}{4} + \frac{7}{8} + 1\frac{1}{16} =$

7. $\frac{5}{32} + \frac{7}{64} + \frac{7}{8} =$

8. $1\frac{3}{8} + \frac{3}{32} + \frac{7}{16} =$

9. $3\frac{1}{16} + \frac{9}{16} + \frac{1}{2} =$

10. $5\frac{1}{5} + 2\frac{3}{10} + 8\frac{1}{2} =$

11. $4\frac{5}{8} + 20\frac{7}{32} =$

12. $\frac{3}{8} + \frac{7}{64} + \frac{9}{16} =$

13. $12\frac{7}{8} + 25\frac{3}{8} =$

14. $\frac{21}{32} + \frac{9}{64} + \frac{1}{4} =$

15. $\frac{3}{8} + 1\frac{1}{2} + \frac{7}{16} + \frac{7}{8} =$

16. $2\frac{1}{4} + \frac{5}{8} + \frac{5}{16} + \frac{17}{32} =$

Apply and Analyze

Name ____________________ Date ____________ Class ____________

Review Activity A-2

Subtracting Fractions

Solve the following problems. Reduce answers to the lowest form. Show all work in the space provided.

1. $\frac{3}{8} - \frac{1}{4} =$

2. $\frac{3}{4} - \frac{5}{16} =$

3. $1\frac{7}{8} - \frac{3}{16} =$

4. $3\frac{1}{2} - \frac{9}{16} =$

   Note: Change $3\frac{1}{2}$ to $2\frac{24}{16}$

5. $10\frac{3}{8} - 7\frac{3}{32} =$

6. $5 - 2\frac{3}{8} =$

7. $12\frac{1}{16} - 8\frac{1}{2} =$

8. $4\frac{1}{4} - 3\frac{1}{16} =$

9. $20\frac{7}{8} - 11\frac{3}{64} =$

10. $15\frac{5}{8} - 5\frac{1}{2} =$

Name ______________________ Date ______________ Class ______________

Review Activity A-3

Multiplying Fractions

Solve the following problems. Reduce answers to the lowest form. Show all work in the space provided.

1. $\frac{3}{4} \times \frac{1}{2} =$

2. $2\frac{5}{8} \times \frac{1}{4} =$

3. $\frac{7}{8} \times 5 =$

4. $6\frac{3}{4} \times \frac{1}{3} =$

5. $12\frac{1}{2} \times \frac{1}{2} =$

6. $4\frac{3}{4} \times \frac{1}{2} \times \frac{1}{8} =$

7. $16 \times \frac{3}{4} =$

8. $9\frac{5}{8} \times \frac{1}{2} =$

9. $10 \times \frac{4}{5} =$

10. $\frac{14}{3} \times 6 =$

Name ______________________ Date ______________ Class ______________

Review Activity A-4

Dividing Fractions

Solve the following problems. Reduce answers to the lowest form. Show all work in the space provided.

1. $2\frac{3}{4} \div 6 =$

2. $12 \div \frac{3}{4} =$

3. $16\frac{1}{8} \div 2 =$

4. $8\frac{2}{3} \div \frac{1}{3} =$

5. $16\frac{1}{4} \div 20 =$

6. $\frac{7}{8} \div \frac{7}{16} =$

7. $15 \div 1\frac{1}{4} =$

8. $21\frac{3}{8} \div 3\frac{1}{8} =$

9. $5\frac{1}{4} \div \frac{3}{8} =$

10. $3\frac{5}{8} \div 2 =$

Name ____________________ Date ____________ Class ____________

Review Activity A-5

Adding and Subtracting Decimals

Solve the following problems. Show all work in the space provided.

1.
```
    4.5625
     .875
    2.75
  + 5.8137
  --------
```

2. 7.0625 + .125 + 8.0 =

3. .832 + .4375 + .27 =

4.
```
   27.9375
 − 16.937
 ---------
```

5. 4.0 – .0625 =

6. 2.25 – 1.125 =

Name ______________________ Date ______________ Class ________________

Review Activity A-6

Multiplying Decimals

Solve the following problems. Show all work in the space provided.

1. $4.825 \times 1.75 =$

2. $167 \times .25 =$

3. $65.96 \times .37 =$

4. $4.95 \times 1.35 =$

5. $93.18 \times .07 =$

Name ______________________ Date ____________ Class ____________

Review Activity A-7

Dividing Decimals

Solve the following problems. Show all work in the space provided.

1. 9.45 ÷ 2.7 =

2. 654.5 ÷ 35 =

3. 1386.0 ÷ 1.65 =

4. 331.266 ÷ 80.6 =

5. 4401.25 ÷ 503 =

Name ______________________ Date ______________ Class ______________

Review Activity A-8

Adding and Subtracting Metric Numbers

Solve the following problems. Show all work in the space provided.

1.
```
   66.67
    1.42
    3.76
+   1.24
--------
```

2. 41.88 + 89.112 + 8.38 =

3. 4.19 + 49.25 + 2.6 =

4.
```
   66.68
−  41.88
--------
```

5. 26.97 – 7.1 =

6. 102.85 – 16.302 =

Name ______________________ Date ______________ Class ______________

Review Activity A-9

Multiplying and Dividing Metric Numbers

Solve the following problems. Show all work in the space provided.

1. Find the total length of six sections, each 72.5 mm long.

2. What is the total thickness of five spacers, each 1.22 mm thick?

3. Find the length, in millimeters, of each part when a 108 mm long rod is divided into seven equal parts. Round off to the nearest thousandth.

APPENDIX **B**

Measurement Tools

LEARNING OBJECTIVES

After completing this unit, you will be able to:

- Identify the importance of the steel rule.
- Read a fractional steel rule, decimal steel rule, and metric steel rule.
- Identify the common scale devices available to the drafter and print reader.
- Read and use the fractional scale found on the architect's scale.
- Read and use common reduction scales found on the architect's scale.
- Read and use the decimal scale found on a civil or mechanical engineer's scale.
- Read and use the scales found on a civil engineer's scale for reduction scales.
- Read and use a fractional scale designed for reduction scales.
- Discuss the common terms used to identify units within the metric system.
- Read an inch micrometer or a metric micrometer.
- Identify common dial or digital calipers as used in industrial applications.

TECHNICAL TERMS

architect's scale
caliper
civil engineer's scale
index line
mechanical engineer's scale
metric scale
micrometer
scale
steel rule

The ability to make accurate measurements is a fundamental skill needed by all who read and use prints. This unit reviews the basic principles of reading many scales, including the steel rule, architect's scale, civil engineer's scale, and micrometer. These measuring instruments or devices are commonly used by drafters, machinists, and others who work with prints on a regular basis.

Principle of Scale Measurements

A scaled drawing is created by specifying that a unit on the drawing, such as 1/2″, is equal to a unit on the actual object, such as 1″. In this case, rather than stating 1/2″ = 1″, this particular scale can appear on the print as SCALE = 1:2, SCALE: 1 = 2, or simply HALF SCALE. Expressions should always be read as "paper" equals "real." In other words, if the scale is expressed as 1 = 4, then one unit on the paper equals four units in real life. If the scale is 1/4″ = 1′-0″, then 1/4″ on the paper equals one foot in real life.

Scale is a term with multiple meanings. As a noun, a scale is a measuring instrument or device. As a verb, scale means the process of proportionately reducing or enlarging an object. For example, an object is "scaled down" to decrease its size. The term "scale" is used synonymously for the word "size." For example, a plastic model car may be a "1/25 scale" model, which also means it is "1/25 size."

Reading a Steel Rule

The *steel rule* is a common tool for a machinist. It is very compact, fits nicely in a pocket, and provides very accurate measurements. To work with this print reading text, your instructor may want you to have and be able to read a 6″ steel rule. With this measuring tool, you would be able to read fractionally in 64ths of an inch, in a decimal format in 100ths (decimal) of an inch, or in the metric units of millimeters. In **Figure B-1**, an enlarged portion of a rule is shown, with one edge divided into 64 parts to the inch and the other edge with 100 parts to the inch.

Steel rules can be purchased in more than one format. Some have a decimal edge in addition to the fractional edge, while others have a metric edge. If your particular industry has a lot of metric and inch conversion, you may wish to have a scale with fractional inches along one edge and millimeters along the other edge. Some machine tools and equipment are specified in fractional measurements (1/2, 1/4, 1/8, 1/16, etc.). Some standard stock sizes or screw threads are still specified with fractions. However, engineering designs and tolerances (variations in allowed size) are usually provided in decimal inches. Therefore, a decimal-equivalent chart is also a handy tool to have when reading a print.

Fractional Rule

The number 64 printed on the end of the rule means each inch is divided into 64 equal parts, **Figure B-2**. Therefore, each small division on the rule is 1/64″. To read a fractional rule, use the edge divided into 64ths and follow these principles to help you determine a measurement:

- Reduce the fractional reading to its smallest denominator. For example, if you are reading 26, then convert 26/64 to 13/32. If you are reading 48, then convert 48/64 to 24/32, then to 12/16, then to 6/8, and finally to 3/4. Appendix A, *Applied Mathematics*, is available if you need to review these principles.
- The major divisions of each inch are numbered in increments of eight (8, 16, 24, etc.). As there are eight of these major divisions, each one is equal to 1/8 of an inch. Think of these major divisions as 1/8 (8/64), 2/8 (16/64), 3/8 (24/64), and so on. Of course, 2/8 can be reduced to 1/4.
- Each of these numbered divisions is also divided into eight smaller divisions. The middle division mark is longer than the others within that division and is 1/16 (4/64) of an inch away from the labeled marks. For example, the 32 mark is 1/2″ (32/64). Therefore, halfway toward the 40 mark is 9/16″. Halfway toward the 24 mark is 7/16″.

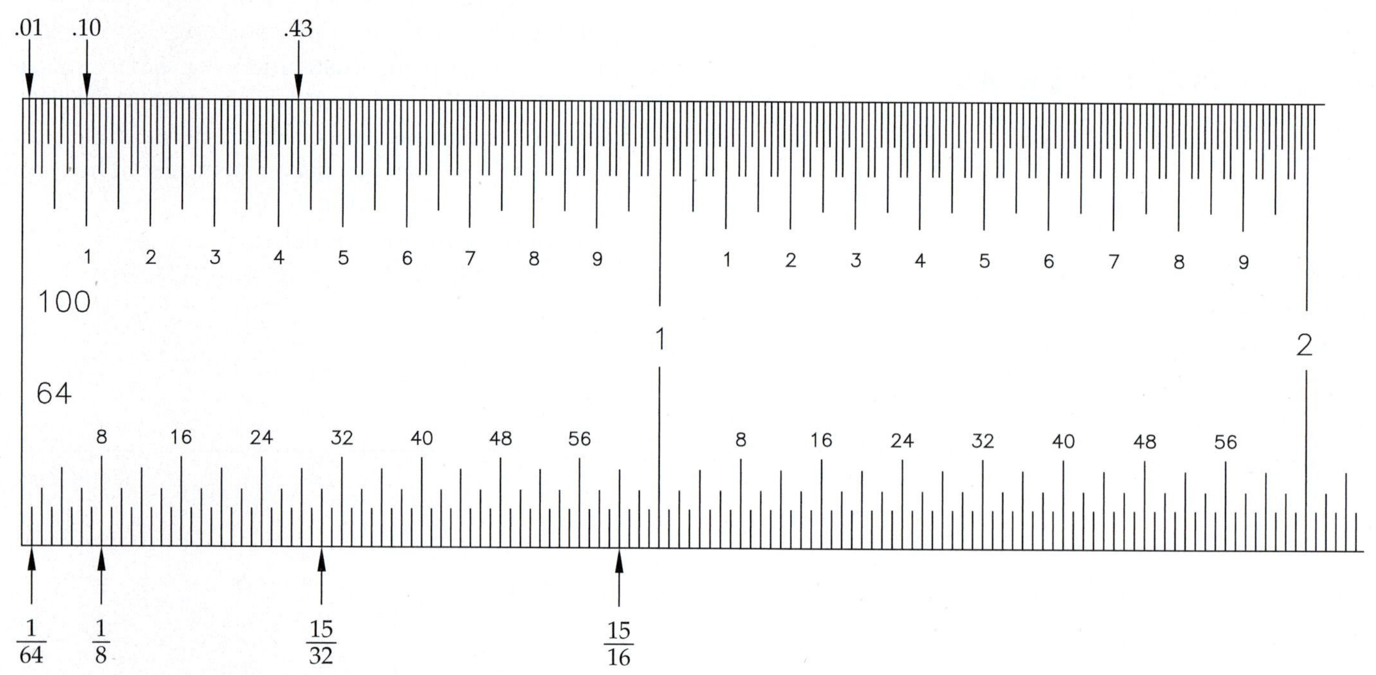

Figure B-1. A typical steel rule has one edge divided into 64 parts to the inch and the other edge with 100 parts to the inch. On some steel rules, one edge is metric.

- Have a decimal equivalent chart or a calculator handy. Some fractional readings make more sense as a decimal. For example, 39/64 may be more meaningful when converted to .609″.

Decimal Rule

Many decimal rules have 100 printed on the end of the rule, **Figure B-3**. This indicates each inch is divided into 100 parts. Therefore, each small division is 1/100″ (.01″). If the scale has a 50 on the end, there are 50 increments to the inch on that scale. To read the decimal rule divided into 100ths, follow these principles to help you determine a measurement:

- Each inch has 100 major divisions. Each major division is divided into groups of 10. Therefore, subdivisions represent 1/10 (10/100), 2/10 (20/100), 3/10 (30/100), and so on.
- Each 1/10 inch division contains 10/100 of an inch. Therefore, these major divisions can be read as .20, .30, .40, and so on. Between each major division is a "half" mark that is longer than the others, representing .25, .35, .45, and so on.
- Each division mark represents .01″. For example, a measurement of .43″ is three small divisions beyond the 4 mark. The smallest divisions have marks of varying lengths to help prevent the lines from all blending together.

Metric Rule

Metric rules are used when reading prints dimensioned in centimeters or millimeters. Machined parts are usually smaller than one meter. One meter is 100 centimeters or 1000 millimeters. Metric dimensions for industrial parts are usually given in millimeters.

Conversion tables for millimeters and inches are shown in Appendix D. A metric rule is shown in **Figure B-4**. A metric rule is read in the same manner as the decimal rule. To read a metric rule, use the edge labeled "mm" and follow these principles to help you determine a measurement:

- Each number represents a centimeter, which is equivalent to 10 millimeters.
- Each small division is 1/10 of a centimeter, or one millimeter. The measurement is expressed as a whole number, such as 1 mm, 9 mm, 17 mm, and so on.
- Between each centimeter division is a "half" mark that is longer than the others. These marks represent 5 mm, 15 mm, 25 mm, etc.

Common Scales Used by Drafters

Traditional instrument drawings have been created by drafters throughout the years, often executed

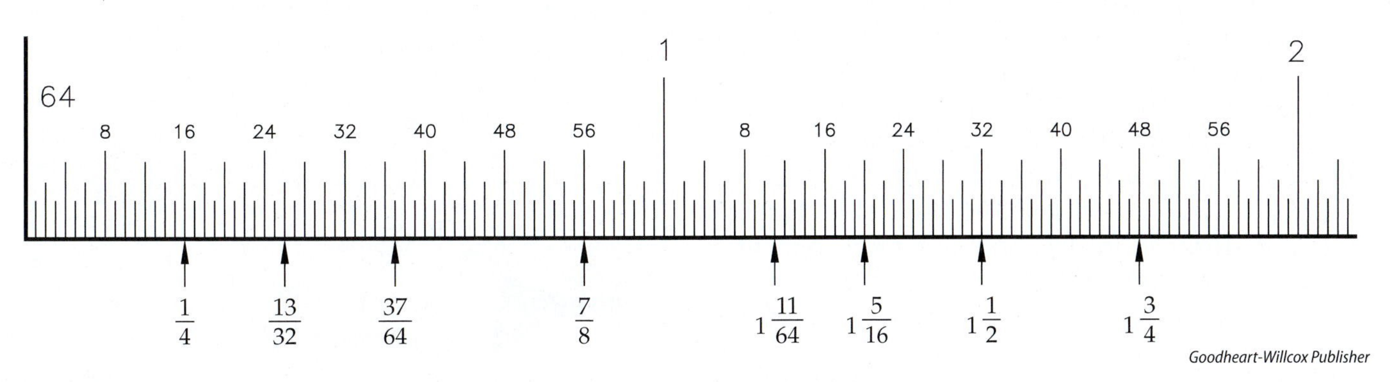

Figure B-2. One edge of a steel rule may be divided into 64 parts to the inch.

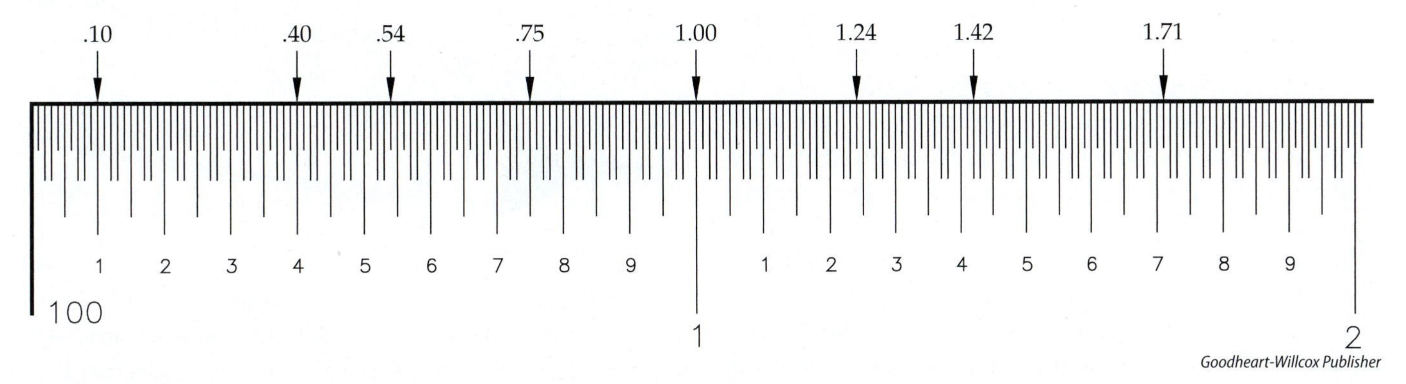

Figure B-3. One edge of a steel rule may be divided into 100 parts to the inch.

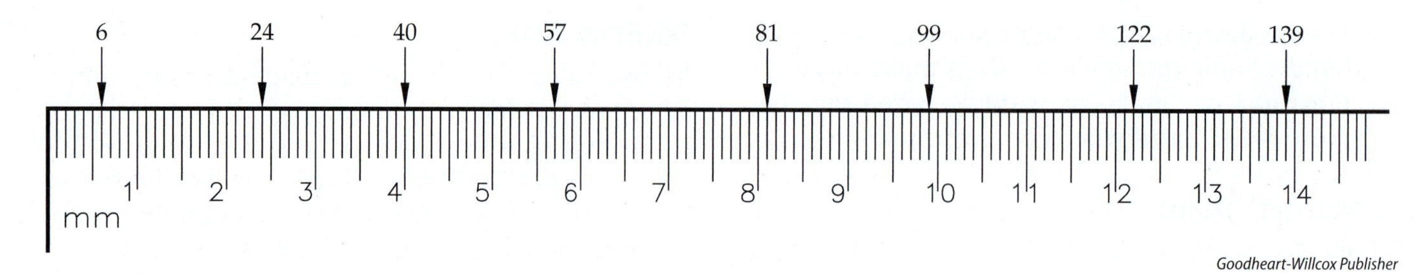

Goodheart-Willcox Publisher

Figure B-4. One edge of a steel rule may be divided into millimeters.

with a variety of scales, depending on the scale of the final drawing. A drafter's ruler is called a "scale" because it has several edges designed for scaling a large object down to paper size. The most common measuring tool of the drafter's collection of instruments has always been the triangular ***architect's scale***, **Figure B-5.** For objects drawn actual size (scale 1 = 1), the architect's scale has an edge of the scale that is fully divided. This "full" edge of the architect's scale is typically 12 inches, with each inch divided into 16ths. Of course, any common ruler would be fine for the drafter to use, but the architect's scale is usually manufactured more precisely than some rulers. While most architect's scales are triangular, there are some flat versions available.

While one edge of the triangular architect's scale has a fully divided inch scale for full-size measurements, the other five edges each contain two scales that share tick marks. Each scale on a given edge originates from a different end of the scale. This provides the drafter with 10 scales for reducing houses and other structures to paper size. Typically, these are expressed as 3/32″ = 1′-0″, 1/8″ = 1′-0″, 3/16″ = 1′-0″, 1/4″ = 1′-0″, 3/8″ = 1′-0″, 1/2″ = 1′-0″, 3/4″ = 1′-0″, 1″ = 1′-0″, 1-1/2″ = 1′-0″, and 3″ = 1′-0″. Architectural scales are designed for feet and inch measurements, such as room sizes of 14′-2″ or 12′-3 1/2″.

With the proper math understanding, some of the edges of the architect's scale can also be used for mechanical design applications. For example, since 3″ = 1′-0″ is a one-to-four ratio, that edge can be used for creating or measuring 1/4 size (1:4 scale) drawings.

Another common scale used by instrumental drafters is the ***civil engineer's scale***, often simply referred to as an *engineer's scale*. See **Figure B-6.** Civil engineers and drafters working on roadway and property drawings often use the civil engineer's scale because it is decimal, with each inch divided into 10ths. Like the architect's scale, the engineer's scale is most commonly manufactured as a triangular device. The full-size edge of an engineer's scale is marked off in divisions of 10 parts to the inch. This scale can be used to lay out drawings based on decimal dimensions.

Goodheart-Willcox Publisher

Figure B-5. An architect's scale is often triangular to allow up to 12 different scales on one instrument.

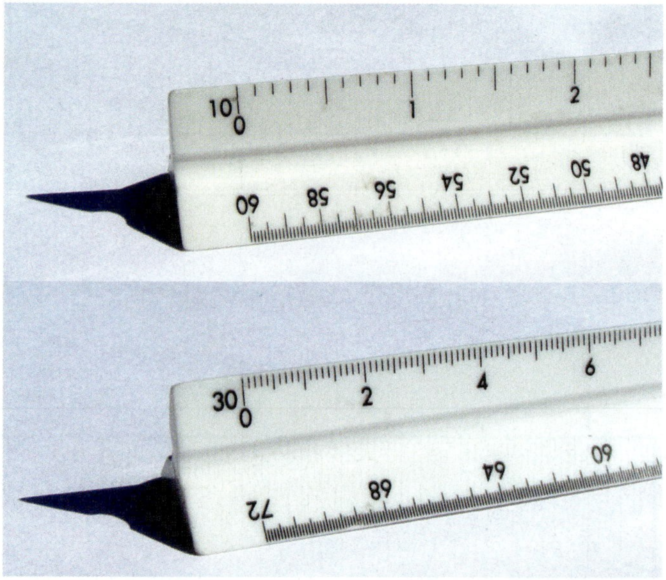

Goodheart-Willcox Publisher

Figure B-6. Civil engineering scales are often triangular, which allows for six different scales on one instrument. Each edge can also have a multiple of scales based on a factor of 10 (1″ = 2′, 1″ = 20′, 1″ = 200′, etc.).

However, the marks can also be used for reduction scales, such as 1″ = 10′, 1″ = 100′, or 1″ = 1000′. In addition, the scale can be used for enlarging drawings at scales such as 1″ = .1″. The other five edges of the triangular civil engineer's scale are divided into 20, 30, 40, 50, and 60 parts and are marked as such. These edges are appropriate for reduction ratios such as 1″ = 20′, 1″ = 30′, 1″ = 40′, 1″ = 50′, and 1″ = 60′. Of course, with the decimal system, the same five edges can be used for 1″ = 200′, 1″ = 300′, and so on. With an understanding of math and decimal places, you can also use the 50 edge of the scale for 1″ = .5″ and, thus, have a DOUBLE SCALE. The 20 edge can be used for 1″ = 2″ and, thus, a HALF SCALE. The review for this appendix provides practice for some of these options.

For many mechanical applications, such as machine part drawings, the scales most frequently applied are quarter size (1/4″ = 1″), half size (1/2″ = 1″), and full size (1″ = 1″), which can also be stated as 1:4, 1:2, and 1:1. While there are scales (measuring instruments) available for these scales, they are not as common. If measuring instruments are produced in this fashion, they are often referred to as ***mechanical engineer's scales***. See **Figure B-7**. Also, some smaller parts of an assembly may need to be enlarged on the drawing for clarity in shape and size. Typical scales for enlargement are twice size (2″ = 1″), triple size (3″ = 1″), and 5X (5″ = 1″). The 5X is read as "five times." For general measurements, a civil engineering scale could be used to measure these drawings using the appropriate edge, although the civil engineering scale is decimal, not fractional.

Metric scales are also triangular in shape. These metric scales are configured in such a way as to provide for reduction or enlargement along each edge. See **Figure B-8**. In metric applications, just as in civil engineering, the divisions are decimal, so 1:10, 1:100, and 1:1000 can all use the same edge.

Combination scales are also available in a variety of configurations. On these scales, common architectural, civil, mechanical, and metric scales may be combined in a variety of ways.

While CAD drafters create drawings using real-world size measurements, regardless of the paper size, the views are printed at a common and standard scale. The final product is the same as drawings created by pencil drafters. Measuring instruments can be used to extract information from the print, if allowed. Remember, however, that measuring a print is not typically allowed.

Reading Common Scales

With an understanding of how scales are used, you must now learn how to read various scales. This section explains how to read the different scales you may use to read prints.

Reading a Fractional-Inch Scale

Study **Figure B-9**. This figure shows the full-size fractional scale on a typical architect's scale. Each inch is divided into 16 parts. The large tick marks indicate whole inches. The next smaller tick marks indicate 1/2 inch increments. The next smaller tick marks are 1/4 inch indicators. The next smaller tick marks are 1/8 inch marks, and the smallest tick marks are 1/16 inch marks.

Reading a Decimal-Inch Scale

Study **Figure B-10**. This figure shows the full-size decimal scale on a typical civil engineer's scale. The tick marks between the whole inches are all the same length and represent 10ths of an inch. To read a decimal dimension of 2.125″ on the civil engineer's scale, start at 0 and move to the right past 2. Continue past the first 10th (.10) to one-fourth (.025) of the next 10th (judged by eye). This represents the decimal .125.

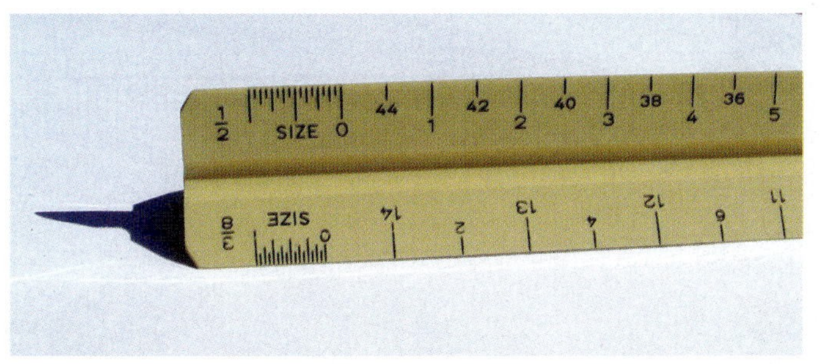

Goodheart-Willcox Publisher

Figure B-7. This triangular scale contains a half-size scale typically found on a mechanical engineer's scale.

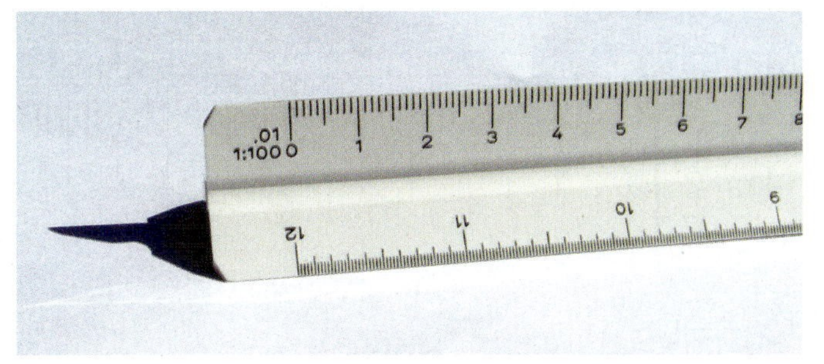

Goodheart-Willcox Publisher

Figure B-8. This metric scale is commonly used if large metric measurements are to be reduced at various scales.

A full-size mechanical engineer's scale typically has 50 parts to the inch. This is also designed for reading in a decimal format. See **Figure B-11**. Notice how 2.125 inches is measured along this scale compared to the previous example.

Reading an Architect's Reduction Scale

The 10 architectural scales are designed for measuring feet and inches. Since industrial prints and topics covered in this textbook are not in feet and inch units, these scales will not be discussed in much detail.

Study **Figure B-12**. This figure shows the edge for 1/8″ = 1′-0″ in combination with the 1/4″ = 1′-0″ on a typical architect's scale. Notice that on the "outside" side of the zero (0) is a 12″ ruler shrunk down to 1/8″, and on the other end of that edge on the outer side of the zero (0) is a 12″ ruler shrunk down to 1/4″. This principle of placing a 12″ ruler at each end of each edge of the architect's scale is consistent on all the edges except for the full-size edge.

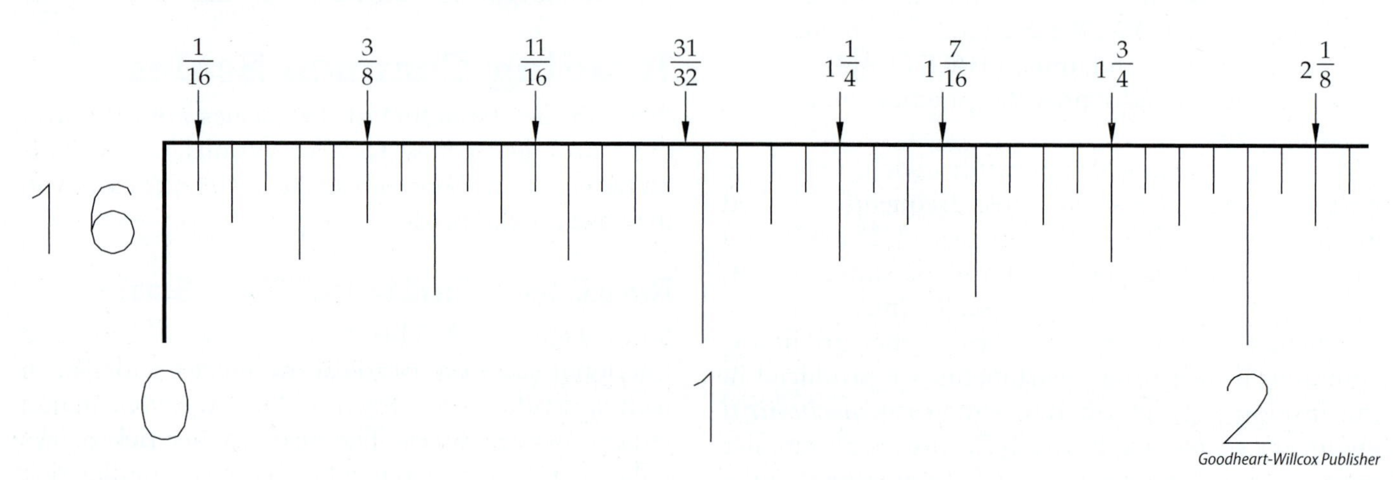

Figure B-9. This full-size fractional scale on a typical architect's scale is divided into 16 parts per inch.

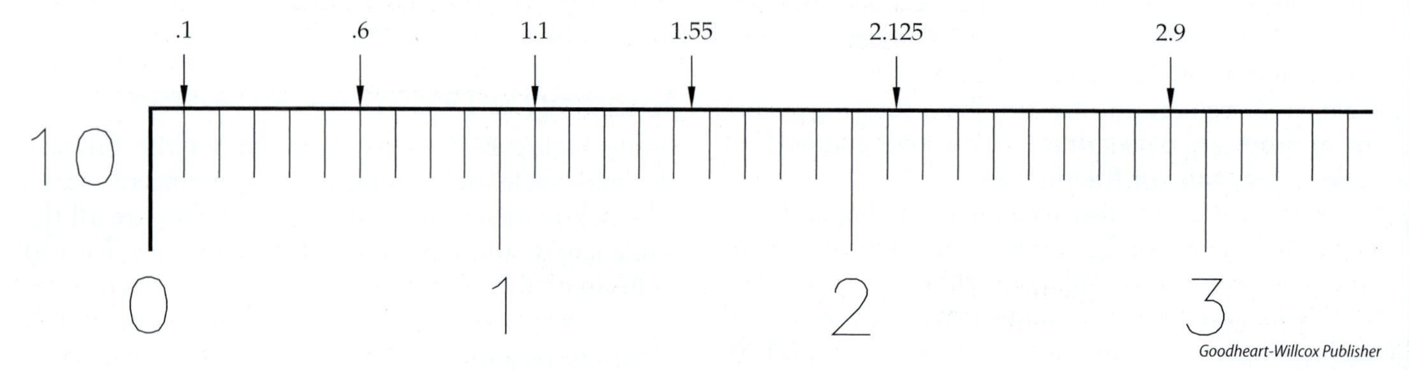

Figure B-10. The full-size edge of a civil engineering scale is marked off in divisions of 10 parts to the inch.

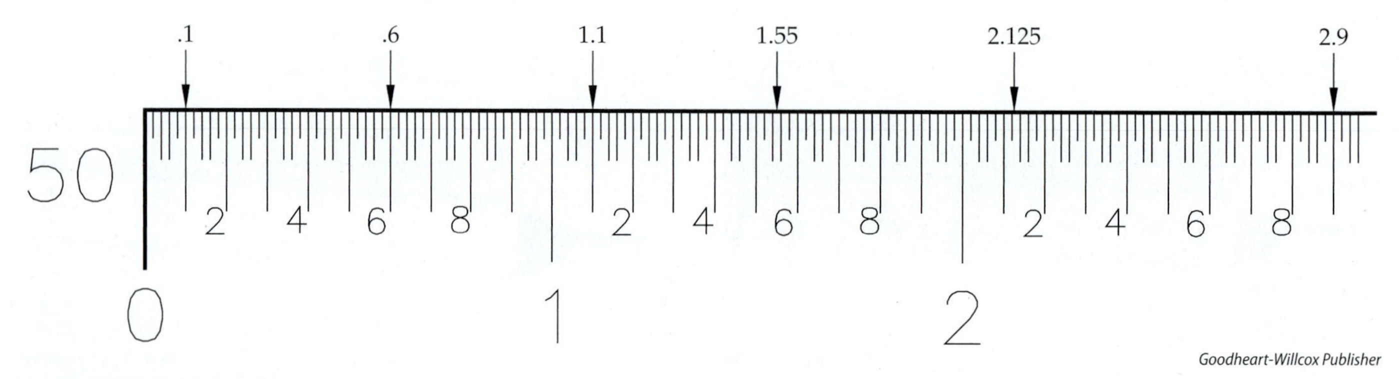

Figure B-11. The full-size decimal scale on a mechanical engineer's scale is divided into 50 parts to the inch. Notice how 2.125″ is measured along this 50 scale compared to the example in the previous illustration.

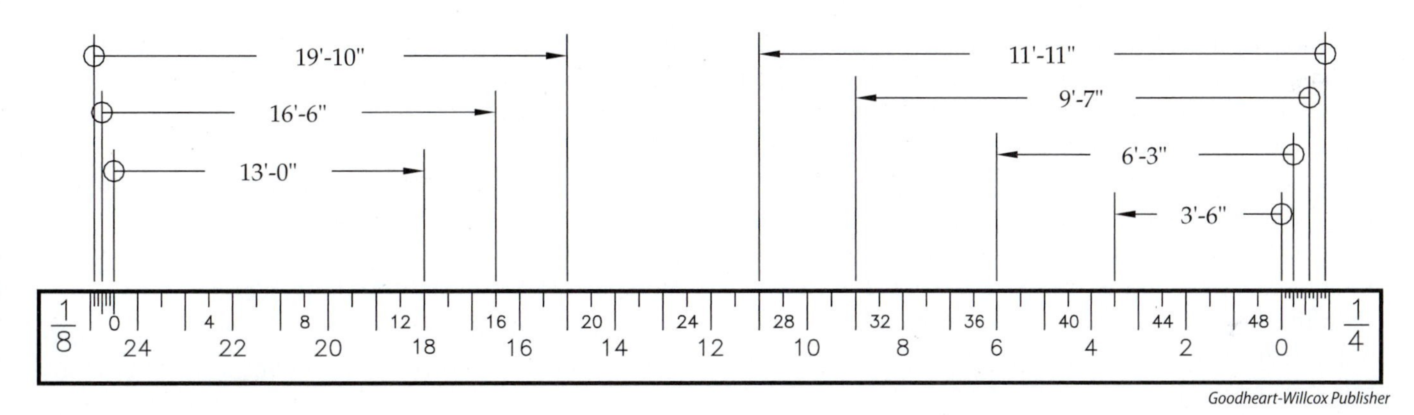

Figure B-12. The architect's scale has five edges, each with two scales that share foot marks. There is also a small 12″ ruler at the end of each edge for that reduction scale.

On the inner side of the zero (0) are tick marks for measuring feet. Since the scale at one end is either one-half or double the scale on the other end, these foot marks can be shared. Some of the numerical labels are for feet increments from the "far" end, while others are for the "near" end. The small 12″ rulers are divided into fractional units, depending on how much room there is for the tick marks to be readable. For example, on the 1/8″ = 1′-0″ end of the scale, there is only room for six tick marks, one for every two inches. On the 1/4″ = 1′-0″ end of the scale, there is room for 12 tick marks, one for each inch, so distances can be measured more precisely.

Reading a Mechanical Engineer's Reduction Scale

Study **Figure B-13.** This figure shows the half-size scale on a mechanical engineer's scale. On instruments that have scales shrunk down for reduced-size drawings, the tick marks become much closer together. Sometimes in these cases, the more-precise tick marks are only placed on one end of the scale. To read a decimal dimension of 2.125″ on the mechanical engineer's scale in **Figure B-13**, locate the 2 on the right side of the zero. Then, move the instrument an additional 1/8″ (.125″) on the scale to obtain the total measurement. The actual distance on the drawing is 1.0625″ (2.125″ at 1/2 scale).

Reading a Civil Engineer's Reduction Scale

Study **Figure B-14.** This figure shows the 20 scale on a civil engineer's scale. The 2 is located one actual inch from zero. Therefore, this edge can be used for any

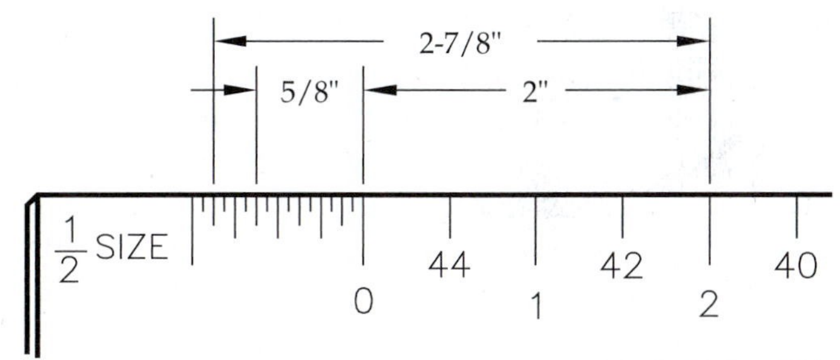

Figure B-13. Similar to the architect's scale, this 1:2 scale is not fully divided. The whole inches are measured on one side of the zero, and the fraction is measured on the other side of the zero.

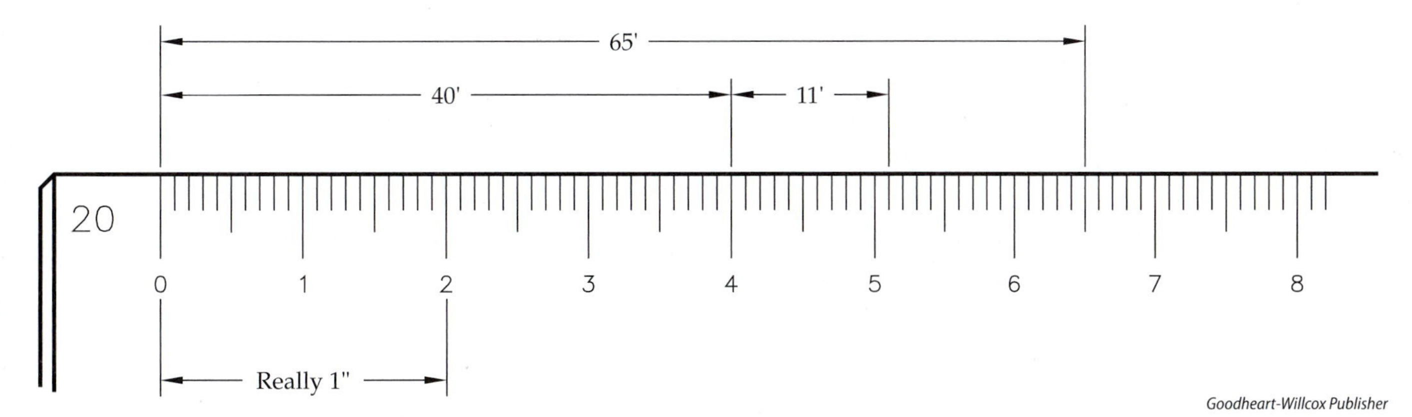

Figure B-14. In this example, the civil engineer's scale is shown being used for measuring at a scale of 1″ = 20′.

reduction of 1 = 2 units, or 20 units, or 200 units. It can be used for 1″ = 2″ or 1″ = 2 cubits. Other scales include 10, 30, 40, 50, and 60.

To read a 1/2 scale (half scale) decimal dimension of 3.75″ on the civil engineer's scale, locate the 3.7 and 3.8 tick marks. Then, estimate a point halfway between those tick marks. Remember, the actual distance on the drawing is 1.875 (3.75 at 1/2 scale).

While the scope of this text does not include civil engineering applications, learning to read this scale can be very useful in your professional career, especially if you are reading and measuring large prints of facility layouts or site drawings. For example, assume the 2, which is really 1″ from 0, represents 20′. Use the tick marks to read measurements just as you would on a full-size scale. The same 2 that represents 20′ in the example above could have represented 2′ or 200′. The measurements you make simply assume the tick marks are for feet, based on the value given to the numeral 2.

Metric Scales

In the workplace, you may encounter both the metric and decimal-inch US customary systems of measurement. Many industries use a system of dual dimensioning where both decimal-inch and metric dimensions are shown on a print. Common International System (SI) metric units are shown in **Figure B-15**. The metric system is somewhat similar to the decimal-inch system. Only the size of the units and terms vary. Both are base-10 number systems, which makes it easy to shift from one multiple or submultiple to another.

The most important aspect to understand in this section is how to measure millimeters. This is the most common unit for mechanical and industrial parts discussed throughout this text. If you are using a triangular metric reduction scale, the 1:100 edge of the scale can be used as a full-size scale when dimensions are in millimeters, since the smallest divisions on the scale are spaced one millimeter apart.

Anyone using the metric system should learn all common units and how to convert between units. Unit conversion is actually a simple shift of the decimal point. See **Figure B-16**.

Metric Units on Industrial Prints

The base unit of the metric system is the meter. Other units derived from the meter are used in various industrial fields as the official unit of measure on metric drawings. See **Figure B-17**. For example, on precision drawings in manufacturing, the millimeter is the standard unit of length. **Figure B-18** shows some typical metric scale designations for drawings.

The ***micrometer*** (mike) is a precision-measurement instrument. It consists of a spindle that rotates in a fixed nut, thus opening or closing the distance between an anvil and spindle. See **Figure B-19**. The spindle screw on an inch micrometer has 40 threads per inch. The spindle is attached to a thimble. As the thimble is turned one revolution, the spindle moves

Quality	Name of Unit	Symbol
Length	meter	m
Mass (weight)	kilogram	kg
Time	second	s
Electric current	ampere	A
Temperature	kelvin	K
Luminous intensity	candela	cd
Amount of substance	mole	mol
Supplementary Units		
Unit	**Name of Unit**	**Symbol**
Plane angle	radian	rad
Solid angle	steradian	sr

Goodheart-Willcox Publisher

Figure B-15. This chart shows common SI metric base units.

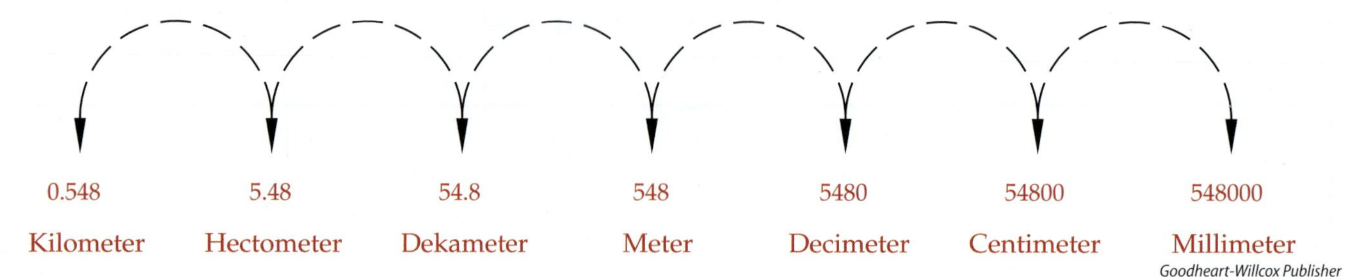

Goodheart-Willcox Publisher

Figure B-16. Converting between metric units involves an understanding of the different prefix terms used in the metric system, but otherwise is as simple as moving the decimal place.

Industry	Unit of Measure	International Symbol	Multiple Factor
Topographical	kilometer	km	10^3 = 1 000m
Building, Construction	meter	m	10^0 = 1 m
Lumber, Cabinet	centimeter	cm	10^2 = 0.01 m
Mechanical Design, Manufacturing	millimeter	mm	10^3 = 0.001 m

Goodheart-Willcox Publisher

Figure B-17. Various industries use different base units of measure.

Reductions		Enlargements
1:2	1:50	2:1
1:2:5	1:100	5:4
1:5	1:200	10:1
1:10	1:500	20:1
1:20	1:1000	50:1

Goodheart-Willcox Publisher

Figure B-18. This chart shows some typical metric scale designations for drawings.

precisely 1/40″ (.025″). The screw on the metric micrometer advances 1/2 mm per revolution of the thimble. Therefore, two revolutions move the spindle 1 mm.

Reading an Inch Micrometer

Notice in **Figure B-19** the line extending the length of the sleeve. This line is called the ***index line*** and is divided into 40 equal parts by vertical lines. Therefore, each vertical line on the sleeve designates 1/40″ (.025″). Two of these lines or spaces are equal to 1/20″ (.025″ + .025″ = .050″). Every fourth line is marked with a number (1, 2, 3, etc.) and designates hundred thousandths (4 × .025″ = .100″). The line marked 1 represents .100″, the line marked 2 represents .200″, and so on.

The beveled edge of the thimble is divided around its circumference into 25 equal parts. Each line is consecutively numbered. One revolution of the thimble moves the spindle .025″. Moving from one line

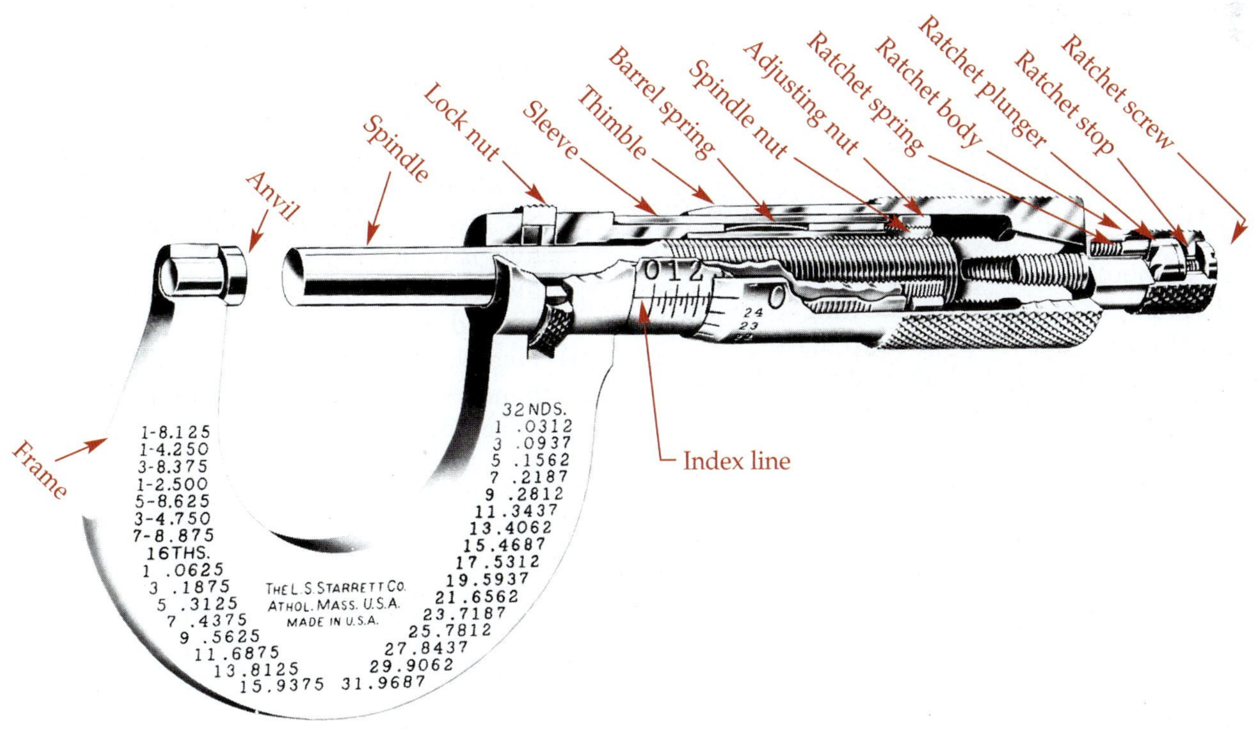

Goodheart-Willcox Publisher

Figure B-19. A micrometer is used to make precise measurements. This is an inch micrometer.

on the beveled edge of the thimble to the next moves the thimble 1/25 of a revolution. Therefore, the spindle is moved 1/25 of .025″, or .001″. Rotating two divisions or lines on the thimble moves the spindle .002″, and so on. Twenty-five divisions indicate a complete revolution of the thimble, or .025″, which is one division on the index line.

To read the micrometer in thousandths, the reading is taken at the edge of the thimble on the index line, as follows:

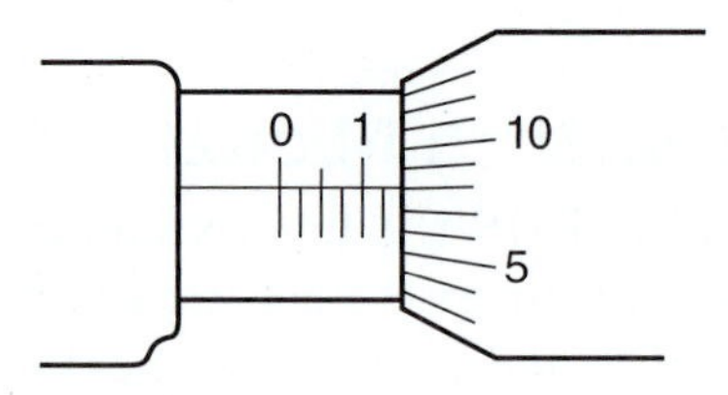

1. Line 1 on the sleeve is visible, representing .100″.
2. Two additional lines are visible, representing .025″ each.
3. Line 8 on the thimble is aligned with the index line, representing .008″.
4. The micrometer reading is .158″, calculated as follows:

$$\begin{array}{rl} .100 & \\ .050 & (2 \times .025'' = .050'') \\ +\ .008 & \\ \hline .158 & \end{array}$$

Try another reading:

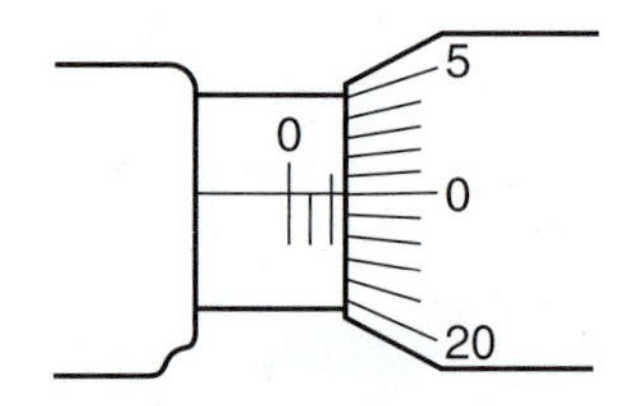

1. No numbered marks are visible.
2. Three lines are visible, representing .025″ each (the thimble edge is on the third mark).
3. Line 0 on the thimble is aligned with the index line, so there is nothing to add.
4. The micrometer reading is .075″, calculated as follows:

$$\begin{array}{rl} .000 & \text{(No numbered marks)} \\ .075 & (3 \times .025'' = .075) \\ +\ .000 & \text{(thimble reads 0)} \\ \hline .075 & \end{array}$$

Tricks of the Trade

An easy way to remember the values of the various micrometer divisions is to think of them as money. Count the figures on the sleeve (1, 2, etc.) as dollars. The extra vertical lines on the sleeve are then quarters. Finally, the divisions on the thimble are pennies. Add up your money and move the decimal point in place of the dollar sign. This is the micrometer reading. In the first example above, line 1 is visible on the sleeve. This is one dollar. Two additional lines are visible. This is two quarters, or 50 cents. Finally, line 8 on the thimble is aligned with the index line. This is eight pennies. Add this currency to get $1.58. Move the decimal point in place of the dollar sign to get a micrometer reading of .158″.

Reading a Metric Micrometer

A metric micrometer measures in hundredths of a millimeter (0.01 mm). Reading this micrometer is quite similar to reading an inch micrometer. The basic difference is the units on the scale are in millimeters.

The sleeve of the micrometer is graduated in millimeters below the index line and in half millimeters above the line. See **Figure B-20**. The thimble is marked in 50 divisions around its circumference. Each small division represents 0.01 mm. Turning the thimble 10 tick marks equals 0.1 mm. Therefore, one complete turn equals 0.5 mm. If you turn the thimble a second full turn, you have added another 0.5 mm. To read the metric micrometer, use the following steps:

1. Note the number of whole millimeter divisions (below index line) on the sleeve. These count as whole millimeters.
2. See if a half millimeter division (above index line) is visible between the whole millimeter division and the thimble. If the thimble is beyond the half mark, add 0.50 mm.
3. Finally, read the thimble for additional hundredths (division on thimble aligning with index line).

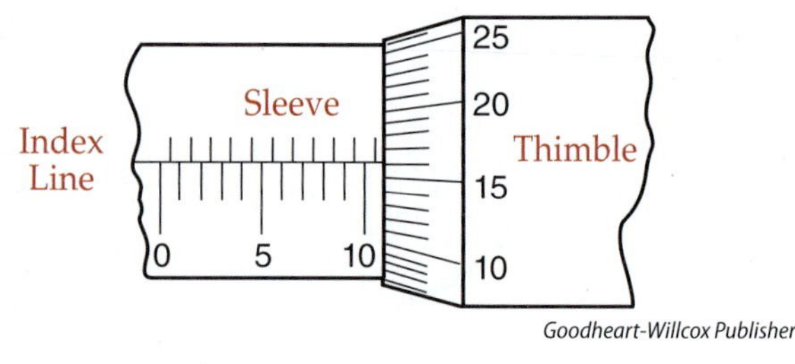

Goodheart-Willcox Publisher

Figure B-20. A metric micrometer is read in a manner similar to reading an inch micrometer.

The reading on the metric micrometer shown in **Figure B-20** is produced as follows:

1. There are 10 whole-millimeter divisions.
2. There is one half-millimeter division.
3. The reading on the thimble is 16.
4. The micrometer reading is 10.66 mm, calculated as follows:

10.00	(10 × 1 mm)
0.50	(1 × 0.50 mm)
+ 0.16	(16 × 0.01 mm)
10.66 mm	

Calipers

Calipers can make internal (inside), external (outside), and depth measurements. A ***caliper*** is a measuring instrument consisting of a main scale with a fixed jaw and a sliding jaw. The sliding jaw has an attached scale, either in the form of a dial or a digital readout. See **Figure B-21**. Each revolution of the dial is 1/10″, with 100 increments around the dial. This allows readings to the nearest 1/1000″. Tick marks along the bar indicate the measurement to the nearest 10th of an inch (for example, between 1.2 and 1.3). The additional thousandths are read from the dial. The two readings are added to get the total measurement. A caliper with a dial indicator may have to be periodically reset to ensure the zero reading is correct when the jaws are closed at a zero distance.

Most state-of-the-art calipers have a digital readout. See **Figure B-22**. Instead of reading the nearest 1/1000″ on a dial, the digital readout displays the *total* measurement. It is also important when using a digital-readout caliper to reset the value to zero when the calipers are closed.

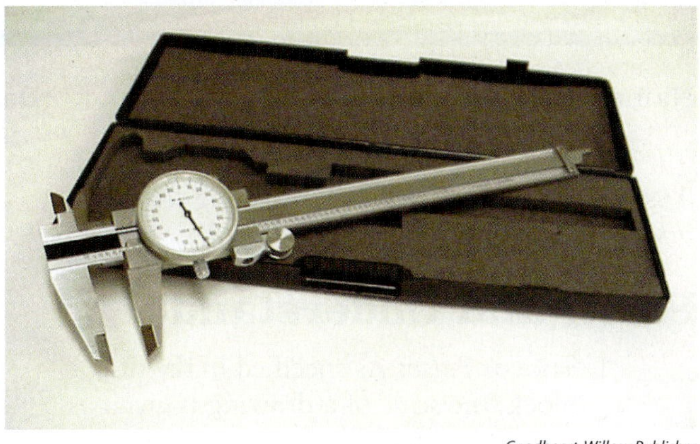

Goodheart-Willcox Publisher

Figure B-21. Dial calipers are used to make precise measurements.

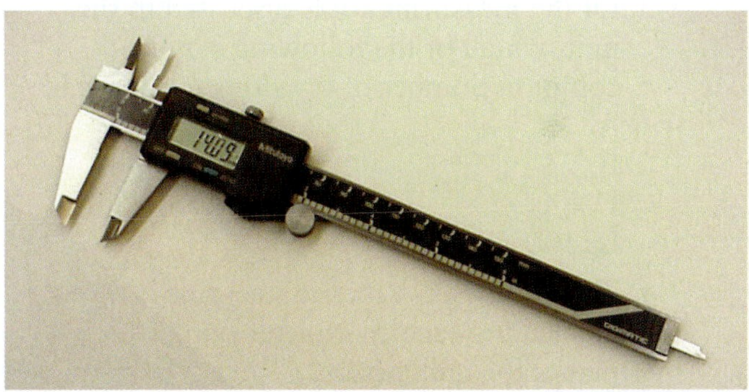

Goodheart-Willcox Publisher

Figure B-22. Calipers may come with digital readouts. These calipers are often referred to as digital calipers.

Appendix Review

Name ______________________ Date ______________ Class ______________

Answer the following questions using the information provided in this appendix.

Know and Understand

____ 1. *True or False?* As specified in the title block, the scale of a drawing is given as "paper" = "real."

____ 2. *True or False?* A steel rule is the most common 12″ measuring device available to a machinist.

____ 3. Of the measuring tools discussed in this unit, which of the following is *not* a common number of divisions for an inch?
A. 16
B. 25
C. 64
D. 100

____ 4. *True or False?* A metric steel rule is often labeled for each centimeter but has tick marks for millimeters.

____ 5. Which description applies to one edge of a triangular architectural scale but not the other five edges?
A. Decimal
B. Civil
C. Enlargement
D. Fully divided

____ 6. For a triangular architect's scale, how many "reduction" scales are there, each featuring foot marks and a 12″ ruler?
A. 5
B. 8
C. 10
D. 12

____ 7. *True or False?* The 20 scale on a civil engineer's scale can be used for reducing measurements at a scale of 1″ = 2″, 1″ = 2′-0″, or 1″ = 20′-0″.

____ 8. Which of the following is *not* a micrometer part?
A. Spindle
B. Throttle
C. Anvil
D. Index line

____ 9. *True or False?* Micrometers measure to the nearest 1/100″ or nearest 1/10 mm.

____ 10. *True or False?* As with micrometers, calipers can only read outside measurements and cannot be used to read internal measurements.

Apply and Analyze

Name ______________________ **Date** ______________ **Class** ______________

Review Activity B-1

Reading a Fractional-64 Steel Rule

Complete the readings indicated below. Write your answers as fractional values in the spaces provided.

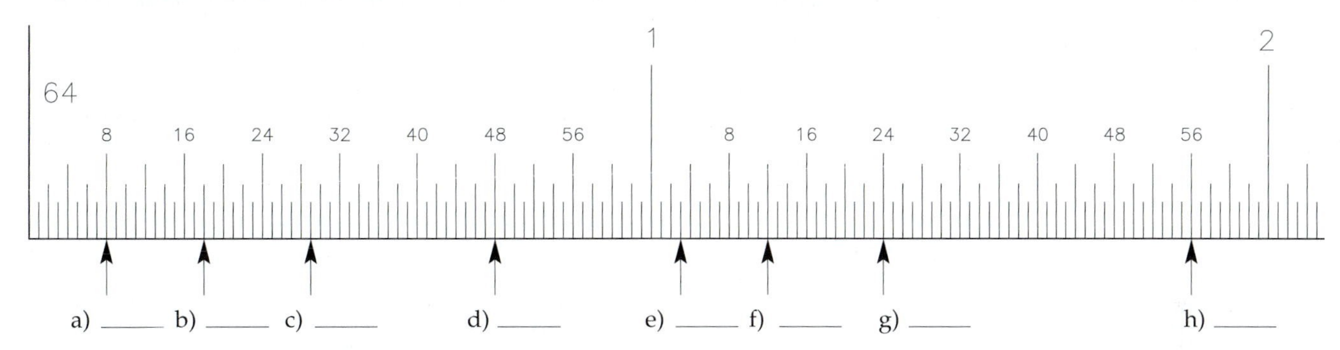

Review Activity B-2

Reading the Decimal-100 Steel Rule

Complete the readings indicated below. Write your answers as two-place decimals in the spaces provided.

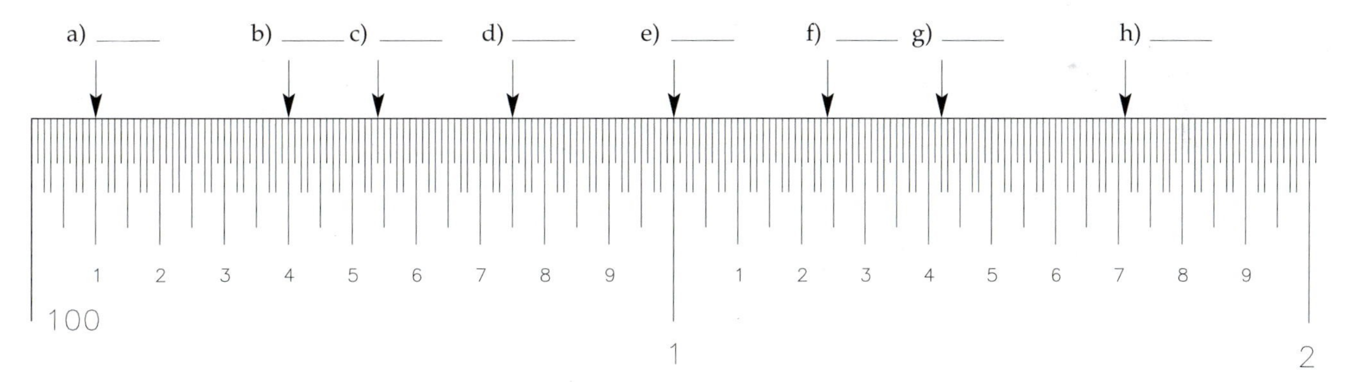

Review Activity B-3

Reading the Metric Steel Rule

Complete the readings indicated below. Write your answers as millimeter measurements in the spaces provided.

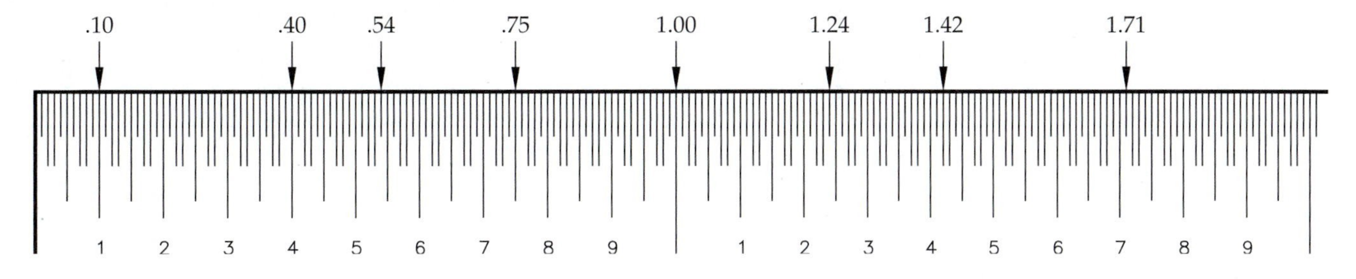

Apply and Analyze

Name ____________________ Date __________ Class __________

Review Activity B-4

Using a Fractional-Inch (16) Full Scale

Measure the length of the following lines using a fractional scale. Dividers can also be used to transfer distances to the scale on the page. Place your reading in the blank as a full scale reading.

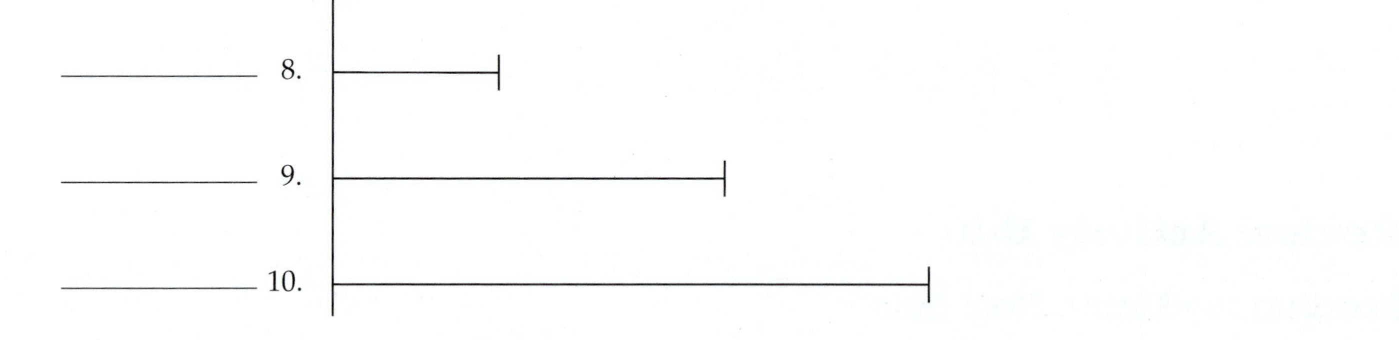

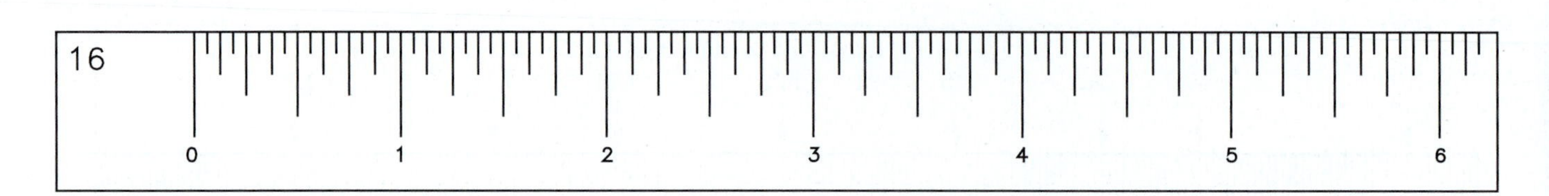

Name ______________ Date ______________ Class ______________

Review Activity B-5

Using a Decimal-Inch (10) Full Scale

Measure the length of the following lines using a decimal scale. Dividers can also be used to transfer distances to the scale on the page. Place your reading in the blank as a full scale reading.

________ 1.

________ 2.

________ 3.

________ 4.

________ 5.

________ 6.

________ 7.

________ 8.

________ 9.

________ 10.

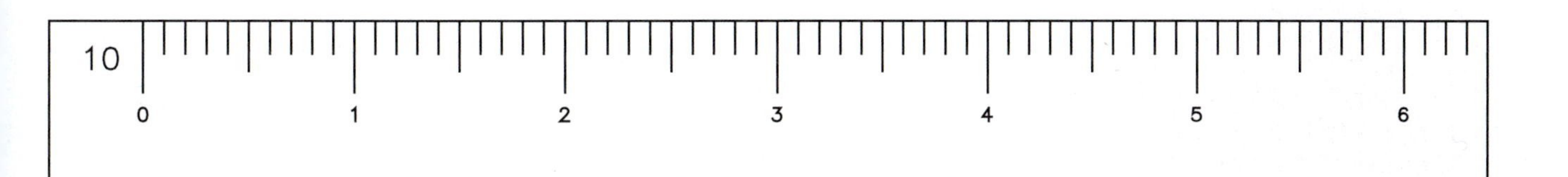

Apply and Analyze

Name ____________________ Date ______________ Class ______________

Review Activity B-6

Using an Architect's Reduction Scale

Shown here are some dimensions along different architect's scales. Locate the zero point on each scale. Inches are to one side of zero and feet to the other side. Evaluate each dimension according to the scale illustrated. Remember, the scale is indicated on the end of the architect's scale. Place your reading in the blank provided.

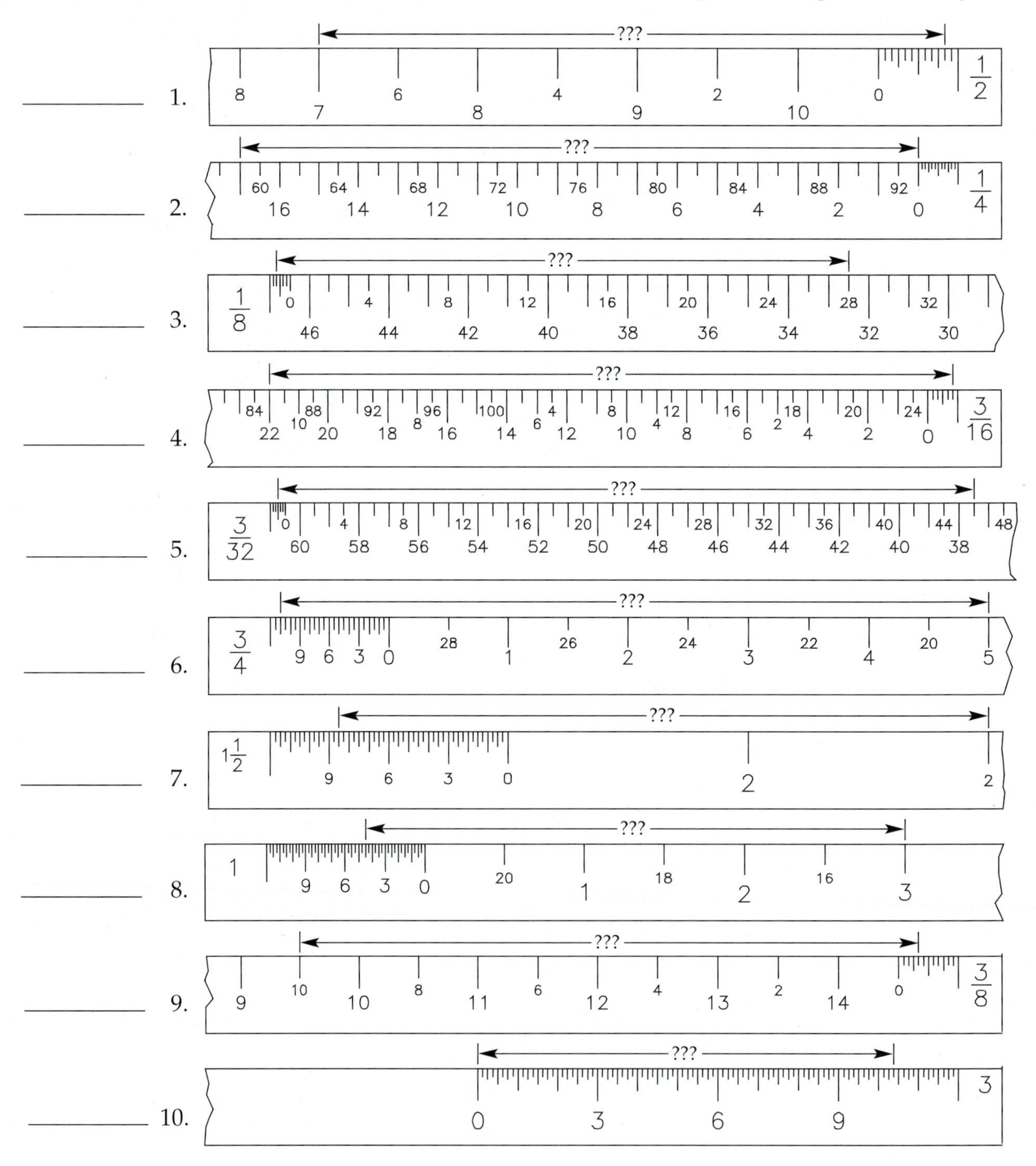

Apply and Analyze

Name ________________ **Date** ________________ **Class** ________________

Review Activity B-7

Using a Civil Engineer's Reduction Scale

Shown are some dimensions along different edges of a civil engineering scale. Evaluate each dimension according to the scale indicated. Place your reading in the blank provided.

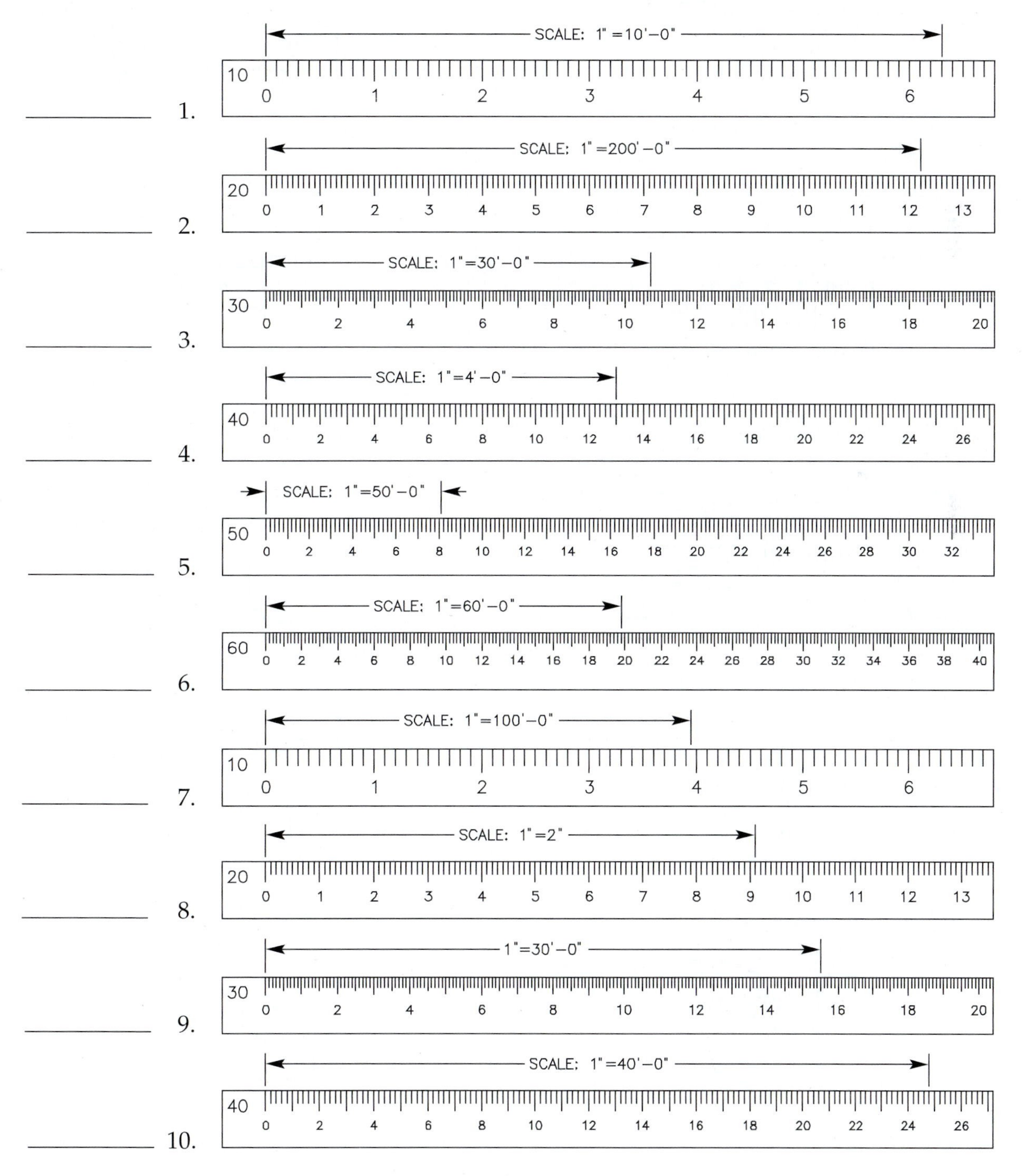

Name ______________________ Date ______________ Class ______________

Review Activity B-8

Reading an Inch Micrometer

Record the readings in the space provided for the following micrometer settings.

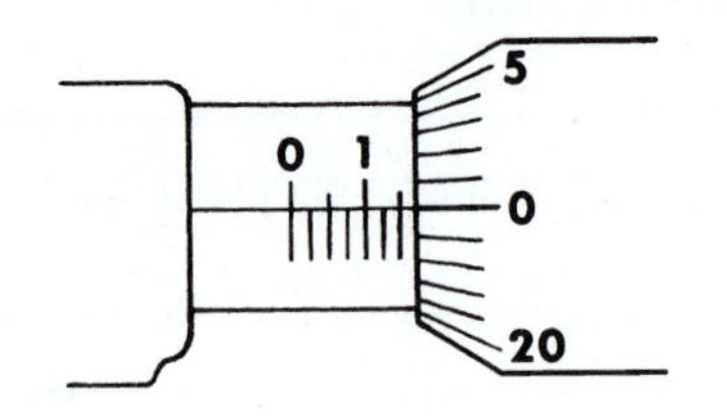

1. ______________________

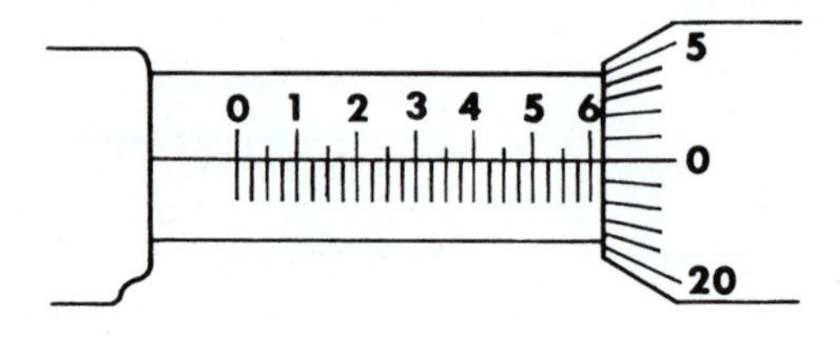

6. ______________________

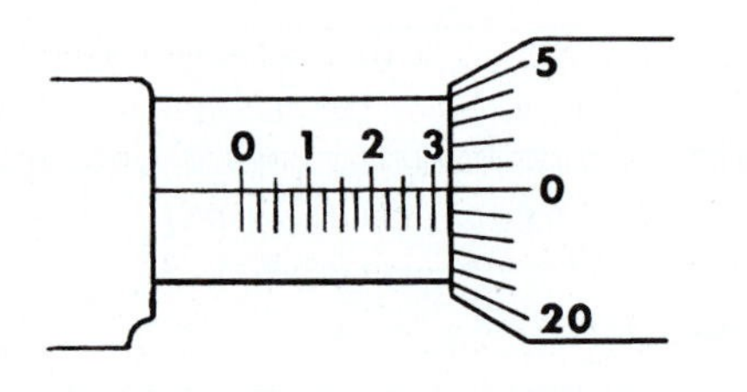

2. ______________________

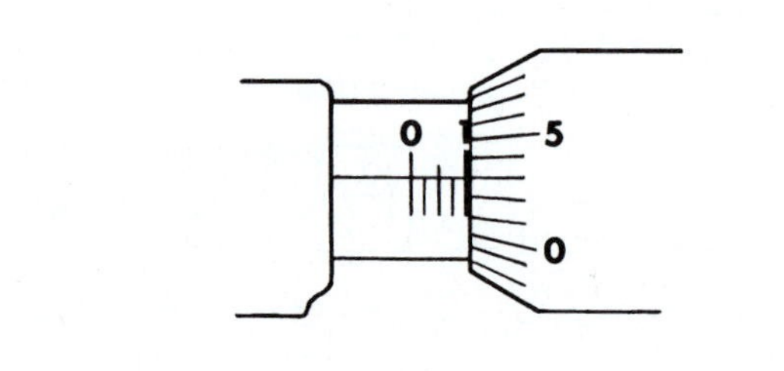

7. ______________________

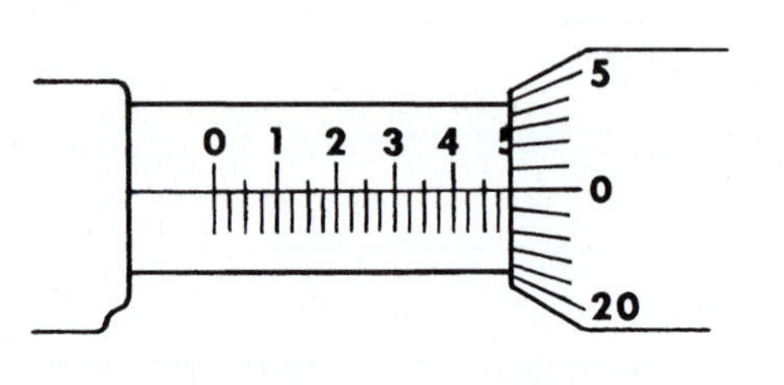

3. ______________________

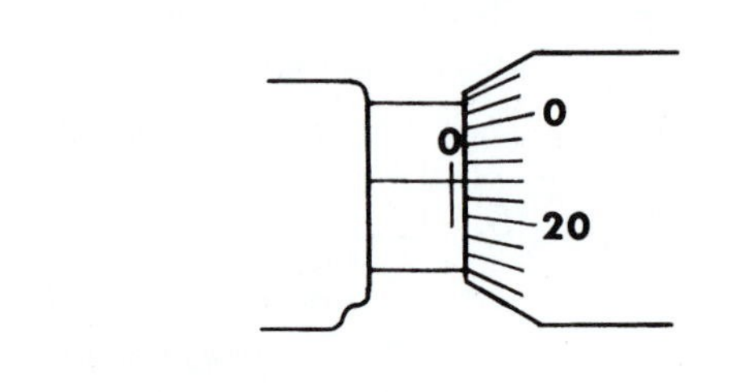

8. ______________________

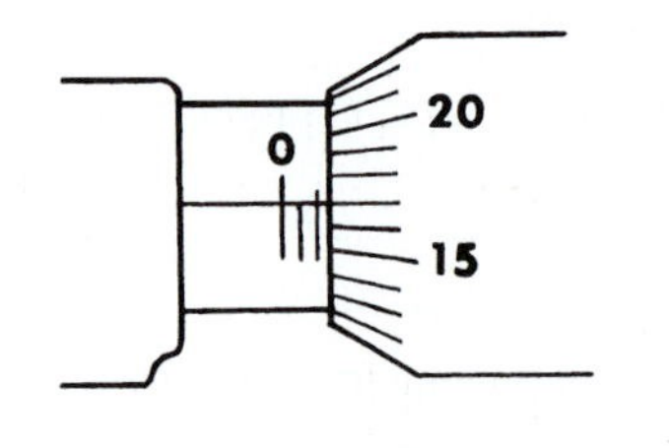

4. ______________________

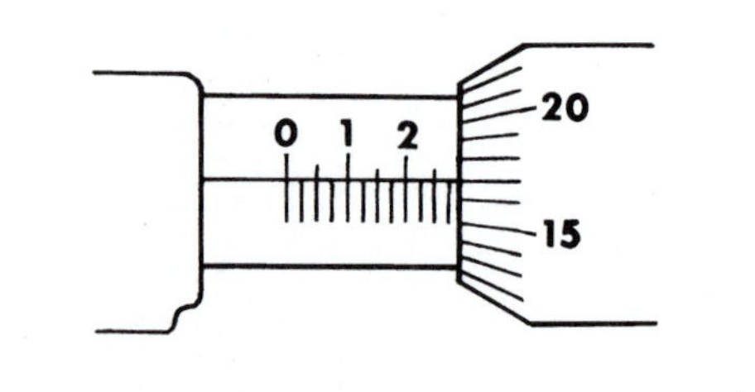

9. ______________________

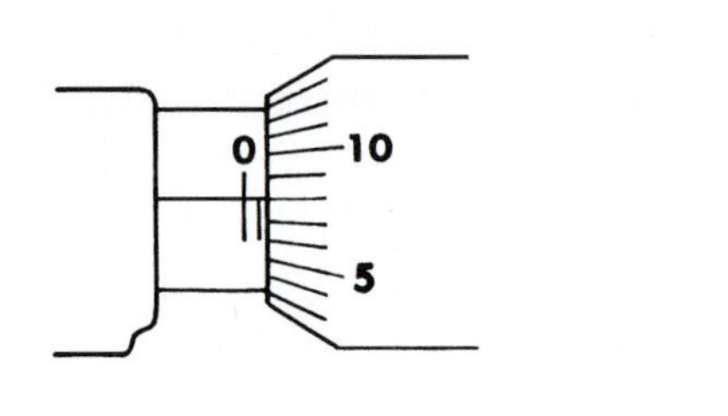

5. ______________________

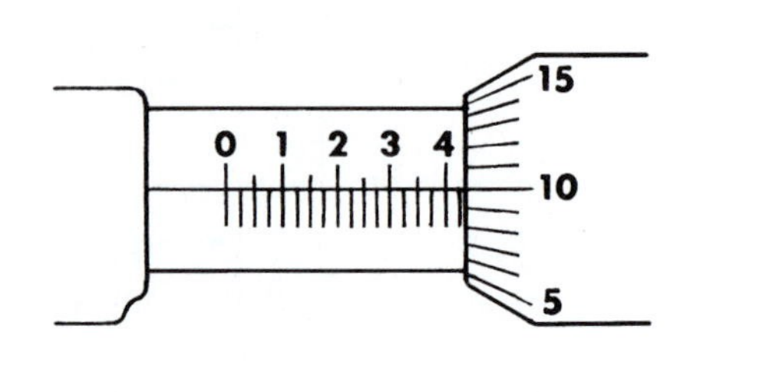

10. ______________________

Name ______________________ Date ______________ Class ______________

Review Activity B-9

Reading a Metric Micrometer

Record the readings in the space provided for the following micrometer settings.

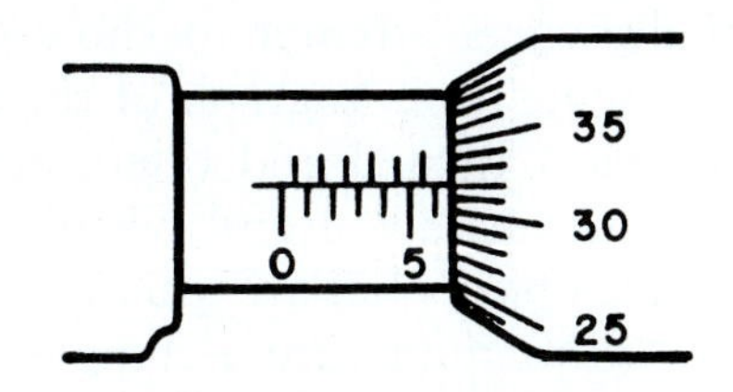

1. ______________________

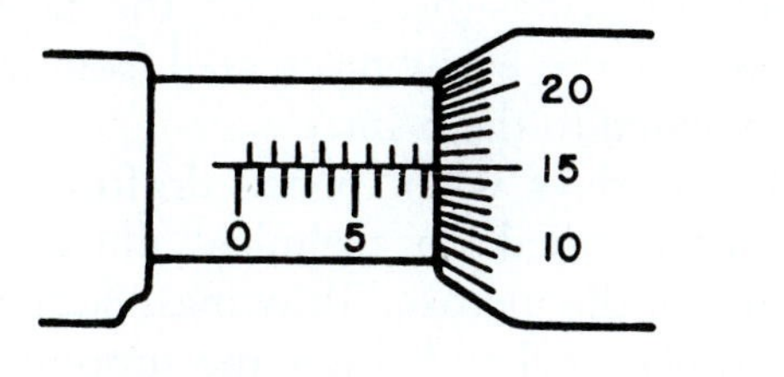

2. ______________________

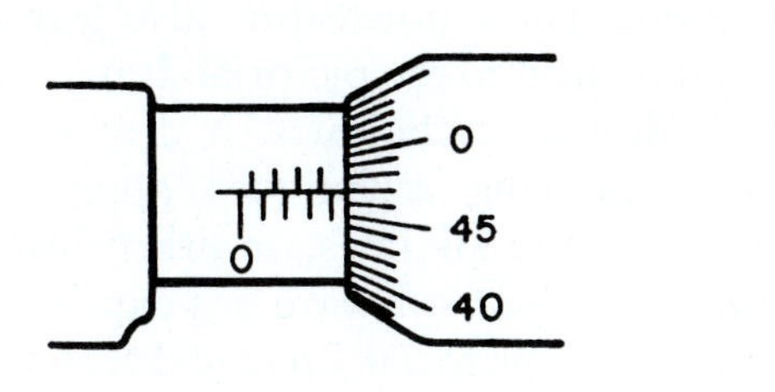

3. ______________________

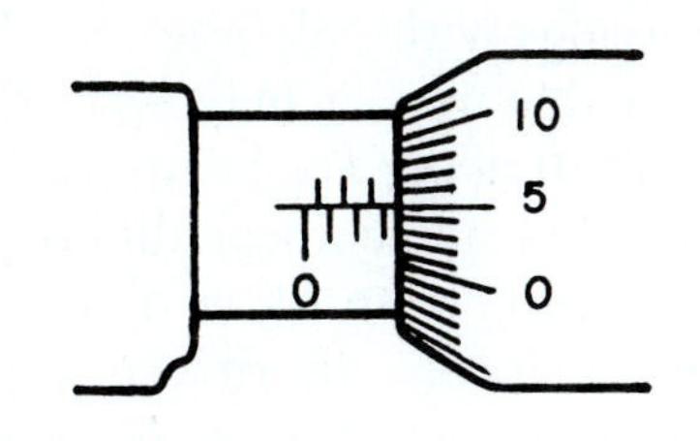

4. ______________________

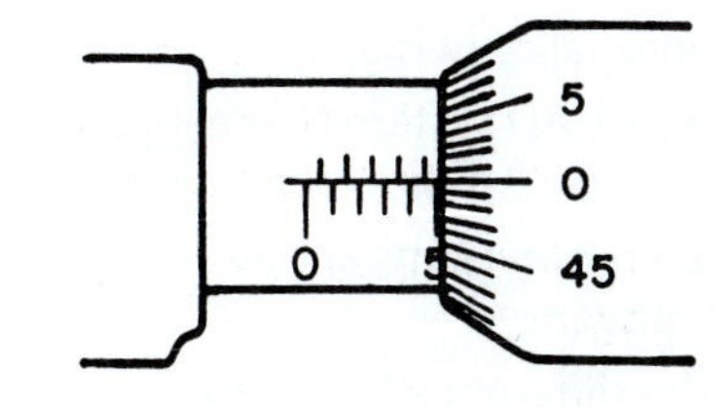

5. ______________________

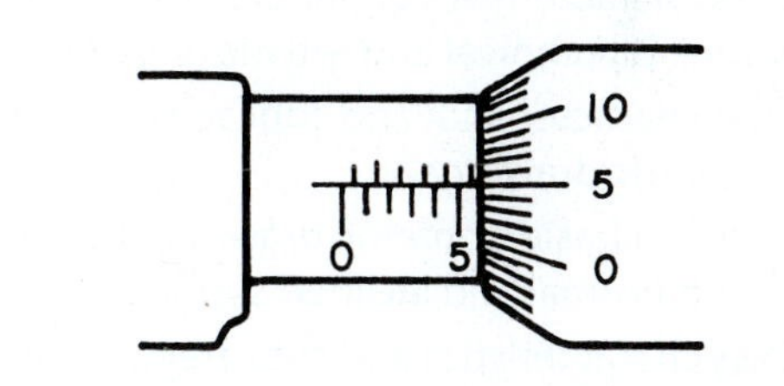

6. ______________________

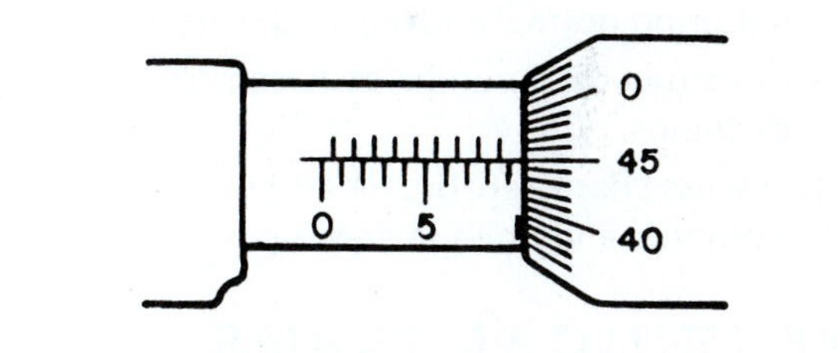

7. ______________________

APPENDIX C

Sketching Pictorial Drawings

LEARNING OBJECTIVES

After completing this unit, you will be able to:

- Define pictorial drawing.
- Discuss how CAD systems have impacted the creation and use of pictorial drawings.
- Identify common terms applied to various types of pictorial drawings.
- Discuss the difference between a projection and a drawing.
- Discuss characteristics of perspective projections and their role in technical and artistic drawings.
- Discuss characteristics and subcategories of axonometric drawings.
- Construct a basic isometric drawing of an object featuring normal and inclined surfaces.
- Discuss characteristics and subcategories of oblique drawings.
- Construct a basic oblique drawing of an object featuring normal and inclined surfaces.
- Construct circles and arcs in isometric and oblique drawings.
- Discuss characteristics for adding dimensions to isometric and oblique drawings.

TECHNICAL TERMS

axonometric projection
cabinet oblique drawing
cavalier oblique drawing
isometric drawing
oblique drawing
perspective drawing
pictorial drawing
rendering
vanishing point

By definition, ***pictorial drawings*** are picture-like representations of objects or products. Pictorial drawings attempt to show the part as if you are viewing a photograph of the actual object, with the object turned and tilted. See **Figure C-1**. Pictorial drawings are useful when making or servicing simple objects but are usually not adequate to show how to manufacture a complex part. In current practice, companies usually build 3D models of a product during the design process and create 2D detail drawings directly from the model. A pictorial view can also easily be created from the same model and added to the print.

In the days when pencil drafting was prevalent, industry relied on technical illustrators to create many of the pictorial drawings. Sometimes these individuals worked in sales and support, creating technical brochures or instructional and maintenance diagrams. These illustrators also learned techniques to add realism to the pictorial drawings with shading or shadowing techniques. A drawing with shading and shadowing techniques applied using pencil tones, pen and ink lines, or airbrushed ink is called a ***rendering***. CAD software has expanded our options for creating pictorial images. Most CAD programs include the capability of producing rendered images that have a photorealistic look and feel, as well as pictorial drawings with only lines. See **Figure C-2**.

Throughout this unit, it may seem that the terms *projection* and *drawing* are being used interchangeably, but there is a technical difference. In drafting practice, a projection of an object is a scientific exercise that involves an object, projectors, and one or more projection planes. With some pictorial projections, lines are mapped out and the final view is projected from other views of the object as tilted or turned in various positions. On the other hand, many of the pictorial views created are simply drawings based on those principles, but no physical projection was constructed. The objects were simply drawn in a manner reflective of the projection method.

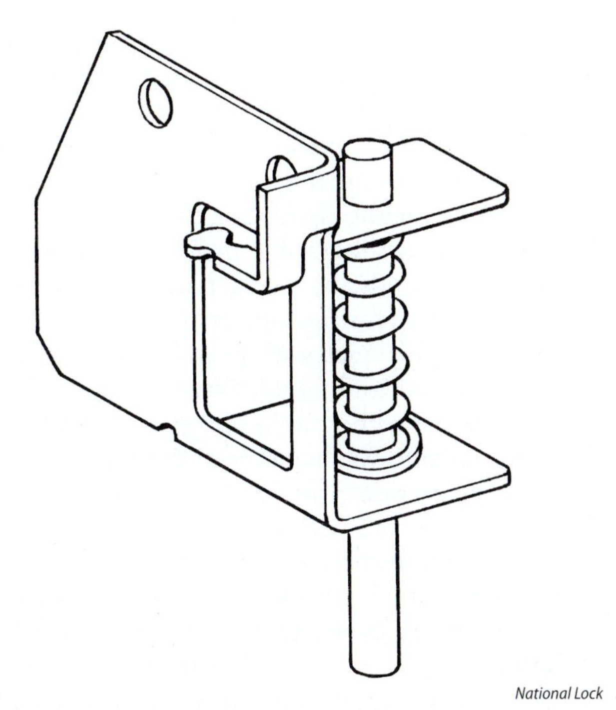

National Lock

Figure C-1. A pictorial drawing is a realistic, picture-like representation of an object.

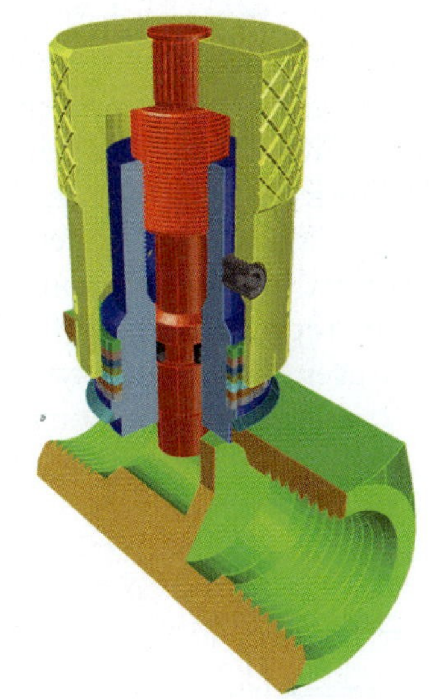

RegO Cryo-Flow Products

Figure C-2. CAD renderings are a form of pictorial drawing.

CAD Applications

As discussed previously, most CAD programs are very effective in creating pictorial drawings or views of a 3D model for use on prints. A clarification related to pictorial terms is also required at this point. With respect to CAD-generated prints, a pictorial drawing can be created with lines and ellipses as a "flat," two-dimensional image based on pictorial projection principles. In other words, the drawing is "drafted" with lines, circles, and other shapes, just like a pencil drawing. A better and more flexible method, however, is to allow the CAD system to generate one or more pictorial *views* from the 3D model. These views are truly projections generated by the CAD system and have the flexibility to be turned and tilted later without erasing lines. Also, as the model changes, the drawing does not have to be updated manually.

A pictorial view of a 3D model is three-dimensional and defined by X-Y-Z coordinates. Of course, once the image is printed onto paper, it too becomes a two-dimensional, "flat" image. Even so, another view can be generated just as quickly if a different orientation or style is desired. As a print reader, you do not need to be concerned with how the image was placed on the paper—you simply need to interpret and visualize the idea. The good news is that there are more options than ever to help print readers visualize and "see" the object.

While CAD has not eliminated the need for manually created pictorial drawings, it has impacted the emphasis placed on constructing pictorial drawings, not only within drafting curricula, but also within industry standards. Even though computer programs can create pictorial drawings and images directly from 3D models, pictorial sketching is still useful in helping visualize multiview drawings. Pictorial sketching is also helpful in communicating your ideas or technical problems to others. Pictorial drawings are very useful in explaining print reading concepts in textbooks, such as this one.

This appendix will explore several terms that define pictorial drawing types, as these terms do appear on some prints or within various 3D modeling and viewing software applications. While this unit will define various types of pictorial projections that can be explained scientifically, the emphasis will be on two of the more common types of drawings that can easily be learned. These drawings can assist the print reader with visualizing an object's features or assist in the communication process.

Perspective Drawings

One type of pictorial representation is ***perspective drawing***, based on a projection system where all projectors converge toward one or more points in the background. This perspective projection system stands in contrast to the parallel projection system discussed in Unit 5. Most industrial drawings are based on the principles of parallel projection wherein object vertices and surfaces are projected onto two-dimensional planes with projectors parallel to each other. Perspective drawings are based on axes that converge toward points referred to as ***vanishing points***, **Figure C-3**. Perspective projections are more realistic to the way in which the human eye works but are more complicated to learn, with projection planes and projectors mapped out in precise and methodical fashion. For that reason, techniques for sketching perspective drawings will not be covered in this appendix. There are many textbooks that deal solely with the construction of perspective drawings, as architects and artists have also had an interest in perspective techniques throughout the years. Be aware that most 3D CAD modeling and viewing programs also allow users to select between parallel and perspective projection modes for on-screen viewing.

Axonometric and Isometric Drawings

Before defining additional pictorial terms, it is appropriate to review orthographic projection methods. As discussed in Unit 5, multiview projections are created by projecting the vertices and surface of an object onto multiple projection planes. The resulting views are based on projections created orthographically, which means parallel lines are projected from the edges and corners of the object perpendicular to the projection planes. If the front surface is parallel to the projection plane, the front surface projects true size and shape, while other surfaces may project as edge view lines. The result is one or more two-dimensional views of the object, with surfaces featured differently in various views. Surfaces on the top portion of the object are featured in the top view, surfaces featured on the right side are featured in the right-side view, and so on. The print reader must visualize the object through multiple views.

One form of pictorial drawing can also be explained with the same orthographic projection principles as multiview methods, by using parallel projectors to project the points of an object onto one projection plane. With this pictorial projection method, the object is turned and tilted, resulting in a pictorial view where the orthographic projectors intersect the projection plane. While pictorial drawings are based on these projection principles, the drawings can be sketched or created without fully understanding the scientific methods underlying the pictorial views.

By definition, the previous paragraph described an ***axonometric projection***, a projection of an object that has been turned and tilted and then projected orthographically onto a projection plane. The three divisions of axonometric projection are *isometric projection*, *dimetric projection*, and *trimetric projection*. See **Figure C-4**. In an isometric projection, all three axes form equal angles. In a dimetric projection, only two axes form equal angles. In a trimetric projection, all axes are oriented at unequal angles. With trimetric projection, each dimension is foreshortened differently, and circular features are projected elliptically, but not the same amount in each direction. With CAD systems, 3D models can easily be rotated in any axonometric direction, with appropriate representation of all dimensions and circular features. With CAD systems, the terms *diametric* and *trimetric* have become unimportant,

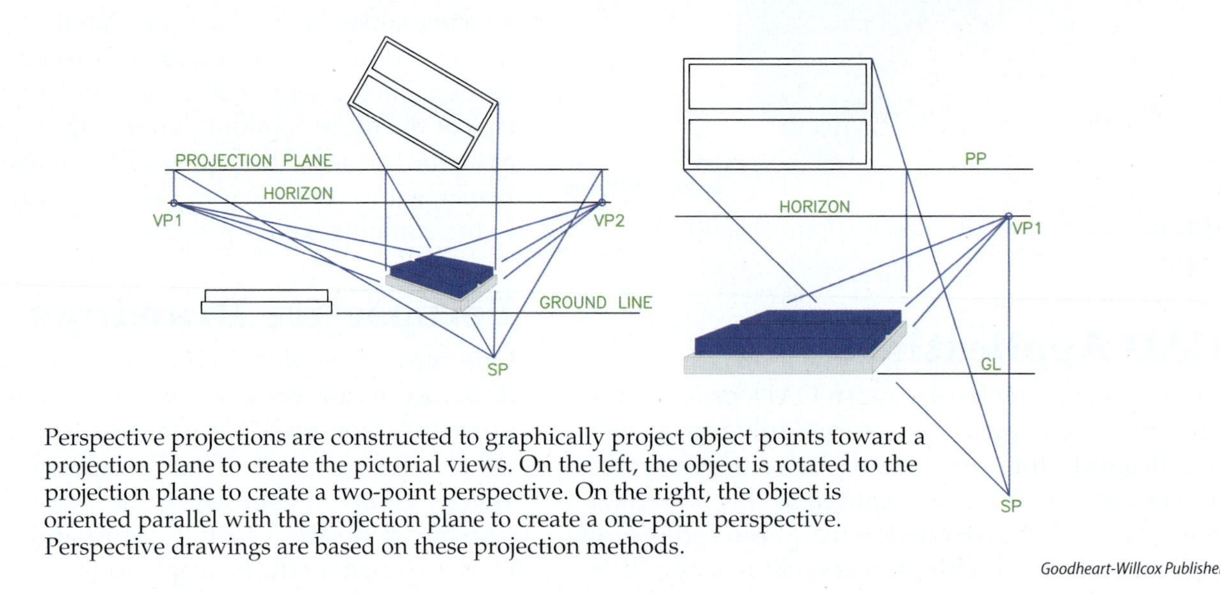

Figure C-3. Perspective drawings feature vanishing points toward which one or more axes converge.

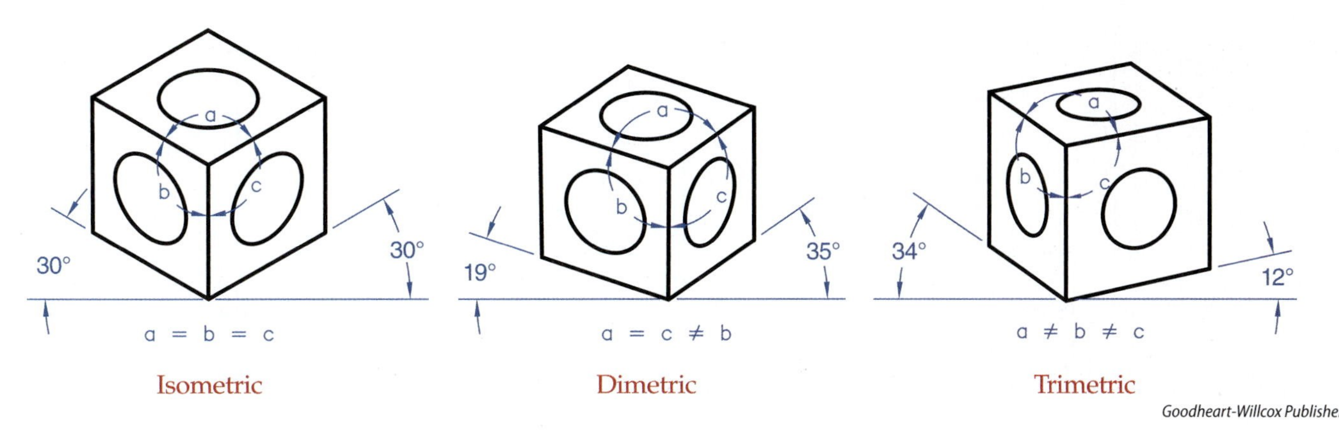

Figure C-4. The three types of axonometric projections are isometric, dimetric, and trimetric.

as drafters no longer need to graphically construct these different angles and proportions.

In drafting education, the most common of the three axonometric choices is isometric, which means "equal measure." ***Isometric drawings*** based on this system of projection are constructed with the X-Y-Z axes projected at angles of 120° to each other. Because of this equal spacing, the same scale of measure can be used along each axis. The width and depth (X and Y) are drawn 30° from horizontal, and the height (Z) is vertical. See **Figure C-5**. The lines forming a Y in the center of the object are equally spaced at 120° to each other and are known as the isometric axes. In an isometric drawing, equal measurements are applied on all axes. In addition, the amount of distortion of circles into ellipses is the same on all surfaces.

While isometric drawings appear less realistic than perspective drawings, they are used more often because direct measurements for all three dimensions can be used in their construction, as the foreshortening is equally projected for all three dimensions. Also, sketching grids are easily created to assist with distances and direction. In isometric drawings, lines appearing horizontal in the front or side views of a multiview drawing are drawn at an angle of 30° in an isometric view. Vertical lines remain vertical in an isometric view. Inclined lines are called *nonisometric lines* and are drawn by locating their endpoints along the isometric axes and connecting the two points. Hidden lines are not shown in isometric views unless required to clarify drawing details.

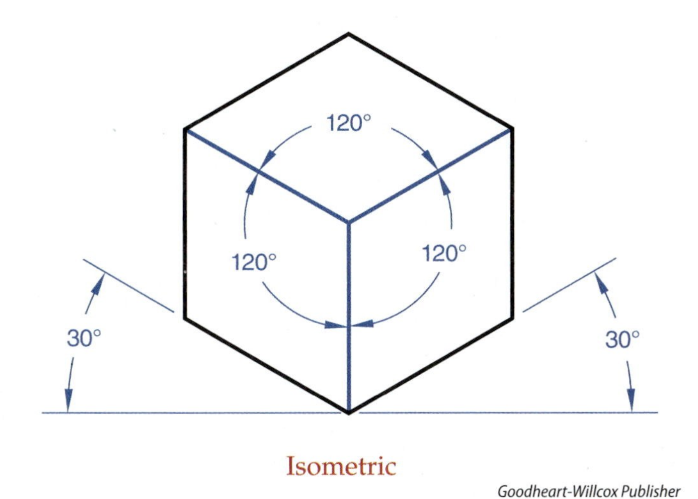

Figure C-5. Isometric drawings are constructed with the X-Y-Z axes projected at angles of 120° to each other.

Constructing an Isometric Sketch

Refer to **Figure C-6** as you go through the procedure for constructing an isometric sketch, which follows:

1. The three orthographic views of a V-block are shown in a print. See **Figure C-6A.**
2. Select a position for the object that best describes its shape, such as a corner of the V-block located at the "front" of the drawing.
3. Sketch the axes for the lower corner. See **Figure C-6B.**
4. Make overall measurements in their true length on the isometric axes or on lines parallel to the axes. See **Figure C-6C.**
5. Construct a box to enclose the object. See **Figure C-6D.** Note: It is more important to make sure lines are parallel than to maintain a precise 30° angle to the horizon.
6. Sketch the isometric lines of the object. See **Figure C-6E.**
7. Locate the endpoints of nonisometric lines. Then, sketch the lines. See **Figure C-6F.**
8. Darken all visible lines and erase the construction lines to complete the isometric sketch. See **Figure C-6G.**

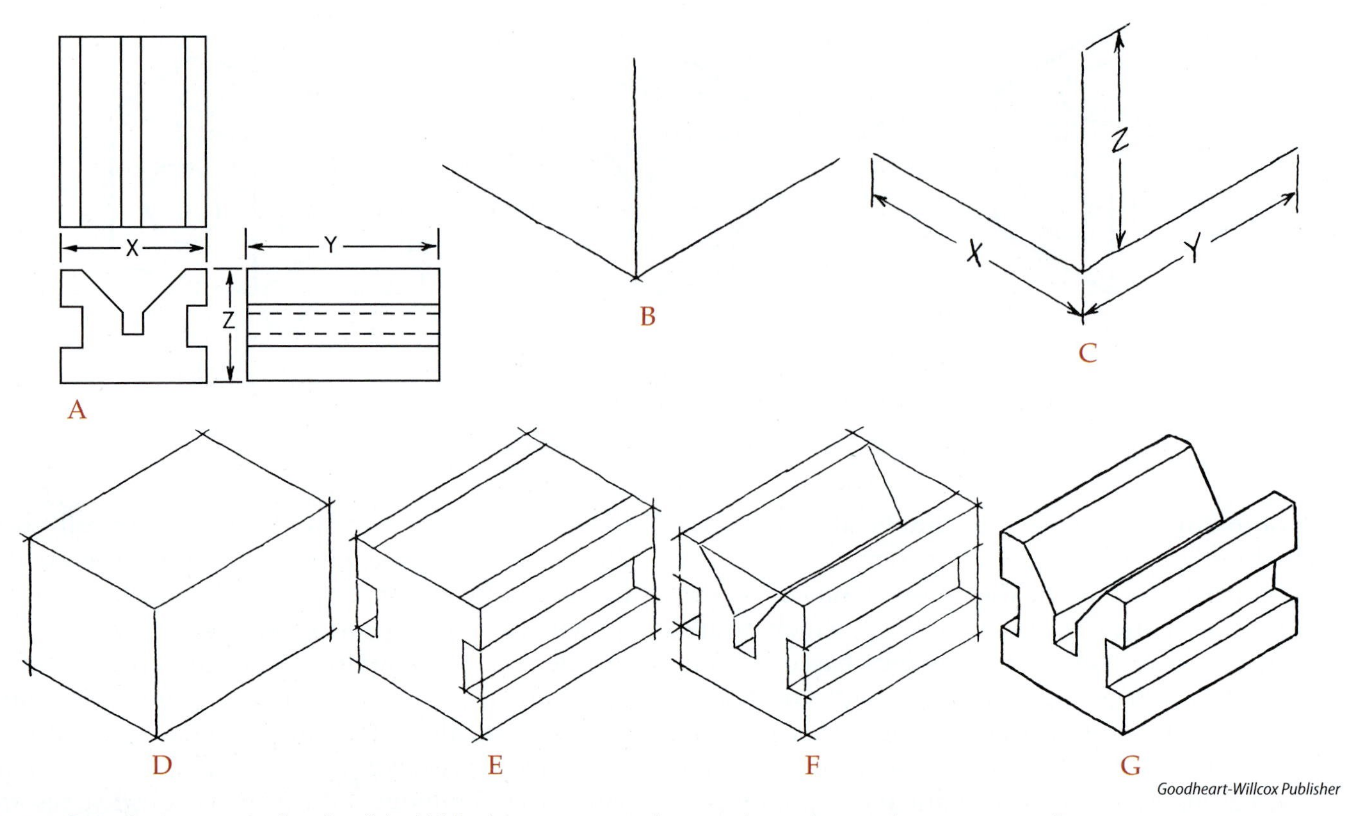

Figure C-6. An isometric sketch of the V-block is constructed using the orthographic views as reference.

Circles and Arcs in Isometric Drawings

Isometric circles and arcs are sketched in a fashion similar to that of regular circles and arcs as discussed in Unit 5. For isometric circles and arcs, you start with an isometric square instead of a normal square. An isometric square resembles a diamond in the isometric view. Circles can be sketched on all three planes of the isometric drawing in the same manner. See **Figure C-7.** The procedure for sketching isometric circles is as follows:

1. Sketch an isometric square enclosing the location of the circle. See **Figure C-8A.**

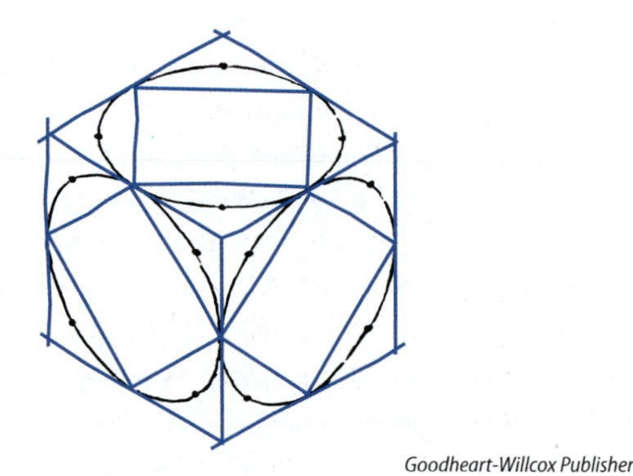

Figure C-7. Circles can be sketched on all three planes of the isometric drawing.

2. Locate and connect the midpoints of the sides of the isometric square. See **Figure C-8B.**
3. Locate the centers of the triangles formed. Then, sketch four isometric arcs, connecting each midpoint to form an ellipse. See **Figure C-8C.**
4. Erase construction lines and darken the ellipse. Notice the ellipse is composed of four arcs, two with a small radius and two with a large radius. See **Figure C-8D.**

The procedure for sketching an isometric arc is as follows:

1. Mark the radius of the arc from the corner. See **Figure C-9A.**
2. Draw a line connecting the two points, forming a triangle. See **Figure C-9B.**
3. Locate the center of the triangle and sketch an arc through this point to smoothly join with the sides. See **Figure C-9C.**
4. Erase construction lines and darken the arc. See **Figure C-9D.**

Isometric Dimensioning

The dimension lines on an isometric drawing or sketch are parallel to the isometric axes. Extension lines are extended aligned with these axes. See **Figure C-10.** This basically creates the dimensions

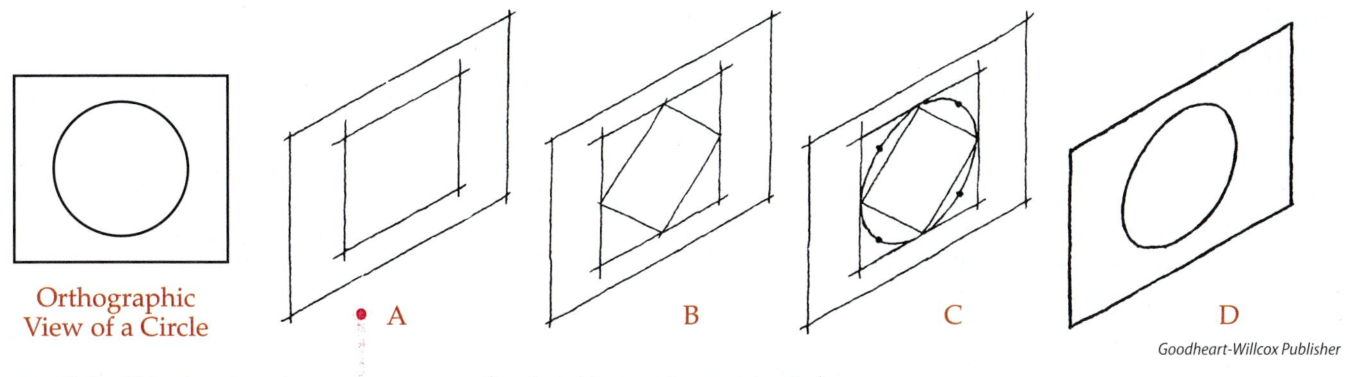

Goodheart-Willcox Publisher

Figure C-8. This drawing shows the process for sketching an isometric circle.

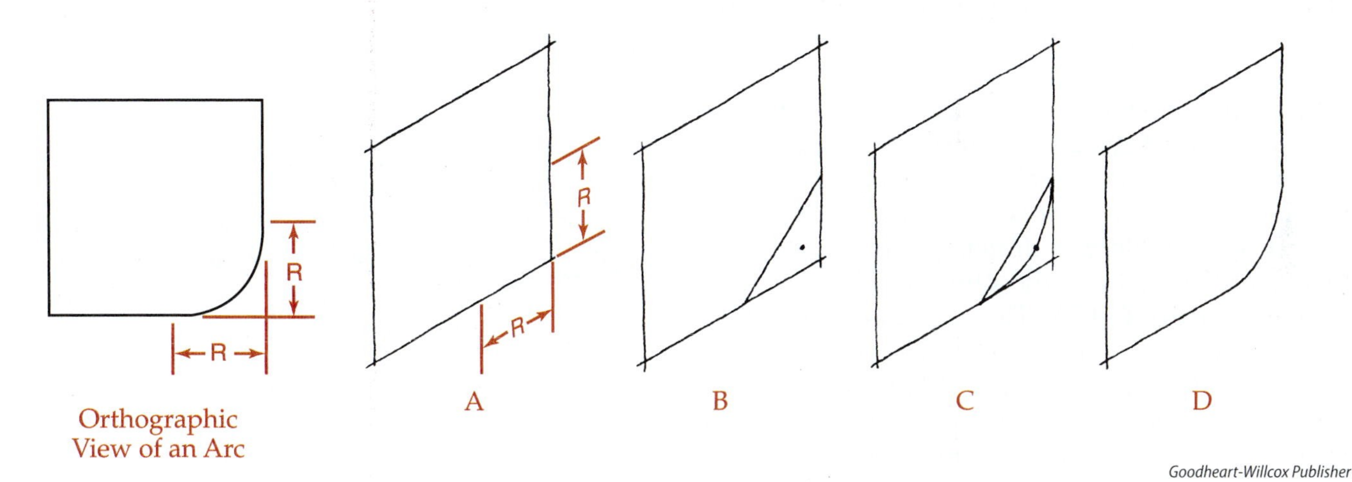

Goodheart-Willcox Publisher

Figure C-9. This drawing shows the process for sketching an isometric arc.

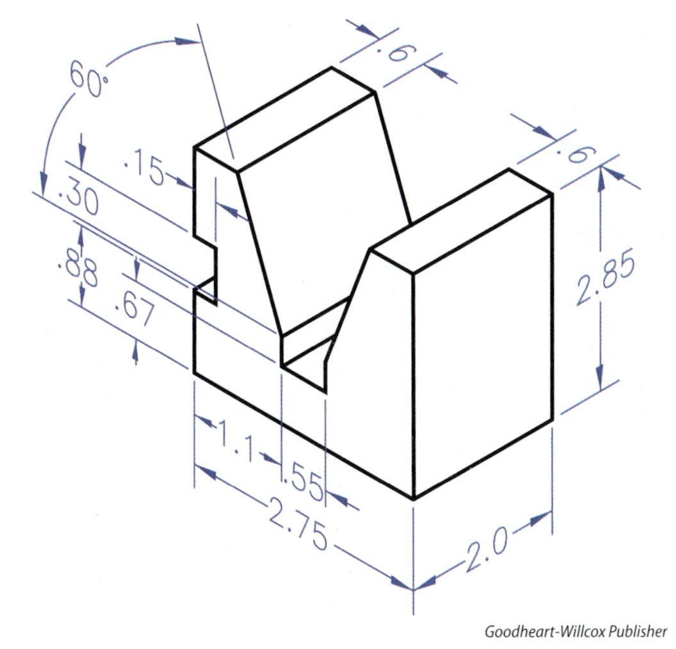

Goodheart-Willcox Publisher

Figure C-10. Dimension text in isometric drawings can be aligned, as shown here, or horizontal.

in the same plane as the surface or features being dimensioned. Notice in **Figure C-10** how the numerals are aligned with the isometric axes. The numeral dimensions can also be horizontally aligned instead. Dimensioning is discussed in detail in Unit 9.

Oblique Drawings

In drafting education, the oblique drawing is another common type of pictorial drawing that can be easily learned. ***Oblique drawings*** are based on oblique projection principles rather than orthographic projection principles. Both types of projection feature parallel projectors, but for oblique projection, the projectors are not perpendicular to the projection plane, resulting in a projection that appears somewhat pictorial. With the object oriented so a front normal face is parallel with the projection plane, the frontal faces project true shape, just as an orthographic projection projects true shape. The front normal faces and rear normal faces will both project true size and shape, but they will be offset from each other. Think of this as drawing two overlapping squares and then connecting the corners. As a result, the depth axis appears slanted up or down and to the left or right. The word oblique means *slanted*, as represented by the treatment of the depth axis. As the front face is shown in true shape and size, this can be an advantage for the person sketching, especially when there are circles or arcs. However, the top and side faces of an oblique sketch are projected at an angle back from the front view, often at 45°. Oblique methods have been used throughout the years with block lettering, when the corners of lettering shapes are projected back at an angle to provide depth.

The oblique method can give a distorted appearance to the drawing, as if the depth measurement is too much. See **Figure C-11**. In drafting education, two principal types of oblique drawings are identified by the measurement scale applied to the depth axis. ***Cavalier oblique drawings*** are drawn with their receding sides to the same scale as the front view. See **Figure C-12A**. This creates an unrealistic appearance. However, the advantage is one scale is used on all three axes. ***Cabinet oblique drawings*** are drawn with their receding depth lines one-half scale to the width and height measurements. This gives a more realistic appearance. See **Figure C-12B**. Most likely, the name of this drawing is derived from the woodworking or cabinetmaking industry. In summary, the terms *cavalier* and *cabinet* are not critical terms for oblique sketches. Oblique drawings work well for regular grid paper sketching, even though depth measurements across the grid are estimated or counted across the corners of the grid. Even on plain paper, depth measurements for oblique sketches are estimated.

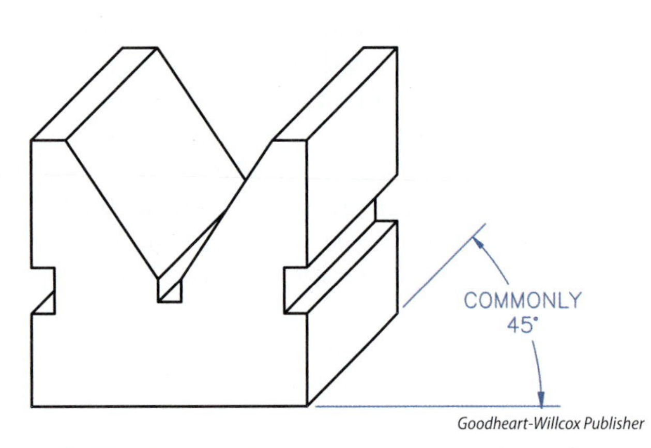

Figure C-11. Oblique drawings are drawn with the front face of an object shown in its true shape.

Circles and Arcs in Oblique Drawings

Circles and arcs are sketched in oblique drawings as true circles or arcs in the front plane. However, when circles and arcs appear on the receding planes, they are sketched in the same manner as isometric circles and arcs. The procedure for sketching oblique circles and arcs is as follows (refer to **Figure C-13**):

1. Sketch an oblique square that will contain the circle.
2. Locate the midpoints of the oblique square's sides. Connect these points. As an option, sketch centerlines for the height, width, and depth directions.

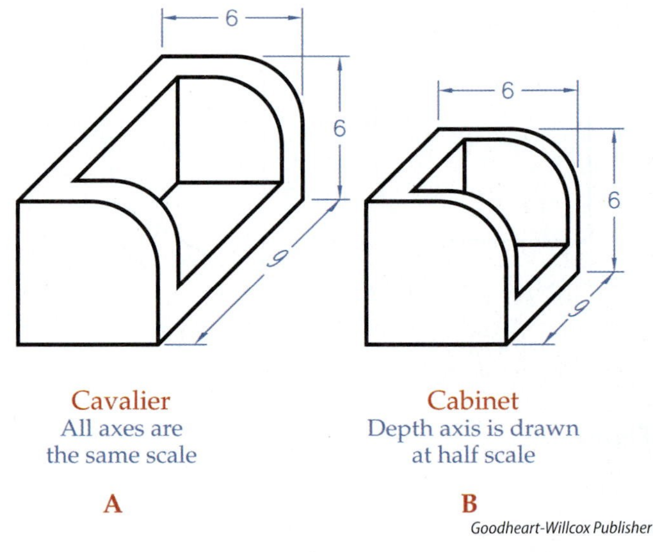

Figure C-12. A—Cavalier oblique drawings are drawn with their receding sides to the same scale as the front view. B—Cabinet oblique drawings are drawn with their receding sides to one-half scale of the front view.

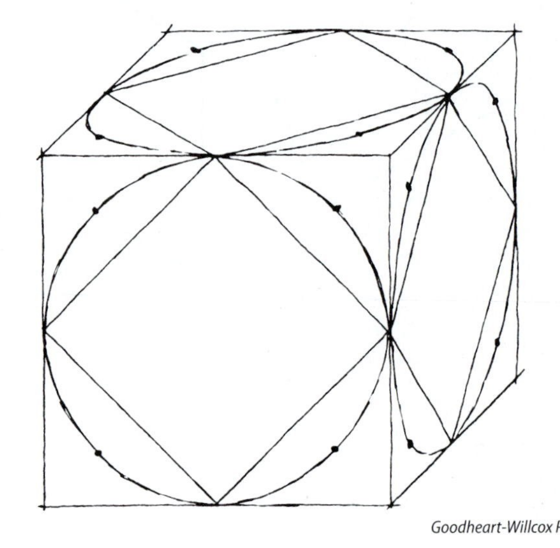

Figure C-13. This drawing illustrates the sketching of oblique circles.

3. Locate the center of the triangles. Sketch oblique circles or ellipses through these points, smoothly joining with the sides of the oblique square.
4. Erase construction lines and darken the circles or ellipses.

Oblique Dimensioning

Oblique dimensioning must be done in the same plane as the surface or feature being dimensioned. This is the same as in isometric dimensioning. **Figure C-14** illustrates the placement of dimensions on an oblique drawing.

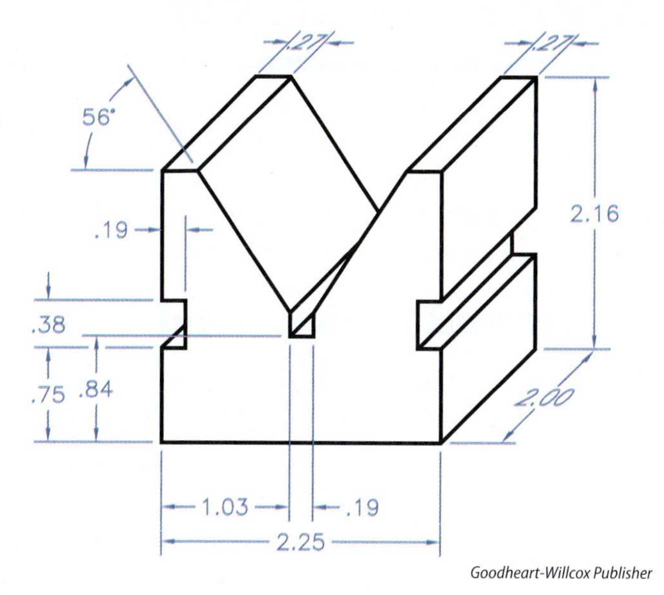

Goodheart-Willcox Publisher

Figure C-14. This drawing demonstrates the dimensioning of an oblique drawing.

Appendix Review

Name ______________________ Date ______________ Class ______________

Answer the following questions using the information provided in this appendix.

Know and Understand

____ 1. Of the choices below, which set of words is descriptive of a pictorial drawing as defined in this appendix?
A. Sales and support
B. Turned and tilted
C. Educational and instructive
D. Shading and shadowing

____ 2. *True or False?* Pictorial *projections* are constructed using views of objects with projection lines and projection planes, while pictorial *drawings* are less formally constructed based on the projection principles.

____ 3. *True or False?* While a CAD system can be used to create a 2D pictorial drawing one line at a time, many CAD systems can also create pictorial views directly from a 3D model.

____ 4. Of the types of pictorial projection or drawing systems, which is more complicated to learn due to the projectors *not* being parallel with each other?
A. Oblique
B. Dimetric
C. Trimetric
D. Perspective

____ 5. Which of the following terms is the umbrella term for the other three?
A. Axonometric
B. Isometric
C. Dimetric
D. Trimetric

____ 6. *True or False?* Isometric drawings are characterized by axes 180° apart, thereby allowing equal measure along each of the three axes.

____ 7. *True or False?* In isometric drawings, square features appear diamond shaped, while round features appear elliptical.

____ 8. *True or False?* Oblique and isometric drawings can be dimensioned, but the dimension values must always be drawn horizontal on the page.

____ 9. The word oblique means slanted, which in oblique pictorial drawings describes the _____ axis.
A. height
B. width
C. length
D. depth

____ 10. *True or False?* One advantage of oblique pictorial drawings is that you will never need to draw an ellipse for a circle.

Name ______________________ **Date** ______________ **Class** ______________

Sketching Activity C-1

Motor Bracket

Create an isometric or oblique sketch of the motor bracket, as assigned by your instructor. If instructed, dimension your sketch.

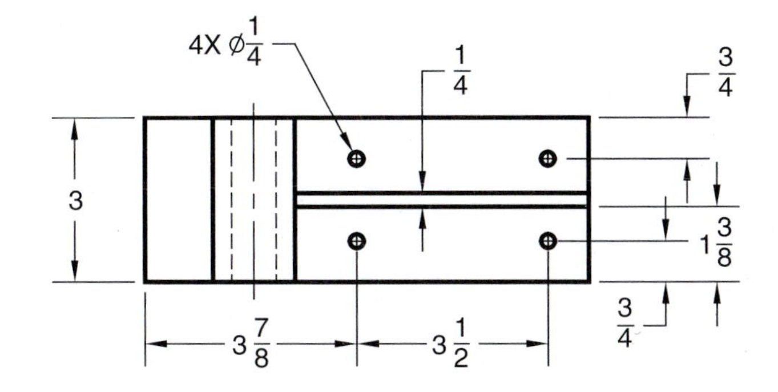

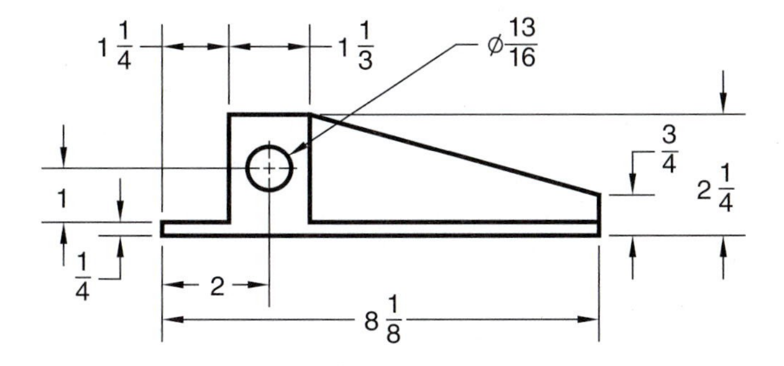

Apply and Analyze

Name ____________________ **Date** ______________ **Class** ______________

Sketching Activity C-2

Bracket Rest

Create an isometric or oblique sketch of the bracket rest, as assigned by your instructor. If instructed, dimension your sketch.

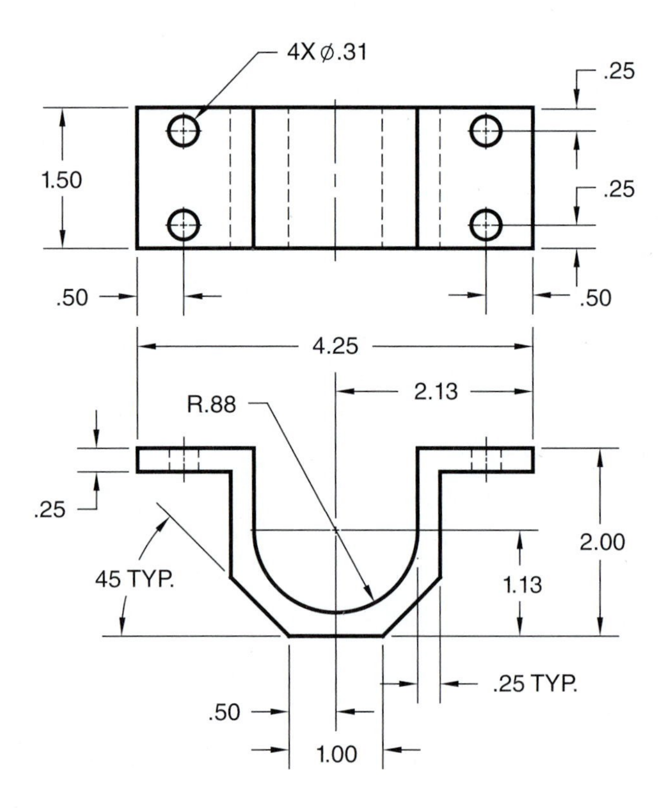

Apply and Analyze

Name ______________________ Date ____________ Class ______________

Sketching Activity C-3

Coupling

Create an isometric or oblique sketch of the coupling, as assigned by your instructor. If instructed, dimension your sketch.

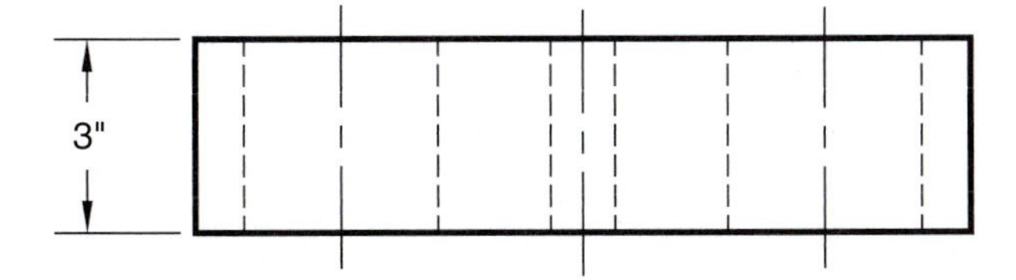

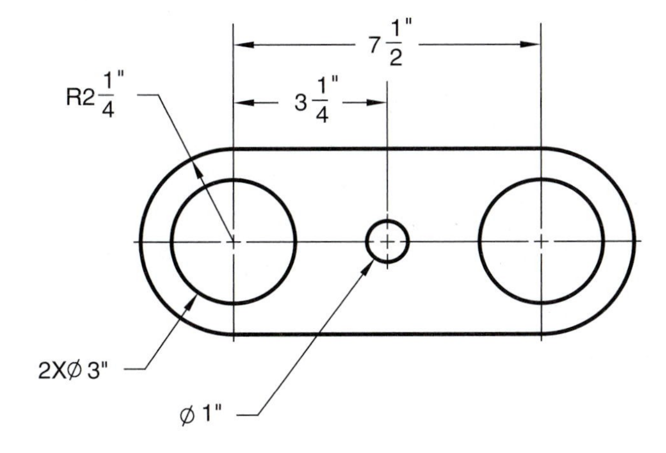

Name ______________________ Date ______________ Class ______________

Sketching Activity C-4

Oil Stones and Holder

Create an isometric or oblique sketch of the oil stones and holder, as assigned by your instructor. If instructed, dimension your sketch.

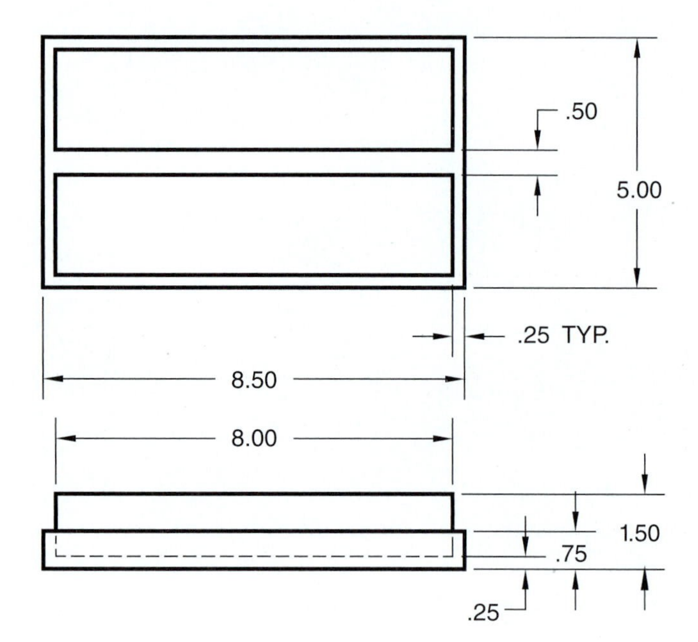

Name ______________________ Date ______________ Class ______________

Sketching Activity C-5

Lock Housing

Create an isometric or oblique sketch of the lock housing, as assigned by your instructor. If instructed, dimension your sketch.

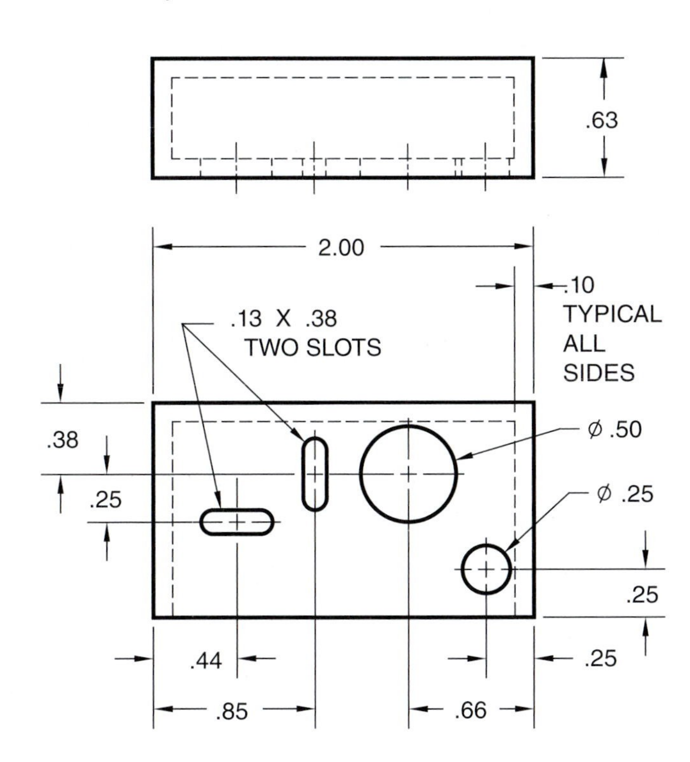

APPENDIX D
Abbreviations and Tables

Standard Abbreviations

A

A	Area
ABRSV	Abrasive
ACCESS	Accessory
ACCUMR	Accumulator
ACET	Acetylene
ACR	Across
ACT	Actual
ACTR	Actuator
ADCN	Advances Document Change Notice
ADD	Addendum
ADH	Adhesive
ADJ	Adjust
ADV	Advance
AERO	Aeronautic
AL	Aluminum
ALIGN	Alignment
ALLOW	Allowance
ALT	Alternation
ALT	Alternate
ALY	Alloy
AMER NATL STD	American National Standard
AMP	Ampere
ANL	Anneal
ANOD	Anodize
ANSI	American National Standards Institute
APPD	Approved
APPROX	Approximate
AR	As Required
ASB	Asbestos
ASME	American Society of Mechanical Engineers
ASSEM	Assemble
ASSY	Assembly
AUTO	Automatic
AUX	Auxiliary
AVG	Average
AWG	American Wire Gage

B

B&S	Brown & Sharpe (Gage)
BAB	Babbitt
BC	Bolt Circle
BEV	Bevel
BHN	Brinell Hardness Number
BL	Base Line
BLT	Bolt
BNH	Burnish
BR	Bend Radius
BRKT	Bracket
BRS	Brass
BRZ	Bronze
BRZG	Brazing
BUSH	Bushing

C

C	Celsius
C TO C	Center to Center
C/R	Chamfer or Radius
CACL	Calculated
CAD	Computer-Aided Drafting
CAM	Computer-Aided Manufacturing
CANC	Canceled
CAP	Capacity
CARB	Carburize
CBORE	Counterbore
CC	Color Code
CDS	Cold-Drawn Steel
CENT	Centrifugal
CFM	Cubic Foot Per Minute
CFS	Cubic Foot Per Second
CH	Case Harden
CHAM	Chamfer
CI	Cast Iron
CIM	Computer-Integrated Manufacturing
CIR	Circular
CIRC	Circumference
CKT	Circuit
CL	Clearance
CLAD	Alclad
CLOS	Closure
CM	Centimeter
CMM	Computer Measuring Machine
CNC	Computer Numeric Control
CONC	Concentric
COND	Condition
CONT	Control
COP	Copper
CPLG	Coupling
CR VAN	Chrome Vanadium
CRS	Cold-Rolled Steel
CSK	Countersink
CTD	Coated
CTR	Center
CTR	Contour
CV	Check Valve
CYL	Cylinder

D

DAT	Datum
DBC	Diameter Bolt Circle
DCN	Document Change Notice
DCN	Drawing Change Notice
DEC	Decimal
DECR	Decrease
DEG	Degree
DET	Detail
DEV	Deviation
DF	Drop Forge
DFT	Draft
DIA or D	Diameter
DIAG	Diagonal
DIAG	Diagram
DIM	Dimension
DISC	Disconnect
DL	Developed Length
DNC	Direct Numerical Control
DP	Diametral Pitch
DR	Drill
DRM	Drafting Room Manual
DUP	Duplicate
DW	Developed Width
DWG	Drawing
DWL	Dowel

E

EA	Each
ECC	Eccentric
ECO	Engineering Change Order
EFF	Effective
ELEC	Electric
ENCL	Enclosure
ENG	Engine
ENG	Engineer
ENGR	Engineering
EO	Engineering Order
EQ	Equal
EQUIV	Equivalent
EST	Estimate

F

F/P	Flat Pattern
FAB	Fabricate
FAO	Finish All Over
FHD	Flat Head
FIL	Fillet
FIM	Full Indicator Movement
FIN	Finish
FIX	Fixture
FL	Fluid
FLEX	Flexible
FLG	Flange
FORG	Forging
FST	Forged Steel
FTG	Fitting
FURN	Furnish
FXD	Fixed

G

GA	Gage
GA	Gauge
GAL	Gallon
GALV	Galvanized
GEN	Generator
GRD	Grind
GRD	Ground
GSKT	Gasket

H

1/2H	Half-Hard
HCS	High-Carbon Steel
HD	Head
HDL	Handle
HDN	Harden
HEX	Hexagon
HF	High Frequency
HOR	Horizontal
HR	Hour
HRS	Hot-Rolled Steel
HS	High Speed
HSG	Housing
HT TR	Heat Treat
HYD	Hydraulic
HYDRO	Hydrostatic

I

ID	Inside Diameter
IDENT	Identification
IMPG	Impregnate
IN	Inch
INCL	Inclined
INCL	Include, Including, Inclusive
INCR	Increase
IND	Indicator
INDEP	Independent
INFO	Information
INSTL	Installation
INTER	Interrupt

J

JCT	Junction
JOG	Joggle

K

KWY	Keyway

L

LAB	Laboratory
LAM	Laminate
LAQ	Lacquer
LC	Low Carbon
LEN	Length
LG	Long
LH	Left Hand
LIN	Linear
LIQ	Liquid
LOM	List of Material
LS	Limit Switch
LTD	Limited
LTR	Letter
LUB	Lubricate

M

M	Magnaflux
MAG	Magnesium
MAINT	Maintenance
MAJ	Major
MALL	Malleable
MAN	Manual
MATL	Material
MAX	Maximum
MEAS	Measure
MECH	Mechanical
MED	Medium
MI	Malleable Iron
MID	Middle
MIL	Military
MIN	Minimum
MISC	Miscellaneous
MK	Mark
ML	Mold Line
MOD	Modification
MOT	Motor
MS	Master Switch
MULT	Multiple

N

NC	Numerical Control
NO	Number
NOM	Nomenclature
NOM	Nominal
NORM	Normalize
NS	Nickel Steel
NTS	Not to Scale

O

OA	Over-All
OBS	Obsolete
OD	Outside Diameter
OPP	Opposite
OZ	Ounce

P

P	Pitch
P	Port
PATT	Pattern
PC	Piece
PC	Pitch Circle
PD	Pitch Diameter
PKG	Package
PL	Parting Line (Castings)
PL	Plate
PLSTC	Plastic
PLT	Pilot
PNEU	Pneumatic
POS	Positive
PRESS	Pressure
PRI	Primary
PROC	Process, Procedure
PROD	Product, Production
PSI	Pounds Per Square Inch
PSIA	Pounds Per Square Inch, Absolute
PSIG	Pounds Per Square Inch, Gage
PV	Plan View

Q

QC	Quality Control
QTY	Quantity
QUAL	Quality
1/4H	Quarter-Hard

R

RAD *or* R	Radius
RD	Round
RECP	Receptacle
REF	Reference
REG	Regular
REG	Regulator
REL	Release
REQD	Required
RH	Right Hand
RH	Rockwell Hardness
RIV	Rivet
RM	Ream

S

SCH	Schedule
SCHEM	Schematic
SCR	Screw
Screw Threads	
NC	American National Coarse
NF	American National Fine
NEF	American National Extra Fine
8N	American National 8 Pitch
NPT	American Standard Taper Pipe
NPSC	American Standard Straight Pipe
NPTF	American Standard Taper (Dryseal)
NPSF	American Standard Straight (Dryseal)
UNC	Unified Screw Thread Coarse
UNF	Unified Screw Thread Fine
UNEF	Unified Screw Thread Extra Fine
8UN	Unified Screw Thread 8 Thread
SECT	Section
SEQ	Sequence
SER	Serial
SERR	Serrate
SF	Spot Face
SH	Sheet
SILS	Silver Solder
SO GR	Soft Grind
SOL	Solenoid
SPEC	Specification
SPL	Special
SST	Stainless Steel
STK	Stock
STL	Steel
SYM	Symbol
SYM	Symmetrical
SYS	System

T

T	Teeth
TAB	Tabulate
TAN	Tangent
TAP	Tapping
THD	Thread
THK	Thick
TIF	True Involute Form
TIR	Total Indicator Reading
TOL	Tolerance
TOR	Torque
TS	Tensile Strength
TS	Tool Steel
TU	Tungsten
TYP	Typical

V

V	Volt
VAC	Vacuum
VAR	Variable
VD	Void
VER	Vernier
VERT	Vertical
VIB	Vibrate
VOL	Volume

W

W	Watt
W	Wide, Width
WASH	Washer
WI	Wrought Iron
WLD	Welded
WP	Weatherproof
WRT	Wrought

Y

YP	Yield Point
YS	Yield Strength

Decimal and Metric Equivalents

INCHES		MILLI-METERS
FRACTIONS	DECIMALS	
	.00394	.1
	.00787	.2
	.01181	.3
1/64	.015625	.3969
	.01575	.4
	.01969	.5
	.02362	.6
	.02756	.7
1/32	.03125	.7938
	.0315	.8
	.03543	.9
	.03937	1.00
3/64	.046875	1.1906
1/16	.0625	1.5875
5/64	.078125	1.9844
	.07874	2.00
3/32	.09375	2.3813
7/64	.109375	2.7781
	.11811	3.00
1/8	.125	3.175
9/64	.140625	3.5719
5/32	.15625	3.9688
	.15748	4.00
11/64	.171875	4.3656
3/16	.1875	4.7625
	.19685	5.00
13/64	.203125	5.1594
7/32	.21875	5.5563
15/64	.234375	5.9531
	.23622	6.00
1/4	.2500	6.35
17/64	.265625	6.7469
	.27559	7.00
9/32	.28125	7.1438
19/64	.296875	7.5406
5/16	.3125	7.9375
	.31496	8.00
21/64	.328125	8.3344
11/32	.34375	8.7313
	.35433	9.00
23/64	.359375	9.1281
3/8	.375	9.525
25/64	.390625	9.9219
	.3937	10.00
13/32	.40625	10.3188
27/64	.421875	10.7156
	.43307	11.00
7/16	.4375	11.1125
29/64	.453125	11.5094
15/32	.46875	11.9063
	.47244	12.00
31/64	.484375	12.3031
1/2	.5000	12.70
	.51181	13.00
33/64	.515625	13.0969
17/32	.53125	13.4938
35/64	.546875	13.8907
	.55118	14.00
9/16	.5625	14.2875
37/64	.578125	14.6844
	.59055	15.00
19/32	.59375	15.0813
39/64	.609375	15.4782
5/8	.625	15.875
	.62992	16.00
41/64	.640625	16.2719
21/32	.65625	16.6688
	.66929	17.00
43/64	.671875	17.0657
11/16	.6875	17.4625
45/64	.703125	17.8594
	.70866	18.00
23/32	.71875	18.2563
47/64	.734375	18.6532
	.74803	19.00
3/4	.7500	19.05
49/64	.765625	19.4469
25/32	.78125	19.8438
	.7874	20.00
51/64	.796875	20.2407
13/16	.8125	20.6375
	.82677	21.00
53/64	.828125	21.0344
27/32	.84375	21.4313
55/64	.859375	21.8282
	.86614	22.00
7/8	.875	22.225
57/64	.890625	22.6219
	.90551	23.00
29/32	.90625	23.0188
59/64	.921875	23.4157
15/16	.9375	23.8125
	.94488	24.00
61/64	.953125	24.2094
31/32	.96875	24.6063
	.98425	25.00
63/64	.984375	25.0032
1	1.0000	25.4000

Number and Letter Size Drills

Number and Letter Size Drills Conversion Chart

Drill No. or Letter		Inch	mm
		.001	0.0254
		.002	0.0508
		.003	0.0762
		.004	0.1016
		.005	0.1270
		.006	0.1524
		.007	0.1778
		.008	0.2032
		.009	0.2286
		.010	0.2540
		.011	0.2794
		.012	0.3048
		.013	0.3302
80	.0135	.014	0.3556
79	.0145	.015	0.3810
	1/64	.0156	0.3969
78		.016	0.4064
		.017	0.4318
77		.018	0.4572
		.019	0.4826
76		.020	0.5080
75		.021	0.5334
		.022	0.5588
74	.0225	.023	0.5842
73		.024	0.6096
72		.025	0.6350
71		.026	0.6604
		.027	0.6858
70		.028	0.7112
		.029	0.7366
69	.0292	.030	0.7620
68		.031	0.7874
	1/32	.0312	0.7937
67		.032	0.8128
66		.033	0.8382
		.034	0.8636
65		.035	0.8890
64		.036	0.9144
63		.037	0.9398
62		.038	0.9652
61		.039	0.9906
		.0394	1.0000
60		.040	1.0160
59		.041	1.0414
58		.042	1.0668
57		.043	1.0922
		.044	1.1176
		.045	1.1430
56	.0465	.046	1.1684
	3/64	.0469	1.1906
		.047	1.1938
		.048	1.2192
		.049	1.2446
		.050	1.2700
		.051	1.2954
55		.052	1.3208
		.053	1.3462
		.054	1.3716
54		.055	1.3970
		.056	1.4224
		.057	1.4478
		.058	1.4732
		.059	1.4986
53	.0595	.060	1.5240
		.061	1.5494
		.062	1.5748
	1/16	.0625	1.5875
52	.0635	.063	1.6002

Drill No. or Letter		Inch	mm
		.064	1.6256
		.065	1.6510
		.066	1.6764
51		.067	1.7018
		.068	1.7272
		.069	1.7526
50		.070	1.7780
		.071	1.8034
		.072	1.8288
49		.073	1.8542
		.074	1.8796
		.075	1.9050
48		.076	1.9304
		.077	1.9558
47	.0785	.078	1.9812
	5/64	.0781	1.9844
		.0787	2.0000
		.079	2.0066
		.080	2.0320
46		.081	2.0574
45		.082	2.0828
		.083	2.1082
		.084	2.1336
		.085	2.1590
44		.086	2.1844
		.087	2.2098
		.088	2.2352
43		.089	2.2606
		.090	2.2860
		.091	2.3114
		.092	2.3368
42	.0935	.093	2.3622
	3/32	.0937	2.3812
		.094	2.3876
		.095	2.4130
41		.096	2.4384
		.097	2.4638
40		.098	2.4892
39	.0995	.099	2.5146
		.100	2.5400
		.101	2.5654
38	.1015	.102	2.5908
		.103	2.6162
37		.104	2.6416
		.105	2.6670
36	1065	.106	2.6924
		.107	2.7178
		.108	2.7432
		.109	2.7686
	7/64	.1094	2.7781
35		.110	2.7940
34		.111	2.8194
		.112	2.8448
33		.113	2.8702
		.114	2.8956
		.115	2.9210
32		.116	2.9464
		.117	2.9718
		.118	2.9972
		.1181	3.0000
		.119	3.0226
31		.120	3.0480
		.121	3.0734
		.122	3.0988
		.123	3.1242
		.124	3.1496
	1/8	.125	3.1750
		.126	3.2004

Drill No. or Letter		Inch	mm
		.127	3.2258
		.128	3.2512
30	.1285	.129	3.2766
		.130	3.3020
		.131	3.3274
		.132	3.3528
		.133	3.3782
		.134	3.4036
		.135	3.4290
29		.136	3.4544
		.137	3.4798
		.138	3.5052
		.139	3.5306
28	.1405	.140	3.5560
	9/64	.1406	3.5519
		.141	3.5814
		.142	3.6068
		.143	3.6322
27		.144	3.6576
		.145	3.6830
		.146	3.7084
26		.147	3.7338
		.148	3.7592
25	.1495	.149	3.7846
		.150	3.8100
		.151	3.8354
24		.152	3.8608
		.153	3.8162
23		.154	3.9116
		.155	3.9370
		.156	3.9624
	5/32	.1562	3.9687
22		.157	3.9878
		.1575	4.0000
		.158	4.0132
21		.159	4.0386
		.160	4.0640
20		.161	4.0894
		.162	4.1148
		.163	4.1402
		.164	4.1656
		.165	4.1910
19		.166	4.2164
		.167	4.2418
		.168	4.2672
		.169	4.2926
18	.1635	.170	4.3180
		.171	4.3434
	11/64	.1719	4.3656
		.172	4.3688
17		.173	4.3942
		.174	4.4196
		.175	4.4450
		.176	4.4704
16		.177	4.4958
		.178	4.5212
		.179	4.5466
15		.180	4.5720
		.181	4.5974
14		.182	4.6228
		.183	4.6482
		.184	4.6736
13		.185	4.6990
		.186	4.7244
		.187	4.7498
	3/16	.1875	4.7625
		.188	4.7752
12		.189	4.8006

Drill No. or Letter		Inch	mm
		.190	4.8260
11		.191	4.8514
		.192	4.8768
		.193	4.9022
10	.1935	.194	4.9278
		.195	4.9530
9		.196	4.9784
		.1969	5.0000
		.197	5.0038
		.198	5.0292
		.199	5.0546
8		.200	5.0800
7		.201	5.1054
		.202	5.1308
		.203	5.1562
	13/64	.2031	5.1594
6		.204	5.1816
5	.2055	.205	5.2070
		.206	5.2324
		.207	5.2578
		.208	5.2832
4		.209	5.3086
		.210	5.3340
		.211	5.3594
		.212	5.3848
3		.213	5.4102
		.214	5.4356
		.215	5.4610
		.216	5.4864
		.217	5.5118
		.218	5.5372
	7/32	.2187	5.5562
		.219	5.5626
		.220	5.5880
2		.221	5.6134
		.222	5.6388
		.223	5.6642
		.224	5.6896
		.225	5.7150
		.226	5.7404
		.227	5.7658
1		.228	5.7912
		.229	5.8166
		.230	5.8410
		.231	5.8674
		.232	5.8928
		.233	5.9182
A		.234	5.9436
	15/64	.2344	5.9531
		.235	5.9690
		.236	5.9944
		.2362	6.0000
		.237	6.0198
B		.238	6.0452
		.239	6.0706
		.240	6.0960
		.241	6.1214
C		.242	6.1468
		.243	6.1722
		.244	6.1976
		.245	6.2230
D		.246	6.2484
		.247	6.2738
		.248	6.2992
		.249	6.3246
8	1/4	.250	6.3500
		.251	6.3754
		.252	6.4008

Drill No. or Letter		Inch	mm
		.253	6.4262
		.254	6.4516
		.255	6.4770
		.256	6.5024
F		.257	6.5278
		.258	6.5532
		.259	6.5786
		.260	6.6040
G		.261	6.6294
		.262	6.6548
		.263	6.6802
		.264	6.7056
		.265	6.7310
	17/64	.2656	6.7469
H		.266	6.7564
		.267	6.7818
		.268	6.8072
		.269	6.8326
		.270	6.8580
		.271	6.8834
		.272	6.9088
		.273	6.9342
		.274	6.9596
		.275	6.9850
		.2756	7.0000
		.276	7.0104
		.277	7.0358
		.278	7.0612
		.279	7.0866
		.280	7.1120
K		.281	7.1374
	9/32	.2812	7.1437
		.282	7.1628
		.283	7.1882
		.284	7.2136
		.285	7.2390
		.286	7.2644
		.287	7.2898
		.288	7.3152
		.289	7.3406
L		.290	7.3660
		.291	7.3914
		.292	7.4168
		.293	7.4422
		.294	7.4676
M		.295	7.4930
		.296	7.5184
	19/64	.2969	7.5406
		.297	7.5438
		.298	7.5692
		.299	7.5946
		.300	7.6200
		.301	7.6454
N		.302	7.6708
		.303	7.6962
		.304	7.7216
		.305	7.7470
		.306	7.7724
		.307	7.7978
		.308	7.8232
		.309	7.8486
		.310	7.8740
		.311	7.8994
		.312	7.9248
	5/16	.3125	7.9375
		.313	7.9502
		.314	7.9756
		.3150	8.0000
		.315	8.0010

(continued)

Number and Letter Size Drills

Number and Letter Size Drills Conversion Chart

Drill No. or Letter	Inch	mm	Drill No. or Letter	Inch	mm	Drill No. or Letter	Inch	mm	Drill No. or Letter	Inch	mm	Drill No. or Letter	Inch	mm
O	.316	8.0264		.354	8.9916		.391	9.9314		.428	10.8712		.465	11.8110
	.317	8.0518		.3543	9.0000		.392	9.9568		.429	10.8966		.466	11.8364
	.318	8.0772		.355	9.0170		.393	9.9822		.430	10.9220		.467	11.8618
	.319	8.1026		.356	9.0424		.3937	10.0000		.431	10.9474		.468	11.8872
	.320	8.1280		.357	9.0678		.394	10.0076		.432	10.9728	15/32	.4687	11.9062
	.321	8.1534	T	.358	9.0932		.395	10.0330		.433	10.9982		.469	11.9126
	.322	8.1788		.359	9.1186		.396	10.0584		.4331	11.0000		.470	11.9380
P	.323	8.2042	23/64	.3594	9.1281	X	.397	10.0838		.434	11.0236		.471	11.9634
	.324	8.2296		.360	9.1440		.398	10.1092		.435	11.0490		.472	11.9888
	.325	8.2550		.361	9.1694		.399	10.1346		.436	11.0744		.4724	12.0000
	.326	8.2804		.362	9.1948		.400	10.1600		.437	11.0998		.473	12.0142
	.327	8.3058		.363	9.2202		.401	10.1854	7/16	.4375	11.1125		.474	12.0396
	.328	8.3312		.364	9.2456		.402	10.2108		.438	11.1252		.475	12.0650
21/64	.3281	8.3344		.365	9.2710		.403	10.2362		.439	11.1506		.476	12.0904
	.329	8.3566		.366	9.2964	Y	.404	10.2616		.440	11.1760		.477	12.1158
	.330	8.3820		.367	9.3218		.405	10.2870		.441	11.2014		.478	12.1412
	.331	8.4074	U	.368	9.3472		.406	10.3124		.442	11.2268		.479	12.1666
Q	.332	8.4328		.369	9.3726	13/32	.4062	10.3187		.443	11.2522		.480	12.1920
	.333	8.4582		.370	9.3980		.407	10.3378		.444	11.2776		.481	12.2174
	.334	8.4836		.371	9.4234		.408	10.3632		.445	11.3030		.482	12.2428
	.335	8.5090		.372	9.4488		.409	10.3886		.446	11.3284		.483	12.2682
	.336	8.5344		.373	9.4742		.410	10.4140		.447	11.3538		.484	12.2936
	.337	8.5598		.374	9.4996		.411	10.4394		.448	11.3792	31/64	.4844	12.3031
	.338	8.5852	3/8	.375	9.5250		.412	10.4648		.449	11.4046		.485	12.3190
R	.339	8.6106		.376	9.5504	Z	.413	10.4902		.450	11.4300		.486	12.3444
	.340	8.6360	V	.377	9.5758		.414	10.5156		.451	11.4554		.487	12.3698
	.341	8.6614		.378	9.6012		.415	10.5410		.452	11.4808		.488	12.3952
	.342	8.6868		.379	9.6266		.416	10.5664		.453	11.5062		.489	12.4206
	.343	8.7122		.380	9.6520		.417	10.5918	29/64	.4531	11.5094		.490	12.4460
11/32	.3437	8.7312		.381	9.6774		.418	10.6172		.454	11.5316		.491	12.4714
	.344	8.7376		.382	9.7028		.419	10.6426		.455	11.5570		.492	12.4968
	.345	8.7630		.383	9.7282		.420	10.6680		.456	11.5824		.493	12.5222
	.346	8.7884		.384	9.7536		.421	10.6934		.457	11.6078		.494	12.5476
	.347	8.8138		.385	9.7790	27/64	.4219	10.7156		.458	11.6332		.495	12.5730
S	.348	8.8392	W	.386	9.8044		.422	10.7188		.459	11.6586		.496	12.5984
	.349	8.8646		.387	9.8298		.423	10.7442		.460	11.6840		.497	12.6238
	.350	8.8900		.388	9.8552		.424	10.7696		.461	11.7094		.498	12.6492
	.351	8.9154		.389	9.8806		.425	10.7950		.462	11.7348		.499	12.6746
	.352	8.9408		.390	9.9060		.426	10.8204		.463	11.7602	1/2	.500	12.7000
	.353	8.9662	25/64	.3906	9.9219		.427	10.8458		.464	11.7856			

Metric Twist Drill Sizes

Metric Twist Drill Sizes					
Metric Drill Sizes (mm)[1]		Decimal Equivalent in Inches (Ref)	Metric Drill Sizes (mm)[1]		Decimal Equivalent in Inches (Ref)
Preferred	Available		Preferred	Available	
	.40	.0157	1.70		.0669
	.42	.0165		1.75	.0689
	.45	.0177	1.80		.0709
	.48	.0189		1.85	.0728
.50		.0197	1.90		.0748
	.52	.0205		1.95	.0768
.55		.0217	2.00		.0787
	.58	.0228		2.05	.0807
.60		.0236	2.10		.0827
	.62	.0244		2.15	.0846
.65		.0256	2.20		.0866
	.68	.0268		2.30	.0906
.70		.0276	2.40		.0945
	.72	.0283	2.50		.0984
.75		.0295	2.60		.1024
	.78	.0307		2.70	.1063
.80		.0315	2.80		.1102
	.82	.0323		2.90	.1142
.85		.0335	3.00		.1181
	.88	.0346		3.10	.1220
.90		.0354	3.20		.1260
	.92	.0362		3.30	.1299
.95		.0374	3.40		.1339
	.98	.0386		3.50	.1378
1.00		.0394	3.60		.1417
	1.03	.0406		3.70	.1457
1.05		.0413	3.80		.1496
	1.08	.0425		3.90	.1535
1.10		.0433	4.00		.1575
	1.15	.0453		4.10	.1614
1.20		.0472	4.20		.1654
1.25		.0492		4.40	.1732
1.30		.0512	4.50		.1772
	1.35	.0531		4.60	.1811
1.40		.0551	4.80		.1890
	1.45	.0571	5.00		.1969
1.50		.0591		5.20	.2047
	1.55	.0610	5.30		.2087
1.60		.0630		5.40	.2126
	1.65	.0650	5.60		.2205
				5.80	.2283

[1] Metric drill sizes listed in the "Preferred" column are based on the R40 series of preferred numbers shown in the ISO Standard R497. Those listed in the "Available" column are based on the R80 series from the same document.

(continued)

Metric Twist Drill Sizes

Metric Drill Sizes (mm)[1]		Decimal Equivalent in Inches (Ref)	Metric Drill Sizes (mm)[1]		Decimal Equivalent in Inches (Ref)
Preferred	Available		Preferred	Available	
6.00		.2362		19.50	.7677
	6.20	.2441	20.00		.7874
6.30		.2480		20.50	.8071
	6.50	.2559	21.00		.8268
6.70		.2638		21.50	.8465
	6.80[2]	.2677	22.00		.8661
	6.90	.2717		23.00	.9055
7.10		.2795	24.00		.9449
	7.30	.2874	25.00		.9843
7.50		.2953	26.00		1.0236
	7.80	.3071		27.00	1.0630
8.00		.3150	28.00		1.1024
	8.20	.3228		29.00	1.1417
8.50		.3346	30.00		1.1811
	8.80	.3465		31.00	1.2205
9.00		.3543	32.00		1.2598
	9.20	.3622		33.00	1.2992
9.50		.3740	34.00		1.3386
	9.80	.3858		35.00	1.3780
10.00		.3937	36.00		1.4173
	10.30	.4055		37.00	1.4567
10.50		.4134	38.00		1.4961
	10.80	.4252		39.00	1.5354
11.00		.4331	40.00		1.5748
	11.50	.4528		41.00	1.6142
12.00		.4724	42.00		1.6535
12.50		.4921		43.50	1.7126
13.00		.5118	45.00		1.7717
	13.50	.5315		46.50	1.8307
14.00		.5512	48.00		1.8898
	14.50	.5709	50.00		1.9685
15.00		.5906		51.50	2.0276
	15.50	.6102	53.00		2.0866
16.00		.6299		54.00	2.1260
	16.50	.6496	56.00		2.2047
17.00		.6693		58.00	2.2835
	17.50	.6890	60.00		2.3622
18.00		.7087			
	18.50	.7283			
19.00		.7480			

[1] Metric drill sizes listed in the "Preferred" column are based on the R40 series of preferred numbers shown in the ISO Standard R497. Those listed in the "Available" column are based on the R80 series from the same document.
[2] Recommended only for use as a tap drill size.

Unified Thread Series and Tap and Clearance Drills

Unified Coarse and Unified Fine Thread Series and Tap and Clearance Drills							
Size	Threads Per Inch	Major Dia.	Pitch Dia.	Tap Drill (75% Max. Thread)	Decimal Equivalent	Clearance Drill	Decimal Equivalent
2	56	.0860	.0744	50	.0700	42	.0935
	64	.0860	.0759	50	.0700	42	.0935
3	48	.099	.0855	47	.0785	36	.1065
	56	.099	.0874	45	.0820	36	.1065
4	40	.112	.0958	43	.0890	31	.1200
	48	.112	.0985	42	.0935	31	.1200
6	32	.138	.1177	36	.1065	26	.1470
	40	.138	.1218	33	.1130	26	.1470
8	32	.164	.1437	29	.1360	17	.1730
	36	.164	.1460	29	.1360	17	.1730
10	24	.190	.1629	25	.1495	8	.1990
	32	.190	.1697	21	.1590	8	.1990
12	24	.216	.1889	16	.1770	1	.2280
	28	.216	.1928	14	.1820	2	.2210
1/4	20	.250	.2175	7	.2010	G	.2610
	28	.250	.2268	3	.2130	G	.2610
5/16	18	.3125	.2764	F	.2570	21/64	.3281
	24	.3125	.2854	I	.2720	21/64	.3281
3/8	16	.3750	.3344	5/16	.3125	25/64	.3906
	24	.3750	.3479	Q	.3320	25/64	.3906
7/16	14	.4375	.3911	U	.3680	15/32	.4687
	20	.4375	.4050	25/64	.3906	29/64	.4531
1/2	13	.5000	.4500	27/64	.4219	17/32	.5312
	20	.5000	.4675	29/64	.4531	33/64	.5156
9/16	12	.5625	.5084	31/64	.4844	19/32	.5937
	18	.5625	.5264	33/64	.5156	37/64	.5781
5/8	11	.6250	.5660	17/32	.5312	21/32	.6562
	18	.6250	.5889	37/64	.5781	41/64	.6406
3/4	10	.7500	.6850	21/32	.6562	25/32	.7812
	16	.7500	.7094	11/16	.6875	49/64	.7656
7/8	9	.8750	.8028	49/64	.7656	29/32	.9062
	14	.8750	.8286	13/16	.8125	57/64	.8906
1	8	1.0000	.9188	7/8	.8750	1-1/32	1.0312
	14	1.0000	.9536	15/16	.9375	1-1/64	1.0156
1-1/8	7	1.1250	1.0322	63/64	.9844	1-5/32	1.1562
	12	1.1250	1.0709	1-3/64	1.0469	1-5/32	1.1562
1-1/4	7	1.2500	1.1572	1-7/64	1.1094	1-9/32	1.2812
	12	1.2500	1.1959	1-11/64	1.1719	1-9/32	1.2812
1-1/2	6	1.5000	1.3917	1-11/32	1.3437	1-17/32	1.5312
	12	1.5000	1.4459	1-27/64	1.4219	1-17/32	1.5312

Metric Tap Drills

Tap Drill Sizes for ISO Metric Threads				
	Series			
	Coarse		Fine	
Nominal Size mm	Pitch mm	Tap Drill mm	Pitch mm	Tap Drill mm
1	0.25	0.75	—	—
1.1	0.25	0.85	—	—
1.2	0.25	0.95	—	—
1.4	0.3	1.1	—	—
1.6	0.35	1.25	—	—
1.7	0.35	1.3	—	—
1.8	0.35	1.45	—	—
2	0.4	1.6	—	—
2.2	0.45	1.75	—	—
2.5	0.45	2.05	—	—
3	0.5	2.5	—	—
3.5	0.6	2.9	—	—
4	0.7	3.3	0.35	3.6
4	—	—	0.5	3.5
4.5	0.75	3.7	—	—
5	0.8	4.2	0.5	4.5
6	1	5	0.5	5.5
6	—	—	0.75	5.25
7	1	6	0.75	6.25
8	1.25	6.8	0.5	7.5
8	—	—	0.75	7.25
8	—	—	1	7
9	1.25	7.8	1	8
10	1.5	8.5	0.75	9.25
10	—	—	1	9
10	—	—	1.25	8.8
11	1.5	9.5	1	10
12	1.75	10.2	0.75	11.25
12	—	—	1	11
12	—	—	1.5	10.5
14	2	12	1	13
14	—	—	1.25	12.8
14	—	—	1.5	12.5
16	2	14	1	15
16	—	—	1.5	14.5
18	2.5	15.5	1	17
18	—	—	2	16
20	2.5	17.5	1	19
20	—	—	1.5	18.5
20	—	—	2	18

(continued)

Metric Tap Drills

Tap Drill Sizes for ISO Metric Threads				
	Series			
	Coarse		Fine	
Nominal Size mm	Pitch mm	Tap Drill mm	Pitch mm	Tap Drill mm
22	2.5	19.5	1	21
22	—	—	1.5	20.5
22	—	—	2	20
24	3	21	1.5	22.5
24	—	—	2	22
26	—	—	1.5	24.5
27	3	24	1.5	25.5
27	—	—	2	25
28	—	—	1.5	26.5
30	3.5	26.5	1.5	28.5
30	—	—	2	28
33	3.5	29.5	2	31
36	4	32	3	33
39	4	35	3	36

Unified Standard Screw Thread Series

Sizes		Basic Major Diameter	Threads Per Inch											Sizes
			Series with Graded Pitches			Series with Constant Pitches								
Primary	Secondary		Coarse UNC	Fine UNF	Extra Fine UNEF	4UN	6UN	8UN	12UN	16UN	20UN	28UN	32UN	
0		0.0600	—	80	—	—	—	—	—	—	—	—	—	0
	1	0.0730	64	72	—	—	—	—	—	—	—	—	—	1
2		0.0860	56	64	—	—	—	—	—	—	—	—	—	2
	3	0.0990	48	56	—	—	—	—	—	—	—	—	—	3
4		0.1120	40	48	—	—	—	—	—	—	—	—	—	4
5		0.1250	40	44	—	—	—	—	—	—	—	—	—	5
6		0.1380	32	40	—	—	—	—	—	—	—	—	UNC	6
8		0.1640	32	36	—	—	—	—	—	—	—	—	UNC	8
10		0.1900	24	32	—	—	—	—	—	—	—	—	UNF	10
	12	0.2160	24	28	32	—	—	—	—	—	—	UNF	UNEF	12
1/4		0.2500	20	28	32	—	—	—	—	—	UNC	UNF	UNEF	1/4
5/16		0.3125	18	24	32	—	—	—	—	—	20	28	UNEF	5/16
3/8		0.3750	16	24	32	—	—	—	—	UNC	20	28	UNEF	3/8
7/16		0.4375	14	20	28	—	—	—	—	16	UNF	UNEF	32	7/16
1/2		0.5000	13	20	28	—	—	—	—	16	UNF	UNEF	32	1/2
9/16		0.5625	12	18	24	—	—	—	UNC	16	20	28	32	9/16
5/8		0.6250	11	18	24	—	—	—	12	16	20	28	32	5/8
	11/16	0.6875	—	—	24	—	—	—	12	16	20	28	32	11/16
3/4		0.7500	10	16	20	—	—	—	12	UNF	UNEF	28	32	3/4
	13/16	0.8125	—	—	20	—	—	—	12	16	UNEF	28	32	13/16
7/8		0.8750	9	14	20	—	—	—	12	16	UNEF	28	32	7/8
	15/16	0.9375	—	—	20	—	—	—	12	16	UNEF	28	32	15/16
1		1.0000	8	12	20	—	—	UNC	UNF	16	UNEF	28	32	1
	1 1/16	1.0625	—	—	18	—	—	8	12	16	20	28	—	1 1/16
1 1/8		1.1250	7	12	18	—	—	8	UNF	16	20	28	—	1 1/8
	1 3/16	1.1875	—	—	18	—	—	8	12	16	20	28	—	1 3/16
1 1/4		1.2500	7	12	18	—	—	8	UNF	16	20	28	—	1 1/4
	1 5/16	1.3125	—	—	18	—	—	8	12	16	20	28	—	1 5/16
1 3/8		1.3750	6	12	18	—	UNC	8	UNF	16	20	28	—	1 3/8
	1 7/16	1.4375	—	—	18	—	6	8	12	16	20	28	—	1 7/16
1 1/2		1.5000	6	12	18	—	UNC	8	UNF	16	20	28	—	1 1/2
	1 9/16	1.5625	—	—	18	—	6	8	12	16	20	—	—	1 9/16
1 5/8		1.6250	—	—	18	—	6	8	12	16	20	—	—	1 5/8
	1 11/16	1.6875	—	—	18	—	6	8	12	16	20	—	—	1 11/16
1 3/4		1.7500	5	—	—	—	6	8	12	16	20	—	—	1 3/4
	1 13/16	1.8125	—	—	—	—	6	8	12	16	20	—	—	1 13/16
1 7/8		1.8750	—	—	—	—	6	8	12	16	20	—	—	1 7/8
	1 15/16	1.9375	—	—	—	—	6	8	12	16	20	—	—	1 15/16
2		2.0000	4 1/2	—	—	—	6	8	12	16	20	—	—	2
	2 1/8	2.1250	—	—	—	—	6	8	12	16	20	—	—	2 1/8
2 1/4		2.2500	4 1/2	—	—	—	6	8	12	16	20	—	—	2 1/4
	2 3/8	2.3750	—	—	—	—	6	8	12	16	20	—	—	2 3/8
2 1/2		2.5000	4	—	—	UNC	6	8	12	16	20	—	—	2 1/2
	2 5/8	2.6250	—	—	—	4	6	8	12	16	20	—	—	2 5/8
2 3/4		2.7500	4	—	—	UNC	6	8	12	16	20	—	—	2 3/4
	2 7/8	2.8750	—	—	—	4	6	8	12	16	20	—	—	2 7/8
3		3.0000	4	—	—	UNC	6	8	12	16	20	—	—	3
	3 1/8	3.1250	—	—	—	4	6	8	12	16	—	—	—	3 1/8
3 1/4		3.2500	4	—	—	UNC	6	8	12	16	—	—	—	3 1/4
	3 3/8	3.3750	—	—	—	4	6	8	12	16	—	—	—	3 3/8

(continued)

Unified Standard Screw Thread Series

Sizes		Basic Major Diameter	Threads Per Inch											Sizes
			Series with Graded Pitches			Series with Constant Pitches								
Primary	Secondary		Coarse UNC	Fine UNF	Extra Fine UNEF	4UN	6UN	8UN	12UN	16UN	20UN	28UN	32UN	
3½		3.5000	4	—	—	UNC	6	8	12	16	—	—	—	3½
	3⅝	3.6250	—	—	—	4	6	8	12	16	—	—	—	3⅝
3¾		3.7500	4	—	—	UNC	6	8	12	16	—	—	—	3¾
	3⅞	3.8750	—	—	—	4	6	8	12	16	—	—	—	3⅞
4		4.0000	4	—	—	UNC	6	8	12	16	—	—	—	4
	4⅛	4.1250	—	—	—	4	6	8	12	16	—	—	—	4⅛
4¼		4.2500	—	—	—	4	6	8	12	16	—	—	—	4¼
	4⅜	4.3750	—	—	—	4	6	8	12	16	—	—	—	4⅜
4½		4.5000	—	—	—	4	6	8	12	16	—	—	—	4½
	4⅝	4.6250	—	—	—	4	6	8	12	16	—	—	—	4⅝
4¾		4.7500	—	—	—	4	6	8	12	16	—	—	—	4¾
	4⅞	4.8750	—	—	—	4	6	8	12	16	—	—	—	4⅞
5		5.0000	—	—	—	4	6	8	12	16	—	—	—	5
	5⅛	5.1250	—	—	—	4	6	8	12	16	—	—	—	5⅛
5¼		5.2500	—	—	—	4	6	8	12	16	—	—	—	5¼
	5⅜	5.3750	—	—	—	4	6	8	12	16	—	—	—	5⅜
5½		5.5000	—	—	—	4	6	8	12	16	—	—	—	5½
	5⅝	5.6250	—	—	—	4	6	8	12	16	—	—	—	5⅝
5¾		5.7500	—	—	—	4	6	8	12	16	—	—	—	5¾
	5⅞	5.8750	—	—	—	4	6	8	12	16	—	—	—	5⅞
6		6.0000	—	—	—	4	6	8	12	16	—	—	—	6

ISO Metric Standard Screw Thread Series

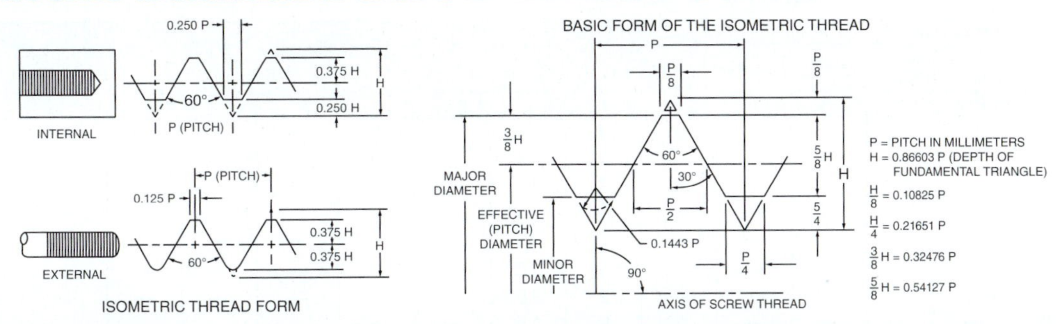

Nominal Size Diam. (mm)			Pitches (mm)														Nominal Size Diam. (mm)
Column[a]			Series with Graded Pitches		Series with Constant Pitches												
1	2	3	Coarse	Fine	6	4	3	2	1.5	1.25	1	0.75	0.5	0.25	0.25	0.2	
0.25			0.075	—	—	—	—	—	—	—	—	—	—	—	—	—	0.25
0.3			0.08	—	—	—	—	—	—	—	—	—	—	—	—	—	0.3
	0.35		0.09	—	—	—	—	—	—	—	—	—	—	—	—	—	0.35
0.4			0.1	—	—	—	—	—	—	—	—	—	—	—	—	—	0.4
	0.45		0.1	—	—	—	—	—	—	—	—	—	—	—	—	—	0.45
0.5			0.125	—	—	—	—	—	—	—	—	—	—	—	—	—	0.5
	0.55		0.125	—	—	—	—	—	—	—	—	—	—	—	—	—	0.55
0.6			0.15	—	—	—	—	—	—	—	—	—	—	—	—	—	0.6
	0.7		0.175	—	—	—	—	—	—	—	—	—	—	—	—	—	0.7
0.8			0.2	—	—	—	—	—	—	—	—	—	—	—	—	—	0.8
	0.9		0.225	—	—	—	—	—	—	—	—	—	—	—	—	—	0.9
1			0.25	—	—	—	—	—	—	—	—	—	—	—	—	0.2	1
	1.1		0.25	—	—	—	—	—	—	—	—	—	—	—	—	0.2	1.1
1.2			0.25	—	—	—	—	—	—	—	—	—	—	—	—	0.2	1.2
	1.4		0.3	—	—	—	—	—	—	—	—	—	—	—	—	0.2	1.4
1.6			0.35	—	—	—	—	—	—	—	—	—	—	—	—	0.2	1.6
	1.8		0.35	—	—	—	—	—	—	—	—	—	—	—	—	0.2	1.8
2			0.4	—	—	—	—	—	—	—	—	—	—	—	0.25	—	2
	2.2		0.45	—	—	—	—	—	—	—	—	—	—	—	0.25	—	2.2
2.5			0.45	—	—	—	—	—	—	—	—	—	—	0.35	—	—	2.5
3			0.5	—	—	—	—	—	—	—	—	—	—	0.35	—	—	3
	3.5		0.6	—	—	—	—	—	—	—	—	—	—	0.35	—	—	3.5
4			0.7	—	—	—	—	—	—	—	—	—	0.5	—	—	—	4
	4.5		0.75	—	—	—	—	—	—	—	—	—	0.5	—	—	—	4.5
5			0.8	—	—	—	—	—	—	—	—	—	0.5	—	—	—	5
		5.5	—	—	—	—	—	—	—	—	—	—	0.5	—	—	—	5.5
6			1	—	—	—	—	—	—	—	—	0.75	—	—	—	—	6
		7	1	—	—	—	—	—	—	—	—	0.75	—	—	—	—	7
8			1.25	1	—	—	—	—	—	—	1	0.75	—	—	—	—	8
		9	1.25	—	—	—	—	—	—	—	1	0.75	—	—	—	—	9
10			1.5	1.25	—	—	—	—	—	1.25	1	0.75	—	—	—	—	10
		11	1.5	—	—	—	—	—	—	—	1	0.75	—	—	—	—	11
12			1.75	1.25	—	—	—	—	1.5	1.25	1	—	—	—	—	—	12
	14		2	1.5	—	—	—	—	1.5	1.25[b]	1	—	—	—	—	—	14
		15	—	—	—	—	—	—	1.5	—	1	—	—	—	—	—	15
16			2	1.5	—	—	—	—	1.5	—	1	—	—	—	—	—	16
		17	—	—	—	—	—	—	1.5	—	1	—	—	—	—	—	17
	18		2.5	1.5	—	—	—	2	1.5	—	1	—	—	—	—	—	18
20			2.5	1.5	—	—	—	2	1.5	—	1	—	—	—	—	—	20
	22		2.5	1.5	—	—	—	2	1.5	—	1	—	—	—	—	—	22
24			3	2	—	—	—	2	1.5	—	1	—	—	—	—	—	24
		25	—	—	—	—	—	2	1.5	—	1	—	—	—	—	—	25
		26	—	—	—	—	—	—	1.5	—	1	—	—	—	—	—	26
	27		3	2	—	—	—	2	1.5	—	1	—	—	—	—	—	27
		28	—	—	—	—	—	2	1.5	—	1	—	—	—	—	—	28

(continued)

a Thread diameter should be selected from columns 1, 2 or 3; with preference being given in that order.
b Pitch 1.25 mm in combination with diameter 14 mm has been included for spark plug applications.
c Diameter 35 mm has been included for bearing locknut applications.
The use of pitches shown in parentheses should be avoided wherever possible.
The pitches enclosed in the bold frame, together with corresponding nominal diameters in Column 1 and 2, are those combinations which have been established by the ISO Recommendations as a selected "coarse" and "fine" series for commercial fasteners. Sizes 0.25 mm through 1.4 mm are covered in ISO Recommendation R 68 and, except for the 0.25 mm size, in AN Standards ANSI B1.10. (ANSI)

ISO Metric Standard Screw Thread Series

Nominal Size Diam. (mm)			Pitches (mm)														Nominal Size Diam. (mm)
Column[a]			Series with Graded Pitches		Series with Constant Pitches												
1	2	3	Coarse	Fine	6	4	3	2	1.5	1.25	1	0.75	0.5	0.25	0.25	0.2	
30			3.5	2	—	—	(3)	2	1.5	—	1	—	—	—	—	—	30
		32	—	—	—	—		2	1.5	—	—	—	—	—	—	—	32
	33		3.5	2	—	—	(3)	2	1.5	—	—	—	—	—	—	—	33
		35[c]	—	—	—	—	—	—	1.5	—	—	—	—	—	—	—	35c
36			4	3	—	—	—	2	1.5	—	—	—	—	—	—	—	36
		38	—	—	—	—	—	—	1.5	—	—	—	—	—	—	—	38
	39		4	3	—	—	—	2	1.5	—	—	—	—	—	—	—	39
		40	—	—	—	—	3	2	1.5	—	—	—	—	—	—	—	40
42			4.5	3	—	4	3	2	1.5	—	—	—	—	—	—	—	42
	45		4.5	3	—	4	3	2	1.5	—	—	—	—	—	—	—	45

a Thread diameter should be selected from columns 1, 2 or 3; with preference being given in that order.
b Pitch 1.25 mm in combination with diameter 14 mm has been included for spark plug applications.
c Diameter 35 mm has been included for bearing locknut applications.
The use of pitches shown in parentheses should be avoided wherever possible.
The pitches enclosed in the bold frame, together with corresponding nominal diameters in Column 1 and 2, are those combinations which have been established by the ISO Recommendations as a selected "coarse" and "fine" series for commercial fasteners. Sizes 0.25 mm through 1.4 mm are covered in ISO Recommendation R 68 and, except for the 0.25 mm size, in AN Standards ANSI B1.10. (ANSI)

Sheet Metal Gages

Gage No.	Sheet Steel		Galvanized Steel		Stainless Steel		Aluminum		Brass		Copper	
	Thick. (in)	lb/ft²	Thick. (in)	lb/ft²	Thick. (in)	lb/ft²	Thick. (in)	lb/ft²	Thick. (in)	lb/ft²	Thick. (in)	lb/ft²
0000000	—	—	—	—	.5000	20.808	—	—	—	—	—	—
000000	—	—	—	—	.4686	19.501	.5800	8.185	—	—	—	—
00000	—	—	—	—	.4375	18.207	.5165	7.289	—	—	—	—
0000	—	—	—	—	.4063	16.909	.4600	6.492	—	—	—	—
000	—	—	—	—	.3750	15.606	.4096	5.780	—	—	—	—
00	—	—	—	—	.3438	14.308	.3648	5.148	—	—	—	—
0	—	—	—	—	.3125	13.005	.3249	4.585	—	—	—	—
1	—	—	—	—	.2813	11.707	.2893	4.083	—	—	—	—
2	—	—	—	—	.2656	11.053	.2576	3.635	—	—	—	—
3	.2391	9.754	—	—	.2500	10.404	.2294	3.237	.2294	9.819	.2294	10.392
4	.2242	9.146	—	—	.2344	9.755	.2043	2.883	.2043	8.745	.2043	9.255
5	.2092	8.534	—	—	.2187	9.101	.1819	2.567	.1819	7.788	.1819	8.2420
6	.1943	7.927	—	—	.2031	8.452	.1620	2.286	.1620	7.185	.1620	7.340
7	.1793	7.315	—	—	.1875	7.803	.1443	2.036	.1443	6.400	.1443	6.536
8	.1644	6.707	.1681	6.858	.1719	7.154	.1285	1.813	.1285	5.699	.1285	5.821
9	.1495	6.099	.1532	6.250	.1562	6.500	.1144	1.614	.1144	5.074	.1144	5.183
10	.1345	5.487	.1382	5.638	.1406	5.851	.1019	1.438	.1019	4.520	.1019	4.616
11	.1196	4.879	.1233	5.030	.1250	5.202	.0907	1.280	.0907	4.023	.0907	4.110
12	.1046	4.267	.1084	4.422	.1094	4.553	.0808	1.140	.0808	3.584	.0808	3.650
13	.0897	3.659	.0934	3.810	.0937	3.899	.0720	1.016	.0720	3.193	.0720	3.250
14	.0747	3.047	.0785	3.202	.0781	3.250	.0641	.905	.0641	2.843	.0641	2.900
15	.0673	2.746	.0710	2.896	.0703	2.926	.0571	.806	.0571	2.532	.0571	2.585
16	.0598	2.440	.0635	2.590	.0625	2.601	.0508	.717	.0508	2.253	.0508	2.302
17	.0538	2.195	.0575	2.346	.0562	2.339	.0453	.639	.0453	2.009	.0453	2.050
18	.0478	1.950	.0516	2.105	.0500	2.081	.0403	.569	.0403	1.787	.0403	1.825
19	.0418	1.705	.0456	1.860	.0437	1.819	.0359	.507	.0359	1.592	.0359	1.626
20	.0359	1.465	.0396	1.615	.0375	1.561	.0320	.452	.0320	1.419	.0320	1.448
21	.0329	1.342	.0366	1.493	.0344	1.432	.0285	.402	.0285	1.264	.0285	1.289
22	.0299	1.220	.0336	1.371	.0312	1.298	.0253	.357	.0253	1.122	.0253	1.148
23	.0269	1.097	.0306	1.248	.0281	1.169	.0226	.319	.0226	1.002	.0226	1.023
24	.0239	.975	.0276	1.126	.0250	1.040	.0201	.284	.0201	.892	.0201	0.910
25	.0209	.853	.0247	1.008	.0219	.911	.0179	.253	.0179	.794	.0179	0.811
26	.0179	.730	.0217	.885	.0187	.778	.0159	.224	.0159	.705	.0159	0.722
27	.0164	.669	.0202	.824	.0172	.716	.0142	.200	.0142	.630	.0142	0.643
28	.0149	.608	.0187	.763	.0156	.649	.0126	.178	.0126	—	.0126	0.573
29	.0135	.551	.0172	.702	.0141	.587	.0113	.159	.0113	—	.0113	0.510
30	.0120	.490	.0157	.640	.0125	.520	.0100	.141	.0100	—	.0100	0.454
31	.0105	.428	.0142	.579	.0109	.454	.0089	.126	.0089	—	.0089	0.404
32	.0097	.396	.0134	.547	.0102	.424	.0080	.113	.0080	—	.0080	0.350
33	.0090	.367	—	—	.0094	.391	.0071	.100	.0071	—	.0071	0.321
34	.0082	.335	—	—	.0086	.358	.0063	.089	.0063	—	.0063	0.286
35	.0075	.306	—	—	.0078	.325	.0056	.079	.0056	—	.0056	0.234
36	.0067	.273	—	—	.0070	.291	.0050	.071	.0050	—	.0050	0.225
37	.0064	.261	—	—	.0066	.275	.0045	.064	.0045	—	.0045	0.202
38	.0060	.245	—	—	.0062	.258	.0040	.056	.0040	—	.0040	0.180
39	—	—	—	—	—	—	.0035	.049	—	—	—	—
40	—	—	—	—	—	—	.0031	.044	—	—	—	—

Precision Sheet Metal Setback Chart

Material Thickness

90° Bend Radius	.016	.020	.025	.032	.040	.051	.064	.072	.078	.081	.091	.102	.125	.129	.156	.162	.187	.250
1/32	.034	.039	.046	.05	.065	.081	.102	.113	.121	.125	.139							
3/64	.041	.046	.053	.062	.072	.090	.108	.119	.127	.131	.145							
1/16	.048	.053	.059	.068	.079	.093	.110	.122	.134	.138	.152							
5/64	.054	.060	.066	.075	.086	.100	.117	.127	.138	.144	.158							
3/32	.061	.066	.073	.082	.092	.107	.124	.134	.142	.146	.160							
7/64	.068	.073	.080	.08	.099	.113	.130	.141	.148	.153	.167	.181						
1/8	.075	.080	.086	.095	.106	.120	.137	.147	.155	.159	.172	.186	.216	.221				
9/64	.081	.087	.093	.102	.113	.127	.144	.154	.162	.166	.179	.193	.223	.228	.263			
5/32	.088	.093	.100	.109	.119	.134	.150	.161	.169	.173	.186	.200	.230	.235	.270	.278		
11/64	.095	.100	.107	.116	.126	.140	.157	.168	.175	.179	.192	.207	.236	.242	.277	.284	.317	
3/16	.102	.107	.113	.122	.133	.147	.164	.174	.182	.186	.199	.213	.243	.248	.283	.291	.324	.405
13/64	.108	.114	.120	.129	.140	.154	.171	.181	.189	.193	.206	.220	.250	.255	.290	.298	.330	.412
7/32	.115	.120	.127	.136	.146	.161	.177	.188	.196	.199	.212	.227	.257	.262	.297	.305	.337	.419
15/64	.122	.127	.134	.143	.153	.167	.184	.195	.202	.206	.219	.233	.263	.269	.304	.311	.344	.426
1/4	.129	.134	.140	.149	.160	.174	.191	.201	.209	.213	.226	.240	.270	.275	.310	.318	.351	.432
17/64	.135	.141	.147	.156	.166	.181	.198	.208	.216	.220	.233	.247	.277	.282	.317	.325	.357	.439
9/32	.142	.147	.154	.163	.173	.187	.204	.215	.223	.226	.239	.254	.284	.289	.324	.332	.364	.446

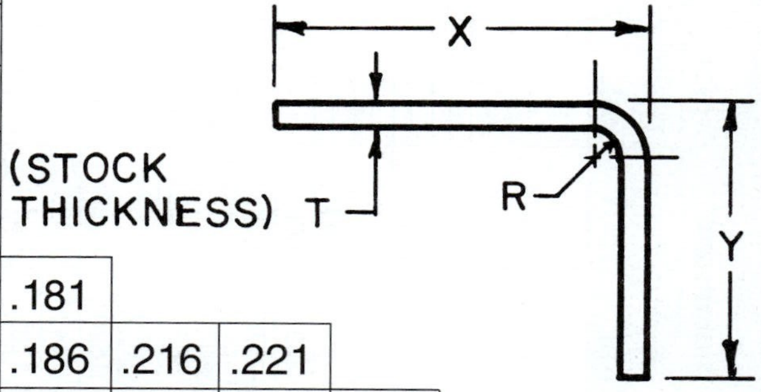

Developed Length = X + Y – Z

Z = Setback Allowance from the chart.

Geometric Dimensioning and Tolerancing Symbols

Characteristic	ASME Y14.5	ISO
STRAIGHTNESS	⏤	⏤
FLATNESS	⏥	⏥
CIRCULARITY	○	○
CYLINDRICITY	⌭	⌭
PROFILE OF A LINE	⌒	⌒
PROFILE OF A SURFACE	⌓	⌓
DYNAMIC PROFILE	△	NONE
ALL AROUND	←○	←○
ALL OVER	←◎	←◎
ANGULARITY	∠	∠
PERPENDICULARITY	⊥	⊥
PARALLELISM	//	//
POSITION	⌖	⌖
CONCENTRICITY	NONE	◎
SYMMETRY	NONE	⌯
CIRCULAR RUNOUT	↗	↗
TOTAL RUNOUT	⌰	⌰
AT MAXIMUM MATERIAL CONDITION	Ⓜ	Ⓜ
AT MAXIMUM MATERIAL BOUNDARY	Ⓜ	NONE
AT LEAST MATERIAL CONDITION	Ⓛ	Ⓛ
AT LEAST MATERIAL BOUNDARY	Ⓛ	NONE
PROJECTED TOLERANCE ZONE	Ⓟ	Ⓟ
TANGENT PLANE	Ⓣ	Ⓣ
FREE STATE	Ⓕ	Ⓕ
UNEQUALLY DISPOSED PROFILE	Ⓤ	UZ
TRANSLATION	▷	NONE
INDEPENDENCY	Ⓘ	NONE
DIAMETER	∅	∅
BASIC DIMENSION	[50]	[50]
REFERENCE DIMENSION	(50)	(50)
DATUM FEATURE	◀[A]	◀[A]

Other Dimensioning and Tolerancing Symbols

Feature	ASME Y14.5	ISO
CONICAL TAPER	⌲	⌲
SLOPE	⌳	⌳
COUNTERBORE	⌴	NONE
SPOTFACE	⌊SF⌋	NONE
COUNTERSINK	⌵	NONE
DEPTH/DEEP	↧	NONE
SQUARE	□	□
DIMENSION NOT TO SCALE	15	15
NUMBER OF TIMES/PLACES	8X	8X
ARC LENGTH	⌒105	⌒105
RADIUS	R	R
SPHERICAL RADIUS	SR	SR
SPHERICAL DIAMETER	SØ	SØ
CONTROLLED RADIUS	CR	NONE
BETWEEN	◄──►	◄──►
FROM-TO	─► or ─▷	NONE
DIMENSION ORIGIN	◄─⌖	NONE
STATISTICAL TOLERANCE	⟨ST⟩	NONE
CONTINUOUS FEATURE	⟨CF⟩	NONE

Symbols for Materials in Section

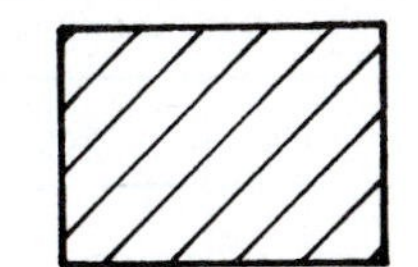

Cast iron and malleable iron. Also for use of all materials.

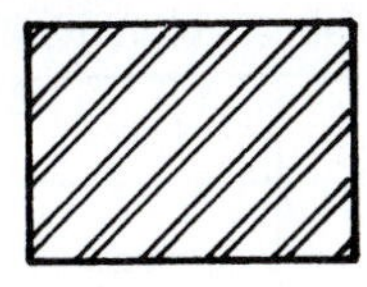

Steel

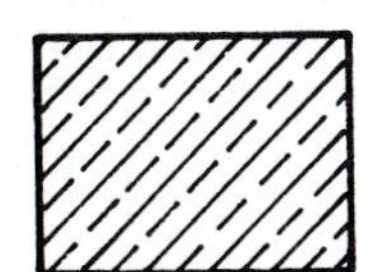

Brass, bronze, and compositions

White metal, zinc, lead, babbitt, and alloys

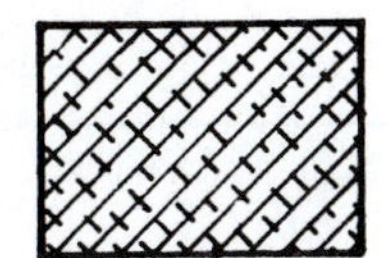

Magnesium, aluminum, and aluminum alloys

Rubber, plastic, electrical insulation

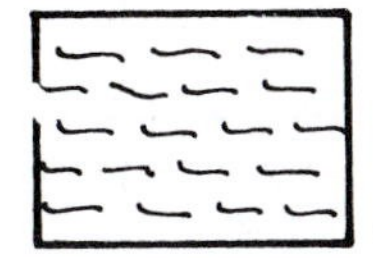

Cork, felt, fabric, leather, and fiber

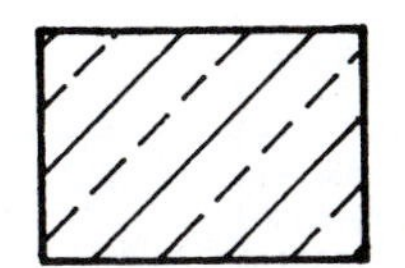

Firebrick and refractory material

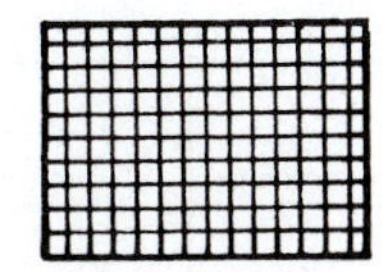

Electric windings, electromagnets, resistance, etc.

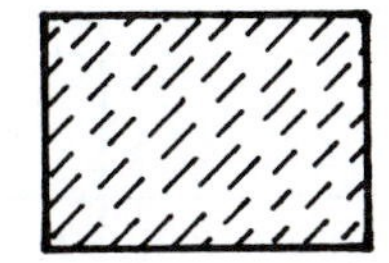

Marble, slate, glass, porcelain, etc.

Water and other liquids

Across grain
With grain
Wood

Standard Welding Symbols

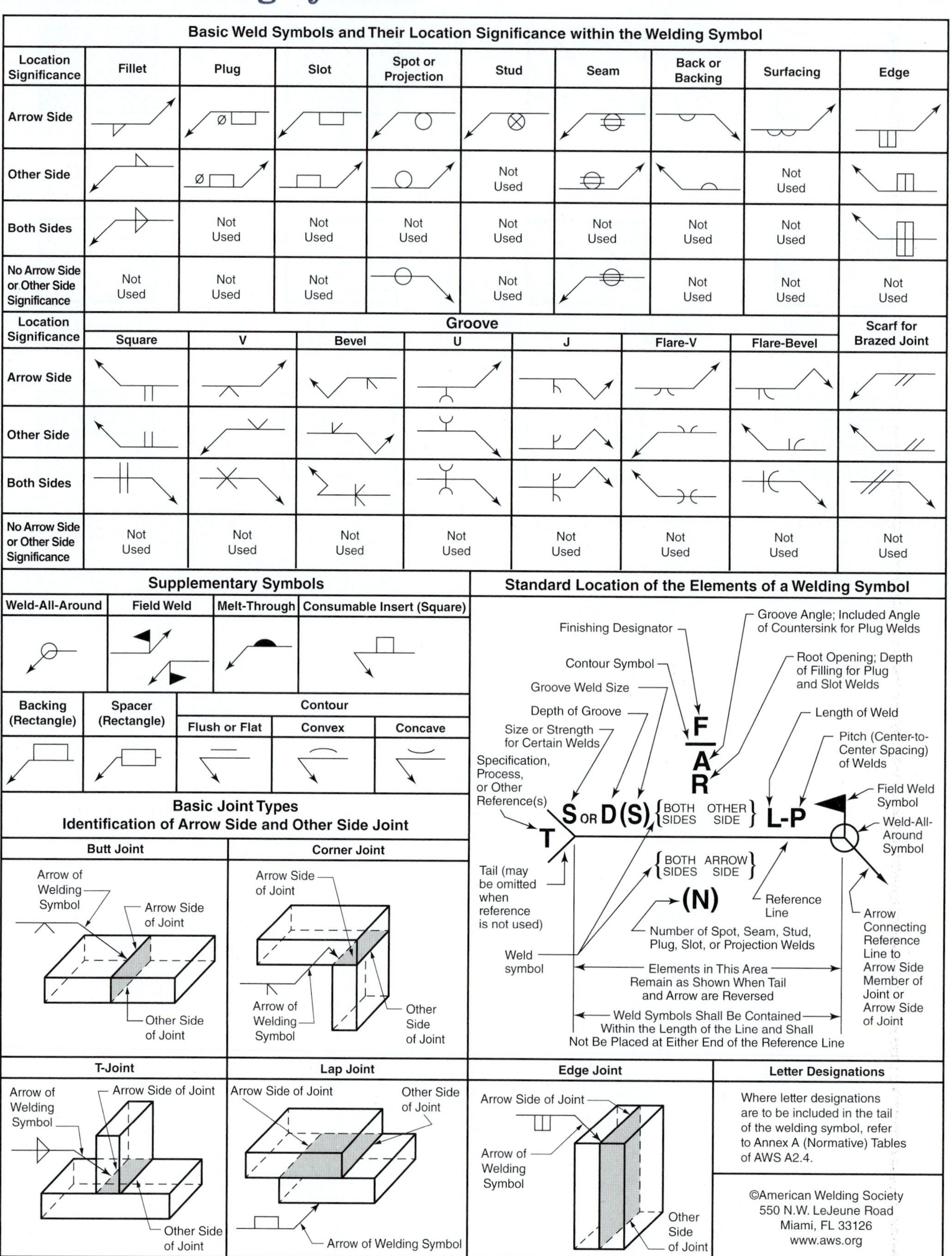

AWS A2.4, Standard Symbols for Welding, Brazing, and Nondestructive Examination, reproduced with permission from the American Welding Society (AWS), Miami, FL USA

(continued)

Standard Welding Symbols

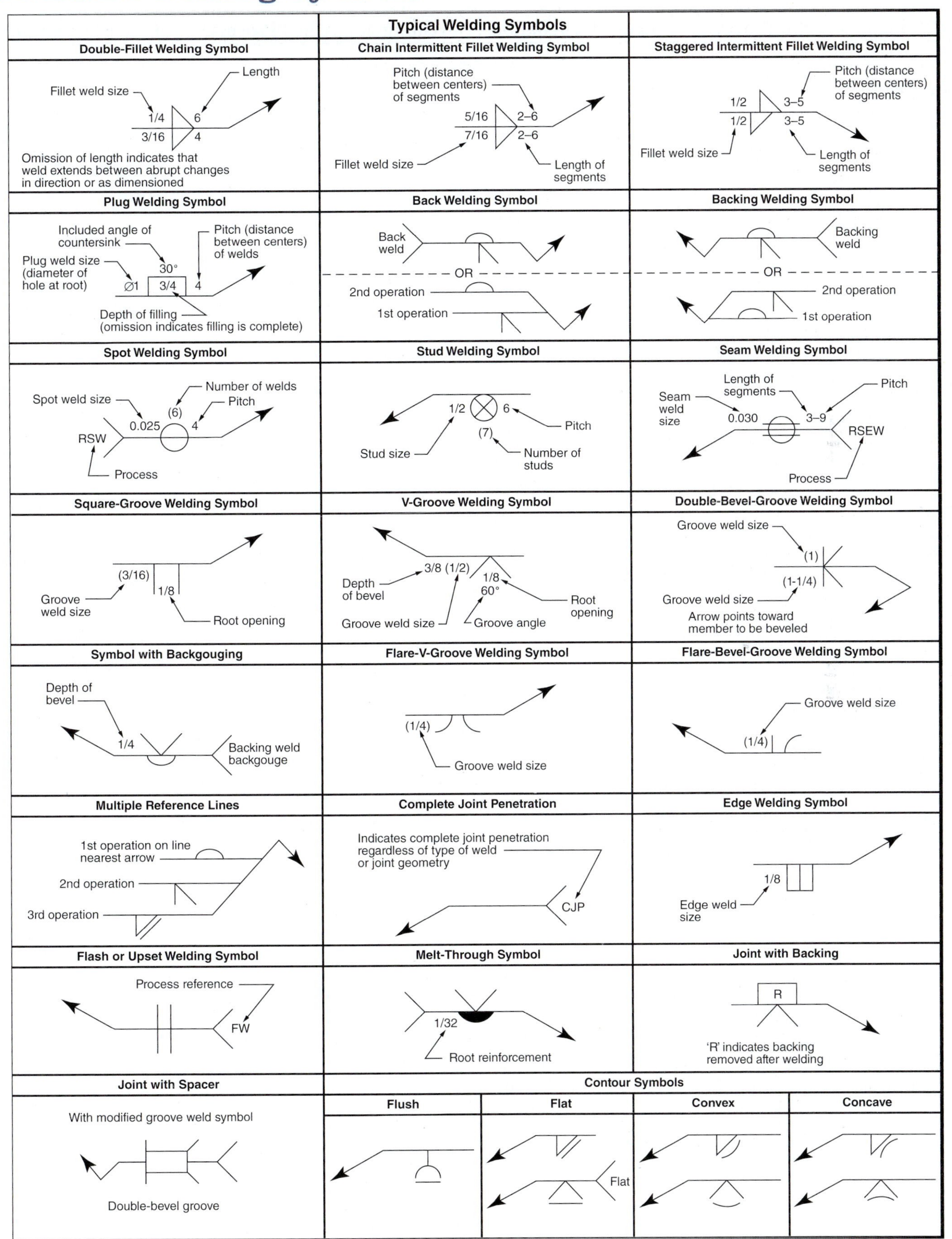

AWS A2.4, Standard Symbols for Welding, Brazing, and Nondestructive Examination, reproduced with permission from the American Welding Society (AWS), Miami, FL USA

Designation of Welding and Allied Processes by Letters

Welding and Allied Processes	Letter Designation
adhesive bonding	AB
arc welding	AW
arc stud welding	SW
atomic hydrogen welding	AHW
bare metal arc welding	BMAW
carbon arc welding	CAW
gas	CAW-G
shielded	CAW-S
twin	CAW-T
electrogas welding	EGW
flux cored arc welding	FCAW
gas metal arc welding	GMAW
pulsed	GMAW-P
short circuit	GMAW-S
gas tungsten arc welding	GTAW
pulsed	GTAW-P
plasma arc welding	PAW
shielded metal arc welding	SMAW
submerged arc welding	SAW
series	SAW-S
brazing	B
block brazing	BB
carbon arc brazing	CAB
diffusion brazing	DFB
dip brazing	DB
furnace brazing	FB
induction brazing	IB
infrared brazing	IRB
resistance brazing	RB
torch brazing	TB
braze welding	BW
arc braze welding	ABW
flow brazing	FLB
flow welding	FLOW
high energy beam welding	HEBW
electron beam welding	EBW
high vacuum	EBW-HV
medium vacuum	EBW-MV
nonvacuum	EBW-NV
laser beam welding	LBW
other welding processes	
electroslag welding	ESW
induction welding	IW
percussion welding	PEW
thermite welding	TW
oxyfuel gas welding	OFW
air acetylene welding	AAW
oxyacetylene welding	OAW
oxyhydrogen welding	OHW
pressure gas welding	PGW
resistance welding	RW
flash welding	FW
projection welding	PW
resistance seam welding	RSEW
high frequency	RSEW-HF
induction	RSEW-I
resistance spot welding	RSW
upset welding	UW
high frequency	UW-HF
induction	UW-I
soldering	S
dip soldering	DS
furnace soldering	FS
induction soldering	IS
infrared soldering	IRS
iron soldering	INS
resistance soldering	RS
torch soldering	TS
wave soldering	WS
solid-state welding	SSW
coextrusion welding	CEW
cold welding	CW
diffusion welding	DFW
explosion welding	EXW
forge welding	FOW
friction welding	FRW
hot pressure welding	HPW
roll welding	ROW
ultrasonic welding	USW
thermal cutting	TC
arc cutting	AC
carbon arc cutting	CAC
air carbon arc cutting	CAC-A
gas metal arc cutting	GMAC
gas tungsten arc cutting	GTAC
plasma arc cutting	PAC
shielded metal arc cutting	SMAC
electron beam cutting	EBC
laser beam cutting	LBC
air	LBC-A
evaporative	LBC-EV
inert gas	LBC-IG
oxygen	LBC-O
oxygen cutting	OC
flux cutting	OC-F
metal powder cutting	OC-P
oxyfuel gas cutting	OFC
oxyacetylene cutting	OFC-A
oxyhydrogen cutting	OFC-H
oxynatural gas cutting	OFC-N
oxypropane cutting	OFC-P
oxygen arc cutting	OAC
oxygen lance cutting	OLC
thermal spraying	THSP
arc spraying	ASP
flame spraying	FLSP
plasma spraying	PSP

Standard Fluid Power Graphic Symbols

THE SYMBOLS SHOWN CONFORM TO THE AMERICAN NATIONAL STANDARDS INSTITUTE (ANSI) SPECIFICATIONS. BASIC SYMBOLS CAN BE COMBINED IN ANY COMBINATION. NO ATTEMPT IS MADE TO SHOW ALL COMBINATIONS.

LINES AND LINE FUNCTIONS	
LINE, WORKING	
LINE, PILOT (L>20W)	
LINE, DRAIN (L<5W)	
CONNECTOR	
LINE, FLEXIBLE	
LINE, JOINING	
LINE, PASSING	
DIRECTION OF FLOW, HYDRAULIC PNEUMATIC	
LINE TO RESERVOIR ABOVE FLUID LEVEL BELOW FLUID LEVEL	
LINE TO VENTED MANIFOLD	
PLUG OR PLUGGED CONNECTION	
RESTRICTION, FIXED	
RESTRICTION, VARIABLE	

METHODS OF OPERATION	
PRESSURE COMPENSATOR	
DETENT	
MANUAL	
MECHANICAL	
PEDAL OR TREADLE	
PUSH BUTTON	

PUMPS	
PUMP, SINGLE FIXED DISPLACEMENT	
PUMP, SINGLE VARIABLE DISPLACEMENT	

MOTORS AND CYLINDERS	
MOTOR, ROTARY, FIXED DISPLACEMENT	
MOTOR, ROTARY, VARIABLE DISPLACEMENT	
MOTOR, OSCILLATING	
CYLINDER, SINGLE ACTING	
CYLINDER, DOUBLE ACTING	
CYLINDER, DIFFERENTIAL ROD	
CYLINDER, DOUBLE END ROD	
CYLINDER, CUSHIONS BOTH ENDS	

METHODS OF OPERATION	
LEVER	
PILOT PRESSURE	
SOLENOID	
SOLENOID CONTROLLED, PILOT PRESSURE OPERATED	
SPRING	
SERVO	

Provided by Vickers Industrial Division

(continued)

Standard Fluid Power Graphic Symbols

MISCELLANEOUS UNITS	
DIRECTION OF ROTATION (ARROW IN FRONT OF SHAFT)	
COMPONENT ENCLOSURE	
RESERVOIR, VENTED	
RESERVOIR, PRESSURIZED	
PRESSURE GAUGE	
TEMPERATURE GAUGE	
FLOW METER (FLOW RATE)	
ELECTRIC MOTOR	M
ACCUMULATOR, SPRING LOADED	
ACCUMULATOR, GAS CHARGED	
FILTER OR STRAINER	
HEATER	
COOLER	
TEMPERATURE CONTROLLER	
INTENSIFIER	
PRESSURE SWITCH	W
BASIC VALVE SYMBOLS	
CHECK VALVE	
MANUAL SHUT OFF VALVE	
BASIC VALVE ENVELOPE	
VALVE, SINGLE FLOW PATH, NORMALLY CLOSED	

BASIC VALVE SYMBOLS (CONT.)	
VALVE, SINGLE FLOW PATH, NORMALLY OPEN	
VALVE, MAXIMUM PRESSURE (RELIEF)	
BASIC VALVE SYMBOL, MULTIPLE FLOW PATHS	
FLOW PATHS BLOCKED IN CENTER POSITION	
MULTIPLE FLOW PATHS (ARROW SHOWS FLOW DIRECTION)	
VALVE EXAMPLES	
UNLOADING VALVE, INTERNAL DRAIN, REMOTELY OPERATED	
DECELERATION VALVE, NORMALLY OPEN	
SEQUENCE VALVE, DIRECTLY OPERATED, EXTERNALLY DRAINED	
PRESSURE REDUCING VALVE	
COUNTER BALANCE VALVE WITH INTEGRAL CHECK	
TEMPERATURE AND PRESSURE COMPENSATED FLOW CONTROL WITH INTEGRAL CHECK	
DIRECTIONAL VALVE, TWO POSITION, THREE CONNECTION	
DIRECTIONAL VALVE, THREE POSITION, FOUR CONNECTION	
VALVE, INFINITE POSITIONING (INDICATED BY HORIZONTAL BARS)	

Provided by Vickers Industrial Division

Glossary

A

accumulator. A container in which fluid is stored under pressure as a source of fluid power. (23)

actuate. To put devices and circuits into action. (23)

actuator. A device for converting hydraulic or pneumatic energy into mechanical energy. (23)

addendum (a). The radial distance between the pitch circle and the top of the tooth for a gear. (18)

addendum circle diameter. The outside diameter of the gear, or the pitch circle diameter plus twice the addendum. (18)

additive manufacturing. Technology that builds parts by adding material in a series of layers. Also known as *three-dimensional (3D) printing.* (1)

aligned dimensioning. A dimensioning system that features all lettering for the dimensions aligned with the dimension lines. See also *unidirectional dimensioning.* (9)

aligned section. A section view wherein the cutting plane is offset, with the result showing an odd number of features; usually used on a cylindrical part aligned with a vertical or horizontal plane. (6)

allowance. The intentional difference in the dimensions of mating parts to provide for different classes of fits. The tightest fit between two mating parts. (10)

all sheets same. A multisheet revision history method in which each sheet is independently revised but, when a revision affects a sheet, the same letter is used for all other sheets affected by the same revision. (14)

alphabet of lines. The list of types of lines and line weights standardized for use in industrial prints. (2)

altered item drawing. A detail drawing that describes an alteration procedure that transforms one part into another part. (15)

American Society of Mechanical Engineers (ASME). An independent, not-for-profit organization that defines standards for engineering drawings. (1)

angle of projection block. A block that identifies whether the drawing is a first-angle or third-angle projection. (3)

angularity. A geometric control applied to a surface or axis to maintain a given angle (other than 90°) of the surface or axis in reference to a datum surface or axis. A basic dimension is used, and no tolerance of degrees is needed. (13)

application block. A block that identifies the larger unit, subassembly, or assembly in which a part or subassembly is used. (3)

arc. Any curved edge with a constant radius and an angle of less than 360°. A portion of a circle. (4)

architect's scale. A triangular scale with one fully divided edge and five edges each containing two scales that share tick marks. (B)

arc welding. A method of fusion welding that uses an electric current via electrodes to generate the necessary heat to melt the metal. (22)

assembly drawing. A drawing showing the working relationship of the various parts of a machine or structure as they fit together. (16)

auxiliary view. A view projected at an angle inclined to the six principal views. (7)

average roughness (R_a). The average deviation of the measurements taken by a profilometer. For a contact-based profilometer, the stylus tip traverses the surface, moving up and down along the tiny ridges, and the R_a simply averages the readings. (12)

axonometric projection. A term applied to the orthographic projection of a rotated and tilted object to form a pictorial view. Types include isometric, dimetric, and trimetric. (C)

B

back gouging. The process of opening or cleaning the root on the other side of a partially welded joint, often accomplished with welding torch equipment or a grinder. (22)

backing symbol. A supplementary symbol in the form of a rectangle that indicates a metal material is to be placed behind the weld during the welding process. (22)

backing weld. A weld run along the back side of a groove weld to create root reinforcement, created *before* the groove weld has been created. (22)

Note: The number in parentheses following each definition indicates the unit in which the term is introduced.

backlash. The play (lost motion or extra space) between moving parts, such as a threaded shaft and nut or the teeth of meshing gears. (18)

back weld. A weld run along the back side of a groove weld to create root reinforcement, created *after* the groove weld has been created. (22)

balloon. A circle, usually connected to a leader line, containing a number or letter used to identify parts on a drawing. (3)

base circle. 1. As applied to gears, the diameter from which an involute tooth curve is generated or developed. 2. As applied to cams, the imaginary circle representing the diameter that would cause no change in the follower from its home position. (18, 19)

baseline dimensioning. A system of dimensioning wherein the features are dimensioned from a common datum surface or feature. (9)

basic dimension. A theoretically exact value used to describe the size, shape, or location of a feature. For geometric dimensioning and tolerancing, a basic dimension value is placed in a box. (10)

basic hole system. A standardized system wherein the calculation of limits for mating features uses the nominal design size at MMC, or smallest size, applied to the hole or internal feature. In the metric system, the basic hole system is referred to as the *hole basis system.* (10)

basic size. The size from which the limits of size are derived by the application of allowances and tolerances. (10)

bend. A uniformly curved section of material that serves as the edge between two sides or legs. (21)

bend allowance. The amount of sheet metal required to make a bend over a specified radius. (21)

bend angle. The number of degrees between flat surfaces on each side of a bend. (21)

bend axis. The theoretical center axis of the bend. (21)

bend deduction. The amount of material to deduct from the blank length to accommodate a bend. Also known as *setback*. (21)

bend instruction. A notation on the drawing giving the number of degrees and direction for forming a flange relative to the surface to which the note is associated. (21)

bend radius. The desired amount of the curvature measured on the inside of the bend, usually identified as R. (21)

bend tangent line. The line of transition between the flat surface leg and the curved portion of the bend. (21)

bevel gear. A gear resembling a cone in appearance with teeth grooves on the conical sides. (18)

bilateral tolerance. A toleranced dimension expressed in such a way as to give the amount of deviation permitted in two directions from the design size. (10)

blank length. The total distance across all legs and the bend allowance. (21)

block diagram. A diagram that maps the components of a system as boxes in a basic flow-chart manner. (23)

blueprint. An early, no longer common form of print featuring a blue background and white lines. Popularly, any plan of action or detailed procedure. (1)

bolt. The general term for a uniform-diameter cylinder with external threads on one end and a head on the other. (17)

bolt circle. A centerline circle used to locate a pattern of holes about an axis. Sometimes referred to as a *circle of centers.* (9)

boss. A small local thickening of the body of a casting, forging, or plastic part to allow more thickness for a bearing area or to support threads from a fastener. (20)

brake. A metalworking tool designed to bend or fold sheet metal. (21)

brake process. A method of forming using a punch and die to bend the surface up and around the shape of the punch. (21)

brazing. The joining of two close-fitting metal parts with a nonferrous filler metal with a melting point lower than that of the base metal. The filler material flows between the parts by capillary action. (22)

break line. A line used to help explain how a feature might look if part of it were broken out, as in section views, or if the feature were shortened to fit on the drawing. (2)

broaching. A manufacturing process for producing irregular hole shapes, such as squares, hexagons, ovals, or splines, by pushing or pulling a machine tool called a broach across the work. (11)

broken arrow. With reference to a welding symbol, an arrow leader that has a bend between the arrow and the reference line, used to indicate on which side of the joint the beveled pieces reside. (22)

broken-out section. A section wherein a cutting plane is started through a part and the part is presumed to be broken-out to show a small portion of internal detail. A break line distinguishes the break. (6)

by drawing method. A multisheet revision history method in which the revision history block is on sheet one and applies to the whole drawing set. (14)

by sheet method. A multisheet revision history method in which each sheet of a multisheet revision is treated independently and has its own revision history block. A revision status of sheets block identifies the revision status of each sheet. (14)

C

cabinet oblique drawing. An oblique drawing wherein the depth measurement is foreshortened, usually by half, to present a more realistic view. (C)

CAGE code. Acronym for Commercial and Government Entity. A five-digit alphanumeric code identifying companies that do business with the federal government. (3)

caliper. A measuring instrument consisting of a main scale with a fixed jaw and a sliding jaw. (B)

cam. A rotating or sliding device used to convert rotary motion into intermittent or reciprocating motion. (19)

cam profile. For a plate cam, the shape of the surface upon which the follower has contact. For a disk cam, the shape of the groove. (19)

cap screw. A finished bolt, which may have a hexagonal head and a built-in washer surface. (17)

casting drawing. A detail drawing showing the dimensions necessary to create a mold. (15)

casting (plastic). The process of forming a plastic object by pouring a fluid solution into an open mold where polymerization (curing) takes place. Because this process requires no pressure, it is one of the simplest techniques for molding plastics. (20)

cavalier oblique drawing. An oblique drawing wherein the depth measurement is not foreshortened. (C)

cavity. A depression, or set of matching depressions, in a mold that shapes the surfaces of the molded object. (20)

center distance. The distance between the center axes of mating gears. (18)

center line. A thin line comprised of a medium dash followed by a longer dash and used to designate symmetrical features or paths of motion. (2)

center line of bend (CLB). A line midway between bend tangent lines that indicates a bend contact line for the punch in a brake process. (21)

chain dimensioning. A system of dimensioning wherein dimensions are linked together end to end. (9)

chain line. A line that notes a special treatment or specification about a specific surface of a part. (2)

chamfer. A beveled edge. (11)

chord. Line segment within a circle connecting two points other than through the center point. (4)

chordal addendum. For gears, the distance from the top of the gear tooth to the chord at the pitch circle. (18)

chordal tooth thickness. For gears, the length of the chord between the intersection of the pitch circle and the two sides of the gear tooth. (18)

circle. Closed curve consisting of an edge that loops 360° around a center point at a fixed distance. (4)

circuit. The complete path of flow in a fluid power system, including the flow-generating device, such as a pump or compressor. (23)

circularity. A geometric control applied to circular elements of round parts, regardless of the axis or datums. Also known as *roundness.* (13)

circular pitch (P_c). The length of the arc along the pitch circle from the center of one gear tooth to the center of the next gear tooth. (18)

circular runout. A geometric control providing control of individual circular elements of a surface as the part is rotated about a datum axis. (13)

circular tooth thickness. The length of the arc along the pitch circle between the two sides of a gear tooth. (18)

circumscribe. To construct a polygon shape around the outside of a circle. (4)

civil engineer's scale. A triangular scale with individual scales based on the decimal system and marked off in divisions of 10, 20, 30, 40, 50, and 60 parts. Also known as an *engineer's scale.* (B)

clearance. For gears, the radial distance between the top of a tooth on one gear and the bottom of the tooth space on the mating gear. (18)

clevis pin. Non-threaded fastening device with a head on one end and a hole designed to accept a cotter pin on the other end. (17)

closed end. The end finish style of drawing compression springs where the pitch changes on the last coil so that the end of the spring is perpendicular to the axis. Also called *square end.* (17)

coaxial. General term applied to the condition in which two features, each having an axis, share a common center. The three-dimensional extension of concentric. (4)

component. A single unit or part. (23)

component list. A tabular chart or form providing the name, model number, quantity, and supplier or manufacturer of all the components in a circuit. Also called a *parts list* or *bill of materials.* (23)
composite tolerancing. In geometric dimensioning and tolerancing, an application of a geometric control in more than one manner to a feature or pattern of features using a multilayer feature control frame. (13)
compression spring. A spring designed to counteract a compressive force in line with the axis of the coil of wire. Characterized by coils that do not touch in the free position. (17)
computer-aided engineering (CAE). A system in which computer programs are used to generate data about manufacturing processes and the properties of parts. (1)
concentric. Having a common center, as circles or diameters. (4)
concentricity. A discontinued geometric control that was formerly applied to the elements of a surface to maintain equal midpoint distance of all elements on each side of a datum axis, regardless of feature shape or size. (13)
conduit. The tubing or piping within the fluid power system via which the fluid is transmitted between components. (23)
conical taper. The ratio of the diameters on each end of a cone-shaped surface, which can be dimensioned in a variety of ways, including a standard conical taper symbol. (10)
conic section. One of four shapes resulting from slicing a cone with a plane: circle, ellipse, hyperbola, or parabola. (4)
conjugate cam. A cam featuring two dual plate cams mounted together, transferring motion to two wheels on the same follower. (19)
consumable insert. A specially designed metal filler placed between the metal parts to be welded. (22)
continuous feature. Two surfaces interrupted by a groove or other feature that are treated as one feature, and formally identified as such with the assistance of a standard symbol. (10)
contour dimensioning. A system of dimensioning wherein the dimensions are placed in the view that best represents the shape of the feature. (9)
control. A device used to regulate the function of a component or system. (23)
control drawing. A drawing of a purchased part to document various engineering requirements to assure the interchangeability of items each time they are purchased. (15)
convention. A generally accepted way of doing something. (2)
conventional practice. The breaking of orthographic projection rules or other scientific explanations for the sake of clarity. Conventional practices are published in standards. (6)
coordinate dimensioning without dimension lines. An arrowless system of dimensioning wherein the horizontal and vertical dimension values are placed next to the feature and are referenced from an origin or datum plane. Also referred to as *ordinate dimensioning*, *arrowless dimensioning*, and *datum dimensioning.* (9)
core. 1. The central member of a laminate to which surface laminates are attached. 2. A channel in a mold for circulation of heat-transfer media. 3. Part of a complex mold that forms undercut parts; usually withdrawn to one side before the main sections of the mold are opened. Also called a *core pin.* 4. The central conductor in coaxial cables. (20)
cotter pin. A non-threaded fastening device made of a folded strip of metal used to prevent parts from slipping or rotating off an assembly. (17)
counterbore. The cylindrical enlargement of the end of a hole to a specified diameter and depth. (11)
counterdrill. A hole with a conical transition to a smaller-diameter hole. The transition is usually shown at the same angle as a drill bit tip (120°). (11)
countersink. The conical enlargement, or chamfered end, of a hole to a specified angle and diameter or depth. (11)
crest. The top of a screw thread ridge. (8)
cutaway diagram. A diagram depicting symbols of components and piping as cutaways. (23)
cutting-plane line. A thick line made of dashes used to indicate the cutting plane used in a section view. Arrows indicating the direction in which the section is to be viewed usually accompany the line. (2, 6)
cylinder. Within the context of fluid power, a device that features a piston rod that moves from one position to another due to applied fluid pressure. (23)
cylindrical cam. A cam with a follower groove cut into the surface of a cylinder. Also known as a *barrel cam.* (19)
cylindricity. A geometric control applied to a cylindrical surface to maintain the form of the surface elements, regardless of the axis or a datum. (13)

D

datum. A point, axis, or plane assumed to be exact for purposes of computation from which the location or geometric control of other features is established. (9)

datum feature. An actual geometric feature of a part used to establish a datum. (13)

datum feature symbol. A standardized symbol used to identify a feature or features of the object as a datum plane or axis. (10)

datum precedence. The order in which an object is placed into the datum reference frame. (13)

datum reference frame. The term applied to the establishment of a part in three-dimensional space for measuring features in all directions. This is especially important for positional tolerancing. (13)

datum target. A specific point, line, or area of contact on a part that can be specified when establishing a theoretical datum. (13)

dedendum (b). For gears, the radial distance between the pitch circle and the bottom of the gear tooth. (18)

dedendum circle diameter. For gears, the pitch circle minus twice the dedendum. (18)

denominator. The bottom number of a fraction, indicating the number of equal parts into which the unit is divided. (A)

depth. The front-to-back measurement of an object as defined from the front view. (5)

depth auxiliary view. An auxiliary view projected from the front view. (7)

depth of fusion. The distance the fusion face of the weld penetrates the parent piece. (22)

design process. Five-step process, with cycling and looping between steps, used to bring an idea to market or to improve a current product. (1)

design web format (DWF). An electronic file format developed by Autodesk and designed for the efficient distribution and communication of design data, such as drawings and models, including the ability to view, review comments, and print. (1)

detail assembly drawing. An assembly drawing in multiview form that serves as a detail drawing for the parts in an assembly. (16)

detail drawing. A drawing that provides all of the information necessary to produce a single part. (15)

detailed representation. One of three methods for representing screw threads on a drawing. This representation appears the most lifelike but is not a true projection. See also *schematic representation* and *simplified representation.* (8)

developed length. The dimensions of the flat part that result in the correct dimensions for the folded part. (21)

diagram assembly drawing. An assembly drawing using conventional symbols joined with single lines to show how parts are assembled. (16)

diameter. Distance of the center axis of a circle from outer edge to outer edge. (4)

diametral pitch (P_d). The number of gear teeth per inch of pitch diameter. Mating gears must have identical diametral pitches. (18)

diazo. An organic compound that evaporates when exposed to light and turns blue when exposed to ammonia. Prints made using a diazo compound are referred to as whiteprints or blue-line prints. (1)

die. A precision tool, usually metal in form, that is used in manufacturing equipment to cut, shape, or form metal and other materials. (21)

dihedral angle. The angle between two plane surfaces, shown true when observing the point view of the intersection between the planes. (7)

dimension. The distance between two points or the size description of an object, indicated as an annotation. (2)

dimensioning. The process of providing the size description of an object. Dimensions are comprised of dimension lines, extension lines, and leader lines. (9)

dimensioning mechanics. Instructions or guidelines describing the size and spacing of all the components of a dimension. (9)

dimension line. The line used in dimensioning that shows the extent and direction of the measurement. (2)

dimension origin symbol. A symbol that indicates which surface acts as the origin for measurements. It replaces one of the arrowheads on the dimension line. (9, 10)

displacement. In fluid power applications, the quantity of fluid that can pass through a pump, motor, or cylinder in a single revolution or stroke. For cam design, the amount the follower moves as displaced by the cam surface contacting the follower. (19)

displacement diagram. A diagram that charts the movement of a cam follower, in a linear fashion, during a cycle of time. (19)

dowel pin. A pin on a part that fits into a hole in a mating part to prevent motion or slipping, or to ensure accurate location of assembly. (17)

draft. A slight taper in a mold wall designed to facilitate removal of a molded object. When the taper is in the opposite direction, it tends to impede removal of the articles and is called a *back draft.* (20)

drafting. The process of creating drawings of objects for the technical fields. (1)

drawing deviation (DD). A formal description of a temporary revision that authorizes a deviation from the print. (14)

drawing graphic sheet. A drawing consisting of orthographic or axonometric views of a three-dimensional model that is not required to provide a complete product definition. (9)

drawing number. Unique number assigned to a print for identification and archival purposes. (3)

drawing revision. Any change made to a drawing after final approval has been given to the drawing. (14)

drawing title. Brief and descriptive name that clearly identifies a part or assembly. (3)

dual dimensioning. Dimensioning using both US customary and SI metric units. (9)

dwell. The period of time for which a cam follower does not move or change position. (19)

E

eccentric. Not having a common center. A device that converts rotary motion into reciprocating (back-and-forth) motion. (4)

edge. In computer modeling, the connecting line between two vertices, often represented as the intersection of two faces. (4)

ejector pin. A hardened steel pin designed to push a plastic product out of a mold. (20)

elastomer. A material having rubber-like characteristics. Elastomers stretch to at least twice their length and return rapidly to their original length. (20)

electron beam welding (EBW). A form of intense energy beam fusion welding that bombards the workpiece with a focused stream of electrons traveling at high speed within a vacuum. (22)

enclosure. A rectangle drawn around a component or components to indicate the limits of an assembly. Port connections are shown on the enclosure line. (23)

endface cam. A cylindrical cam with only one cam surface, allowing the follower to ride on the rim of the cylinder. Also known as a *side cam.* (19)

engineering change order (ECO). One of several possible terms for the documentation accompanying the change of an industrial print. A number is usually assigned to any change made to a drawing once the part is in production. Also called *alteration notice (AN)*, *change in design (CID)*, *engineering notice (EN)*, *engineering change notice (ECN)*, and *advance drawing change notice (ADCN).* (14)

envelope. The representative squares, in singular or multiple combination, that symbolize a valve and its characteristics in various positions. (23)

evaluation length. The entire scope of the surface profile being evaluated, versus the sampling length or roughness cutoff, which defines segments within the profile. (12)

exploded assembly drawing. A term given to an assembly drawing utilizing a technique for spreading the parts away from each other but in-line with each other as assembled. (16)

extension line. The line used in dimensioning to extend the object out away from the view so the dimension lines do not block the visualization of the shape. Sometimes referred to as witness lines. (2)

extension spring. A spring designed to counteract a pulling force in line with the axis of the coil of wire. Characterized by coils that touch in the free position. (17)

external thread. Screw thread cut on an external surface. (8)

extrusion (plastic). The continuous forming of primarily thermoplastic materials by forcing the material through a die at the end of an extruder. (20)

F

face. 1. In computer modeling, a closed set of edges, which may be limited by the system to three or four edges, which in turn, if coplanar, can form surfaces. 2. In welding, the surface of the weld bead that is exposed and can be contoured as flat, concave, or convex. (4, 22)

face cam. A cam with a follower groove in the flat surface of a disc. Also known as a *groove cam.* (19)

face gear. A circular disc with a ring of teeth cut in its side face that can be driven by a spur gear, similar in function to a 90° bevel gear set. (18)

faceted. In computer modeling, a term that describes the representation of a model as defined by faces that, in combination with each other, form a polygonal mesh. (4)

faying surfaces. The surfaces that contact each other for the purpose of bonding by methods such as adhesives, bolts, rivets, or weldment. (22)

feature. A portion of a part, such as a diameter, hole, keyway, or flat surface. (13)

feature control frame. The rectangular box within geometric dimensioning and tolerancing practices containing such items as the geometric control, tolerance amount, and, if applicable, modifiers and datum references. (13)

feature of size. In geometric dimensioning and tolerancing, a feature with a center axis or center plane. (13)

filler. The material that is added to the weld, perhaps in the form of consumable electrode welding rods, inserts, or spacers. (22)

fillet. Interior rounded corner or edge. See also *round* and *runout.* (5)

fillet weld. A weld created in a triangular fashion, often within a right-angle corner joint or on one or both sides of a T-joint. (22)

finish. General finish requirements, such as paint, chemical, or electroplating, rather than surface texture or roughness. See *surface texture.* (12)

first-angle projection. A system that serves as the foundation for orthographic projection and is used primarily in countries *other than* the United States and Canada. See *third-angle projection.* (5)

flag. A local note consisting of a number enclosed in an equilateral triangle. The note text appears elsewhere on the print, with other flagged local note text. (11)

flange. An edge or collar fixed at an angle to the main part or web, as in an I-beam. In sheet metal design, a portion of an item being bent, such as a side or leg. (21)

flank. The surface area joining the crest and root that could also be considered the "side" of the ridge or groove. (8)

flank angle. The angle between the flank and a line perpendicular to the axis. (8)

flash. The thin web of excess material that is forced into crevices between mating mold surfaces and remains attached to the molded part. (20)

flatness. A geometric control applied to a surface or center plane to maintain flatness of the surface elements, regardless of other surfaces or datum references. (13)

flaw. An unintentional interruption in the surface, such as a crack, pit, or dent. (12)

fluid. A liquid or gas. (23)

fluid conditioner. A component whose function is to control the physical characteristics of the fluid, such as heating, cooling, or filtering. (23)

fluid power. Power that is generated, controlled, and transmitted using fluids under pressure. (23)

folding line. Reference line marking the "hinge" between projection planes. It should be a phantom line and can be marked with projection labels. Also known as a *bend line.* (7)

follower. In cam design, the machine or mechanism that moves with reciprocating movement by following the cam as it rotates or oscillates. (19)

follower constraint. The constraining force that keeps the follower in contact with the cam, such as gravity, spring tension, or the positive control of a groove. (19)

form block line (FBL). A line representing the surface of a forming tool about which the metal will be bent. (21)

form tolerance. The tolerances of straightness, flatness, circularity, and cylindricity, each controlling the form of individual features without respect to datums or other features. (13)

freeform surface. A three-dimensional surface defined *not* by rigidly defined dimensions such as is common to planes and cylindrical or conical surfaces, but rather by equation-driven curves, such as NURBS-based surfaces. (4)

frontal plane. In multiview projection theory, a projection plane onto which the front view is projected. (5)

full indicator movement (FIM). The total movement of the dial indicator needle in measuring the variance of a surface. Also known as *full indicator reading (FIR)* and *total indicator reading (TIR).* (13)

full section. A section view wherein the object has been fully cut by a flat plane. For example, a wheel cut in half by a plane. (6)

fusion face. The portion of surface along the joint that will experience a phase change, joining the parent metal to the weld bead. (22)

fusion welding. A main category of welding processes wherein the pieces are joined using a process that melts the metals of each piece as well as any filler material. (22)

G

gage. The thickness of sheet metal. The higher the gage, the thinner the material. Also spelled *gauge.* (21)

garter spring. A long extension spring designed to encompass a circular feature in a belt-like manner. (17)

gas welding. A type of fusion welding wherein a gas fuel mixed with pure oxygen creates the intense heat needed to create the weld. Also referred to as *oxyfuel welding.* (22)

gate. In injection and transfer molding, a channel through which the molten resin flows from the runner into the cavity. It may have the same diameter or cross section as the runner but more often is restricted to 1/8″ or less. (20)

gate vestige. The residual material that remains after a gate is torn or cut away from a part. (20)

gear blank. A common starting point for manufacturing gears that primarily is a gear with the teeth still unformed. (18)

general note. A note that applies to the entire drawing or model. (11)

geometric dimensioning and tolerancing (GD&T). A system of dimensioning and tolerancing a part with respect to the actual function or relationship of part features that can be most economically produced. (13)

globoidal cam. A more complex type of cam that may be designed in a concave or convex barrel fashion that features a groove that drives an oscillating follower, or as a thread-like cam surface that drives a turret wheel follower in an intermittent rotating fashion. (19)

graphic diagram. A diagram consisting of graphic symbols joined by lines. (23)

groove weld. A weld made into a groove-shaped joint. Often used if larger pieces of metal are to be welded. (22)

gusset. A small plate used to reinforce assemblies. Also, in plastics, an angular piece of material added to strengthen two adjoining walls. (20)

H

half section. A section view wherein the object is only cut halfway through to the center. In essence, one-fourth of the object is cut, and the result is a "double exposure." On one side of the center line, you see external detail with no hidden lines. On the other side of the center line, you see internal detail as a section. (6)

heat affected zone (HAZ). The area around the weld bead that has been impacted by the welding process and may be subject to change in properties or weakness. (22)

height. The top-to-bottom measurement of an object as defined from the front view. (5)

height auxiliary view. An auxiliary view projected from a top view. (7)

helical gear. A gear with teeth at an angle to the axis of revolution and therefore curved in a helical manner, providing for a smoother operation but adding to the difficulty of manufacture. (18)

helix. A geometric shape generated by a point moving around an axis while progressing along that axis, like a barber pole or candy cane stripe. (4)

herringbone gear. A gear that combines helical gear teeth of opposing directions, featuring a V-shape appearance and allowing it to operate without the axial load that a helical gear incurs. Also known as a *double helical gear.* (18)

hidden line. A thin series of short dashes spaced closely together representing a hidden feature. (2)

hobbing machine. A specialized milling machine designed specifically to form gears, using a specialized cutting tool known as a hob. Also called a *hobber.* (18)

honing. A method of finishing a hole or other surface to a precise tolerance using an abrasive block and rotary motion. (12)

horizontal plane. In multiview projection theory, a projection plane onto which the top view is projected. (5)

hydraulic. Those processes using pressurized liquid, channeled through tubing or pipes, to perform mechanical work. (23)

hydroforming. A method of forming characterized by bending thin metal material over a form block or into a specially shaped die with high-pressure hydraulics, in conjunction with a rubber bladder or diaphragm, helping to shape the product. (21)

hypoid gear. A complex variation of the spiral bevel gear that allows for axes to be offset. (18)

I

improper fraction. A fraction where the numerator is greater than the denominator. (A)

inclined surface. In multiview projection theory, a surface perpendicular to one principal projection plane but inclined to the other two. (5)

independency. A specification for a feature of size that indicates perfect form is not required at the maximum material condition. (13)

index line. A line extending the length of the sleeve of a micrometer and divided into 40 equal parts by vertical lines, each designating 1/40″ (.025″). (B)

inscribe. To construct a polygon around the inside of a circle. (4)

inseparable assembly drawing. A drawing composed of two or more parts that, once assembled, are permanently joined and thus become one unit. (16)

insert. A material placed into a mold cavity prior to molding for the purpose of surrounding the material by the molding material. (20)

inside mold line (IML). The line representing the intersection of inside surfaces if projected to a sharp edge. (21)

installation assembly drawing. An assembly drawing providing the instructions to install or erect a piece of equipment. Often shown in pictorial form. (16)

interchangeable manufacturing. The process of manufacturing wherein all parts of the same designation or part number can be interchanged regardless of place or date of manufacture. (10)

internal diameter. For an internal gear, the diameter of a circle coinciding with the tops of the gear teeth. (18)

internal gear. A gear that features teeth cut on the inside of a cylinder, meshing with a spur gear, with both gears turning in the same direction but at different speeds. (18)

internal thread. Screw thread cut on an internal surface. (8)

International Organization for Standardization (ISO). The agency responsible for establishing and publishing manufacturing standards worldwide. (1)

International System of Units (SI). The modern form of the metric system used widely outside the United States, and not just for linear distances. Linear distances are based on the meter, with the millimeter being common on industrial prints. (9)

interpretation. That aspect of print reading that involves interpreting the meaning of lines, symbols, dimensions, notes, and other information on a print. (1)

involute. A spiral curve generated by a point on a chord as the chord unwinds from a circle or a polygon. (4)

isometric drawing. A common form of pictorial drawing in which the three-dimensional axes are all equally spaced to each other. (C)

J

joggle. An offset displacement of material from its original plane, often to allow for smooth material overlap, as in a mating sheet metal flange for welding overlap. (21)

K

key. A section of metal that prevents a gear or pulley from rotating on its shaft. (11)

keyseat. A recess in a shaft that holds the key. (11)

keyway. A slot in the hub of a pulley or gear that allows the hub to slide over the key. (11)

K-factor. A factor used in formulas to allow for the neutral plane offset. (21)

knockout. Any part or mechanism of a mold used to eject the molded article. (20)

knurling. The process and result of machining a diamond or straight-line pattern on a cylindrical part. It is often used on a gripping surface, to increase the diameter of the part, or to increase friction between mating parts. The rolls are called knurls. (11)

L

lapping. A method of finishing a surface with a very fine abrasive. (12)

laser beam welding (LBW). A method of welding that uses laser beams optically focused on the workpiece surfaces, generating an energy beam that brings the solid metal to a liquid phase change. (22)

lay. Direction of the predominant surface pattern. (12)

lead. For screw threads, the amount a threaded part advances when revolved one turn. (8)

lead angle. For worm gearing, the angle the lead makes with a perpendicular line to the worm axis. (18)

leader line. A line with an arrowhead on one end and a shoulder on the other end leading to the text of a local note. Also known as a *leader.* (2)

least material boundary (LMB). With respect to datum establishment, the boundary represented by the least material condition of the datum feature. (13)

least material condition (LMC). When a feature contains the least amount of material allowed by the toleranced dimension, such as the largest hole diameter or the smallest shaft diameter. (10)

left-hand thread. Thread with the helical shape in a reversed direction, allowing for advancement of the threaded part when rotated counterclockwise. Also known as *reverse thread*. (8)

leg. 1. The straight segment of stock on either side of a bend. 2. In welding, the size of the weld, which for a fillet weld is expressed with two values for each side of the weld bead triangle. (21, 22)

lettering. Freehand sketching alphanumeric shapes according to standard practices. (2)

limit dimensioning. Method of expressing a tolerance with two limit dimensions, the maximum and minimum size. (10)

limits. The extreme permissible dimensions of a part resulting from the application of a tolerance. (10)

linear cam. A cam that moves in a straight line. (19)

linear pitch. For worm gearing, the distance from a given point on one worm thread to the next. (18)

local note. A note located directly on or near a particular feature of the object, and usually accompanied by a leader line. (11)

location dimension. Indicates the location of features such as holes, slots, and grooves. (9)

lowest common denominator (LCD). The smallest number to which each denominator can be converted. (A)

M

machine screw. A threaded fastener of 1/4″ diameter or less, similar to a cap screw. (17)

machining center. Hole drilled and countersunk for the purpose of holding a part between lathe centers or in a machining fixture. (11)

machining drawing. A detail drawing providing the necessary dimensions and information to manufacture a finished part from an existing base part such as a casting. (15)

major diameter. Largest diameter of the thread. (8)

material boundary modifier. A modifier applied to a datum feature reference specifying the applicable material boundary. (13)

material condition modifier. A modifier applied to a tolerance specification indicating the applicable material condition. (13)

maximum material boundary (MMB). With respect to datum establishment, the boundary represented by the maximum material condition of the datum feature (13).

maximum material condition (MMC). When a feature contains the maximum amount of material allowed by the toleranced dimension, such as the smallest hole diameter or the largest shaft diameter. (10)

mechanical engineer's scale. A scale used in mechanical engineering applications, such as machine part drawings. Typical mechanical engineer's scales are quarter size, half size, and full size. (B)

mechatronics. A field of study combining a wide array of mechanical and electrical applications in high-tech systems such as robotics and automation. (23)

metric scale. A triangular scale with decimal divisions, configured to provide for reduction or enlargement along each edge. (B)

micrometer. A precision measuring device with a spindle that rotates in a fixed nut and opens or closes the distance between an anvil and the spindle. Also known as a *mike*. (B)

minor diameter. Smallest diameter of the thread. (8)

miter gears. Bevel gears in a 90° relationship and having the same size and number of teeth, designed to transfer motion without changing speed. (18)

mixed number. A number comprised of a whole number and a proper fraction. (A)

model and drawing method. A hybrid method of preparing a product definition data set that includes both a drawing graphic sheet and a 3D model. (1)

model only method. A method for preparing a product definition data set by attaching annotations to an individual 3D model file. (1)

modification drawing. A detail drawing describing the necessary steps to modify one part into another. (15)

modifying drawing. A detail drawing describing objects that are made from other parts. (15)

module (*m*). In metric-based gear applications, the ratio defined as the pitch diameter divided by the number of teeth. Mating metric-based gears must have identical modules. (18)

mold. A hollow form or matrix into which a material is placed and that determines the final shape of the material. (20)

molding. The process of forming a plastic object by forcing molten plastic into a hollow mold using pressure. (20)

mold shrinkage. The immediate shrinkage of a plastic part after it has been removed from the mold. Both the mold and molded part are measured at normal room temperature. (20)

monodetail drawing. A detail drawing showing only one part. (15)

monomer. A molecule of low molecular weight capable of reacting with identical or different monomers to form a polymer. (20)

multidetail drawing. A detail drawing showing more than one part on one sheet. (15)

multiple-start thread. Special-application thread comprised of two or more adjacent helical ridges wrapped around the surface of a cylinder. Formerly referred to as multiple threads, or more specifically, double threads or triple threads. (8)

multiview drawing. Drawing consisting of multiple two-dimensional projection views of an object arranged systematically. (5)

N

neutral arc bend radius. The bend radius plus a percentage (K-factor) of the metal thickness. (21)

neutral plane. The area within a flat blank that will neither compress nor stretch as bending occurs. (21)

nominal size. A general classification term used to designate the size of a commercial product. (8, 10)

normal surface. In multiview projection theory, a surface that is parallel with one principal projection plane but perpendicular to the other two. (5)

not-to-scale method. A method used to indicate a change is made to a dimension value but the feature is not redrawn to scale. A heavy line is drawn below the revised dimension value to indicate that it is not to scale. (14)

now condition. In the revision process, the print after a change is made. (14)

numerator. The top number of a fraction, indicating the total number of equal parts given; for example, 3/16 means 3 of 16 equal parts are indicated. (A)

nut. An internally threaded fastening device designed to be screwed onto a bolt or stud. (17)

O

oblique drawing. A type of pictorial drawing wherein the depth of an object is "slanted" to one side, while the front-facing surfaces remain true shape. See *cavalier oblique drawing* and *cabinet oblique drawing.* (C)

oblique surface. In multiview projection theory, a surface inclined to all three principal projection planes. (5)

offset section. A section view wherein the cutting plane is bent, or offset, through features that are not aligned. (6)

open end. The end finish style of drawing compression springs that simply features the wire ending after a specified number of coils, spaced evenly based on the pitch. Also called *plain end.* (17)

orientation tolerance. The permitted variation of the orientation of a feature specified with an angularity, parallelism, or perpendicularity tolerance. (13)

orthographic projection. System of projection using parallel lines to project the view of a three-dimensional object onto a perpendicular two-dimensional plane. (5)

outline section. A section view of a section with large surface areas wherein the section lines are drawn as segments along the visible object lines only. Large parts are often drawn with outline sections. (6)

outside diameter. The diameter of a circle coinciding with the tops of the teeth of an external gear. (18)

outside mold line (OML). The line representing the intersection of outside surfaces if projected to a sharp edge. (21)

overbending. Bending an extra increment to compensate for springback. (21)

overmold. The process of injection molding a plastic over an existing plastic part. This term also refers to the material forming the overmolding. (20)

P

parallel. A term describing a geometric relationship wherein two features, such as lines, do not intersect, even if extended. (4)

parallelism. A geometric control applied to a surface or axis to maintain the feature's parallel orientation in reference to a datum surface or axis. (13)

partial section. A section view wherein a removed section shows details of an object without showing the complete view. (6)

partial view. A view that is incomplete in the sense that part of the object is not drawn. A short break line will usually indicate the division between what is shown and what is not shown, although a center line could also be used for one-half of a symmetrical object. (5)

parting line. The edge of the mold cavity where both mold halves touch. (20)

parts list. A tabular chart or form listing the quantity and description of all the individual parts of a machine, structure, or assembly. (3)

pattern development drawing. A detail drawing that applies dimensions to the shape of the part while it is a flat piece of metal or other stock. (15)

perpendicular. A term describing a geometric relationship wherein two features, such as lines, form a 90° angle with each other. (4)

perpendicularity. A geometric control applied to a surface or axis to maintain the feature's perpendicular orientation in reference to a datum surface or axis. (13)

perspective drawing. A form of pictorial drawing based on the projection of points toward one or more vanishing points. This drawing more closely represents what the human eye sees than oblique or isometric drawings. (C)

phantom line. A thin line comprised of long dashes separated by two shorter dashes and used to indicate alternate positions, repeated details, or adjacent positions of related parts. (2)

pictorial diagram. A diagram depicting components in a pictorial manner and the piping or conduits as lines. (23)

pictorial drawing. A picture-like representation of an object or product. (C)

pin diameter. The diameter of the measuring pin or ball used between the teeth of a gear. (18)

pinion. In gear sets, the spur gear that is smaller and often considered the driving gear; especially common as the term for the spur gear that meshes with a rack. (18)

pin measurement. A measurement made over pins for an external gear and between pins for an internal gear. (18)

pitch (P). The distance from a point on one thread to a corresponding point on the next thread. (8)

pitch circle. For gears, the circle with a diameter representing the mating diameter with the other gear, as if the mating gears were friction rollers. It is an imaginary circle located approximately halfway between the roots and tops of the gear teeth. (18)

pitch diameter. 1. For threads, an imaginary diameter at the point wherein the width of the thread ridge is equal to the thread groove. 2. For gears, the diameter of the pitch circle (D). (8)

plastic. A synthetic material formed by chemical reaction and capable of being formed into a finished shape. (20)

plotter. An output device that prints on large-format paper, individual sheets, and, most often, roll-fed stock. Originally a vector-based output device that drew prints with pens. Modern plotters use ink-cartridge systems. (1)

plug weld. A weld made through a hole in one of the pieces of metal. (22)

plus and minus dimensioning. Method of expressing tolerances with a basic size and a plus and minus value. (10)

pneumatic. Those processes using pressurized air or gas to perform mechanical work. (23)

point of tangency. The single point shared by tangent features, such as a line and a circle or two circles. (4)

polygon. Closed shape composed of straight lines. (4)

polygon mesh. In computer modeling, a collection of vertices, edges, and faces that define the shape of a polyhedral object. (4)

polymer. The basic molecular structure of a plastic material. The product of *polymerization.* (20)

polymerization. The joining of two or more molecules to form a new and more complex molecule whose physical properties are different. (20)

port. An internal or external terminus of a passage in a component. (23)

portable document format (PDF). An electronic file format developed by Adobe Inc. and designed to allow users to easily share printable documents regardless of computer platform. (1)

positional tolerance. The permitted variation of a feature from the exact or true position indicated on the drawing. (13)

pressure angle. For gears, the angle between the tooth profile and a radial line at the intersection of mating gear teeth on the pitch circle. (18)

primary auxiliary view. An auxiliary view projected from, and adjacent to, a principal view. (7)

primitive. Basic form used to create complex models in CAD programs. (4)

print. The generic term for a copy of an original drawing. (1)

print reading. The process of analyzing a drawing or print to obtain information. (1)

product definition data set. The total collection of information required to completely define a component part or an assembly of parts. (1)
profile of a line. A two-dimensional geometric control of the elements of an irregular surface. (13)
profile of a surface. A three-dimensional geometric control that extends throughout an irregular surface. (13)
profile plane. In multiview projection theory, a projection plane onto which the right-side view is projected. (5)
profile tolerance. A geometric control that maintains a surface or elements of a surface to a designated design shape, as defined by basic dimensions; often referenced to a datum surface or axis. (13)
profilometer. A device that can measure a surface's profile and report the data. The original profilometer was a device that had a stylus, or needle, that moved relative to the surface profile. (12)
projectors. Parallel lines used to project the points of a three-dimensional object onto a two-dimensional plane. (5)
proper fraction. A fraction where the numerator is less than the denominator. (A)

R

rack. A straight bar with gear teeth that, in conjunction with a pinion, creates a mechanism for rotary motion to be transferred to linear motion. (18)
radial cam. A cam consisting of a wheel with an irregular shape. Also known as a *plate cam*. (19)
radius. Distance from the center point to the outer edge of a circle or arc. (4)
rapid manufacturing. The application of additive manufacturing technologies to the full-scale production of parts and products. (1)
rapid prototyping. Technology capable of producing a prototype directly from a 3D CAD model, usually made of plastic or metal. (1)
raster image. An image composed of tiny pixels or dots. (1)
reaming. Finishing a drilled hole to a close tolerance using a machine tool called a reamer. (11)
reference dimension. Dimension used only for information purposes and not for governing production or inspection operations. The value is given in parentheses in current standards. (9)
regardless of feature size (RFS). The condition where a geometric tolerance must be met no matter where the feature lies within its size tolerance. (13)
regardless of material boundary (RMB). The condition wherein a datum feature must be established by maximum contact with a feature regardless of the material boundary. (13)
removed section. A section view or set of views "removed" and placed out of projection with the regular views of the object. (6)
removed view. A view that has been moved out of alignment with the main views that are arranged in the usual manner. A viewing-plane line will be given, with alphabetic labels to identify the removed view as VIEW A-A, or a similar label. (5)
rendering. A photorealistic illustration of a design idea, object, or architectural project that takes the pictorial drawing to another level. Rendering techniques range from simple line shading to complex airbrushed effects or computer-generated renderings. (C)
reservoir. A container for storing fluid in a fluid power system. (23)
resin. The basic material from which plastics are formed. (20)
resistance welding. A method of welding wherein phase change occurs due to resistance to electric current channeled through two pieces of metal held together under pressure for a defined amount of time. (22)
restriction. A reduced cross-sectional area in a line or passage that produces a pressure drop. (23)
retaining ring. A non-threaded fastening device typically inserted into a groove and used in both internal and external applications. (17)
revision. Any change made to the original drawing. (3)
revision authorization document. A document that contains a complete record of the changes that have been made to an original drawing. This document and the information it contains varies among industries. (14)
revision history block. A block that is a record of changes made to a drawing. (3)
revision status of sheets block. A block that is a record of the revision status for each sheet of a multiple-sheet drawing. The block is located next to the revision history block or is a separate document. (3)

revolved section. A section view wherein the cutting plane is presumed to be cutting parallel with the line of sight for a view, but the sectional shape is revolved 90° and drawn directly on the same view. (6)

right-hand thread. Thread with a helical shape allowing for advancement of the threaded part when rotated clockwise. The most common form of thread. (8)

rivet. A non-threaded fastening device used in permanent assemblies. (17)

root. 1. For a screw thread, the bottom edge or surface between two ridges. 2. For welding, the bottom portion of a weld bead, which for a fillet weld is the corner. (8, 22)

root opening. The empty space between two pieces to be welded. (22)

rotational molding. A casting process that creates hollow objects by rotating a powdered thermoplastic in a closed, heated mold until it melts and fuses into a solid mass. (20)

round. Exterior rounded corner or edge. See also *fillet* and *runout.* (5)

runner. In an injection or transfer mold, the usually circular channel connecting the sprue with the gate to the mold cavity. (20)

runout. Filleted or rounded edge that intersects with a curved surface and tails out. (5)

runout tolerance. A geometric control applied to circular elements (circular runout) or surface elements (total runout), as determined by measuring surface points while revolving the part about a designated datum axis. (13)

S

sampling length. An extracted length of the overall surface profile for collecting roughness data in segments. Also called *roughness cutoff.* (12)

scale. 1. A measuring device. 2. A process of reducing or enlarging an object proportionately. 3. The ratio between the size of the part as drawn and the actual size of the part. 4. The size; e.g., full scale is equal to the expression "full size." (3, B)

schematic representation. A method of representing screw threads in a simplified form using alternating thick and thin lines for root and crest lines. See also *detailed representation* and *simplified representation.* (8)

screw gear. An application of two helical gears in a nonparallel and nonintersecting arrangement, wherein the helical teeth are crossed and mesh with one another. Also called *crossed helical gear.* (18)

screw thread. A ridge or groove of a particular shape that follows a helical path around the surface of an external or internal cylindrical feature. Also known as a *thread.* (8)

secondary auxiliary view. An auxiliary view projected from a primary auxiliary view in order to depict an oblique surface in true size and shape. (7)

section line. A thin line in a pattern that fills in, or "hatches," an area presumed to be cut by a cutting plane in a section view. (2, 6)

section view. A cross-sectional view at a specified point of a part or assembly. (6)

sector. Area bounded by two radii and the included arc. (4)

sequence of operations. The order of a series of processes or movements. (23)

serration. An internal notch or sharp tooth on a surface or edge. (18)

setback. The amount of material to deduct from the blank length to accommodate a bend. Also known as *bend deduction.* (21)

set screw. A threaded fastener used to prevent pulleys from slipping on shafts, hold collars in place on shafts, and hold shafts in place on assemblies. (17)

simplified representation. A method of representing screw threads in a simplified form using hidden lines for the root diameter. See also *detailed representation* and *schematic representation.* (8)

single limit. Method of expressing a tolerance value with a MIN or MAX note, for use in a few special cases such as edge break, thread length, or hole depth. (10)

single-start thread. A screw thread comprised of a single ridge wrapped around the cylindrical surface, as opposed to multiple-start threads that are comprised of two or more adjacent ridges. (8)

sink mark. A shallow depression or dimple on the surface of an injection-molded part caused by a short shot or local internal shrinkage after the gate seals. Also known as a *sink.* (20)

sinter. The technique used to shape parts from metal powders by heating the powder until the particles melt and fuse together. (1)

size dimension. Indicates the size of the part and the size of its various geometric features, such as holes, fillets, and slots. (9)

slot weld. A weld made through a slot in one of the pieces of metal. (22)

soldering. A method of joining metals with a nonferrous filler metal without having to heat the base metals to their melting point. (22)

solenoid. An electrical switch incorporating a metal rod that can be extended or retracted based on the application of electricity. (23)

solid-state welding. A method of welding that joins the metals without phase change, as pieces are heated to an extent but not to the point of melting. (22)

spatial visualization. The ability to mentally visualize and manipulate two-dimensional and three-dimensional figures. (5)

spline. 1. A curved "line" fit smoothly through a series of points. 2. A raised area on a shaft designed to fit into a recessed area of a mating part. (4)

spotface. A machined circular spot on the surface of a part to provide a flat bearing surface for a screw, bolt, nut, washer, or rivet head. (11)

spot weld. A type of resistance weld joining pieces of metal at separate spots, rather than with a continuous weld. (22)

spring. A device designed to either compress or expand due to a force and counteract that force with an equivalent force. (17)

springback. The opening of a bend after forming due to the elasticity of the material, usually specified in degrees or a percentage. (21)

spring washer. A flat spring designed to exert a force if another part, such as a threaded fastener, attempts to flatten it. (17)

sprue. In an injection or transfer mold, the main feed channel connecting the mold orifice with the runners leading to each cavity gate. The term is also given to the piece of plastic material formed in this channel. (20)

spur gear. A commonly used gear represented generally by a cylinder, or wheel, with equally spaced teeth. (18)

standard. A voluntary guideline. (1)

statistical process control (SPC). A system used in industry to monitor a manufacturing process using data collection within a particular analysis method for quality control purposes. (10)

statistical tolerance. A tolerance value flagged with a symbol that indicates the dimension is to be produced with statistical control. (10)

steel rule. A measuring device, usually 6″ long and made of steel, with a fractional, decimal inch, or metric scale on one or both edges. (B)

stitch line. A line that indicates a sewing process on a part or an assembly. (2)

straightness. A geometric control applied to an element, axis, or elements of a surface to maintain straightness of the elements, regardless of a datum. (13)

stud. A threaded fastener with external threads on both ends. (17)

stud welding. A type of welding wherein studs or nuts are welded onto a base metal using equipment and fasteners especially designed for that purpose. (22)

subassembly. Any assembly that fits within a larger assembly. (16)

subassembly drawing. A drawing that describes a group of related parts composing a subunit of a larger mechanism. Subassemblies often are given their own part numbers and are stored in inventory as if they are single parts. (16)

supplementary symbol. A symbol used along with the welding symbol to further specify the characteristics of the weld. (22)

surface roughness. Fine irregularities in the surface texture. (12)

surface texture. The lay, roughness, waviness, and flaws of a surface. (12)

surface waviness. Widely spaced component of surface texture due to such factors as machine chatter, vibrations, work deflections, warpage, and heat treatment. (12)

surfacing weld. A weld buildup created on a flat or cylindrical surface for the purpose of creating a surface with different properties or dimensions. (22)

symmetry. A discontinued geometric control formerly applied to maintain the elements of two surfaces equidistant about a center plane, regardless of feature size. (13)

symmetry line. A line that shows the center axis of a part where both sides are symmetrical. (2)

T

tabular dimension. A dimension measured from mutually perpendicular datums and listed in a table on the drawing instead of on the views. (9)

tangent. A condition wherein two geometric features share a single point—for example, a line drawn so it intersects a circle only once, even if extended. See also *point of tangency.* (4)

thermoforming. The process of forming a plastic object by forcing a softened sheet of thermoplastic onto a mold using vacuum or positive pressure. (20)

thermoplastics. Resins or plastic compounds that can be repeatedly softened by temperature increase and hardened by temperature decrease. (20)

thermoset plastics. Resins or plastic compounds that cannot be softened after they have cured. (20)

thickness. The stock thickness of the sheet metal, usually identified in formula as T. (21)

third-angle projection. A system that serves as the foundation for orthographic projection used primarily in the United States and Canada. See *first-angle projection.* (5)

thread. A ridge or groove of a particular shape that follows a helical path around the surface of an external or internal cylindrical feature. Also known as a *screw thread.* (8)

thread angle. The angle formed between two adjacent flanks. (8)

thread class. Definition of the fit between mating threads. (8)

thread form. A description of the shape of the ridge (or groove) forming the thread. (8)

three-dimensional (3D) printing. Technology that builds parts by adding material in a series of layers. Also known as *additive manufacturing.* (1)

throat. The depth of the weld bead from the face to the root, which for a fillet weld is measured in actual, theoretical, or effective terms. (22)

throat diameter. The diameter of a circle coinciding with the tops of the worm gear teeth at their center plane. (18)

title block. Boxed area normally located in the lower-right corner of a print that provides general information and aids in identification and filing of the print. (3)

toe. The top portion of a weld bead as measured along the pieces, which for a fillet weld is the leg distance away from the root. (22)

tolerance. The total amount of variation permitted for a feature's location or size based on the dimensional difference between limits. (3, 10)

tolerance block. Boxed area on a drawing that indicates the general tolerance limits specified for the drawing. (3)

tool and die maker. A machinist who specializes in precision metalworking, including the creation of tools and dies used in sheet metal forming. (21)

tooth face. The curved surface of a gear tooth that lies outside of the pitch circle. (18)

tooth flank. The curved surface of a gear tooth that lies inside of the pitch circle. (18)

tooth space. The distance at the pitch circle between two adjacent gear teeth. (18)

torsion spring. A spring designed to counteract a force that is at a right angle to the axis of the coil of wire. (17)

total runout. A geometric control providing a composite control of all surface elements as the part is rotated about a datum axis. (13)

true geometry view. True profile of a part. (9)

true position. Theoretically exact location of the axis or median plane of a feature. (13)

typical (TYP). When associated with a dimension or feature, this term means the dimension or feature applies to all locations appearing to be identical in size and configuration. (11)

U

undimensioned drawing. Full scale drawing that defines a part with a true geometry view. (9)

unidirectional dimensioning. A dimensioning system where all lettering for dimensions is horizontal on the page. See also *aligned dimensioning.* (9)

Unified Thread Standard (UTS). The ASME standard thread form and series defined for screw threads in the United States. It is the main standard for bolts, nuts, and a wide variety of threaded fasteners. Often referred to simply as *Unified threads.* (8)

unilateral tolerance. A toleranced dimension that allows deviation from the design size in one direction only. (10)

US customary units. The unit system used primarily within the United States, based on the inch, foot, yard, and mile. Linear distances on industrial prints are often given in inches with decimal fractions. (9)

V

vanishing point. An imaginary point in a perspective drawing toward which one or more receding axis lines converge. (C)

vellum. A paper made of 100% rag content and impregnated with a synthetic resin to provide high transparency. (1)

vertex. In computer modeling, a single 2D or 3D point within the geometric shape, containing X, Y, and Z coordinate position values. The plural of vertex is vertices. (4)

viewer program. A piece of software that allows a user to open, view, and print the electronic file of a 2D drawing or 3D model. (1)

viewing-plane line. A line accompanied by arrows used to indicate a special view from a particular direction that may be confusing if not accompanied by the line. See *cutting-plane line.* (2)

visible line. The thick, continuous line representing all edges of an object visible in a particular view. Sometimes referred to as an object line. (2)

visualization. An element of print reading described as the ability to "see" the shape of an object from prints showing various views. (1)

W

warpage. The distortion of a part caused by nonuniform shrinkage. (12, 20)

was condition. In the revision process, the print before a change is made. (14)

washer. A device used to increase the bearing surface for bolt heads and nuts and to prevent surface marring. (17)

weld. The joint formed between two parts that have been joined by melting. (22)

weld bead. The material deposited along the joint that is formed as the melted metal solidifies. (22)

welding. A fabrication process that involves joining metal pieces by melting them along a joint, with a filler material being added in some cases. (22)

welding drawing. A detail drawing showing parts that need to be welded together. (15)

welding symbol. A symbolic notation used to indicate the type and size of weld to be applied at a seam, joint, or location on the assembled pieces. (22)

weld joint. The location where two pieces of metal come together to be welded. (22)

weld symbol. The symbol attached to the reference line within a welding symbol that indicates the type of weld. (22)

whole depth. The total depth of a gear tooth space, or the addendum plus the dedendum. (18)

wide-format printer. A printer designed to accommodate wide media (paper, vinyl, or textiles), available in widths of two feet to several feet, and also available using various ink technologies. (1)

width. The left-to-right measurement of an object as defined from the front view. (5)

width auxiliary view. An auxiliary view projected from a right-side view. (7)

working depth. The sum of the addendums of two mating gears. (18)

working drawings. The general term applied to a set of drawings providing details for the manufacture of a product or correct assembly of all parts. (15)

worm. A threaded, screw-like device that drives a worm gear. (18)

worm gear. A gear with teeth cut on an angle to be driven by a worm (screw). (18)

Z

zone. Area on a print identified by numbers and letters marked along the border of the drawing. (3)

Index

D

E

F

G

H

I

J

K

L

M

N

O

P

R

S

T

U

V

W

Z